High-Temperature Superconductivity

Physical Properties, Microscopic Theory, and Mechanisms

High-Temperature Superconductivity

Physical Properties, Microscopic Theory, and Mechanisms

Edited by

Josef Ashkenazi
Stewart E. Barnes
Fulin Zuo
University of Miami
Coral Gables, Florida

Gary C. Vezzoli
U.S. Army Materials Technology Laboratory
Watertown, Massachusetts

and

Barry M. Klein
Naval Research Laboratory
Washington, D.C.

PLENUM PRESS • NEW YORK AND LONDON

Library of Congress Cataloging-in-Publication Data

High-temperature superconductivity : physical properties, microscopic
  theory, and mechanisms / edited by Josef Ashkenazi ... [et al.].
      p.   cm.
   Proceedings of a workshop held Jan. 3-9, 1991, at the University
 of Miami, Coral Gables.
   Includes bibliographical references and index.
   ISBN 0-306-44117-9 (hardbound)
   1. High temperature superconductivity--Congresses. I. Ashkenazi,
 Josef.
 QC611.98.H54H543453   1992
 537.6'23--dc20   91-43401
                                                                 CIP

Proceedings of the University of Miami Workshop on Electronic Structure
and Mechanisms of High-Temperature Superconductivity, held January 3-9, 1991,
in Coral Gables, Florida

ISBN 0-306-44117-9

© 1991 Plenum Press, New York
A division of Plenum Publishing Corporation
233 Spring Street, New York, N.Y. 10013

All rights reserved

No part of this book may be reproduced, stored in a retrieval system, or transmitted
in any form or by any means, electronic, mechanical, photocopying, microfilming,
 recording, or otherwise, without written permission from the Publisher

Printed in the United States of America

PREFACE

This volume contains the proceedings of the University of Miami Workshop on the subject of "Electronic Structure and Mechanisms for High Temperature Superconductivity". The workshop was held at the James L. Knight Physics Building on the campus of the University of Miami, Coral Gables, 3-9 January 1991. Some 106 scientists from 12 countries attended this workshop, most of whom presented either invited or contributed papers.

The reader will find in this volume a series of papers discussing the most important experimental and theoretical developments as of winter/spring 1990/1991. Despite more than four years of intensive research on high-T_c materials, there has been considerable controversy both with respect to the interpretation of experiment and even more so in connection with the construction of an appropriate theory. In this regard, workshops such as this, gathering scientists with many viewpoints, and varying specialization, and fostering constructive discussions, are important in the development of a common ground. Of major concern in the present context were the basic physical processes involved in high-temperature superconductivity.

Experimental work in this volume includes reports of photoemission spectra, positron annihilation, the de Haas-van Alphen effect, nuclear magnetic resonance (NMR), and various transport and other measurements. The first three of these have generally been interpreted in terms of fairly conventional Fermi liquid band-structure models. However, in particular, the width of the photoemission peaks as they approach the Fermi surface has been interpreted by some as suggesting non-Fermi liquid behaviour. The transfer of spectral intensity upon entering the superconducting state has also been cited as evidence for behaviour not consistent with a normal Fermi liquid.

NMR experiments are often interpreted as being in accord with a single band model in which magnetic correlation effects are important, $i.e.$, as suggestive of theoretical approaches based upon the single band Hubbard model. The remarkable recent de Haas-van Alphen experiments seem consistent with the existence of a Fermi surface not unlike that predicted by band structure calculations. Clearly, in connection with superconductors, transport measurements and the dependence of such basic properties as the critical temperature T_c upon doping and upon other chemical factors remain a basic frame of reference within which all theories must fit.

A wide range of theoretical approaches are described. Those theories based upon Luttinger liquids, marginal Fermi liquids, or anyons might be classified as being "exotic" or highly innovative. Others based upon the predictions of Fermi liquid theory and local-density approximation (LDA) type band structure calculations, including first-principles calculations of phonon frequencies rely on more classical concepts, while theories based upon anomalous Fermi-liquid type states, Kondo lattices, defects, anharmonicity, polarons, or bipolarons, may be considered to be in a domain somewhere between the two extremes. P. W. Anderson, is perhaps the person who most forcefully suggests that, when viewed as a whole, the experimental evidence is simply not consistent with a picture based upon conventional Fermi liquid ideas. A contrasting point of view is offered by those who point to the remarkable agreement between the parameter-free predictions of LDA electronic structure calculations and the details of measured electronic and phonon properties.

The organizing committee of this workshop wishes to thank all of the participants for their contributions to what was an animated, enjoyable, and very informative experience. The organizers are indebted to the University of Miami and, in particular, the Office of Naval Research, through the Department of the Navy Grant N00014-91-J-1224, for their financial support.

These proceedings correspond to the first full scale topical condensed matter physics workshop to be hosted by the Physics Department, University of Miami, Coral Gables. Given sufficient support, it is hoped to hold similar topical conferences on a regular basis. The organizing committee comprised Drs. Josef Ashkenazi, Stewart E. Barnes, Fulin Zuo, Barry M. Klein and Gary C. Vezzoli. The help of the student assistants Dan Vacaru, Jose Alexandro Orta, Yuntao Ge, Xuemei Song, Sharon Zane and Lai Leong, during the workshop, is gratefully acknowledged, as is the assistance of Mrs. Judy Mallery. The assistance of Drs. C. Balseiro, F. de la Cruz and M. Tovar in providing coordination with Latin American scientists is also acknowledged.

The Editors

CONTENTS

The Theory of High T_c Superconductivity....................1
   Philip W. Anderson

Electronic Structure, Lattice Dynamics, and Electron-Phonon
     Interaction in High T_c Superconductors...............7
   R.E. Cohen, H. Krakauer, and W.E. Pickett

Marginal Fermi-Liquid Theory of the High T_c Materials.....19
   C.M. Varma

Recent Progress in Semionic Pairing Theory of High
     Temperature Superconductivity........................25
   Z. Zou

Mean Field Analysis of the CuO_2 Lattice: Comparison to
     Monte Carlo Simulations..............................37
   A.J. Fedro, Y. Zhou, T.C. Leung, and B.N. Harmon

On the Inequivalence of the Three Band, Kondo Heisenberg and
     t-J Models...45
   O.F. de Alcantara Bonfim and G. Reiter

Holes in the Quantum Antiferromagnet: The Theory of Small
     Polarons...53
   A. Auerbach and B.E. Larson

Propagator-Renormalized Perturbation Theory for the Hubbard
     Model..61
   J.W. Serene and D.W. Hess

Slave Boson Approach to the One Dimensional t-J Model.....69
   T.K. Lee and Z. Wang

Universality of Behavior of Electrons on a Ring...........77
   F.V. Kusmartsev

Vector Mean Field Theory of Statistics....................89
   M.D. Johnson, C. Gros, S.M. Girvin, and G.S. Canright

Spinon-Holon Statistics, and Broken Statistical Symmetry for
     the t-J and Hubbard Models in 2D....................95
   S.E. Barnes

Frequency Dependent Conductivity of High T_c Oxide
  Superconductors..107
  H. Kim and P.S. Riseborough

Electronic Structure and the Superconducting Gap of
  $Bi_2Sr_2CaCu_2O_{8+\delta}$...113
  Z.-X. Shen, D.S. Dessau, and B.O. Wells

Superconducting and Magnetic Phase Boundaries in Doped Cu-
  Based Superconducting Oxides: Contrasting Charging
  and Pair Breaking Mechanisms.........................119
  Y. Gao, A. Kebede, P. Pernambuco-Wise, M. Kuric,
  J.E. Crow, R.P. Guertin, T. Mihalisin, N.D. Spencer, and
  D.W. Cooke

Carrier Scattering Rates Measured by Hall Effect and
  Magnetoresistance in the High-T_c Oxides............131
  N.P. Ong, T.R. Chien, T.W. Jing, T.V. Ramakrishnan,
  Z.Z. Wang, J.M. Tarascon, and K. Remschnig

Superconductive Tunneling in $YBa_2Cu_3O_7$ Thin Films:
  Dependence Upon Crystallographic Orientation.......137
  L. H. Greene, J. Lesueur, W.L. Feldmann, and A. Inam

Infrared Properties: The Normal State, The Energy Gap, and
  The Temperature Dependence of the Gap..............147
  Z. Schlesinger, R.T. Collins, L.D. Rotter, F. Holtzberg,
  C. Feild, U. Welp, G.W. Crabtree, J.Z. Liu, and Y. Fang

Infrared Studies of High-T_c Superconductors: Where's the
  Gap?..159
  D.B. Tanner, D.B. Romero, K. Kamaras, G.L. Carr,
  L. Forro, D. Mandrus, L. Mihaly, and G.P. Williams

Fermi Surface of YBCO by dHvA............................177
  J.L. Smith, C.M. Fowler, B.L. Freeman, W.L. Hults,
  J.C. King, and F.M. Mueller

Direct, Experimental Evidence of the Fermi Surface in
  $YBa_2Cu_3O_{7-x}$..183
  H. Haghighi, J.H. Kaiser, S.L. Rayner, R.N. West,
  J.Z. Liu, R. Shelton, R.H. Howell, P.A. Sterne, F. Solal,
  and M.J. Fluss

The Acute Spectral Structure of Single-Domain
  $YBa_2Cu_3O_{6.9}$...189
  J.G. Tobin, C.G. Olson, F.R. Solal, C. Gu, J.Z. Liu,
  and M.J. Fluss

Infrared Study of the Order Parameter in $YBa_2Cu_3O_7$: Analogy
  to Short Range Ordering in Itinerant
  Ferromagnets..197
  D. van der Marel, A. Wittlin, H.-U. Habermeier, and
  D. Heitmann

Thermoelectric Power and Resistivity Measurements of
  $Pr_xY_{1-x}Ba_2Cu_3O_{7-\delta}$ Up to 1200 K and a Theoretical
  Analysis..205
  B. Fisher, J. Genossar, L. Patlagan, G. Koren,
  J. Ashkenazi, and C.G. Kuper

Determination of the Coulomb Parameters of $La_{1.85}(Sr,Ba)_{0.15}CuO_4$
from Auger Spectroscopy and Electronic Band
Structure...215
R. Bar-Deroma, J. Felsteiner, R. Brener, J. Ashkenazi,
and D. van de Marel

Some Transport Properties in $YBa_2Cu_3O_7$......................225
A.T. Seshadri and B. Subrahmanyam

On Transport Properties of HTSC Oxides.......................231
M. Affronte and D. Pavuna

Normal-State Transport Properties of $YBa_2Cu_3O_{7-\delta}$:
Conventional Metallic Picture............................235
J.L. Cohn, S.A. Wolf, V. Selvamanickam, and K. Salama

Electron Tunneling and Acoustic Mode in High-T_c
Superconductors..241
A. Kallio

Plasmons in High-T_c Cuprate Superconductors................251
J.H. Kim, I. Bozovic, J.S. Harris, Jr., W.Y. Lee,
C-B. Eom, T.H. Geballe

The Role of Magnetism in High-T_c Superconductivity.........257
G.C. Vezzoli, B.M. Moon, M.F. Chen, T. Burke, and
F. Craver

A Unified Approach to A Description of the Cuprate
Superconductors: Major Normal and Superconducting
Parameters and The Mechanism Responsible for the
High-T_c..275
V.Z. Kresin, H. Morawitz, and S.A. Wolf

Electron-Hole Asymmetry: The Key to Superconductivity....295
J.E. Hirsch

Non-BCS Features of Tunneling and Point-Contact Spectroscopy
Data of Organic Superconductors and High-T_c
Cuprates...309
M. Weger and R.F. Wehrhahn

Properties and Superconductivity of Large Bipolarons.....319
David Emin

Anomalous Dynamics and High T_c.............................331
J.W. Flocken, J.R. Hardy, and H.M. Lu

Dynamic Jahn-Teller Theory of High T_c
Superconductivity..341
K.H. Johnson, D.P. Clougherty, and M.E. McHenry

μSR Measurements of Penetration Depth in Exotic
Superconductors..353
Y.J. Uemura

On the Orbital Angular Momentum of Cu 3d Holes in High T_c
    Superconductors. Is a Cu $3d_{3z^2-r^2}$ Bipolaron the
    Superconducting Pair?................................363
  A. Bianconi, A.M. Flank, P. Lagarde, C. Li, I. Pettiti,
  M. Pompa, and D. Udron

Kosterlitz-Thouless-Like Behavior over Extended Ranges of
    Temperature and Layer Thickness in Crystalline
    $YBa_2Cu_3O_{7-x}/PrBa_2Cu_3O_{7-x}$ Superlattices...................377
  D.H. Lowndes and D.P. Norton

Local Structural Anomaly at T_c Observed by Neutron
    Scattering...389
  T. Egami, B.H. Toby, S.J.L. Billinge, Chr. Janot,
  J.D. Jorgensen, D.G. Hinks, M.A. Subramanian,
  M.K. Crawford, W.E. Farneth, and E.M. McCarron

Anomalous Magnetic Properties of High Temperature
    Superconductors......................................401
  R.E. Walstedt

The Oxygen Isotope Effect in $(Y_{1-x-y}Pr_xCa_y)$
    $Ba_2Cu_3O_{7-\delta}$...411
  J.P. Franck, S. Gygax, J. Jung, M.A-K. Mohamed,
  and G.I. Sproule

Bosonic Mechanism for High Temperature Superconductors...417
  F.M. Mueller, G.B. Arnold, and J.C. Swihart

The Role of the Axial Oxygen in High-T_c Materials........425
  J. Mustre de Leon, S.D. Conradson, P.G. Allen,
  I. Batistic, and A.R. Bishop

Thermally Activated Depinning of Vortices in High-T_c
    Superconductors......................................437
  E.H. Brandt

Recent μSR Results on Magnetic Properties of Cuprate
    Materials..447
  E.J. Ansaldo

Topological Effects in Disordered Phase of Two-Dimesional
    Magnet...453
  A.M. Tsvelik and M. Yu Reizer

Specific Heat of $YBa_2Cu_3O_7$: Volume Fraction of Super-
    conductivity; Parameters Characteristic of the
    "Ideal" Superconductive State........................459
  N.E. Phillips and R.A. Fisher

Specific Heat and Thermal Expansion of the Bi and Tl High
    Temperature Superconductors near T_c................469
  D. Wohlleben, W. Schnelle, E. Braun, and H. Broicher

Current Status of Fermi Liquid Based Approaches to the
    Cuprates...481
  K. Levin, Q. Si, J.H. Kim, and J.P. Lu

x

Van Hove Scenario for HITC Superconductivity............493
    C.L. Kane, D.M. Newns, P.C. Pattnaik, C.C. Tsuei, and
    C.C. Chi

The t-J Model at Small t/J: Numerical, Perturbative and
    Supersymmetric Results................................503
    T. Barnes

Plasmon Exchange Model For the High-T_c Superconductors...515
    S.M. Bose and P. Longe

Properties of Superlattices Made of High-T_c
    Superconductors.......................................523
    P. Kumar, R.A. Guyer, S. Obukhov, and Y.S. Sun

Electronic Structure Fermi Liquid Theory of High T_c
    Superconductors.......................................529
    J. Yu and A.J. Freeman

Dynamical Spiral State in Two-Dimensional Hubbard
    Model...541
    Z.Y. Weng and C.S. Ting

Valence-Fluctuation Scenario for Cuprate Supercon-
    ductivity: The Finite-U Pairing Mechanism............547
    B. Brandow

The Van Hove Singularity and High-T_c Superconductivity:
    The Role of (Nanoscopic) Disorder.....................555
    R.S. Markiewicz

Two-Component Theory and Dynamic Structural
    Correlations..561
    Y. Bar-Yam

Search for the Correct Microscopic Theory for the High
    Temperature Cuprate Superconductors...................569
    J. Ashkenazi, D. Vacaru, and C.G. Kuper

Re-Analysis of Photoemission Data for CuO: Revision of
    the Configuration-Energy Scheme for Cuprate
    Materials...583
    B.H. Brandow

Self-Energy Corrections for NiO...........................591
    B. Szpunar and V.H. Smith Jr.

2D One-Band Hubbard Model for the Cuprates................597
    S.B. Bacci, E.R. Gagliano, and R.M. Martin

Interlayer Pairing and c-axis Versus ab-Plane Gap
    Anisotropy in High-T_c Superconductors...............609
    R.A. Klemm and S.H. Liu

Muon Spin Relaxation Studies of the Layered Copper
    Oxides..621
    G. Aeppli and D.R. Harshman

Neutron Scattering Studies of Spin Correlations in
    Metallic $YBa_2Cu_3O_{6+x}$............................629
  J.M. Tranquada

Recent Studies of Chemical Doping in High-T_c
    Superconductors....................................641
  Y.H. Kao

Participants...653

Author Index 659

Subject Index 663

THE THEORY OF HIGH T_c SUPERCONDUCTIVITY

Philip W. Anderson

Joseph Henry Laboratories of Physics
Princeton University
Princeton, NJ 08544

I would like to talk here about several developments in the continuing story of the theory of high T_c. As many of you will have read, or heard, I believe that the outline of the correct theory is, at this point, reasonably complete, as expressed in a sequence of statements, each supported both by theoretical and by experimental evidence, which I have called "Central Dogma" because they play the role of structuring our choices at each of a sequence of check points along the way to a complete theory. A theory which does not deal with the weight of experimental evidence and theoretical constraint which lies behind this overall structure has no chance of being correct. One of the most serious problems in this field is what I call "myopia", the tendency both of theorists and experimentalists to ignore unequivocal findings outside their own narrow specialty. Most serious of all, it seems that many theorists, especially, are determinedly not subjecting either their own work, or even that of others, to critical scrutiny. I myself welcome criticism, but find a very disturbing lack of serious criticism. I think people have a duty to examine other theories critically and explain why they reject straightforward theoretical or experimental arguments. There is also the tendency to mouth conventional wisdom just because it is the conventional wisdom. It is not physics, for instance, to pretend that the normal state is a conventional Fermi liquid in any sense; especially, there is no excuse for using Eliashberg-Anderson-Schrieffer theory (now called MacMillan-Rowell, or MMR) which assumes Migdal's theorem— that all vertex corrections may be ignored—to be valid for this peculiar metal; there is no excuse for continuing to discuss as relevant to cuprates theories of short-range RVB's, anyons, and flux phases, all of which depend crucially on the existence of a spin gap, when experiment unequivocally shows that there is no spin gap in the normal state and there is a Fermi surface with a straightforward Pauli spin susceptibility, a metallic, if peculiar, Hall effect, clear photoemission traces, and perhaps even a de Haas – van Alphen effect. On a more chemical level, there is no excuse for deviating from the basic simple, one-band Hubbard model, given the clear evidence that the carriers of spin are the same electrons which carry the charge, and that both the basic exchange interactions and the kinetic energy parameters of the underlying model are easily estimated.

After summarizing very briefly my overall picture, I would like to discuss some new developments, especially:

(a) New photoemission evidence and the theory of photoemission in the interlayer tunneling mechanism.

(b) New ideas, somewhat speculative, about the nature of the gap function in the superconducting state, supported by some new ideas, not speculative at all, about the superconducting pair susceptibility of the one-dimensional Hubbard model.

(c) Some new ideas completing the arguments I have given that the Fermi liquid theory breaks down for all two-dimensional Hubbard models, especially repulsive ones.

Brief summary: the dogmas I have called I, II, III lead to these conclusions:

(1) The normal state is the key problem

(2) The model is a one-band repulsive Hubbard model in 2 dimensions — one band in that the charge and spin carriers are the same electrons, although different admixtures of d and p for different excitations. The scale of the interactions is large in the 2d model.

Dogma (IV) is not really one dogma — it leads to several conclusions.

(3) The normal state is not a Fermi liquid but a new, correlated state, the properties of which closely resemble those of the 1D Hubbard Model.

(4) The normal state has a Fermi surface but the particles which define the Fermi surface are not ordinary electrons but spinons: $Z = 0$ Fermions carrying spin but no charge. Charge is carried by a separate excitation.

(5) This 2d state is not superconducting but has a large pair susceptibility: it is like a $T_c = 0$ superconductor (and the $\sigma \propto 1/T$ is the fluctuation conductivity.)

(6) The best-kept secret: superconductivity arises only from interplanar tunneling. Phonons are probably pair-breaking if anything.

The evidence for all this I have summarized at length elsewhere.

I think one of the main avenues to an understanding of these materials is photoemission. Two of the superconductors at least, BiSrCaCuO and David Mitzi's crystals, are easily cleaved to a clean surface, and recent improvement in the resolution of photoemission, plus the strict 2-dimensionality of the bands, make excellent data available both in normal and superconducting states.

I have spoken elsewhere about the normal state data, and I will just show my version of a heuristic fit to them (Fig. 1). The critical point I want to make is that the backgrounds are incompatible with the sharpness of the features if the background is to be ascribed either to bad resolution or to Lorentzian Fermi liquid broadening. In particular, I believe the programs used to fit the data are often erroneous in that Lorentzians with fixed breadths Γ are used, whereas in any theory $\Gamma(E - E_F) \to 0$ and also $\Gamma(E < E_g) \equiv 0$. Where Γ is comparable with the quasiparticle energy, this effect makes a difference: real Fermi Liquid peaks do not look like this.

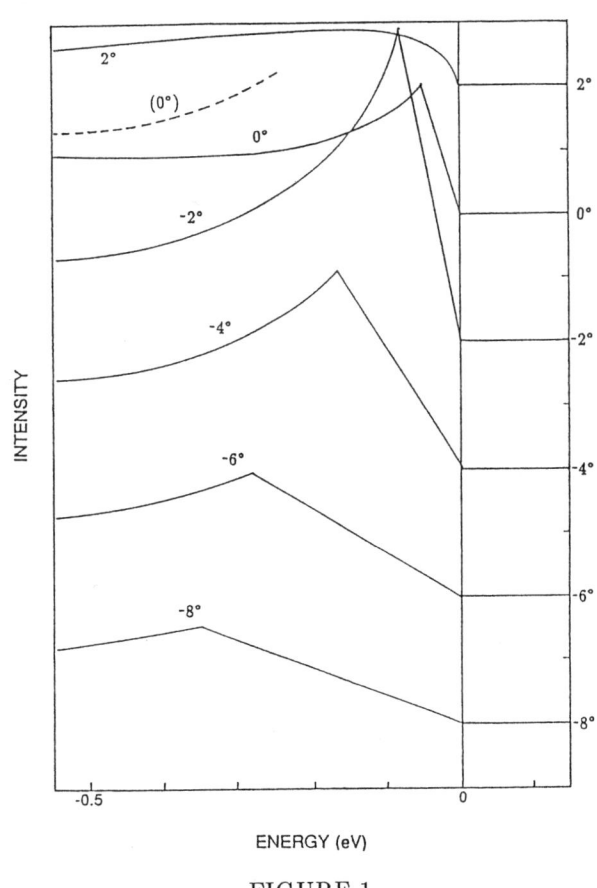

FIGURE 1

Much more interesting is what happens when the sample goes superconducting. A most extraordinarily strong and sharp peak arises at the energy gap. I have tried to fit these peaks within BCS theory and the fact is this cannot be done: it is always true that the growth of intensity in the peak is $>$ than the loss in the gap while it is a rigorous sum rule that in BCS the gain in I in the peak $\leq$ the loss in I in the gap. This phenomenon is shown in some recent data of Spicer's group (Fig. 2) which just confirms the earlier data of Arko et al (Fig. 3) and Petroff et al (Fig. 4) which I show here—here for the Spicer data I take the $\Gamma - X$ scan since an even more interesting result occurs on the $\Gamma - M$ scan.

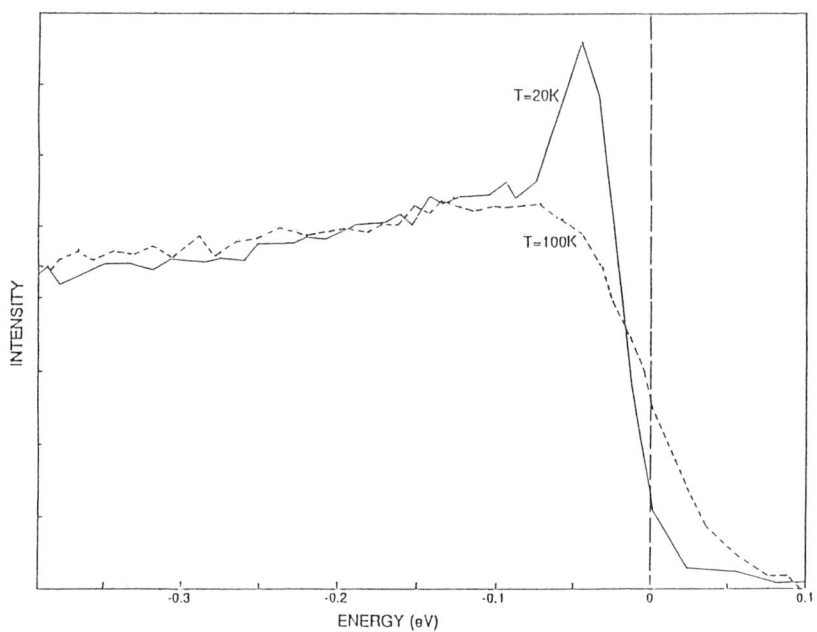

FIGURE 2

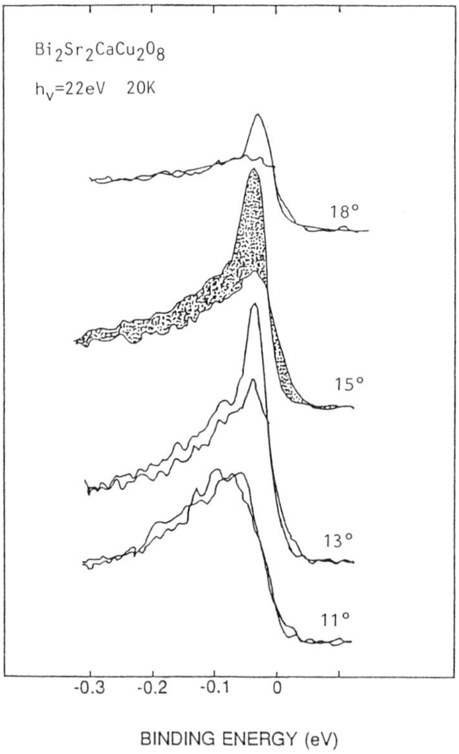

FIGURE 3

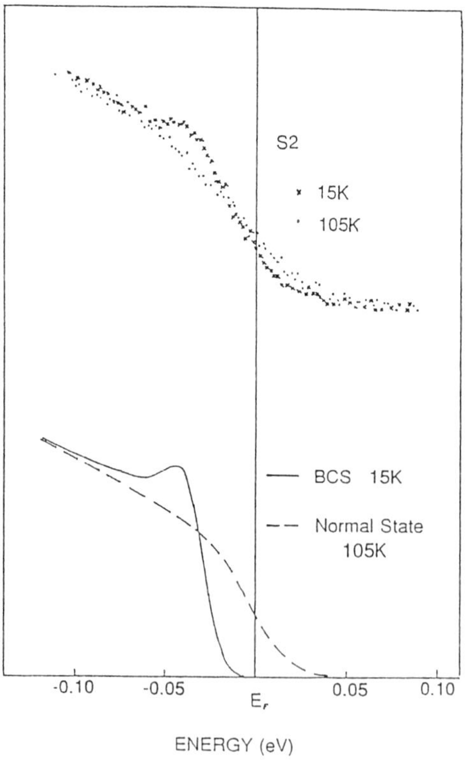

FIGURE 4

In the talk I gave a brief summary of the theory behind the photoelectric effect, emphasizing that all background comes from scattering downwards in energy (as one sees by the absence of background above E_F) so that the "background", which does not rise when the giant superconducting peak appears near E_F, but actually decreases, must primarily be intrinsic—what you see is what is there. In figure (5) I show an actual calculated shape for the normal state which closely resembles the observed spectrum with a smooth, smallish background added.

In the Dessau et al data along ΓM a new phenomenon appears. In most directions the Fermi surface should be split by $\sigma - \pi$ interaction due to hopping between the two close planes. In the normal state this splitting is never seen, because the Luttinger liquid "confines" electrons and is strictly two-dimensional. In this data we see this splitting <u>reappearing</u> in the superconducting state, because now the excitations have taken on true quasiparticle character and are three-dimensional. I reproduce Dessau et al's figure (Fig. 6). In essence, this figure is an explicit "smoking gun" showing the source of the superconducting condensate energy.

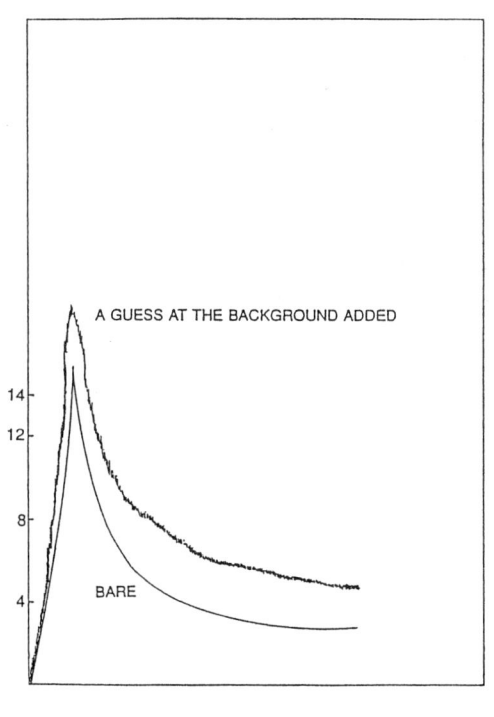

FIGURE 5

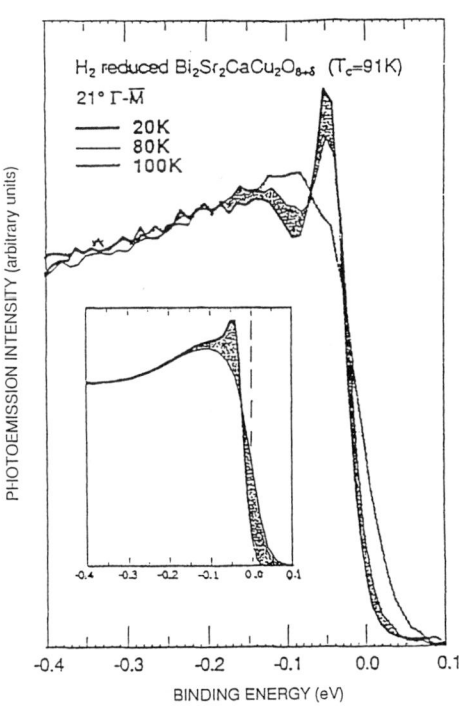

FIGURE 6

Finally, I discuss new ideas on the shape of the energy gap. From the photoemission results, the dimensional crossover idea is experimentally proven: interlayer tunneling appears at T_c and the resulting energy is adequate to cause condensation. This "Josephson" mechanism is new and very different from that of BCS, especially in that the main interaction <u>conserves momentum</u>: $\Delta(k)$ in layer 1 depends primarily on $\Delta(k)$ in layer 2 and not on $\Delta(k')$ anywhere. I have set up a heuristic gap equation which includes this effect but also takes into account that the electron-electron interaction is truly <u>repulsive</u>, which means that in real space $\Delta(r_1 - r_2 = 0) \equiv 0$. This condition requires that $\int \Delta(k)dk \equiv 0$. If we stick to singlet pairing, in the spirit of RVB, this means that Δ is not a strong function of the direction $\hat{k}$ but must have a node as a function of $|k-k_F|$, and by symmetry this node will be at $k = k_F$ itself: there is a node everywhere exactly at the Fermi surface, a node which becomes very steep as $T \to 0$. The proposed behavior of the gaps as a function of $k - k_F$ and temperature is sketched in Figure (7). Many strange properties of these superconductors fall into place if one understand the implications of this gap function: notably, the "two-fluid model" behavior, the appearance of large gap near T_c, the sensitivity of specific heat to magnetic field and structure, etc.

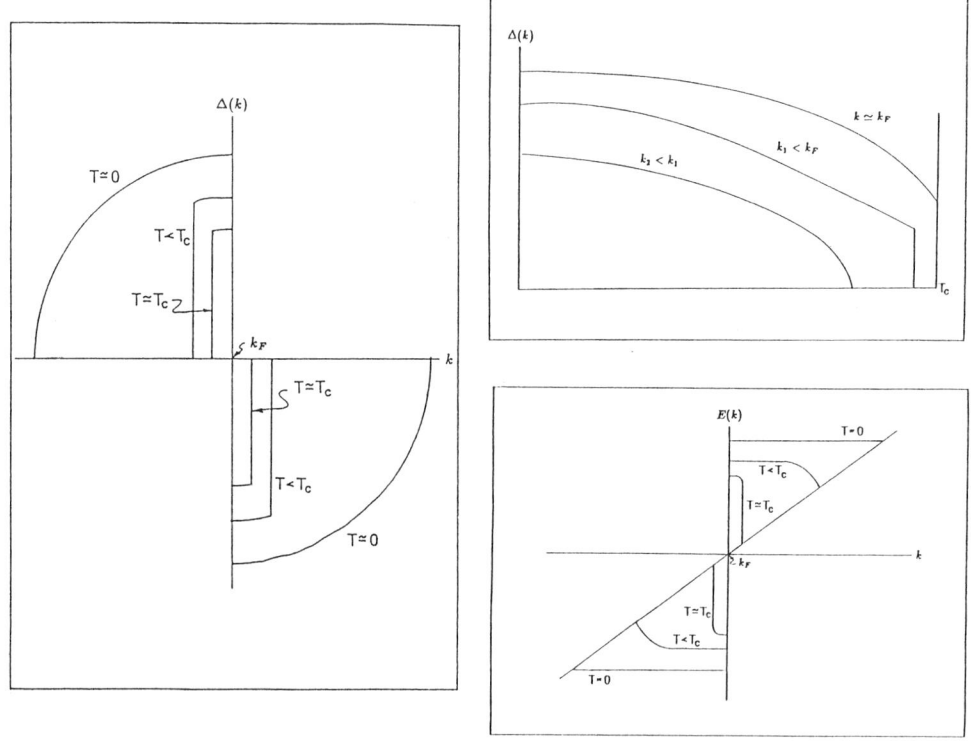

FIGURE 7

ELECTRONIC STRUCTURE, LATTICE DYNAMICS, AND ELECTRON-PHONON INTERACTION IN HIGH T_c SUPERCONDUCTORS

R.E. Cohen

Geophysical Laboratory
Carnegie Institution of Washington
5251 Broad Branch Rd., N.W.
Washington, D.C. 20015

H. Krakauer

Department of Physics
College of William and Mary
Williamsburg VA 23185

W.E. Pickett

Complex Systems Theory Branch
Naval Research Laboratory
Washington, D.C. 20375

ABSTRACT

Density functional calculations within the local density theory (LDA) have successfully predicted and reproduced electronic, structural, and vibrational properties of high T_c superconductors. Successful predictions include Fermi surfaces, phonon frequencies, structural instabilities, etc. If Fermi liquid theory is not applicable to high T_c superconductors, it is difficult to explain why first-principles calculations within the LDA give as good agreement with properties of the metallic high T_c oxides as is found in other metals. This is not to say that the antiferromagnetic insulator relatives of the high T_c superconductors can be quantitatively described within the local spin density approximation (LSDA). It appears, however, that the doped metals are represented well by conventional band theory. Experiments have now demonstrated strong electron-phonon coupling in the high T_c superconductors. The question now is not whether the electron-phonon interaction is strong in the high T_c superconductors, but whether it is strong enough to give T_c's above 100K in itself. The electron-phonon interaction is enhanced in the high T_c oxides by poor screening and ionicity that leads to strong non-local electron-lattice interactions. We also find evidence of strong anharmonicity. Calculations support the possibility that the electron-phonon interaction may be responsible for high T_c.

ELECTRONIC STRUCTURE

One of the most important basic questions about the High T_c superconductors is the appropriate framework to describe the normal state. If the normal state can be described as a Fermi liquid much of the conventional theory of metals can be carried over, many types of experiments can be interpreted in a straightforward way, and many properties can be calculated from first principles using band theoretical methods. On the other hand, if the high T_c oxides are exotic, totally new types of metals, it would appear to be unlikely that many conventional approaches would be at all applicable. Here we address the doped metals that actually become superconducting on cooling, rather than the insulating "parent" phases such as undoped pure La_2CuO_4 which is an antiferromagnetic insulator. The origin of the insulating state, where band structure predicts metallic behavior, is an interesting question in its own right, but there is no a priori reason to suppose that the insulating state is directly related to the superconducting behavior. In fact, the highest T_c in $YBa_2Cu_3O_{7-\delta}$ is two electrons away from the insulating $\delta=1$ parent compound. T_c increases as doping increases and antiferromagnetism decreases which implies that magnetism is not the origin of high T_c, and if anything decreases superconductivity.

The discovery of the 40K $(Ba,K)BiO_3$ superconductor should clarify the likely origin of high T_c. Though $(Ba,K)BiO_3$ has T_c of "only" 40K, it is hard to believe that a different mechanism acts in it than in the cuprates which have T_c's which range from 0K to over 120K. After all, $(Ba,K)BiO_3$ is an oxide which shares many of the features of the higher T_c cuprates, such as a perovskite framework and a nearby insulating parent compound. $(Ba,K)BiO_3$ not only has no copper and no antiferromagnetism, but it is cubic, and not an anisotropic layered structure like most of the cuprates. This suggests that high T_c is not "caused" by two-dimensionality or by antiferromagnetism. There are several conventional explanations for the apparent higher peak T_c's in the cuprates than the bismuthates. For example, the cuprates uniquely have Cu d states in the same energy range as the O 2p states which causes greater hybridization than in the bismuthates. Also, the layered structures may increase T_c in a conventional electron-phonon picture, because it allows the strongest non-local ionic coupling as is described below.

In a way it is fortunate that conventional band theory incorrectly gives a metallic state for the undoped superconductors such as La_2CuO_4. If the calculations were exact, we would have to "dope" the material in our calculations in order to calculate properties of the metals which become superconducting. Fortunately, we are already in the correct phase, and can calculate properties of hypothetical pure metallic La_2CuO_4, for example, and use a rigid band picture to understand the effects of doping on the metals. A reasonable way to consider the effects of doping in the real system is that there is a phase boundary between the insulators and the metals. Disorder and inhomogeneity in the samples, however, smears out the boundary to give complicated behavior between the undoped insulator and the doped metallic phase. This is consistent with Mott's picture of there always being a first-order phase transition between metallic and insulating states.

Fermi Surface

One of the greatest successes of band theory has been the prediction of the Fermi surfaces of the high T_c oxides long before experimental measurements gave convincing evidence that a Fermi surface even existed in the doped high T_c oxides. Only very recently have these experiments begun to be successful, and so far agreement with the band calculations appears to be as good as is found in other, conventional, metals. Several experimental studies appear in this volume. So far it appears that the Fermi surfaces are in essentially the positions predicted by the best calculations. It appears from photoemission studies that the main discrepancies may be effective masses, which in the calculations are about a factor of two smaller than the experimental effective masses. This is consistent with other transition metals, such as Fe. It must also be borne in mind that none of the conventional metals have been studied with the resolution that has been used to study the high T_c oxides, so any revolutionary conclusions based on deviations from the simplest Fermi liquid theory must be withheld until normal metals have been studied with the same resolution. For example, much has been made of the shapes and widths of the photoemission peaks as a function of energy away from E_F. However, the Fermi liquid theory only predicts the variation in lifetime with energy in the neighborhood of E_F. Furthermore, all of these discussions are based on simple single-band Fermi liquids, but all of the high T_c oxides have multiple bands crossing E_F, with touching bands at E_F that will lead to electronic scattering down to zero energy. [In $(La,Sr,Ba)_2CuO_4$ it is necessary to consider the tilts that lead to the orthorhombic structure in order to get multiple bands.] In fact, it is possible that many of the features of "marginal Fermi liquid" theory, such as the Raman background, can be explained by interband scattering.

Most of the interpretations of photoemission data in the high T_c oxides assume strong two dimensionality in the electronic structure. Though the layered oxides are certainly highly anisotropic, band calculations show significant three-dimensional dispersion. Recently, special effort has been taken to accurately calculate the three-dimensional Fermi surface in $YBa_2Cu_3O_7$ (Fig. 1) and it was found that the dispersion along c varied significantly among different parts of the Brillouin zone. Fig. 2 shows the width (in k space) that would arise from three-dimensional dispersion alone. Though the bands would be sharp at some points in the zone, in other places, such as near X, c dispersion should cause significant broadening of photoemission spectra. Thus the photoemission linewidths must be interpreted with much caution, since they are projections along c.

Resistivity

Another issue that has received much attention is the near linear resistivity of many of the cuprate superconductors. This is not as unique as many have suggested. For example, NbO, which is a cubic, low T_c oxide (T_c=1.5 K), has a linear resistivity. $YBa_2Cu_4O_8$, which has T_c up to 90 K when doped, has a nonlinear resistivity with saturation at high temperatures.[1] $YBa_2Cu_4O_8$ has a similar structure to $YBa_2Cu_3O_7$ but has double, rather than single chains and a tetragonal structure with no twinning. The occurrence of a non-linear resistivity in such a similar high T_c oxide demonstrates that the linear resistivity, whatever its origin, is not intrinsic to high T_c. Linear resistivity is obtained in any case for $La_{2-x}Sr_xCuO_4$ and $YBa_2Cu_3O_7$ with a straightforward application of Bloch-Boltzmann theory.[2]

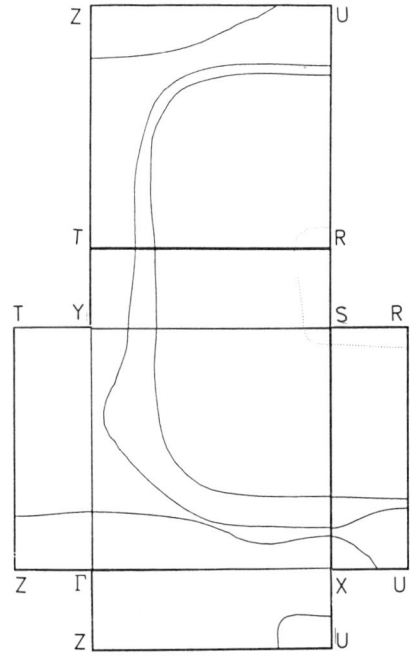

Figure 1. Calculated Fermi surface cross sections in $YBa_2Cu_3O_7$.

Recently it has been suggested that resistivity at constant volume in $YBa_2Cu_3O_7$ is indeed nonlinear and exhibits saturation, whereas the resistivity at constant pressure is linear.[3] This analysis requires a large change in resistivity with volume, which was measured by the same group. If indeed these data are correct, then the linear resistivity at room pressure is just a fluke, and the transport at constant volume (which is what is predicted in various theories) is typical of metals with strong electron-phonon interaction.[4]

Several analyses have been published which suggest that resistivity constrains the electron-phonon interaction to be small because of the absence of saturation at high temperatures.[2,5] These conclusions may have to be modified if the above volume effect on resistivity is verified. Furthermore, the multiple bands and anisotropy of these

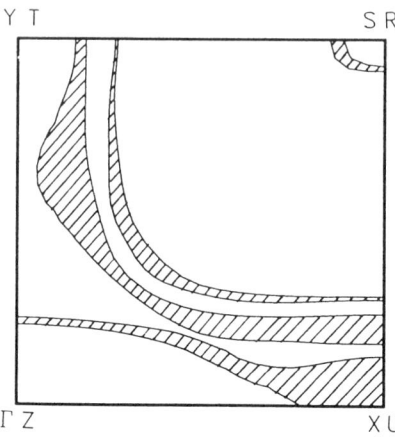

Figure 2. Fermi surface for all k_z projected onto the $k_z=0$ plane in $YBa_2Cu_3O_7$. The shaded regions indicate the dispersion of the Fermi surface along z.

structures needs to be taken into account in order to properly calculate bounds on λ_{tr}. Finally, there is no a priori reason to assume that $\lambda_{tr}=\lambda_{e\text{-}p}$. Recently Mazin and Dolgov[6] have investigated the difference between λ_{tr} and $\lambda_{e\text{-}p}$ within the rigid muffin tin approximation in $YBa_2Cu_3O_7$ and suggest that the difference may be considerable. However, they were not able to estimate the difference quantitatively since the phonons, electronic structure, and electronic phonon coupling need to be known in detail before the differences in λ_{tr} and $\lambda_{e\text{-}p}$ can be calculated explicitly, which is beyond present capabilities. Furthermore, the rigid muffin tin approximation has been demonstrated to be qualitatively incorrect for the high T_c oxides, as is discussed further below.

Another important question is the two-dimensional versus three-dimensional character of the electronic structure. Many high T_c theories are based on and require two-dimensionality. However, though the crystal and electronic structures are quite anisotropic, we believe the materials are three-dimensional metals. One critical property to understand the dimensionality of the metals is the c-axis resistivity. Many experiments show semiconductor c-axis resistivity in the normal state. However, a few experiments show linear c-axis resistivity. It is not clear how often this linear resistivity is observed. DC conductivity measurements require electrons to travel all the way through the sample from one electrode to another. Since stacking faults are common along c, it is easy to understand extrinsic non-metallic resistivity due to these defects.[7] Optical and AC measurements should may help unravel whether the c-axis transport is metallic or not.

In summary, the transport properties are complicated in these materials, and it should not be surprising that they are not completely understood. The complexity in itself is however no justification for considering these materials to be non-Fermi liquids or exotic superconductors.

LATTICE DYNAMICS

Frozen Phonon Calculations

Shortly after the discovery of $La_{2-x}Sr_xCuO_4$,[8] tight-binding calculations which neglected ionicity predicted incorrectly that the breathing mode is unstable, rather than the octahedral tilt that leads to the low-temperature orthorhombic (LTO) structure.[9] This result was used as evidence of the failure of band theory. Simple non-empirical ionic calculations, however, gave the breathing mode to be the highest frequency mode at X, and correctly gave the tilt mode to be unstable.[10] This result was verified by full-potential LAPW calculations, which make no assumptions about ionicity or the charge density response. Not only were qualitatively correct results obtained for the breathing and tilt modes, but calculated phonon frequencies agreed as closely with experiment as is found with any other material.[11] Since then still more modes have been calculated within the harmonic approximation, and are generally in good agreement with experiment, except possibly in the frequency region where an extra modes are observed,[12] probably due to strong anharmonicity (discussed below).

LAPW calculations for all fifteen $YBa_2Cu_3O_7$ Raman modes also show excellent agreement with experiment.[13] The only significant disagreement was for the 440 cm^{-1}, in phase O2-O3 mode, which is found 18% too low. Rodriguez et al.[14] found very similar results using the FLMTO method, but obtained only a 9% error for the 440 cm^{-1} mode by including some anharmonic third order terms in their fit of the energies. This suggests to us that the greater error for this mode is due to anharmonicity.

Strong anharmonicity is not limited to the tilt that leads to the LTO structure in $La_{2-x}(Ba,Sr)_xCuO_4$. Other anharmonic modes include the E_u O sliding mode at Γ,[11] and the B_{2u} sliding mode at X which also involves x-y motions of La and O_z.[15] Experimentally, octahedral tilts have been found to lead to another phase transition in $La_{2-x}Ba_xCuO_4$ from the LTO phase to a low temperature tetragonal (LTT) structure for x~0.12.[16] This transition has strong effects on transport and superconductivity and will be discussed next.

Octahedral Tilts

The phase transition from the high temperature tetragonal (HTT) to LTO structures appears to have little or no effect on transport or superconductivity in $La_{2-x}(Sr,Ba)_xCuO_4$. This is consistent with band structure calculations that show little effect on the states at E_F for the LTO structure compared with HTT.[11] The phase transition from LTO to LTT, however, leads to a change in the resistivity,[17] and most dramatically, a sharp suppression of T_c to 0K.[16] Furthermore, anomalous isotope effects are observed in the region of the phase transition, with α_O greater than 0.5.[18] We have performed a large number of self-consistent calculations using the full-potential LAPW method in order to understand whether band theory makes any predictions that can explain this behavior.

The LTO and LTT structures differ in the directions of the octahedral tilts. Following Axe,[16] we can describe the tilts in the La_2CuO_4 structure in terms of a two-dimensional order parameter $Q=(Q_1,Q_2)$ where $Q=(0,0)$ gives the HTT structure, $Q=(a,0)$ describes the LTO distortion along X, $Q=(0,a)$ describes LTO along X', $Q=(a,a)$ gives the LTT structure, and $Q=(a,b)$ gives a structure between LTT and LTO with symmetry Pccn. We have investigated the potential surface for the tilts for the HTT, LTO, LTT, and Pccn structures. The LTT and Pccn structures have 28 atoms in the primitive unit cell, and we found it necessary to use matrix sizes up to 3300 (RK_{max}=6.5) in order to obtain fully converged total energy differences. These calculations, we believe, are the largest all-electron full potential calculations yet done on any system.

Fig. 3 shows the calculated total energy differences for the various octahedral tilts. These points were fit to tetragonal invariants to 12th order. The curves in Fig. 3 are the fitted function along the directions of the calculated points. The fitted potential surface is shown in Fig. 4. The static potential surface is an octuple well, with minima along the four symmetry related LTO directions, and deeper minima along the four LTT directions. The calculated potential surface is consistent with the LTT phase being the lowest energy, and thus lowest energy, structure. These calculations were done with the lattice parameters of the x=0.12 Ba-doped phase (a=5.362 Å, c=13.24 Å), without any orthorhombic lattice strain for the LTO phase, and without any "virtual crystal" or other type of doping away from the pure La_2CuO_4 composition. Orthorhombic lattice strain would further stabilize the LTO structure and "doping" would increase the stability of the LTT phase, as discussed below. The exact shape and relative depths of the potential wells are likely to be strong functions of volume, as was found in $BaTiO_3$, and the experimental lattice constants are functions of x. Most likely increasing the volume, as occurs with the addition of Ba, increases the relative depth of the LTT wells relative to the LTO wells. Thus qualitatively we can understand the compositional dependence of the LTT phase transition, but it would be very difficult to calculate exactly the phase transition temperature as a function of composition.

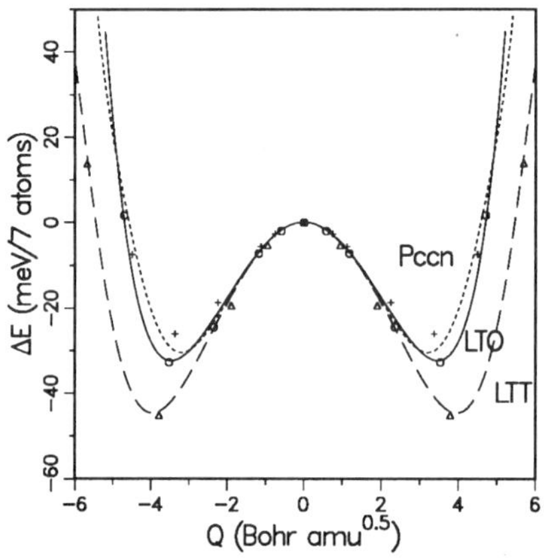

We have studied the effects of the octahedral tilts on the electronic structure, and we find the electronic structure to be a strong function of the LTT tilt. Fig. 5 shows the density of states of LTT versus LTO. There is a large dip in the density of states almost exactly at E_F due to the LTT distortion. This dip is not due to a self-consistency or Peierls type effect because the actual calculations were performed with E_F in the undoped position (E~0.1 eV). Self-consistency

Figure 3. Change in energy versus normal mode amplitude for LTO, LTT, and Pccn distortions. The curves are those given by the equation in the text.

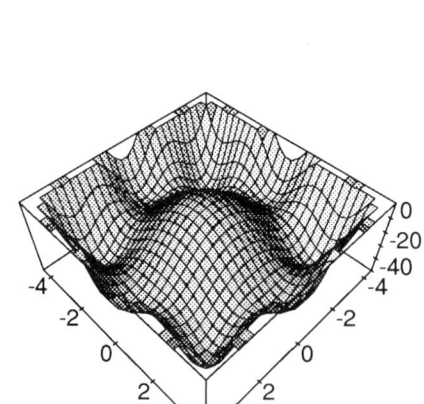

Figure 4. Fitted potential surface for tilts. The axes have the same units as in Fig. 3. The HTT structure is at the center, LTT at the four corner minima, and LTO at the four minima along ±x and ±y.

should further decrease the density of states at E_F if E_F sat directly on the dip. Fig. 6 shows the calculated band structure of the LTT structure. A gap opens up from M to A and a large portion of the Fermi surface is lost. This resulting decrease in the density of states is sufficient to explain the change in resistivity and the drop in T_c associated with the LTT phase transition. The reduction in the density of states is also the right size to account for the reduction in linear specific heat coefficient measured by Okajima et al.[19] for the LTT phase.

The reason such a gap doesn't open up in the LTO structure is that the bands are forced to stick together at E_F because of symmetry. In the LTT structure the folded bands are no longer forced to stick together everywhere. It is possible that in the Pccn phase, a gap opens on even more of the Fermi surface. Although the potential surface does not have a local minimum for Pccn, the fact that tilt librations will affect the bands at E_F implies strong electron-phonon coupling for these anharmonic modes.

The calculated potential surface implies local tilting of the octahedra, even in the high temperature phase. This tilting can change frequencies of other modes through off-diagonal anharmonic coupling. For example, if one regards the tilts as quasi-static, they lower the local symmetry and can split degenerate modes or change mode frequencies. Similar tilts of the Cu-O pyramids may occur in $YBa_2Cu_3O_7$, and may be responsible for the EXAFS results that have been interpreted in terms of double well potentials for the O_4 ion.[20] In fact, a double well of the type proposed in that study is inconsistent with Raman data.[21] O4 displacements in x, y, and z directions are Raman active and have frequencies that are in excellent agreement with LAPW frozen phonon calculations, which show harmonic potential wells for all three O4 symmetry modes at Γ in $YBa_2Cu_3O_7$.[13] Another possible source of the EXAFS data is the buckling of the Cu-O chains predicted by the LAPW computations.

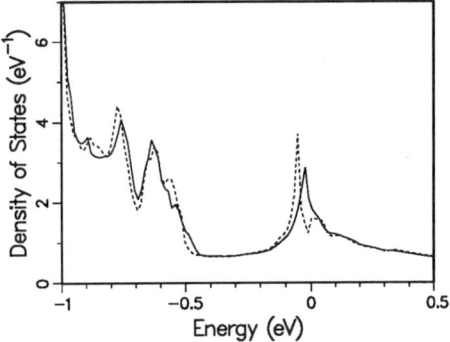

Figure 5. Density of states for in the LTO (solid) and LTT (dashed) structures. The Fermi level (E=0) is for x=0.12 doped. Note the large decrease in the density of states exactly at E_F.

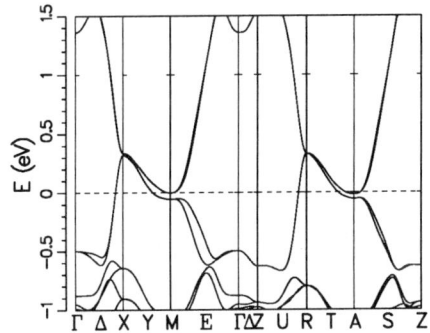

Figure 6. Calculated band structure for $La_{2-x}(Ba,Sr)_xCuO_4$ in the LTT structure. The Fermi level (E=0) is for x=0.12. Note the gap that opens at E_F.

The anharmonicity displayed in Fig. 4 may also lead to the unusual isotope effects near x=0.12[22] observed by Crawford et al.[18] Due to the extreme anharmonicity, the coupling strength λ can become mass dependent because the structure itself becomes mass dependent. If the atomic masses were infinite, or, in other words, in the static limit, only the lowest energy points on the static potential surface would be occupied, which is the LTT structure. At finite temperatures the low oxygen mass allows the octahedra to tilt dynamically. However, the heavier oxygen isotope will prefer the LTT tilt relative to the lighter isotope.

This affect appears to be large only in the immediate region near the LTT phase transition. If 5% more of the O^{18} sample is transformed to LTT relative to the O^{16} sample, it would account for increases of α_O of 0.4. This is large enough to account for the isotope shifts as large as 0.7 observed by Crawford et al.

Electron-phonon interaction

Superconductivity in all conventional superconductors is generally believed to be a result of the electron-phonon interaction. This belief derives from many observations, including the presence of an isotope effect in many (but not all) superconductors, the observation of phonon anomalies that correlate with T_c, the similarity of tunneling structure with phonon density of states, the similarity of λ_{e-p} with λ_{tr}, the general success of BCS theory and its strong coupling extensions in Eliashberg theory, and the success in calculating superconducting properties from first principles for many simple superconductors based on the electron-phonon interaction. There is no maximum T_c in the conventional strong coupling theory, so high T_c in itself is no reason to discount the electron-phonon interaction as the cause of high T_c.

Indeed, many experiments have indicated strong electron-phonon coupling in the high T_c oxides. Toby et al.[23] have found a local structural rearrangement at T_c in $Tl_2Ba_2CaCu_2O_8$. Such a structural change accompanying the onset of superconductivity is a completely new one, and implies unusually strong coupling between the carriers and the distortion. Raman[24,25] and IR[26] data indicate frequency and linewidth shifts below T_c that indicate strong coupling. Fano antiresonances in the Raman spectra also indicate strong coupling between phonons and electronic states.[27] Femtosecond spectroscopy gives an estimate of $\lambda <\omega^2>$[28] and Brorson et al.[29] find $\lambda=0.9$ for $YBa_2Cu_3O_7$ if $<\omega^2>^{1/2}=23$ meV=267 K is assumed. This λ in itself cannot account for $T_c>90$ K, but it is still fairly large. The current experimental status is that though there is much evidence that there is a strong electron-phonon interaction in the high T_c superconductors, no evidence is currently available that suggests a λ between 2 and 3, which is necessary in order to reach T_c's above 90 K,[30] other than possibly T_c itself.

Calculations that assume only a local electron-phonon interaction, which often works well for conventional superconductors, cannot give T_c of greater than 90K in the high T_c oxides. We would like to understand if there are reasons to suspect that the usual approximations are not applicable to the high T_c oxides and to investigate whether such high T_c can possibly be understood within an electron-phonon framework. We have concluded that the conventional approximations, such as the rigid muffin tin approximation (RMTA), are not applicable, and that an electron-phonon interaction mechanism is plausible.[13,31]

Most conventional superconductors have a high density of states at E_F, so that the electron-phonon interaction is local; when an atom is displaced, only electrons near the displaced atom feel the change in potential since the potential change is strongly screened by the rest of the carriers. In the high T_c oxides this is not true for three reasons: Firstly, the density of states is much lower, than, say in Nb or other conventional high T_c materials. Secondly, the structures are generally anisotropic and are such that the carriers are concentrated in only part of the structure. Thus screening is quite low between the metallic layers, for instance. Thirdly, the chemistry is such that the atoms are more properly thought of as charged ions, such as La^{3+}, O^{2-}, etc. These three facts lead to strong non-local electron-phonon coupling, where carriers interact with atomic motions at other sites. This effect has been demonstrated in first-principles frozen phonon calculations for both La_2CuO_4[11,31] and $YBa_2Cu_3O_7$.[13] For example, if the Ba ion is displaced along c in $YBa_2Cu_3O_7$, it changes the potential in the Cu-O planes and chains, and thus has a significant electron-phonon coupling strength. In the so-called rigid muffin tin approximation (RMTA) or other local approximations the Ba motion would not contribute at all to superconductivity, since there are few carriers at the Ba site.

Ideally, we would like to able to calculate, without approximations, the electron-phonon coupling for enough phonons to accurately obtain the $\alpha^2F(\omega)$ electron-phonon spectral function, calculate T_c, and compare with experiment to "prove" an electron-phonon mechanism for $YBa_2Cu_3O_7$ or $La_{2-x}(Ba,Sr)_xCuO_4$. Even if this program were currently possible, a host of difficulties would arise. For example, the coulomb repulsion parameter μ^* may be quite different in the high T_c oxides than in simpler materials where it is believed to be about 0.1. Also, there is no rigorous theory for how to treat the strong anharmonicity in the high T_c oxides, such as the unstable tilts in $La_{2-x}(Ba,Sr)_xCuO_4$ and the unstable chain mode in $YBa_2Cu_3O_7$.[13] The problem of anharmonicity is even more severe than it appears on first consideration, because the anharmonic modes in general interact with all other modes in the crystal. An example of this was demonstrated recently by Lichtenstein et al.[32] for $BaBiO_3$ where it was shown that the octahedral tilt strongly interacts with the strongly coupling oxygen breathing mode. This is important because the tilt mode is not believed to interact strongly with carrier states at E_F whereas the breathing mode does.

In spite of the technical and theoretical difficulties in calculating T_c from first principles, we have estimated the electron-phonon coupling strength λ_v for a number of modes in $La_{2-x}(Ba,Sr)_xCuO_4$ and $YBa_2Cu_3O_7$ in order to ascertain whether there is any possibility of obtaining electron-phonon coupling strengths between 2-3 that are required to achieve T_c's of greater than 90K. In other words, we are investigating the theoretical plausibility of the electron-phonon interaction being responsible for high T_c. These calculations are also a strong test of band theory in general and the LDA in particular since the results are sensitive to details in the electronic structure at E_F as well as the lattice dynamics and energetics.

The electron-phonon coupling strength has been estimated for the 15 branches that are Raman active at Γ in $YBa_2Cu_3O_7$.[13] Several assumptions were made. It was assumed that the electron-phonon matrix element for each branch is constant and that the bands are dispersionless. The latter approximation is relatively minor, but the former approximation probably severely underestimates the coupling strengths of the B_{2g} and B_{3g} modes. The deformation potentials were calculated for all fifteen zone center modes. The results, expressed as $39\lambda_v$ are 1.9, 1.7, 1.0, 0.8 and 0.2 for the A_g modes at calculated frequencies of 105, 127, 312, 361, and 513 cm^{-1}. Strictly speaking, these probably cannot be compared directly with Raman linewidths since these results represent approximations to averages over each phonon branch. By studying the temperature dependencies of the linewidths of Raman active modes in $YBa_2Cu_3O_7$, Friedl et al.[33] estimate $39\lambda_v$ to be 0.39 for the 500 cm^{-1} mode. Note that both the estimated and measured linewidths for the 500 cm^{-1} mode are quite high for such a high frequency mode, and that it is a quantity between $\omega\lambda$ and $\omega^2\lambda$ that gives the contribution of a mode to T_c.

Very recently we have calculated $\lambda_v(X)$ for the four fully symmetric X-point modes (those with the symmetry of the planar breathing mode) in $La_{2-x}(Ba,Sr)_xCuO_4$.[34] Unlike the calculations for $YBa_2Cu_3O_7$ discussed above, the calculations for La_2CuO_4 make no approximations other than the LDA and are highly converged with respect to k-point sampling. The electron-phonon matrix elements were calculated directly by integrating the eigenfunctions at the Fermi level with the changes in self-consistent potential with displacement.[35] Table I shows the results for the symmetry modes and eigenmodes. The coupling strengths at the X-point are huge. This is because the phase space for scattering at X, given by $\xi(q) = \Sigma \, \delta(E_k - E_F)\delta(E_{k+q}-E_F)$ is very large due to the strong nesting of the Fermi surface.[36] If we assume constant matrix elements, the average coupling would be $\xi(q)/N(E_F)^2 = 7.8$ times smaller, giving

Table I. Calculated electron-phonon coupling constants and linewidths for fully symmetric X-point modes in La_2CuO_4.

Eigenmodes:				Symmetry Modes:			
ω (cm^{-1})	$\lambda(q,v)$	$2!*\lambda$	γ/ω	ω (cm^{-1})	$\lambda(q,v)$	$2!*\lambda$	γ/ω
156	0.531	11.144	0.0	155	0.003	0.056	0.00034
299	0.269	5.648	0.03	339	0.769	16.151	0.1063
475	0.025	0.515	0.00	404	0.014	0.297	0.0025
731	0.278	5.841	0.08	609	0.238	4.998	0.0593
	sum	avg			sum	avg	
	1.102	5.787			1.024	5.375	

an average over the four branches of $2!*\lambda=0.74$ which is in the correct ball park for explaining T_c in $La_{2-x}(Ba,Sr)_xCuO_4$.

We conclude that the electron-phonon coupling strength is undoubtedly strong for several modes in $YBa_2Cu_3O_7$ and $La_{2-x}(Ba,Sr)_xCuO_4$, and thus that an electron-phonon mechanism for high T_c is plausible. On the other hand, there are some non-conventional alternatives that are also consistent with our calculations. For example, we find some electronic states with are much more strongly coupled than others with given modes, and the modes at X in $La_{2-x}(Ba,Sr)_xCuO_4$ are so strongly coupled that it is possible that strong-coupling theory breaks down.

Another possibility is that anharmonicity enhances T_c. One model for anharmonic enhancement is based essentially on the divergence in the (vibrational) susceptibility in a quantum double well potential due to tunneling or "two-level-states."[37,38] However, higher order terms must drastically reduce the susceptibility in a double well, because in the limit of separated wells the susceptibility should be that of near-harmonic wells expanded around the well minima, rather than that of a double well expanded around the maximum. Nevertheless, the presence of multiple well potentials leads to larger than normal r.m.s. displacements which may enhance T_c somewhat. Unfortunately, there is no rigorous non-perturbative procedure for estimating the effects of strong anharmonicity, and perturbation theory is not applicable to strong anharmonicity in multiple well potentials. Even the perturbative approach is extremely complicated and has not been tractable even for simpler superconductors.[39] However, the electron-lattice system with carriers strongly coupled to a mode involved in a phase transition is an extremely complex system, and it will be some time before it is clear to what extent anharmonicity can enhance T_c. Flach et al.[40] recently presented a self-consistent phonon analysis of the effects of double well potentials and phase transitions on T_c, and find in their model an enhancement in T_c when the structural phase transition coincides with the onset of superconductivity. This should be a fruitful area to explore given the observations of atomic displacement changes at T_c.[23]

CONCLUSIONS

In spite of early intimations to the contrary, band theory has surpassed every expectation in predicting electronic and energetic properties of the doped high T_c oxides. Nevertheless, an understanding of the doped insulators is of interest in its own right, and an understanding of the complicated combination of magnetic, dynamical, and compositional properties in intermediate phases between the undoped "parent compounds" and the doped superconductors is probably one of the most difficult problems in physics. We have chosen to concentrate on the well-doped superconductors, and have found that conventional band theory, when applied with sufficiently accurate methods, does as good a job for the doped high T_c superconductors as for any other material.

We find that the high temperature tetragonal structure of $La_{2-x}(Ba,Sr)_xCuO_4$ is unstable with respect to octahedral tilts in any direction. Furthermore, tilts in certain directions, such as the LTT tilts, strongly influence the carrier states at the Fermi level, where as tilts in other directions, such as the LTO tilts, do not affect the states at E_F. The unusual isotope effects observed in $La_{2-x}(Ba,Sr)_xCuO_4$ can be understood in terms of the calculated potential surface for octahedral tilts.

Finally, we find evidence for very large electron-phonon interactions in $La_{2-x}(Ba,Sr)_xCuO_4$ and $YBa_2Cu_3O_7$. Whether the electron-phonon interaction is strong enough to be completely responsible for high T_c remains to be seen. In fact, it is possible that coupling to between certain modes and certain electron states that the ordinary strong coupling theory breaks down.

ACKNOWLEDGEMENTS

We would like to thank P.B. Allen, L.L. Boyer, B.M. Klein, M.V. Klein, I. Mazin, D. Papaconstantopoulos, D. Singh, and B. Szpunar for helpful discussion. Calculations were performed on the Cray 2 at the National Center for Supercomputing Applications (REC) and on the IBM 3090 at the Cornell National Supercomputing Facility (HK). The work was supported in part by National Science Foundation Grant No. DMR-87-19535 to HK.

REFERENCES

1. S. Martin et al., Phys. Rev. B 39:9611 (1989).
2. P. B. Allen, W. E. Pickett, and H. Krakauer, Phys. Rev. B 37:7482 (1988).
3. B. Sundqvist and B. M. Andersson, Solid State Commun. 76:1019 (1990).
4. Z. Fisk and G. W. Webb, Phys. Rev. Lett. 36:1084 (1976).
5. M. Gurvitch and A. T. Fiory, in *Novel Mechanisms of Superconductivity*, S. A. Wolf and V.Z. Kresin, eds. Plenum, New York (1987).
6. I. I. Mazin and O. V. Dolgov, On the estimation of the electron-phonon coupling from the resistivity, unpublished.
7. S. Martin, A. T. Fiory, R. M. Fleming, L. F. Schneemeyer, and J. V. Waszczak, Phys. Rev. Lett. 60:2194 (1988).
8. J. G. Bednorz and K. A. Mueller, Z. Phys. 64:189 (1986).
9. W. Weber, Phys. Rev. Lett. 58:1371 (1987).
10. R. E. Cohen, W. E. Pickett, L. L. Boyer, and H. Krakauer, Phys. Rev. Lett. 60:817 (1988).
11. R. E. Cohen, W. E. Pickett, and H. Krakauer, Phys. Rev. Lett. 62:831 (1989).
12. W. Reichardt et al., Proc. Intl. Conf. on the Manifestations of Electron-Phonon Interaction in Cu-O Superconductors, Oaxtepec, Mexico, Dec. 1990.
13. R. E. Cohen, W. E. Pickett, and H. Krakauer, Phys. Rev. Lett. 64:2575 (1990).
14. Rodriguez et al., Phys. Rev. B 42:2692 (1990).
15. W. E. Pickett, R. E. Cohen, and H. Krakauer, in: "Proc. Intl. Symp. of High T_c Superconductors," Dubna, USSR (1989).

16. J. D. Axe et al., Phys. Rev. Lett. 62:2751 (1989).
17. Sera et al., Solid State Commun. 69:851 (1989).
18. M. K. Crawford et al., Phys. Rev. B 41:282 (1990).
19. Y. Okajima et al., Physica B 165&166:1349 (1990).
20. J. Mustre de Leon, S. D. Conradson, I. Batistic, and A. R. Bishop, Phys. Rev. Lett. 65:1675 (1990).
21. K. F. McCarty, J. Z. Liu, R. N. Shelton, and H. B. Radousky, Phys. Rev. B 41:8792 (1990).
22. W. E. Pickettt, R. E. Cohen, and H. Krakauer, Lattice instabilities, isotope effect, and high T_c superconductivity in $La_{2-x}Ba_xCuO_4$, unpublished.
23. B. H. Toby, T. Egami, J. D. Jorgensen, and M. A. Subramanian, Phys. Rev. Lett. 64:2414 (1990).
24. B. Friedl, C. Thomsen, and M. Cardona, Phys. Rev. Lett. 65:915 (1990).
25. K. F. McCarty, H. B. Radousky, J. Z. Liu, and R. N. Shelton, Temperature dependence of the linewidths of the Raman-active phonons of $YBa_2Cu_3O_7$: Evidence for a superconducting gap between 440 and 500 cm^{-1}, unpublished.
26. B. Guttler et al., J. Phys.: Condens. Matter 2:8977 (1990).
27. M. V. Klein, Physica C 162-164:1701 (1988).
28. P. B. Allen, Phys. Rev. Lett. 59:1460 (1987).
29. Brorson et al., Solid State Commun. 74:1305 (1990).
30. D. A. Papaconstantopoulos, J. Superconduc. 2:317 (1989).
31. R. E. Cohen, W. E. Pickett, H. Krakauer, and D. A. Papaconstantopoulos, Phase Transitions 22:167 (1990).
32. A. I. Lichtenstein et al., Structural phase diagram and electron-phonon interaction in $Ba_{1-x}K_xBiO_3$, unpublished.
33. B. Friedl, C. Thomsen, E. Schonherr, and M. Cardona, Solid State Comm. 76:1107 (1990).
34. H. Krakauer, W. E. Pickett, and R. E. Cohen, unpublished.
35. H. Krakauer, R. E. Cohen, and W. E. Pickett, Mat. Res. Soc. Symp. Proc. 141:165 (1989).
36. W. E. Pickett, H. Krakauer, and R. E. Cohen, Physica B 165&166:1055 (1990).
37. J. R. Hardy and J. W. Flocken, Phys. Rev. Lett. 60:2191 (1988).
38. N. M. Plakida, V. L. Aksenov, and S. L. Dreschler, Europhys. Lett. 4:1309 (1987).
39. A. E. Karakozov and E. G. Maksimov, Sov. Phys. JETP 47:358 (1978).
40. S. Flach, P. Hartwich, and J. Schreiber, Solid State Comm. 75:647 (1990).

MARGINAL FERMI-LIQUID THEORY OF THE HIGH T_c MATERIALS

C. M. Varma

AT&T Bell Laboratories
600 Mountain avenue
P.O. Box 636
Murray Hill, NJ 07974

INTRODUCTION

My reading of the experimental data in high T_c materials is that the normal state is not a Landau Fermi-liquid, but the superconductive state, despite some of its peculiarities, is of the generic BCS-s state variety. The anomalous nature of the normal state has been extensively described. I agree with Anderson and others in the general view point that the key to the problem is the normal state. The superconductive state (SS) and its peculiarities should follow. The peculiarities of the SS, I have in mind are primarily connected with the disrespect of BCS coherence factors in various properties (no peak in T_1^{-1} near T_c but a peak in electromagnetic absorption etc.), possibly large ratio of $2\Delta(0)/T_c$ etc.

In an effort to constrain the fruitful theoretical approaches to the high T_c problem, my colleagues and I advanced a phenomenological hypothesis[1] for the excitation spectrum in the normal state of high T_c superconductors. The hypothesis is that of the "marginal Fermi liquid" (MFL) phenomenology over a wide range of momenta the charge (ρ) and spin (σ) polarizabilities have a low energy scale given by the temperature of measurement rather than any parameter in the Hamiltonian:

$$\mathrm{Im}\tilde{P}_{\rho,\sigma}(q,\omega) \sim \begin{cases} -N(0)\,\omega/T, & \omega \ll T \\ -N(0), & \omega_c \gg \omega \gg T. \end{cases} \tag{1}$$

Here ω_c is a cut-off energy. Equation (1) implies a contribution,

$$\mathrm{Re}\,\tilde{P}(q,\omega) \sim N(0) \ln \frac{\omega_c}{T}\,; \quad \omega \leq T. \tag{2}$$

As a result of Eq. (1), the frequency dependence of the electron self-energy is given by

$$\Sigma(\omega) = \lambda\omega \left(\log \frac{x}{\omega_c} + i \frac{\pi}{2} x\,\text{sgn}\,\omega\right), \qquad (3)$$

where $x = \max(|\omega|, T)$ and λ is a coupling constant. In turn, this leads to a quasiparticle spectral weight, z, which depends logarithmically on energy or temperature, namely

$$1/z = (1 + \lambda \ln \frac{y}{\omega_c}), \qquad (4)$$

where $y = \max[(E_k - \mu), T]$ with E_k the real part of the quasiparticle energy. It is the weakest imaginable way in which the quasiparticle concept of the conventional Fermi liquid theory could be violated (suggesting the term "marginal" in MFL).

The MFL hypothesis reconciles the experimental observation of a Fermi-surface similar to the band-structure calculations with the anomalous scattering observed in transport measurements. Angle resolved photoemission results[2] have been found consistent with the prediction of Eq. (3). Equations (1) in the charge channel provides a retarded attractive interaction in the Cooper channel and has been used to calculate several properties of the superconductive state. Recent neutron scattering results[3] have been observed to be directly consistent with Eq. (1) in the spin-channel.

In Eq. (1), we have emphasized the frequency dependence; any momentum dependence of $P_{\rho,\sigma}$ with a frequency dependence still of the form (1) over most of the range of q-space leaves most of the phenomenological conclusions unaltered. Equation (1) cannot be correct at long wavelengths $v_F q \ll \omega$; elsewhere we have presented the phenomenological expression which properly obeys the hydrodynamic requirements.

Physical Content of the Marginal Fermi-liquid Hypothesis

The high density of low energy excitations represented by Eq. (1) is accompanied by another experimental fact:[4] the chemical potential μ in the metallic phase for electron density less than 1 (per CuO_2 cell) or more than 1 appears close to the middle of the insulating gap. This indicates a resonance which draws its spectral weight from states below and states above the insulating gap on the passage from the insulator to the metal.

We find[5] that in certain highly correlated models of relevance to the Cu-O materials the breakdown of a Landau Fermi-liquid to a MFL state occurs through a divergence in certain low energy effective interactions, which self-consistently pin μ to be near the middle of the gap of the insulating phase. The singularity in Eqs. (1) and (2) mirrors this effective interaction.

Our physical idea is the following: At 1/2 filling, the Cu-O materials are antiferromagnetic insulators with a gap for one particle (charge) *incoherent* excitations on Cu of order U, and (particle-hole) spin and change transfer excitations (these will both be referred to as excitons) of much lower energy. The latter are excitations between the lower hybrid Cu-O states and the upper (Hubbard) set of states. For large enough interactions and electron density not too far from 1/2-filling, a "ghost of the upper Hubbard band" still remains in the metallic paramagnetic state, and should be thought of the incoherent part of the single particle spectrum, which now also have a low energy "coherent" part. For appropriate strongly correlated models the excitons between the ghost and the low energy single particle states bind to the low energy one particle excitations leading to an almost dispersionless three-body resonance at the self-consistently determined chemical potential. Such resonances lead to singular two-body vertices for low energy properties which are essential for a breakdown of Landau theory.

Ruckenstein and I have discussed the detailed theory based on these ideas elsewhere. Let me briefly discuss here the choice of the model for discussing Cu-O materials.

Model for Cu-O Materials

The Cu-O materials at 1/2 filling are antiferromagnetic insulators well describable by a one band repulsive Hubbard model. The insulating gap is due to charge transfer excitation $C^{++}O^{--} \rightarrow Cu^+O^-$, so the parameters in this model are highly renormalized parameters dependent on Cu-O hopping energy, on-site repulsions and quite importantly the Madelung energy of these highly ionic solids. Given this Hubbard model, the experimental properties at 1/2 filling are such that no new physics is required to understand them. The physics is simply, that of a Néel state with zero-point spin-wave fluctuations. No unusual quantum fluctuations are involved.

An important question is — Does it follow from the fact that the 1/2 filled case is well described by a one-band Hubbard model that the metallic state is also described by a one-band Hubbard model. Physical arguments were given that the Cu-O materials have special features in their solid state chemistry that a three-band model with on-site and intra-site interactions (to represent the Madelung energy) is in general the minimum necessary model.[6] This model reduces to the one-band Hubbard model at 1/2 filling. The only way to be sure if it can or can not be so reduced in the metallic phase is to do calculations in the more complicated model and to see if it has instabilities other than the one band Hubbard model. Two kinds of calculations have been done — generalized RPA calculations and slave Boson calculations. They both give, unlike the one band Hubbard model, s-wave superconductivity in an appropriate range of parameters, electron concentration etc., besides other new features. This is not the right place to describe at any length the extra physics in the three-band Hubbard model. Briefly, "Cu^{++}" and "O^{--}" in the solid state have their ionization and affinity energies within the hopping energy between them. Low energy states in the metal cause fluctuation of charge both on Cu and on oxygen. The Madelung energy term then begins to play a dynamical role mixing charge transfer resonances in the low energy excitations. These lead to an effective s-wave attraction.

There is a rather pointless discussion going about as to whether a one-band or a three-band model is necessary and sufficient for the description of the crucial physics. In band structure calculations one finds only a single band within about 1 eV of the chemical potential of the hybrid Cu-O variety. So a one-band model would appear enough to discuss the low energy physics. The question is, what are the interaction between the states of this band. In calculating these interactions in a strong-coupling situation, one may have essential contributions from intermediate states, which belong to other bands. One may then work either with multiple bands in the model Hamiltonian or eliminate all except one in favor of new effective interactions in the Hilbert space kept. The essential message of well defined calculations is that such new interactions are relevant, the one band Hubbard model is inadequate.

Outline of the Microscopic Calculation

Ruckenstein and I see this in our calculations as well. Our calculations proceed as follows. We divide the self-consistent single particle Green's function into two parts:

$$G = G^L + G^H .$$

G^L describes the low energy physics and G^H represents the incoherent, nearly momentum independent features away from the chemical potential, i.e. the ghost bands. We write the exact low energy scattering vertex Γ in terms of a vertex Γ^{LH} which is irreducible with respect to $G^L G^L$. Γ^{LH} is in turn expressed in terms of Γ^{HH}, a vertex which is irreducible with respect to both $G^L G^L$ and $G^L G^H$. The large on-site Cu interaction U is renormalized in Γ^{HH} so that Γ^{LH} is determined by the exchange energy J and the Cu-O Coulomb interactions V.

The determination of Γ^{LH} has strong similarities to the problem of threshold singularities. For V of order the charge transfer gap Γ^{LH} has an excitonic singularity $(\omega - \omega_{ex}) - (1 - \delta^2/\pi^2)$, with $\delta \to 0$ for large V/E_G. The low energy vertex Γ and the low energy self-energy $\Sigma(\omega)$ are next calculated. For the indicated V, the self-consistent chemical potential is pinned to ω_{ex} within energies of O(kT). The behavior of the polarizabilities in that case is of the marginal Fermi-liquid form. An important point in the analysis is that in the pinned regime, antiferromagnetic interactions among the Cu-moments and the conduction electrons disappears.

In conclusion, in an appropriate model and in the strong-coupling limit, the transfer of spectral weight from the incoherent bands to those near the chemical potential, in the passage from an insulator to a metal, occurs via a singular resonance near the chemical potential. This violates a fundamental condition for the validity of Landau Fermi-liquid theory. Marginal behavior then ensues. It ought to be remembered however that marginal behavior is in turn unstable at low temperatures either to superconductivity or localization due to disorder.

REFERENCES

1. C. M. Varma, P. B. Littlewood, S. Schmitt-Rink, E. Abrahams and A. E. Ruckenstein, Phys. Rev. Letters 63, 1996 (1989); G. Kotliar et. al. Europhysis Letters (1991) to be published.

2. C. G. Olson et. al., Science 45, 731 (1989); T. Takahashi et. al., Phys. Rev. B39, 6636 (1989).

3. S. M. Holden, G. Aeppli, H. Mool, D. Rytz, M. F. Huntley and Z. Fisk, Phys. Rev. Letters, 66, 821 (1991).

4. Fukuda et. al., Solid State Comm. 72, 1183 (1989); J. W. Allen et. al., Phys. Rev. Letters, 64, 595 (1990).

5. A. E. Ruckenstein and C. M. Varma, (preprint).

6. C. M. Varma, S. Schmitt-Rink and E. Abrahams, Solid State Comm. 62, 681 (1987).

RECENT PROGRESS IN SEMIONIC PAIRING THEORY OF HIGH TEMPERATURE SUPERCONDUCTIVITY

Z. Zou

Department of Physics
Stanford University
Stanford, CA 94305

and

Lawrence Livermore National Laboratory
P.O. Box 808
Livermore, CA 94550

1. Introduction

It is a pleasure to speak to you about some recent theoretical developments of semionic pairing mechanism of high-temperature superconductivity. The concept of fractional statistics has become familiar to condensed matter physicists since the discovery of the fractional quantum Hall effect. I would like to remind you of two facts-that particles obeying fractional statistics exist in nature and that 1/2 fractional statistics has the ability to cause superconductivity. The first one has its precedent in the fractional quantum Hall effect and the second one was established by the work of Laughlin et al.[1,2] The more difficult question of whether fractional statistics accounts for superconductivity in the cuprite oxides is, however, unclear. Substantial progress towards answering that question has been made. This includes the establishment of complete equivalence of the original Baskaran-Zou-Anderson[3,4] Gutzwiller projection and gauge field[5,6] approach to the Kalmeyer-Laughlin[7] quantum Hall analogy. We have shown[8,9,10] that Anderson's procedure[3] of Gutzwiller projecting a particle-hole pair makes excitations equivalent to the quantum Hall quasiparticles. These holons have surprisingly high mobility and thus form a good variational basis set for the t-J Hamiltonian. It has also been shown that the commensurate flux state, first suggested by Anderson,[11] maps exactly onto the chiral spin-liquid state of Wen, Wilczek and Zee[12] with a macroscopic number of holons. A consequence of this is that the work of Liang and Trivedi[13] amounts to a variational study of semion gas on a lattice, which shows that the chiral spin-liquid state can be energetically competitive. My plan of this talk is to review these developments and to derive an effective Hamiltonian for the t-J model in the long wavelength limit. I'll also speculate on the direction of possible future work.

2. Superconductivity due to Semionic Pairing

A gas of semions may be described as a gas of spinless fermions whose equation of motion is governed by the Hamiltonian

$$\mathcal{H} = \sum_j \frac{1}{2m}(\vec{P}_j + \vec{A}_j)^2 , \qquad (1)$$

where

$$\vec{A}_j(\vec{r}_j) = \frac{1}{2}\sum_{i\neq j}\frac{\hat{z}\times(\vec{r}_i - \vec{r}_j)}{|\vec{r}_i - \vec{r}_j|^2} . \qquad (2)$$

This problem was studied by Hanna, Fetter and Laughlin[2] who found that its ground state superconducts. Several authors[14] have later confirmed this conclusion. Thus, the ability of semionic pairing to cause superconductivity is established as a matter of principle. It still remains to understand how and why semions described by Eq.(1) and (2) would occur in real materials. The rest of my talk will be focused on this issue. I will derive an effective Hamiltonian similar to Eq.(1) starting from the t-J Hamiltonian. This can be done in a particular saddle point approximation and in the long wavelength limit.

3. The Chiral Spin-Liquid State

Particles obeying 1/2 fractional statistics can be constructed from the chiral spin-liquid state of the form[8,9]

$$|\Psi\rangle = P_G |\chi\rangle \qquad (3)$$

where $P_G = \prod_j (1 - n_{j\uparrow}n_{j\downarrow})$ is the Gutzwiller[18] projection operator, and $|\chi\rangle$ is a single Slater determinant formed by completely filling a magnetic band of flux π per plaquette. We have specifically

$$|\chi\rangle = \prod_{q,\sigma} c^\dagger_{q\sigma}|0\rangle \qquad (4)$$

where $c^\dagger_{q\sigma}$ creates an electron of spin σ in the state

$$\phi_q(\vec{r}_j) = e^{i\vec{q}\cdot\vec{r}_j}\{2E_q[E_q - \cos q_x]\}^{-1/2}$$
$$\times \begin{pmatrix} \cos q_x - E_q; & \text{j even} \\ -\cos q_y + im\sin q_x \sin q_y; & \text{j odd} \end{pmatrix} \qquad (5)$$

"Even" and "odd" in this expression refer to the two different sites in the magnetic unit cell, as shown in Fig. 1, m is a gap parameter characterizing the amount of chirality, and

$$E_q = -\sqrt{\cos^2 q_x + \cos^2 q_y + m^2 \sin^2 q_x \sin^2 q_y}\,. \tag{6}$$

The band structure is plotted in Fig. 2. The properties of this class of wave functions and the reasoning leading to them were discussed at length elsewhere.[9] The key properties are that the state has a finite spin-spin correlation length which depends on m, that this length diverges as $m \to 0$, and that the spin-spin exchange energy as $m \to 0$ compares[15] favorably with that of the exact ground state, which is thought to be ordered antiferromagnetically. One of this class of states is, in fact, equal to the Kalmeyer-Laughlin quantum Hall ground state. This class of states violate both the parity and time reversal symmetries. The order parameter characterizing the broken symmetries is the expectation value of the chiral operator $\langle \vec{S}_i \cdot \vec{S}_j \times \vec{S}_k \rangle$, where sites i, j and k form a triangle.

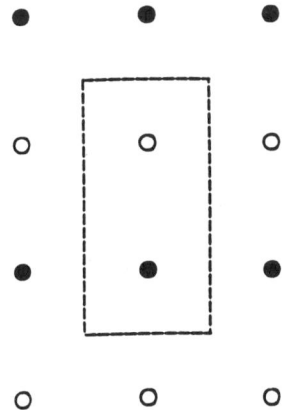

FIG. 1. Illustration of magnetic unit cell. "Even" and "odd" sites are depicted as empty and solid circles, respectively.

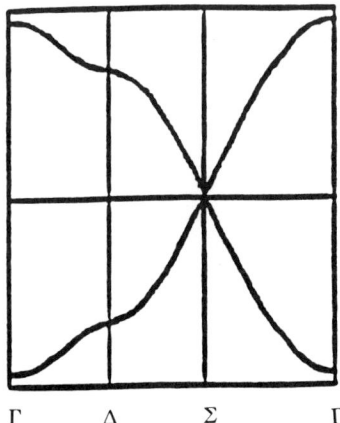

FIG. 2. Band Structure defined by Eq.(6).

4. Spinons and Holons

The spin excitation of the the chiral spin liquid state is a neutral spin 1/2 particle, a spinon, which may be created by Anderson's procedure[3] of Gutzwiller projecting a particle-hole pair. That is, we shall represent the spinons of spin σ and σ' at site j_A and j_B by the wavefunction

$$P_G\, c^\dagger_{j_A\sigma} c_{j_B\bar{\sigma}'} |\chi\rangle = P_G\, c_{j_A\bar{\sigma}} c^\dagger_{j_B\sigma'} |\chi\rangle \,, \tag{7}$$

where $\bar{\sigma}$ stands for the opposite of σ. In published work,[8,9] we have shown that these two expressions are equal and are equivalent to the fractional quantum Hall quasiparticle wavefunctions.

The charged excitations, holons, can be made by removing the electron at the spinon center

$$|j_A, j_B\rangle = c_{j_A\sigma} c_{j_B\sigma'} P_G\, c^\dagger_{j_A\sigma} c_{j_B\bar{\sigma}'} |\chi\rangle \,, \tag{8}$$

regardless of the choice of σ and σ'. Note that this wave function is different from the one obtained by removing two electrons from the magnetic band before projecting. To verify that the holons created above obey 1/2 fractional statistics, we calculate the phase

$$\theta = \mathrm{Im}\, \ln[\langle j'_A, j_B | \mathcal{H}_{tJ} | j_A, j_B\rangle] \,, \tag{9}$$

where $\mathcal{H}_{tJ}$ is the t-J Hamiltonian.[3] The sum of the phases over a path of hops encircling the stationary holon equals $\pm\pi$. This is expected to be exact only in the limit that the two holons are far apart. However, the tendency to equal to π should be present even in small encircling paths, particularly whem m is large. We have calculated[8] the total phase along the sequence of near-neighbor hops

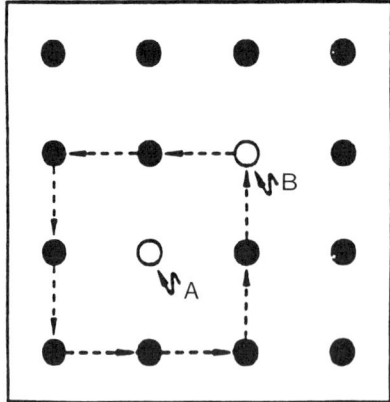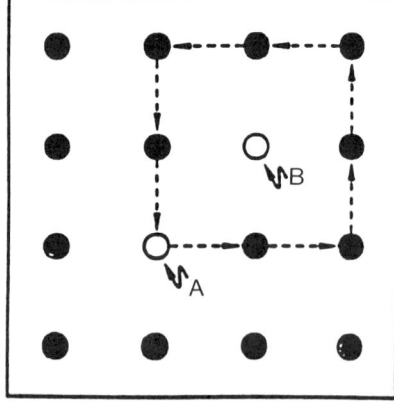

FIG. 3. Sequence of near-neighbor hops on the 4X4 torus used in computing the phases defined by Eq.(9). The empty circles represent the holon being moved and the holon being encircled.

on a 4×4 torus as shown in Fig. 3. For the typical case of $m = 1$, we found that the total phase is roughly 3.66, a number amazingly close to π, considering the proximity of the two holons and the small sample size. We found the total phase to be the same for all values of m studied, no matter how small, and to change sign abruptly as m changes sign. It follows that the fractional statistics at large separation must be exact. There is no other way the total phase could so closely approximate π and be so insensitive to $|m|$. One can also show the fractional statistics of holons by means of analytic continuation methods. This is possible because one of the holon states is simply the Kalmeyer-Laughlin holon wave function which exhibits 1/2 fractional statistics exactly. So long as the spin gap remains, it is always possible to evolve the holon wave function (8) into the Kalmeyer-Laughlin holon wave function adiabatically.

The minimum kinetic energy of an isolated holon is also calculated[8] using the wavefunction (8). We found $T_{\text{holon}} = -2.3|t|$, which is considerably more negative than the kinetic energy of an ordinary hole. A large value of the holon kinetic energy not only supports the idea that holons can stabilize the chiral spin-liquid state, it indicates that decay of a holon into a spinon and an ordinary hole is energetically unfavorable, and thus that our specific choice of holon wavefunction is a good one. The conclusion we draw from the study of isolated holons is that the holons probably have the capacity to stabilize the chiral spin state but that a macroscopic number of them is required to do so. Therefore, the quantum mechanics of a gas of holons obeying 1/2 fractional statistics becomes the key problem.

5. Multi-holon Wave Functions

Anderson's procedure of making holons described above can, in principle, be repeated to generate multi-holon wave functions. This turns out to be completely equivalent[16] to the Gutzwiller projected commensurate flux state. Let us denote the coordinates of M holons by a set of complex variables $(h_1, \ldots, h_M)$, the coordinates of N up-spin electrons by $(z_1, \ldots, z_N)$ and the coordinates of N down-spin electrons by $(\eta_1, \ldots, \eta_N)$. Since every lattice site is occupied by either a spin or a holon, $M + 2N = N_s$ is the total number of sites. The multi-holon basis set is of the form

$$|h_1, \ldots, h_M\rangle = P_G |\chi\rangle , \qquad (10)$$

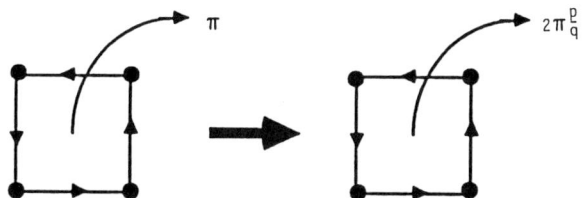

FIG. 4. The key idea in constructing multi-holon basis set. The flux is renormalized to be commensurate with the density of electrons.

where $|\chi\rangle$ is a Slater determinant of filled magnetic bands. The magnetic bands are solutions of the Hofstadter[17] Hamiltonian, a tight-binding Hamiltonian with a uniform magnetic field. The magnetic flux per plaquette is commensurate with the density of electrons; i.e. as shown in Fig. 4, the flux per plaquette is $2\pi N/N_s = 2\pi p/q$, where p, q are relative primes. The well known results for the Hofstadter model relevant to our discussions are that there are q distinct magnetic bands, that each band contains N_s/q states, and that the lower p bands are separated by a large gap from the rest of the spectrum. Evidently, the lower p bands can just accommodate the $2N$ electrons. The Slater determinant of the p filled bands is taken to be the amplitude of an up and down spin arrangement for a particular holon configuration $(h_1, \ldots, h_M)$. The basis set (10) explicitly violates the time reversal symmetry for $p/q \neq 1/2$.

The fractional statistics of the holon basis set manifest itself as an extra phase in the holon hopping integral

$$\theta = \text{Im } \ln[\langle h'_1, \ldots h_M | \mathcal{H}_{tJ} | h_1, \ldots, h_M \rangle] \; , \tag{11}$$

the sum of which over a path of hops encircling one stationary holon equals $\pm \pi$. We've done monte carlo calculations,[16] using the above wave functions, on a 6×6 lattice for $p/q = 1/3$. The total phase around a loop varies typically from 2.4 to 2.8, very close to π. In addition to the numerical evidence, we have also demonstrated the 1/2 statistics by means of analytic continuation methods. Since the analytic proof involving the theta functions is rather complicated, I refer you to the original paper[16] for details.

In summary, the multi-holon basis set consists of spin singlet wave functions because they are Slater determinants and the Gutzwiller projector commutes with the spin operator. The basis set exhibits 1/2 fractional statistics exactly. Although they have excellent spin-spin exchange energy[19,20] due to the large energy gap of the Hofstadter spectrum, they do not automatically give rise to good kinetic energy. Thus, a formidable task is to find the optimum combination of the basis set[16]

$$|\Psi_h\rangle = \sum_{h_1,\ldots,h_M} a_{h_1,\ldots,h_M} |h_1, \ldots, h_M\rangle \; , \tag{12}$$

satisfying the variational condition

$$\delta \left[\frac{\langle \Psi_h | \mathcal{H}_{tJ} | \Psi_h \rangle}{\langle \Psi_h | \Psi_h \rangle} \right] = 0 \; . \tag{13}$$

We note that this variational problem is essentially equivalent to the semion gas problem defined in (1) and (2) on a lattice, since it amounts to solving Schrodinger's equation with the fractional statistics basis.

We've solved the variational equation (13) in the Hartree-Fock approximation, and we found that the holon kinetic energy is much too high as compared to other variational states, indicating that Hartree-Fock is a bad approximation for the holons on the lattice. Fortunately, an upper bound for the ground state energy of the t-J Hamiltonian within our basis set can be obtained by examining the work of Liang and Trivedi.[13] This upper bound suggests that the macroscopically doped chiral spin-liquid state can be energetically competitive.[16] Specifically, at small dopings (< 10%), the chiral spin state is stable against the Gutzwiller projected free fermi sea for J/t as small as 0.2. It is important to recognize that our simple basis set does not have sufficient degrees of freedom in the large doping limit for small J/t.

6. Effective Action

I would like to use the remaining time to discuss the fractional statistics from the point of view of a gauge theory.[5,6] I shall review the gauge theory for the flux saddle point in accordance with the variational basis set. A Chern-Simons term in the effective action will be derived[16] based on a pertubation theory. The coefficient of the Chern-Simons term is exactly quantized[21,16] because the vacuum state, consisting of p filled Hofstadter bands, possesses a quantized Hall conductance, ensuring the exactness of the 1/2 fractional statistics of the p/q basis functions. Presently, the validity of neither the flux saddle point nor the perturbation theory is justified, because the gauge theory is strongly coupled. Nevertheless, the following exercise serves the purpose of relating the variational description to the gauge theory and serves as a starting point for future work.

The t-J Hamiltonian may be represented in the slave boson formalism in which we write[22] $c_{i\sigma} = b_i^\dagger f_{i\sigma}$, where the f's and b's are fermi and bose operators, respectively. The t-J Hamiltonian can thus be written

$$\mathcal{H}_{tJ} = -t \sum_{\langle jk \rangle \sigma} b_j^\dagger b_k f_{k\sigma}^\dagger f_{j\sigma} - \frac{J}{2} \sum_{\langle jk \rangle \sigma\sigma'} f_{j\sigma}^\dagger f_{k\sigma} f_{k\sigma'}^\dagger f_{j\sigma'} - \frac{J}{4} \sum_{\langle jk \rangle} b_j^\dagger b_j b_k^\dagger b_k , \quad (14)$$

where we have omitted constant terms, and the bose and fermi fields must satisfy the local constraint $\sum_\sigma f_{i\sigma}^\dagger f_{i\sigma} + b_i^\dagger b_i = 1$. Rearranging terms in (14), we have

$$\mathcal{H}_{tJ} = -\frac{J}{2} \sum_{\langle jk \rangle} \left(\sum_\sigma f_{j\sigma}^\dagger f_{k\sigma} + \frac{t}{J} b_j^\dagger b_k \right) \left(\sum_\sigma f_{k\sigma}^\dagger f_{j\sigma} + \frac{t}{J} b_k^\dagger b_j \right)$$
$$+ \frac{t^2}{2J} \sum_{\langle jk \rangle} (b_j^\dagger b_k)(b_k^\dagger b_j) - \frac{J}{4} \sum_{\langle jk \rangle} b_j^\dagger b_j b_k^\dagger b_k . \quad (15)$$

The corresponding Lagrangian is

$$\mathcal{L}(\tau) = \sum_i b_i^\dagger \partial_\tau b_i + \sum_{i\sigma} f_{i\sigma}^\dagger \partial_\tau f_{i\sigma}$$
$$+ i \sum_i \lambda_i \left(\sum_\sigma f_{i\sigma}^\dagger f_{i\sigma} + b_i^\dagger b_i - 1 \right) + \mathcal{H}_{tJ}(\tau) , \quad (16)$$

where τ is the Euclidean time variable and λ_i is a Lagrange multiplier, integration over which in a path integral implements the constraint exactly. We note that $\mathcal{L}(\tau)$ is invariant under a $U(1)$ gauge transformation:[5] $f_{i\sigma} \to f_{i\sigma} e^{i\theta_i(\tau)}$, $b_i \to$

$b_i e^{i\theta_i(\tau)}$, and $\lambda_i \to \lambda_i - \partial_\tau \theta_i(\tau)$. Thus, the Lagrange multiplier λ_i acts like the temporal component of a vector potential. We can replace $\lambda_i(\tau)$ with $A_{0i}(\tau)$. $\mathcal{L}(\tau)$ may be linearized by a Hubbard-Stratonovich transformation:

$$\mathcal{L}(\tau) = \sum_i b_i^\dagger \partial_\tau b_i + \sum_{i\sigma} f_{i\sigma}^\dagger \partial_\tau f_{i\sigma} + \frac{J}{2} \sum_{jk} |\chi_{jk}|^2$$

$$+ \sum_i i A_{0i} \left(\sum_\sigma f_{i\sigma}^\dagger f_{i\sigma} + b_i^\dagger b_i - 1 \right) + \frac{t^2}{2J} \sum_{jk} (b_j^\dagger b_k)(b_k^\dagger b_j) \quad (17)$$

$$- t \sum_{jk} \chi_{jk} b_j^\dagger b_k - J \sum_{jk} \chi_{jk} f_{j\sigma}^\dagger f_{k\sigma} \ ,$$

where we have introduced a complex Hubbard-Stratonovich field

$$\chi_{jk} = \langle \sum_\sigma f_{j\sigma}^\dagger f_{k\sigma} + \frac{t}{J} b_j^\dagger b_k \rangle = |\chi| e^{i\mathcal{A}_{jk}} \quad (18)$$

Under the gauge transformation, the linearized Lagrangian remains invariant, provided that the phases of χ transform accoording to $\mathcal{A}_{ij} \to \mathcal{A}_{ij} - (\theta_i - \theta_j)$.

Our p/q basis set corresponds to a saddle point solution to (17) in which the Hubbard-Stratonovich field $\chi = |\chi| \exp i\mathcal{A}_{ij}$ is chosen to be constant in amplitude and its phase is $\mathcal{A}_{ij} = \int_i^j d\vec{s} \cdot \vec{A}(s)$, with the total flux enclosed per plaquette $2\pi \frac{p}{q}$. Thus, if we ignore the fluctuations of the amplitude $|\chi|$, $\mathcal{L}(\tau)$ in (17) is a U(1) gauge theory in the saddle point approximation. The bosons, the holon degrees of freedom, interact with the fermions, the spinon degrees of freedom, via the U(1) gauge field. The gauge theory is strongly coupled because the usual "electromagnetic" energy density term is absent from the Lagrangian, which implies that the gauge potential can have arbitrary fluctuations with little energy cost, at least in the zeroth order. The fluctuations of the gauge field will polarize the vacuum, affecting the excitation spectrum. To see this, we perform a perturbation calculation[16] around the saddle point and derive a long wavelength effective action for the bosons. As remarked before, the validity of the pertubation theory is not clear because of the strong coupling nature of the gauge field. Still, we wish to point out the consequences of the perturbation theory. Since the mean field fermion spectrum has a large energy gap, we may integrate out the fermion degrees of freedom.[16] We expand the hopping phases $e^{i\mathcal{A}_{jk}}$ in the long wavelength limit by $e^{i\mathcal{A}_{jk}} \to 1 + i\mathcal{A}_{jk} - \mathcal{A}^2/2$. The part of the effective action depending on A_μ is

$$\frac{1}{N_s} S_f[A] = \frac{1}{2\beta} \int \frac{d\omega}{2\pi} \sum_k A_\mu(\vec{k}, \omega) \Pi^{(f)}_{\mu\nu}(\vec{k}, \omega) A_\nu(-\vec{k}, \omega) \ , \quad (19)$$

where $\beta = 1/T$ is the inverse of the temperature, ω is the frequency variable, $A_\mu(\vec{k}, \omega)$ is the Fourier component of the vector potential A_μ, and $\Pi^{(f)}_{\mu\nu}(\vec{k}, \omega)$ can be calculated[16] from the fermion vacuum polarization and the diamgnetic response depicted in Fig. 5. We know that the p filled Hofstadter bands have an integer Hall conductance[23] σ_H whose value is determined by solving the Diophantine equation[23]

$$\sigma_H = 1 + s\frac{q}{p} \quad \text{with} \quad |\sigma_H| \leq \frac{q}{2} \tag{20}$$

where s is an integer. Therefore $\sigma_H = 1$ is the unique solution. Thus, the effective action after integrating out the fermions must contain a Chern-Simons term whose coefficient is proportional to the integer Hall conductance. The effective Lagrangian to the linear order in ω and $\vec{q}$ is thus

$$\delta\mathcal{L} = \frac{i}{2\pi}\varepsilon_{\alpha\beta\gamma} A_\alpha(q_\beta A_\gamma - q_\gamma A_\beta) \;. \tag{21}$$

This is simply the momentum space representation of the Chern-Simons term

$$\delta\mathcal{L} = \frac{1}{2\pi}\varepsilon_{\alpha\beta\gamma} A_\alpha F_{\beta\gamma} \;, \tag{22}$$

where $F_{\beta\gamma} = \partial_\beta A_\gamma - \partial_\gamma A_\beta$ is the field strength of the gauge field.

For small dopings, the effective mass approximation is appropriate for the bosons. Let us denote the boson field by $\psi(\vec{r})$ in the long wavelength limit and the effective mass of boson by m^*. Neglecting the short range holon-holon interactions, the effective Lagrangian for the bosons and the gauge field is then given by,

$$\mathcal{L}_{\text{eff}}(\tau) = \psi^\dagger(\partial_0 + iA_0)\psi + \frac{1}{2m^*}\psi^\dagger(\vec{\partial} + i\vec{A})^2\psi \pm \frac{1}{2\pi}\varepsilon_{\mu\nu\gamma}A_\mu F_{\nu\gamma} + \ldots, \tag{23}$$

where the "$\pm$" signs correspond to the two degenerate vacua (time-reversal partners), respectively. The omitted terms contain higher orders of derivatives. The fractional statistics of the ψ-field, the holons, follows from the coupling to the Chern-Simons term.[6] The fractional statistics is exact bacause the coefficient of the Chern-Simons term is a topological quantity, the Hall conductance of the p filled magnetic bands. The effective action (23) is the field theoretical equivalence of the semion gas problem defined by Eq.(1) and (2).

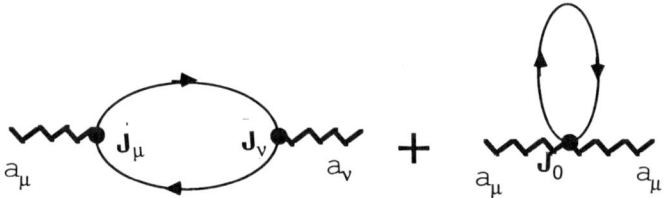

FIG. 5. The vacuum polarization and diamagnetic response of spinons. A Chern-Simons term arises in the long wavelength limit.

7. CONCLUSIONS

The key question of whether mobile carriers have the capacity to stabilize the chiral spin liquid state in the context of the t-J Hamiltonian and thus account for high temperature superconductivity is only partially answered. The variational calculations show that the macroscopically doped chiral spin liquid state is energetically competitive, especially for small dopings or large J/t. While we recognize that the matter of stability of the chiral spin-liquid state cannot be resolved variationally, it is significant that two very different lines of reasoning, the Baskaran-Zou-Anderson analogy with the fermi sea plus the gauge field and the Kalmeyer-Laughlin analogy with the fractional quantum Hall state, have now led to the same conclusions. The next step is to do realistic calculations for real materials and compare them with experiments. Reliable experimental predictions must be made so that experimentalists are able to provide a final resolution of the high-T_c problem. The gauge theory approach seems, in my opinion, a promising way to accomplish that goal.

8. ACKNOWLEDGMENTS

Much of the work presented here was done in collaboration with R.B. Laughlin and J.L. Levy. I would also like to thank S.D. Liang and S.B. Libby for many useful discussions. This work was supported by the National Science Foundation under Grant No. DMR-88-16217. Additional support was provided by the U.S. Department of Energy through the Lawrence Livermore National Laboratory under contract No. W-7405-Eng-48.

REFERENCES

1. R.B. Laughlin, Phys. Rev. Lett. 60: 2677(1988); Science, 242: 525(1988).
2. A.L. Fetter, C.B. Hanna and R.B. Laughlin, Phys. Rev. B 39: 9679(1989); *ibid*, 40: 8745(1989).
3. P.W. Anderson, Science 235: 1196(1987); and in: "Frontiers and borderlines of many-particle physics", edited by J.R. Schrieffer and R.A. Broglia, North Holland, Amsterdam 1988.
4. G. Baskaran, Z. Zou and Anderson, Solid State Comm. 63: 973(1987).
5. G. Baskaran and P.W. Anderson, Phys. Rev. B 37: 580(1988).
6. Z. Zou, Phys. Rev. B 40: 2262(1989).
7. V. Kalmeyer and R.B. Laughlin, Phys. Rev. Lett. 59: 2095(1987); Phys. Rev. B 39: 11879 (1989).
8. Z. Zou and R.B. Laughlin, Phys. Rev. B 42: 4073(1990).
9. R.B. Laughlin and Z. Zou, Phys. Rev. B 41: 664(1990).
10. Z. Zou, B. Doucot and B.S. Shastry, Phys. Rev. B 39: 11424(1989).
11. P.W. Anderson, B.S. Shastry and D. Hristopulos, Phys. Rev. B 40: 8939(1989).

12. X.G. Wen, F. Wilczek, and A.Zee, Phys. Rev. B 39: 11413(1989).
13. S.D. Liang and N. Trivedi, Phys. Rev. Lett. 64: 232(1990).
14. Y.H. Chen, F. Wilczek, E. Witten and B.I. Halperin, Int. J. Mod. Phys. B 3: 1001(1989).
15. I. Affleck and B. Marston, Phys. Rev. B37: 3774(1988).
16. Z.Zou, J.L. Levy, and R.B. Laughlin, Stanford preprint (1991).
17. D.R. Hofstadter, Phys. Rev. B 14: 2239(1976).
18. M.C. Gutzwiller, Phys. Rev. Lett. 10: 159(1963); W.F. Brinkman and T.M. Rice, Phys. Rev. B 2: 4302(1970).
19. Y. Hasegawa, P. Lederer, T.M. Rice and P. Wiegman, Phys. Rev. Lett. 63: 907(1989); P.Lederer, D. Poiblanc and T.M. Rice, Phys. Rev. Lett. 63: 1519(1989).
20. B. Doucot, M. Ogata and T.M. Rice, Preprint 1990
21. Y. Hasegana, O.Narikiyo, K. Kuboki, and H. Fukuyama, J. Phys. Soc. Jpn, 59: 822(1990). Z. Zou, unpublished.
22. Z. Zou and P.W. Anderson, Phys. Rev. B 37: 627(1988).
23. D. Thouless, in: "The quantized Hall effect", edited by R.E. Prange and S.M. Girvin, Springer-Verlag (1990).

MEAN FIELD ANALYSIS OF THE CUO₂ LATTICE:

COMPARISON TO MONTE CARLO SIMULATIONS

A. J. Fedro[*,#], Yu Zhou[*], T. C. Leung[§] and B. N. Harmon[§]

[*] Materials Science Division, Argonne National Laboratory
[#] Department of Physics, Northern Illinois University
[§] Ames Laboratory - USDOE and Iowa State University

INTRODUCTION

The 2D multi-band Hubbard models have been attracting much attention recently since they provide one of the simplest models for the high T_c superconductors where one assumes the relevant motion is confined to the Cu-O planes. Recent Monte Carlo (MC) studies[1,2], using the three-band Hubbard model to describe this motion, have been done in a wide parameter range for various values of doping δ away from half-filling ($\delta = 0$ is defined as one hole per Cu site). We summarize their results as follows: We define the on-site O energy, ε_p, and the Cu on-site energy, ε_d. Then there are two basic regimes depending on whether the on-site energy difference ε between the O and Cu sites ($\varepsilon = \varepsilon_p - \varepsilon_d$) is greater or less than the on-site Cu Coulomb repulsion U_{dd}. If $U_{dd} \gg \varepsilon$ the behavior of the system is controlled by ε (charge transfer limit). In this case, at half-filling, they find strong antiferrromagnetic correlations and evidence of a charge transfer gap ($\approx \varepsilon$). The antiferromagnetic correlations decrease rapidly as one dopes away from half-filling. In the other case $\varepsilon \gg U_{dd}$ the behavior is controlled by U_{dd}. Here the O occupation is always small and we essentially have an effective single band model with a Mott-Hubbard gap which depends on U_{dd}.

In this short paper we outline a mean field (MF) calculation of the multi-band Hubbard model of the CuO₂ lattice based on a projection operator scheme[3,4]. We then make a detailed comparison of our MF results to the above-mentioned MC simulations which we will regard as "exact". In this approach, the two usual Hubbard projection operators are defined (involving creation/destruction of a fermion at site j with spin σ either in the presence or absence of the fermion at the same site j with spin $-\sigma$), and the model is solved with an equation of motion method for these operators using a well-defined truncation procedure which treats the complicated statistics of these operators properly and reduces to well-established decoupling in the case of pure fermions. In this way we generate an exact Dyson-like equation for the Green's functions where the self energy is found to contain both static and dynamic contributions. We will ignore the effects due to the dynamic terms (in this context termed memory functions) in this paper thus defining a MF set of equations for the needed Green's functions. These equations are then solved self-consistently for all band fillings and for possible antiferromagnetic (AF) order. We point out that this formulation deals with the physical fermion directly. There is no need to separate the spin and charge degrees of freedom as is done in slave theories and thus no need for additional constraint equations which make the number of coupled MF equations unwieldy. Also it has been shown by Ruckenstein and Schmitt-Rink[4] that, in the large spin degeneracy (large N) limit, the present approach and the slave boson theory are identical. We finally note that this method is essentially equivalent to the method of irreducible Green's functions[5] which has also been used recently in the studies of the single-band Hubbard and t-J models. The paper is organized as follows: In Section II, our MF formalism is outlined. Section III contains the results of our calculations and the comparisons to the MC simulations. The conclusions are presented in Section IV.

MEAN FIELD FORMALISM

In this Section we present the results of a projection operator method for determining the needed Green's functions. The derivation will be given elsewhere[6]. We are interested in the general Hamiltonian used in the theory of the 2D CuO_2 lattice which can be written in *hole notation* as follows[7,8]:

$$H = \varepsilon_d \sum_{j,\sigma} n_{j,\sigma} + U_{dd} \sum_{j} n_{j,+} n_{j,-} + \varepsilon_p \sum_{j,\lambda,\sigma} n_{j+\lambda/2,\sigma} + U_{pp} \sum_{j,\lambda} n_{j+\lambda/2,+} n_{j+\lambda/2,-}$$

$$+ U_{dp} \sum_{j,\alpha,\lambda,\sigma,\sigma'} n_{j,\sigma} n_{j+\alpha\lambda/2,\sigma'} - t_{dp} \sum_{j,\alpha,\sigma} [d^\dagger_{j\sigma} (p_{j+\alpha x/2,\sigma} - p_{j+\alpha y/2,\sigma}) + H.c.]$$

$$+ t_{pp} \sum_{j,\sigma} [(p^\dagger_{j+x/2,\sigma} - p^\dagger_{j-x/2,\sigma})(p_{j+y/2,\sigma} - p_{j-y/2,\sigma}) + H.c.] \quad (1)$$

where $\lambda = x,y$ and $\alpha = \pm$. Here the Cu d-hole operator for site j and spin σ is given by $d_{j\sigma}$ with the corresponding number operator $n_{j\sigma} = d^+_{j\sigma} d_{j\sigma}$. The oxygen p-hole operators surrounding the j-th Cu site are defined by $p_{j+\alpha\lambda/2,\sigma}$ with the corresponding number operators given by $n_{j+\alpha\lambda/2,\sigma} = p^+_{j+\alpha\lambda/2,\sigma} p_{j+\alpha\lambda/2,\sigma}$. ε_p and ε_d are the on-site O and Cu energies respectively, U_{pp} and U_{dd} the corresponding on-site Coulomb repulsions, and U_{dp} the near neighbor Cu-O repulsion. t_{dp} is the near neighbor Cu-O hopping matrix element and t_{pp} the O-O near neighbor hopping matrix element. The orbital sign convention is such that the LDA values for both t_{dp} and $t_{pp} > 0$. We define the space lattice to be that of the Cu's, i.e., a 2D square lattice of spacing a. The four surrounding oxygens are then at a distance a/2 from the central Cu atom. The LDA numbers for the parameters are given by[9]

$$U_{dd} = 10.5 \text{ eV} \quad ; \quad U_{pp} = 4.0 \text{ eV} \quad ; \quad U_{dp} = 1.2 \text{ eV}$$
$$t_{dp} = 1.3 \text{ eV} \quad ; \quad t_{pp} = 0.65 \text{ eV} \quad ; \quad \varepsilon_p - \varepsilon_d = 3.6 \text{ eV} \quad (2)$$

The available MC results at the present time set $U_{pp} = U_{pd} = 0$ as well as $t_{pp} = 0$, and the parameters used in the MC calculations are not the above LDA parameters. Since we would like to include the effects due to finite t_{pp} in our discussion, the relevant multi-band Hamiltonian in Eqn. (1) we need for the MC comparison has the form

$$H = \sum_{j,\nu,\sigma} \varepsilon_\nu n_{j\nu\sigma} + \sum_{j,\nu} U_{\nu\nu} n_{j\nu,+} n_{j\nu,-} + \sum_{j,j',\nu,\nu',\sigma} t_{j,\nu,j',\nu'} c^\dagger_{j\nu\sigma} c_{j'\nu'\sigma} \quad (3)$$

where j denotes the unit cell position, and ν the atoms in the unit cell. In this framework we can easily incorporate possible long range antiferromagnetic (AF) order. The hopping matrix elements are real and satisfy $t_{j\nu;j'\nu'} = t_{j'\nu';j\nu}$ with $t_{j\nu;j\nu} = 0$. The site number operators are defined by $n_{j\nu\sigma} = c^+_{j\nu\sigma} c_{j\nu\sigma}$. ε_ν are the single level energies and $U_{\nu\nu}$ the on-site repulsions.

We now separate the pure fermion excitation operators $c_{jv\sigma}$ into two operators (as done in the original work of Hubbard) $f^\alpha_{jv\sigma}$, with $\alpha = \pm$, defined as following:

$$f^\alpha_{jv\sigma} = n^\alpha_{jv,-\sigma} c_{jv\sigma} \quad ; \quad c_{jv\sigma} = \sum_\alpha f^\alpha_{jv\sigma} \quad \text{etc...} \tag{4a}$$

where we define

$$n^\alpha_{jv,-\sigma} = \begin{cases} n_{jv-\sigma} & \alpha = + \\ 1 - n_{jv-\sigma} & \alpha = - \end{cases} \tag{4b}$$

We then form the needed retarded Green's functions $G^{\alpha\alpha'}_{jv;j'v'\sigma}(t)$ as follows:

$$G^{\alpha\alpha'}_{jv;j'v'\sigma}(t) = -i < [f^\alpha_{jv\sigma}(t), f^{\alpha' \dagger}_{j'v'\sigma}]_+ > \quad ; \quad t \geq 0 \tag{5}$$

where the $f^\alpha_{jv\sigma}(t)$ are the ordinary Heisenberg operators

$$f^\alpha_{jv\sigma}(t) = e^{iHt} f^\alpha_{jv\sigma} e^{-iHt} \quad ; \quad f^\alpha_{jv\sigma}(0) = f^\alpha_{jv\sigma} \tag{6}$$

To solve for these Green's functions we introduce the projection operator $P^2 = P$ as follows:

$$PX = \sum_{j'',v'',\alpha'',\sigma''} f^{\alpha'' \dagger}_{j''v''\sigma''} <[f^{\alpha''}_{j''v''\sigma''}, X]_+> / <n^{\alpha''}_{j''v'',-\sigma''}> \tag{7}$$

Since the projections defined in Eqns. (7) involve the averages $<n^{\alpha''}_{j''v'',-\sigma''}>$, it is convenient to write the equations of motion for the rescaled Green's functions defined as follows:

$$\widetilde{G}^{\alpha\alpha'}_{jv;j'v'\sigma}(t) = (<n^\alpha_{jv,-\sigma}><n^{\alpha'}_{j'v',-\sigma}>)^{-1/2} G^{\alpha\alpha'}_{jv;j'v'\sigma}(t) \tag{8}$$

Use of Eqns. (5) to (8), we find, in standard Mori fashion[6], the exact equations for the Laplace and spatial transform of the Green's functions in Eqn. (8), $\widetilde{G}^{\alpha\alpha'}_{kvv'\sigma}(\varpi)$, the following:

$$\sum_{\alpha''v''} \{ \varpi \delta_{vv''}\delta_{\alpha\alpha''} - E^{\alpha\alpha''}_{kvv''\sigma}(\varpi) \} \widetilde{G}^{\alpha''\alpha'}_{kv''v'\sigma}(\varpi) = \delta_{vv'}\delta_{\alpha\alpha'} \tag{9a}$$

with

$$E^{\alpha\alpha''}_{kvv''\sigma}(\varpi) = \delta_{vv''}\delta_{\alpha\alpha''}\varepsilon^\alpha_v + t_{kvv''}(<n^\alpha_{v,-\sigma}><n^{\alpha''}_{v'',-\sigma}>)^{1/2}$$

$$+ \alpha\alpha'' [\Delta_{kvv''\sigma} + M_{kvv''\sigma}(\varpi)] (<n^\alpha_{v,-\sigma}><n^{\alpha''}_{v'',-\sigma}>)^{-1/2} \tag{9b}$$

where we assume that the "n" averages are independent of the unit cell and

$$\varepsilon_v^\alpha = \varepsilon_v + U_{vv}\delta_{\alpha,+} \tag{10}$$

The static mean field corrections $\Delta_{kvv''\sigma}$ are the spatial transforms of

$$\Delta_{jv;\,j''v''\sigma} = \langle[f_{jv\sigma}^-, L_t(f_{j''v''\sigma''}^{-\dagger})]_+\rangle - t_{jv;\,j''v''}\langle n_{v,-\sigma}^-\rangle\langle n_{v'',-\sigma}^-\rangle \tag{11}$$

where the Liouville operator $L = [H,]_-$ and L_t is defined as that part of the commutator with respect to H of Eqn. (3) which is proportional to the $t_{jv;\,j''v''}$'s. Finally, the dynamic memory functions $M_{kvv''\sigma}(\varpi)$ are the Laplace and spatial transforms of

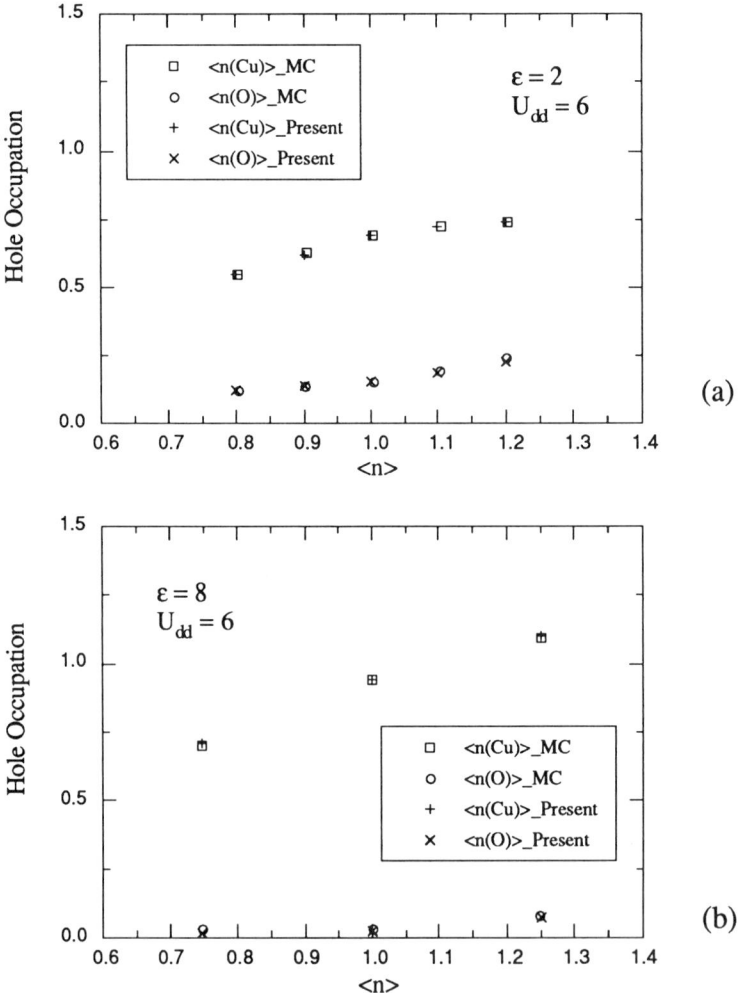

Fig.1 Hole occupation of the Cu and O sites versus doping <n> for two different regimes: (a) Charge transfer limit and (b) Mott-Hubbard limit. Here $t_{pp} = 0$ and the inverse temperature $\beta = 8$. Excellent agreement with MC results are seen.

$$M_{jv;\,j''v''\sigma}(t) = -i \langle [f^-_{jv\sigma}, L\exp[-it(1-P)L](1-P)L(f^{-\dagger}_{j''v''\sigma})]_+ \rangle \qquad (12)$$

In the energy matrix of Eqn. (9b), if we set both the $\Delta_{kvv''\sigma}$'s and $M_{kvv''\sigma}(\varpi)$'s equal to zero, we have the Hubbard I solution. Notice that the Hubbard I solution misses both these static and dynamic terms. This is due to the naive truncation which essentially treats the "f" operators as if they were pure fermions. If one truncates the equations of motion properly by using the the projection operator given in Eqn. (7), one automatically handles the complicated statistics of the $f^\alpha_{jv\sigma}$'s correctly, and obtains the result given in Eqns. (9). Now our mean field (MF) solution is that generated by setting the memory functions $M_{kvv''\sigma}(\varpi)$ to zero in Eqns. (9).

RESULTS

We now compare our MF results with the MC results of Scalettar et.al.[1]. We will show that our mean field analysis gives remarkable agreement with these results in all parameter ranges. The comparison with the work of Dopf et al[2] gives similar agreement, but, due to lack of space, we will not give this comparison here. In all results shown, the unit of energy is defined by $t_{dp} = 1$ and we set $U_{pp} = U_{dp} = 0$ throughout.

In Figures (1a) and (1b) we plot the Cu and O hole occupation versus filling $\langle n \rangle$ in two regimes: $U_{dd} = 6$ and $\varepsilon = 2$, where the insulating state is characterized by a charge transfer gap and the Mott-Hubbard limit with $U_{dd} = 6$ and $\varepsilon = 8$. As observed by Scalettar et.al., in the charge transfer limit given in Fig. 1a, when we move away from half-filling, $\langle n \rangle = 1$, the added holes tend to go to the O sites while added electrons tend to go to the Cu sites. However, in the Mott-Hubbard limit in Fig. 1b, one sees little change in the hole occupation on the O sites with doping since the O p-states have higher energy than the doubly-occupied Cu d-states, so that added holes tend to fill the Cu sites. Notice that the MC and MF points are basically indistinguishable.

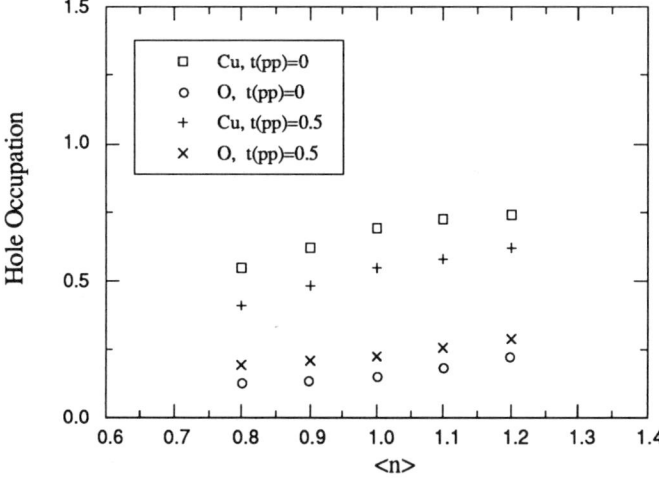

Fig.2 Effect of t_{pp} on the hole occupation at the Cu and O sites in the charge transfer limit. All parameters are the same as those given in Fig.1a except for finite t_{pp}.

In Figure 2 we show the effects of t_{pp} on the hole occupations at the Cu and O sites in the charge transfer case of $U_{dd} = 6$ and $\varepsilon = 2$ given in Fig. 1a. With finite t_{pp}, the p-band is broadened, resulting in increased hybridization between Cu and O and the O hole-occupation increases followed by a corresponding decrease of the Cu occupation.

In Fig. 3 we plot the Cu hole-occupation number $\langle n_d \rangle$ and Cu local-moment $\langle m_z^2 \rangle$ as functions of the charge transfer energy ε for the half-filled case with $U_{dd} = 6$ and $\beta = 10$. For $t_{pp} = 0$, where MC results are available, our MF calculation again yields good agreement. When t_{pp} is finite, the hole occupation on the Cu sites is suppressed dramatically in the charge-transfer limit while the suppression is much smaller in the Mott-Hubbard limit as shown in Fig.3a. Similar results are found for the local-moment of Cu in Fig.3b.

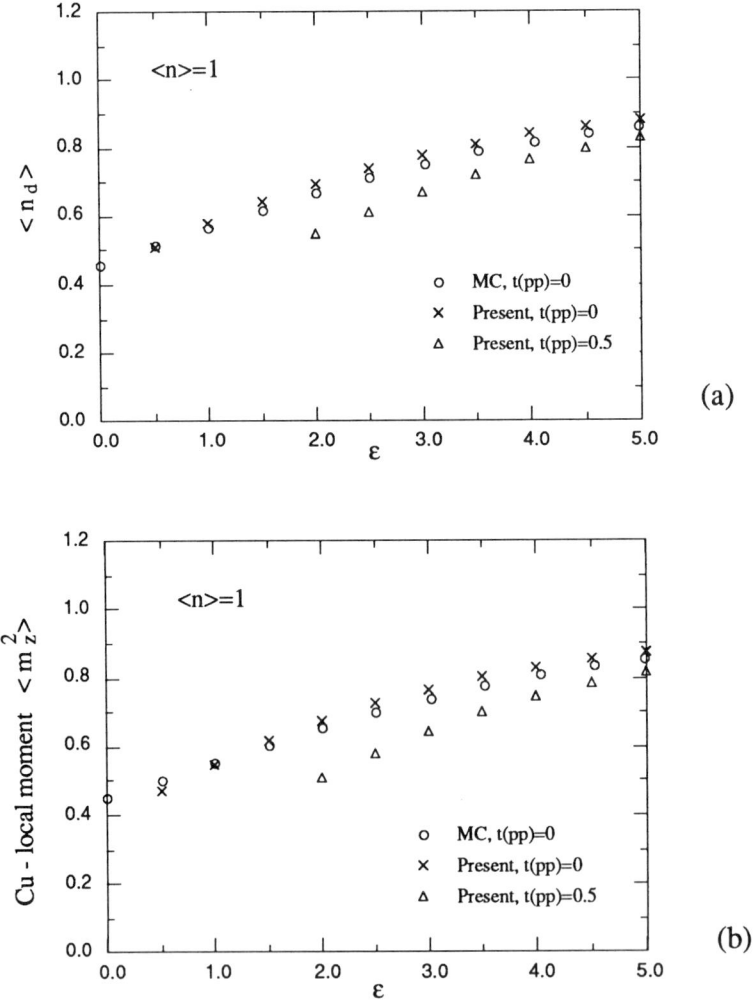

Fig.3 Hole-occupation number and local-moment on the Cu sites as functions of the charge transfer energy for half-filled case with $U_{dd} = 6$ and inverse temperature $\beta = 10$.

Finally, in Fig.4, we plot the staggered magnetization $<m_z>$ on the Cu sites versus band-filling $<n>$ for the charge-transfer limit case: $U_{dd} = 6$ and $\varepsilon = 2$. The system has its maximum magnetization at half-filling. As we dope the system away from half-filling, with either holes or electrons, $<m_z>$ decreases and vanishes eventually. It is also seen that turning on t_{pp} suppresses not only the value of $<m_z>$ but it vanishes much more rapidly with doping.

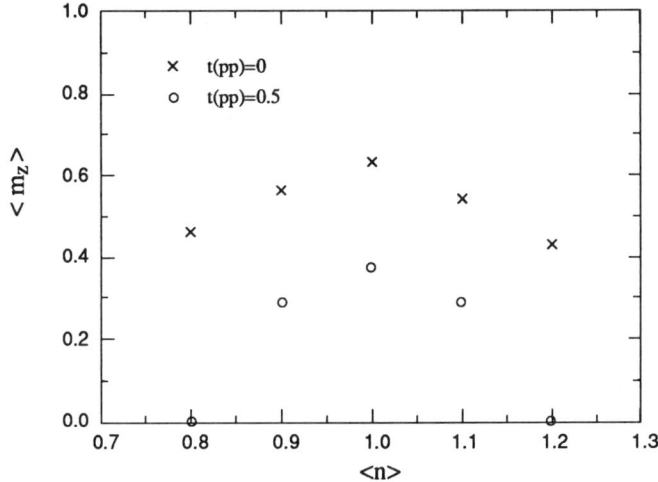

Fig.4   Cu - staggered magnetization versus the band filling $<n>$. Parameters are the same as those in Fig.1a except for finite t_{pp}.

CONCLUSIONS

We have developed a projection operator-based MF formalism to calculate various static properties of the 3-band Hubbard model. The results of this MF calculation agree remarkably well with the Monte Carlo simulation results in the parameter range where MC results are available. In addition, we have examined the effect of hopping between Oxygen sites, and found that t_{pp} suppresses the hole occupation at the Cu sites and the AF order while moving away from half-filling, especially in the charge transfer limit. Additional comparisons between the present MF calculation, other MF theories and Monte Carlo simulations have been done and these results will be presented elsewhere. It is obvious that the MF results shown here indicate that our formalism provides a relatively accurate description of the statics in the CuO_2 lattice.

ACKNOWLEDGMENTS

This work is supported by the U.S. Department of Energy, BES-Materials Sciences, under Contract #W-31-109-ENG-38, and by the U.S. Department of Energy at Iowa State University under contract #W-74005-ENG-82.

REFERENCES

1. R. T. Scalettar, D. J. Scalapino, R. L. Sugar and S. R. White, Preprint.
2. G. Dopf, A. Muramatsu and W. Hanke, Phys. Rev. B 41, 9264 (1990).

3. A. J. Fedro and S. K. Sinha, Valence Instabilities, ed. P. Wachter and H. Boppart (North-Holland, Amsterdam, 1982), p 371; A. J. Fedro and B. D. Dunlap, Jpn. J. Appl. Phys., 26 Supplement 26-3, 463, (1987).
4. Andrei E. Ruckenstein and Stefan Schmitt-Rink, Phys. Rev. B 38, 7188 (1988).
5. E.G. Goryachev, E.V. Kuzmin and S. G. Ovchinnikov, J. Phys. C. 15, 1481 (1982); V. Yu. Yushankhai, N.M. Plakida and P. Kalinay, Physica C 174, 401 (1991).
6. Yu Zhou, A. J. Fedro, T. C. Leung and B. N. Harmon, in preparation.
7. C. M. Varma, S. Schmitt-Rink and E. Abrahams, Solid State Commun. 62, 681 (1987).
8. V. Emery, Phys. Rev. Lett. 58, 2794 (1987).
9. M. S. Hybertsen, M. Schluter and N. E. Christensen, Phys. Rev. B 39, 9028 (1989); A. K.McMahan, R. M. Martin and S. Satpathy, Phys. Rev. B 38, 6650 (1988).

ON THE INEQUIVALENCE OF THE THREE BAND, KONDO HEISENBERG AND t-J MODELS

O. F. de Alcantara Bonfim and George Reiter

Texas Center for Superconductivity and Physics Department
University of Houston
Houston, TX 77204-5504

Abstract

We compare the spectrum for the three band (Emery) Model, the Kondo Heisenberg model and the t-J model within the lowest order self consistent approximation. The band minima for the three band model is at $(\pi/2,\pi/2)$ and its bandwidth is proportional to J, as in the case with the other two models. The quasiparticle wavefunction is calculated to the same level of approximation, and used to demonstrate that the quasiparticle in the three band model cannot be described as a moving singlet.

The claim by Zhang and Rice [1] of the equivalence of the three band (Emery) model in the strong coupling limit with the t-J model, due to the formation of bound singlet pairs between oxygen and copper holes on small systems [2] together with numerical simulations supporting that claim, have served to focus attention on the t-J model as the "simplest" model for the CuO_2 planes that was thought to contain the essential physics. Indeed, it seems the natural model to describe the electron doped systems, so that a principle of using the minimum necessary model would commend it. In a series of papers, however, Emery and Reiter [3] showed that the ZR truncation of the original strong coupling Hamiltonian was an approximation, and that there was no small parameter available to justify the approximation. We wish to answer the question of how good is the ZR approximation for the hole doped system, by treating in detail the single hole in an antiferromagnetic background for the strong coupling limit of the Emery model and the t-J model. We will use the approximation scheme, introduced in this context by Kane, Lee and Read [4], and Schmitt, Rink, Varma and Ruckenstein [5].

The approximation consists of treating the Cu spin background using spin wave theory, and calculating the lowest order self energy self consistently. The first approximation is likely to be excellent [6], and the second is found empirically to be very good. The quasiparticle dispersion relations derived in this way are an excellent approximation to exact results on small systems. [7] Ramsak and Prelovsek (RP) [8] observing that the means of deriving the coupling between the charge and the background Cu spins using Schwinger boson methods was suspect, used the approximation for the Kondo Heisenberg (KH) model, which goes over to the t-J model in the limit of strong Kondo coupling. They showed that different methods of coupling the spin to the charge gave very similar dispersion relations for the quasiparticle, but that there was a non-analyticity introduced by scattering from low energy magnons in the KH model regularization of the t-J model that was not present in the Schwinger boson treatment. While we share RP's suspicions of the Schwinger boson treatment of the coupling, we find that it reproduces the bandwidths as a function of J obtained by exact diagonalization of small systems much better than the KH treatment. We will make comparison with the KH model anyway, as it will enable us to make clear the differences between the t-J and Emery models, and in particular, why there is a non-zero value for the average spin on the oxygen hole in the latter case. We find the non-analyticity in the Emery model as well.

We will take as our starting point the strong coupling limit of the three band model, [3] with two of the parameters $t_1 = t^2/\varepsilon$, $t_2 = t^2/u-\varepsilon$ set equal to one another and the Cu-O and O-O Coulomb interactions neglected. The Hamiltonian is then

$$H = \frac{4\bar{t}}{N} \sum_{\substack{k,k' \\ \sigma,\sigma'}} \rho_k \rho_{k'} c^+_{k'\sigma} c_{k\sigma'} \vec{S}_{\sigma\sigma'} \cdot \vec{S}_{k-k'} + J \sum_{i,\Delta} \vec{S}_i \cdot \vec{S}_{i+\Delta} \qquad (1)$$

where $\bar{t} = t_1 = t_2$, $\rho_k^2 = 2(2 + \cos k_x + \cos k_y)$, $\vec{s} = 1/2\vec{\sigma}$, and $\vec{S}_k$ is the Fourier transform of the copper spin operators $\sum_m \vec{S}_m e^{i\vec{k}\cdot\vec{m}}$ and $\vec{\Delta}$ denotes nearest neighbors. It should be noted that $c^+_{k_0\sigma}|0>$, where $\vec{k}_0 = (\pi,\pi)$ is an exact eigenstate, with eigenvalue 0. This is not a result of the strong coupling limit, and may be seen to be a feature of the original Emery model [9]. The Wannier state associated with $c^+_{k\sigma}$ is primarily localized on the four oxygens surrounding a given copper site. The existence of a zero eigenvalue is a consequence of the symmetry of the position of the oxygen on the bond between two copper ions. This symmetry is lost in the closely related KH model, where the oxygen Wannier state is exchange coupled only to a single copper ion. We may recover the KH model by expanding the ρ_k in $\cos k_x + \cos k_y$, i.e. replace

$$\rho_k \rho_{k'} \to 4(1 + 1/4 (\cos k_x + \cos k_y) + 1/4 (\cos k_{x'} + \cos k_{y'})) \qquad (2)$$

The Hamiltonian (1) then becomes

$$H = \frac{16\bar{t}}{N} \sum_i \vec{S}_i \cdot \vec{s}_i + 2\bar{t} \sum_{\substack{i,\Delta \\ \sigma,\sigma'}} \vec{S}_i \cdot \left[c^+_{i\sigma} \vec{S}_{\sigma\sigma'} c_{i+\Delta,\sigma'} + c^+_{i+\Delta,\sigma} \vec{S}_{\sigma\sigma'} c_{i\sigma} \right] + J \sum_{i,\Delta} \vec{S}_i \vec{S}_{i+\Delta} \qquad (3)$$

and $\vec{s}_i = \sum_{\sigma,\sigma'} c^+_{i\sigma} \vec{S}_{\sigma\sigma'} c_{i\sigma}$

This is the Kondo-Heisenberg model, with a value for the Kondo coupling, in RP's notation of $V = 8t_1$, $t_1 = 2\bar{t}$. If we allow V and t_1 to be arbitrary, then for large V/t_1 one expects to be able to treat the problem in terms of a singlet hopping with an effective t. One might think that $V/t_1 = 8$ would make an excellent approximation, but the violation of the symmetry of the oxygen coupling implied by (2) is critical. Indeed, if one retained the Kondo coupling to both adjacent copper spins for the oxygen, and took the limit V/t_1 large, one would obtain a model with three spin polarons, i.e. the ground state of $V(\vec{S}_i + \vec{S}_{i+1}) \cdot \vec{s}_{i+1/2}$, which are spin 1/2 entities, hopping on a lattice. The actual result is intermediate between the two pictures, but always has some spin on the oxygen.

We will treat the copper spins within the spin wave approximation. It is convenient also to introduce, following RP, staggered fermion operators

$$\tilde{c}^+_{i\sigma} = \sum_s 1/2 \,(1 + 4\sigma s e^{i\vec{k}_0 \vec{r}_i})c^+_{is} \tag{4}$$

where $s,\sigma = \pm 1/2$. This corresponds to the spin on the oxygen Wannier state at the site i being in the same direction as the copper spin at that site, assuming a Neel ordering, for $\tilde{c}^+_{i+}$, and in the opposite direction for $\tilde{c}^+_{i-}$.

The Hamiltonian is then, after the usual spin wave transformation

$$H = \sum_q \omega_q \alpha^+_q \alpha_q + \frac{1}{\sqrt{N}} \sum_{\substack{k,q \\ \sigma,\sigma'}} U_{\sigma\sigma'}(k,q) \left[\tilde{c}^+_{k-q,\sigma}\tilde{c}_{k,\sigma'}\alpha^+_q + \tilde{c}^+_{k\sigma'}\tilde{c}_{k-q,\sigma}\alpha_q\right]$$

$$+ \sum_{k,\sigma,\sigma'} \varepsilon_k \sigma z \tilde{c}^+_{k\sigma}\tilde{c}_{k\sigma'} \tag{5}$$

where $\omega_q = 4JS(1 - 1/4(\cos q_x + \cos q_y)^2)^{1/2}$, $\varepsilon_k = t\langle S_i\rangle \rho_k \rho_{k+k_0}$, and the α_q are the spin wave operators, defined for q throughout the original zone. $\langle S_i \rangle = 1/2$ to lowest order in the spin wave expansion. The matrix $U(k,q)$ is

$$U(K,q) = \frac{t}{2}\begin{pmatrix} u_q(\rho_k - \bar{\rho}_k)(\rho_{k-q} + \bar{\rho}_{k-q}) + v_q(\rho_k + \bar{\rho}_k)(\rho_{k-q} - \bar{\rho}_{k-q}) & u_q(\rho_k + \bar{\rho}_k)(\rho_{k-q} + \bar{\rho}_{k-q}) + v_q(\rho_k - \bar{\rho}_k)(\rho_{k-q} - \bar{\rho}_{k-q}) \\ u_q(\rho_k - \bar{\rho}_k)(\rho_{k-q} - \bar{\rho}_{k-q}) + v_q(\rho_k + \bar{\rho}_k)(\rho_{k-q} + \bar{\rho}_{k-q}) & u_q(\rho_k + \bar{\rho}_k)(\rho_{k-q} - \bar{\rho}_{k-q}) + v_q(\rho_k - \bar{\rho}_k)(\rho_{k-q} + \bar{\rho}_{k-q}) \end{pmatrix}$$

$$(6)$$

where u_q, v_q are the usual Bogoliubov coefficients, satisfying $u_q + v_q = [1 - 1/2(\cos q_x + \cos q_y)/(1 + 1/2(\cos q_x + \cos q_y))]^{1/4}$, $u_q^2 - v_q^2 = 1$, and $\bar{\rho}_k = \rho_{k+k_0}$.

The approximation we will use for calculating the Green's function

$$G(k,\omega) = [G_0^{-1}(k,\omega) - \Sigma(k,\omega)]^{-1} = \begin{pmatrix} G_+ & G_{+-} \\ G_{-+} & G_- \end{pmatrix} \tag{7}$$

where $G_0(k,\omega) = \begin{pmatrix} \omega-\varepsilon_k & 0 \\ 0 & \omega+\varepsilon_k \end{pmatrix}$, is

$$\Sigma_{\sigma\sigma'}(k,\omega) = \frac{1}{N}\sum_{\tau\tau'}\sum_q U_{\tau\sigma}(k,q)G_{\tau\tau'}(k-q,\omega-\omega_q)U_{\tau'\sigma'}(k,q) \tag{8}$$

and Σ and G are calculated self consistently [4,5,8]. It is also equivalent to the retracable path approximation of Brinkman and Rice, [10] and corresponds to a definite approximation for the wavefunction of the quasiparticles, as we now show. Let

$$|\psi\rangle = a^0(k) \begin{pmatrix} c^+_{k\uparrow} \\ c^+_{k\downarrow} \end{pmatrix} |0\rangle + \frac{1}{\sqrt{N}} \sum_q a^1(k,q) \begin{pmatrix} c^+_{k-q\uparrow} \\ c^+_{k-q\downarrow} \end{pmatrix} \alpha^+_q |0\rangle$$

$$+ \frac{1}{N} \sum_{q_1 q_2} a^2(k,q_1,q_2) \begin{pmatrix} c^+_{k-q_1-q_2\uparrow} \\ c^+_{k-q_1-q_2\downarrow} \end{pmatrix} \alpha^+_{q_1} \alpha^+_{q_2} |0\rangle + \ldots \tag{9}$$

where the $a^n(k, q_1 \ldots q_n)$ are row vectors, and the vacuum is the product of the spin wave ground state and the hole vacuum.

Then the Schroedinger equation, $(\lambda - H)|\psi\rangle = 0$ corresponds to an infinite set of equations, the first two of which are

$$a^0(k) \begin{pmatrix} \lambda - \varepsilon_k & 0 \\ 0 & \lambda + \varepsilon_k \end{pmatrix} - \frac{1}{N} \sum_q a^1(k,q) U(k,q) = 0 \tag{10}$$

$$a^1(k,q) \begin{pmatrix} \lambda - \varepsilon_{k-q} - \omega_q & 0 \\ 0 & \lambda + \varepsilon_{k-q} - \omega_q \end{pmatrix} - \frac{1}{N} \sum_{q_2} a^2(k,q,q_2) U(k-q,q_2)$$

$$- \frac{1}{N} \sum_{q_1} a^2(k,q_1,q) U(k-q,q_1) = a^0(k) U^T(k,q) \tag{11}$$

Let us assume that

$$a^2(k,q_1,q_2) = a^1(k,q_1) U^T(k-q_1,q_2) G(k-q_1-q_2, \lambda - \omega_{q_1} - \omega_{q_2})$$

and neglect the second sum appearing in Eq.(11). If we require the self consistency (8) to hold

$$\Sigma(k-q_1; \lambda-\omega_{q_1}) = \frac{1}{N} \sum_{q_2} U^T(k-q_1,q_2) G(k-q_1-q_2, \lambda-\omega_{q_1}-\omega_{q_2}) U(k-q_1,q_2) \tag{12}$$

we then have

$$a^1(k,q) = a^0(k) U^T(k,q) G(k-q, \lambda-\omega_q) \tag{13}$$

and λ is determined by

$$a^0(k) G^{-1}(k,\lambda) = 0 \tag{14}$$

If λ is a pole of G, then $a^0(k)$, $a^1(k,q)$, $a^2(k-q_1-q_2,q_1,q_2)$ will be the wavefunction of the quasiparticle in this approximation. the procedure outlined above can be continued to all orders, and we will obtain a solution

$$a^n(k,q_1 \ldots q_n) = a^{n-1}(k,q_1 \ldots q_{n-1}) U^T(k-q_1 \ldots q_{n-1},q_n) G(k-q_1 \ldots q_n; \omega-\omega_{q_1} \ldots \omega_{q_n}) \tag{15}$$

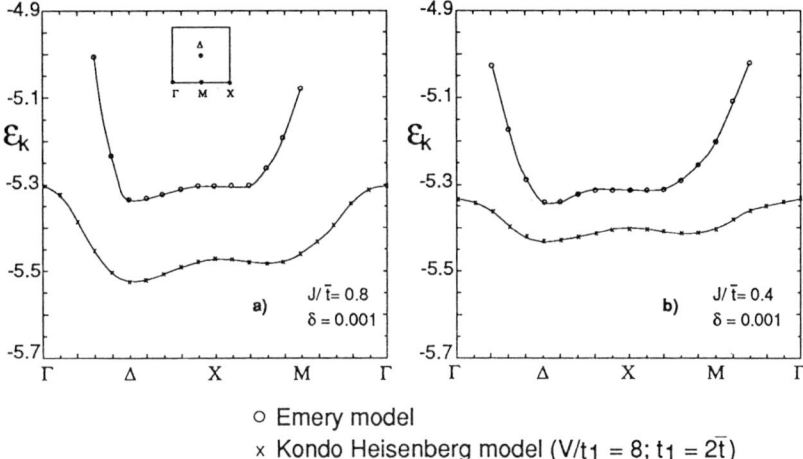

○ Emery model
× Kondo Heisenberg model (V/t_1 = 8; t_1 = $2\bar{t}$)

Fig. 1 Comparison of the quasiparticle dispersion relation for the Kondo-Heisenberg model, V/$\bar{t}$=16, with the three band (Emery) model dispersion relation having the same value of $\bar{t}$=1/2, for two values of J/$\bar{t}$, a) J/$\bar{t}$ = .8 b) J/$\bar{t}$ = .4

by neglecting in the Schroedinger equation the annihilation of all but the last magnon created. This is the essence of Brinkman and Rice's [10] retracable path approximation. The wavefunction obtained is then obviously on the same level of approximation as the Green's function obtained by solving (7,8) self consistently.

The dispersion relation of the two models defined as the location of the quasiparticle pole for two values of J/$\bar{t}$ are shown in Fig. 1. Note that the minimum in the band occurs at ($\pi/2$, $\pi/2$) in both cases. The dispersion along the zone boundary direction (Δ-x) is very similar, but the Emery model dispersion is much larger in the (Γ-Δ) direction.

One can show from the symmetry of

$$U(k,q) = \begin{pmatrix} U_{++} & U_{+-} \\ U_{-+} & U_{++} \end{pmatrix}_{k+k_0,q} = \begin{pmatrix} -U_{++} & U_{+-} \\ U_{-+} & -U_{--} \end{pmatrix}_{k,q}$$

that

$$\begin{pmatrix} \Sigma_{++} & \Sigma_{+-} \\ \Sigma_{-+} & \Sigma_{--} \end{pmatrix}_{k+k_0} = \begin{pmatrix} \Sigma_{++} & -\Sigma_{+-} \\ -\Sigma_{-+} & \Sigma_{--} \end{pmatrix}_{k}$$

and

$$\begin{pmatrix} G_{++} & G_{+-} \\ G_{-+} & G_{--} \end{pmatrix}_{k+k_0} = \begin{pmatrix} G_{++} & -G_{+-} \\ -G_{-+} & G_{--} \end{pmatrix}_{k},$$

and further, that at k=0, U_{++} = U_{+-}, U_{--} = U_{-+}, which suffices to make all the entries of $\Sigma_{\alpha\beta}$ identical. Since $\varepsilon_{k=0}$=0, this implies that the poles of the Green's function are located at ω=0 and ω=$2\Sigma_{++}(0,\omega)$. The pole at ω=0 is at the lower limit of the upper band. Its wave function is $1/\sqrt{2}\,(c^+_{k_0\uparrow} - c^+_{k_0\downarrow})|0\rangle$, and there is no coupling of the fermion to the spin background. The remaining pole is the upper limit of the lower band. Its wavefunction has $a_0 = 1/\sqrt{2}(1,1)$. Thus the wavefunction contains equal amplitudes for the fermion to be aligned or antialigned with the Cu spin on the same sublattice and is manifestly not a spin singlet. The situation improves for the singlet picture at the band minimum at $(\pi/2, \pi/2)$. From symmetry, $\Sigma_{+-}=\Sigma_{-+}=0$, so that the wave function $a^o(\pi/2,\pi/2)$ is $(0,1)$, and the oxygen spin is indeed antialigned with the copper spin on the same sublattice (since only c^+_{k-} appears in the wavefunction). Whether or not the flipped copper spin is sufficiently localized to complete the singlet wavefunction can only be determined numerically. A detailed comparison of the spin patterns for the models will be presented elsewhere. It is clear, however, that the spin on the oxygen hole varies continuously with k.

Finally, we show a comparison of the bandwidths on a 4x4 system, calculated as above with the numerical exact diagonalization results of Dagotto et al. [11] on 4x4 clusters, for the t-J model, and Frenkel et al. [2,12] for the Emery model. The theoretical bandwidth is defined to be the difference between the $\vec{k}=0$ and $\vec{k}=(\pi/2,\pi/2)$ points on the dispersion curve.

The KH model is clearly a poor approximation to the Emery model, indicating the need to retain the Kondo coupling of the oxygen hole to both

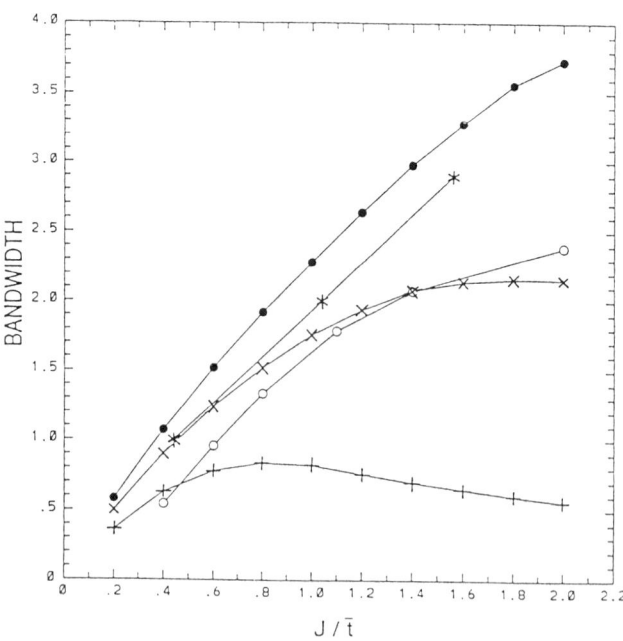

Fig. 2 Variation of the bandwidth as a function of of $J/\bar{t}$ for the Emery model (·), the Kondo Heisenberg model [8] (+) and the t-J model (x) computed within lowest order self consistent approximation. The open circles are the exact results for the t-J model on a 4x4 lattice. [11] The (*) are the exact results for the Emery model by Frenkel et al. [2], on a 4x4 lattice.

neighboring copper spins. It is surprisingly, also a poor approximation to the t-J model, even for $V/t_1=8$, within the approximation used here. The off-diagonal matrix elements of $U(k,q)$ in the Emery model have the same dependence on q for small q and $|q-q_0|$ as do those of the KH model, and the same non-analyticity appears in the self energy. The t-J model self energy is regular, however, so the models are inequivalent in that respect, as well as the appearance of spin of the quasiparticle. Both differences arise from the coupling of the singlet and triplet bands in the Emery and KH models.

ACKNOWLEDGEMENTS

GR would like to acknowledge the hospitality of the Laboratoire de Physique des Solides, Universite de Paris during the preparation of this manuscript, and Vic Emery, for valuable conversations. Most calculations presented in this paper were made possible by a generous grant of computer time from Cray Research. This work was partially supported by Grant No. MDS 972-88-G-0002 from U. S. Defense Advanced Research Projects Agency and State of Texas.

REFERENCES

1. F. C. Zhang and T. M. Rice, Phys. Rev. B 37, 3759 (1988); Phys. Rev. B 41, 7243 (1990); Phys. Rev. B 41, 2560 (1990).
2. C. X. Chen, H. B. Schuttler and A. J. Pedro, Phys. Rev. B. 41, 2582 (1990); D. M. Frenkel, R. J. Gooding, B. I. Shraiman, E. D. Siggia, Phys. Rev. B 41 350 (1990); M. S. Hybertsen, E. B. Stechel, M. Schluter and D. R. Jennison, Phys. Rev. B 41, 11068 (1990).
3. V. Emery and G. Reiter, Phys. Rev. B 38, 11938; Phys. Rev. B 41, 7247 (1990).
4. C. Kane, P. Lee and N. Read, Phys. Rev. B. 39, 6880 (1989).
5. S. Schmitt-Rink, C. Varma and A. Ruckenstein, Phys. Rev. Lett 60, 2793 (1988).; M. Marsiglio, A. Ruckenstein, S. Schmitt-Rink and C. Varma, Phys. Rev. B 43, 10882 (1990).
6. T. Becher and G. Reiter, Phys. Rev. Letts. 63, 1004 (1989): erratum Phys. Rev. Letts. 64 109 (1990).
7. Z. Liu and E. Manousakis, preprint.
8. A. Ramsak and P. Prelovsek, Phys. Rev. B. 42, 10415 (1990).
9. V. J. Emery, Phys. Rev. Lett. 58, 2794 (1987).
10. W. F. Brinkman and T. M. Rice, Phys. Rev. B 2, 1324 (1970).
11. E. Dagotto, R. Joynt, A. Moreo, S. Bacci and E. Gagliano, Phys Rev B 41, 9049 (1990).
12. The results of Frenkel et al. [2] in Fig.4 of their paper have been adjusted by multiplying the abscissa by four to compensate for their use of 4J in the Hamiltonian, Eq.2 of Ref.2.

HOLES IN THE QUANTUM ANTIFERROMAGNET: THE THEORY OF SMALL POLARONS

Assa Auerbach and B.E. Larson
Physics Department
Boston University
Boston, Massachusetts 02215

ABSTRACT: The spin-hole coherent states path integral is used to generate a systematic large-spin expansion of the t-J model on the square lattice. The single hole's classical energy is minimized by small polarons with short ranged interactions. We derive the polarons' low energy Lagrangian in which inter-sublattice hopping has been eliminated. The polarons' interaction with the Néel gauge field was argued to produce superconductivity in the magnetically disordered phase. We discuss the relevance of this model to experiments in slightly doped copper oxides.

1. Introduction

The discovery of high temperature superconductivity has spurred intense investigations of the two dimensional doped antiferromagnet. In the strong coupling limit, the t-J Hamiltonian, derived from the large-U Hubbard model[1], is often used to describe the low lying excitations. At zero doping, it directly reduces to the quantum antiferromagnetic Heisenberg model (QHM). Substantial progress has been recently achieved in understanding the Heisenberg limit, both theoretically and experimentally[2]. The effects of doping, however, are still highly controversial.

Continuum theories[3] and the Schwinger Boson- Slave Fermion mean field theory[4] predict spiral magnetic phases at finite doping concentrations. Several approaches have found an instability of the t-J model towards phase separation into hole-rich and hole-poor domains[5]. This instability is also detected by the mean field negative compressibility[6]. It could be argued that in the real systems, phase separation is suppressed by longer range Coulomb interactions. Recently however, the RPA determinant has also been found to be unstable (negative) in a *range of momenta*[6]. The offending fluctuations can be identified with local enhancements of the spiral distortion. Clearly, the holes drive strong perturbations of the spins on the lattice constant scale, which are difficult to treat within the continuum and mean field approaches.

The path integral of spin coherent states has been fruitfully used by Haldane to map the QHM onto the non linear sigma model and to derive the topological Berry phases[7]. It provided a unified semiclassical treatment of the ordered and disordered phases of the quantum antiferromagnet. In this paper we generalize this path integral to represent the t-J model by defining "spin-hole coherent states". This will lead to a semiclassical expansion of the ground state and low excitations in the presence of holes. Although the expansion is formally controlled by the large spin size S we have learned that (at least for the undoped case[2]) it can work well even for $S = \frac{1}{2}$.

2. The Semiclassical Formulation of the t-J Model

The t-J Hamiltonian is given by[6]

$$\mathcal{H}^{tJ} = t \sum_{\langle i;j \rangle} f_i^\dagger f_j \mathcal{F}_{ij} - \frac{J}{4} \sum_{\langle i;jk \rangle} (\delta_{ik} - f_k^\dagger f_i) \mathcal{A}_{ij}^\dagger \mathcal{A}_{kj} (1 - f_j^\dagger f_j) \quad (1)$$

where $\mathcal{A}_{ij}^\dagger \equiv (a_i^\dagger b_j^\dagger - b_i^\dagger a_j^\dagger)$ and $\mathcal{F}_{ij}^\dagger \equiv (a_j^\dagger a_i + b_j^\dagger b_i)$. $\langle i;j \rangle$ ($\langle i;jk \rangle$) denote summation over sites i and their first (second) nearest neighbors on the square lattice. $J = 4t^2/U$ is the Heisenberg superexchange constant. t and U are the hopping and interaction parameters of the parent Hubbard model. Eq. (1) includes *all the terms* to second order in t/U. The operators a_i, b_i (f_i) are Schwinger Bosons (Slave Fermions), and the Hilbert space is subjected to the constraint $a^\dagger a + b^\dagger b + f^\dagger f = 2S$ at each site. This constraint generalizes the original Hubbard $S = \frac{1}{2}$ states to arbitrary spin S.

The spin-hole coherent states are defined as follows:

$$|\hat{\Omega}, \xi\rangle_S = |\hat{\Omega}_S\rangle |0\rangle + |\hat{\Omega}_{S-\frac{1}{2}}\rangle \xi f^\dagger |0\rangle . \quad (2)$$

$|\hat{\Omega}_S(\theta, \phi)\rangle = (2S!)^{-1/2} (ua^\dagger + vb^\dagger)^{2S} |0,0>$ are the standard spin coherent states, where $u = \cos(\theta/2) e^{-i\phi/2}$ and $v = \sin(\theta/2) e^{i\phi/2}$. ξ is a Grassman variable. The states (2) allow a resolution of the identity in the S-sector:

$$\frac{2S}{4\pi} \int d\phi \, d\cos\theta \, d\xi^* d\xi \exp[-\alpha_S \, \xi^* \xi] \, |\hat{\Omega}, \xi\rangle\langle\hat{\Omega}, \xi| = I \quad (3)$$

Where the factor $\alpha_S = (2S+1)/(2S)$ is required for normalizing the matrix elements to unity. In the grand canonical partition function, α_S is replaced by unity by renormalizing the chemical potential μ.

Following standard procedure[7] we use (3) to construct the path integral for the partition function:

$$Z = \int \mathcal{D}\hat{\Omega} \, \mathcal{D}\xi^* \, \mathcal{D}\xi \exp\left[\int_0^\beta d\tau \sum_i \left(i(2S - \xi_i^*\xi_i)\mathbf{A}(\hat{\Omega}_i) \cdot \dot{\hat{\Omega}}_i + \xi_i^*\dot{\xi}_i\right) - H^{tJ}[\hat{\Omega}, \xi^*, \xi]\right] \quad (4)$$

H^{tJ} in (4) is given by Eq. (1) where $a, b, f \rightarrow u, v, \xi$. $\mathbf{A} \cdot \dot{\hat{\Omega}}_i$ is the spin-kinetic term, where $\mathbf{A}(\Omega)$ is the vector potential of a unit magnetic monopole at the origin $\Omega = 0$. The Fermion "time derivatives" denote the discrete form $\dot{\xi} = (\xi(\tau) - \xi(\tau - \epsilon))/\epsilon$, where ϵ is the timestep.

H^{tJ} has quadratic and quartic Fermion terms. We decouple the four-fermion terms by the Hartree-Fock approximation. For our purposes this approximation is justified by the following arguments: (i) hole-correlation corrections are higher order in hole density and (ii) the quartic terms vanish in the ferromagnetically correlated regions, where the hole density is high. We define $\rho_{ij}[\hat{\Omega}] = \langle f_i^\dagger f_j \rangle$ to be determined self-consistently, and write $H^{tJ} \approx \bar{H}^J + \bar{H}^f - \mu \sum_i \xi_i^* \xi_i$ where

$$\bar{H}^J = -\frac{\bar{J}}{8} \sum_{\langle i;jk\rangle} (\delta_{ik} - e^{i\psi_{ik}}\rho_{ik}\rho_{jj})\sqrt{(1-\hat{\Omega}_j\cdot\hat{\Omega}_k)(1-\hat{\Omega}_i\cdot\hat{\Omega}_j)}$$

$$\bar{H}^f = \frac{\bar{t}}{\sqrt{2}} \sum_{\langle i;j\rangle} \sqrt{1+\hat{\Omega}_i\cdot\hat{\Omega}_j}\, e^{i\gamma_{ij}}\, \xi_i^*\xi_j + \frac{\bar{J}}{8} \sum_{\langle i;jk\rangle} \sqrt{(1-\hat{\Omega}_j\cdot\hat{\Omega}_k)(1-\hat{\Omega}_i\cdot\hat{\Omega}_j)}$$

$$\times \left((1-\rho_{jj})e^{i\psi_{ik}}\,\xi_i^*\xi_k + (\delta_{ik}-e^{i\psi_{ik}}\rho_{ik})\,\xi_j^*\xi_j\right)$$

(5)

Here we define the *classical* parameters $\bar{J} \equiv 4JS^2$ and $\bar{t} \equiv 2tS$. The sums in $\bar{H}^f$ represent two distinct hopping processes: intersublattice hopping ("t-terms") and intra-sublattice hopping ("J-terms"). γ_{ij} and ψ_{ik} are the phases of $u_i^*u_j + v_i^*v_j$ and $(u_i^*v_j^* - v_i^*u_j^*)(u_kv_j - v_ku_j)$ respectively. When the spins $\hat{\Omega}_i$ have short range antiferromagnetic order the γ phases in the t-terms fluctuate wildly, while $\psi_{ik} = \eta_i \mathbf{A}^N \cdot (\mathbf{x}_i - \mathbf{x}_k)$ represents a slowly varying Néel gauge field $\mathbf{A}^N(\mathbf{x})$ whose curl is the tolopological density of the staggered magnetization[8]. $\eta_i = +1, (-1)$ on sublattice A (B), is the corresponding "sublattice charge". Weigmann, Wen, Shankar and Lee have studied Lagrangians which contain similar intra-sublattice $\mathbf{A}^N$-coupled hopping terms in the context of high T_c superconductivity[9]. Returning to the t-J model, we see that the t-terms, are not $\mathbf{A}^N$-gauge invariant, *and do not conserve the sublattice charges*. Although the t-terms cannot be justifiably ignored, especially in the $\bar{t}/\bar{J} > 1$ regime, we shall soon see how they are effectively eliminated from the low energy Lagrangian.

We begin by integrating out the fermions to obtain a spin partition function

$$Z^s = \int \mathcal{D}\hat{\Omega}\, \exp\left[\int_0^\beta d\tau\left(i\sum_i (2S-\rho_i)\mathbf{A}(\hat{\Omega}_i)\cdot\dot{\hat{\Omega}}_i - \bar{H}^J[\hat{\Omega}] - E^f[\hat{\Omega}]\right)\right],\quad (6)$$

where $E^f[\hat{\Omega}]$ is the time-retarded action (free energy) of the hamiltonian $\bar{H}^f$. Here we concentrate on the zero temperature case $\beta = \infty$. Eq. (6) is a useful starting point for the semiclassical approximation. In the classical limit, $S \to \infty$, the spins are frozen, i.e. $\langle \dot{\hat{\Omega}} \rangle = 0$. The first step is to minimize $\bar{H}^J + E^f$ for a given number of holes. The second step is include the semiclassical fluctuations whose dynamics are given by the kinetic terms. We discuss the single hole and the many hole cases seperately.

3. Results

3.a) The single hole

In the regime $\bar{t}/\bar{J} > 0.87$, the "polaron" which is a *local* alignement of spins, yields a lower energy than any of the possible uniform states, including: the Néel state, spiral states and canted states. This result helps to explain the instability in the RPA fluctuations about the uniform states[6].

We used a Lanzcos algorithm on the Connection Machine to minimize the energy for 128×128 spins. The polaron variational parameters were chosen to describe a ferromagnetic core, a transition region, and a far field antiferromagnetic tail. The latter is completely determined by the boundary condition $\delta\theta$ and the pure Heisenberg model (i.e. the Laplace equation).

Our results are quite simple. For $1 < \bar{t}/\bar{J} < 4.1$ the single hole energy is minimized by the five-site polaron (one flipped spin), depicted in Fig. 1. The hole density is approximately 1/2 and 1/8 on the central and neighboring sites respectively, with a small amount of leakage (due the J-terms in (5)) to sites further away. For $4.1 < \bar{t}/\bar{J} < 6.6$, the polaron has two flipped spins (diagonally across a plaquette), and at larger values the core radius increases slowly as $R_c \propto (\bar{t}/\bar{J})^{1/4}$. The most important fact is that the small polarons *do not distort the Néel background*. In particular, the configurations centered on a bond[10] are considerably higher in energy. We also find that the polarons have no tails, i.e. $\delta\theta = 0$, throughout the regime discussed above. This follows from competing contributions of order $\pm \bar{J}(\delta\theta)^2$ of $\bar{H}^J$ and E^f. Since in addition, the density ρ is exponentially localized near the polaron sites, we conclude that *the classical interactions between polarons are short ranged*.

The spin dynamics produce leading order quantum effects on two scales: harmonic spin wave fluctuations, of order $1/S$, and polaron intersite tunneling, of order $\exp[-\alpha S]$.

The spin waves at zero temperature reduce the staggered magnetization by their zero-point motion. Perturbative spin wave analysis of the average moment, yields an additional reduction which goes as $\Delta m_0 \propto -(\bar{t}/\bar{J})x$, for small hole density x. More dramatic reduction of the ordered moment can arise from correlations between polarons. This is investigated elsewhere[6].

The polaron breaks the lattice translational symmetry. This symmetry is restored by tunneling events, where two spins flip their orientation (see Fig. 1). The tunneling matrix element can be computed as follows: The azimuthal coordinates are analytically continued to $i\phi \to \tilde{\phi}_i$, while their cannonical momenta $S_i^z = (2S - \rho_i)\cos\theta_i$ are kept real. It can be readily verified that $\sum_i S_i^z$, and $\bar{H}^J + E^f$ are conserved along the tunneling path $\bar{S}_i^z(\tilde{\phi})$ which minimizes the action. As a result of these conservation laws, we obtain a selection rule: *tunneling can only take place between sites on the same sublattice!* Thus, for polarons there are no effective "t-terms". The tunneling rate Γ_{ik} between flipped spins at i and k is:

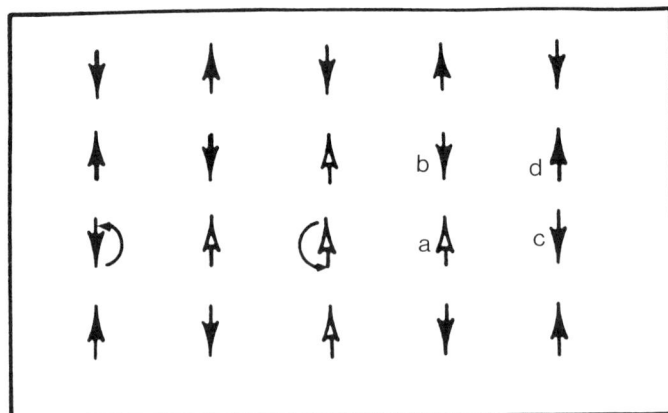

Fig. 1. The five site polaron. The hole density is primarily concentrated on the sites of the unfilled arrows. The circular arrows represent an allowed tunneling path, where the polaron hops two lattice constants to the left.

$$\Gamma_{ik} = \Gamma_0 \exp\left(-\sum_{i'} \int d\tilde{\phi}_{i'} \bar{S}^z_{i'}\right) \approx S^{1/2} \beta_{ik} \bar{t} \exp(-S\alpha_{ik}) , \qquad (7)$$

where α_{ik}, β_{ik} are slowly varying dimensionless functions of $\bar{t}$ and $\bar{J}$. For five site polarons and $S = \frac{1}{2}$ we estimate the exponent to be roughly unity, but a fuller treatment of the multidimensional tunneling problem is required for a quantitative determination of the polaron's effective mass. The single polaron occupies a Bloch wave of dispersion $\epsilon_{\mathbf{k}} = 2\Gamma_c(\cos(2k_x) + \cos(2k_y)) + 2\Gamma_b(\cos(k_x + k_y) + \cos(k_x - k_y))$, where c, b denote the site of the other flipped spin as labelled in Fig. 1. By energetic arguments, $\Gamma_b \leq \Gamma_c$. Thus the single polaron energy is minimized at $\mathbf{k} = (\pi/2, \pi/2)$. This results agrees with other studies of the single hole spectral function in the t-J model[11]. In the presence of long wavelength background spin fluctuations, the tunneling rate Γ_{ik} is multiplied by the overlap between the aligned spins and the distorted background given by $\exp[i\eta_i \mathbf{A}^N(\mathbf{x}_i - \mathbf{x}_k)]$, where A^N and η_i are the familiar Néel gauge field and sublattice charge respectively. Thus the polaron hopping conserves the sublattice charge and is gauge invariant with respect to $\mathbf{A}^N$, as there are no effective "t-terms".

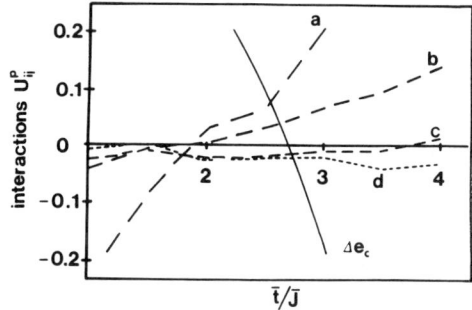

Fig. 2. Classical interactions between polarons, in units of $\bar{J}$. Lines (a)-(d) represent the second polaron positions as labelled in Fig. 1. The solid line represents the relative condensation energy per hole of the hole-rich phase (see text).

3.b) Interactions

The interactions between two polarons were computed in the regime $\bar{t}/\bar{J} = 1-4$. We define $U_{ij}^p = \frac{1}{2}e_{ij} - 2e_h$, where e_{ij} is the relaxed energy of a two-hole polaron with flipped spins at sites i and j. U_{ii}^p is repulsive, and of order $0.6\bar{J} - 2.6\bar{J}$. The intersite interactions, for neighboring polarons at sites (a)-(d) (see Fig. 1.) are plotted in Fig. 2. We find both attractive and repulsive interactions, and it is interesting to note that for $\bar{t}/\bar{J} < 1.8$ there is a near neighbour attraction of antiferromagnetically correlated spins. We also consider the possibility of polaron condensation into hole-rich domains[1]. The condensation energy per hole is determined by minimizing it with respect to the spin configuration, and the density. The spins in the hole-rich domains align ferromagnetically, and the energy per hole is given by $e_{fm} = -4\bar{t} + 4\sqrt{B\bar{J}\bar{t}\pi}$.

This results coincides with that of Emery, Kivelson and Lin[5], except that their quantum correction factor B=0.584 is here set to 1/2. In Fig. 2., the condensation energy $\Delta e_c = e_{fm} - e_h$ is plotted. We find that it becomes negative at $\bar{t}/\bar{J} = 2.7$, *above which phase-separation will occur for large S.*

Attractive interactions and negative condensation energies may result in charge density waves or superconductivity in the ground state of the quantized model. However, if realistic intersite Coulomb repulsions are included in the t-J model, the attractive interactions may change sign. In particular, phase separation will be supressed, or pushed to higher values of $\bar{t}/\bar{J}$.

4. The effective Lagrangian of Small Polarons

The information given above allows us to write the effective Lagrangian for a dilute system of small polarons:

$$\mathcal{L}^{s-p} = \sum_i \left[i(2S - p_i^* p_i)\mathbf{A}(\hat{\Omega}_i) \cdot \dot{\hat{\Omega}}_i + p_i^* \dot{p}_i \right] + \frac{\bar{J}}{8} \sum_{\langle i;j \rangle} \hat{\Omega}_i \cdot \hat{\Omega}_j + \sum_i (e_h - \mu) \, p_i^* p_i$$
$$+ \sum_{\langle i;jk \rangle} \Gamma_{ik} e^{\eta_i \mathbf{A}^N \cdot \mathbf{x}_{ik}} \, p_i^\dagger p_k + \sum_{ij} U_{ij}^p \, p_i^* p_i \, p_j^* p_j$$

(8)

Eq. (8) is the main result of this paper. $\mathcal{L}^{s-p}$ describes spinless Fermions p_i with short range interactions U_{ij}^p coupled to Heisenberg spins. The formation of polarons can be viewed as a strong short-wavelength renormalization of the original f-holes by the spins. As a consequence, the uncomfortable t-terms have been conveniently eliminated. In the dilute density and long wavelength limit, the spin terms can be relaced by the quantum non linear sigma model[7]. We recall that previous investigations of continuum versions of Eq. (8) have proposed that superconductivity will be obtained in the magnetically disordered phase[9]. The pairing interaction between charges on opposite sublattices depends on the spin wave gap. Further investigations of Eq. (8) might determine whether this gap can be generated within $\mathcal{L}^{s-p}$ with physically acceptable parameters.

Eq. (8) has experimental consequences, relevant to the slightly doped cuprate systems: (i) The Hall coefficient is inversely proportional to the number of holes. (ii) Disorder gives rise to a temperaure dependent conductivity typical of weakly localized semiconductors[12]. (iii) The polaron internal excitations could be probed by optical absorption. (iv) Recent NMR studies[13] of $YBa_2Cu_3O_{6+x}$, have shown a dramatic decrease in intensity of the NMR resonance under doping. The NMR frequency, however, has varied very slightly. This indicates that the changes in local spin correlations due to presence of holes is abrupt, as would be given by polarons with *sharp* boundaries. If the polarons had long range effects on the spins, the NMR frequency would vary smoothly with dopant concentration. We thank H. Alloul for this observation, which was made during our poster representation in this workshop.

Acknowledgements

The computations were performed in part on the Connection Machine at Boston University. This work has been supported by a grant from NSF, DMR-8914045. A. Auerbach acknowledges Alfred P. Sloan Foundation for a fellowship.

REFERENCES

[1] See e.g. J.E. Hirsch, *Phys. Rev. Lett.* **54**, 1317 (1985).

[2] S. Chakravarty, Proceedings of the 1989 Symposium on High Tc Superconductivity, Eds. K.S. Bedell *et. al.*, (Addison-Wesley, 1990) and references therein.

[3] B.I. Schraiman and E.D. Siggia, *Phys. Rev. Lett.* **62**, 1564 (1989).

[4] C. Jayaprakash, H.R. Krishnamurthy and S. Sarker, *Phys. Rev. B* **40**, 2610 (1989); C.L. Kane, P.A. Lee, T.K. Ng, B. Chakraborty and N. Read, *Phys. Rev. B* **41**, 2653 (1990).

[5] L.B. Ioffe and A.I. Larkin, *Phys. Rev. B* **37**, 5730 (1988); V.J. Emery, S.A. Kivelson and H.Q. Lin, *Phys. Rev. Lett.* **64**, 475 (1990); M. Marder, N. Papanicolau and G.C. Psaltakis, *Phys. Rev. B* **41**, 6920 (1990).

[6] A. Auerbach and B.E. Larson, *Bull. Am. Phys. Soc.* **35**, 379-380, (1990), BU-PHY preprint; A. Auerbach, B.E. Larson and G. Murthy, unpublished.

[7] F.D.M. Haldane, "Two Dimensional Strongly Correlated Electron Systems", Eds. Z.Z. Gan and Z.B. Su, (Gordon and Breach, 1988), pp. 249-261; *Phys. Rev. Lett.* **61**, 1029 (1988).

[8] A.M. Polyakov, "Gauge Field And Strings", (Harwood, 1987), P. 140.

[9] P.B. Wiegmann, *Phys. Rev. Lett.* **60**, 821 (1988); X-G. Wen, *Phys. Rev. B* **39**, 7223 (1989); R. Shankar, *Phys. Rev. Lett.* **63**, 203 (1989); P.A. Lee, *Phys. Rev. Lett.* **63**, 690 (1989).

[10] A. Aharony, R.J. Birgeneau, A. Coniglio, M.A. Kastner and H.E. Stanley, *Phys. Rev. Lett.* **60**, 1330 (1988).

[11] S. Schmitt-Rink, C.M. Varma and A.E. Ruckenstein, *Phys. Rev. Lett.* **60**, 2793 (1988); C. Kane, P.A. Lee and N. Read, *Phys. Rev. B* **39**, 6880 (1989); E. Dagotto, A. Moreo, R. Joynt, S. Bacci and E. Gagliano, *Phys. Rev. B* **41**, 2585 (1990).

[12] C.Y. Chen, N.W. Preyer, P.J. Picone, M.A. Kastner, H.P. Jenssen, D.R. Gabbe, A. Cassanho and R.J. Birgeneau, *Phys. Rev. Lett.* **63**, 2307 (1989).

[13] P. Mendels, H. Alloul, J.F. Marucco, J. Arabski and G. Collin, Physica C **171**, 429 (1990).

PROPAGATOR-RENORMALIZED PERTURBATION THEORY

FOR THE HUBBARD MODEL

J.W. Serene and D.W. Hess

Complex Systems Theory Branch
Naval Research Laboratory
Washington, D.C. 20375-5000

Introduction

Anderson [1] has argued forcefully that the unusual properties of the high temperature superconductors for $T > T_c$ place these materials in the universality class described by Haldane's Luttinger liquid theory (LL) [2] instead of the more familiar class of metals governed by Landau's Fermi liquid theory. The proven examples of Luttinger liquids are all one dimensional systems where the breakdown of the Fermi liquid picture can be traced to susceptibilities which diverge for $T \to 0$ on account of the 1D analog of perfect Fermi surface nesting. The low-energy elementary excitations of the LL ground state exhibit separation of spin and charge, and if one trys to cast the one-electron Green's function into standard Fermi liquid form, the resulting quasiparticle renormalization factor a_k retains a strong frequency and temperature dependence, and vanishes as a power law when these variables tend to zero.

A quasiparticle renormalization factor that vanishes as the inverse logarithm of the frequency and temperature has been proposed on phenomenological grounds by Varma et al. [3], who show that a number of the anomalous normal state properties observed in the superconducting cuprates follow from a self energy of the marginal Fermi liquid (MFL) form,

$$\Sigma(\varepsilon) = \frac{2\alpha}{\pi} \left[\varepsilon \ln \frac{x}{\omega_c} - i\frac{\pi}{2}x \right], \qquad (1)$$

where $x = \max(|\varepsilon|, T)$ and ω_c is a cutoff of order the bandwidth W.

While most of the (to date unsuccessful) attempts to derive a LL or MFL from a microscopic model have started from a strong-coupling limit, several groups have proposed that special features of a two-dimensional band structure could enhance the phase space for electron-electron scattering and thereby yield a LL or MFL-like self energy in weak coupling. Virosztek and Ruvalds [4] consider electron-electron scattering on a nearly nested Fermi surface in a 2D Hubbard model, and present an approximate analytic derivation of a MFL self energy for $U \ll W$. Newns and collaborators [5] argue that a LL or MFL self energy can occur if the Fermi surface lies close to a van Hove singularity, even without nesting.

We present self consistent numerical calculations of the one-electron Green's function and self energy for the 1D and 2D Hubbard models, using the fluctuation exchange approximation (FEA) of Bickers and co-workers [6]. Within this framework, and with the help of a massively parallel algorithm implemented on NRL's Connection Machine, we have been able to work with very large lattices (128 × 128 in 2D) and very low temperatures ($T/W \approx 0.002$), which are essential if one hopes to identify low-energy anomalies characteristic of a LL or MFL. While preferable to the FEA, exact methods such as diagonalization or quantum Monte Carlo are practical only for small lattices and are thus unable to obtain true low energy properties. The FEA should in any case be investigated thoroughly, because it contains at least the seeds of all known electron gas instabilities or anomalies. Anderson has argued that a 2D Luttinger liquid will be invisible to any formalism ultimately derived from perturbation theory, including fully renormalized methods such as ours, because the breakdown of Fermi liquid theory results from a pathology of the scattering amplitude for two electrons in the zero-density limit (that is, with an otherwise empty Fermi sea) [7]. We do not find this argument completely persuasive, because it seems to us that the applicability of perturbation theory always depends essentially on the state being perturbed about, and the pathology identified by Anderson has been shown to be rendered harmless by the presence of a nonzero electron density [8].

Renormalized Perturbation Theory

Long ago, Luttinger and Ward [9] showed that the thermodynamic potential can be expressed as a functional of the exact temperature Green's function $G(\mathbf{k}, \varepsilon_n)$ and self energy $\Sigma(\mathbf{k}, \varepsilon_n)$, as

$$\Omega(T, \mu) = -2\text{Tr}\left[\Sigma G + \ln(-G_0^{-1} + \Sigma)\right] + \Phi[G], \quad (2)$$

where $\Phi[G]$ and Σ are related by

$$\Sigma(\mathbf{k}, \varepsilon_n) = \frac{1}{2} \frac{\delta \Phi[G]}{\delta G(\mathbf{k}, \varepsilon_n)}, \quad (3)$$

and G and Σ satisfy Dyson's equation,

$$G(\mathbf{k}, \varepsilon_n) = [G_0^{-1}(\mathbf{k}, \varepsilon_n) - \Sigma(\mathbf{k}, \varepsilon_n)]^{-1}. \quad (4)$$

We have applied this formalism to the single-band Hubbard model, with Hamiltonian

$$H = -t \sum_{<i,j>,\sigma} (c_{i\sigma}^\dagger c_{j\sigma} + c_{j\sigma}^\dagger c_{i\sigma}) + U \sum_i n_{i\uparrow} n_{i\downarrow}. \quad (5)$$

In the fluctuation exchange approximation, the functional $\Phi[G]$ is the sum of a single second order graph, plus particle-hole and particle–particle bubble chains describing exchanged density, spin-density, and (singlet) pair fluctuations,

$$\Phi = \Phi_2 + \Phi_{ph}^{df} + \Phi_{ph}^{sf} + \Phi_{pp} \quad (6)$$

$$\Phi_2 = -\frac{1}{2}\text{Tr}\left[\chi_{ph}^2\right] \quad (7)$$

$$\Phi_{ph}^{df} = \frac{1}{2}\text{Tr}\left[\ln(1 + \chi_{ph}) - \chi_{ph} + \frac{1}{2}\chi_{ph}^2\right] \quad (8)$$

$$\Phi_{ph}^{sf} = \frac{3}{2}\text{Tr}\left[\ln(1 - \chi_{ph}) + \chi_{ph} + \frac{1}{2}\chi_{ph}^2\right] \quad (9)$$

$$\Phi_{pp} = \text{Tr}\left[\ln(1 + \chi_{pp}) - \chi_{pp} + \frac{1}{2}\chi_{pp}^2\right], \quad (10)$$

where the particle-hole and particle-particle susceptibility bubbles are

$$\chi_{pp}(\mathbf{q}, \omega_m) = U(T/N) \sum_{\mathbf{k}} \sum_n G(\mathbf{k} + \mathbf{q}, \varepsilon_n + \omega_m) G(-\mathbf{k}, -\varepsilon_n) \quad (11)$$

$$\chi_{ph}(\mathbf{q}, \omega_m) = -U(T/N) \sum_{\mathbf{k}} \sum_n G(\mathbf{k} + \mathbf{q}, \varepsilon_n + \omega_m) G(\mathbf{k}, \varepsilon_n). \quad (12)$$

For this $\Phi[G]$ we have solved Eqs. (3) and (4) numerically for the Hubbard model in 1D and 2D. Because the real part of the LL or MFL self energy goes through $\varepsilon = 0$ continuously but with infinite slope at $T = 0$, we focus not on the self energy itself, but instead on

$$a_M(\mathbf{k}, T)^{-1} = 1 - \frac{\text{Im } \Sigma(\mathbf{k}, \varepsilon_0)}{\varepsilon_0}.$$

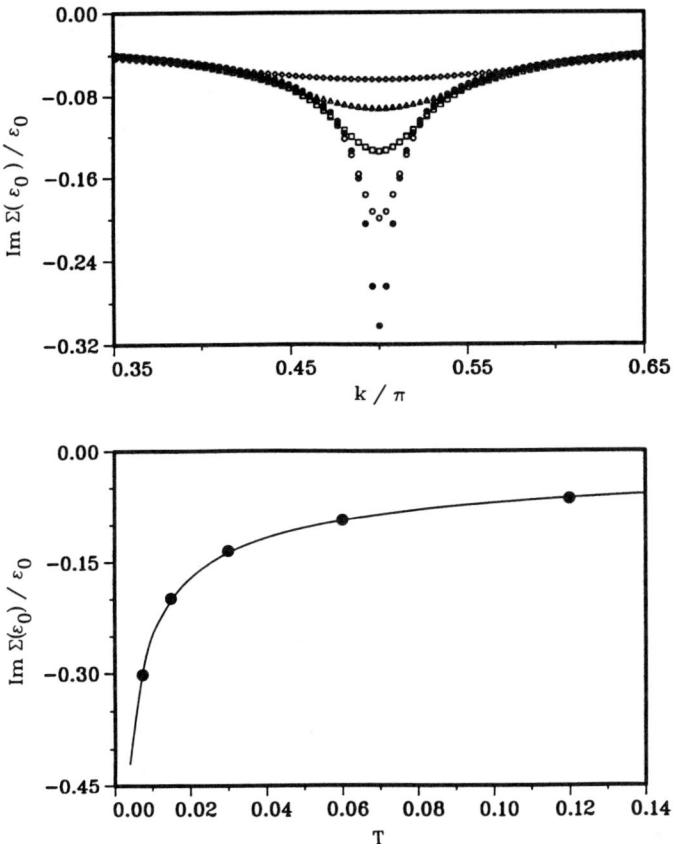

Figure 1. Top: Im $\Sigma(\varepsilon_0)/\varepsilon_0$ plotted as a function of k for $T/t = 0.0075$ (•), 0.015 (○), 0.03 (□), 0.06 (△), 0.12 (◇) for the Hubbard model in 1D with $U = t$ and $n = 1$. Bottom: Im $\Sigma(\varepsilon_0)/\varepsilon_0$ for $k = k_F$ as a function of temperature. The solid line is a power law fit to our numerical results (see text).

For a normal Fermi liquid, $a_M(\mathbf{k},T)^{-1}$ is weakly temperature dependent and approaches the inverse quasiparticle renormalization factor $a_\mathbf{k}^{-1}$ when $T \to 0$, but if the renormalization factor vanishes, then $a_M(\mathbf{k},T)^{-1}$ must diverge in this limit. For example, using the MFL model self energy of Eq. (1) as the spectral function for the retarded self energy, one easily generates an analytic result for $\Sigma(\mathbf{k},\varepsilon_0)$, for which $a_M(\mathbf{k},T)^{-1}$ diverges as $\ln(T/\omega_c)$.

1D Hubbard Model

In 1D we use 512 k points with periodic boundary conditions and up to 2048 Matsubara frequencies, at temperatures $T/t = 0.0075, 0.015, 0.03, 0.06$ and 0.12. The factor of two between successively increasing temperatures reflects the need for a constant cutoff energy for the Matsubara frequencies, together with our use of fast Fourier transforms to perform frequency convolutions. Fig. 1 shows results at half-filling with $U = t$ for the k-dependence of Im $\Sigma(k,\varepsilon_0)/\varepsilon_0$ and for the T-dependence of Im $\Sigma(k_F,\varepsilon_0)/\varepsilon_0$. The latter is well described by the function $-0.019T^{-0.56}$ as also shown in Fig. 1. The apparent divergence of $a_M(\mathbf{k},T)^{-1}$ as $T \to 0$ and the sharp notch in the self energy as a function of k are not consistent with a Fermi liquid. Although the FEA is completely inadequate in 1D at $T = 0$, it may be reasonable at high enough temperatures or weak enough coupling so that it yields $a_M(\mathbf{k}, T)$ not too much less than 1; at the lowest temperature shown in Fig. 1, $a_M(\mathbf{k}, T) \approx 0.75$. In any case, our major motivation for these 1D calculations was to investigate how anomalies that ultimately lead to the Luttinger liquid are manifest in the self-consistent FEA.

2D Hubbard Model

In 2D we use a 128×128 mesh of $\mathbf{k}$ points with periodic boundary conditions. This fine mesh is required to avoid finite-size effects at the lowest temperatures; with a 64×64 lattice, deviations from the correct T^2–dependence of the noninteracting thermodynamic potential are significant at $T/t = 0.015$. We used Matsubara frequency cutoffs of up to twelve times the bandwidth, corresponding to 1024 frequencies at $T/t = 0.03$. The self energy was calculated for $T/t = 0.015, 0.03, 0.06$ and 0.12, where the factor of 2 between successively increasing temperatures has the same origin as in 1D.

Half Filling

At half-filling, the Hubbard model with nearest-neighbor hopping has a familiar diamond-shaped Fermi surface with perfect nesting along the $\{11\}$ directions. Virosztek and Ruvalds have constructed an approximate self-consistent analytic treatment of scattering across a nested Fermi surface and found MFL-like behavior in weak coupling, using a model equivalent to taking $\Phi = \Phi_2$. The Fermi surface also passes through van Hove singularities at the corners of the diamond in the $\{10\}$ directions. The van Hove singularities have been suggested as independent sources of non-Fermi-liquid behavior [5]. In a separate paper [10] we give a detailed comparison between our results and the model of Virosztek and Ruvalds; here we focus on the manifestations of non-Fermi-liquid behavior and compare the perfectly nested 2D model and the 1D model. Since the characteristic features of 1D models also originate from perfect nesting (which occurs automatically, at all fillings), one can think of perfect nesting as a mechanism for converting a 2D model into an effectively 1D model.

Fig. 2 shows results, analogous to those in Fig. 1, for $U = 0.5t$ at half filling. As a function of k along (11), Im $\Sigma(\mathbf{k}, \varepsilon_0)/\varepsilon_0$ has the same notched structure as in 1D, with the depth of the notch growing with decreasing temperature. The temperature dependence on the Fermi surface is again well-described by a power law. The expression $0.0049\,T^{-0.31}$ provides an excellent fit; the prefactor varies by less than $\sim 6\%$ over the Fermi surface, but the exponent varies significantly with direction, reaching ~ -0.47 for k along (10). In Fig. 2 we also show the temperature dependence of Im $\Sigma(\mathbf{k}_F, \varepsilon_0)/\varepsilon_0$ along (11) on a log-log plot, together with the 1D result for the same U and same filling; the 1D and 2D results are strikingly similar, with the only apparent difference being in the prefactor of the corresponding power laws. The MFL model self energy gives a qualitatively similar temperature dependence, but the quality of the fit is not as good as a power law, for any reasonable choice of the cutoff ω_c and coupling strength α. In addition, both the phenomenological MFL self energy and the analytic model of Virosztek and Ruvalds assume that the momentum dependence of the self energy is smooth, in marked contrast to our results.

For nearly perfect Fermi surface nesting we expect that, just as in 1D, the results of the fluctuation exchange approximation will become unreliable when $a_M(\mathbf{k}_F, T)$ approaches 0. In the case of perfect nesting, this will occur at sufficiently low temperature, no matter how weak the bare coupling U, while for imperfect nesting a minimum coupling is required, depending on the deviation from half filling. In either case, the 1D analogy makes it plausible that the strong-coupling crossover is to Anderson's conjectured 2D Luttinger liquid, although we cannot rule out other possibilities, including a highly-renormalized Fermi liquid. With imperfect nesting it is also possible (for small enough U) that the nesting-induced growth of the susceptibilities with decreasing temperature cuts off before the renormalization factor becomes significantly smaller than 1. In this case the low temperature crossover is almost certainly to an ordinary

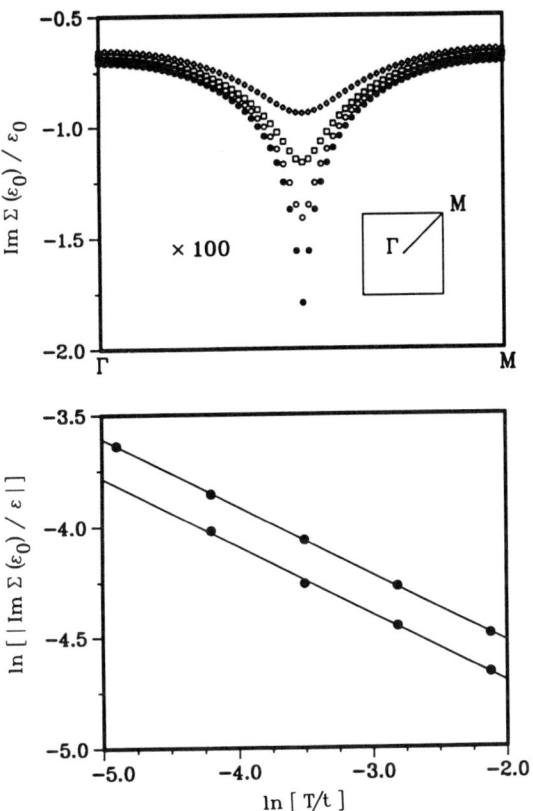

Figure 2. Top: Im $\Sigma(\varepsilon_0)/\varepsilon_0$ as a function of k for k along the (11) direction and $T/t = 0.015$ (•), 0.03 (∘), 0.06 (□), and 0.12 (⋄). Bottom: $\ln|\text{Im }\Sigma(\varepsilon_0)/\varepsilon_0|$ plotted against $\ln T/t$ for the nested 2D Hubbard model (upper) compared with the 1D Hubbard model (lower), with $U = 0.5t$ and $n = 0.5$. The straight lines are power law fits (see text).

2D Fermi liquid. From a calculation with $\Phi = \Phi_2$, we find that for $U = t$ the signs of deviations from Fermi liquid behavior have essentially disappeared by $n = 0.89$ [10].

Quarter Filling

Well away from half filling, divergences of the susceptibilities due to nesting are no longer present and one might at first glance expect the FEA to produce a Fermi liquid without question. However, as pointed out by Schmitt-Rink, Varma, and Ruckenstein [11] for negative U and by Randeria and Engelbrecht [12] for positive U, the particle-particle susceptibility bubble of the noninteracting isotropic Fermi gas exhibits singular behavior at nonzero frequencies for $q \approx 2k_F$. This singular behavior follows from the nonzero limit of the density of states at zero energy in 2D, and leads to a new branch of collective two-electron (two-hole) excitations for $U < 0$ ($U > 0$). In the Hubbard model the density of states is nonzero at both the top and bottom of the band, and so we expect the same sort of singular behavior here as occurs for the free Fermi gas. An important question is then whether these singularities (and associated collective excitations) of the noninteracting particle-particle susceptibility can lead to a marginal Fermi liquid or Luttinger liquid in 2D away from half filling.

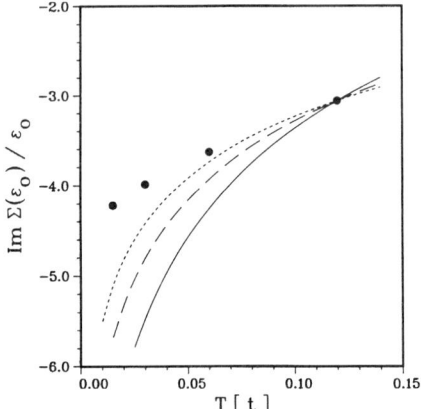

Figure 3. Im $\Sigma(\varepsilon_0)/\varepsilon_0$ plotted against T/t for $U = 8$, $n \sim 0.53$ and **k** on the Fermi surface in the (11) direction (●). The MFL prediction for cutoff energies $\omega_c = 8t$ (short dash), $4t$ (dash), and $2t$ (solid) is shown for comparison. At each cutoff the coupling strength was adjusted to fit the fluctuation exchange approximation at $T/t = 0.12$.

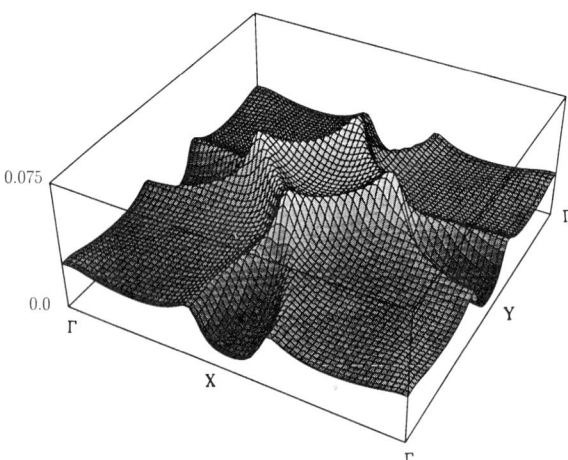

Figure 4. The particle-particle susceptibility, $\chi_{pp}(\mathbf{q}, \omega_m = \omega_1)$ for $U = 8t$, $n = 0.53$ and $T/t = 0.015$ plotted for an extended zone with equivalent Γ points at each corner and M at the center.

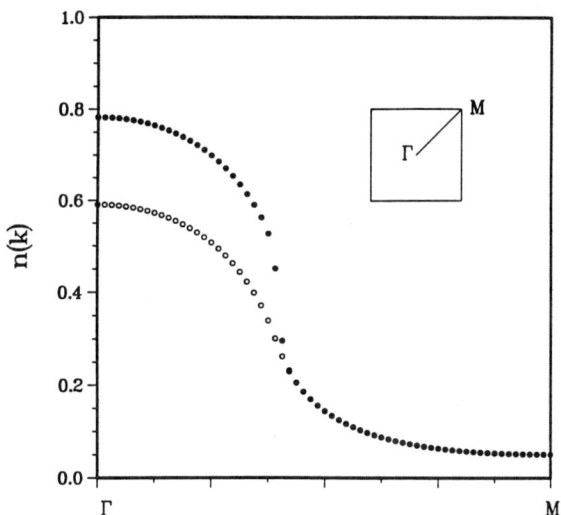

Figure 5. The momentum distribution function $n(\mathbf{k})$ plotted for $\mathbf{k}$ along the (11) direction and $T/t = 0.015$. The open circles are the incoherent part.

Fig. 3 shows the temperature dependence of Im $\Sigma(\mathbf{k}, \varepsilon_0)/\varepsilon_0$ with $\mathbf{k}$ on the Fermi surface in the (11) direction for $U = 8t$ and $n = 0.53$, together with the same quantity calculated from the MFL model for $\omega_c = 8t, 4t, 2t$ and α adjusted to agree with the FEA at one temperature ($\alpha = 0.49, 0.63, 0.88$). The temperature dependence in the (10) direction is similar. Over this temperature range the results from the FEA are inconsistent with the MFL theory for any reasonable choice of parameters. The weak temperature dependence found in the FEA instead suggests a Landau Fermi liquid description. The particle-particle susceptibility bubble in this approximation shows interesting structure as a function of both Matsubara frequency and momentum in the vicinity of $q = 2k_F$. In Fig. 4 we plot the momentum dependence of Im $\chi_{pp}(\mathbf{q}, \omega_m)$ for $\omega_m = 2\pi T$. Apparently this structure is too weak to alter the character of the self-consistent solutions, which probably reflects the circumstance that the collective modes lie so close to the neighboring continuum that they are either not excited or else "ionized" with probability exponentially close to one. We also found no deviations from Fermi liquid behavior in the thermodynamic potential and number density of the negative-U Hubbard model, except for those clearly associated with true superconducting fluctuations (with $q \approx 0$).

To test the Fermi-liquid interpretation of the FEA, we have examined the temperature-dependence of the thermodynamic potential obtained from Eq. (2). For a Fermi liquid, $\Omega(T, \mu)$ should have the form

$$\Omega_{FL}(T, \mu) = \Omega(0, \mu) + \frac{1}{2}\gamma(\mu)T^2 + \cdots, \qquad (13)$$

where the coefficient $\gamma(\mu)$ is given by the same expression as for a noninteracting Fermi gas, but with the noninteracting single particle energy replaced by the quasiparticle energy [13]; this is equivalent to the Fermi surface integral,

$$\gamma(\mu) = (1/6) \int_{FS} ds \frac{a_{\mathbf{k}}^{-1}}{\left|\nabla[\xi_{\mathbf{k}}^0 + \text{Re}\,\Sigma^R(\mathbf{k}, 0)]\right|}. \qquad (14)$$

We have evaluated this expression using a discrete-lattice approximation to the Fermi surface integral and approximating $a_{\mathbf{k}}$ by $a_M(\mathbf{k}, T)$ and Re $\Sigma^R(\mathbf{k}, 0)$ by Re $\Sigma(\mathbf{k}, \varepsilon_0)$. At $T/t = 0.03$, $U = 8t$, and $n \approx 0.53$, we find $\gamma = 1.67/t$, while the two lowest temperature results for $\Omega(T, \mu)$ give $\gamma = 1.62/t$.

In Fig. 5, we show the momentum distribution $n(\mathbf{k})$ at $T/t = 0.015$ with $\mathbf{k}$ along the (11) direction. If the quasiparticle interpretation of our results is correct, then $n(\mathbf{k})$ is the sum of a continuous incoherent background and a quasiparticle part described by a Fermi distribution with weight $a_{\mathbf{k}}$,

$$n(\mathbf{k}) = n_{inc}(\mathbf{k}) + a_{\mathbf{k}} f(\xi_{\mathbf{k}}). \tag{15}$$

Using the same approximations as used for Eq. (14), we evaluate the quasiparticle contribution and obtain the corresponding approximation for $n_{inc}(\mathbf{k})$, which appears to be continuous at k_F. It is interesting to observe that the strongly interacting Fermi liquid momentum distribution function shown in Fig. 5 is essentially indistinguishable from that of a MFL.

Summary

In the framework of the self-consistent fluctuation exchange approximation for the Hubbard model, we find closely analogous deviations from Landau Fermi liquid theory in one and two dimensions in weak coupling ($U \ll W$). In 1D these Luttinger-liquid-like deviations are known to presage the appearance of a true Luttinger liquid ground state, but in 2D the ultimate fate of these anomalies as $T \to 0$ is not obvious to us. Whether the weak-coupling anomalies can explain the MFL phenomenology of the cuprate superconductors (independent of the appropriate zero temperature limit) clearly requires further investigation, although the apparent need for special Fermi surface properties gives cause for some skepticism. Calculations near quarter filling but with relatively strong coupling ($U = 8t$) appear to fit within the Fermi liquid picture, and thus cast considerable doubt on the idea that repeated electron-electron scattering necessarily invalidates Fermi liquid theory in two dimensions.

Acknowledgments

We thank P.W. Anderson and D.M. Newns for stimulating discussions.

References

[1] P.W. Anderson, Phys. Rev. Lett. **64**, 1839 (1990); **65**, 2306 (1990).

[2] F.D.M. Haldane, J. Phys. C: Solid State Phys. **14**, 2585 (1981).

[3] C.M. Varma, P.B. Littlewood, S. Schmitt-Rink, E. Abrahams, and A.E. Ruckenstein, Phys. Rev. Lett. **63**, 1996 (1989).

[4] A. Virosztek and J. Ruvalds, Phys. Rev. B **42**, 4064 (1990).

[5] D.M. Newns, Bull. Am. Phys. Soc. **36**, 386 (1991); C.L. Kane, D.M. Newns, P.C. Pattnaik, C.C. Tsuei and C.C. Chi, these proceedings.

[6] N.E. Bickers, D.J. Scalapino and S.R. White, Phys. Rev. Lett. **62**, 961 (1989).

[7] P.W. Anderson, private communication.

[8] M. Randeria and J.R. Engelbrecht, preprint.

[9] J.M. Luttinger and J.C. Ward, Phys. Rev. **118**, 1417 (1960).

[10] D.W. Hess and J.W. Serene, J. Phys. Chem. Solids, to be published, (1991).

[11] S. Schmitt-Rink, C.M. Varma and A.E. Ruckenstein, Phys. Rev. Lett. **63**, 445 (1989).

[12] J.R. Engelbrecht and M. Randeria, Phys. Rev. Lett. **65**, 1032 (1990) and to be published.

[13] J.M. Luttinger, Phys. Rev. **119**, 1153 (1960).

SLAVE BOSON APPROACH TO THE ONE DIMENSIONAL t-J MODEL

T.K. Lee* and Ziqiang Wang[†]

*Department of Physics
Virginia Polytechnic Institute and State University
Blacksburg, Virginia 24061

[†]Department of Physics
Rutgers University
Piscataway, New Jersy 08854

Abstract: The one-dimensional t-J model is studied using functional integral approach. The separation of spin and charge degrees of freedom is treated in the slave boson representation. The density response function is calculated when holes are described by bosons as well as by spinless fermions à la Jordan-Wigner transformation. We show that the long wavelength limit of the t-J model can be mapped onto the Tomonaga-Luttinger model. The forward scattering amplitudes g_2 and g_4, and the electron momentum distribution exponent α are obtained in terms of the t-J model parameters.

I. Introduction

It is commonly believed that the normal state properties of the high-T_c superconductors may be described by the two-dimensional repulsive large-U Hubbard model or the t-J model [1]. The large on-site Coulomb repulsion between electrons has imposed a very strong constraint on the low energy states of this model. To treat the constraint forbidding double occupancy in a straightforward way, a useful approach is the so-called slave boson (SB) method. In the SB representation, electronic spin degrees of freedom are described by fermionic operators while the charge degrees of freedom by spinless bosons. The constraint that total number of bosons and fermions on each site equals to one is imposed only on average by the introduction of Lagrange multiplier. Due to the strong correlated nature of the system, the success of the SB approach to the t-J model is still limited and little progress has been made beyond the mean-field theory. There exists no exact results on the two-dimensional t-J model that can be used as references to guide intuitions and approximations.

In one dimension, the t-J model may be solved exactly for $J = 2t$ using Bethe ansatz [2]. It can also be studied numerically [3] with less difficulty. Anderson has argued [4] that the correct treatment of the two-dimensional t-J model will ultimately produce a ground state with charge and spin separation. The electron momentum distribution $n(k)$ does not have a discontinuity at the Fermi surface as in Landau Fermi liquid, but rather exhibits power-law singularity. This metallic state is called the Luttinger liquid. The low energy excitations in this case are argued to provide a coherent explanation of the anomalous normal state properties of the copper-oxide superconductors. Up to now, Luttinger liquid has only been found in one-dimensional models. In particular the well known and exactly solubale Tomonaga-Luttinger (TL) model describes the low energy Luttinger liquid fixed point.

In this work we apply the slave boson method to the one-dimensional t-J model. Many recent numerical studies [3,6,7] and Bethe ansatz solutions [2] at $J \ll t$ and $J = 2t$ show indications that the physical properties of the 1D t-J model, and in particular of its ground state is essentially the same as that of the TL model [5] with only forward scattering between electrons. This result is consistent with the renormalization group analysis of one-dimensional Fermi gas model [5] where the backward scattering amplitude renormalizes to zero whereas that of the forward scattering to a finite value. Our goal is to identify the "bare" forward and backward scattering amplitudes in terms of the parameters t and J and then carry out the renormalization group analysis. In the present paper, we consider only the long wavelength or forward scattering with emphasis on the methodology. The effect of backward scattering and the renormalization group analysis will be reported elsewhere.

This paper is organized as follows. Section II contains a brief review on the TL model. In particular, the density response function is shown to be related to the exponent α appeared in the electronic momentum distribution $n(k)$. In Section III, the one-dimensional t-J model is studied using the functional integral formalism in the SB representation. It is shown that including fluctuations to one-loop order, the density response function in the long wavelength and low energy regime has the same form as that of the TL model. The exponent α has the value 0.077 for small hole concentration in the limit of $J \ll t$. The discrepancy with the exact value 1/8 of the Hubbard model is expected since the renormalization due to backward scattering is not included at one-loop order. In Section IV, we consider the slave fermion (SF) approach. The charged boson in the SB approach is transformed into a spinless fermion by a Jordan-Wigner transformation to better treat the hard core nature of the boson. We show that the density response function is again of the same form as that of the TL model. The exponent α is calculated to be 0.26 at $J \ll t$ and $\delta \ll 1$. This result is in very good agreement with the numerical result of Ref.[8] based on variational wavefunction. The results are summarized in Section V.

II. The Tomonaga Luttinger Model

The TL model was reviewed in detail by Sólym [5]. The electrons are separated into left and right moving branches and the Hamiltonian is

$$H_0 = \sum_{k\sigma} v_F(k - k_F)a^\dagger_{k\sigma}a_{k\sigma} + \sum_{k\sigma} v_F(-k - k_F)b^\dagger_{k\sigma}b_{k\sigma}. \tag{1}$$

The scattering takes place only in the forward direction according to

$$H_1 = \sum_{k_1 k_2 q \sigma \sigma'} g_2(a^\dagger_{k_1\sigma} a_{k_1-q\sigma} b^\dagger_{k_2\sigma'} b_{k_2+q\sigma'})$$
$$+ \frac{1}{2} \sum_{k_1 k_2 q \sigma \sigma'} g_4(a^\dagger_{k_1\sigma} a^\dagger_{k_2\sigma'} a_{k_2+q\sigma'} a_{k_1-q\sigma} + b^\dagger_{k_1\sigma} b^\dagger_{k_2\sigma'} b_{k_2+q\sigma'} b_{k_1-q\sigma}) \quad (2)$$

This model can be solved exactly. The response function of the density operator $\rho_q = \sum_{k\sigma}(a^\dagger_{k\sigma}a_{k+q\sigma} + b^\dagger_{k\sigma}b_{k+q\sigma})$ has the form

$$\langle \rho_q \rho_{-q} \rangle = -\frac{2}{\pi}\gamma_\rho \frac{u_\rho q^2}{\omega^2 - u_\rho^2 q^2}, \quad (3)$$

where ω is the frequency. The sound velocity u_ρ and the parameter γ_ρ are given by

$$u_\rho^2 = (v_F + g_4/\pi)^2 - g_2^2/\pi^2 \quad (4)$$
$$\gamma_\rho = \left(1 + \frac{g_4 - g_2}{\pi v_F}\right)^{1/2} / \left(1 + \frac{g_4 + g_2}{\pi v_F}\right)^{1/2}. \quad (5)$$

The exact Green's function for the electrons can be evaluated [5]. Its momentum distribution $n(k)$ has a power law type behavior near Fermi surface

$$n(k) = \frac{1}{2} - \text{const.}|k \mp k_F|^\alpha \text{sgn}(\pm k - k_F) \quad (6)$$

where

$$\alpha = \frac{(\gamma_\rho - 1)^2}{4\gamma_\rho}. \quad (7)$$

The density-density correlation function at $2k_F$ and $4k_F$ has the form

$$\langle \rho(r)\rho(0) \rangle = \text{const.} + A_0 r^{-2} + A_2 r^{-(1+\gamma_\rho)} \cos(2k_F r) + A_4 r^{-4\gamma_\rho} \cos(4k_F r), \quad (8)$$

which shows the typical $2k_F$ and $4k_F$ low energy oscillations. Detailed analysis can be found in Ref.[5].

III. Slave boson (SB) approach

The t-J model Hamiltonian is given by

$$H = -t \sum_{\langle ij \rangle \sigma}(d^\dagger_{i\sigma} d_{j\sigma} e^\dagger_j e_i + h.c) + J \sum_{\langle ij \rangle}(\vec{S}_i \cdot \vec{S}_j - \frac{1}{4}n_i n_j) \quad (9)$$

where $\langle ij \rangle$ denotes a nearest neighbor pair and the second term is the spin-1/2 Heisenberg interaction. The density operator $n_i = \sum_\sigma d^\dagger_{i\sigma} d_{i\sigma}$. In this Hamiltonian the fermion operators $d^\dagger_{i\sigma}$ describe the spin degrees of freedom whereas the boson operators $e^\dagger_i$ represent the charge degrees of freedom and keep track of empty sites. An important part of the Hamiltonian (9) is the holonomic constraint $\sum_\sigma d^\dagger_{i\sigma} d_{i\sigma} + e^\dagger_i e_i = 1$, which will be treated on average by introducing Lagrange multiplier in the imaginary time functional integral form of the partition function. Following Affleck and Marston [9], we rewrite the magnetic term

$$J\sum_{\langle ij \rangle}(\vec{S}_i \cdot \vec{S}_j - \frac{1}{4}n_i n_j) = -\frac{J}{2}\sum_{\langle ij \rangle}(\sum_{\sigma} d^\dagger_{i\sigma}d_{j\sigma})(\sum_{\sigma'} d^\dagger_{j\sigma'}d_{i\sigma'}) - \frac{J}{2}\sum_{\langle ij \rangle} n_i n_j.$$

The second term will be neglected in the following discussion whereas the first term can be decoupled by introducing link variables $D_{ij} = J/2\langle \sum_\sigma d^\dagger_{i\sigma}d_{j\sigma}\rangle$ through a Hubbard-Stratanovich transformation leading to the following Lagrangian

$$\begin{aligned}L = &\sum_{i\sigma} d^\dagger_{i\sigma}\frac{\partial}{\partial \tau}d_{i\sigma} + \sum_i e^\dagger_i \frac{\partial}{\partial \tau}e_i \\ &+ \frac{2}{J}\sum_{\langle ij \rangle}|D_{ij}|^2 + \sum_{\langle ij \rangle}(D^*_{ij}\sum_\sigma d^\dagger_{i\sigma}d_{j\sigma} + h.c.) \\ &- t\sum_{\langle ij\sigma \rangle}(d^\dagger_{i\sigma}d_{j\sigma}e^\dagger_j e_i + h.c.) - \mu\sum_{i\sigma}d^\dagger_{i\sigma}d_{i\sigma} \\ &+ i\sum_i \lambda_{i\sigma}(d^\dagger_{i\sigma}d_{i\sigma} + e^\dagger_i e_i - 1) \end{aligned} \qquad (10)$$

where μ is the chemical potential. The field $d_{i\sigma}$ becomes now a Grassmann variable and e_i and D_{ij} are complex scalar fields. λ_i is the Lagrange multiplier. The partition function at temperature β^{-1} is given by the functional integral

$$Z = \int \mathcal{D}d\mathcal{D}d^\dagger \mathcal{D}D\mathcal{D}D^* \mathcal{D}e\mathcal{D}e^\dagger \mathcal{D}\lambda \exp(-\int_0^\beta d\tau L). \qquad (11)$$

Define the scalar field $e_i \equiv \sqrt{\rho_i}e^{i\theta_i}$ and make the following gauge transformation: $d_{i\sigma} \to d_{i\sigma}e^{i\theta_i}$; $D_{ij} \to D_{ij}e^{-i\theta_i}$ and $\lambda_i \to \lambda_i - \partial_\tau \theta_i$, we see that the phase of the boson field θ_i can be absorbed into λ_i through gauge transformations and therefore does not appear in the Lagrangian. We will work in the unitary gauge with real scalar field ρ_i and λ_i which becomes now dynamical.

The uniform mean field theory is obtained by choosing $\langle \rho_i \rangle = \rho_0$, $\langle D_{ij}\rangle = D_0$ and $\langle i\lambda_i \rangle = \lambda_0$. The Green's function for the fermion field $d_{k\sigma}$ is

$$G_\sigma(k, i\omega_n) = \frac{1}{i\omega_n - \epsilon(k)}$$

where $\epsilon(k) = -2\Delta \cos k$ and $\Delta = D_0 + \rho_0 t$. The mean field parameters are obtained at zero temperature as

$$D_0 = J\frac{\sin k_F}{\pi}, \qquad \rho_0 = 1 - \frac{2k_F}{\pi}, \qquad \lambda_0 = \frac{4t}{J}D_0,$$

which determines ρ_0 to be the hole concentration δ. This mean field state is formally justified in the limit of large spin degeneracy [9], and would be simply a one-dimensional Fermi liquid. However inelastic scattering effect is enhanced as only forward and backward channels are available due to one dimensionality. We will show below that including one-loop fluctuations the theory is identical to the TL model in the low energy and long wavelength limit, and thus the system is in fact a Luttinger liquid.

Integrating out the Grassmann field $d_{i\sigma}$ leaves one with four coupled fluctuating scalar fields: $\delta\rho_i$, δR_{ij}=real part of δD_{ij}^*, δA_{ij}=imaginary part of δD_{ij}^* and $\delta\lambda_i$. On the level of one-loop calculation, which is equivalent to the random phase approximation, modes of different wavelengths do not couple. The propagators of these fields can be calculated analytically. Retaining only the long wavelength and low frequency fluctuations, we obtain the density response function

$$\langle \frac{1}{N_s}\sum_{k\sigma} d^\dagger_{k\sigma}d_{k+q\sigma}, \frac{1}{N_s}\sum_{k'\sigma'} d^\dagger_{k'\sigma'}d_{k'-q'\sigma'}\rangle = -\frac{\rho_0 t}{\pi\Delta}\frac{2v_F q^2}{\omega^2 - v_s^2 q^2} \quad (12)$$

where the Fermi velocity $v_F = 2\Delta\sin k_F$ and the sound velocity v_s is given by

$$v_s^2 = \frac{\rho_0 t}{\Delta}\left[v_F^2 + \frac{2v_F}{\pi}(4t\cos k_F - J\cos^2 k_F)\right]. \quad (13)$$

This form is identical to the density correlation of the TL model (Eq.3) with $u_\rho = v_s$ and

$$\gamma_\rho = v_F \rho_0 t / v_s \Delta. \quad (14)$$

The forward scattering amplitudes g_2 and g_4 defined in equation (2) can be extracted as

$$g_2 = 4t\cos k_F - J\cos^2 k_F, \quad g_4 = 4t\cos k_F - J. \quad (15)$$

In the limit $J \ll t$ and for small hole concentration, we have $\gamma_\rho = 1/\sqrt{3}$. From equation (7) we obtain the exponent $\alpha \simeq 0.077$. This result deviates from the exact result of the Hubbard model [7] which has $\gamma_\rho = 1/2$ and $\alpha = 1/8$. We expect that the backward scattering will renormalize the values of g_2 and g_4. This work will be discussed elsewhere.

In the SB approach, the effect of $4k_F$ correlation (Eq.8) is not easily seen. However this effect is easily obtained in the slave fermion (SF) representation where the boson e_i is turned into a fermion by a Jordan-Wigner transformation. The latter approach is apparently better suited to incorporate the hard-core nature of the e_i bosons.

IV Slave fermion (SF) approach

The boson operators e_i representing holes in equation (9) can be transformed by

$$e_i = f_i \exp\left(i\pi \sum_{l<i} f_l^\dagger f_l\right) \quad (16)$$

where f_i is a spinless fermion operator. It is easy to verify, using Eq.(16), $e_j^\dagger e_i = f_j^\dagger f_i$ for nearest neighbors i and j such that the Hamiltonian (9) is formally unchanged. The corresponding Lagrangian has the form

$$L = \sum_{i\sigma} d_{i\sigma}^\dagger \frac{\partial}{\partial \tau} d_{i\sigma} + \sum_i f_i^\dagger \frac{\partial}{\partial \tau} f_i$$
$$+ \frac{2}{J}\sum_{\langle ij\rangle} |D_{ij}|^2 + \sum_{\langle ij\rangle}[(D_{ij}^* \sum_\sigma d_{i\sigma}^\dagger d_{j\sigma} + \frac{2t}{J} f_i^\dagger f_j) + h.c.]$$
$$+ \frac{J}{2t^2}\sum_{\langle ij\rangle}[F_{ij}^2 + iF_{ij}(f_i^\dagger f_j + f_j^\dagger f_i)]$$
$$+ \frac{J}{2t^2}\sum_{\langle ij\rangle}[A_{ij}^2 + A_{ij}(f_i^\dagger f_j - f_j^\dagger f_i)]$$
$$- \mu \sum_{i\sigma} d_{i\sigma}^\dagger d_{i\sigma} + i \sum_{i\sigma} \lambda_i (d_{i\sigma}^\dagger d_{i\sigma} + f_i^\dagger f_i - 1) \tag{17}$$

where F_{ij} and A_{ij} are real scalar fields. Since the Lagrangian is invariant under the local gauge transformation discussed under equation (10), we may choose the gauge such that D_{ij} is real.

In the mean field theory, we have $\langle D_{ij}\rangle = D_0$, $\langle iF_{ij}\rangle = F_0$, $\langle A_{ij}\rangle = 0$ and $\langle i\lambda_i\rangle = \lambda_0$. The Green's function for the spin-1/2 fermions (spinons) $d_{i\sigma}$ and the spinless fermions (holons) are

$$G_\sigma(k, i\omega_n) = \frac{1}{i\omega_n - \epsilon_\sigma(k)}, \qquad G_0(k, i\omega_n) = \frac{1}{i\omega_n - \epsilon_0(k)}$$

respectively, where the energies are $\epsilon_\sigma(k) = -2\Delta_\sigma \cos k$ and $\epsilon_0(k) = -2\Delta_0 \cos k$. The mean field parameters at zero temperature are given by

$$\Delta_\sigma = J \frac{\sin k_F}{\pi} + t\frac{\sin k_F'}{\pi}, \qquad \Delta_0 = 2t\frac{\sin k_F}{\pi}$$

and $F_0 = 2t\Delta_\sigma/J - \Delta_0$. The Fermi wave vector for holons is $k_F' = \pi\delta = \pi - 2k_F$, where δ is the hole concentration.

The advantage of the SF method is that the low energy excitations at wave vector $q = 4k_F = 2\pi - 2k_F'$ are explicitly included. Hence all low energy excitations are already present even in the mean field theory. The oscillator strength of these modes may be renormalized by including higher order corrections. The other merit of the SF approach lies in the ground state energy. Using variational Monte Carlo method, Hellberg and Mele [8] showed that the mean field state formulated above, where both spinon and holon wavefunctions are Slater determinants, has lower variational energy then the mean field state of the SB approach with only the spinon described by an occupied Fermi-sea wavefunction. The numerical calculations were done with the constraint that there are no two particles on the same site at a given time.

Integrating out the fermion fields $d_{i\sigma}$ and f_i, we obtain the propagators of the four coupled real scalar fields $(\delta D_{ij}, \delta F_{ij}, \delta A_{ij}, \delta \lambda_i)$ at one-loop order. In particular, the hole density response function in the long wavelength limit is obtained as

$$\langle \frac{1}{N_s}\sum_k f_k^\dagger f_{k+q}, \frac{1}{N_s}\sum_{k'} f_{k'}^\dagger f_{k'-q'}\rangle = \langle \frac{1}{N_s}\sum_{k\sigma} d_{k\sigma}^\dagger d_{k+q\sigma}, \frac{1}{N_s}\sum_{k'\sigma'} d_{k'\sigma'}^\dagger d_{k'-q'\sigma'}\rangle$$
$$= -\frac{J}{(2t\cos k_F')^2}[1 + \frac{4}{J}\langle \delta F_q, \delta F_{-q}\rangle] = -\frac{1}{\pi}\frac{v_F' q^2}{\omega^2 - v_s^2 q^2} \tag{18}$$

where
$$v_s^2 = v_F'^2 + \frac{v_F'}{\pi}\left(4t\cos k_F \cos k_F' - J\cos^2 k_F + \frac{\pi}{2}v_F\right). \tag{19}$$

Here the holon and spinon Fermi velocities are given by $v_F' = 2\Delta_0 \sin k_F'$ and $v_F = 2\Delta_\sigma \sin k_F$ respectively. Neglecting the renormalization effect due to backward scattering and comparing equations (18) and (3) we find $\gamma_\rho = v_F'/2v_s$ and $u_\rho = v_s$. This result can also be obtained by integrating out holons f_i, giving rise to an effective Lagrangian for the spin-1/2 spinons $d_{i\sigma}$ whose long wavelength part has a form consistent with the TL model with γ_ρ and u_ρ given above. Alternatively, we could integrate out the spinon fields $d_{i\sigma}$. This results in an effective Lagrangian for the holons which is formally identical to the spinless fermion version of the TL model. In the latter case of spinless fermions, equation (3) should be reduced by a factor of 2. Thus using equation (18), we obtain for holons $\gamma_\rho' = v_F'/v_s = 2\gamma_\rho$. Instead of giving by equations (6) and (7), the momentum distribution function $n'(k) = \langle f_k^\dagger f_k \rangle$ has the form

$$n'(k) = \frac{1}{2} - \text{const.}|k \mp k_F'|^{\alpha'} \text{sgn}(\pm k - k_F') \tag{20}$$

where
$$\alpha' = \frac{(\gamma_\rho' - 1)^2}{2\gamma_\rho'}. \tag{21}$$

The density-density correlation function near $2k_F'$ will be of the form $r^{-2\gamma_\rho'}\cos(2k_F' r)$. Since $\gamma_\rho' = 2\gamma_\rho$ and $2k_F' = 2\pi - 4k_F$, this result agrees with that of the TL model in equation (8).

In the limit of $J \ll t$ and for small hole concentration $\delta \ll 1$, we have $\gamma_\rho = 1/\sqrt{7}$, and $\alpha \simeq 0.256$ for spinons, and $\gamma_\rho' = 1$ and $\alpha' = 0.04$ for holons. These values can be compared with exact results in this limit, i.e. $\gamma_\rho = 1/2$, $\alpha = 1/8$, $\gamma_\rho' = 1$ and $\alpha' = 0$. It is interesting to note that the numerical result on $\langle c_{k\sigma}^\dagger c_{k\sigma} \rangle$ obtained by Hellberg and Mele [8] using trial wavefunction of the mean field state discussed in this section gives a power law exponent $\simeq 0.27 \pm 0.01$ which is very close to the value of α derived above.

V. Summary and Conclusions

The main results of the TL model was briefly reviewed. The one-dimensional t-J model is first studied using the slave boson approach. The long wavelength density response function obtained in the one-loop calculation has a similar form as that of the TL model, although the power law exponent α of the momentum distribution $n(k)$ near k_F thus obtained does not agree with the exact results. This indicates the importance of the renormalization effects due to backward scattering. The low energy excitations near $4k_F$ are not included in the mean field theory of the SB approach.

The slave fermion approach based on Jordan-Wigner transformation is favorable in several aspects. It is better suited to treat the hard-core nature of the bosons imposed by the constraint. As in the exact solution, the mean field theory already produces low energy excitations at wave vectors $q \simeq 0$, $2k_F$ and $4k_F$. This mean field state also has very good variational energies [8]. After a one-loop calculation is carried out, the density response function for small q is again similar to that of the TL model. The momentum distribution exponent obtained in this state is very close to the numerical result based on variational Monte Carlo method.

What we have reported here is a first step in setting up the formalism for a comprehensive study of the one-dimension t-J model. Both the exact results from Bethe ansatz and results based on the Tomonaga-Luttinger model can be used as references for carefully examining approximations used in the calculations. It also helps one to better understand the slave boson method applied to strongly correlated electron systems in general. The effect of renormalization due to backward scattering and higher order calculations based on the results presented here are in progress.

Acknowledgments: We would like to thank professors S.E. Barnes, J. Ashkenazi and the University of Miami for hospitality. We are grateful to D.H. Lee, G. Kotliar and X.F. Wang for useful discussions. T.K. wishes to acknowledge a partial support of this work from Thomas F. and Kate Miller Jeffress Memorial Trust. Z.W. was supported in part by National Science Foundation Grant No. DMR-8915895.

References

[1] "High Temperature Superconductivity", edited by K. Bedell, D. Coffey,

D. Mettzer, D. Pines and J.R. Schrieffer, (Addison-Wesley 1990).

[2] N. Kawakami and S.K. Yang, Phys. Rev. Lett. **65**, 2309 (1990);

P.A. Bares and G. Blatter, Phys. Rev. Lett. **64**, 2567 (1990).

[3] F.F. Assaad and D. Würtz, preprint;

M. Ogata, M. Luchini, S. Sorella and F.F. Assaad, preprint.

[4] Y. Ren and P.W. Anderson, paper in this proceeding.

[5] J. Sólym, Adv. in Phys. **26**, 201 (1979).

[6] A. Parola and S. Sorella, Phys. Rev. Lett. **64**, 1831 (1990).

[7] M. Ogata and H. Shiba, Phys. Rev. **B41**, 2326 (1990).

[8] C.S. Hellberg and E.J. Mele, preprint.

[9] I. Affleck and B. Marston, Phys. Rev. **B37**, 3774 (1988).

UNIVERSALITY OF BEHAVIOR OF ELECTRONS ON A RING

F.V. Kusmartsev

Department of Physics, University of California
Los Angeles, California 90024-1547
Institut für Theoretische Physik
der Universität zu Köln
D-5000 Köln 41,Germany

and

L.D. Landau Institute for Theoretical Physics
Moscow,117940, GSP-1, Kosygina 2, V-334, USSR

Using the Bethe-ansatz method we have calculated exactly the spectrum of fermions located on a ring in a transverse magnetic field within the framework of models associated with Heisenberg-Ising spin-chains and the Hubbard Hamiltonian. We find a universality in the behavior of spinless fermions on the ring which proves the existence of the flux phase state. We show that, with an even number of spinless fermions on the ring, there always exists a flux phase state. This conclusion does not depend on the interactions between spinless fermions. On the other hand, the spin degrees of freedom may strongly change the magnetic properties of the ring. We have also calculated, exactly, the spectrum of electrons on a ring in a magnetic field in a model which includes the Hubbard interaction. We have found that, for a ring, there exist very unusual resonances when the flux through the ring is changed. The appearance of these resonances depends critically upon the total number of electrons N. They appear when the flux is changed by $\Delta f = 1/N$. In analogy with two-particle bound state, one can appreciate that, on the ring with strong interaction between electrons, there appears a N-particle bound state. Such resonances might be found in aromatic molecules.

PACS numbers: 71.10, 72.15.Lh, 73.6 A.q

INTRODUCTION

The discovery of the Quantum Hall Effect and of high T-c superconductivity has motivated the intense interest in strongly-correlated two-dimensional electronic systems. There are a lot of attempts to find an analogy between these two phenomena. In the search for this analogy a flux phase state for the ground state of such systems has been constructed. This state of the condensed many-fermion system on a two dimensional lattice is characterized by spontaneous orbital currents and violates the fundamental symmetries C,P and T[1-5].

There exists a numerical proof of the possibility of the existence of flux phase states on a lattice [6,7]. In these works it has been shown that the energy of the electrons on the lattice in magnetic field is lower than the energy of free electrons without field. The presence of the magnetic field means that the hopping amplitude is a complex number and the hamiltonian breaks time reversal symmetry. If the hopping amplitude is a complex number it should be that there is some physical reason for the time reversal symmetry breaking of the hamiltonian. If such a cause is the magnetic field it is not clear that the energy of the total system, consisting of the energy of electrons in magnetic field and of the energy of a magnetic field, can be lower than the energy of electrons without field. In other words it is not clear which energy is lower, the energy of the electrons in a magnetic flux, including the pure magnetic energy or the energy of free electrons without field. In order to show that the total energy of fermions and a field may be lower than the energy of free fermions without field in, our previous work we have considered a simple example: a quasi-two-dimensional system which is a ring in magnetic field [8], where by direct diagonalization we calculated exactly the spectrum of electrons located on a ring in the tight-binding approximation in a transverse magnetic field. We have shown that, for an even number of electrons on the ring, there exist a flux phase state. That is we have shown that for an even number of electrons on the ring the total electron energy decreases linearly with the flux, and therefore is larger than the energy of the pure magnetic field which increases quadratic. The analogous ideas have been considered in quasiclassical approximation in Ref.[9]. In present work we show that this phenomenon has universal character and qualitatively does not depend on the interaction between fermions. For an odd number of fermions such phenomenon does not exist.

1. MODEL 1

In order to show that the existence of the flux phase state, on a ring with interactions between fermions depends on the even of odd nature of the number of particles, we consider the Heisenberg-Ising spin-chain model (see [10] and reference therein) governed by the Hamiltonian

$$H = -\frac{1}{2}\sum_{j=1}^{N}(\sigma_j^x\sigma_{j+1}^x + \sigma_j^y\sigma_{j+1}^y + \delta\sigma_j^z\sigma_{j+1}^z) \tag{1}$$

The eigenfunction of this Hamiltonian may be given by Bethe's ansatz in the form [11]:

$$\psi(x_1,\ldots,x_M) = \sum_P A(P)exp[i\sum_{j=1}^{M}P_{Pj}x_j] \tag{2}$$

where P is a permutation of integers from 1 to M and $A(P)$ are $M!$ coefficients which may be determined from the Bethe equations[11,10]. These equations define the variables P_j which corresponds to each of the M fermions or pseudofermions. Although the spins can be described by fermions, as it is well known, there are two different cases which are usually called the "c-cyclic" problem and "a-cyclic" problem[12]. The "a-cyclic" problem describes directly the chain of spins on a ring. The "c-cyclic" problem describes real fermions on a ring. The Hamiltonian, describing real fermions and associated with Hamiltonian (1) is (see, for details [12]):

$$H = -\frac{1}{2}\sum_{j=1}^{N}(c_j^+c_{j+1} + \delta(c_j^+c_j - \frac{1}{2})(c_{j+1}^+c_{j+1} - \frac{1}{2})). \tag{3}$$

Where c_i^+ and c_j are the fermion creation and annihilation operators, respectively; the sum is over nearest-neighbor bonds and N is the number of sites on the ring.

The behavior of these two systems in magnetic field is quite different. Generally speaking the "a-cyclic" problem and "c-cyclic" problem are different because of the boundary conditions. For "a-cyclic" problem the boundary conditions are always periodic. For "c-cyclic" problem the choice depends on the total number of fermions (even or odd number). For an even number of fermions they are aperiodic, while for an odd number of particles they are again periodic. Therefore in the "c-cyclic" problem the behavior of the system strongly depends on the filling factor.

This problem has been studied, in detail, in the thermodynamic limit [13-17], where solutions have been found with the aid of Wiener-Hopf techniques of Yang and Yang [13].

Let us consider the discrete case when on the ring there exist N sites with M particles. Such a case has already been considered for "a-cyclic" problem with some values of the parameter δ [18]. Since the effect of the magnetic flux may be described by imposing twisted boundary condition (see, for details [18-20]), the familiar transcendental equations for the wavevectors of the Bethe-ansatz solution are modified only by the addition of the phase $2\pi f$ to the usual boundary conditions:

$$NP_j = 2\pi I_j + 2\pi f - \sum_{l=1, l \neq j}^{M} \theta(P_j, P_l) \tag{4}$$

where the phase shift $\theta(P_i, P_j) = -\theta(P_j, P_i)$ is equal to

$$\theta(p,q) = 2\arctan\left[\frac{\delta \sin(\frac{p-q}{2})}{\cos(\frac{p+q}{2}) - \delta\cos(\frac{p-q}{2})}\right] \tag{5}$$

The values I_j are integer quantum numbers which correspond to the location of the particles on the levels of the quantization of the Hamiltonian (1). The distribution of the I_j is different for even or odd numbers of particles and also depends upon the flux $2\pi f$. For the ground state and the case of $2\pi f = 0$ with an even number of particles we have (see[12])

$$I_1, \ldots, I_{2K} = -K, -(K-1), \ldots, -1, 0, 1, \ldots, K-1 \tag{6}$$

In the case of an odd number of particles, the distribution is different from (6), and has a symmetrical form

$$I_1, \ldots, I_{2K+1} = -K, -(K-1), \ldots, -1, 0, 1, \ldots, K-1, K \tag{7}$$

For $2\pi f \neq 0$ the set of I_j strongly depends on the value of $2\pi f$. Let us consider two different cases: a) the case of zero interaction between particles ($\delta = 0$) and b) the case when there are infinitely strong interaction ($\delta \to \infty$).

The first case, of free fermions on a ring, we have already considered by direct diagonalization of the Hamiltonian (3) [8]. Here we obtain the same result by the solution of Bethe equations, which at $\delta = 0$, have the simple form

$$P_j = \frac{2\pi I_j}{N} + \frac{2\pi f}{N}, \tag{8}$$

where the set of values I_j depends on the value of the flux $2\pi f$. From (8) we obtain the expression for the total energy:

$$E = \sum_j E_j = -2 \sum_j \cos(P_j), \tag{9}$$

where $E_j = -2\cos P_j$ are one particle energies. Here we neglect the constant shift in energy which does not depend on the flux. Using (8), (6) and (7) we obtain the following expressions for the ground state energy:

1) For an even number of particles ($M = 2K$)

$$E_{gr}^e = -2D[\alpha cos(\frac{2\pi f}{N}) + \beta sin(\frac{2\pi f}{N})] \qquad (10)$$

where D is the positive constant: $D = sin(\frac{\pi M}{N})/sin(\pi/N)$; $\alpha = cos(\pi/N)$ and $\beta = sin(\pi/N)$ are also positive constants and $0 \leq f \leq 1$;

2) For an odd number of particles ($M = 2K + 1$):

$$E_{gr}^o = -2Dcos(\frac{2\pi f}{N}) \quad , \quad \frac{-1}{2} \leq f \leq \frac{1}{2} \quad . \qquad (11)$$

The the ground state energy is a periodic function of the flux with a period $f_T = 1$. For the interval $1 \leq f \leq 2$ we should choose another set of the quantum numbers I_j. Instead of (6) we should take

$$I_1, \ldots, I_{2K} = -K - 1, -K, \ldots, -1, 0, 1, \ldots, K - 2 \qquad (12)$$

and so on.

Let us consider now the second case of an infinitely strong interaction, i.e., $\delta \to \pm\infty$. In this case the Bethe equations are also simplified and instead of (4) and (5) we have,

$$P_1 + P_2 + \ldots + P_j(N - M + 1) + \ldots + P_M = 2\pi I_j + 2\pi f. \qquad (13)$$

This equation has no solutions when $M = N$. Let $M < N$, then we have the following solution,

$$P_i = \frac{2\pi}{NM} \sum_j I_j + \frac{2\pi f}{N} + \frac{2\pi}{M(N-M)} \sum_j (I_i - I_j). \qquad (14)$$

For this solution, one must choose, in an appropriate way, the quantum numbers I_j. By this one means that this choice should correspond to the ground state. This case is absolutely equivalent to a free particle state, that is we should consider independently the cases of the even and odd number of particles. Then for a even number of fermions we should choose the quantum number I_j for each region of f, $k < f < (k+1)$, where k is arbitrary integer number. The same procedure should be followed for an odd number of particles. In this case a different sets of quantum numbers I_j corresponds to the regions $\frac{1}{2}(2k - 1) < f < \frac{1}{2}(2k + 1)$, where k is arbitrary integer. Using the formula for P_i one obtains, for the ground state energy, the expressions similar to (10) and (11): For even number of fermions

$$E_{gr}^e = -2D_1[\alpha cos(\frac{2\pi f}{N}) + \beta sin(\frac{2\pi f}{N})] \qquad (15)$$

where D_1 is a positive or a negative constant which characterizes the system with interactions:

$$D_1 = \frac{sin(\frac{\pi M}{N-M})}{sin(\frac{\pi}{N-M})}. \qquad (16)$$

Below we consider the case when D_1 is positive.

For odd number of fermions $M = 2K + 1$ the expression for the ground state energy has another universal form:

$$E^o_{gr} = -2D_1 \cos(\frac{2\pi f}{N}) \quad , \quad \frac{-1}{2} \le f \le \frac{1}{2} \quad , \tag{17}$$

For other values of f these functions should be periodically continued. As one can see from the results obtained, for even number of particles the energy always decreases linearly with increasing of flux f starting from zero (see, (10), (15)). This property of fermions on a ring is very important and plays a key role in the existence of the flux phase state on the ring. For a homogeneous distribution of the magnetic field on the ring the energy of this field is quadratically dependent on the flux. Therefore the minimum of total energy will correspond to the spontaneous nonzero value of the flux f_{sp}. $(f_{sp} \neq 0)$

That is, if $\frac{2\pi f}{N} \ll 1$, then from (10) or (15) we have

$$E^e_{gr} = -2D_i(\alpha + f\frac{2\pi\beta}{N}), \tag{18}$$

where D_i may be D or D_1. This property of the ground state energy for an even number of the electrons leads to the existence of a flux phase state. The flux for this state f_{sp} may be determined by a minimization of the total energy, E_Σ, consisting of the energy E^e_{gr} and the energy of magnetic field $E_M = Cf^2$:

$$E_\Sigma = Cf^2 - 2D_i\alpha - \frac{4\pi\beta D_i}{N}f \tag{19}$$

where C is a some positive constant which may be a self-inductance of a some cylinder of cyclic polymer molecules. In this case the magnetic field f will be the average field generated by orbital currents of $L - 1$ rings which acts on L-th ring. From this expression we find the value of the spontaneously generated magnetic flux:

$$f_{sp} = \frac{2\pi D_i \beta}{CN} \tag{20}$$

with energy $E_{\Sigma sp} = -1/C(2\pi D_i\beta/N)^2 - 2D_i\alpha$ which is lower than the energy of the free electrons without flux $E_{free} = -2D_i\alpha$ (ref. [8]).

For an odd number of fermions on a ring this property of ground state energy is absent. At small f the ground state energy is quadratically depend on f with the minimum at $f = 0$ (see formula (11) and (17)). On the other hand when D_1 is negative, still for a the ring, there may exist an anionic state, i.e., in this case, with an odd number of particles, the state with half-integer flux corresponds also to cusp like energy maximum and which will always be present.

The other "a-cyclic" problem, associated with Hamiltonian (1), has already been considered in Ref. [18], for some values of the parameter δ. We want to point out several details which have been missed in Ref. [18]. For the "a-cyclic" problem the expression of the ground state energy always has the form (11) or (17). The interaction changes only the coefficient D. In contrast to the result obtained in [18], the ground state energy is a periodic function of the flux f with period $f_T = 1$, it has cusp like maxima at half-odd values of flux and smooth minima at integer values of flux. For the "a-cyclic" problem, this result does not depend on the parity nor on having an odd number of particles in the system.

On the other hand the behavior of the fermionic system (the "c-cyclic" problem) is quite different. The location of the cusp maxima and smooth minima correspond to there being on even or odd number of fermions. For even number of electrons these cusp like maxima are located at integer values of the flux. For an odd number of fermions, cusp like maxima correspond to a half-odd quantum of the flux f. The cusps correspond to phase transitions to a state with a new set of quantum numbers I_j.

Generally speaking each quantum number I_j corresponds to an energy level on which one fermion is localized. The total energy is the sum of these one fermion energies (see (9) and also in Ref. [8]). The total current is the sum of one fermion currents J_{I_j}. The sum of these one fermion currents is the total current on the ring. It is obvious from (15) that for zero magnetic field and for a even number of fermions the total current is not equal to zero. It is equal to

$$J(0) = \frac{4\pi D_i}{N} sin(\frac{\pi}{N}). \qquad (21)$$

The appearance of this persistent current can be explained by means of group theory [21]. Without a magnetic field the system has a C_{NV} symmetry group. This group has one dimensional and two dimensional irreducible representations. The first level corresponds to one dimensional irreducible representation A_1. The other levels (except the highest one) are two-fold degenerated and therefore correspond to the E_2 representation of the C_{NV} symmetry group. In magnetic field, the irreducible representation E_2 transform into reducible ones which decay into irreducible representations A_1. Thus all degenerate levels split into nondegenerate ones and the symmetry of the system is broken. In such a way there is a generation of a spontaneous magnetic flux which is equivalent to the Jahn-Teller phenomenon [21]. This phenomenon has a general character and may occur in any ring system with degenerate levels.

Thus, for an even number of electrons, upon increasing the magnetic field on the ring from zero the total current decreases. At the value of the magnetic flux $f_c = \frac{1}{2}$ the total current equals zero. This value of the flux corresponds to the absolute minimum of the total energy. For other values of f this function should be periodically continued. This function has a cusp like maxima at $f = (2l-1)/2$ and smooth minima at $f = l$ where l is an arbitrary integer. At these points, $f = l$, and the total current is equal to zero. The expression describing the total current at arbitrary f has a form

$$J^o = \frac{4\pi D_i}{N} sin(\frac{2\pi}{N}(f - \frac{2l-1}{2})) \quad , \quad \frac{2l-3}{2} \leq f \leq \frac{2l-1}{2}, \qquad (22)$$

and one can see from this formula that at $f = (2l-1)/2$ the current makes a jump from $J_{max} = (4\pi D_i/N)\sin(\pi/N)$ to $J_{min} = -(4\pi D_i/N)\sin(\frac{\pi}{N})$. At this point the total current changes sign, but conserves its absolute value. This jump has a deep meaning in that it characterizes a phase transition. The abrupt changing of the diamagnetic current at the point of the transition helps the ring to trap a new quantum of the total flux. In this case the diamagnetic current screens the magnetic flux only before the transition and after the transition it does not screen the external magnetic flux, but, rather, the diamagnetic current generates the additional quantum of the flux.

The dependence of the diamagnetic current on the magnetic flux behaves qualitatively like the dependence of the current associated with single particle in the ground state on the flux through the hole in a superconducting ring [22]. The diamagnetic current does not have a direction which brings the total flux to zero. The direction of the orbital current is such that it brings the total flux closer to the nearest integer number of flux quanta. The direction of the diamagnetic current alters suddenly at each transition.

Thus, the general conclusions that can be drawn here are that 1) at even number of electrons the system has a phase transition at integer values of the flux $f = l$ and 2) at an odd number of electrons the system has a transitions when $f = (2l-1)/2$ (l is arbitrary integer). At each of these transitions the system traps one quantum of flux per ring.

Using these results one can explain the experiments [23] which have observed the oscillation of the magnetization on the scale of half a flux quantum. If we have a system with 10^7 rings, half of the rings may have an odd number of electrons in each ring the other half of rings may have an even number of electrons. The total energy of this system will have cusp like maxima at each half of the flux quantum, i.e. it is a periodical function with the period $f = 1/2$. The same is true for the magnetization and the magnetic susceptibility. This explanation is consistent with the beautiful results of Bouchiat and Montambaux [24]. In contrast to [25], at each half of the flux quantum our theory predicts cusp like maxima in the magnetic susceptibility, which may be observed at low temperatures. For the even number of fermions the expression for the magnetic susceptibility is given by the formula:

$$\chi = -2D_i(\frac{2\pi}{N})^2 \cos\left[\frac{2\pi}{N}(f - \frac{2l-1}{2})\right] \tag{23}$$

where $l - 1 \leq f \leq l$. For an odd number of fermions, l should be half odd integer number. It is worthy of note that this function is always negative. This means that this system is always diamagnetic, but makes the paramagnetic jumps at the points of transition.

Moreover one can connect this phenomena with Aharonov-Bohm resistance oscillations with period $1/2$ predicted in [26] and observed experimentally in [27]. If we consider a ring with an even number of electrons, and add one fermion, then the change in the Aharonov-Bohm phase will not be f. This additional fermion will have a direction such that its current is opposite that with an even number of electrons. Due to these two motions, in opposite directions, the change of the Aharonov-Bohm phase will be $2f$. This is the reason for the periodicity, with the period $1/2$, of the total energy of the system of rings with even and odd numbers of electrons. This change of the Aharonov-Bohm phase is also the cause of the oscillation of the magnetoresistance [26,27]. The oscillation phenomena of the magnetoresistance, of the magnetization, and of the magnetic susceptibility, as well as persistent currents, should exist for an array of quantum dots of the type which can be produced by modern technology [28].

As we described above, the two particle interaction between fermions does not change the character of the generation of a spontaneous magnetic flux. We should mention effects which can destroy the effect. Impurities, as well as a finite temperature, will disperse the cusp like maxima. If the disorder, or the temperature, increases this dispersion also increases. Therefore there exists a critical amount of disorder and a temperature at which the spontaneous flux disappears. These questions will be considered in details in other publications [10].

2. HUBBARD HAMILTONIAN ON A RING

We have shown that, for even number of fermions on a ring, there exist a flux phase state. On the other hand, interacting electrons can have a different behavior. Specifically, spin-flip processes may destroy nonzero orbital moment, appeared in the case of spinless fermions. We show here that, such processes do exist and have an amusing character. The orbital moment is completely destroyed. In order to demonstrate this phenomenon we will solve, by the Bethe ansatz method, the Hubbard Hamiltonian having a form,

$$H = -t \sum_{<i,j>,\sigma} a_{i\sigma}^+ a_{j\sigma} + U \sum_{i=1}^{L} n_{i+} n_{i-}, \tag{24}$$

which involves as parameters, the electron hopping integral t and the on-site repulsive Coulomb potential U. The operator $a^+_{i\sigma}(a_{i\sigma})$ creates (destroys) an electron with spin projection σ ($\sigma = +$ or $-$) at a ring site i, and $n_{i\sigma}$ is the occupation number operator $a^+_{i\sigma}a_{i\sigma}$. The summations in Eq. (24) extend over the ring sites i or $-$ as indicated by $<i,j>, \sigma -$ over all distinct pairs of nearest-neighbor sites, along the ring with the spin projection σ.

The Hamiltonian (24) models a system of M electrons with spin projection up $\sigma = +$ and $N - M$ electrons with down spin $\sigma = -$.

For the case with a magnetic field, we will use the same form of the wave function as has been proposed by Bethe [11,29],

$$\psi(x_1,\ldots,x_N) = \sum_P [Q,P] exp[i\sum_{j=1}^N k_{Pj} x_{Qj}]. \tag{25}$$

where $P = (P_1,\ldots,P_N)$ and $Q = (Q_1,\ldots,Q_N)$ are two permutations of $(1,2,\ldots,N)$. The coefficients $[Q,P]$ as well as $(K_1,\ldots K_N)$ are determined from the Bethe equations, which because of the magnetic field, are changed by the addition of the flux phase $2\pi f$,

$$e^{i(k_j L - 2\pi f)} = \prod_{\beta=1}^{M}\left(\frac{it\sin k_j - i\lambda_\beta - U/4}{it\sin k_j - i\lambda_\beta + U/4}\right) \tag{26}$$

$$-\prod_{j=1}^{N}\left(\frac{it\sin k_j - i\lambda_\beta - U/4}{it\sin k_j - i\lambda_\beta + U/4}\right) = \prod_{\alpha=1}^{M}\left(\frac{i\lambda_\alpha - i\lambda_\beta + U/2}{i\lambda_\alpha - i\lambda_\beta - U/2}\right). \tag{27}$$

These equations are greatly simplified in the limit $U \to \infty$. As a result we have the following equations:

$$Lk_j = 2\pi(I_j + \frac{1}{N}\sum_\alpha J_\alpha) + 2\pi f \tag{28}$$

where the quantum numbers I_j and J_α are connected with charge and spin degrees of the freedom, respectively. The sets of these quantum numbers of course strongly depend on the magnetic flux f. They are different for an even and for an odd numbers of particles. One has the following classification for the sets (I_j) and (J_α).

For an even number of the electrons N and for an even number of the up spins M the numbers I_j should be integer and the numbers J_α should be half-odd integers. For an even number of the electrons N and for an odd number of up spins M the numbers I_j should be half-odd integer and J_α should be integers. For an odd number of the electrons N and for an even number of the up spins M, both the numbers I_j and J_α should be integers. For an odd number of electrons N and an odd number of up spins M, the numbers I_j and J_α should be half-odd integers.

For example, at zero magnetic field they have the following form [30]: when $N = 2K$ is an even number, then

$$I_1,\ldots,I_{2K} = -K, -(K-1),\ldots,-1,0,1,\ldots,K-1, \tag{29}$$

$$J_1,\ldots,J_M = -\frac{M-1}{2},\ldots,-\frac{1}{2},\frac{1}{2},\ldots,\frac{M-1}{2}, \tag{30}$$

when M is also an even number; and

$$I_1,\ldots,I_{2K} = -K+\frac{1}{2},\ldots,-\frac{1}{2},\frac{1}{2},\ldots,K-\frac{1}{2}, \qquad (31)$$

$$J_1,\ldots,J_M = -\frac{M-1}{2},\ldots,-1,0,1,\ldots,\frac{M-1}{2}, \qquad (32)$$

when M is an odd number.

If $N = 2K+1$ is an odd number then

$$I_1,\ldots,I_{2K+1} = -K,\ldots,-1,0,1,\ldots,K, \qquad (33)$$

$$J_1,\ldots,J_M = -\frac{M}{2},\ldots,-1,0,1,\ldots,\frac{M}{2}-1, \qquad (34)$$

when M is an even number;

$$I_1,\ldots,I_N = -\frac{N}{2},\ldots,-\frac{1}{2},\frac{1}{2},\ldots,\frac{N}{2}, \qquad (35)$$

$$J_1,\ldots,J_M = -\frac{M}{2},\ldots,-\frac{1}{2},\frac{1}{2},\ldots,\frac{M}{2}, \qquad (36)$$

when M is an odd number.

In a magnetic field these sets of numbers are shifted. The optimal set is determined by minimizing the total energy of N electrons with M up spins. Using this principle for an even number of electrons one may obtain the following expressions of the total energy:

$$E^{evenM}_{evenN} = -Dcos(\frac{2\pi}{L}(f - \frac{M}{2N} + \frac{N-M}{2N} + l + \frac{M}{N}k)) , \qquad (37)$$

and

$$E^{oddM}_{evenN} = -Dcos(\frac{2\pi}{L}(f + l + \frac{M}{N}k)) , \qquad (38)$$

where l and k are arbitrary integers, and D is the positive constant:

$$D = 2t\frac{sin(\frac{\pi N}{L})}{sin(\frac{\pi}{L})}. \qquad (39)$$

For an odd number of electrons, the expressions of the total energy have another form:

$$E^{evenM}_{oddN} = -Dcos(\frac{2\pi}{L}(f - \frac{M}{2N} + l + \frac{M}{N}k)) , \qquad (40)$$

and

$$E^{oddM}_{oddN} = -Dcos(\frac{2\pi}{L}(f - \frac{N-M}{2N} + l + \frac{M}{N}k)) , \qquad (41)$$

when the values of M are even and odd, respectively. It is worthy of note that all these expressions are periodic functions of f with periods $f_T = L, f_T = LM/N$ and, of course, $f_T = L(N-M)/N$. The quantum numbers l and k describe the charge and spin vortices.

Let us consider the case of an odd number of particles N. For the ground state the quantum numbers k, l and M are not independent and are governed by the equation which can be obtained from eq. (40):

$$l + \frac{M}{N}k = \frac{M}{2N} \tag{42}$$

Let all spins be down, i.e. $M = 0$ then a simple solution of (42) is $k = l = 0$. Then from (40) the ground state energy is given by

$$E^0_{oddN} = -D\cos(\frac{2\pi}{L}f), \tag{43}$$

With numbers k,l zero, at the point $f = 1/2N$, there will exist a transition to the state with $M = 2$, where for the ground state energy, from (40), we have:

$$E^2_{oddN} = -D\cos(\frac{2\pi}{L}(f - \frac{1}{N})). \tag{44}$$

Upon further increasing the magnetic flux f, at the point $f = 3/2N$ there exists the transition to the state with $M = 4$ and so on. That is, at the point $f = (2p-1)/2N$ there is the transition to the state with $M = 2p$, where here k and l are equal to zero and p is an integer. The ground state energy in this state has the form:

$$E^{2p}_{oddN} = -D\cos(\frac{2\pi}{L}(f - \frac{p}{N})) \quad , \tag{45}$$

This expression describes the ground state energy only in the following interval of the flux f: $(2p-1)/2N \leq f \leq (2p+1)/2N$. If in the initial state all spins are up, then the energy should be described by the equation (41). With $k = l = 0$ we again have transitions with $\Delta M = 2$. Here the ground state energy has smooth minima when

$$f_{min} = \frac{p}{N}, \tag{46}$$

where p is an arbitrary integer. The ground state energy has cusp like maxima at

$$f_{max} = \frac{p}{N} + \frac{1}{2N} \tag{47}$$

Upon increasing the magnetic flux from zero the total current decreases. At the point $f = 1/2N$ this diamagnetic current changes sign, but conserves its absolute value. The expression describing the total current for an arbitrary f has the form

$$J = -\frac{2\pi D}{L}\sin(\frac{2\pi}{L}(f - \frac{p}{N})) \quad , \quad \frac{2p-1}{2N} \leq f \leq \frac{2p+1}{2N} \quad . \tag{48}$$

One can see from this formula that, at $f = (2p+1)/2N$, the current makes a jump from $J_{min} = -(2\pi D/L)\sin(\pi/L)$ to $J_{max} = (2\pi D)/L\sin(\pi/L)$. This jump has a significant meaning since it characterizes the phase transition. The abrupt change of the diamagnetic current at the point of the transition helps the ring to trap a new fractional quantum of the total flux $f = 1/N$ which characterizes the spin-charge vortex. In this case the diamagnetic current screens the magnetic flux only before the transition while after the transition, rather than screening the external magnetic flux, the diamagnetic current generates the additional fractional quantum of the flux $f = 1/N$.

The dependence of the diamagnetic current on the magnetic flux behaves qualitatively like the dependence of the current associated with a single particle with the effective charge $e^* = Ne$ in the ground state of a superconducting ring with a flux through the hole. The diamagnetic current does not have a direction which brings the total flux to zero, rather the direction of the orbital current is such that it brings the total flux closer to the nearest fractional number of flux quantum $f = p/N$. The direction of the diamagnetic current changes suddenly at each transition. It is amusing that at each of these transitions there is the generation of the spin-charge vortex, described by quantum numbers k, l and M. These quantum numbers are not independent. With $k = l = 0$ the transition is simply the flip of two spins which can even be in the opposite direction to the magnetic field. This can be explained by the observation that the changing of the sign of the current at the transitions points implies a flip of the orbital momentum. As a result we have spin-orbital momentum flip or, in other words, the generation of a spin charge vortex. If we take into account the Zeeman interaction, then in high magnetic field $f >> 1$ this idealized physical picture may be changed.

The same behavior can be obtain with an even number of particles. In this case the ground state is also strongly degenerate. The quantum numbers l, k and M for ground state energy are governed by an equation analogous to (42):

$$l + \frac{M}{N}k = 0. \tag{49}$$

With $k = l = 0$, and without a field, the ground state is a singlet. Thus, we have obtained results which are equivalent to the "Nagaoka" theorem. That is, for a ring with the quantum numbers $l = k = 0$ and with an odd number of particles, the system favors the ferromagnetic state, while with an even number of particles it favors an antiferromagnetic singlet state. However, it is worthy of note that all states with different M may give the same energy dependence with an appropriate choice of of the quantum numbers l and k, i.e., the ground state energy is strongly degenerate.

Let, for example, $l = k = 0$ then with an even number of particles, in the region $-1/2N \le f \le 1/2N$, the ground state energy is described by equations (37) with $M = N/2$ (it coincides with (42)); in the region $1/2N \le f \le 3/2N$, it is described by equations (37) with $M = (N/2) + 2$ (it coincides with (43)) and so on. One general conclusion which can be drawn here is that the ground state energy is a periodic function of the flux f through the ring with period $f_T = 1/N$ (see, for comparison [31]).

This phenomenon has a simple explanation which is related to the multivalued nature of a wave function of a hole on a ring. If we take the hole around the ring once, we have, in effect, changed the entire ring from one ground state phase to the other (all electrons have moved only by one site). Only by taking the hole around N times do we restore the system to its original state. The total change of the Aharonov-Bohm phase due to this process is Nf_T. For the periodicity of the total wave function this change, according to gauge invariance, must equal unity. Whence the ground state energy as a function of the flux f has the period $f_T = 1/N$. The strong Coulomb on site-site repulsion prevents free one electron motion on the ring ("Coulomb blockade") but it does allowed the motion of all N electrons together .

The magnetization and susceptibility of this ring are also periodic functions with this universal period $f_T = 1/N$.

For the ring system of singly connected quantum dots [28] such a phenomenon may also readily observed. As there is an experiment in which the oscillation of the magnetization with period $f = 1/2$ for a copper rings has been observed [23]. If instead of a copper ring we take a ring consisting of some transition metal (Mn,Fe,Co or Ni) then, probably, for such rings it would be possible to observe oscillations of the magnetization and magnetic susceptibility with period $f_T = 1/N$, i.e., the period of the oscillation is inversely proportional to the density of electrons on the ring. At intermediate values of $U \sim t$ the period of the oscillation may be equal to an arbitrary fractional number. The analogous system, where we have found the similar "$1/N$" oscillations, is that of free electrons on a ring with one magnetic "Kondo" impurity. The phenomenon found in present work may be also observed for aromatic molecules in transverse magnetic field. Of special importance may be optical experiments.

ACKNOWLEDGMENTS: I wish to thank F.Woynorovich and S.Kivelson for fruitful discussions, Stewart Barnes and M.D.Johnson for many very useful remarks, and also M.L.Ristig and S.Feng for their hospitality. This work was supported by Alexander von Humboldt Foundation.

REFERENCES

[1] V.Kalmeyer, R.B.Laughlin, Phys.Rev.Lett. 59,2095 (1987)
[2] J.Affleck, J.B.Manston, Phys. Rev.B37 (1988), 3774
[3] X.G.Wen, Fr. Wilczek, A.Zee, Phys.Rev. B39, (1989) 11413
[4] P.W.Anderson, Physica Scripta T27,60 (1989)
[5] P.Wiegmann, Phys.Scripta T27,(1989) 160
[6] Y.Hasegawa, P.Lederer, T.M.Rice, P.B.Wiegmann Phys.Rev.Lett 63 (1989) 907
[7] G.Montambaux, Phys.Rev.Lett.63 (1989) 1657
[8] F.V.Kusmartsev "Flux phase states on a ring" Preprint, Cologne 10-11-1990 (1990) and Pisma Eksp. Teor. Fiz 53, 27 (1991)
[9] D.Wohlleben, E.Esser, E.Zipper and M.Szopa. Possibility of Orbital Magnetic Phase Transitions in Mesoscopic Metallic Rings. Preprint, Cologne (1990).
[10] F.V.Kusmartsev, Preprint, Cologne (1990) and J. Phys. Cond. Matt. 3 (1991) in Press.
[11] H.Bethe Z.Physik 71,205 (1931)
[12] E.Lieb, T.Schultz and D.Mattis, Ann. of Phys. 16,407 (1961)
[13] C.N.Yang and C.P.Yang. Phys.Rev. 147,303(1966), 150,321 (1966) 150,321 (1966), 150,327 (1966), 151,258 (1966)
[14] J.des Cloizeaux and M.Gaudin J.Math Phys. 7,1384, (1966)
[15] C.J.Hamer, G.R.W.Quispel and M.T.Batchelor J.Phys.A 20,5677 (1987)
[16] F.Woynarovich and H.P.Eckle J.Phys.A 20,L97 (1987)
[17] F.Woynarovich Phys.Rev.Lett. 59,259 (1987); 59,1264(E) (1987)
[18] B.Sutherland and B.S.Shastry Phys.Rev.Lett.65,1833 (1990)
[19] N.Bayers, C.N.Yang, Phys.Rev.Lett. 46,7 (1961)
[20] C.N.Yang, Rev.Mod.Phys.34,694 (1962)
[21] L.D.Landau and I.M.Lifschitz, Quantenmechanic, Akademie-Verlag, Berlin, p.370
[22] J.M.Blatt. Theory Superconductivity, Academic Press.New York 325 (1964)
[23] L.P.Levy, G.Dolan, J.Dunsmuir and H.Bouchiat, Phys.Rev.Lett. 64,2074 (1990)
[24] H.Bouchiat, G.Montambaux J.Phys.(Paris)50,2695 (1989)
[25] V.Ambegaokar, U.Eckern, Phys.Rev.Lett. 65,381 (1990)
[26] B.L.Altshuler, A.G.Aronov, B.Spivak, Pis'ma Zh.Eksp.Teor.Fiz. 33,(101) 1981
[27] D.Y.Sharvin and Y.D.Sharvin, Pis'ma Zh.Eksp.Teor.Fiz. 33,104 (1981)
[28] T.Demel, D.Heitman, P.Grambow and K.Ploog, Phys. Rev. Lett. 64, 788 (1990)
[29] E.Lieb, F.Wu, Phys.Rev.Lett. 20,1445 (1968)
[30] F.Woynarovich, J.Phys.A: Math.Gen., 22,4243 (1989)
[31] G.S.Canright and M.D.Johnson. Orbital ferromagnetism of anions. Preprint (1990)

VECTOR MEAN FIELD THEORY OF STATISTICS

M.D. Johnson[1], C. Gros[2], S.M. Girvin[3], and G.S. Canright[4]

[1]Department of Physics, University of Central Florida, Orlando, FL 32816
[2]Institut für Physik, Universität Dortmund, 4600 Dortmund 50, Germany
[3]Department of Physics, SW-117, Indiana University, Bloomington, IN 47405
[4]Department of Physics, University of Tennessee, Knoxville, TN 37996-1200

INTRODUCTION

In 2 + 1 dimensions, a particle can be not just a boson or a fermion, but something in between. Such particles or quasiparticles are said to possess fractional statistics.[1] One way to think of statistics is by reference to the phase $e^{i\theta}$ which develops when two identical particles are exchanged. For bosons $\theta = 0$ and for fermions $\theta = \pi$. In 2 + 1 dimensions it turns out that θ can take any value, and so particles with general θ have been called anyons.[2]

Perhaps the simplest way to understand fractional statistics is in terms of the anyon model of Wilczek.[2] Imagine taking fermions or hard-core bosons and adding to each an infinitesimal solenoid, or flux tube. The flux in each tube and charge on each particle are presumed to be identical. When two particles are exchanged, an Aharonov-Bohm phase develops. By appropriate choice of flux and charge, any value of θ can be obtained. This model can only describe point particles in 2+1 dimensions, where the flux tubes do not entangle. A deeper reason for the dimensional restriction is given by considering path integrals on multiply connected manifolds.[3] The basic idea is that in two dimensions, but not three, it is possible to keep track of the number of times hard-core particles wind around one another.

The anyon model is an instance of what has come to be called *statistical transmutation*.[1] By adding a long-ranged gauge interaction, ordinary bosons or fermions can be turned into noninteracting particles with exotic statistics. In particular, it is possible to convert fermions into hard-core bosons, or vice-versa, in this way.

Anyon models are difficult to study, naturally so since they correspond to fermions or bosons with long-range interactions. One popular approach we call vector mean-field theory (VMFT).[4] Here one removes the flux from the particles and spreads it uniformly through the system, and studies the simpler problem of fermions or hard-core bosons moving in a uniform magnetic field B. One then hopes that the approximate Hamiltonian might still yield the basic physics, or at least be a good starting point for better calculations.

Improbable as VMFT may appear, there is some evidence in its favor. The ground state of fermions in a uniform magnetic field turns out to be the exact Hartree-Fock ground state wavefunction for hard-core bosons treated as interacting fermions. Adding RPA corrections to this then restores the gaplessness which must be correct.[5] Moreover, in numerical studies conducted on rather small systems, VMFT proved to be remarkably accurate.[6] These studies found the exact ground state $|ex\rangle$ for some choice of statistics, and also the ground state $|vmf\rangle$ for hard-core bosons in a fixed uniform B. The energy of $|vmf\rangle$ turns out to be very close to the exact energy of $|ex\rangle$ only when B is very close to the value given by spreading out the flux tubes. Even more remarkably, in that case there also turns out to be a very large overlap $\langle ex|vmf\rangle$ between the two wavefunctions. A converse of VMFT—mapping fermions in a magnetic field to free bosons—led to the discovery of algebraic off-diagonal long-range order in the fractional quantum Hall effect.[7] (See also Ref. 8.)

Let us then think more seriously about VMFT, and explore in detail one test case: hard-core bosons represented as fermions plus flux tubes. This can be done exactly. We want to study VMFT, in which the bosons are approximated by fermions moving in a uniform magnetic field. These results have been discussed in a more comprehensive paper (Ref. 9) to which we will refer for details. Here we will try to highlight one point: small-system results notwithstanding, VMFT should be expected to fail on the lattice (except in the dilute limit), and to be reasonably accurate in the continuum. This can be explained as a result of phase fluctuations on the lattice, or, from a perhaps complementary viewpoint, as a result of the point zeros in the continuum versus the line zeros on the lattice.

There are two reasons for using the VMFT fermion representation of hard-core bosons as our test case. It represents the most stringent test, since this is the largest statistical transmutation possible. And it is nonetheless analytically tractable, since we need only deal with fermions and bosons.

VARIATIONAL ANALYSIS OF VMFT

We consider bosons moving on a square lattice, with the nearest-neighbor hopping Hamiltonian

$$H_B = -t \sum_{\langle ij \rangle} \left(b_i^\dagger b_j + b_j^\dagger b_i \right).$$

If the bosons have hard cores, this becomes equivalent to the spin-1/2 xy model. (One can identify $\downarrow$ as an empty site and $\uparrow$ as a site occupied by a boson.)

The hard-core bosons can be replaced by spinless fermions plus appropriate flux tubes. The latter are described by a Hamiltonian H_F. This mapping is exact. Every observable of the fermion system subject to H_F is identical to the corresponding observable of the boson system described by H_B. This includes, for example, the energy spectrum and the density of each eigenstate.

Next we want to exercise VMFT by removing the flux tubes from the fermions and replacing them by an appropriate uniform magnetic field. This approximation, fermions moving on a lattice subject to a uniform field, is the Hofstadter problem, described by the mean-field Hamiltonian H_{vmf}. We consider two cases, a density $n = 1/2$ per site and $n = 1/4$. These are particularly simple cases for which the Hofstadter problem can be solved analytically. For details of this calculation, and the others discussed below, see Ref. 9.

What we would like to do is compare $|ex\rangle$, the exact ground-state of hard-core bosons, with $|vmf\rangle$, the approximate ground state for fermions in a uniform magnetic field. For example, one might consider

$$E_{ex} = \langle ex|H_B|ex\rangle \quad (exact)$$
$$E_{var} = \langle vmf|H_F|vmf\rangle \quad (variational)$$
$$E_{vmf} = \langle vmf|H_{vmf}|vmf\rangle \quad (VMFT).$$

Here E_{ex} is the exact ground state energy. This could equally well have been calculated in the fermion representation, using H_F and the fermionic equivalent (i.e., the singular-gauge transformation) of $|ex\rangle$. E_{var} is the variational energy of the VMFT ground-state. This obeys the usual variational principle $E_{var} \geq E_{ex}$. (Remember that H_F is the fermion-representation equivalent of H_B.) On the other hand, E_{vmf} is not a variational estimate at all, and ordinarily would not be bounded by a variational principle (although this case is an exception[9]). E_{vmf} is the VMFT estimate of the ground state energy, computed using the VMFT Hamiltonian instead of the true Hamiltonian.

The difficulty with this plan, of course, is that the exact ground state $|ex\rangle$ can only be calculated for very small systems. For these, as mentioned, VMFT works well.

What can be learned using only E_{vmf} and E_{var}? We can view the corresponding Hamiltonians H_{vmf} and H_F as probes to study the properties of the VMFT wavefunction. If the two energies are the same, perhaps it is because the wavefunction looks the same from both viewpoints. If this occurs it is presumably because VMFT gets the physics basically right.

More generally, one should look for a systematic connection between E_{var} and E_{vmf} as a function of some system parameters (such as system size or a systematically varying choice of gauge). If a

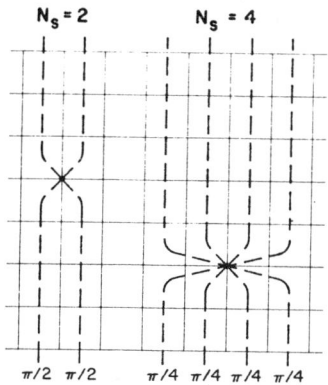

Fig. 1. Two gauge choices for the vector potential of flux tubes attached to particles.

systematic connection exists, it might be possible to construct a renormalized mean-field theory in the spirit of the Gutzwiller approximation.

An important point has been glossed over above: E_{var} is a mixed gauge quantity. The flux tubes enter H_F as a vector potential, with a choice of gauge. Similarly, H_{vmf} and hence $|vmf\rangle$ depend on the gauge chosen for the uniform magnetic field. E_{var} depends on two separate gauge choices, and hence is not gauge invariant. For lattice calculations, the vector potential is conveniently bunched into δ functions. Two possible gauge choices for the flux tubes are shown in Fig. 1, where the vector potential is concentrated into $2N_s$ strings (shown for N_s equal to 2 or 4). A particle moving counterclockwise across a string experiences a phase jump of π/N_s. Similar choices can be made for the uniform field. Experience shows that the best agreement comes from matching the gauges.[6,9]

We have computed E_{var} and E_{vmf} on finite cylinders of dimensions $L_x \times L_x$ for $L_x = 4$ to 20, for $n = 1/2$ and $1/4$. We also have varied the gauge by choosing $N_s = 2$ to 16. Because H_F incorporates the flux tubes, the evaluation of E_{var} is nontrivial for large systems; we have used a variational Monte Carlo approach.[10] Results for the half-filled system are given in Fig. 2, for a fixed gauge choice $N_s = 2$. There we show the total energies and also the energies corresponding to the horizontal (x) bonds only. For $N_s = 2$ the gauge strings do not cut any vertical (y) bonds (see Fig. 1), so for this case $E_{vmf,y} = E_{var,y}$.

First consider the total energies. For small systems, $E_{var} \approx E_{vmf}$, in agreement with exact diagonalizations.[6] But for larger systems it is evident from Fig. 2 that the two diverge considerably. Similar behavior is found for $n = 1/4$. This leads to our first conclusion: for large, densely occupied lattices, VMFT does not work well.

LATTICE VS. CONTINUUM

There are two ways to understand the apparent failure of VMFT on a lattice. They lead to the conclusion that VMFT should be satisfactory for the continuum or a dilute lattice.

Phase Fluctuations

The Hamiltonian H_F which describes hard-core bosons in the fermion representation (i.e., which is the statistical transmutation of H_B) is

$$H_F = -t \sum_{\langle ij \rangle} e^{i\phi_{ij}} f_i^\dagger f_j + H.c.$$

(Here $f, f^\dagger$ are fermion operators.) That is, the hop from j to i has a phase ϕ_{ij} which depends on the positions of all of the other particles in the system via the gauge phase Θ_{ik}.[9]

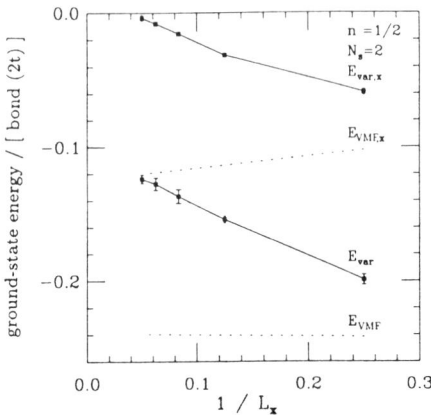

Fig. 2. Estimates of the ground state energy per bond for half-filling and $N_s = 2$.

$$\phi_{ij} = \sum_{k \neq i,j} (\Theta_{ik} - \Theta_{jk}) f_k^\dagger f_k .$$

VMFT pretends that ϕ_{ij} is independent of particle positions when, in fact, it fluctuates as particles move about. $E_{vmf,x}$ measures the energy of horizontal bonds, accepting this fiction. Hence the horizontal bonds add with little cancellation. This is visible in Fig. 2, where $E_{vmf,x}$ is tending to a constant nonzero value. In fact, the particles do move about, and consequently the phases fluctuate more or less randomly. The result is that the x energies, as measured by $E_{var,x}$, have considerable cancellation, and tend to zero for large system sizes.

We can make a simple estimate of how the phases should fluctuate on a horizontal bond. Consider a particular horizontal bond. Let $N_\pm$ be the number of particles above (below) this bond whose strings cross it. The phase for hopping with the others fixed is $\phi_x = \pi(N_+ - N_-)/N_s$. For a typical bond, $N_\pm \sim nN_sL_y/2$. Fluctuations of order $\Delta N_\pm \sim \sqrt{N_\pm}$ then give phase fluctuations of order $\Delta\phi_x \sim 2\pi\Delta N_\pm/N_s$, or

$$\Delta\phi_x \sim \pi\sqrt{2nL_y/N_s}.$$

There are two points to notice here. Increasing N_s can decrease the fluctuations, and so reduce the cancellation in $E_{var,x}$. Indeed this is visible if N_s is changed from 2 to 4.[9] On the other hand, N_s cannot be increased past the system size, and so the minimum possible fluctuations are of order

$$\Delta\phi_x \sim \pi\sqrt{2n}.$$

In the dense limit, the fluctuations are of order π. The fluctuations get smaller as the density of particles on the lattice is reduced. A similar result holds for y fluctuations.

The chief lack of VMFT is that it ignores the fluctuations in the gauge fields, and hence in the phases. This will not be a severe failure if the fluctuations are in truth small—that is, if the density is small. So this argument suggests that VMFT should be at its best in the continuum limit.

Point vs. Line Zeros

Let us consider the fractional quantum Hall effect, and an explanation for the off-diagonal long-range order found there.[7] The ground state for $\nu = 1/m$, with m odd, is very nearly Laughlin's wavefunction, a Gaussian multiplied by

$$\psi = \prod_{i<j}(z_i - z_j)^m .$$

Here for each fermion (at coordinate $z_j = x_j + iy_j$) there are m quanta of flux. If we model a fermion as a boson plus a flux tube of $-m$ quanta, and perform VMFT, we get free bosons. Ignoring the hard core, one expects $\psi_{vmf} \sim 1$, and hence long-range order. Now transforming back to the fermion representation gives

$$\psi_{vmf} = \prod_{i<j} \frac{(z_i - z_j)^m}{|z_i - z_j|^m}.$$

Here we have "solved" the problem by statistical transmutation, followed by VMFT in the new (boson) representation, and then transformation back. The result is a wavefunction which correctly reproduces the complicated phase structure of Laughlin's wavefunction. Adding RPA-like fluctuations to VMFT will get the amplitude correct, as well.

The lesson here is that VMFT works in this case (i.e., gives something close to the Laughlin wavefunction) because the true state has zeros at the positions of other particles. The inverse singular gauge transformation will inevitably introduce point phase windings which had better match up with the wavefunction's nodes.

But this was for the continuum. On the lattice, generically one expects plane-wave-type solutions (and gets them, for example, for $|vmf\rangle$). These have nodes along *lines*, but the singular gauge transformation gives *point* phase windings.

CONCLUSIONS

We have looked at the simplest case of statistical transmutation, where hard-core bosons are represented equivalently as fermions plus flux tubes. On the lattice, with a simple hopping Hamiltonian, this is equivalent to the xy model. Vector mean-field theory amounts to removing the flux tubes and spreading them uniformly. For large systems and high densities VMFT is rather poor.

The failure of VMFT can be traced to its neglect of the phase fluctuations which result from fluctuations in particle position. In the continuum limit these fluctuations are minimized, so VMFT should then be a sensible starting point. This is supported by the nodal structure. On the lattice nodes tend to form lines, whereas the singular gauge transformations generate point phase windings. In the continuum the nodes can be points.

The concept of statistical transmutation gained currency in the study of fractional statistics. It is worth bearing in mind that these ideas can also be useful for understanding the properties of ordinary fermions and bosons.

This work was supported by the National Science Foundation under Grant No. DMR-8802383, and by the National Center for Supercomputer Applications under Grant No. PHY890017N.

REFERENCES

1. For a general overview of fractional statistics, see G.S. Canright and S.M. Girvin, *Science* **247**, 1197 (1990).
2. F. Wilczek, *Phys. Rev. Lett.* **49**, 957 (1982); **48**, 1144 (1982).
3. J.M. Leinaas and J. Myrheim, *Nuovo Cimento* **37B**, 1 (1977).
4. D.P. Arovas, R. Schrieffer, F. Wilczek and A. Zee, *Nucl. Phys.* B **251**, 117 (1985).
5. R.B. Laughlin, *Phys. Rev. Lett.* **60**, 2677 (1988); A.L. Fetter, C.B. Hanna and R.B. Laughlin, *Phys. Rev. B* **39**, 9679 (1989); C.B. Hanna, R.B. Laughlin and A.L. Fetter, *Phys. Rev. B* **40**, 8745 (1989).
6. G.S. Canright, S.M. Girvin and A. Brass, *Phys. Rev. Lett.* **63**, 2291,2295 (1989).
7. S.M. Girvin and A.H. Macdonald, *Phys. Rev. Lett.* **58**, 1252 (1987).
8. M.D. Johnson and G.S. Canright, *Phys. Rev. B* **41**, 6870 (1990); G.S. Canright and M.D. Johnson, *Comments Cond. Matt. Phys.* **15**, 77 (1990).
9. C. Gros, S.M. Girvin, G.S. Canright and M.D. Johnson, *Phys. Rev. B* **43**, 5883 (1991).
10. See, e.g., C. Gros, *Ann. Phys. (N.Y.)* **189**, 53 (1989), and references therein.

SPINON-HOLON STATISTICS, AND
BROKEN STATISTICAL SYMMETRY FOR THE
t-J AND HUBBARD MODELS IN 2D

S. E. Barnes

Physics Department, University of Miami, Coral Gables, FL 33124, U.S.A.

Départment de Physique de la Matière Condensée, Université de Genève, 24 quai Ernest Ansermet, 1211 Geneva 4, Switzerland

Theoretische Physik, ETH-Hönggerberg, CH 8093 Zurich, Switzerland

I. INTRODUCTION

The mean-field approach to the t-J model, or equivalently to the large U Hubbard model, has been repeatedly[1,2,3,4] suggested as a paradigm for theories in which the spin and charge degrees of freedom are decoupled; the excitations associated with the former are called spinons, and with the latter holons. The usual formulation for this model involves the author's[5] auxiliary particles (slave bosons or fermions) and a mean-field approximation based upon a certain large N expansion[6]. Within such a formulation the physical value is $N = 2$ and the holons and spinons retain the statistics assign them within the auxiliary particle method, i.e., for the slave boson technique, the spinons are fermions and holons bosons while for the slave fermion technique the statistics are interchanged. However it is not at all clear that the expansion about the large N saddle point converges. For this and some other, more technical, reasons the present author questions the appropriateness of the large N technique.

In addition, Zhang et al.[7], suggest that the Gutzwiller projected wavefunctions lead to an exchange energy which is four times as large as that given by the free particle averages which occur in the large N mean-field approaches. Performing the Gutzwiller projected is roughly equivalent to enforcing *exactly* the constraint, $Q_i = 1$, that there be only a single auxiliary particle at a given site. The exchange energy within large N technique is smaller because this constraint is only obeyed on average, i.e. only the average $< Q_i >$ is equal to unity.

Here it is suggested a mean-field approach be based upon the Ritz variational principle, i.e., that it correspond to the best wavefunction in which the spin and charge degrees of freedom are factored. It is shown that such an approach can be made to satisfy the requirement that $Q_i = 1$ exactly (at least in the thermodynamic limit), thereby leading to a larger magnitude for the exchange energy.

As the author has stated elsewhere[8], this latter formulation leads to the conclusion that fermionic statistics are appropriate for *both* the holons and spinons; this claim has been misunderstood. The author has also shown that, alone, the kinetic energy term of the t-J model is unstable towards the formation of a certain flux phase. That is for $J = 0$ or $U = \infty$ models the holons experience an effective flux generated by the spin wavefunction, and which might be represented by a vector potential $\vec{A}_k$, and vice versa. The existence of such an average flux, for a mean-field theory, in turn, can be interpreted as mean-field theory of holons (and spinons) with fractional statistics, i.e., it is inferred that the holons and spinons are anyons. The *real* claim of the author is that within a mean-field theory in which $Q_i = 1$, *and* which is consistent with the commutation rules for the physical conduction electron field, i.e., that $\{c_{i\sigma}^\dagger, c_{j\sigma'}\} = \delta_{ij}\delta_{\sigma\sigma'}$, the nominal statistics for both the spinons and holons are fermionic. However, deviations from these nominal statistics are indicated by the formation of a finite flux in the ground state. Within such a ground state, this statistical flux experienced by the spinons is equal to that experienced by the holons and because of this relationship the nominal fermion statistics represent a reference which remains relevant.

The generation of this effective flux represents a strong, albeit average, residual interaction between the sectors. It is natural to ask if there is a representation of the physical field in which this interaction vanishes. The search for this representation leads to the idea[8] of a *broken statistical symmetry*; this concept will be elaborated upon below.

The spinons and holons correspond to the decomposition of the physical field into a product. Usually, for the t-J model (i.e., for U large Hubbard model), this is written as

$$c_{i\sigma} = b_{i0}^\dagger f_{i\sigma} \tag{1}$$

where $b_{i0}^\dagger$ is the boson associated with the empty site and where the $f_{i\sigma}$ are fermions corresponding to sites occupied by a single electron of spin σ. (Technically, this does not have the correct, on site, commutation rules unless there is added the term $\sigma f_{i-\sigma}^\dagger b_{i\uparrow\downarrow}$, where $b_{i\uparrow\downarrow}$ represents the doubly occupied site.) However, more generally in two dimensions it is possible to write

$$c_{i\sigma} = \mathcal{A}_{i0}^\dagger a_{i\sigma} \tag{2}$$

where $\mathcal{A}_{i0}^\dagger$ and $a_{i\sigma}$ are complementary anyons such that the physical field obeys the correct (off-site) commutation rules. Clearly there is a free choice of statistics for one of these particles; this can be thought of as a gauge symmetry. Writing the physical field in such a fashion implies that the anyons are treated in a singular gauge. The (practical) alternative is to revert to the decomposition Eq. 1 and attach an equal magnitude flux tube to each auxiliary particle. There will be a fictitious magnetic field

$$\vec{A}_h + \vec{A}_s \tag{3}$$

generated by the holon and spinon flux tubes respectively. (Note: the physical vacuum corresponds to a state in which there is a $b_{i0}^\dagger$ for each site; the associated flux must be canceled by a uniform background field.) Now, e.g., when a mean-field theory is constructed for the holon sector, there are two contributions to the fictitious average flux generated by the spinon sector, first that generated by the mean-field treatment of $\vec{A}_s$ and, second, the effective flux corresponding to $-\vec{A}_k$ and generated by the spin wavefunction. With a suitable choice for the magnitude of the flux tube attached to each auxiliary particle these two mean-field contributions will cancel whence the holons see only the their own flux. In this fashion the mean-field interaction between the spinons and holons is eliminated, i.e., the holons become *free anyons*. That there is only a single such choice for the magnitude of the flux tubes, and hence for the *free* anyon statistics is what is being called *broken statistical symmetry*.

The rest of this contribution describes the mathematics which can be used to put the above statements on a more or less sound basis.

II. COHERENT STATES

One of the principle objectives in constructing a mean-field trial wavefunction is to maintain the constraint that

$$Q_i = \sum_\sigma f_{i\sigma}^\dagger f_{i\sigma} + b_{i0}^\dagger b_{i0} = 1. \tag{4}$$

There is a simple means by which to accomplish this objective. Assume that $|1/2>$ is an eigenstate appropriate to half filling (one electron per atom). A basis set which satisfies this constraint may be obtained by the action of $c_{i\sigma}$ on $|1/2>$ without the need for an explicit Gutzwiller projection. The idea is to construct a trial wavefunction using only this physical set of states.

A mean-field theory, based upon the Ritz variational principle, corresponds to the best product wavefunction. If the trial wavefunction is in the form,

$$|\text{spinon}>|\text{holon}>, \tag{5}$$

then the effective Hamiltonian for, e.g., the holon sector is $\mathcal{H}_h = <\text{spinon}|\mathcal{H}|\text{spinon}>$. It is a trivial result that, if $|\text{holon}>$ is the ground state for $\mathcal{H}_h$ and $|\text{spinon}>$ that for $\mathcal{H}_s$, then the product corresponds to a minimal wavefunction for the original Hamiltonian $\mathcal{H}$.

Clearly, the mapping $\mathcal{H} \rightarrow \mathcal{H}_h$ is a projection. The full Hamiltonian $\mathcal{H}$ is associated with a space spanned by product vectors of the form $|\text{spinon}> \otimes |\text{holon}>$, while $\mathcal{H}_h$ is defined in the space spanned by a set of $|\text{holon}>$ vectors. One mapping is evidently unique, i.e., the holon sector vacuum $|0>$ necessarily corresponds to the state $|1/2>$. An *overcomplete* basis for the holon mean-field theory is generated by the action of the $b_{i0}^\dagger$ on $|0>$. This basis definitely does not conform with the constraint $Q_i = 1$ since it permits any number of $b_{i0}^\dagger$ particles at a given site. However of more importance is a more subtle point which has to do with statistics. It would appear that if $i \neq j$ the state $b_{i0}^\dagger b_{j0}^\dagger |0>$ is an acceptable $Q_i = 1$ state which is the equivalent of the state $c_{i\sigma} c_{j\sigma} |1/2>$ (any value of spin). However there is an evident problem since $c_{i\sigma} c_{j\sigma} |1/2> = -c_{j\sigma} c_{i\sigma} |1/2>$, while $b_{i0}^\dagger b_{j0}^\dagger |0> = +b_{j0}^\dagger b_{i0}^\dagger |0>$. It is implied, e.g., that the single particle density matrix calculated within the mean-field theory does not have the correct symmetry.

Both this statistics-symmetry problem, and the problem of how to maintain $Q_i = 1$, can be resolved if it is possible to determine the appropriate representation of the physical field $c_{i\sigma} = b_{i0}^\dagger f_{i\sigma}$ within the holon mean-field theory. To almost any level of rigor, it is clear that this is simply

$$<\text{spinon}|c_{i\sigma}|\text{spinon}> = b_{i0}^\dagger <\text{spinon}|f_{i\sigma}|\text{spinon}>. \tag{6}$$

That the expectation value, $<\text{spinon}|f_{i\sigma}|\text{spinon}>$, of a single fermion operator exists requires the introduction of *coherent states*[9]. Evidently, such coherent states cannot correspond to a fixed number of the $f_{i\sigma}^\dagger$, particles. A slightly non-standard[9] form for such a state is:

$$|s> = e^{\sum_{i\sigma} <f_{i\sigma}> \xi_i f_{i\sigma}^\dagger} |0> \tag{7}$$

where it is understood that $<f_{i\sigma}>$ is a c-number and where here $|0>$ is, in this context, the vacuum for the spinon sector. The quantities ξ_i are Grassmann variables, they obey anti-commutation rules $\{\xi_i, \xi_j\} = \{\xi_i^*, \xi_j\} = \{\xi_i, f_{j\sigma}\} = \{\xi_i, f_{j\sigma}^\dagger\} = 0$, etc., for all i and j, e.g., $\{\xi_i, \xi_i\} = 0$, which implies that $(\xi_i)^2 = 0$. The normalization for $|s>$ is $1/\mathcal{N}_s = <s|s>^{-1} = \exp\{-\sum_i <f_{i\sigma}^\dagger><f_{i\sigma}> \xi_i^* \xi_i\}$. Notice there is only a single ξ_i associated with a given site i. It would be more normal to have two $\xi_{i\sigma}$, one each for the two $f_{i\sigma}^\dagger$; the present choice implies there can only be a single $f_{i\sigma}^\dagger$ acting at a given site (with a finite result). This is a necessary condition if the constraint is to be satisfied. The state $|s>$ is an eigenstate of the $f_{i\sigma}$, i.e., $f_{i\sigma}|s> = <f_{i\sigma}> \xi_i |s>$. The Grassmann variables reflect the commutation properties of the fermion field, i.e., $f_{i\sigma} f_{j\sigma'} |s> = <f_{i\sigma}><f_{j\sigma'}> \xi_i \xi_j |s> = -<f_{i\sigma}><f_{j\sigma'}> \xi_j \xi_i |s> = -f_{j\sigma'} f_{i\sigma} |s>$.

With $|s>$ as the trial wavefunction the projection of the physical field is,

$$c_{i\sigma} \rightarrow b_{i0}^\dagger <f_{i\sigma}> \xi_i. \tag{8}$$

Since $<f_{i\sigma}>$ is a c-number the relevant operator is $F_{i0}^\dagger \equiv b_{i0}^\dagger \xi_i$, and corresponds to an effective spinless fermion. Specifically, the equivalent of $c_{i\sigma} c_{j\sigma} |1/2>$ is $F_{i0}^\dagger F_{j0}^\dagger |0>$ which is, correctly, equal to $-F_{j0}^\dagger F_{i0}^\dagger |0>$. Also, because $\xi_i^2 = 0$, it follows that $(F_{i0}^\dagger)^2 = 0$, i.e., there can be no more that a single holon sector particle at a given site (again a necessary condition in order that the constraint be satisfied). It is trivial show that there is an exclusion principle for $\vec{k}$-states constructed from the $F_{i0}^\dagger$ operators.

Consider next the mean-field theory for the spinon sector. Now

$$c_{i\sigma}^\dagger \to f_{i\sigma}^\dagger <b_{i0}> \qquad (9)$$

where, again, the expectation value $<b_{i0}>$ is a c-number. The $Q_i = 1$ states for the spinon sector are obtained by the action of $f_{i\sigma}$ on the half-filled, spinon sector, state $|1/2>$, i.e., the $f_{i\sigma}$ are the appropriate projection of $c_{i\sigma}$ for the spinon sector.

It perhaps needs to be emphasized, a finite value of $<b_{i0}>$ *does not* imply Bose-Einstein condensation. An evident counter example[9] is the original application of coherent states to the simple harmonic oscillator problem!

In fact, that the expectation value $<b_{i0}>$ is finite follows automatically from the requirement that $Q_i = 1$ and that the coherent state $|s>$ does not correspond to a fixed number of the $f_{i\sigma}^\dagger$ particles; the number of $b_{i0}^\dagger$ must have an equal but opposite variance.

A state with $<Q_i> \le 1$ is obtained by writing

$$|h> = \prod_i (1+ <b_{i0}> \xi_i b_{i0}^\dagger)|0>, \qquad (10)$$

where it is noted that all the commutators of the form $[\xi_i, b_{i0}^\dagger] = 0$ and where, since the terms in the product do not commute, some specific, but unimportant, order is implied. This state *does not* have the property that $b_{i0}|h> = <b_{i0}> \xi_i |h>$, i.e., it is not an eigenstate of b_{i0} (although there does exist a definition involving such a coherent state). The normalization factor is $1/\mathcal{N}_h = <h|h>^{-1} = \exp\{-\sum_i <b_{i0}^\dagger><b_{i0}> \xi_i^* \xi_i\}$. This state *does have* the desired property that $<h|b_{i0}|h>/\mathcal{N}_h = <b_{i0}> \xi_i$. Because the b_{i0} operators commute with the ξ_i variables, it follows that, despite the appearance of the ξ_i, there is no conflict with the requirement that, e.g., $b_{i0}(b_{j0}|h>) = +b_{j0}(b_{i0}|h>)$.

There is one important complication which has been ignored up until now; the Grassmann variables are of the nature of operators and it is necessary to extend the physical Hilbert space in order to accommodate them. By the same token it is necessary, at some point, to project the wavefunction $|h>|s>$ of the extended space back onto the original physical Hilbert space. As usual[9], this is accomplished by including the projection integral $\mathcal{P}_G = \prod_i \int d\xi_i d\xi_i^* e^{-\sum \xi_i \xi_i^*}$ to the right of *each* physical matrix element. However there is only a single such projection for a given matrix element and when this latter is factored the integral $\mathcal{P}_G$ must be associated with *either* the spinon or holon part; clearly it cannot be associated with both. In order that the expectation value of $f_{i\sigma}^\dagger$ be a Grassmann variable and that of $b_{i0}^\dagger$ be a c-number, it must be that $\mathcal{P}_G$ is included in the holon averages.

Since only the combinations $\xi_i b_{i0}^\dagger$ or $\xi_i f_{i\sigma}^\dagger$ occur, it follows that the product $|h>|s>$, is the sum of states of the form $\prod_i \xi_i b_{i0}^\dagger \prod_{i'} \xi_{i'} f_{i'\sigma}^\dagger |0>|0>$ for which it is easy to show that $Q_i^2 - Q_i = 0$, i.e., that each $Q_i = 1$ or 0; it follows that $<Q_i> \le 1$. If the normalized product $|h>|s>/\mathcal{N}_h \mathcal{N}_s$ is required to satisfy $<Q_i> = 1$ then it follows that $Q_i = 1$ without any variance. Since one has, e.g., that $<f_{i\sigma}^\dagger f_{i\sigma}> = |<f_{i\sigma}>|^2$, it follows that what is required is,

$$<Q_i> = \sum_\sigma |<f_{i\sigma}>|^2 + |<b_{i0}>|^2 = 1. \qquad (11)$$

Consider, e.g., the *projected* value of $<c_{i\sigma}> \equiv <f_{i\sigma} b_{i0}^\dagger> = \mathcal{P}_G[<s|f_{i\sigma}|s><h|b_{i0}^\dagger|h>/\mathcal{N}_h\mathcal{N}_s] = \mathcal{P}_G[<f_{i\sigma}> \xi_i <h|b_{i0}^\dagger|h>/\mathcal{N}_h] = <f_{i\sigma}> \mathcal{P}_G[<h|F_{i0}^\dagger|h>/\mathcal{N}_h] \equiv <f_{i\sigma}><F_{i0}^\dagger> = <f_{i\sigma}><b_{i0}>^*$ where $<f_{i\sigma}>$ and $<b_{i0}>$ are the c-numbers introduced with the definitions of the wavefunctions. The difference between $<b_{i0}>^*$ and $<F_{i0}^\dagger> = \mathcal{P}_G[<h|F_{i0}^\dagger|h>/\mathcal{N}_h]$ is that the former is a matrix element between wavefunctions in Fock space while the latter is defined in terms of the larger Hilbert space which accommodates the Grassmann variables.

III. CONSTRUCTION OF THE GROUND STATE

So far, the trial wavefunction $|h>|s>$ is merely the skeleton of a ground state wavefunction since almost the entire physical content is contained in the two averages $<f_{i\sigma}>$ and $<b_{i0}>$.

A somewhat pedagogical route for the determination of the former goes as follows. Consider, e.g., an average of the form,

$$<s|f_{i\sigma}^\dagger f_{i\sigma}|s> = |<f_{i\sigma}>|^2 = \sum_{\vec{k}} <s|f_{i\sigma}^\dagger|\vec{k}><\vec{k}|f_{i\sigma}|s>. \quad (12)$$

where the $|\vec{k}>$ are a complete set of kets which differ from $|s>$ by one particle and the momentum $\hbar\vec{k}$. An attempt at factoring this expression might be written as, $<f_{i\sigma}> = \sum_{\vec{k}} <\vec{k}|f_{i\sigma}|s>$. However this gives

$$|<f_{i\sigma}>|^2 = <s|f_{i\sigma}^\dagger f_{i\sigma}|s> = \sum_{\vec{k}}\sum_{\vec{k}'} <s|f_{i\sigma}^\dagger|\vec{k}><\vec{k}'|f_{i\sigma}|s>. \quad (13)$$

The double sum is symptomatic of the fact that the matrix elements of $|s>$ conserve neither, number, momentum nor spin. A similar, but less severe problem occurs for the BCS wavefunction since it does not conserve particle number. A version of the BCS theory in which particle number is conserved is obtained by performing certain phase averages. The same strategy is applied here. The decomposition of a given expectation value is written as

$$<f_{i\sigma}> = \sum_{\vec{k}} e^{-i\theta_{n\sigma}} <\vec{k}|f_{i\sigma}|s>, \quad (14)$$

where now it is understood, when evaluating a physical matrix element, an average

$$\mathcal{P}_s = \prod_{n\sigma} \int_{-\pi}^{\pi} \frac{d\theta_{n\sigma}}{2\pi} \quad (15)$$

is to be taken. With this projection, the decomposition Eq. (14) leads to the correct result for the average of Eq. (12).

If it first assumed that strictly $U = \infty$, i.e., that $J = 0$, then after the projection onto the spinon space the effective mean-field Hamiltonian is inevitably $\mathcal{H}_s = -t_s \sum_{<ij>\sigma} f_{i\sigma}^\dagger f_{j\sigma}$. The minimum $<\mathcal{H}_s>$ corresponds to the free Fermi sea value for the near neighbour density matrix $<f_{i\sigma}^\dagger f_{j\sigma}>$. In turn this corresponds to,

$$<f_{i\sigma}> = \sum_{\vec{k}\leq \vec{k}_F^s} \frac{1}{\sqrt{N}} e^{-i\theta_{\vec{k}\sigma}} e^{i\vec{k}\cdot\vec{r}_i}, \quad (16)$$

where $\vec{r}_i$ is the location of the site i, $\hbar\vec{k}_F^s$ is *spinon* the Fermi momentum for a given direction in momentum space, and N is the number of sites. Notice that the normalization is chosen so that $\mathcal{P}_s \sum_\sigma |<f_{i\sigma}>|^2 = 1 - \delta$. Trivially,

$$<f_{i\sigma}^\dagger f_{j\sigma}> = [\mathcal{P}_s \sum_{\vec{k}}\sum_{\vec{k}'} <s|f_{i\sigma}^\dagger|\vec{k}><\vec{k}'|f_{j\sigma}|s>] = \frac{1}{N}\sum_{\vec{k}\leq \vec{k}_F^s} e^{i\vec{k}\cdot(\vec{r}_i-\vec{r}_j)}, \quad (17)$$

which is the free particle expression for the density matrix.

The corresponding detail for the holon sector is a little more complicated because of the need to perform the Grassmann variable projection $\mathcal{P}_G$ for each physical matrix element. Beginning with,

$$< F_{i0}^\dagger F_{i0} >= \sum_{\vec{k}}[\mathcal{P}_G < h|F_{i0}^\dagger|\vec{k}>][\mathcal{P}_G < \vec{k}|F_{i0}|h>] = |<b_{i0}>|^2 = |\mathcal{P}_G < F_{i0}^\dagger >|^2, \quad (18)$$

the attempt to factor the sum takes the form,

$$< F_{i0} >= \sum_{\vec{k}} e^{-i\Theta_{\vec{k}0}} < \vec{k}|F_{i0}|h>, \quad (19)$$

where again when evaluating an average the projection,

$$\mathcal{P}_h = \prod_{\vec{k}} \int_{-\pi}^{\pi} \frac{d\Theta_{\vec{k}0}}{2\pi} \quad (20)$$

must be included. It follows, e.g., that,

$$< F_{i0}^\dagger F_{i0} >= \mathcal{P}_h \sum_{\vec{k}} \sum_{\vec{k}'} [\mathcal{P}_G < h|F_{i0}^\dagger|\vec{k}>][\mathcal{P}_G < \vec{k}'|F_{i0}|h>]. \quad (21)$$

For $U = \infty$, i.e., $J = 0$, the effective, $\mathcal{H}_h = -t_h \sum_{<ij>} F_{i0}^\dagger F_{j0}$ and again the ground state must correspond to a free Fermi sea, i.e.,

$$< F_{i0} >= \sum_{\vec{k} \leq \vec{k}_F^h} e^{-i\Theta_{\vec{k}0}} \frac{1}{\sqrt{N}} e^{-i\vec{k}\cdot\vec{r}_i} \xi_i \xi_i^*. \quad (22)$$

where $\hbar \vec{k}_F^h$ is the *holon* Fermi momentum for a given direction in momentum space. Again this gives correctly the normalized free fermion density matrix, i.e.,

$$< F_{i0}^\dagger F_{j0} >= \frac{1}{N} \sum_{\vec{k} \leq \vec{k}_F^h} e^{i\vec{k}\cdot(\vec{r}_i - \vec{r}_j)}. \quad (23)$$

It is easy to check, with appropriate values for $\vec{k}_F^s$ and $\vec{k}_F^h$, that indeed,

$$< Q_i >= \mathcal{P}_h \mathcal{P}_s [\sum_\sigma |<f_{i\sigma}>|^2 + |<b_{i0}>|^2] = 1, \quad (24)$$

as it must if $Q_i = 1$ exactly.

It must be emphasized that the phase averages contained in $\mathcal{P}_h$ and $\mathcal{P}_s$ are quite different from those introduced, but never executed, by others in order to maintain the constraint $Q_i = 1$. Because of the Grassmann variables, the present states satisfy $Q_i = 1$ exactly when the requirement $<Q_i>=1$ is enforced and these phase averages, which here contract double sums, are in fact performed.

It should also be noted that the present trial wavefunction corresponds to the best *uniform* factorable such function for the *kinetic energy*. When factored, the kinetic energy reduces to $-t\sum_{<ij>\sigma} <f_{i\sigma}^\dagger f_{j\sigma}><F_{i0}F_{j0}^\dagger>$ and, given that both the $f_{i\sigma}^\dagger$ and $F_{i0}^\dagger$ particles are fermions, the most negative kinetic energy corresponds to a Fermi sea for both sectors. As has been shown above, the proposed trial wavefunction gives such free fermion expectation values for all elements of the density matrices $<f_{i\sigma}^\dagger f_{j\sigma}>$ and $<F_{i0}F_{j0}^\dagger>$. The word *uniform* is important;

if, e.g., the phase of $< f_{i\sigma}^\dagger f_{j\sigma} >$ is different for different bonds, then it is possible to construct a flux state ground state which has a lower energy. This will be discussed below.

IV. MEAN-FIELD OPERATOR EXPANSIONS

Having constructed the be best uniform trial wavefunction, it is possible to perform expansions about this proposed ground state. This technique is really quite standard, e.g., in the context of spinwave theory, the Holstein-Primakoff theory[10] of spinwaves corresponds to an expansion about the mean-field value $S_z = S$ corresponding to the (classical) ground state. Here is made an expansion of the individual auxiliary operators about their mean-field values, i.e., written are, e.g.,

$$b_{i0} = <b_{i0}> + h_{i0} \quad \text{and,} \quad f_{i\sigma} = <f_{i\sigma}> \xi_i + s_{i\sigma}. \tag{25}$$

It is to be noted, given that b_{i0} is a Bose, and $f_{i\sigma}$ a Fermi, operator, then the corresponding operators h_{i0} and $s_{i\sigma}$ which occur in these expansions have, respectively, the same statistics. In this context it is important that $f_{i\sigma}$ is expanded about a Grassmann variable $<f_{i\sigma}> \xi_i$.

It is important to check the expansion of the physical field, i.e.,

$$c_{i\sigma}^\dagger = <f_{i\sigma}^\dagger><b_{i0}>\xi_i^* + <f_{i\sigma}^\dagger>\xi_i^* h_{i0} + <b_{i0}> s_{i\sigma}^\dagger \\ + \sigma <f_{i-\sigma}> \xi_i e_{i\uparrow\downarrow}^\dagger + \sigma <b_{i\uparrow\downarrow}^\dagger> s_{i-\sigma}, \tag{26}$$

where it *is* the case that mean-field value $<b_{i\uparrow\downarrow}^\dagger>$ but *not* expansion $e_{i\uparrow\downarrow}^\dagger$ of $b_{i\uparrow\downarrow}^\dagger$, associated with the doubly occupied sites, is negligible.

It must be that the commutation relation, $\{c_{i\sigma}, c_{i\sigma}^\dagger\} = 1$ for the physical field is satisfied. However, since $\{\xi_i^* h_{i0}, \xi_i h_{i0}^\dagger\} = \xi_i^* \xi_i$, and, $\{\xi_i^* e_{i\uparrow\downarrow}, \xi_i e_{i\uparrow\downarrow}^\dagger\} = \xi_i^* \xi_i$, the result involves Grassmann variables. This is a consequence of working in the large, spinon plus holon plus Grassmann variable, Hilbert space. Explicitly when (diagonal) matrix elements of these commutation rules are taken in the physical sub-space, a projection $\mathcal{P}_G$ must be effected and the result involves $\mathcal{P}_G \xi_i^* \xi_i = 1$, i.e., the combinations such as $F_{i0}^\dagger \equiv \xi_i b_{i0}^\dagger$ or $H_{i0}^\dagger \equiv \xi_i h_{i0}^\dagger$ do obey fermion commutation rules in the physical sub-space.

Explicitly taking (diagonal) matrix elements of the physical commutation rule $\{c_{i\sigma}, c_{i\sigma}^\dagger\} = 1$ *in the physical sub-space*, leads to the mean-field expression for the constraint, i.e., $\mathcal{P}_G\{c_{i\sigma}, c_{i\sigma}^\dagger\} = <Q_i> = \sum_\sigma |<f_{i\sigma}>|^2 + |<b_{i0}>|^2 = 1$, i.e., the fermionic commutation rules for the physical field *are* obeyed.

It might be useful to note, for the holon sector, all finite physical matrix elements can be reduced to $\mathcal{P}_G <0|F_{i0} F_{i0}^\dagger|0> = \mathcal{P}_G \xi^* \xi = 1$.

The full $t - J$ model, written in terms of auxiliary operators, is[5],

$$\mathcal{H} = -t \sum_{<ij>\sigma} f_{j\sigma}^\dagger b_{j0} b_{i0}^\dagger f_{i\sigma} + J \sum_{<ij>} \vec{S}_i \cdot \vec{S}_j, \tag{27}$$

where $q_i = e\left[b_{i0}^\dagger b_{i0} - 1\right]$ is the charge, and $\vec{S}_i$ is the spin operator, for the site i.

At least in the context of mean-field theories, some care is required if $\vec{S}_i \cdot \vec{S}_j$ is to be correctly expressed in terms of auxiliary operators. To understand the problem, consider, e.g., $(J/2)S_i^+ S_j^-$. Within a theory which admits a physical mean-field pair amplitude, $<c_{i\downarrow}c_{j\uparrow}>$, the *pair field part* of this interaction is $-(J/2)c_{i\uparrow}^\dagger c_{j\downarrow}^\dagger <c_{i\downarrow}c_{j\uparrow}> + \ldots$, which is biquadratic and must be written in terms auxiliary operators as $-(J/2)f_{i\uparrow}^\dagger b_{i0} f_{j\downarrow}^\dagger b_{j0} <c_{i\downarrow}c_{j\uparrow}>$. It therefore follows, within a theory with a finite pair amplitude, the auxiliary particle mapping should be,

$$\frac{J}{2}S_i^+ S_j^- \rightarrow \frac{J}{2} f_{i\uparrow}^\dagger b_{i0} b_{i0}^\dagger f_{i\downarrow} f_{j\downarrow}^\dagger b_{j0} b_{j0}^\dagger f_{j\uparrow}, \tag{28}$$

and $not \to (J/2)f_{i\uparrow}^\dagger f_{i\downarrow} f_{j\downarrow}^\dagger f_{j\uparrow}$, which is the simpler form adopted by all previous authors[1,2,3,4].

It is important not to confuse the physical pair amplitude $<c_{i\downarrow}c_{j\uparrow}>$ and the similar RVB amplitude $<f_{i\downarrow}f_{j\uparrow}>$. That the latter is finite is necessary for the former to be non-zero but is not a sufficient condition. Within the present factorable theory $<c_{i\downarrow}c_{j\uparrow}> = <F_{i0}^\dagger F_{j0}^\dagger>$ $<f_{i\downarrow}f_{j\uparrow}>$ and hence *both* a spinon *and* a holon pair amplitude are needed in order to have a physical pair amplitude.

To obtain a theory of spinons and holons with physical, RVB and holon pair amplitudes, it is merely necessary to substitute the expansion for $f_{i\sigma}^\dagger$ into the t-J Hamiltonian and factor out these pair amplitudes. After diagonalization of the kinetic energy, such a theory comprises the following effective Hamiltonian:

$$\mathcal{H}_{\text{eff}} = \sum_{\vec{k}\sigma} |\mathcal{E}_{\vec{k}\sigma} - \mu_1| s_{\vec{k}\sigma}^\dagger s_{\vec{k}\sigma} + \sum_{\vec{k}\sigma}(\epsilon_{\vec{k}} - \mu_2) F_{\vec{k}0}^\dagger F_{\vec{k}0}$$
$$- \sum_{\vec{k}\vec{k}'} \mathcal{V}_{\vec{k},\vec{k}'}\left(\Delta_{\vec{k}'}^* s_{\vec{k}\uparrow}^\dagger s_{-\vec{k}\downarrow}^\dagger + H.c.\right) - \sum_{\vec{k}\vec{k}'} V_{\vec{k},\vec{k}'}\left(D_{\vec{k}'}^* F_{\vec{k}0}^\dagger F_{-\vec{k}0}^\dagger + H.c.\right), \quad (29)$$

where, μ_1 and μ_2 are the chemical potentials for the two sectors, the order parameters are $\Delta_{\vec{k}} = <s_{\vec{k}\uparrow}^\dagger s_{-\vec{k}\downarrow}^\dagger>$ and $D_{\vec{k}} = <F_{\vec{k}0}^\dagger F_{-\vec{k}0}^\dagger>$, the energy $\mathcal{E}_{\vec{k}\sigma} = 2t\tilde{\delta}$; $\tilde{\delta} = (1/N)\sum_{\vec{k}} \gamma_{\vec{k}} N_{\vec{k}}$ contains the density of holes, and is small (assuming δ is small), while $\epsilon_{\vec{k}} = 2\tilde{t}\gamma_{\vec{k}}$; $\tilde{t} = (t/N)\sum_{\vec{k}\sigma} \gamma_{\vec{k}} n_{\vec{k}\sigma}$ is only weakly dependent on δ and implies a bandwidth of the order of $8t$. This reduced model reflects the possibility of both spinon and holon pairing, via the interactions,

$$\mathcal{V}_{\vec{k}\vec{k}'} = (1-\delta)^2 \frac{3Jg_s}{4}\gamma_{\vec{k}-\vec{k}'} + \frac{3Jg_s}{4}\sum_{\vec{k}''\vec{k}'''}\left(D_{\vec{k}''}^* \gamma_{\vec{k}-\vec{k}'+\vec{k}''-\vec{k}'''} D_{\vec{k}'''}\right), \quad (30)$$

and

$$V_{\vec{k}\vec{k}'} = \frac{3Jg_s}{4}\sum_{\vec{k}''\vec{k}'''}\left(\Delta_{\vec{k}''}\gamma_{\vec{k}-\vec{k}'+\vec{k}''-\vec{k}'''}\Delta_{\vec{k}'''}^*\right). \quad (31)$$

The factor of g_s occurs because of correlations effects induced by Gutzwiller projection enforced by the Grassmann variables. The preliminary value is $g_s = 4$ following the claim of Zhang et al.[7] that Gutzwiller projected ground states correspond to a lower energy than do the conventional mean-field theories based upon the large N expansion.

V. FLUX STATES

In the above has been discussed the construction of a trial wavefunction and a pairing theory assuming that the searched for separable mean-field ansatz which is uniform. It turns out that this uniform solution does not have the lowest energy, i.e., there occurs what can be though as an all electronic Peierls instability towards a ground state in which the phases of the effective hopping matrix elements are non-uniform. This can also be thought of as a flux phase in which the flux seen by the holon sector is generated by the spinon sector and vice versa. In order to exhibit the phenomenon more clearly, in this section, it will again be assumed that $J = 0$. This is done to isolate the spinon-holon interaction via the kinetic energy rather than to make specific claims about the globally stable solution in this limit.

What has been shown elsewhere[8], is that if, instead of taking, $<f_{i\sigma}^\dagger f_{j\sigma}> = $ const, for all near neighbour bonds, it is assumed that,

$$<f_{i\sigma}^\dagger f_{j\sigma}> = |<f_{i\sigma}^\dagger f_{j\sigma}>|e^{i(e/\hbar c)\int_j^i \vec{A}_k \cdot d\vec{s}} \quad (32)$$

where $\vec{A}_k$ is the vector potential corresponding to a uniform magnetic field $\vec{B}$ perpendicular to the assumed two dimensional plane, and if it is assumed that the value of B is such that the flux is $\Phi = (p/q)\Phi_0$ per plaquette of the lattice, where p/q is an irreducible proper fraction and Φ_0 is the flux quantum, then it is found that the similar expectation value for the holon sector is of the form,

$$< b_{i0} b_{j0}^\dagger > = | < b_{i0} b_{j0}^\dagger > | e^{-i(e/\hbar c) \int_j^i \vec{A}_k \cdot d\vec{s}}. \tag{33}$$

The effect of the fictitious field cancels in the physical density matrix element

$$< c_{i\sigma}^\dagger c_{j\sigma} > = < f_{i\sigma}^\dagger f_{j\sigma} > < b_{i0} b_{j0}^\dagger > = | < f_{i\sigma}^\dagger f_{j\sigma} > | | < b_{i0} b_{j0}^\dagger > |, \tag{34}$$

although the effect contained in the averages $| < f_{i\sigma}^\dagger f_{j\sigma} > |$ and $| < b_{i0} b_{j0}^\dagger > |$ does not.

As is usual for flux states, when p/q is chosen correctly there are gaps which appear at the Fermi surface and the net effect is to lower the energy as compared with the zero flux state. For a giving filling n the optimal flux state is one with $n = p/q$. For a given value of q the original single band splits into q sub-bands and the lower p bands are filled. It is also the case that the largest gap occurs between the p and $p+1$ bands. This leads to a problem since the optimal flux for the holon sector corresponds to $\delta = p/q$ while for the spinon sector $(1/2)(1-\delta) = p/q$ for both of the spinon ($\sigma = \uparrow$ or $\downarrow$) bands. Roughly[8], the splitting between the bands corresponds to that between Landau levels, i.e., $\sim B$. As is usual, for gaps in a fermionic density of states the associated energy gain relative to the gapless, i.e., zero flux state, should $\sim B^2 \sim (p/q)^2$ or $\sim B^2 \ln B \sim (p/q)^2 \ln(p/q)$. The total kinetic energy for both the spinons or holons are the same, i.e., $\sim \delta t$. It follows that the largest energy to be gained is that associated with the spinon sector, since appropriate value of p/q is larger (for small δ), and so the optimal flux state has

$$\frac{p}{q} = \frac{1}{2}(1-\delta), \tag{35}$$

in this limit.

Notice that the holons will have $(q-2p)$ sub-bands filled and that this corresponds to a cusp like minimum energy for the holon filling $\delta = (q-2p)/q$; it follows that there is a cusp like minimum of the *total* energy.

VI. BROKEN STATISTICAL SYMMETRY

The aim of any many-body theory is to determine the nature of the elementary excitations. Often these can be understood in terms of a set of non-interacting (or possibly weakly interacting) effective particles, i.e., the "quasi-particles". Usually these are fermion and/or bosons, as, e.g., for a "Landau Fermi liquid". It is of considerable interest[11] to know if, in two dimensions, it is possible to have quasi-particles which are non-interacting, or weakly interacting, "anyons". That the ground state of the kinetic energy is a flux state, within mean-field theory, is suggestive that the elementary excitations are indeed anyons. For this finite flux ground state it is inappropriate to think of the fermionic spinons and holons as elementary excitations since they are rather strongly interacting via the fictitious field $\vec{A}_k$.

As was mentioned in the introduction, at least for two dimensions, the auxiliary particle decompositions known as "slave bosons" or "slave fermions" are not the only two possibilities, rather there exists a continuum of possibilities of the form,

$$c_{i\sigma} = \mathcal{A}_{i0}^\dagger a_{i\sigma} \tag{36}$$

where $\mathcal{A}_{i0}^\dagger$ and $a_{i\sigma}$ are complimentary anyons such that the physical field is anti-symmetric for the interchange of physical electrons. This freedom of statistics can be thought of as a gauge symmetry.

As usual, the anyons in this decomposition can be replaced by fermions and/or bosons with flux tubes attached. It will be assumed that each auxiliary particle has the same type, and magnitude, of fictitious charge, e, and that there is only a single fictitious magnetic field, i.e., the anyon $\mathcal{A}_{i0}^\dagger$ is represented by a boson $b_{i0}^\dagger$, with a certain flux attached, while $a_{i\sigma}^\dagger$ (not $a_{i\sigma}$) is reflected by a fermion $f_{i\sigma}^\dagger$ carrying the same flux. Clearly the physical vacuum with a $b_{i0}^\dagger$ particle at each site has a uniform distribution of flux tubes the effect of which must be canceled by a uniform background field. With this, the destruction operator $f_{i\sigma}$ creates a flux tube which cancels that due $b_{i0}^\dagger$ and the physical electron created by $c_{i\sigma}^\dagger = b_{i0}^\dagger f_{i\sigma}$ carries no net flux. The total magnetic field corresponds to $\vec{A} = \vec{A}_s + \vec{A}_h$.

The the spinon-holon interactions are treated in a mean-field approximation whence, the spinon field becomes replaced by a uniform average $<\vec{A}_s>$. To this must be added a second uniform fictitious flux $-\vec{A}_k$ which reflects the non-uniform effective hopping matrix elements discussed in the previous section. Total fictitious field generated by the spinon sector is now,

$$<\vec{A}_s> -\vec{A}_k. \tag{37}$$

and it is possible to set $<\vec{A}_s> = \vec{A}_k$ so that this field is null. With this condition the only field seen by the holons is $\vec{A}_h$ and this sector is described by free anyons. Of course, since the mean-field theory is constructed using Grassmann variables these free anyons are effective fermions, reflected by the $F_{i0}^\dagger$, to which are attached the flux tubes. The same considerations lead to free anyons for the spinon sector with the same value for the magnitude of the average flux, i.e., with $<\vec{A}_h> = - <\vec{A}_s>$.

The magnitude of, e.g., $<\vec{A}_s>$ is determined if a second mean-field approximation is made for the spinon sector. In this approximation the flux of the spinons themselves is replaced by the mean-field value $<\vec{A}_s>$. Since the other two contributions to the average field have been made to cancel this field alone which must satisfy the flux condition $\Phi = (p/q)\Phi_0$ with $p/q = (1/2)(1-\delta)$ for the reasons explained in connection with the flux states. The corresponding relationship between flux per particle and the filling, to the authors knowledge, has not yet been determined exactly for free anyons and it can be only speculated that these relationships remain exact.

That there is a special decomposition of the physical field which leads to both the spinons and holons being free anyons (demonstrated here with $J = 0$) can reasonably called *broken statistical symmetry*, i.e., all choices for this decomposition are *not* equivalent.

VII. CONCLUSION

Using the usual slave boson version of the Barnes auxiliary particle method for strongly correlated problems, it is found that within a mean-field theory which maintains, (i) the physical field commutation rules $\{c_{i\sigma}, c_{j\sigma'}^\dagger\} = 0; i \neq j$ and (ii) the constraint $Q_i = 1$, both the spinons and holons are fermions *but* that the ground state, even for the $U = \infty$, i.e., $J = 0$, limit, is a flux state. The fictitious flux seen, e.g., by the spinons is generated by the holons and vice versa. This fictitious field therefore represents, an albeit average, large residual interaction between the two sectors.

With a more general version of the auxiliary particle method in which both the spinons and holons are anyons, this average residual interaction can be removed whence each sector is described by free anyons. Both the holons and spinons have the same statistics, i.e., they are both fermions (or possibly hard core bosons) with the same magnitude of statistical flux attached. The spinons have $S = 1/2$. Near to half filling both particles will be close to, but at least within a mean-field approximation for the statistical flux, but not exactly semions[11].

REFERENCES

[1] See P. W. Anderson in *Frontiers and Borderlines in Many Body Physics* (North-Holland, Amsterdam, 1987).

[2] Z. Zou, and P. W. Anderson, Phys. Rev. **37**, 627 (1988).

[3] See, for different points of view e.g., K. Flensberg et al., Phys. Rev. B**40**, 850 (1989); F. Nori, E. Abrahams, and G. T. Zimanyi, Phys. Rev. B**41**, 7277 (1990); T. Kopp, F. J. Seco, S. Schiller, P. Wölfle, Phys. Rev. B**38**, 11 835 (1988).

[4] See also several contributions in this volume which use the slave boson mean-field approximation.

[5] S. E. Barnes, J. Phys. F **6**, 115 (1976); J. Phys. F **6**, 1375 (1976); J. Phys. F **7**, 2637 (1977); Adv. Phys. **30**, 801 (1980); Phys Rev. B**40**, 723 (1989); Rapid Communication, Phys. Rev. B**41**, 11 701 (1990).

[6] P. Coleman, Phys. Rev. B**29**, 3035 (1984).

[7] F. C. Zhang, C. Gros, T. M. Rice and H. Shiba, Supercond. Sci. Technol. **1** 36 (1988).

[8] S. E. Barnes, Proc. of the 19th Int. Conf. on Low Temperature Physics, D. S. Betts, ed., Physica B **165** & **166** (North-Holland, Amsterdam, 1990) p. 987; Phys. Rev. B, to be published, 1991.

[9] See, J. W. Negele and H. Orland, *Quantum Many-Particle Systems*, (Addison-Wesley, Redwood City, California, 1988).

[10] See, e.g., C. Kittel *Quantum Theory of Solids*, (John-Wiley & Sons, NY, 1966).

[11] For an interesting discussion, see, Z. Zou, this volume.

FREQUENCY DEPENDENT CONDUCTIVITY OF HIGH T_c OXIDE SUPERCONDUCTORS

Hayoung Kim and Peter S. Riseborough

Department of Physics, Polytechnic University
333 Jay Street, Brooklyn, New York 11201

INTRODUCTION

The normal state properties of the cuprate superconductors are quite unusual, and depend strongly on the degree of doping. In the absence of electron-electron interactions, one would expect that the undoped systems would be simple metals, however it was found that the undoped systems are antiferromagnetic and insulating[1,2]. The antiferromagnetic insulating phase can be viewed either as an example of a Slater antiferromagnet or of a Mott-Hubbard insulator, which occurs whenever each ion is occupied by an integer number of electrons. When the systems are doped, an excess number of electrons or holes is introduced and the system becomes metallic and non-magnetic[3,4]. In this note, we shall calculate the frequency dependent electrical conductivity, taking into account the effect of both charge and spin fluctuations[5,6], and compare the results with recent experiments[7-9].

The normal state of the doped cuprate systems is to be described in terms of a Hubbard model, with N fold spin and orbital degeneracy. The electrons are described by a density of states with total band width 2W, that has the characteristic discontinuities at the band edges associated with the two dimensional anisotropy of the Copper-Oxide planes. The electrons occupying the d orbitals on the same ion are assumed to interact via a highly screened coulomb interaction U_{dd}. The Hamiltonian H governing the electronic system can be written as

$$H = H_0 + H_I \qquad (1.a),$$

where H_0 describes the band of non-interacting electrons

$$H_0 = \sum_{k\,\alpha} e(\underline{k})\, a^+_{\alpha,\underline{k}}\, a_{\alpha,\underline{k}} \qquad (1.b),$$

in which $a^+_{\alpha,\underline{k}}$, $a_{\alpha,\underline{k}}$ respectively create and annihilate and electron in the αth spin-orbit sub-band state labeled by the Bloch wave vector $\underline{k}$. The coulomb interaction between the electrons occupying the orbitals on the same lattice is described by H_I, where

$$H_I = \sum_{i\alpha\beta} U_{dd}/2 \, a^+_{\alpha,i} \, a^+_{\beta,i} \, a_{\beta,i} \, a_{\alpha,i} \quad (1.c),$$

where $a^+_{\alpha,i}$ and $a_{\alpha,i}$ are the creation and annihilation operators for the Wannier orbitals on site $\underline{R}_i$.

The electronic bands have an electron filling fraction of less than half, corresponding to the case of hole doping. The electronic self-energy is calculated as an expansion in descending powers of N the degeneracy[6], where $u = N U_{dd}/W$ is a constant coupling strength. The lowest order term corresponds to the unrestricted Hartree-Fock approximation. The lowest order term indicates that the paramagnetic phase will undergo a ferromagnetic instability at a critical value of $u = 2(N-1)$, and an antiferromagnetic instability at a smaller value of u.

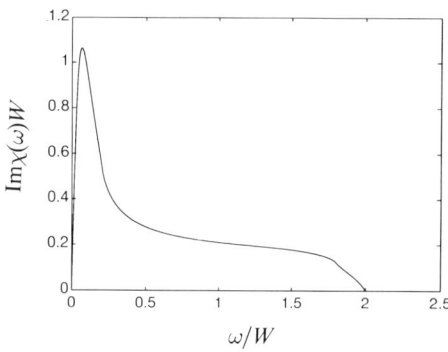

Fig. 1. The spectral density for the collective excitations of the system. The spectral density $\mathrm{Im}\chi(\underline{q},\omega)$ is plotted in units of $1/W$, versus the dimensionless energy ω/W. The following parameters where chosen; $u/u_c = 0.85$, $\mu/W = -0.8$.

The next order term in the expansion yields the Random-Phase Approximation to the self-energy, with coupled charge and spin fluctuations[6]. This non-trivial irreducible self-energy, due to electron-electron interactions, is given by the expression

$$\hat{\Sigma}(i\omega_n) = W \sum_{\underline{q}} \int dx \, \frac{u^2}{\pi} \, \mathrm{Im}\chi(\underline{q},x)$$

$$\left\{ \frac{[1 - f(e(\underline{k}-\underline{q})) + N(x)]}{i\omega_n + \mu - e(\underline{k}-\underline{q}) - x} + \frac{[f(e(\underline{k}-\underline{q})) + N(x)]}{i\omega_n + \mu - e(\underline{k}-\underline{q}) + x} \right\}$$

(2.a),

where $i\omega_n$ are the odd Matsubara frequencies and the spectral density for the coupled spin and charge boson collective excitations $\mathrm{Im}\chi(\underline{q},x)$ is given by

$$\chi(\underline{q},x) = \chi^{(0)}(\underline{q},x) \, [1 + u\chi^{(0)}(\underline{q},x)]^{-1} \cdot [1 - u\chi^{(0)}(\underline{q},x)/(N-1)]^{-1}$$

(2.b),

and the non-interacting polarizability is $\chi^{(0)}(\underline{q},x)$. This self energy also has an additional part due to weak inelastic, isotropic, scattering from a low density of impurities.

The impurity contribution gives a small shift to the energy and an energy independent elastic scattering contribution $i/2\tau$ to the imaginary part of the self-energy. The irreducible part of the electronic self-energy is evaluated in the local approximation[10,11], which is expected to be reasonable if the collective fluctuations are local in character. The spectral density for these coupled boson excitations are shown in figure 1. The low energy peak in this spectrum can be identified with spin fluctuations, whereas the high energy tail is associated with the charge fluctuations. For values of u close to uc the spin fluctuation energy scale is low and the electronic bands experience significant renormalization effects, leading to large mass enhancement factors, $Z = [1 - \partial/\partial\omega\ \Sigma(\omega)\ ]|_{\omega=0}$, for the quasi-particle masses, and the quasi-particle lifetimes.

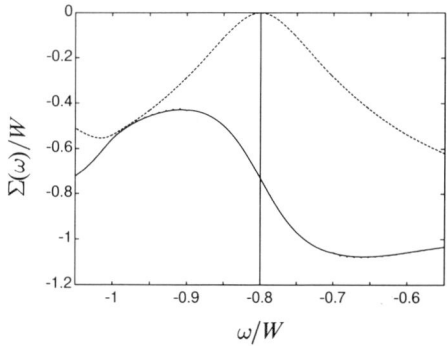

Fig. 2. The real and imaginary parts of the irreducible self energy $\Sigma(\omega)$ in units of W, as a function of ω/W.

The real and imaginary parts of the self-energy are shown in figure 2. The imaginary part of the self-energy shows the characteristic ω^2 variation near the fermi-energy, expected for a fermi liquid at zero temperature, but rapidly changes over to an approximate linear variation off the fermi-energy similar to that postulated for marginal fermi-liquids[12]. The total angle integrated density of states shows a narrow peak at the fermi energy, and the upper and lower Hubbard split bands residing above and below the fermi-energy, respectively. Due to the $\underline{k}$ independence of the self-energy, Luttinger's theorem[13,14] results in the value of the density of states at the fermi energy being equal in magnitude to the, bare, non-interacting density of states. The intensity of the upper and lower Hubbard split bands are controlled by the high energy charge fluctuations.

RESULTS

The frequency dependent conductivity is given, by the expression[15]

$$\sigma_{x,x'}(\underline{q},\omega) = 1/i\omega \, [\, F_{x,x'}(\underline{q},\omega+i\delta) - \delta_{x,x'} \, ne^2/m \,]$$
(3.a),

where $F_{x,x'}(\underline{q},\omega+i\delta)$ is the analytic continuation of the function

$$F_{x,x'}(\underline{q},i\omega_m) = k_B T \int_0^\beta d\tau_1 \int_0^\beta d\tau_2 \, \exp[-\hbar\omega_m(\tau_1 - \tau_2)] \, F_{x,x'}(\tau_2,\tau_1)$$
(3.b).

The time ordered, thermally averaged correlation function $F_{x,x'}(\tau_2,\tau_1)$ is defined as

$$F_{x,x'}(\tau_2,\tau_1) = \, < \, T \, j_x(-\underline{q},\tau_2) \, j_{x'}(\underline{q},\tau_1) \, >$$
(3.c),

in which the j_x are the components of the paramagnetic current operator and $i\omega_m$ are the even Matsubara frequencies. The conductivity can be re-written in terms of the full thermal one electron Green's functions and a Vertex Function as

$$F_{x,x'}(i\omega_m) = -e^2 \, k_B T \sum_n \sum_{\underline{k},\alpha,\beta} v_{x,\alpha,\beta,\underline{k}} \, G_\beta(\underline{k}-\underline{q}/2, i\omega_n) \cdot$$
$$\cdot G_\alpha(\underline{k}+\underline{q}/2, i\omega_n+i\omega_m) \, \Lambda_{x',\alpha,\beta,\underline{k}}(i\omega_n, i\omega_n+i\omega_m)$$
(4),

where the Vertex function is given by $\Lambda_{x,\alpha,\beta,\underline{k}} = v_{x,\alpha,\beta,\underline{k}}$, to lowest order in $1/N$. Following Holstein's discussion of the non-degenerate band case[15], we shall take the $\underline{q} \to 0$ limit, in which the conductivity is composed primarily from the intra-sub band contribution as the f sum rule shows that the inter-sub band term vanish for sufficiently small ω. The conductivity will be evaluated with the above lowest order approximation for the vertex function, and with the full Green's functions replaced by the expressions obtained with the approximate self-energy. For low temperatures, and low external frequencies, the resulting expression for the conductivity can be expanded in powers of ω and $1/\tau$, the leading order contribution is

$$\sigma(\omega) = - \int \frac{dz}{2mW} \, e^2 \, (\partial/\partial z \, f(z)) \frac{[1/\tau]}{[1/\tau^2 + (Z\omega)^2]} \, (\, z - \Sigma(z) + W \,)$$
(5).

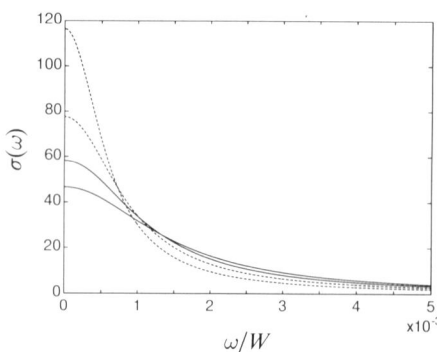

Fig. 3. The low frequency variation of the conductivity $\sigma(\omega)$, for various inelastic scattering rates $1/\tau$ in units of W/h.

This produces a Drude like conductivity,

$$\sigma(\omega) = \frac{n e^2/m\tau}{[(Z\omega)^2 + 1/\tau^2]} \quad (6),$$

where n is the electron density given[13,14] as

$$n = (\mu - \Sigma(\mu) + W)N/2W = (e(\underline{k}_f) + W)N/2W \quad (7).$$

where $e(\underline{k}_f)$ is the fermi-energy and N/2W is the non-interacting one electron density of states. The width of the Drude peak is renormalized by the factor of Z which reduces the single impurity elastic scattering rate[16,17] to $1/\tau Z$, giving a narrow Drude peak centered at $\omega = 0$. This shows that the renormalizations cancel in the d.c. limit. The low frequency Drude peaks shown in figure 3, were calculated for different impurity elastic scattering rates. At higher frequencies the simple Drude formulae breaks down. Near the frequency where the imaginary part of the self energy changes from a fermi-liquid like ω^2 to a linear ω variation, the conductivity crosses over to yield a flatter variation with ω, an almost $1/\omega$ variation, in contrast to the $1/\omega^2$ variation associated with the Drude tail. This cross-over can be clearly seen in figure 4. The conductivity in the intervening frequency range shows a maximum, as shown in figure 4. The position of this maximum roughly corresponds to the energy for excitations between the fermi-level peak and the precursor of the lower Hubbard band, which is slightly larger than ($\mu + W - \Sigma(\mu)$). Therefore, this peak in the frequency dependent conductivity may be ascribed to intra-band transitions. In the extreme high frequency limit, the many body enhancements drop out and result in an ordinary, unrenormalized, $1/\omega^2$ Drude tail to the conductivity[18].

The frequency dependent conductivity has been calculated in by using a 1/N expansion technique that maintains the fermi-liquid behavior for states close to the fermi-level. The conductivity has a d.c. limit which is dominated by elastic scattering from impurities and is totally devoid of many body

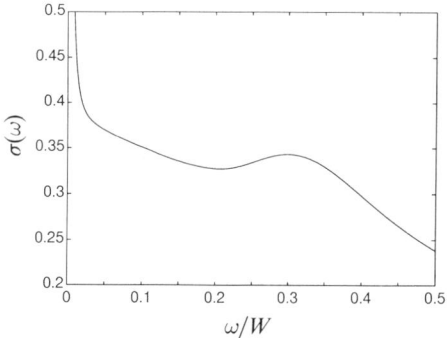

Fig. 4. The higher frequency behavior of the conductivity $\sigma(\omega)$, versus ω/W.

enhancements. At low frequencies, the conductivity has a Drude form with a width which is given by the impurity inelastic scattering rate reduced by the quasi-particle mass enhancement factor Z. The experimental results on high T_c superconductors [7-9], are consistent with these results and give many body enhancement factors of roughly 2 to 3. At higher frequencies, elastic scattering dominates over the inelastic scattering and there is a cross-over to a frequency dependence characteristic of a marginal fermi-liquid, similar to that observed in experiment[7]. At frequencies of the order of $(W + e(k_f))/h$ the conductivity shows a weak maximum. This maximum is due to transitions between the incoherent states of the lower Hubbard split band to the quasi-particle states near the fermi-level. This maximum is close to the energy of the inter-band transition obtained from an analysis of the experimental data[8].

ACKNOWLEDGEMENTS

This work was supported by the U.S. Department of Energy, Office of Basic Energy Sciences through DE-FG01-84ER45127.

REFERENCES

(1) P.B. Allen, Z. Fisk and A. Migliori, **Physical Properties of High Temperature Superconductors**, Donald Ginsburg Ed., World Scientific Publishing Co., Singapore (1989).
(2) R.J. Birgenau and G. Shirane, **Physical Properties of High Temperature Superconductors**, Donald Ginsburg Ed., World Scientific Publishing, Singapore (1989).
(3) A.T. Fiory, S. Martin, R.M. Fleming, L.F. Schneemeyer, J.V. Waszak, A.F. Hebard and S.A. Sunshine, Physica C $162-164$, 1195 (1989).
(4) R.A. Walstead, W.W. Warren Jr. , R.F. Bell and G.P. Espinosa, Phys. Rev. B 40, 2572 (1989).
(5) A. Kampf and J.R. Schrieffer, Phys. Rev. B 42, 7639 (1990).
(6) P.S. Riseborough and Hayoung Kim, Phys. Rev. B 42, 7675 (1990).
(7) J. Orenstein, G. Thomas, A.J. Millis, S.L. Cooper, D.H. Rapkine, T. Timusk, L.F. Schneemeyer and J.V. Waszak, Phys. Rev. B (1990).
(8) T. Timusk and D. Tanner, **Physical Properties of High Temperature Superconductors**, Donald M. Ginsburg Ed., World Scientific Publishing Co., Singapore (1989).
(9) Z. Schlesinger, R.T. Collins, F. Holtzberg, C. Feild, S.H. Blanton, U. Welp, G.W. Crabtree, Y. Fang and J.Z. Liu, Phys. Rev. Lett. 65, 801 (1990)
(10) J. Freidel and C.M. Sayers, J. Phys. (Paris) 38, 697 (1977).
(11) W. Metzner and D. Vollhardt, Phys. Rev. Lett. 86, 324 (1989).
(12) C.M. Varma, P.B. Littlewood, S. Schmidt-Rink, E. Abrahams and A.E. Ruckenstein, Phys. Rev. Lett. 63, 1996 (1989).
(13) J.M. Luttinger, Phys. Rev. 119, 1153 (1960).
(14) J.M. Luttinger, Phys. Rev. 121, 942 (1961).
(15) T. Holstein, Ann. Phys. 29, 410 (1964).
(16) J.H. Kim, K. Levin, R. Wentzcovitch and A. Auerbach, Phys. Rev. B 40, 11378 (1989).
(17) G. Kotliar, P.A. Lee and N. Read, Physica C $153-155$, 538 (1988).
(18) P.S. Riseborough, Phys. Rev. B 27, 5775 (1983).

ELECTRONIC STRUCTURE AND THE SUPERCONDUCTING GAP OF $Bi_2Sr_2CaCu_2O_{8+\delta}$

Z.-X. Shen, D.S. Dessau, and B.O. Wells

Stanford Electronics Laboratory
Stanford University
Stanford, California 94305

Introduction

During the last several years, extensive photoemission studies have been performed on the cuprate superconductors[1]. These studies revealed, among other things, details near the Fermi level which are sensitive to the shape of the Fermi surface, the highly correlated nature of these oxides, as well as the superconducting gap states. These are arguably the most fundamental questions raised by the novel oxide superconductors. Most of the angle-resolved photoemission efforts were concentrated on $Bi_2Sr_2CaCu_2O_{8+\delta}$ (referred as Bi2212 in the rest of the paper) because of its superior material properties (stability in vacuum and ability to cleave)[2-4]. In this short paper, we give a brief summary of the experimental work on Bi2212 involving the Stanford group in collaboration with groups from Los Alamos, Ames and IBM. We intend to make this paper a "road map" so that the readers can get a general picture of our results, and can find more detailed information about specific issues in the references listed.

Summary of Early Results on Bi2212

Bi2212 was the first cuprate compound where a clear Fermi edge was observed in the photoemission spectra above T_c[5,6]. Furthermore, the O1s core level spectrum was very clean, without extra peaks near -532 eV, as previously observed in the earlier discovered cuprates[7]. The crystal cleaves between the two weakly bonded Bi-O planes[8]. The terminating surface Bi-O layer has the same incommensurate superstructure of the bulk Bi-O layers as determined by LEED and STM studies[9,10]. These points are important since they give confidence on results taken from Bi2212, as summarized in the following paragraphs.

Normal State Properties

(i) The experimental Fermi surface is very consistent with the LDA Fermi surface, which implies that the Luttinger Theorem is obeyed. Among others, the excellent work of Olson et al. gave the most convincing evidence of this point[2-4]. For the band along Γ-X direction, they found that the experimental band crossed the Fermi level at the same k value where the theoretical band intercepted the Fermi level. Recently, we provided more information about the experimental Fermi surface[11]. The issue we addressed is whether there is a Bi-O electron pocket near the M point in the Brillouin zone, as band calculations have suggested[12]. Our result showed that, at least for the samples prepared under certain conditions, the band near the M point has substantial Bi-O character. This is consistent with the prediction of the band calculation. Further studies will attempt to address the issue of whether a separate Bi-O electron pocket exists.

(ii) The doping behavior of metallic Bi2212 can be well described by the renormalized band approaches using various forms of the Anderson Hamiltonian[13-18]. The T_c of the Bi2212 decreases as we doped it with more holes by annealing it at higher oxygen partial pressure. Based on this, we concluded that the doping levels of Bi2212 crystals are higher than the doping level corresponding to the maximum T_c. In this high doping region, the valence band features shift in energy position with doping. This could be best interpreted as a result of a chemical potential shift due to the filling and emptying of the bands at E_F. At the same time, we saw indications that the lowest energy band approached the Fermi level at different k values in differently doped samples which was consistent with the chemical potential shift[19]. The spectral weight near E_F increases together with the hole doping, which can be interpreted as a manifestation of the spin and charge fluctuations within the framework of the renormalized band approach[13-18]. This doping behavior of the Bi2212 compound is quite different from the picture obtained near the insulating regime[20], suggesting that some aspects of the Fermi liquid behavior have been recovered at such high doping levels.

(iii) Correlation effects persist in the Bi2212 system despite the fact that it is far away from the antiferromagnetic insulating regime. It is important to point out that even though the one-electron band calculation appears to give a good description of the Fermi surface, the correlated nature of the material still persists. This is most clearly indicated by the existence of the valence band satellite near -12 eV[21,22]. Such satellite structures cannot be explained within the context of the one-electron picture. It is a manifestation of the correlated many-body nature of the material. To the lowest order approximation of the Anderson Hamiltonian, this satellite could be interpreted as the lower Hubbard band. It's energy position gives an estimate of the Coulomb interaction among the two electrons on the same Cu site to be about 6-8 eV. Therefore, Bi2212 is a strong positive U system.

(iv) Several other issues should be pointed out about the Bi2212 compound, as listed in the following paragraphs.

The first is that of the quasiparticle lifetime. Earlier angle-resolved photoemission work by Olson et al. showed that the quasiparticle lifetime was inversely proportional to the quasiparticle energy[3]. This is not consistent with the Fermi liquid picture, but is consistent with other hypotheses[23]. We have very similar experimental data. However, we caution that uncertainties do exist. This includes the complications of the background subtraction, k_z broadening (due to the incomplete two dimensionality), the final state effects and so on. Furthermore, the photoemission peaks used to obtain information of the quasiparticle lifetime had their energies higher than 40 meV(500K in temperature). The nature of the low temperature physics may be very different.

The second issue that needs to be discussed is the uniformity of the doping in the Bi2212 compound. Even though oxygen doping is much better than cation doping in terms of doping uniformity[24], we still have to worry about where the oxygen goes in our interpretation of the photoemission data. Indeed, we have found recently that oxygen goes to different sites as we increase the partial pressure of oxygen under which the Bi2212 crystal is annealed[25].

The third issue is that of the core level spectra. The recent core level data with improved energy resolution showed interesting new information. e.g. The Cu2p core line obtained from Bi2212 with higher resolution (0.3 eV) showed details that were very different from that of CuO, suggesting that the cuprates and CuO are more different than we used to think. Similar conclusions were drawn based on a comparison of experimental data and the Anderson impurity model calculation[26], and the band calculation[27].

The fourth issue is related to the extremely short probing depth of the photoelectron spectroscopy (less than 5 Å in some cases). Since the unit cell along the c-axis is about 12 Å, this means that the signal obtained in photoemission is not even an average of one unit cell. It is amazing that so much information obtained from photoemission is so consistent with predictions of the bulk properties such as the LDA Fermi surface. To what extent the photoemission data can be regarded as reflecting the bulk properties should always be a question in our mind. On the other hand, the fact that the length scale of photoemission is shorter than the dimension of the crystal also provides opportunities for us to extract information related with only part of the crystal structure[11,28].

The last issue is related to the fact that an interesting property of the Bi cuprate system is that T_c is related with the number of the Cu-O planes in the compound[29]. Preliminary studies showed that the density of states at E_F is higher in the Bi2212 compound than the Bi2201 compound and this is probably related to the higher T_c in Bi2212[22]. A systematic investigation of Bi2201, Bi2212 and Bi2223 may help us to understand the role of interlayer coupling between the Cu-O planes, which is important to some of the theoretical models[30].

Therefore, we summarize the normal state properties of the Bi2212 with the following remarks. This system shows aspects of correlation (as a consequence of the large positive U) and some one-electron like features. The Fermi surface is very consistent with the LDA Fermi surface. The doping behavior in this system is explainable within the context of the renormalized band approach.

The Superconducting State properties

Bi2212 is presently the only cuprate materials where a superconducting gap was observed in the photoemission spectra below T_c[4,11,31,32]. This was made possible by the high energy resolution of the spectrometer, the high transition temperature and the large superconducting gap of the material. An advantage of angle-resolved photoemission is that it provides us the opportunity to study the superconducting gap at different parts of the Brillouin zone. We give a brief account of our results in the following paragraphs.

(i) For the cuprate superconductors, because the coherence length is comparable with the lattice constant (ζ_0~a), the conventional Ginzberg-Landau approach predicted a clear reduction of the gap parameter at the surface[33]. However, we found experimental evidence which showed that the superconducting gap parameter at the surface was in fact equal to or even larger than that in the deeper layers[34]. This result is in contrast to the conventional wisdom, and is consistent with a calculation by Valls et al. who suggested an enhancement of the surface gap parameter due to Friedel type of oscillations[35]. This surface effect may be important in interpreting experimental data of the superconducting state obtained by surface sensitive techniques such as photoemission, EELS, and tunnelling.

(ii) We observed anomalous spectral weight transfer at the superconducting transition of Bi2212[36]. In the conventional BCS picture, the spectral weight in the superconducting gap region is depleted and transferred to a peak below the gap as the sample goes through the superconducting transition (see inset of Fig.1). For Bi2212, we observed something that is very different as shown in the figure. Along the Γ-M direction, a dip appears at higher binding energies(~ -0.09 eV). Along the Γ-X direction, on the other hand, no such clear dip is observed. Furthermore, we found that the photoemission spectral weight transfer clearly varies in k space. For example, the photoemission spectral weight increases along the Γ-X direction but decreases along the Γ-M direction. These changes of spectral weight are strongly dependent on the doping level of the crystals. As we have pointed out in our paper, these data may provide important information about the nature of the high temperature superconducting state.

(iii) As shown in Fig.1, the spectral weight transfer at the superconducting transition is very different along Γ-M and Γ-X directions. Then a natural question to ask is whether the supercondcuting gap parameter along the two directions are the same or not. This turns out to be a very tricky issue. As discussed in detail elsewhere, we found evidence for the gap parameter to be larger along the Γ-M direction than Γ-X direction[34]. This result is different from the finding by Olson et al.[32]. We are not sure about the reasons behind this difference. However, a possible explanation could be found in the calculation by Valls et al who suggested a Friedel type of oscillations for the gap parameter near the surface[35]. As indicated earlier, the gap parameters along the two directions may be associated with different layers (the gap parameter along Γ-M is more closely related with the Bi-O layer). The physical surface (of the electron cloud) near the Bi-O layer may be very different in different samples and thus lead to different answers.

It is also appropriate to point out two possible alternative explanations of our experimental data. The first is that by Anderson who suggests that the interlayer coupling is a key to the superconductivity. The interlayer coupling along Γ-M may be stronger so that one sees a larger gap. The second is the Van Hove singularity scenario. There is a possibility of a Van Hove singularity near M (If not completely rounded out by the mixing of the Bi-O states). Hence the gap near M may be larger due to stronger coupling to the higher density of states there.

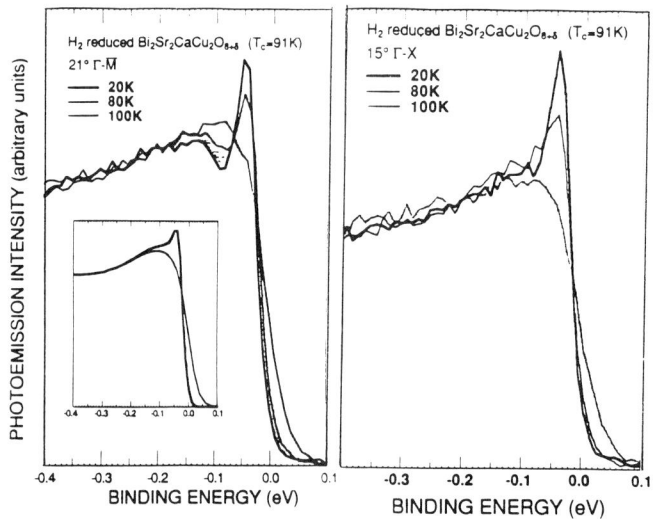

Fig.1. High resolution PES spectra along two different directions in k space recorded above and below the superconducting transition temperature.

(iv) We have used high resolution photoemission to search for the existence of the proximity effect between the cuprate and a conventional metal in two junctions made with the best possible conditions available at this time[32]. The first junction is that of Au/c-axis Bi2212. We have cleaved a Bi2212 crystal in UHV (2×10^{-10} torr) at low temperature (20K), and evaporated 5 Å Au overlayer. We saw 5Å Au forms a metallic overlayer, but we did not see a proximity induced gap which is equal or larger than 5 meV. Since the high temperature superconductors are highly two dimensional in the sense that superconducting carries reside mainly in the important Cu-O sheets, putting down a metallic overlayer along the direction which is perpendicular to the Cu-O plane may make the proximity effect difficult to occur. The natural solution to the problem is to try to form junctions along the a or b axes, where the coherence length is much longer. To do this, we studied a junction made of Au/a-aixs YBCO film. A 15 Å Au overlayer was deposited in situ after the growth of the a-axis YBCO film in 10^{-6} torr vacuum. Again, our data put an upper limit of the proximity effect induced gap size to be 5 meV or smaller in the junctions specified above.

Therefore, we can see that PES provides much new information about the high temperature superconducting states. Using this technique, we can address important scientific issues such as the superconducting gap state, the surface superconductivity, the possible gap anisotropy, and the proximity effect.

Acknowledgement: We wish to express our thanks to following colleagues for their helpful discussions and collaborations: Dr. W.E. Spicer, Dr. A.J. Arko, Dr. R.S. List, Dr. C.G. Olson, Dr. D.B. Mitzi, and Dr. A.Kapitulnik. This work is supported by NSF grant DMR-8913478, NSF support through the Center for Materials Research at Stanford, the JSEP contract DAAD 29-85-K-0048.

References

1. See, for example, P.A.P. Lindberg, Z.-X. Shen, I. Lindau, W.E. Spicer, Surface Science Reports, Vol. 11, 1-138 (1990), and references therein
2. T. Takahashi, H. Matsuyama, H. Katayama-Yoshida, Y. Okabe, S. Hosoya, K. Seki, H. Fujimoto, M. Sato, and H. Inokuchi, Nature 334, 691 (1988)
3. C.G. Olson, R. Liu, D.W. Lynch, R.S. List, A.J. Arko, B.W. Veal, Y.C. Chang, P.Z. Jiang, and A.P. Paulikas, Phys. Rev. B 42, 381 (1990); C.G. Olson, R. Liu, A.-B. Yang, D.W. Lynch, A.J. Arko, R.S. List, B.W. Veal, Y.C. Chang, P.Z. Jiang, and A.P. Paulikas, Science 245, 731 (1989)
4. R. Manzke, T. Buslaps, R. Claessen, and J. Fink, Europhysics Lett. 9, 477 (1989)
5. M. Onellion, M. Tang, Y. Chang, G. Margaritondo, J.M. Tarascon, P.A. Morris, W.A. Bonner, and N.G. Stoffel, Phys. Rev. B 38, 881 (1988)
6. Z.-X. Shen, P.A.P. Lindberg, I. Lindau, W.E. Spicer, C.B. Eom, and T.H. Geballe, Phys. Rev. B38, 7152 (1988)
7. Göran Wendin, J. Phys. (Paris), Colloq. 48, C9-1157 (1987), and references therein
8. P.A.P. Lindberg, Z.-X. Shen, B.O. Wells, D.S. Dessau, D.B. Mitzi, I. Lindau, W.E. Spicer, and A.Kapitulnik, Phys. Rev. B 39, 2890 (1989)
9. P.A.P. Lindberg, Z.-X. Shen, B.O. Wells, D. Mitzi, I. Lindau, W.E. Spicer, A. Kapitulnik, Appl. Phys. Lett. 53, 2563 (1988)
10. M.D. Kirk, J. Nogami, A.A. Baski, D.B. Mitzi, A. Kapitulnik, T.H. Geballe, and C.F. Quate, Science 242, 1673 (1988)
11. B.O. Wells, D.S. Dessau, W.E. Spicer, C.G. Olson, D.B. Mitzi, A. Kapitulnik, R.S. List, and A.J. Arko, Phys. Rev. Lett., 65, 3056 (1990)
12. S. Massidda, J. Yu, and A.J. Freeman, Physica C 52, 251(1988)
13. Z.-X. Shen et al., preprint
14. D.M. Newns, M. Rasolt, and P.C. Pattnaik, Phys. Rev. B 38, 6650 (1988)
15. C.A.R. Sa' de Melo and S. Doniach, Phys. Rev. B 41, 6633 (1989)
16. Qimiao Si, Jian Ping Lu, and K. Levin, Physica C, 162-164, 1467 (1989); K. Levin, see her article in this proceeding
17. M. Tachiki, Spinger Series in Solid State Sciences, Vol. 89, 138 (1989); S. Ishihara, H. Matsumoto, and M. Tachiki, Phys. Rev. B 42, 10041 (1990)
18. J. Zanaan, and Olle Gunnarsson, Physica C, 162-164, 821 (1989)
19. D.S. Dessau, Preprint
20. J.W. Allen, C.G. Olson, M.B. Maple, J.-S. Kang, L.Z. Liu, J.H. Park, R. Anderson, Phys. Rev. Lett. 64, 595 (1990)
21. Z.-X. Shen, P.A.P. Lindberg, B.O. Wells, D.B. Mitzi, I. Lindau, W.E. Spicer and A. Kapitulnik, Phys. Rev. B 38, 11820 (1988)
22. Z.-X. Shen, P.A.P. Lindberg, P. Soukiassian, I. Lindau, W.E. Spicer, C.B. Eom, T.H. Geballe, Phys. Rev. B39, 823 (1989)
23. C.M. Varma, see his article in this proceeding
24. D.B. Mitzi, L.W. Lombardo, A. Kapitulnik, S.S. Laderman, and R.D. Jacowitz, Phys. Rev. B 41, 6564 (1990)
25. F. Parmijiani, Z.-X. Shen, D.B. Mitzi, I. Lindau, W.E. Spicer, and A. Kapitulnik, Physr. Rev. B, in press
26. O. Gunnarsson, O. Jepsen, and Z.-X. Shen, Phys. Rev. B 42, 8707 (1990)

27. The LDA-band structure calculation for the cuprates does a better job of predicting experimental results than the LDA calculation for CuO does[R. Cohen, private communication]
28. Z.-X. Shen, P.A.P. Lindberg, D.S. Dessau, I. Lindau, W.E. Spicer, D.B. Mitzi, I. Bozvic, and A. Kapitulnik, Phys. Rev. B 39, 4295 (1989)
29. J.M. Tarascon, W.R. Mckinnon, P. Barboux, D.M. Hwang, B.G. Bagley, L.H. Greene, G.W. Hull, Y. LePage, N. Stoffel, and M. Giroud, Phys. Rev. B 38, 8885 (1988)
30. P.W. Anderson, see his article in this proceeding
31. J.-M. Imer, F. Patthey, B. Dardel, W.-D. Scheider, Y. Baer, Y. Petroff, and A. Zettl. Phys. Rev. Lett. 62, 336 (1989)
32. C.G. Olson, R. Liu, And D.W. Lynch, R.S. List, A.J. Akro, B.W. Veal, Y.C. Chang, P.Z. Jiang, and A.P. Paulikas, Solid State Comm. 76, 411 (1990)
33. G. Deutscher and K.A. Müller, Phys. Rev. Lett. 59, 1745(1987)
34. B.O. Wells et al., preprint
35. O.T. Valls, M.T. Beal-Monod, T. Giamarchi, S. W. Pierson, Physica B 165&166, 1081(1990)
36. D.S. Dessau et al, submitted to Phys. Rev. Lett.
37. D.S. Dessau, B.O. Wells, Z.-X. Shen, W.E. Spicer, R.S. List, A.J. Arko, C.G. Olson, D.B. Mitzi, A. Kapitulnik, Appl. Phys. Lett., to be published

SUPERCONDUCTING AND MAGNETIC PHASE BOUNDARIES IN DOPED Cu-BASED SUPERCONDUCTING OXIDES: CONTRASTING CHARGING AND PAIR BREAKING MECHANISMS

Y. Gao, A. Kebede,[a] P. Pernambuco-Wise, M. Kuric,[b] J. E. Crow, R. P. Guertin,[b] T. Mihalisin,[a] N. D. Spencer[c] and D. W. Cooke[d]

Physics Department and Center for Materials Development and Technology
Florida State University
Tallahassee, FL 32306

INTRODUCTION

After several years of intensive study of the high transition temperature superconducting oxides (HTS), the nature and origin of the pairing mechanism has eluded investigators. Theories with local, real space pairing and more conventional models with k-space pairing have evolved and a variety of mechanisms have been proposed.[1-5] A useful and now well known method of probing the relevant interactions in both low and high temperature superconductivity is through the use of impurities selected to investigate specific aspects of the response of the system to controlled perturbations, with particular emphasis given to changes of the superconducting properties.

There are two common features associated with most of the known HTS: First, the charged carriers are paired in the quasi-2-dimensional Cu-O planes giving rise to conductivity in those planes with a weak pair coupling perpendicular to those planes. Second, strong electron correlations can lead to long range magnetically ordered phases (normally but not exclusively Cu-ordering) in materials where the chemical parameters are adjusted so that the material is near or just beyond the superconducting/normal

---

[a] Physics Department, Temple University, Philadelphia, PA 19122
[b] Physics Department, Tufts University, Medford, MA 02155
[c] Reaserch Division, W. R. Grace & Co.-Conn., 7379 Route 32, Columbia, MD 21044
[d] Los Alamos National Laboratory, Los Alamos, NM 87545

phase boundary. Examples are the antiferromagnetism in $La_{2-y}M_yCuO_4$ for M = Ba or Sr and $y \leq 0.02$ and $YBa_2Cu_3O_{7-d}$ for $\delta \geq 0.6$.[6]

Most Cu-based impurity studies have focused on the effects of doping on the superconducting and normal state phase boundaries and are primarily directed at the effects of impurity substitution at the Cu-sites, e.g., substitution of Zn and Ni in $(La, Ba)_2CuO_4$ (LaBCO) and Zn, Ni, Co in $YBa_2Cu_3O_7$ (YBCO7).[6,7] These studies have led to evidence for charge (valence) mechanisms and d-hole localization, both with attendant pair breaking effects. Others have looked at the more indirect probes of rare earth (RE) doping on the cation Y site located midway between the Cu-O planes in YBCO7.[9,10] With the exception of RE = Ce, Pr, Pm and Tb, the surprising result is that rare earth substitution has little or no effect on the transition temperature T_c and on other superconducting properties. This is attributed to a vanishingly small exchange interaction J_{df}, between the charged carriers (normally d-holes) in the Cu-O planes and the 4f electrons. It should be noted that in all these cases the RE is 3+. For fully RE substituted YBCO7, viz REBCO7, the RE moments order antiferromagnetically in the superconducting state with Néel temperatures $T_N \leq 2.5K$. Of the exceptions noted, REBCO7 does not form in the orthorhombic structure for RE = Ce and Tb, leaving PrBCO7 as an anomaly. In $Y_{1-x}Pr_xBa_2Cu_3O_7$ [(Y,Pr)BCO7], Pr replaces the Y cation (valence 3+) and causes a depression of T_c with $T_c \rightarrow 0$ for $x \approx 0.54$.[9,10] In addition, the Pr moments in PrBCO7, i.e., $x = 1.0$, antiferromagnetically order with $T_{N2} = 17K$, a temperature fully two orders of magnitude higher than that derived from scaling T_N from other REBCO7 systems[10,12,13]. For the $x = 1.0$ normal system, the Cu moments antiferromagnetically order with $T_{N1} = 280K$.[14] Several measurements including specific heat[12], magnetic susceptibility[12], neutron scattering[13] and NMR[15] seem to suggest that Pr as a cation dopant in YBCO7 is tetravalent, or very nearly tetravalent and that the f-electrons strongly hybridize with the O p-holes within the Cu-O planes. This hybridization is responsible for the depression of T_c through possible changes in the carrier density within the Cu-O planes and/or depairing due to the enhanced exchange interaction between the charged carriers in the Cu-O planes and the Pr moments at the Y site[12,15]. As both of these effects will cause degradation of the superconducting properties, it would be useful to examine systems that would allow a clearer separation of the relative strengths of these two perturbations. Studies of $(Y_{1-x}Pr_x)_{1-y}Ca_yBa_2Cu_3O_7$ have led to a better understanding of the effects of both valence charging and pair breaking contributions to the depression of T_c observed in the (Y,Pr)BCO system.[16] Studies of oxygen depleted YBCO7 allow examination of the effect of non-magnetic perturbations on HTS, but interpretations are complicated by a second superconducting phase at reduced O content and a change in structure near the critical O concentration where $T_c \rightarrow 0$. A much cleaner system is $Bi_2Sr_2Ca_{1-x}Y_xCu_2O_8$ [BS(Ca,Y)CO].[17] Trivalent Y does not carry a magnetic moment and replaces divalent Ca midway between the superconductivity-active Cu-O planes, resulting in changes in the carrier density without causing any magnetic depairing. In Y substituted BSCCO, T_c is depressed due the substitution of Y^{3+} for Ca^{2+} with $T_c \rightarrow 0$ for $x \approx 0.6$ and with an additional (presumably Cu) ordering occurring at $x > 0.6$.[18]

In this paper, we review the results of our studies of (Y,Pr)BCO7 and BS(Ca,Y)CO and focus on a comparison of the superconducting and magnetic phase boundaries and the very interesting region where these phases approach each other. We include measurements of the x-dependence of the room temperature lattice constants and the x- and T-dependence of the low temperature specific heat, magnetic susceptibility, resistivity and muon spin relaxation (μSR).

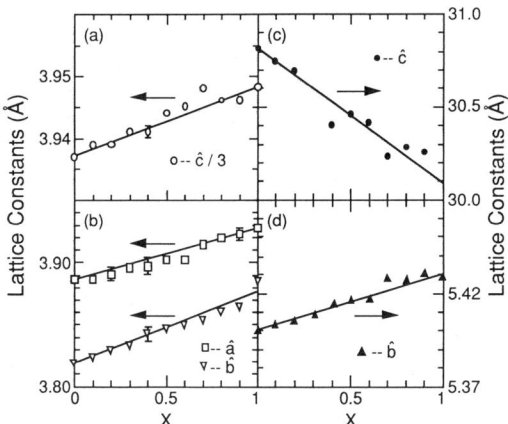

Fig. 1. The concentration dependence of the room temperature unit cell parameters for $Y_{1-x}Pr_xBa_2Cu_3O_7$ (a,b) and $Bi_2Sr_2Ca_{1-x}Y_xCu_2O_8$ (c,d).

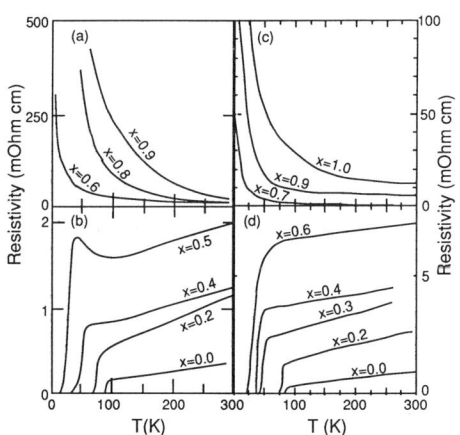

Fig. 2. The temperature and concentration dependences of the electrical resistivity of $Y_{1-x}Pr_xBa_2Cu_3O_7$ (a,b) and $Bi_2Sr_2Ca_{1-x}Y_xCu_2O_8$ (c,d).

EXPERIMENTAL DETAIL

The polycrystalline samples of (Y,Pr)BCO used in this experiment were prepared using standard solid state reaction techniques. Stoichiometric amounts of high purity Y_2O_3, Pr_6O_{11}, $BaCO_3$, and CuO were thoroughly mixed and initially calcined at 950 °C for several hours with the exception that for Pr concentration greater than x = 0.5 the calcining temperature and other processing temperatures did not exceed 900 °C. The reacted products were powdered and reheated for 12 h at the calcining temperature several additional times. During the final sintering process, the samples were heated in flowing oxygen, cooled and held at 600 °C for 10 h before removing from the furnace. Based on thermogravimetric measurements and neutron diffraction studies, the oxygen content is relatively independent of the Pr-doping. The polycrystalline samples of BS(Ca,Y)CO were prepared by the co-precipitation of bismuth, strontium, calcium, yttrium and copper ions as carbonates and hydroxycarbonates using tetramethylammonium carbonate, followed by filtration, drying, heating in air at 540 °C for 5 h and then at 800 °C for 20 h. After the initial reaction, the powders were pressed into pellets and sintered in CO_2-free air for 96 h at 875 °C with a slow cooling to 360 °C and furnace cooled to room temperature. The pellets were then ground, pressed in pellets and fired, this procedure being repeated a number of times to improve homogeneity.

X-ray diffraction patterns were obtained using automated Rigaku and/or Siemens diffractometers with $CuK\alpha$ radiation. All the samples used in this study were single phase, i.e., second phases where near or less than the detectable limits using X-rays. For (Y,Pr)BCO7, the system remains orthorhombic for all x. Shown in Fig. 1a and 1b is the x-dependence of the room temperature, orthorhombic lattice constants. The cell volume for PrBCO is $181.1 A^3$, this being nearly equal to the value expected from an extrapolation of the cell volumes for the other rare earth substituted YBCO vs rare earth trivalent metallic radii, ie. assuming Pr enters the lattice as 3+. This extrapolation may be fortuitous however, as other structural studies indicate that the Pr-Pr ion separation in Pr doped YBCO is inconsistent with that anticipated based on similar measurements on RBCO[19]. Those measurements would suggest that Pr in YBCO is mixed valent. Other non-structural measurements to be presented later, also point to mixed valent or tetravalent behavior for Pr in YBCO7.

The Y-doping dependence of the room temperature lattice constants for BS(Ca,Y)CO is shown in Fig. 1c and 1d. The X-ray diffraction data were fit assuming a tetragonal structure, and as shown in Fig. 1c, the c-axis lattice constant decreases whereas the a-axis constant increases with increasing Y-doping. For both systems, i.e., (Y,Pr)BCO7 and BS(Ca,Y)CO, there is no evidence for a change in structure or break in the x-dependence of the lattice constants in the vicinity of the x-concentration required for complete suppression of superconductivity which occurs in the vicinity of $x \approx 0.6$ for both systems.

The resistivity of small rectangular bars cut from the sintered pellets was measured for $1.8 \leq T \leq 300K$ using a standard four probe dc method. The specific heat was measured for $1.2 \leq T \leq 20K$ using a quasi-adiabatic method. The magnetic susceptibility was measured for $1.8 \leq T \leq 300K$ and $0 \leq H \leq 5.5T$ using a Quantum Design SQUID magnetometer and for $2 \leq T \leq 300K$ and $0 \leq H \leq 9T$ using a vibrating sample magnetometer.

RESULTS AND DISCUSSION

Shown in Fig. 2 is the x-and T-dependence of the resistivity for (Y,Pr)BCO7 and BS(Ca,Y)CO. For both systems, the resistivity is metallic-like for most of the samples in the superconducting regime, switching to resistivity characterized by a negative temperature coefficient for samples with a concentration beyond that necessary to depress $T_c \to 0$. The switching of the resistivity from metallic to semiconductor-like at the critical concentration x_{cr} for $T_c \to 0$ is a common feature of impurity doped HTS. However, the magnitude of the resistivity change in the vicinity of x_{cr} for (Y,Pr)BCO7[8] and BS(Ca,Y)CO is much less than that displayed by dopants that enter the Cu-site, e.g., Zn substituted YBCO7. For the Zn-doped YBCO7 system, there is good evidence for a strong localization of the Cu d-holes within the Cu-O planes as reflected by the development of a strong g = 2 electron spin resonance signal for $x > x_{cr}$.[8] In (Y,Pr)BCO7,however, the dopant only appears to change carrier density without strong evidence for disorder induced localization as occurs in the Zn-doped system. Attempts to fit the resistivity data shown in Fig. 2 to localization models were not successful. For (Y,Pr)BCO7, $\rho(T)$ for $x < x_{cr}$ shows a weak maximum for $T > T_c$ not seen in the BS(Ca,Y)CO system, a difference that is not yet understood.

The squares in Fig. 3 show the x-dependence of T_c for (Y,Pr)BCO7 and BS(Ca,Y)CO. There are clear similarities between these two systems in that x_{cr} for both systems is in the vicinity of x = 0.6, i.e., $x_{cr} \approx 0.54$ and 0.65 for (Y,Pr)BCO7 and BS(Ca,Y)O respectively. However there are noteworthy differences. Shown in the insert of Fig. 3a is T_c vs x for

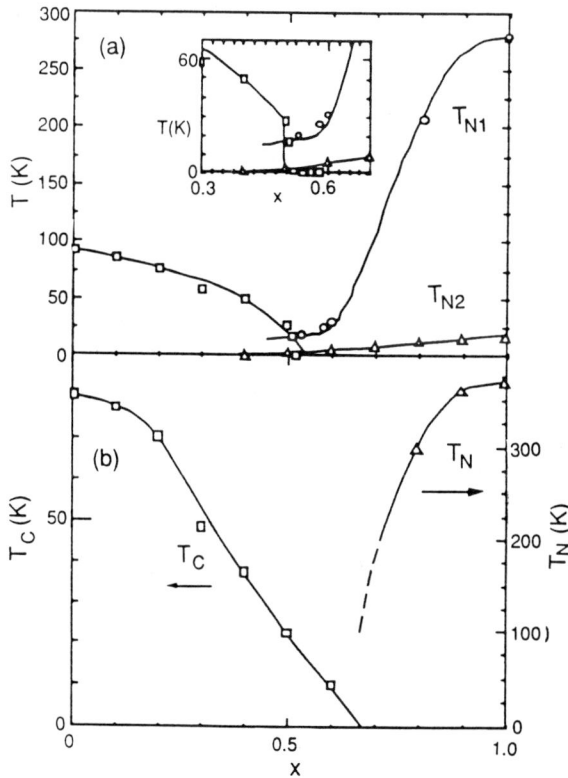

Fig. 3. The magnetic phase diagrams of $Y_{1-x}Pr_xBa_2Cu_3O_7$ (a) and $Bi_2Sr_2Ca_{1-x}Y_xCu_2O_8$ (b). T_{N1} and T_{N2} in (a) refer to the Néel temperatures for ordering of the Cu-moments and Pr moments, respectively. In both cases the squares show T_c as a function of impurity concentration. The insert in (a) shows the detail of the crossover region between superconductivity and anitferromagnetic phases.

(Y,Pr)BCO7 and the x dependence of the Néel temperatures for ordering of the Cu-moments within the Cu-O planes and the ordering of the Pr moments between those planes, T_{N1} and T_{N2} respectively. Determination of these ordering temperatures will be discussed later. As is the case for some conventional magnetically doped superconductors with localized magnetic depairing, there is a more rapid depression of T_c in the vicinity of the overlap between superconducting and magnetic phases.[20] The more rapid depression of T_c for $x \approx x_{cr}$ in (Y,Pr)BCO7 is evidence for a 4f-conduction electron-hybridization driven renormalization of T_c vs x due to magnetic exchange field effects and is indicative of magnetic depairing. However more detailed studies of the T_c vs x for $x \approx x_{cr}$ in the BS(Ca,Y)CO system are required before one can conclude this unequivocally.

Additional evidence of strong hybridization between the 4f and conduction electrons in (Y,Pr)BCO7 is apparent in the x- and T-dependence of the low temperature specific heat, C(T). Shown in Fig. 4b is C(T)/T vs T^2 at low temperatures for (Y,Pr)BCO7. C(T)/T is linear in T^2 for $T^2 > 40K^2$ and a fit of these data yields an estimate of γ, the electronic specific heat coefficient and Θ_D, the Debye temperature. The temperature dependence of C(T) shown in Fig. 4b and the γ-values obtained are similar to the behavior seen in highly correlated f-electron systems.[21] Shown in Fig. 5 is γ obtained from fitting C(T) to $\gamma T + \beta T^3$ vs x for (Y,Pr)BCO7. For x = 0.4, $\gamma \approx 400$mJ/mol-Pr K^2 which indicates that there must be considerable 4f electron weight in the density of states at the Fermi energy. Such large γ-values are consistent with theories of the mixed valent state in the limit that the valence is near an integral value.[22] For strongly mixed valent systems, i.e., systems with valences well away from integral values, the electronic contribution is not expected to be enhanced. An indication that these heavy-fermion-like γ-values are due to f-electron character and not localized d-states is found in the C(T) for the BS(Ca,Y)CO systems. By comparison, C(T) for BS(Ca,Y)CO does not display any strong enhancements and the γ-values are very low compared to those obtained for (Y,Pr)BCO7, see Fig. 4a and Fig. 5. Based on these two measurements, the similarities between the two systems and nature of the doping, it is natural to associate the large γ-values measured for (Y,Pr)BCO7 with strong hybridization between the f-electrons and the charged carriers in the Cu-O planes and to conclude that this hybridization contributes to the depression in T_c.

The magnetic susceptibility $\chi(T)$ and temperature derivative of the magnetic susceptibility $\chi'(T)$ for (Y,Pr)BCO7 are shown in Fig. 6a and 6b, respectively. $\chi(T)$ for (T_c or 20K) $\leq T \leq$ 400K (not shown in Fig.6) can be fitted to a modified Curie-Weiss relationship, i.e., $\chi(T) = \chi_0 + C/(T + T^*)$, with an effective paramagnetic moment $\mu_{eff} \approx 2.50 \pm 0.6$ μ_B which is close to the anticipated value for Pr^{4+} ($\mu_{eff} = 2.54\mu_B$) and much smaller than the value for Pr^{3+} ($\mu_{eff} = 3.58\mu_B$). The reduced μ_{eff} (as compared to that anticipated for Pr^{3+}) could be due to crystal electric field (CEF) effects which would split the 6-fold degenerate angular momentum states and lead to a reduced μ_{eff}. However, recent measurements of μ_{eff} for BS(Ca,Pr)CO yield a value of $\mu_{eff} = 3.52 \pm 0.04\mu_B$ which is close to the value expected for Pr^{3+} and it would be logical to anticipate that the CEF environment at the Pr-site in BSPrCO would be somewhat similar to that experienced in (Y,Pr)BCO7.

The susceptibility for PrBCO7 has a small anomaly in the vicinity of 17K which can be seen much clearer in $\chi'(T)$, see Fig 6b. An anomaly in C(T) for PrBCO7 is also seen at 17K. Neutron diffraction studies have established that this ordering is due to an antiferromagnetic ordering of the Pr-moments located midway between the Cu-O planes.[13] The size of the ordered magnetic moment is also consistent with a Pr^{4+} CEF split ground state.[13] From C(T) and $\chi(T)$ measurements, the Néel temperature

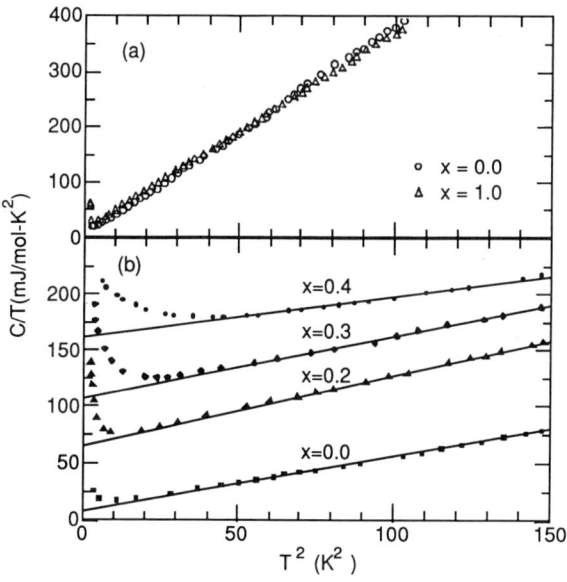

Fig. 4. Specific heat data defined by temperature for $Bi_2Sr_2Ca_{1-x}Y_xCu_2O_8$ (a) and $Y_{1-x}Pr_xBa_2Cu_3O_7$ (b) vs temperature squared.

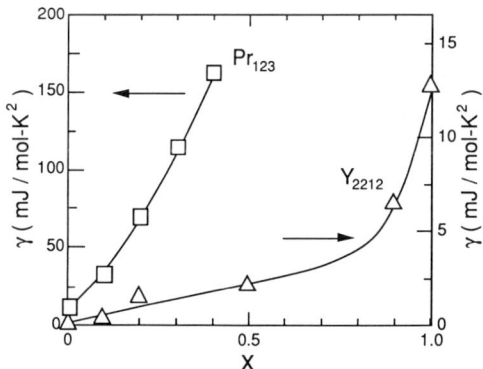

Fig. 5. The electronic specific heat coefficient obtained from fits to the data in Fig. 4 ($T^2 > 40K^2$) vs impurity concentration for $Y_{1-x}Pr_xBa_2Cu_3O_7$ (□) and $Bi_2Sr_2Ca_{1-x}Y_xCu_2O_8$ (△).

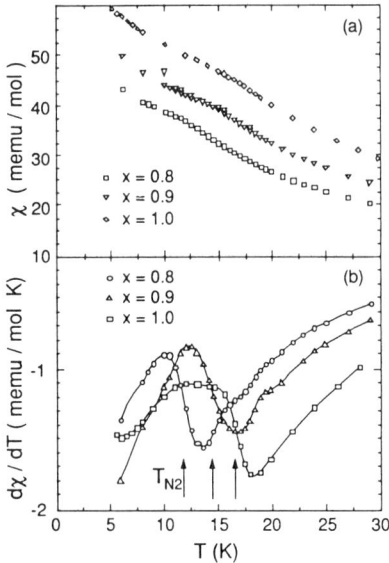

Fig. 6. The magnetic susceptibility $\chi(T)$ and first temperature derivative of the magnetic susceptibility $\chi'(T)$ for $Y_{1-x}Pr_xBa_2Cu_3O_7$ vs temperature. The arrows in (b) indicate T_{N2}, the magnetic ordering temperature of the Pr moments for each of the concentrations.

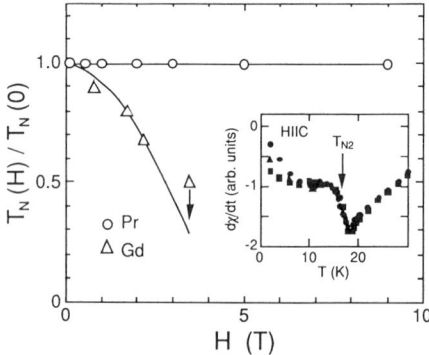

Fig. 7. The magnetic field dependence of the renormalized Néel temperature showing the lack of dependence of Pr doped 123 as compared to Gd doped 123. The insert shows the temperature dependence of the first temperature derivation of the susceptibility for $H = 1$, 3 and 5T.

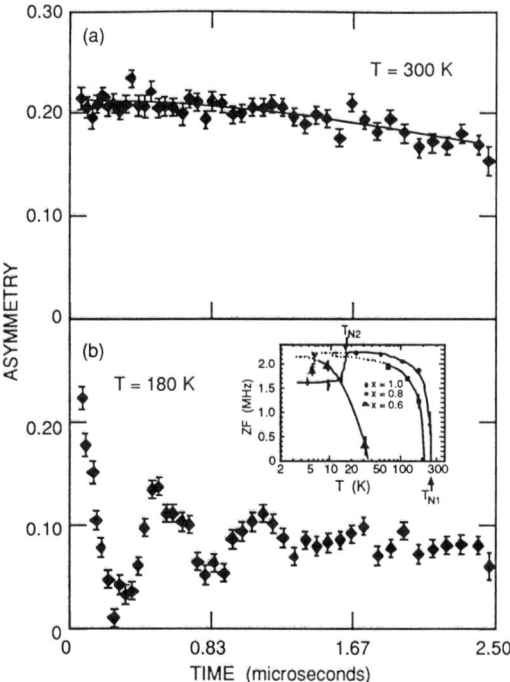

Fig. 8. Representative zero field muon spin resonance spectra for $Y_{1-x}Pr_xBa_2Cu_3O_7$ at T = 300K (a) and T = 180K (b). The insert in (b) shows the zero field precessional frequency as a function of temperature.

T_{N2} due to Pr ordering can be determined and leads to the antiferromagnetic phase boundary shown in Fig. 3a by the triangles. There is evidence for a coexistence region of magnetic order and superconductivity for $x \leq x_{cr}$. An unusual feature of the T_{N2} is its independence of strong magnetic fields as shown in Fig. 7. Shown in the insert of Fig. 7 is $\chi'(T)$ measured in H = 1, 3 and 5T vs T for PrBCO7. Shown in Fig. 6 is the field dependence of T_N for both GdBCO7 and PrBCO7. Whereas T_N for GdBCO7 is depressed by a magnetic field as expected for a conventional antiferromagnet with localized spins, T_N for PrBCO7 is unaffected by magnetic fields up to H = 9T. This is a very surprising result considering the energy scale associated with T_N and the applied magnetic field.

For (Y,Pr)BCO7, a second magnetic phase transition occurs at $T \leq T_{N2}$ which is attributed to the magnetic ordering of Cu-moments within the Cu-O planes.[14] Shown in Fig. 8 are representative zero field μSR spectra for PrBCO7 at T = 300K and 180K.[14] The spectrum shown in Fig. 8a is that expected for a paramagnetic system whereas that in Fig. 8b shows an oscillatory response representative of a well defined precessional frequency due to an internal hyperfine field at the muon site. The dependence of this zero field precessional frequency on temperature for several (Y,Pr)BCO7 samples is shown in the insert of Fig.8b.[14] The Néel temperature T_{N1} due to Cu ordering is defined by the temperature where the zero field precessional frequency goes to zero. Also seen in the insert in Fig. 8b is the lower temperature magnetic transition due to the ordering of the Pr-moments. T_{N1} vs x for (Y,Pr)BCO7 is shown in Fig. 3a. A spin glass like magnetic state is seen by μSR for $T \leq T_c$ but it is not clear at this time if this phase transition is due to a continuation of the Cu ordering phase boundary, as shown in Fig. 7b, or due to Pr ordering which was seen in C(T) data.[14] Combining the data from resistivity, magnetic susceptibility, specific heat and muon spin relaxation, one obtains the very rich superconducting-magnetic phase diagram shown in Fig. 3a. In Fig. 9, $\chi(T)$ vs T is shown for several Y-concentrations in BS(Ca,Y)CO. The most remarkable features of this data are the abrupt change in $\chi(T)$ for $x \geq x_{cr}$ and the evidence of magnetic ordering at high temperatures as reflected by the small anomaly seen in $\chi(T)$, see insert in Fig. 9. This anomaly corresponds to features seen in μSR experiments and has been attributed to an antiferromagnetic ordering of the Cu-moments in the Cu-O planes.[18] Attempts to observe this ordering in neutron diffraction studies have been unsuccessful possibly because of the small size of the Cu-moments and the lower Cu density in this system as compared to YBCO. Using the anomaly apparent in $\chi(T)$ and associating this with Cu ordering, a more complete phase diagram for BS(Ca,Y)CO is obtained, this being shown in Fig. 3b. Measurements are underway to extend the magnetic phase boundary into the vicinity of x_{cr}.

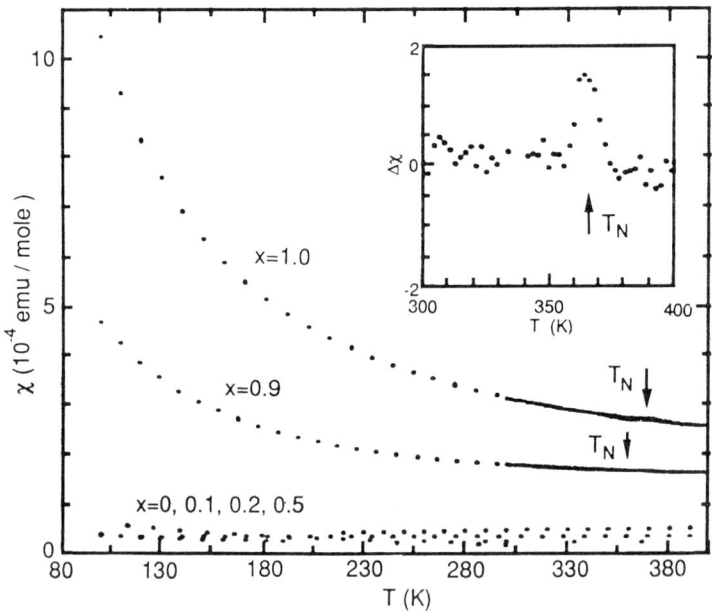

Fig. 9. The temperature dependence of the magnetic susceptibility $\chi(T)$ for $Bi_2Sr_2Ca_{1-x}Y_xCu_2O_8$. The insert shows an expansion of the data at T_N.

CONCLUSIONS

In this paper we have demonstrated for the first time clearly distinguishable cases of HTS pairbreaking. Although superconducting-normal phase diagrams for (Y,Pr)BCO7 and BS(Y,Pr)CO have similarities, and indeed the critical concentration ($x_{cr} = 0.6$) is similar for both systems, the low temperature heat capacities are quite dissimilar, with (Y,Pr)BCO7 showing "heavy fermion"-like behavior and BS(Y,Pr)CO showing conventional metallic low temperature behavior. For (Y,Pr)BCO7 this suggests f-electron participation in the low temperature properties and probably driving the pairbreaking mechanism, but for BS(Ca,Y) simple hole valence filling effects probably cause suppression of the superconductivity. Furthermore, the susceptibility data puts the Pr valence in (Y,Pr)BCO7 near 4+ but in the analogous experiments for P doped BS(C,Pr)CO it is close to 3+. Finally, in this paper we have examined two quite separable magnetic orderings in concentrated (Y,Pr)BCO7, namely, a CU ordering at high temperature and a Pr ordering at low temperature. There are slow indications of a similar Cu ordering at high temperatures in BS(Ca,Y)CO. Further experimentation should investigate the nature of the high temperature ordering in BS(Ca,Y)CO with a similar effort to find a low temperature ordering in the Pr-based counterpart.

Studies such as outlined in this paper share a goal of seeking to understand superconductivity degradation through judicious chemical altering of HTS lattices, thereby moving towards an understanding of the underlying pairing mechanisms that give rise to high temperature superconductivity.

ACKNOWLEDGMENTS

The technical assistance of R. Lichti, C. Boekema and S. F. J. Cox in obtaining the μSR data is acknowledged. Work at Los Alamos is supported by the U. S. D. O. E. We would also like to acknowledge the skillful preparative work of Ted Pedors and the support provided by J. O'Rielly.

REFERENCES

1) See articles in *High Temperature Superconductivity: The Los Alamos Symposium—1989*, ed. by K. S. Bedell, D. Coffey, E. Meltzer, D. Pines and J. R. Schrieffer, (Addison-Wesley, Redwood City, California: 1990).
2) V. J. Emery, Phys. Rev. Lett. **58**, 2794 (1987).
3) J. R. Schrieffer, X. G. Wen and S. C. Zhang, Phys. Rev. Lett. **60**, 944 (1988).
4) P. W. Anderson, Science **235**, 1196 (1987).
5) S. Massidda, J. Yu, A. J. Freeman and D. D. Koelling, Phys. Lett. **A122**, 198 (1987).
6) R. J. Birgeneau and G. Shirane, *Physical Properties of High Temperature Superconductors*, ed. D. M. Ginsberg (World Scientific Publishing: 1989).
7) K. Sekizawa, Y.Takano and T.Inaba, Physica C **162–164**, 1277 (1990).
8) S. T. Ting, Y. Gao, C. S. Jee, S. Rahman, J. E. Crow, T.Mihalisin, G. H. Myer, I. Perez, R. E. Saloman, P. Schlottmann and J. Schwegler, Physica B **163**, 227 (1990).
9) Z. Fisk, J. D. Thompson, E. Zirngiebl, J. L. Smith and S.W. Cheong, Sol. St. Comm. **62**, 743 (1987); P. H. Hor, R. L. Meng, Y. Q. Wang, L. Gao, Z. J. Huang, J. Bechtold, K. Forster and C. W. Chu, Phys. Rev. Lett. **58**, 1891 (1987).

10) For a review of rare earth substitutions in HTS, see J. T. Markert, Y. Dalichouch and M. B. Maple in *Physical Properties of High Temperature Superconductors*, ed. by D. M. Ginzberg (World Scientific, Teaneck, N.J.: 1990) Vol I, 265.
11) L. Soderholm, K. Zhang, D. G. Hinks, M. A. Beno, J. D. Jorgenson, C. U. Segre and I. K. Schuller, Nature **328**, 604 (1987).
12) A. Kebede, C. S. Jee, J. Schwegler, J. E. Crow, T. Mihalisin, G. H. Myer, R. E. Saloman, P. Schlottmann, M. V. Kuric, S. H. Bloom and R. P. Guertin, Phys. Rev. **B40**, 4453 (1989); C. S. Jee, A. Kebede, D. Nichols, J. E. Crow, T. Mihalisin, G. H. Myer, I. Perez, R. E. Saloman and P. Schlottmann, Sol. St. Comm. **69** 379 (1989).
13) W. H. Li, J. W. Lynn, S. Skanthakumar, T. W. Clinton, A.Kebede, C. S. Jee, J. E. Crow and T. Mihalisin, Phys. Rev. **B40**, 5300 (1989); S. Skanthakumar, W. H. Li, J.W. Lynn, A. Kebede, J. E. Crow and T. Mihalisin, Physica B **163**, 239 (1990).
14) D. W. Cooke, R. S. Kwok, R. L. Licthi, T. R. Adams, C. Boekema, W. K. Dawson, A. Kebebe, J. Schwegler, J. E. Crow and T. Mihalisin, Phys. Rev. **B41**, 4801 (1990); D. W. Cooke, M. S. Jahan, R. S. Kwok, R. L. Licthi, T. R. Adams, C. Boekema, W. K. Dawson, A. Kebebe, J. Schwegler, J. E. Crow and T. Mihalisin, Hyperfine Int. **63**, 213 (1990).
15) A. P. Reyes, D. E. Maclaughlin, M. Takigawa, P. C. Hammel, R. H. Heffner, J. D. Thompson, J. E. Crow, Phys. Rev. **B43**, 2989 (1991); D. E. Maclaughlin, A. P. Reyes, M. Tokigawa, P. C. Hammel, R. H. Heffner, J. D. Thompson and J. E. Crow (to be published in Proc. of Valence Fluctuation, Rio de Janeiro, Brazil: 1990).
16) J. J. Neumeier, T. Bjornholm, M. B. Maple and Ivan K.Schuller, Phys. Rev. Lett. **63**, 2516 (1989).
17) T. Tamegai, K. Koga, K. Suzuki, M. Ichihara, F. Sakai and Y. Iye, Jpn. J. Appl. Phys. **28** L112 (1989); T. Fuiita and T. Tomita, Physica C **162–164**, 985 (1989).
18) N. Nishida, S. Okuma, H. Miyatake, T. Tamegai, Y. Iye, R. Yoshizaki, K. Nishiyama, K. Nagamine, R. Kadono and J. H. Brewer, Physica C **168**, 23 (1990); R. De Renzi, C. Bucci, P. Carretta, G. Guidi, R. Tedeschi, C. Calestani and S. F. J. Cox, Physica C **162–164**, 155 (1989).
19) M. E. Lopez-Morales, A. Bezinge, P. M. Grant and D. Riso-Jara, Physica C **162–164**, 61 (1989).
20) Y. Dalichaouch, M. S. Torikachvili, E. A. early, B. W. Lee, C. L. Seaman, K. N. Yang, H. Zhou and M. B. Maple, Solid State Commun. **65**, 1001 (1988); M. B. Maple, Y. Dalichaouch, E. A. Early, B. W. Lee, J. T. Markert, M. W. McElfresh, J. J. Neumeier, C. L. Seaman, M. S. Torikachvili, K. N. Yang and H. Zhou, Proc. Int. Discussion Meeting on HTS, Austria (Plenum, New York).
21) G. R. Stewart, Rev. Mod.Phys. **56**, 755 (1984).
22) P. Schlottmann, Physics Reports **181**, 1–119 (1989).

CARRIER SCATTERING RATES MEASURED BY HALL EFFECT AND MAGNETORESISTANCE IN THE HIGH-T_c OXIDES

N.P. Ong, T.R. Chien, T.W. Jing, and T.V. Ramakrishnan*
Department of Physics, Princeton University, Princeton, New Jersey 08544

Z.Z. Wang
Lab. 2M/CNRS, 196 Ave Henri Ravera, 92220 Bagneux, France

J.M. Tarascon and K. Remschnig
Bell Communication Research, Red Bank, New Jersey 07701
*On leave from the Indian Institute of Science, Bangalore

Introduction

Transport measurements are a useful way to probe the electronic properties of the oxide superconductors because they couple directly to the relevant charge excitations. The unusual linear-temperature variation of the resistivity has received much attention. However, other transport properties are also anomalous when compared with conventional metal behavior. In this report, we survey some recent Hall effect and magnetoresistance measurements on single crystals of $YBa_2Cu_3O_7$ (YBCO) and $Bi_2Sr_2CuO_6$ (Bi 2201).

Hall effect

It is known that the Hall coefficient R_H is strongly temperature dependent in YBCO.[1] The Hall resistivity ρ_{xy} varies roughly as $1/T$ between room temperature and T_c. A similar temperature dependence has been observed in single crystal Bi 2212 and 2223, Tl 2212 and 2223, and in 214 ($La_{2-x}Sr_xCuO_4$). It is instructive to compare these observations with the R_H vs. T curves in the elements Cu, Ag, Mg, Ca, W.[2] Although in these elements R_H shows a strong temperature dependence at low temperatures, it becomes temperature-independent above $\zeta\Theta_D$, where Θ_D is the Debye temperature and ζ is a fraction between 0.2 and 0.3. In conventional metals, it is rare for R_H to display strong variation at temperatures above about half Θ_D (except where the Fermi Surface changes because of nesting instabilities). In contrast, an R_H that is positive and falls monotonically even at temperatures above Θ_D appears to be generic to the layered "hole-type" cuprates. To further investigate this interesting anomaly, we have examined how the Hall conductivity σ_{xy} and the Hall angle Θ_H vary with temperature.[3] By measuring ρ_{xy} and ρ_{xx} simultaneously in microtwinned crystals of "90K" YBCO, we have computed both σ_{xy} and tan

Θ_H. We find that ρ_{xy} is not strictly linear in $1/T$. Above ~250 K, it starts to show negative curvature. However, σ_{xy} follows a strict power law up to our highest temperature 370 K, viz.[3]

$$\sigma_{xy} = A/T^3 \qquad (1)$$

We remark that the Boltzmann expression for the Hall conductivity involves the relaxation time squared, i.e. $\sigma_{xy} \sim \tau^2$. If we assume τ to be the transport scattering time τ_{tr} entering σ_{xx}, we expect σ_{xy} to vary as $1/T^2$. (From infrared reflectivity experiments,[4] the transport lifetime is observed to vary as

$$\hbar/\tau_{tr} = \eta\, k_B T \qquad (2)$$

where $\eta \approx 2\text{-}3$.) An interesting way to analyze Eq. 1 is to plot the reciprocal of Θ_H against the square of the temperature, viz. $\cot\Theta_H$ vs. T^2. In this form, we obtain[3] a straight line which passes close to the origin, i.e.

$$\cot\Theta_H = DT^2 + C \qquad (3)$$

where $D \approx 5.7 \times 10^{-3}$ K^{-2}, and $C \approx 4$, is a constant associated with impurity scattering (negligible in YBCO). Writing $\tan\Theta_H = \omega_c \tau_c$ (ω_c is the cyclotron frequency), Eq. 3 suggests that the "cyclotron" relaxation time τ_c has a temperature dependence $1/T^2$, in contrast with the linear variation for τ_{tr} (Eq 2). In the Boltzmann approach, the Hall current may be viewed as a two-step process: First, the displacement of the Fermi surface (FS) by the electric field **E** introduces the change in distribution function $\delta f_E = -e\mathbf{E}\cdot\mathbf{v_k}\tau_{tr}(\partial f_\mathbf{k}/\partial\varepsilon_\mathbf{k})$. Secondly, the **B** field rotates each state **k** through an angle Θ_H, and alters the distribution function by $\delta f_B = e\tau_c\, \mathbf{v_k}\times\mathbf{B}\cdot\nabla\delta f_E$. The Hall current J_{xy}, equal to the sum of $e\delta f_B v_y$ over the FS (if **E**//x), involves the product of the two relaxation times τ_{tr} and τ_c. Although standard discussions assume that τ_{tr} and τ_c are identical, there appears to be no intrinsic reason that this must be so. We interpret Eq. 3 (which involves τ_c alone) as evidence that the two time scales are, in fact, distinct in the cuprates. An energy scale T_H is implied by the T^2 dependence. Writing

$$\hbar/\tau_c = k_B T^2/T_H, \qquad (4)$$

we find that T_H is ~16 K, a rather low energy scale (assuming $m^* = 1$). Comparing Eqs. 3 and 4, we find that $\tau_{tr}/\tau_c = T/T_H\eta$ exceeds 1 above ~ 40 K. Thus, in the normal state of YBCO, the Hall-angle relaxation rate increases more rapidly with temperature than the transport relaxation rate. *This enhanced scattering of the Hall response suppresses the observed R_H at high temperatures, and gives the temperature variation of R_H its characteristic 1/T profile.* Recently, some interesting theories have been advanced to describe the enhanced Hall scattering. Anderson has proposed that spin excitations alone are involved in Hall angle relaxation, whereas the longitudinal relaxation $1/\tau_{tr}$ involves both spin and charge degrees of freedom.[5] Two groups recently reported calculations of the Hall response in large-U gauge models. Nagaosa and Lee[6] find such a high temperature suppression of R_H. Ioffe, Kalmeyer and Weigmann[7] computed the temperature dependence of R_H and find strong curvature in its temperature dependence.

Experiments are in progress to extend these measurements to 500 K. The variation of the C term in Eq. 3 with impurity concentration is also being investigated in YBCO crystals doped with Ni and Zn. In ceramic samples of 214 (doped with Sr at $x = 0.16$), cot Θ_H also obeys Eq. 3, but with significantly larger C (≈ 500), and D ($\approx 7 \times 10^{-2}$ K^{-2}). The larger C reflects the strong disorder introduced by Sr doping.

Magnetoresistance

At low temperatures, quantum corrections to the conductivity of a normal metal become observable. From extensive research in the last decade, a good understanding of these corrections in conventional metals has emerged.[8,9] Such measurements are invaluable for what they tell us about the dephasing time of carriers. The application of these techniques, especially magnetoresistance (MR), to the layerd cuprates has been hindered by the appearance of superconducting fluctuations at high temperatures, and by the very large field scales encountered. However, some members of the cuprates do not show a superconducting transition down to 4.2 K, so that it is feasible to search for such corrections. We have carried out a detailed program to study the MR in Bi 2201 single crystals.[10] The samples have a metallic parameter $k_F l$ of ≈ 25, and display linear variation of ρ vs. T. Near 20 K, ρ changes rather abruptly switches to localization behavior (without an intervening T-independent range). For the samples used in the MR experiments, the "localization corrections" at 0.3 K constitute only 0.1 of the value of ρ at 20 K, i.e. we may apply the weak-localization calculations.

In these experiments, it is important to disentangle various contributions to the MR. Comparing the MR with field **H** parallel to the current ("longitudinal MR") and with **H** normal to the CuO$_2$ plane, **H**//c ("transverse"), is very useful in this regard. The classical MR arising from "cyclotron" bending of the orbits is difficult to observe in these samples because of the poor mobility ($\Theta_H \sim 10^{-2}$ to 10^{-3} at 8 T). It turns out that the quantum corrections dominate. The most important terms are as follows.[8,9] 1) Field-suppression of weak-localization ("coherent backscattering") effects leads to a negative MR in the transverse MR. (This effect exists in the absence of interaction. Strong spin-orbit scattering can alter the sign of this term at low temperatures. However, no spin-orbit effects are observed in our experiments.), 2) Interaction effects in the presence of disorder, combined with Zeeman "spin-splitting" in a field, leads to a positive MR (Lee-Ramakrishnan effect[11]). This MR term is isotropic, so it appears in both field orientations, 3) Field suppression of the Aslamasov-Larkin and Maki-Thompson (ALMT)

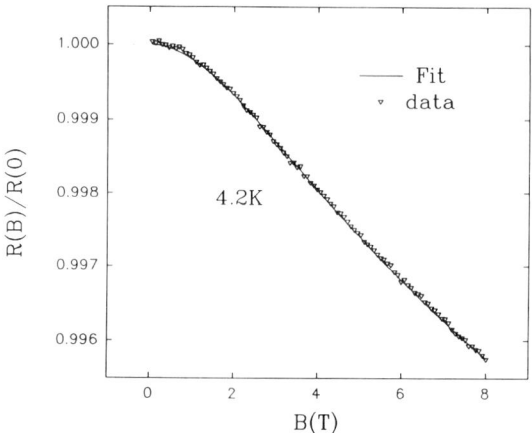

Figure 1 The orbital component of the transverse MR in Bi 2201 crystal at 4.2 K compared with the predictions of weak-localization theory. The parameters derived from the fit are α (5.23), l_0 (101 Å), and l_{in} (130 Å). [Jing et al, Ref. 10]

(superconducting fluctuation) terms leads to a large, anisotropic positive MR in both field orientations.[12]

In order to eliminate the ALMT terms, we selected crystals that show no transition down to 0.3 K. In these samples, highly anisotropic MR is observed.[10] With **H//c**, the MR is negative below 20 K, whereas in the longitudinal geometry, the MR is positive at all temperatures above 2 K (see below). We find that the transverse MR is very well fitted to the weak localization expression involving the digamma function (Fig. 1). Surprisingly, the field scale B_ϕ is about 10 to 50 times larger than in conventional metals, which implies that the carriers diffuse a relatively short distance before dephasing. (l_{in} is the distance each carrier diffuses before it encounters enough inelastic events to dephase. The field scale B_ϕ derived from fitting to the weak-localization expression is related to l_{in} by[8]

$$B_\phi l_{in}^2 \approx \phi_0/4, \qquad (5)$$

where ϕ_0 is the flux quantum, $l_{in}^2 = D\tau_\phi$, D is the diffusion constant, and τ_ϕ the dephasing time.) From our measurements, the value of l_{in} varies from ~250 Å at 0.5 K to 100 Å at 20 K. This relatively short diffusion length implies that there is intense Coulomb scattering between the holes at 0.5 K. A more surprising finding is that the dephasing rate varies as $\sqrt{T}$, instead of the more familiar linear-T behavior observed in disordered metals and semiconductors. Interaction theory predicts that the dephasing rate varies as[13]

$$\hbar/\tau_\phi = (k_B T/k_F l) \ln(k_F l/2) \quad (6).$$

Instead of Eq. 6, we find that in Bi 2201, the dephasing rate is given by[10]

$$\hbar/\tau_\phi = k_B T \sqrt{T_\phi/T} \quad (7)$$

where the energy scale T_ϕ is ~ 700 K (if $m^* \approx 1$), i.e. comparable to the spin exchange constant J (1,200 K) for the copper spins. Equation 7 implies that the level widths, as measured by $\hbar/\tau_\phi$ is very broad compared with $k_B T$ (~26 at 1 K). This is a very puzzling result in the context of Fermi liquid theory, since it implies that the quasi-particles are not well defined in the limit $T = 0$.

The longitudinal MR displays a rich behavior. It is convenient to divide the temperature interval into high (above 20 K), medium (2 to 20 K) and low (below 2 K). In the medium range, the MR is positive and varies approximately as H^2. However, the field scale B_s extracted from the quadratic dependence is much larger than the Zeeman field scale based on equating $g\mu B$ to $k_B T$. In order to invoke the Lee-Ramakrishnan effect[11] to account for the positive MR, we need to change the cut-off of the divergence. (The large level widths implied by the data on $\hbar/\tau_\phi$ also requires this rescaling of the Zeeman field.) By equating $g\mu B$ to $\hbar/\tau_\phi$, we find more reasonable agreement with the measured scale B_s, both in magnitude and in the temperature dependence. Thus, there appears to be experimental support for a broad level width, both from the transverse and longitudinal MR. As stated above, a negative anomaly appears below 2 K in the longitudinal MR. This is at present unexplained, but could arise from suppression of spin scattering of the carriers.

The high temperature regime above 20 K, where the resistivity abruptly changes to linear-T behavior, is the most interesting.[14] As the temperature increases above 20 K, the observed MR becomes more and more isotropic, i.e. the transverse MR becomes indistinguishable from the longitudinal MR. All vestiges of weak localization corrections to the MR become unobservable. The MR is now entirely due to spin coupling to the external field in this regime. At present, we are aware of no mechanism (quantum or classical) that explains the high temperature isotropic MR. If the MR arises from scattering of carriers by spin fluctuations, one would expect isotropic, negative MR (the applied field suppresses the scattering). Close to the Mott limit, an applied field enhances carrier hopping via the Nagaoka mechanism, but this also leads to negative, isotropic MR. Classical MR leads to positive MR, but not in the longitudinal geometry, and certainly not isotropic. At present, the high temperature MR appears to be a novel type of MR that may be generic to the cuprates in the linear-T regime. A small, positive MR has also been reported by Matsuda, Hirai, and Komiyama[15] in ceramic YBCO above 140 K, where the ALMT terms do not dominate.

Conclusion

These transport studies show that the various relaxation times of the holes have to be considered separately. Although the transport scattering rate (also observed in IR reflectivity) is linear in T, it is dangerous to assume that the inverse lifetime of the carriers remains linear in T at low temperatures. The MR results on Bi 2201 show that the dephasing rate is, in fact, greatly enhanced over $\hbar/\tau_{tr}$, and is also much larger than $k_B T$. The physical picture suggested is that of charge carriers that interact very strongly with each other. They diffuse only 250 Å at 0.5 K before dephasing. The broad level widths implied are incompatible with Fermi liquid theory. It is important to extend these studies to other cuprates to see if this is a general feature. The Hall studies also show that the "transverse" Hall current in the presence of a field encounters scattering mechanisms that are different from those that limit the longitudinal current. The ubiquitous temperature dependence of R_H in the hole cuprates appears to be caused by the different scattering rates, rather than by the existence of multiple bands.

We acknowledge helpful discussions with P.W. Anderson, H. Fukuyama, G. Kotliar, and P.A. Lee. This research is supported by the U.S. Office of Naval Research and by the Seaver Institute.

References

1. For a survey, see N.P. Ong, in *Physical Properties of High Temperature Superconductors*, edited by D.M. Ginsberg (World Scientific Singapore, 1990), Vol. 2, p. 459.
2. See for e.g., Colin Hurd in *The Hall effect and its Applications*, edited by C.L. Chien, and C.R. Westgate (Plenum, New York, 1980), p. 1.
3. T.R. Chien, D.A. Brawner, Z.Z. Wang, and N.P. Ong, Phys. Rev. B **43**, 6242 (1991).
4. K. Kamaras et al, Phys. Rev. Lett. **64**, 84 (1990); L. Forro et al, Phys. Rev. Lett. **65**, 1941 (1990).
5. P.W. Anderson, private communication.
6. Naoto Nagaosa and Patrick A. Lee, Phys. Rev. Lett. **64**, 2450 (1990), and to appear.
7. L.B. Ioffe, V. Kalmeyer and P.B. Wiegmann, Phys. Rev. B **43**, 1219 (1991).
8. For a review of MR and weak localization, Gerd Bergmann, Phys. Rept. **107**, 1 (1984).
9. For a review, P.A. Lee and T.V. Ramakrishnan, Rev. Mod. Phys. **57**, 287 (1985).
10. T.W. Jing, N.P. Ong, T. V. Ramakrishnan, J.M. Tarason, and K. Remschnig, to be published.
11. P.A. Lee and T.V. Ramakrishnan, Phys. Rev. B **26**, 4009 (1982).
12. A.G. Aronov, S. Hikami, and A.I. Larkin, Phys. Rev. Lett. **62**, 965 (1989).
13. B.L. Altshuler, A.G. Aronov, and D.E. Khmel'nitskii, J. Phys. C **15**, 7367 (1982).
14. T.W. Jing, N.P. Ong, J.M. Tarascon, and K. Remschnig, to be published.
15. Y. Matsuda, T. Hirai, and S. Komiyama, Solid State Commun. **68**, 103 (1988).

SUPERCONDUCTIVE TUNNELING IN $YBa_2Cu_3O_7$ THIN FILMS: DEPENDENCE UPON CRYSTALLOGRAPHIC ORIENTATION

L. H. Greene, J. Lesueur*, W. L. Feldmann and A. Inam

*BELLCORE
Red Bank, NJ 07701*

ABSTRACT

Superconducting thin films of $YBa_2Cu_3O_7$ are reproducibly grown completely *in-situ* by on-axis planar (ONP), inverted cylinder (IC) and off-axis planar (OFFP) magnetron sputtering, as well as laser deposition. Films of orientations (001), (103) and (100) are then used in our tunneling experiments. Gap-like structures at 18meV are routinely and reproducibly seen when there is some **a-b** plane directed towards the film surface, i.e. in the following situations: (100) laser deposited films, (103) OFFP films, (001) OFFP films which are of high crystallographic orientation but exhibit sub-micron **a**-axis grains on the surface, (001) IC films which are grainy and (001) ONP films that have been etched. No gap-like feature is seen in the unetched (001) ONP films, which are single crystal-like. The junctions exhibit zero-bias anomalies (ZBA) with the sign depending upon the crystallographic orientation of the film. When the ZBA is a conductance peak, its magnitude and sign depends upon applied magnetic field (up to 5T) indicating this feature is related to magnetic scattering at the interface. When the films are not **c**-axis oriented, the tunneling conductance shows oscillations at energies ranging from 20 to 80mV, probably due to geometrical interferences and Andreev scattering.

INTRODUCTION AND GOALS

Tunneling has proven to be one of the most important probes of the superconducting state giving important parameters such as details of the gap and phonon density of states. In the case of highly anisotropic materials such as the superconducting cuprates, it is desirable to perform tunneling into different crystallographic orientations: One may probe the superconducting properties in a more "favorable" direction, e.g., along the **a-b** planes where the coherence length is longer than that in the **c**-direction. In addition, any anisotropy in the gap function could be mapped out by directional-dependent tunneling; and the effect of tunneling into different chemical surfaces can be probed. We present here features in the tunneling characteristics that depend upon the crystallographic orientation and growth morphology of $YBa_2Cu_3O_7$ superconducting films.

FILM GROWTH

The crystallographic orientations of the films studied here are (001) or "**c**-axis", (103) or "nearly **a-b**-axis" and (100) or "**a**-axis". Three methods of magnetron sputter-deposition are used to produce superconducting YBa$_2$Cu$_3$O$_7$ thin films; on-axis planar (ONP), off-axis planar (OFFP) and inverted cylinder (IC). In each case, film growth is performed *in-situ* in a UHV-compatible system with multiple sputter sources to allow *in-situ* layering for junction fabrication. Reproducibility of film growth to the quality quoted here is close to 100%. The details of the growth parameters and physical properties of the films are given in Tables I and II, respectively. The ONP films[1,2], grown with a 2 inch diameter compensated target (Y:Ba:Cu = 1.08:1.75:4.5) parallel to the substrate, exhibit slightly depressed T_c's of 83K but show the highest degree of microstructural homogeneity: These are **c**-axis oriented with no evidence of **a**-axis grains as determined by SEM and TEM cross-sectional analysis. These films are epitaxial with a rocking curve of <0.3° as determined by x-ray analysis and finally, the absence of an internal ac-Josephson effect below T_c indicates no grains within a junction area. These ONP films were used in the YBa$_2$Cu$_3$O$_7$/Au/Pb Josephson junction studies[1,2]. The OFFP films are also grown with a 2 inch diameter target, but in this case, a stoichiometric target (1:2:3) is used at an angle of 90° to the substrate[3]. Although these films are predominately **c**-axis oriented, they exhibit ~0.2μm **a**-axis grains or "sticks" and ~0.5μm "stones" separated by a few μm's as seen by SEM, they are of a high crystalline quality and are homogeneous enough over a 1cm^2 area on LaAlO$_3$ to be used in working microwave filters (see next section). Films with (103) orientation and T_c = 91K can also be grown by this OFFP method, and the details will be reported elsewhere[4]. Films grown by the IC method[5] are of **c**-axis orientation with ~0.5° rocking curves and T_c's of 90K, but exhibit a grainy growth morphology, with a roughness on the half-micron scale. The details of the laser-deposited **a**-axis oriented films are given in Ref. 6.

Table I. Magnetron Sputtering Film Growth Parameters

	"ONP" ON-AXIS PLANAR	"OFFP" OFF-AXIS PLANAR	"IC" INVERTED CYLINDER
TARGET COMP (Y:Ba:Cu)	1.08:1.75:4.5	1:2:3	1:2:3
SUBST TEMP (°C)	830	715-735	810
PRESS (O$_2$/Ar in mT)	200/400	35/165	120/340
DEP. RATE (Å/hr)	1500	200	2700
T_c / δT_c (K)	83 / 3	93 / .5	91 / .5

Table II. Structural and Superconducting Properties of Films

MORPHOLOGIES:	"ONP" ON-AXIS PLANAR	"OFFP" OFF-AXIS PLANAR		"IC" INVERTED CYLINDER	LASER
XTAL-ORIENT	(001)	(001)	(103)	(001)	(100)
T_c (K)	83	92	87	90	92 Onset
APPEARANCE	shiny	shiny	shiny	hazy	shiny
GRAIN B'NDYS?	no	yes	yes	yes	twins
ANY a-AXIS ⊥ ?	none obs	grains	yes	grainy	yes
TUNNELING: ? REPRODUCIBLE GAP-LIKE FEATURE	NO; UNLESS ETCHED	YES	YES	YES	YES

PASSIVE MICROWAVE DEVICE

The primary focus of our research is in the area of active devices such as tunneling and Josephson junctions. The criteria on the materials required to fabricate such junctions are usually significantly more stringent than those required to fabricate passive devices. However, the ability to produce microwave devices is an excellent measure of superconducting film quality and our films grown by the OFFP method are of such quality and size to be used in a working microwave device. These films are grown on 0.864x1.17cm^2 LaAlO$_3$ substrates, exhibit T_c =92K as measured both by four-probe resistance and ac-susceptibility, 4-5% ion channeling (He at 2MeV) yield, <0.5° rocking curves and are completely reproducible. Line stripes of 25μm width patterned on these films are continuous, indicating they are pinhole-free to this dimension. A narrow-band 9GHz filter package designed and built by David R. Sarnoff Labs[7] shows 3.9dB loss at 72K and 4.5 dB loss at 77K, which is ~25x lower than that of Cu at the same temperature, allowing their use for space applications.

TUNNEL JUNCTION FABRICATION

The *in-situ* films of YBa$_2$Cu$_3$O$_7$ are removed from the vacuum chamber, an insulator is applied to define the junction area and Pb cross-strips are evaporated at room temperature to provide the counter electrode. The "natural" barrier, probably due to an oxygen-deficiency at the surface, is used as the tunneling insulator. Typically there are two to five junctions on each film, each with an area of ~0.1 mm^2 and resistances varying from 10 to 1000Ω at T=4.2K. We reproducibly observe a well-defined Pb gap of less that 1% leakage at T=1.2K and distinct Pb phonons. The observation of high-quality Pb tunneling characteristics establishes the existence of an elastic tunneling current, and only such junctions are reported here.

TUNNELING RESULTS: GAP-LIKE FEATURE

Before our results are presented, recall related points in previously reported data on planar tunnel junctions. Quasiparticle tunneling into the surface[8,9] or edge[10] of **c**-axis oriented films or crystals[11,12] reveals a gap-like feature (GLF) at ~18meV with conductance within this region on the order of 40-50% of that of the normal state. A zero-bias resistance peak is routinely observed which has been suggested to be a smaller gap at ~5meV perhaps associated with the **c**-axis direction[12].

Films taken as-grown with the ONP method, although exhibiting the highest-quality Pb tunneling curves, have always failed to produce the signature of a $YBa_2Cu_3O_7$ gap. We emphasize that many experiments were performed on these films. Even proximity-electron tunneling spectroscopy[1,2] experiments that have previously been shown to yield a gap on other low-coherence length superconductors[13], also revealed no $YBa_2Cu_3O_7$ GLF. In these particular experiments, we first established good electrical contact between $YBa_2Cu_3O_7$ and Au by demonstrating the existence of an ac-Josephson effect in $YBa_2Cu_3O_7$/Ag/Pb trilayers. However, $YBa_2Cu_3O_7$/Au/Ta-0 trilayers grown *in-situ* yielded only outstanding-quality Pb tunneling curves with approximately the expected parabolic background conductance extending to high biases (~400mV). When these films are exposed to a methanol:1%Br etch[14] for ~10sec, (the etch that had been previously used to obtain reproducible tunneling results on crystals[12]) a reproducible but weak $YBa_2Cu_3O_7$ GLF and a linear conductance above the gap are observed, as shown in Fig. 1a. Note that for this **c**-axis film with a depressed T_c of 82K, a weak feature, also at 18meV, is observed.

Tunneling measurements on films grown by each of the other growth techniques and crystallographic orientations give completely reproducible and similar results. The tunneling conductance for a **c**-axis oriented but grainy film, grown by the IC method is shown in Fig. 1b. In Fig. 2, curves derived from measurements on films of (001) and (103) by OFFP and (100) by laser deposition are displayed. Note that in each case the GLF occurs near 18meV and there is a substantial conductance at low biases, consistent with previous observations.

If this GLF were a BCS-like superconducting gap, then the energy value would indicate strong-coupling, with $2\Delta/kT_c \sim 4.6$. We caution however, that there is always a significant conductance within the "gap" and its energy value does not change with the measured T_c of the film: The films with $T_c=82K$ (ONP grown film) and $T_c=92K$ exhibit the same "gap" value. This is consistent with findings by Geerk et. al[15] who also find that a film with a lower T_c decreases the magnitude of the resistance and not the energy value of the GLF.

ZERO-BIAS ANOMALY (ZBA)

As discussed earlier, previously reported measurements show a peak in the *resistance* around zero bias. Our measurements show a resistance peak only for films whose **c**-axis is normal to the plane of the substrate, or (001) films. As found earlier, this peak is fairly insensitive to an applied magnetic field[12]. In Fig. 3 the low-bias conductances at T=2K at B=1.5T (Pb normal) and 5T for a (001) OFFP

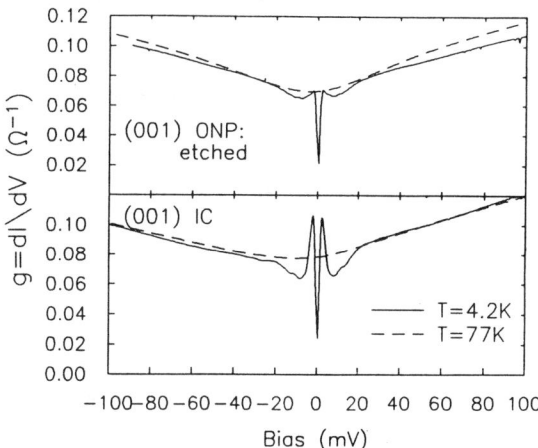

Figure 1. The tunneling conductance as a function of applied voltage is shown at $T=4.2K$ and $T=77K$ by the solid and dashed lines, respectively, for an etched (001) ONP film with $T_c=82K$ and a (001) IC film with $T_c=90K$. At $T=4.2K$ the Pb gap at $\sim 1meV$ and the weak gap-like feature of the $YBa_2Cu_3O_7$ at $\sim 18meV$ is seen.

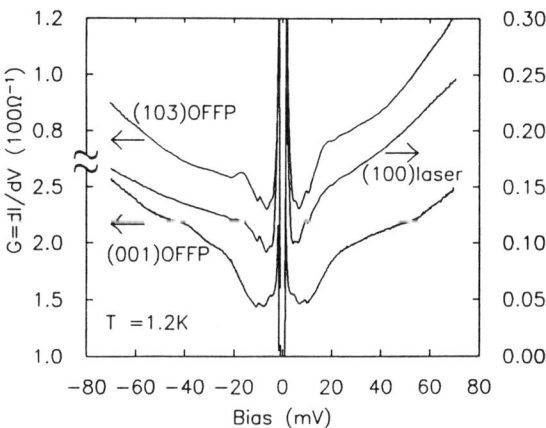

Figure 2. The tunneling conductance as a function of applied voltage is shown for three different films at 1.2K: (100) Laser deposited, (103) OFFP and (001) OFFP.

film are displayed. In contrast, for films that are not of **c**-axis orientation this feature appears as a *conductance* peak with a strong magnetic-field dependence[16]. In Fig. 4, we first show the temperature dependence of an **a**-axis, or (100) laser-deposited film: The magnitude of the ZBA decreases as the temperature is increased from 10K to 38K. The magnetic field dependence of this feature at T=10K is shown in Fig. 5 for a (103) OFFP-deposited film: As the magnetic field is increased from B=0T to 5T, the conductance peak at B=0T changes sign and becomes a resistance peak. Such behavior qualitatively agrees with the magnetic scattering model of a ZBA[17,18].

The sign of the low-bias feature within ±5mV depends upon the crystallographic orientation of the film; positive in resistance only if the **c**-axis is normal to the plane. Otherwise, the sign and magnitude also depends upon the magnitude of the applied magnetic field, indicating that this feature is due to magnetic scattering. The crystallographic and magnetic field dependencies together show that this is a ZBA more likely due to magnetic scattering at the interface than a superconducting gap. Because the structural and transport analyses of film quality, including resistivity, ac-susceptibility, channeling and TEM all show comparably high-quality materials, we do not at present assign the origin the origin of this ZBA to magnetic impurity phases in the as-grown superconducting films. The sign of the ZBA cannot be correlated with magnetic impurities in the vacuum systems or target materials as the sign of the ZBA can be the same for the sputtered and laser films and different for two sputtered films. It is likely that the origin of this ZBA is due to twin orientation, which is a crystallographic effect intrinsic to $YBa_2Cu_3O_7$: Recall that the material forms the O_6 phase at twin boundaries, which is magnetic.

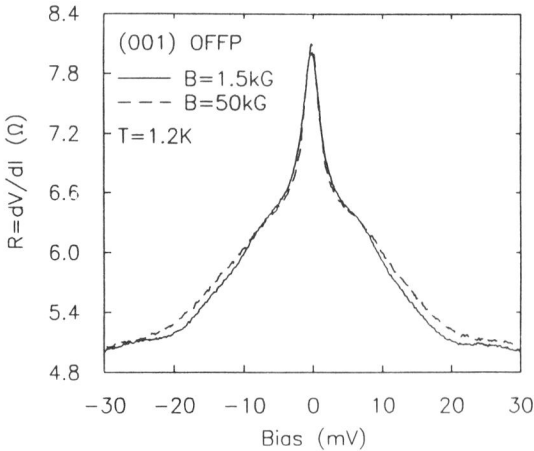

Figure 3. *The ZBA of a **c**-axis normal film, seen as an extra resistance peak of width ±5mV in the dynamic resistance, is shown at T=1.2K under applied magnetic fields of B=1.5T and 5T by the solid and dashed lines, respectively.*

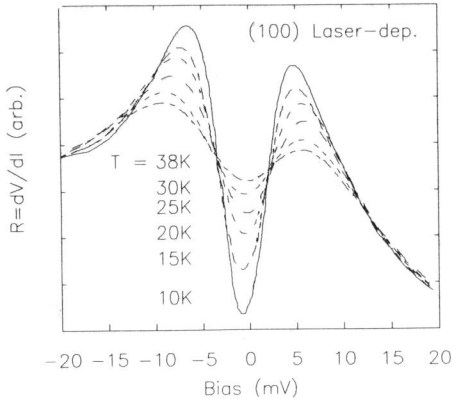

Figure 4. *The temperature dependence of the ZBA, which is a conductance peak in films in which the c-axis is not oriented normal to in the plane is shown for several temperatures from T = 10K to 38K for a (100), laser-deposited film.*

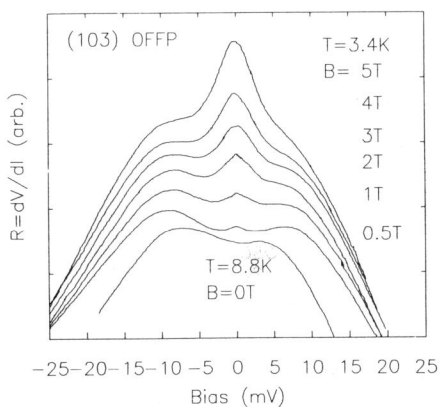

Figure 5. *The magnetic field dependence of the ZBA for a (103) OFFP film is shown for fields applied parallel to the plane of the substrate from 0 to 5T. For B=0T the temperature of the film is 8.8K to remove the Pb gap, otherwise T=3.4K. Note the change in sign from a conductance to a resistance peak with increasing field.*

HIGH-BIAS OSCILLATIONS

Another feature observed in the tunneling conductance that is dependent upon the crystallographic orientation of the film is a set of small oscillations at biases in the 20 to 80mV range[16]. Similar weak oscillations have been previously observed in films of either mixed or predominately **a**-orientation[19]. We find these oscillations are present only in films which are not oriented with the **c**-axis normal to the plane, namely (100) and (103) oriented films.

These oscillations are so small in magnitude that they are not observable in the raw I-V curves but appear after numerical derivatives are taken. A second derivative curve, derived from the dV/dI vs. V data, accentuates these oscillations and is shown in Fig. 6(a) for a (100) laser-deposited film at T=4.2K. In Fig. 6(b), data from the same film taken at 18K (above the T_c of the Pb) is displayed as the solid line. The dashed line is derived from curve (a) by convolving with a Gaussian to simulate a temperature broadening of 18K. Note that the data taken at T=18K and those derived from that taken at 4.2K require approximately a 1meV shift to match. We conclude that these oscillations shift with the disappearance of the Pb gap.

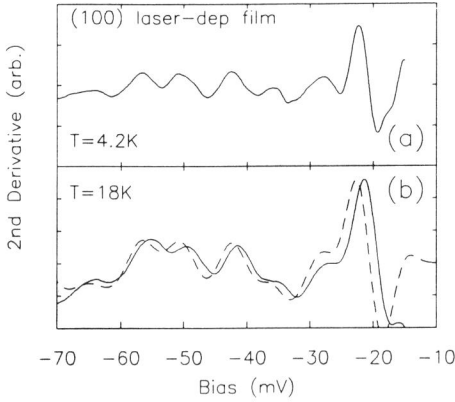

Figure 6. The shift of the high-bias oscillations by the energy of the Pb gap is demonstrated. The second derivative, d^2V/dI^2, is shown for a (100) laser-deposited film at (a) T=4.2K and (b) T=18K by the solid lines. The dashed line is the data taken at T=4.2K convolved with a Gaussian to simulate the thermal smearing at T=18K. Note the ~1meV shift required to match the curves in (b).

These oscillations, only observed in the (100) and (103) films, are consistent with Andreev scattering. The oscillation period is ~6mV and the $YBa_2Cu_3O_7$ film thickness in each case is 1000Å. Following the standard formula for Tomasch oscillations[20]: $E_n^2 = \Delta^2 + (nF)^2$, where $F = hv'_F/2D$, h is Planck's constant, v'_F is the Fermi velocity and D is the film thickness. We derive $v'_F = 3 \times 10^7$, which is in agreement with the value often quoted in the literature[21]. The observation of these weak oscillations only in films that are not **c**-axis oriented is probably due to the much longer coherence length in the direction of the **a-b** planes. Further experiments are in progress to test the film thickness dependence of these oscillations.

CONCLUSIONS

Tunneling experiments on high-quality superconducting thin films of $YBa_2Cu_3O_7$ with crystallographic orientations (001), (103) and (100) which are reproducibly grown completely *in-situ* by magnetron sputtering and laser deposition are reported. Gap-like structures at ~18meV are routinely and reproducibly seen for films with some **a-b** plane directed towards the Pb counter electrode. No gap-like feature is seen in (001) films that are completely **c**-axis oriented. These results suggest the importance of the crystallographic orientation of the film for tunneling to occur: We find that a component of the CuO_2 plane-direction is required to be perpendicular to the counter electrode. A resistance peak, turning on at ~5mV and centered around V=0 in is observed in films with the **c**-axis normal to the plane. Conversely, in the (100) and (103) films, a magnetic-field-dependent conductance peak is observed. We believe this structure to be a simple zero-bias anomaly due to magnetic scattering at the interface and not a superconducting gap. When the films are not **c**-axis oriented, the tunneling conductance shows oscillations at energies ranging from 20 to 80mV, which are consistent with Tomasch oscillations as derived from Andreev scattering.

ACKNOWLEDGMENTS

It is a pleasure to thank J. M. Rowell, J.-M. Tarascon, B. G. Bagley and J. H. Wernick for many stimulating discussions. We also wish to thank F. Shokoohi, B. J. Wilkens, L. A. Farrow, P. F. Miceli, R. Ramesh and J. B. Barner for excellent assistance in film-quality analyses as well as scientific discussions.

REFERENCES

[*] Permanent address: CSNSM-CNRS, Bat108, 91405 Orsay Campus, FR.

1. L. H. Greene, J. B. Barner, W. L. Feldmann, L. A. Farrow, P. F. Miceli, R. Ramesh, B. J. Wilkens, B. G. Bagley, J. M. Tarascon, J. H. Wernick, M. Giroud and J. M. Rowell, Physica C **162-164**, 1069-1070 (1989).

2. L. H. Greene, W. L. Feldmann, J. B. Barner, L. A. Farrow , P. F. Miceli, R. Ramesh, B. J. Wilkens, B. G. Bagley, M. Giroud and J. M. Rowell, in *High-Temperature Superconductors: Fundamental Properties and Novel Materials Processing*, J. Narayan, C. W. Chu and L. F. Schneemeyer, eds. (Materials Research Society, Pittsburgh, 1990) p. 991

3. N. Newman, K. Char, S. M. Garrison, R. W. Barton, R. C. Taber, C. B. Eom, T. H. Geballe and B. Wilkens, Appl. Phys. Lett. **57**, 520 (1990).

4. L. H. Greene, W. L. Feldmann, F. Shokoohi, R. Ramesh, P. F. Miceli, B. J. Wilkens, J. Lesueur and B. G. Bagley, in preparation.

5. X. X. Xi, G. Linker, O. Meyer, E. Nold, B. Obst, F. Ratzel, R. Smithey, B. Strehlau, F. Weschenfelder and J. Geerk, Z. Phys. B **74**, 13 (1989).

6. A. Inam, C. T. Rogers, R. Ramesh, K. Remschnig, L. Farrow, D. Hart, T. Venkatesan and B. Wilkens, Appl. Phys. Lett. **57**, 2484 (1990).

7. D. Kalokitis, A. Fathy, V. Pendrick, E. Belohoubek, A. Findikoglu, A. Inam, X. X. Xi, T. Venkatesan and J. B. Barner, Appl. Phys. Lett. **58**, 537 (1990); L. H. Greene, W. L. Feldmann, B. G. Bagley, J. B. Barner, F. Shokoohi, B. J. Wilkens, P. F. Miceli, D. Kalokitis, A. Fathy and V. Pendrick, in preparation.

8. J. Geerk, X. X. Xi, and G. Linker, Z. Phys B, **73**, 329 (1988);

9. J. Kwo, T. A. Fulton, M. Hong, and P. L. Gammel, Appl. Phys. Lett. **56**, 788 (1990).

10. J. S. Tsai, I. Takeuchi, J. Fujita, S. Miura, T. Terashima. Y. Bando, K. Iljima and Y. Yamamoto Physica C **157**, 537 (1989);

11. A. Fournel, I. Oujia J. P. Sorbier, H. Noel J. C. Levet, M. Potel and P. Gougeon, Europhys. Lett. **6**, 653 (1988);

12. M. Gurvitch, J. M. Valles, A. M. Cucolo, R. C. Dynes, J. P. Garno, L. F. Schneemeyer and J. V. Wasczak, Phys. Rev. Lett. **63**, 1008 (1989).

13. J. J. Hauser, D. D. Bacon and W. H. Haemmerle, Phys. Rev. **151**, 296 (1966); J. Geerk, M. Gurvitch, D. B. McWhan and J. M. Rowell, Physica **109 & 110B**, 1775 (1982); For general review: E. L. Wolf *Principles of Electron Tunneling Spectroscopy* (Oxford University Press, N. Y., 1985) Chapter 5, and references therein.

14. R. P. Vasquez, B. D. Hunt and M. C. Foote, Appl. Phys. Lett. **53**, 2692 (1988).

15. J. Geerk, R.-L. Wang, H. C. Li, G. Linker, O. Meyer, F. Ratzel, R. Smithey and H. Keschtkar, IEEE Trans. on Magn. **27** (1991). in press.

16. J. Lesueur, L. H. Greene, W. L. Feldmann and A. Inam, in preparation.

17. J. Applebaum, Phys. Rev. Lett. **17**, 91 (1966); Phys. Rev. **154**, 633 (1967).

18. L. Y. L. Shen and J. M. Rowell, Phys. Rev. **165**, 566 (1968).

19. Mark Lee, M. Naito, A. Kapitulnik and M. R. Beasley, Solid State Commun. **70**, 449 (1989).

20. W. L. McMillan and P. W. Anderson **16**, 85 (1966); J. M. Rowell, Phys. Rev. Lett. **30**, 167 (1973).

21. Thomas Timusk and David B. Tanner in *Physical Properties of High Temperature Superconductors I*, Donald M. Ginsberg, ed. (World Scientific, Singapore, 1989) p. 363.

INFRARED PROPERTIES: THE NORMAL STATE, THE ENERGY GAP, AND THE TEMPERATURE DEPENDENCE OF THE GAP

Z. Schlesinger, R. T. Collins, L. D. Rotter, F. Holtzberg, C. Feild

IBM Watson Research Center, Yorktown Heights, New York 10598

U. Welp(1,2), G. W. Crabtree(2), J. Z. Liu[*] and Y. Fang(2)

1. Science and Technology Center for Superconductivity, 2. Materials Science Division, Argonne National Lab, Argonne, Illinois 60439
[*]Physics Dept., University of California, Davis, California, 95616

ABSTRACT

Because the CuO_2 planes are the universal element in superconductors with T_c above ~40 K, understanding their fundamental properties is central to the problem of understanding the origin of high T_c. Here we discuss some of the contributions made by infrared measurements in this area. We emphasize reflectivity measurements of single domain $Y_1Ba_2Cu_3O_7$ crystals, where one can distinguish between chain and plane contributions to the infrared conductivity. For the CuO_2 planes at low temperature $\sigma_{1s}(\omega)$ is very small (~0) up to roughly 500 cm^{-1} ($8kT_c$), where it rises rapidly, suggesting a gap of width $8kT_c$ in the in-plane electronic-excitation spectrum. In the normal state $\sigma_1(\omega)$ drops unusually slowly with ω; the $\sim 1/\omega$ dependence of $\sigma_1(\omega)$ can be directly related to the $\sim 1/T$ dependence of σ_{dc} (i.e, to the linear T resistivity), as well as to other spectroscopic properties. (Describing the conductivity in terms of a scattering rate which is linear in both frequency and temperature can help elucidate these relationships.) Examining the temperature dependence of $\sigma(\omega)$ in the superconducting state, one finds that the gap does not collapse as $T \rightarrow T_c$ as expected in BCS theory, but instead fills-in in a highly unconventional manner. Relationships between the infrared conductivity, the penetration depth and nuclear relaxation rates suggest a phenomenology for $Y_1Ba_2Cu_3O_7$ which is fundamentally different from that of BCS superconductors.

TEXT

The use of infrared techniques to study the superconducting state dates to work of Glover, Ginsberg, Richards and Tinkham and others(1-4) in the late 50's and 60's. Their work provided the first spectroscopic evidence for an energy gap, preceding both tunneling spectroscopy and the BCS theory, and is widely recognized as quite important to the emerging understanding of superconductivity. Infrared measurements have also been used to study fundamental properties of a variety of metals in the normal state including highly correlated oxides and heavy Fermion materials(5-7). Here we will briefly discuss some of the applications of infrared techniques to the study of cuprate materials.

High-Temperature Superconductivity
Edited by J. Ashkenazi et al., Plenum Press, New York, 1991

With infrared measurements evidence for a superconducting energy gap can be obtained from either transmission or reflectivity characteristics. In the latter case, one of the signatures of the energy gap is an enhancement of the reflectivity at low frequency in the superconducting state. From the frequency range over which this enhancement occurs, one can estimate the magnitude of the gap. This technique was pioneered by Richards and Tinkham, who used this approach to estimate the magnitude of the gap in a number of elemental superconductors(3). From their data, an example of which is shown in figure 1a, they obtained gap values in the range from about 3.5 to 4.5 kT_c, consistent with the then emerging BCS-Eliashberg theory. More recently the same technique was used on $Ba_{1-x}K_xBiO_3$, and from the reflectivity enhancement (figure 1b) an energy gap of conventional magnitude ($\sim 3.5 kT_c$) was also inferred(8). For both elemental superconductors and $Ba_{1-x}K_xBiO_3$ the infrared estimates of the gap preceeded tunneling results, and were later confirmed by tunneling data.

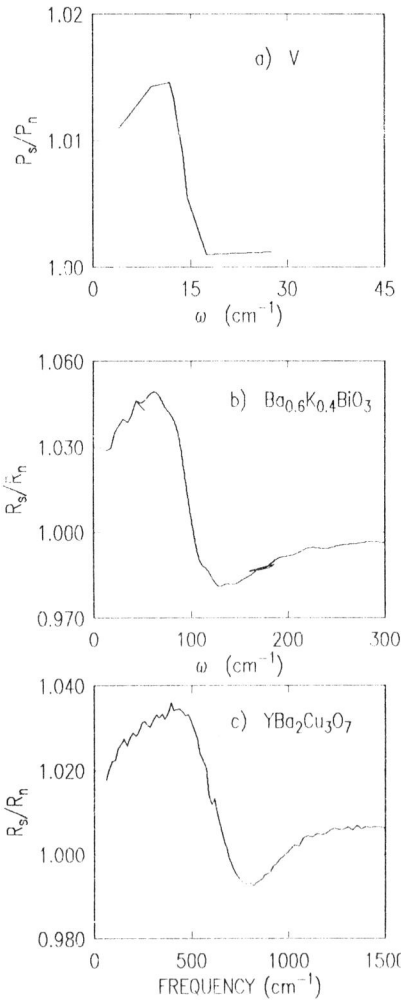

Fig. 1. The ratio of the reflectivity in the superconducting state to that in the normal state is shown for an elemental superconductor(3), a bismuthate(8), and a cuprate(10). For the upper curve (a) the vertical scale is uncertain because the measurements are made in a cavity(3). From the frequency range over which the reflectivity is enhanced the authors have inferred energy gaps of $\sim 3.5 kT_c$, $3.5 kT_c$ and $8 kT_c$, respectively, from these superconductors.

In the cuprates early difficulties in measuring the gap due to the large in-plane/out-of-plane anisotropy were quickly recognized (see, e.g. 9). This problem was avoided by studying crystals with the electric field polarized in the a-b plane. With such measurements of twinned $Y_1Ba_2Cu_3O_7$ crystals an enhancement of the reflectivity, which is quite similar to that found in the elementals and $Ba_{1-x}K_xBiO_3$ was observed(10), as shown in fig 1c. Based on the similarity of this reflectivity enhancement to those observed in other superconductors(3), this curve was first interpreted as evidence for a superconducting energy gap of 500 cm^{-1} in 1987(10). The surprising thing about this inference lies in the observation that this energy scale corresponds to $8kT_c$, a result which is completely beyond the range of any previously reported superconducting gap. This result, along with the absence of an Hebel-Slichter enhancement in $1/T_1$, provided an early indication of unusual behavior in the superconductivity of a cuprate.

Partly because of its unprecedented size, this result was controversial. Evidence for a smaller, conventional gap ($2\Delta/kT_c \simeq 3.5$) was reported based on estimates of the frequency at which R reaches unity(11), however, these results do not appear to be holding up to closer scrutiny(12). At the present time there appears to be only one energy scale showing up reliably in the infrared data (500 cm^{-1}), and although the interpretation of this feature remains controversial, as discussed below, the nature of the data itself is becoming reasonably clear.

Subsequent measurements of infrared properties have focussed on obtaining the absolute reflectivity, from which the conductivity is obtained(13-19) via Kramers-Kronig transform. Most of the physics is contained in the real part of $\sigma(\omega)$, and this is the quantity that is most useful as a guide to or test of theoretical approaches. In figure 2 we show the conductivity (real part) of a twinned $Y_1Ba_2Cu_3O_7$ crystal as a function of frequency at temperatures from 50 to 250 K. Each conductivity spectrum starts at a fre-

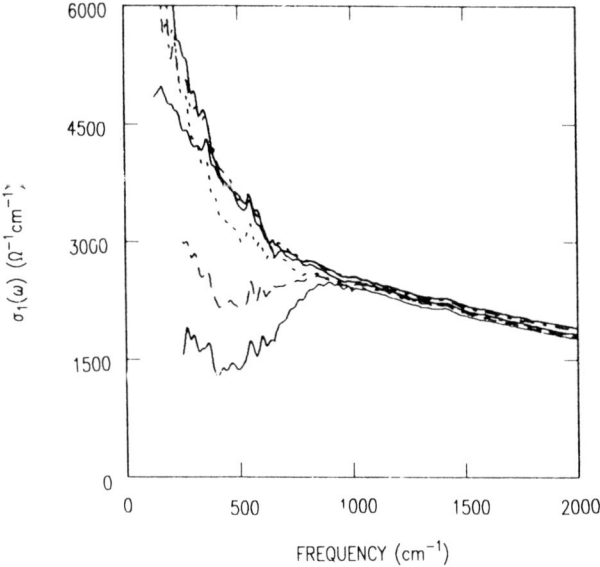

Fig. 2. The conductivity of a twinned $Y_1Ba_2Cu_3O_7$ crystal is shown for T = 50 K (lower solid curve), 80 K (dashed), 100 K (dotted), 150 K (solid), 200 K (dot-dashed), and 250 K (solid). A gap-like feature develops below ~100 K, however, a background conductivity of ~1500 (Ωcm)$^{-1}$ remains below 500 cm^{-1} even at low temperature.

quency at least 20 % above our lowest frequency data point in the reflectivity. In this way one can avoid any significant influence of the low frequency termination used in the Kramers-Kronig transform on the conductivity spectra(18,19). At very low frequency the conductivity in the normal state must increase like $1/T$, to follow the known d.c. temperature dependence of the resistivity. This increase occurs for frequencies roughly less than πkT, and to satisfy the sum rule on $\sigma_1(\omega)$, the conductivity at higher frequencies must decrease. This decrease is spread out over a wide range of frequencies, as discussed in ref. 18, and thus leads to only a very modest temperature dependence in $\sigma_1(\omega)$ at any particular frequency above $\sim kT$. (The sum rule says that the integral of $\sigma_1(\omega,T)$ over all ω must be independent of T.)

Near T_c the conductivity in the vicinity of 500 cm^{-1} and below begins to decrease rapidly. By 50 K most of the conductivity below 500 cm^{-1} is missing, although a background conductivity at a level of about 1500 $(\Omega cm)^{-1}$ remains. The missing conductivity has gone

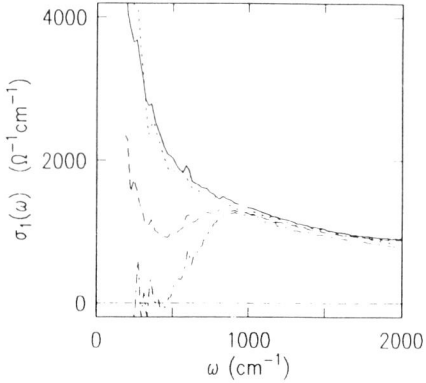

Fig. 3. The conductivity of a twin-free $Y_1Ba_2Cu_3O_7$ crystal measured with the IR electric field perpendicular to the chain axis ($E\|\hat{a}$) is shown for $T = 50$ K (dot-dashed curve), 80 K (dashed), 100 K (dotted) and 125 K (solid). In this geometry only the response of carriers in the CuO_2 planes is measured and the background conductivity observed in the twinned crystals for $\omega \lesssim 500$ cm^{-1} is greatly reduced or absent at low temperature.

into a Dirac-delta function at $\omega = 0$, which represents the disappationless response of the superconducting condensate(20,21). (This is not an assumption, it can be explicitly shown by examining $\sigma_2(\omega)$.) The strength of this delta function is intimately related to the penetration depth, and in fact from the infrared conductivity one estimates a penetration depth of about 150 nm(17,18) in agreement with magnetization and μsr measurements(23,24).

Although the data from twinned crystals contains a great deal of information, it has some built-in ambiguity since contributions to the electrodynamic response from both chains and planes are mixed together. Recent experiments(22) on twin-free crystals have allowed us to resolve this final anisotropy, and thus obtain the pure CuO_2 plane response in the high quality $Y_1Ba_2Cu_3O_7$ material. Since CuO_2 planes appear to be the universal element required for very high T_c, elucidating their properties, as distinct from those of their environment, is of central importance.

Essential aspects of the in-plane (a-b) anisotropy of $Y_1Ba_2Cu_3O_7$ are explored in a recent publication(22). With the infrared electric field along the â axis (i.e., perpendicular to the CuO chains), one obtains essentially the conductivity of just the CuO_2 planes. This CuO_2 plane conductivity is shown in figure 3, for T up to 125 K. (Note that this temperature range is not the same as that of the previous figure.) At low temperature, the main difference between these spectra and those of fig 2 is the absence of the large background conductivity below 500 cm^{-1}. The fact that this large background conductivity is present for E∥b̂ indicates that it is associated with the chains(22).

From an emperical point of view one can clearly say that between about 100 and 50 K most of the area in the conductivity below 500 cm^{-1} is going into the delta-function at $\omega = 0$, and that at low temperature a conductivity threshold is present at about 500 cm^{-1} (8 kT_c). This is clearly the main absorption edge in the infrared data. Gap estimates from tunneling and photoemission consistent with this unusually large value are often reported. We note as well that the evidence from infrared measurements for a ~500 cm^{-1} gap in the CuO_2 planes of $Y_1Ba_2Cu_3O_7$ is essentially equivalent to the best infrared evidence for a gap in a conventional superconductor, as shown in fig 4. It is based on this sort of comparison, as well as that shown in figure 1, and the temperature dependencies shown in figures 2 and 3, that we have associated the 500 cm^{-1} feature with a pair excitation threshold in the superconducting state, i.e., an energy gap(9,17,18,22).

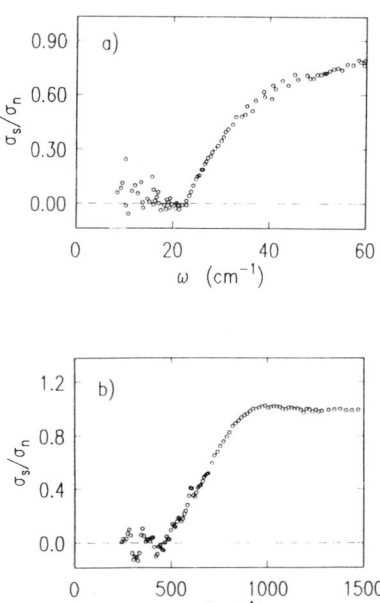

Fig. 4. A comparison of the best infrared evidence for an energy gap in Pb and in $Y_1Ba_2Cu_3O_7$ is shown. These spectra show the ratio of the conductivity in the superconducting state to that in the normal state. Pair excitation thresholds of about 22 cm^{-1}(4) and 500 cm^{-1}(10,17), respectively, have been inferred from these data.

One approach which attempts to interpret the infrared data without using a gap has been proposed by Tanner and coworkers(25). Their approach involves two steps:
1. one fits the normal state conductivity to a sum of contributions from Lorentz oscillators and a Drude term.
2. one then collapses the Drude term to a Dirac-δ function at $\omega = 0$ (reduce its width to 0) while leaving the Lorentz oscillators unchanged(25). This is supposed to fit the superconducting state.

By adjusting the strengths, frequencies and widths of some oscillators (including a Drude contribution) one can obtain a good fit to the normal state conductivity, as shown in figure 5a, where the dashed curve is a fit to the data at 100 K (circles), and the individual contributions of each oscillator are shown by the dotted curves. Here we have fit the normal state CuO_2 plane conductivity with 4 oscillators (11 parameters), following closely the example of Kamaras et al.(25). The attempt to fit the superconducting state by collapsing the Drude term to a δ-function, however, is unsuccesful. This is shown by the solid curve in figure 5b, for which we have collapsed the Drude term (so that is does not contribute to $\sigma_1(\omega)$ at finite ω, but kept the remaining oscillators unchanged. In fact, this procedure also works very poorly even for the twinned data, including that of Karamas et al., who originally suggested otherwise(25a). This is the point of their erratum(25b), which acknowledges that to fit the superconducting state data one must change the Lorentz oscillators in addition to collapsing the Drude term. The observation that this does not provide a reasonable description of the superconducting state, causes us to question whether it is appropriate to divide the conductivity into Drude and non-Drude components in this low frequency range. As we discuss below, the frequency dependence of the normal state conductivity is quite smooth through this range ($\omega \lesssim 2000$ cm^{-1}), thus provided no obvious basis for dividing the conductivity into distinct parts.

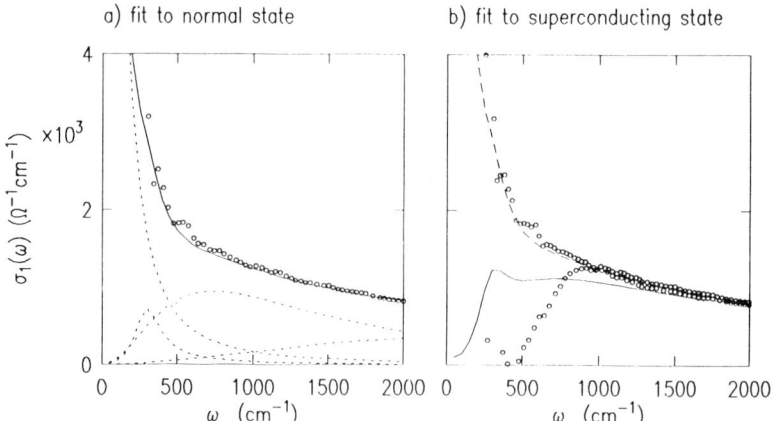

Fig. 5. We examine the procedure suggested by Kamaras et al.(25) to fit the infrared data. The circles show the measured conductivity of the CuO_2 planes in $Y_1Ba_2Cu_3O_7$ (binned on the x axis for clarity) at 100 K and 50 K. In (a) the solid curve shows a fit to the normal conductivity using 4 oscillators, including a Drude term centered at $\omega = 0$. (The dotted curves show the contributions of each oscillator.) In (b), the solid curve is obtained by collapsing the Drude term width to 0. Contrary to the claims of Kamaras et al.(25a), we find that this does not provide a good fit to the superconducting state data (lower circles).

The original motivation for dividing the infrared conductivity into distinct parts came from the early work on ceramic superconductors, where a conductivity with a huge bump in the mid-infrared range was inferred from the Kramers-Kronig transform of reflectivity data. Soon it was realized that the origin of this feature was related to problems associated with anisotropy(9), and that the intrinsic conductivity of the CuO_2 planes falls smoothly and monotonically for $\omega \lesssim 10,000$ cm^{-1}. Data from $Y_1Ba_2Cu_3O_7$ and $Bi_2Sr_2CaCu_2O_{8-y}$ confirm this general behavior. $La_{2-x}Sr_xCuO_4$ is more complicated, since at low doping there is a bump in the mid-infrared range, while at higher doping, near the optimum T_c, the conductivity becomes similar to that of the 123 and 2212. In fig. 6 we show the conductivity of the CuO_2 planes ($\vec{E} \parallel \hat{a}$) as a function of frequency on both linear and log-log plots. (In contrast to the smooth behavior exhibited for the planes alone, we note that the conductivity for $\vec{E} \parallel \hat{b}$, which includes chain contributions, exhibits a bump at about 2000 cm^{-1}(22)). The open diamonds show the conductivity of the quintessential cuprate insulator La_2CuO_4, as recently reported by Uchida et al.(26). In figure 6b one sees that the CuO_2 plane conductivity is linear with a negative slope which is slightly less than one up to about 10,000 cm^{-1}, where it breaks away sharply from this smooth linear behavior. In the same vicinity one sees the conductivity of the cuprate insulator rising sharply, presumably due to excitations from the filled "oxygen" band to the empty upper Hubbard band. These data suggest that for the CuO_2 planes there is no motivation to treat the mid-infrared conductivity ($\omega \sim 2000$ cm^{-1}) as distinct from the low frequency conductivity, since the behavior is quite smooth and continuous in the frequency domain.

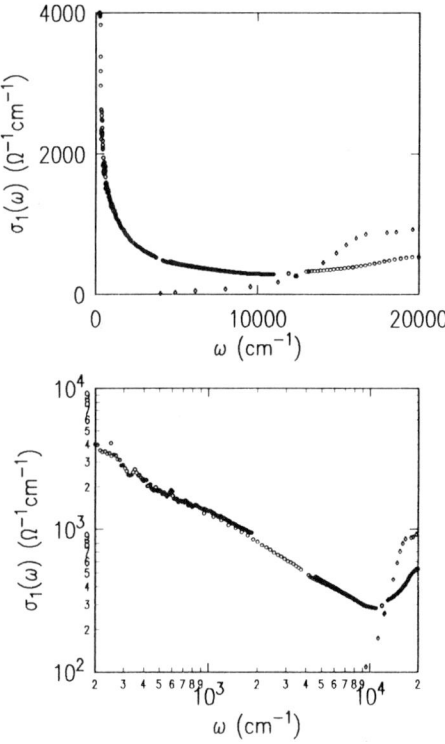

Fig. 6. The normal state conductivity of the CuO_2 planes is shown on both linear and log-log plots. The circles are data from twin-free $Y_1Ba_2Cu_3O_7$; the diamonds show the conductivity of the cuprate insulator, La_2CuO_4, from Uchida et al.(--).

Treating the infrared conductivity as a single entity allows one to make several connections with other measurements(27). For example, the observation that the conductivity falls roughly like $1/\omega$ (instead of the Drude $1/\omega^2$), can be related in a fundamental way (causality) to the $1/T$ temperature dependence of σ_{dc}. One can also describe the conductivity in terms of a scattering rate which is linear in both ω and T (14,22,27-29). This directly connects the infrared behavior to the temperature dependence of the resistivity, and further helps to elucidate relationships between the infrared data(22), the Raman background(28), and the photoemission linewidths, which have been discussed by Anderson(27) and by Varma and co-workers(29).

Thus far we have discussed the normal state conductivity, which drops unusually slowly as a function of ω, and the superconducting state conductivity, which is suggestive of a gap of ~ 500 cm^{-1} ($8kT_c$) in the in-plane electronic excitation spectra. In view of the large size of the gap relative to T_c, it is not unexpected that the temperature dependence is also quite unusual. As shown in fig 7, one finds that as T approaches T_c the conductivity in the gap region ($\omega \lesssim 500$ cm^{-1}) fills-in in a highly unusual manner (no gap collapse), and that some remnant of the gap may be present up to roughly 10 or 20 K above T_c. The persistence of

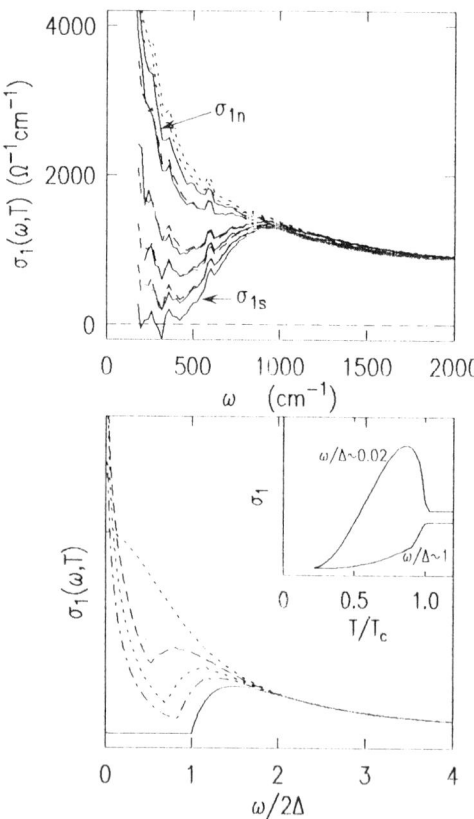

Fig. 7. a) The conductivity of the CuO$_2$ planes in Y$_1$Ba$_2$Cu$_3$O$_7$ is shown at temperatures of 30, 50, 60, 70, 80, 90, 100, 110, and 120 K. The dashed curves show a fit to the intermediated temperature conductivities as a superposition of high and low temperature spectra (Eq. 1), as described in the text. b) Calculated conductivities within the BCS model(22) at T = 0, 0.7, 0.8, 0.9T$_c$ and in the normal state (dotted) are shown. As discussed in ref.31, these BCS curves show gap collapse, and enhancement effects at low ω, which are not seen in Y$_1$Ba$_2$Cu$_3$O$_7$ (a).

a gap-like depression of $\sigma_1(\omega)$ to temperatures somewhat (10-20 K) above T_c, and the absence of gap collapse in $Y_1Ba_2Cu_3O_7$ have been discussed by us previously(17,18). Conclusions regarding the absence of gap collapse in the infrared data, which are quite consistent with our own view, have been reached independently by van der Marel et al.(30). Possible implications of these observations are discussed in more detail recently in a paper in which the relationships between the infrared, NMR and penetration depth data are examined(31). Briefly, we find that:

1. there is no evidence for gap collapse in the infrared conductivity. Instead the infrared data below 100 K can be described using the formula

$$\sigma_1(\omega,T) = f(T)\sigma_{1s}(\omega) + (1 - f(T))\sigma_{1n}(\omega) \quad (1)$$

where $\sigma_{1s}(\omega)$ is the conductivity at 30 K and $\sigma_{1n}(\omega)$ is the conductivity at 100 K ($T_c = 93$ K). The fits are shown as dashed lines in Fig. 7.

2. the area missing from the conductivity is consistent with the penetration depth vs T,

3. the temperature dependence of $\sigma_1(\omega, T)$ in the infrared, and of $1/T_1T$, at NMR frequencies (~0.01 cm^{-1}) are essentially the same (fig 8a).

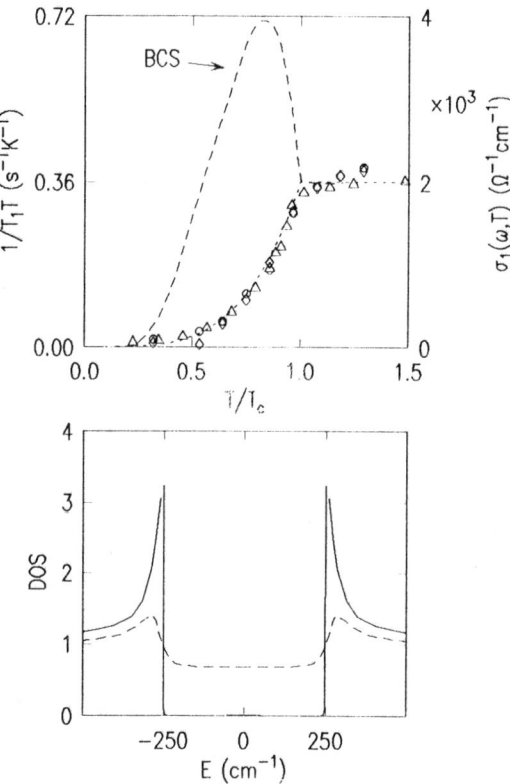

Fig. 8. The temperature dependence of the nuclear Korringa product $1/T_1T$ (planar oxygen), and of the CuO_2 plane conductivity at 220 and 440 cm^{-1} are shown. (The 220 cm^{-1} conductivity is scaled by 0.42 to facilitate comparison.) The observation of very similar temperature dependence at these very different frequencies is inconsistent with conventional BCS gap collapse and coherence factors. In part (b) a schematic density of states evolution which may be consistent with the infrared and NMR data (in (a) and fig 7) is shown. In this picture the gap fills-in, but does not collapse to lower energy as $T \to T_c$.

Regarding the 2nd point, which is discussed in more detail in ref (31), we note that the area missing from the conductivity is a measure of the superfluid density, and the order parameter of the superconducting state. Thus one finds that although the gap stays large, the order parameter collapses gradually as T approaches T_c. The relationship between the infrared conductivity and the penetration depth, which is based on the Glover-Ferrel-Tinkham sum rule, is discussed in (31).

Regarding the 3rd point, typically both $\sigma_1(\omega, T)$ and $1/T_1T$ are related to integrals over products of density of states(20) (displaced by the probing frequency). The observation of similar temperature dependencies at these very different probing frequencies ($\sim$100 MHz vs 10 THz) is inconsistent with ordinary BCS gap collapse near T_c. It may be consistent, however, with a density of states evolution like that illustrated in fig 8b, where the gap goes away not by collapsing, but rather by filling in as $T \to T_c$.

To conclude, we find that:

1. the normal state conductivity drops unusually slowly with frequency in the infrared range. The roughly $1/\omega$ dependence of $\sigma_1(\omega)$ appears to be related to the anomalous behavior of the resistivity, photoemission, and electronic Raman scattering in the normal state.

2. the conductivity at low temperature is very small or zero up to roughly 500 cm^{-1}, suggesting a superconducting energy gap of width $2\Delta = 8kT_c$ in the in-plane electronic excitation spectrum.

3. the temperature dependence of $\sigma_1(\omega)$ in both the normal and superconducting state is quite unusual. In particular, although the development of the superconducting condensate fraction is gradual, the feature which develops into the gap arises from the normal state with its full energy. There is no evidence for gap collapse.

These observations, as well as many others with other techniques, suggest that the cuprates are highly unusual. In particular the temperature dependence near T_c suggests that the physics of the transition to the superconducting state lies somewhere between the familiar paradigms of BCS theory (Fermi surface instability) and Bose condensation.

Acknowledgements: The authors acknowledge valuable conversations with P. W. Anderson, E. Abrahams, H. Fukuyama, D. H. Lee, P. B. Littlewood, D. M. Newns, D. J. Scalapino, C. P. Slichter, M. Tinkham and C. M. Varma

References

1. R. E. Glover and M. Tinkham, Phys. Rev.104, 844 (1956), Phys. Rev.108, 243 (1957)
2. D. M. Ginsberg and M. Tinkham, Phys. Rev.118, 990 (1960)
3. P. L. Richards and M. Tinkham, Phys. Rev.119, 575 (1960)
4. L. H. Palmer and M. Tinkham, Phys. Rev.165, 588 (1968)
5. F. E. Pinkerton, B. C. Webb, A. J. Sievers, J. W. Wilkins and L. J. Sham, Phys. Rev. B30, 3068 (1984)
6. B. C. Webb, A. J. Sievers and T. Mahalisin, Phys. Rev. Lett.57, 1951 (1986)
7. P. E. Sulewski, A. J. Sievers, M. B. Maple, M. S. Torikachvili, J. L. Smith and Z. Fisk, Phys. Rev. B38, 5338 (1988)
8. Z. Schlesinger, R. T. Collins, J. A. Calise, D. J. Hinks, A. W. Mitchell, Y. Zheng, B. Dabrowski, N. E. Bickers, and D. J. Scalapino, Phys. Rev. B40, 6862 (1989)
9. Z. Schlesinger, R. T. Collins, M. W. Shafer and E. M. Engler, Phys. Rev.B36,5275 (1987)

10. Z. Schlesinger, R. T. Collins, D. L. Kaiser, and F. Holtzberg, Phys. Rev. Lett. **59**, 1958 (1987)
11. G. A. Thomas, J. Orenstein, D. H. Rapkine, M. Capizzi, A. J. Millis, R. N. Bhatt, L. F. Schneemeyer and J. V. Waszczak, Phys. Rev. Lett. **61**, 1313 (1988)
12. T. Pham, H. D. Drew, S. H. Mosely and J. Z. Liu, Phys. Rev. B**41**, 11681 (1990)
13. S. Tajima, S. Uchida, H. Ishii, H. Takagi, S. Tanaka, U. Kawabe, H. Hasegawa, T. Aita, and T. Ishiba, Mod. Phys. Lett. B **1**, 353 (1988)
14. R. T. Collins, Z. Schlesinger, F. Holtzberg, P. Chaudari, and C. Field, Phys. Rev.B. **39**, 6571 (1989)
15. S. Tajima, H. Ishii, T. Nakahashi, T. Takagi, S. Uchida, M. Seki, S. Suga, Y. Hidaka, M. Suzuki, T. Murakami, K. Oka and H. Unoki, J. Optical Soc. America B **6**, 475 (1989)
16. J. Schutzmann, W. Ose, J. Keller, K. F. Renk, B. Roas, L. Schultz and G. Saemann-Ischenko, Europhysics Lett. **8**, 679 (1989)
17. R. T. Collins, Z. Schlesinger, F. Holtzberg and C. Field, Phys. Rev. Lett. **63**, 422 (1989)
18. Z. Schlesinger, R. T. Collins, F. Holtzberg, C. Feild, G. Koren and A. Gupta, Phys. Rev. B**41**, 11237 (1990)
19. J. Orenstein, G. A. Thomas, A. J. Millis, S. L. Cooper, D. H. Rapkine, T. Timusk, L. F. Schneemeyer and J. V. Waszczak, Phys. Rev. B**42**, 6342 (1990)
20. M. Tinkham, Introduction to Superconductivity, Robert E. Krieger Publishing, Malabar, Florida (1975)
21. R. A. Ferrel and R. E. Glover, III, Phys. Rev. **109**, 1398 (1958); M. Tinkham and R. A. Ferrel, Phys. Rev. Lett.**2**, 331 (1959)
22. Z. Schlesinger, R. T. Collins, F. Holtzberg, C. Feild, U. Welp, G. W. Crabtree, Y. Fang and J. Z. Liu, Phys. Rev. Lett.**65**, 801 (1990)
23. L. Krusin-Elbaum, R. L. Greene, F. Holtzberg, A. P. Malozemoff and Y. Yeshurun, Phys. Rev. Lett.**62**, 217 (1989)
24. D. R. Harshman, L. F. Schneemeyer et al, Phys. Rev. B**39**, 851 (1989)
25. K. Kamaras, S. L. Herr, C. D. Porter, N Tache, D. B. Tanner, S. Etemad, T. Venkatesan, E. Chase, A. Inam, X. D. Wu, M. S. Hegde, and B. Dutta, a) Phys. Rev. Lett. **64**, 84 (1990); b) Erratum, ibid, 1962 (1990)
26. S. Uchida, T. Ido, H. Takagi, T. Arima, Y Tokura and S. Tajima, Phys. Rev. B**43**, 7942 (1991)
27. P. W. Anderson, in Strong Correlation and Superconductivity, edited by H. Fukuyama, S. Maekawa and A. Malozamoff (Springer-Verlag, Berlin, 1989)
28. M. V. Klein, ibid
29. C. M. Varma, P. B. Littlewood, S. Schmitt-Rink, E. Abrahams and A. E. Ruckenstein, Phys. Rev. Lett. **63**, 1996 (1989); ibid, E, vol. **64**, 497 (1990)
30. D. van der Marel, M. Bauer, E. H. Brandt, H.-U. Habermeier, W. Koenig, and A. Wittlin, Phys. Rev. B**43**, 8606 (1991)
31. R. T. Collins, Z. Schlesinger, F. Holtzberg, C. Feild, U. Welp, G. W. Crabtree, J. Z. Liu and Y. Fang, Phys. Rev. B**43**, 8701 (1991)

INFRARED STUDIES OF HIGH-T_c SUPERCONDUCTORS: WHERE'S THE GAP?

D.B. Tanner, D.B. Romero, K. Kamarás, and G.L. Carr,*
Department of Physics, University of Florida, Gainesville, FL 32611

L. Forro,[†] D. Mandrus, and L. Mihaly
Department of Physics, SUNY, Stony Brook, NY 11794

G. P. Williams
National Synchrotron Light Source, Brookhaven National Laboratory, Upton, NY 11973

Abstract

Infrared measurements in superconductors offer the opportunity to study the superconducting energy gap, the dynamics of the conduction electrons, and the nature of low-energy excitations in the material. The infrared transmittance of single crystals of $Bi_2Sr_2CaCu_2O_8$ has been measured and the optical properties determined by Kramers-Kronig analysis. The normal-state data are compared to both one- and two-component descriptions of the optical conductivity. Both approaches allow for a qualitative description of the T-dependent far-infrared conductivity and the nearly T-independent midinfrared conductivity. Below T_c the samples show almost zero absorption at low frequencies, with an onset around 150 cm^{-1}. At higher frequencies, there is a characteristic maximum in the transmittance at $\sim$ 700 cm^{-1} and corresponding structure in $\sigma_1(\omega)$. Although these features are revealed clearly below T_c, we conclude that neither should be assigned to the superconducting gap.

I. INTRODUCTION

The infrared properties of the high-T_c superconductors remains the object of considerable interest. At the present time it is our opinion that there are two open issues in the interpretation of infrared spectra: the nature of the non-Drude absorption in the mid infrared and the observation of the superconducting energy gap. These issues—which of course are not unrelated—will be addressed in this paper.

---

* Present address: Grumman Corporate Research Center, Bethpage, NY 11714

[†] Permanent address: Institute of Physics of the University, P.O. Box 305, Zagreb, Yugoslavia

A. The midinfrared absorption

All of the high T_c superconductors show non-Drude behavior in the midinfrared region.[1-13] Even though the dc resistance is decreasing substantially with decreased temperature, there is very little temperature dependence to the midinfrared reflectance. At lower frequencies there is a definite temperature dependence, and above T_c the far-infrared conductivity is in good agreement with the dc conductivity. This difference between far-infrared and midinfrared behavior has been explained in two ways, which we call "one-component" and "two-component" pictures.

1. One-component approach

In a one-component approach,[4,6,12,13] all of the midinfrared absorption is viewed as due to the same carriers, namely those carriers which are responsible for the dc/far-infrared conductivity and which become the superconducting condensate below T_c. The different behavior in the two regions is attributed to a strong frequency dependence to the scattering rate and effective mass. The frequency dependence must be rather strong, since the far-infrared data imply a 100 K scattering rate of $1/\tau \sim 100$ cm^{-1} while the midinfrared data would require $1/\tau \sim 5000$ cm^{-1}.

A strong frequency dependence can arise from a Holstein[14] process in a metal, where an electron can absorb a photon, emit a phonon (or some other excitation), and scatter. This Holstein approach has been studied by several workers;[4,6,12] as yet it has not been possible to reconcile the large values of λ implied by the midinfrared data with the relatively small values estimated from transport[15] and far-infrared[9,11] data.

A frequency dependent scattering also arises in the "marginal Fermi liquid" (MFL) described by Varma et al.[16] They suggest that for frequencies smaller than some cutoff, the self-energy of the charge carriers should take the form,

$$\text{Im}\,\Sigma = \begin{cases} \lambda T & \omega < T; \\ \omega & \omega > T. \end{cases} \tag{1}$$

A similar functional form has been put forth by Ruvalds and Virosztek[17] in a theory for the "nested Fermi liquid" (NFL). Once Σ is known, the dielectric function is given by

$$\epsilon(\omega) = \epsilon_\infty - \frac{\omega_p^2}{\omega(\omega - \Sigma)} \tag{2}$$

where $\omega_p = \sqrt{4\pi n e^2/m_b}$ is the plasma frequency of n free carriers with band mass m_b and ϵ_∞ is the contribution from higher frequency excitations (interband transitions, etc.). Both MFL and NFL give a low-frequency resistivity which is T-linear in concert with T-independent, large damping rate, high-frequency behavior. The midinfrared conductivity of YBa$_2$Cu$_3$O$_7$ has been interpreted within this marginal Fermi liquid picture by Schlesinger et al.[13]

2. Two-component approach

In the two-component picture,[1-5,7-10] the oxide superconductors are viewed as containing two types of carriers: free carriers which are responsible for the dc conductivity and which condense to form the superfluid below T_c, and bound carriers which have a semiconductor-like gap. The total dielectric function is written:

$$\epsilon(\omega) = \epsilon_D + \epsilon_{MIR} + \epsilon_\infty \tag{3}$$

where ϵ_∞ is the high-frequency contribution, ϵ_{MIR} the bound-carrier contribution, and ϵ_D the dielectric function contribution from the free carriers. The free-carrier contribution is that of a Drude model,

$$\epsilon_D(\omega) = -\frac{\omega_{pD}^2}{\omega^2 + i\omega/\tau} \qquad (4)$$

where ω_{pD} is the plasma frequency of the free carriers and $1/\tau$ is their (essentially ω-independent) relaxation rate. The T-linear temperature dependence of the resistivity is assumed to come from the temperature dependence of $1/\tau$ since in the Drude model, $\rho = 4\pi/\omega_{pD}^2\tau$. The bound carriers are in a broad, nearly T-independent, band (represented by an overdamped Lorentzian oscillator or a sum of broad oscillators) throughout the mid infrared. It recently has been shown that the structure in this midinfrared band may be attributed to interaction of those carriers with phonons.[18]

The story is obviously more complicated than parallel absorption by bound and free carriers since *both* are installed by the doping process which converts the copper oxide materials from their insulating to superconducting forms. Both are absent in $YBa_2Cu_3O_6$, La_2CuO_4, and other insulating systems. Indeed, both the free and bound carriers must in some sense be the *same* carriers, consisting of holes on the CuO_2 planes. (Some may also be on the chains in $YBa_2Cu_3O_{7-\delta}$, but because very similar behavior is seen in $Bi_2Sr_2CaCu_2O_8$, $Pb_2Sr_2(Y/Ca)Cu_3O_{9-\delta}$, and $La_{2-x}Sr_xCuO_4$, none of which have chains, this is a complication unique to the $YBa_2Cu_3O_{7-\delta}$ system.) Furthermore, both components are found to have essentially the same oscillator strengths: $\omega_{pD} \sim \omega_{pe} \sim 10000$ cm^{-1}.

Most far-infrared measurements show that the carrier concentration (as measured by the plasma frequency) is nearly T-independent, so that the T-linear resistivity must come from a T-linear scattering rate.[9] But, the fact that the dc scattering rate is T-dependent strongly implies that it also must be ω-dependent.[4] The argument goes as follows. The temperature dependence of $1/\tau$ comes because the carriers are scattered from some thermal excitation, for example, phonons, and they are scattered by these excitations because they interact with them. Thus, at finite frequencies, the carriers must also be able to emit these excitations as well. This Holstein[14,19] process leads to a frequency-dependent scattering; as ω increases, $1/\tau$ increases from its dc value $1/\tau(0)$ to $1/\tau(0) + \lambda\langle\Omega\rangle$ at high frequencies, where $\langle\Omega\rangle$ is a properly weighted average over the distribution function of the excitations. The measurements of Gurvitch et al.[15] and Kamarás et al.[9] suggest that $\lambda \approx 0.3$ and $\langle\Omega\rangle < 100$ cm^{-1} (in order to maintain a linear resistivity), so that the frequency dependence of $1/\tau$ is not strong at temperatures above T_c.

B. The superconducting energy gap

An ordinary superconductor has a gap Δ in its excitation spectrum. This gap causes the frequency-dependent conductivity $\sigma_{1s}(\omega)$ to be zero up to $\omega = 2\Delta$; above this frequency $\sigma_{1s}(\omega)$ rises to join the normal-state conductivity at several times 2Δ. The area removed from $\sigma_{1s}(\omega)$ appears under the zero-frequency delta function of the superconductor. The superconductor has zero absorption (100 % reflectivity) at frequencies below 2Δ and reduced absorption up to several times 2Δ.

Far-infrared measurements were important in establishing the existence of a gap in metallic superconductors and in determining its magnitude.[20–23] With the discovery of the high-T_c compounds, there have been many attempts to do the same. At the present time there is a lot of controversy—to say the least—about infrared determinations of the gap. Very similar reflectance spectra for crystals and oriented films have been presented by a number of workers.[24,3–5,7–9,12,25] $\mathcal{R}_s$ shows typically a two-step structure. In YBa$_2$Cu$_3$O$_{7-\delta}$ ab-plane data this takes the form of an absorption onset (a decrease in $\mathcal{R}$ from $\sim 100\%$ to $\sim 98\%$) at 140 cm^{-1} and a second shoulder in the 400–500 cm^{-1} region. Both the 140 cm^{-1} onset and the 400 cm^{-1} shoulder have been assigned to the superconducting gap.

Thomas et al.[4] suggested that the apparent onset of absorption at ≈ 140 cm^{-1} in ab-plane data for reduced T_c samples might be the gap; this would give $2\Delta/k_BT_c = 3.2$. Schützmann et al.[7] also put the gap at 140 cm^{-1} in measurements of 91 K T_c samples, which would make $2\Delta/k_BT_c = 2.1$ Schlesinger et al.[24,25,12] have argued that the 400 cm^{-1} shoulder is the gap, giving $2\Delta/k_BT_c = 8.0$. When the ratio of superconducting to normal state reflectance, $\mathcal{R}_s/\mathcal{R}_n$, is plotted, this shoulder appears as a maximum. A maximum in $\mathcal{R}_s/\mathcal{R}_n$ does occur at 2Δ in ordinary superconductors because $\mathcal{R}_s$ is 100% out to 2Δ and then decreases to join $\mathcal{R}_n$, which has been decreasing like $1 - A\sqrt{\omega}$. This similarity to ordinary superconductors was the motivation for the $8k_BT_c$ gap assignment.[24,25,12]

Schlesinger et al.[13] have presented data on untwinned crystals of YBa$_2$Cu$_3$O$_{7-\delta}$ which shows for $T = 30$ K an apparent 99–100% reflectance out to 450 cm^{-1} ($7k_BT_c$) for $\vec{E} \parallel a$ at which point there is a shoulder and decreasing reflectance. Except for the $\sim 100\%$ rather than 98% reflectance, the data look very much like the ab-plane data. For $\vec{E} \parallel b$, the reflectance is smaller, with the 450 cm^{-1} shoulder less pronounced. Schlesinger et al. assign the gap for ab-plane carriers to this onset and attribute the b-axis absorption below this frequency to carriers on the chains.

In a different type of experiment, Pham et al.[26] measured the direct absorption in YBa$_2$Cu$_3$O$_{7-\delta}$ crystals. They find a finite absorption at all frequencies. The absorption increases like ω^2, reaching 0.5% around 150 cm^{-1}. There is an increase above the initial ω^2 behavior beginning around 150–200cm^{-1}. The direct absorption data are in agreement therefore with reflectance data but the better signal/noise ratio allows absorption to be seen at frequencies where the reflectance appears to reach 100%. Pham et al.[26] assign $\approx 3.5T_c$ to the gap value.

If the two-component picture of the midinfrared absorption is correct, then the presence of the second component can obscure any gap absorption and make its determination difficult. This effect occurred some time ago in early studies of La$_{2-x}$Sr$_x$CuO$_4$ ceramics, which showed a sharp reflectance drop in the 50 cm^{-1} region. This drop, which initially was assigned to the gap, was shown to be caused by a zero-crossing of $\epsilon_1(\omega)$, caused by the interplay between the negative contribution of the superfluid and a strong positive contribution from a phonon.[27,28] Using a similar approach, Timusk et al.[3] argued that the maximum seen in $\mathcal{R}_s/\mathcal{R}_n$ in YBa$_2$Cu$_3$O$_{7-\delta}$ was affected by dispersion as well; in this case $\epsilon_1(\omega)$ does not actually cross zero but its magnitude drops to a small value around 500 cm^{-1} due to the midinfrared absorption, leading to a reduction in the reflectance.

Kamarás et al.[9] showed that various features in the superconducting-state conductivity spectrum of $YBa_2Cu_3O_{7-\delta}$ which have been assigned to the gap can also be seen in the normal-state conductivity. This is illustrated in Fig. 1, which shows the frequency-dependent conductivity (from Kramers-Kronig analysis of the reflectance) for an ab-plane $YBa_2Cu_3O_{7-\delta}$ film above and below T_c. For the data above T_c, the free carrier contribution to $\sigma_1(\omega)$ (represented by a Drude model) has been subtracted in order to reveal the midinfrared term. It can be seen that both the 150 cm^{-1} onset and a conductivity minimum around 420 cm^{-1} are present in both temperature regimes. Because the features persist above T_c, Kamaras et al.[9] concluded that they were from the non-Drude midinfrared absorption and not associated with the superconducting gap. Above T_c this absorption is partially masked by the Drude absorption of the free carriers; below T_c, when the Drude carriers condense into a delta function, the midinfrared absorption becomes fully revealed.

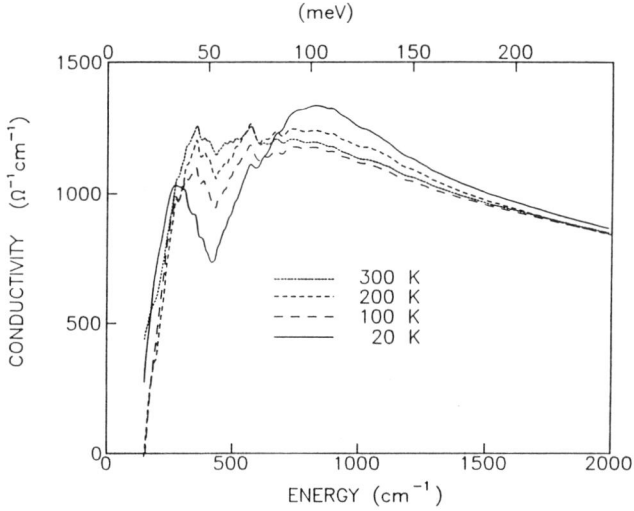

Fig. 1. Frequency-dependent conductivity of a $T_c = 89$ K $YBa_2Cu_3O_{7-\delta}$ film above and below T_c. The free-carrier contribution has been subtracted from the above-T_c data.

The question naturally arises: why should the gap not seen? Kamaras et al.[9] suggested that this is because the high-T_c materials are in the "clean limit," when $2\Delta \gg 1/\tau$. In this limit, all of the Drude oscillator strength exists at low frequencies and goes into the zero-frequency delta function conductivity of the superconductor. None is left for transitions across the gap.

The presence of the midinfrared absorption is a key point in the clean limit argument. Given sufficient sensitivity, the gap could be seen even in the clean limit of ordinary metals with no other low energy excitations. However, in the presence of the midinfrared absorption, especially if there is some temperature dependence in it, picking out the gap becomes more difficult. Intermediate limit calculations by Timusk et al.[29] and Bickers et al.[30] show gap structure, but this structure rather weak and difficult to see in the presence of the much stronger midinfrared conductivity.

The conductivity of Fig. 1 shows a clear minimum or "notch" in the 400–500 cm^{-1} region both above and below T_c. This notch has also been seen by Thomas et al.[4] and by Schützmann et al.[7] In a careful study of oxygen-deficient YBa$_2$Cu$_3$O$_{7-\delta}$ crystals Cooper et al.[8] showed that this notch appears at the same frequency (430 cm^{-1}) in a number of samples, even though their T_c varies between 30 and 90 K. They interpreted this insensitivity to T_c as ruling out a conventional superconducting gap. The conductivity of $T_c = 85$ K Bi$_2$Sr$_2$CaCu$_2$O$_8$ is very similar to the YBa$_2$Cu$_3$O$_{7-\delta}$ conductivity of Fig. 1, except that the onset and notch are at somewhat higher frequencies.[5] It recently has been shown that this structure can be accounted for if coupling of the carriers responsible for the midinfrared absorption to phonons is taken into account.[31,18]

II. EXPERIMENT

A. Samples

Our samples were free-standing crystals of Bi$_2$Sr$_2$CaCu$_2$O$_8$, prepared as described by Forro et al.[32] The samples studied are flakes ca 1×1 mm^2 in area and ~ 1000Å in thickness. Electrical resistance is comparable to many other samples; there is a nearly linear resistance decrease with decreasing temperature and a superconducting transition at 82 K. Extrapolation of the resistance to zero Kelvin leads to a "residual" resistance of around 25% of the 100 K resistance.

X-ray absorption measurements showed that the crystals are extremely uniform in their thickness while electron diffraction revealed that some of the samples are untwinned. Because of the single-domain character of these untwinned crystals, we are able to measure the infrared anisotropy of the Bi$_2$Sr$_2$CaCu$_2$O$_8$ ab plane.

B. Transmittance measurements

Because the samples are free standing, the transmittance may be measured over wide frequency ranges without the complications that would be added by a substrate. We measured the transmittance between 80 and 30,000 cm^{-1}. The far-infrared data were taken using beamline U4IR at the National Synchrotron Light Source. We used a Bruker 113v interferometer in the mid infrared and a Perkin-Elmer 16U monochromator in the near infrared and visible. A continuous-flow cryostat provided cooling to 20 K.

III. RESULTS

A. Transmittance

Fig. 2 shows the (unpolarized) transmittance $\mathscr{T}$ of a Bi$_2$Sr$_2$CaCu$_2$O$_8$ free-standing crystal at temperatures between 20 and 300 K. The transmittance is rather low overall, but increases with increasing frequency. The low frequency $\mathscr{T}$ is rather different above and below the superconducting transition, with a finite intercept for $T > T_c$ contrasting with $\mathscr{T} \propto \omega^2$ for $T < T_c$. In the far infrared, the transmittance of a thin conducting film is well approximated by[20]

$$\mathscr{T} = \frac{1}{(1 + Z_0\sigma_1 d/2)^2 + (Z_0\sigma_2 d/2)^2} \qquad (5)$$

where $Z_0 = 377$ Ω is the impedance of free space, d the film thickness, and $\sigma_1 + i\sigma_2$ the complex conductivity of the film. If $\mathscr{T} \ll 1$, then this further simplifies to $\mathscr{T} \approx$

$4/|Z_0\sigma d|^2$. Above T_c the $\omega = 0$ intercept is then a direct measure of the dc conductivity. (In an ordinary metal, $\sigma_2 << \sigma_1$ at low frequency.) Below T_c the inductive response of the superfluid dominates the absorptive part at low frequencies (i.e., $\sigma_1 << \sigma_2 \propto 1/\omega$) and $\mathscr{T} \propto \omega^2$. Thus the lowest-frequency part of these data is in accord with expectations for a thin superconducting layer.

At higher frequencies, $\mathscr{T}$ increases quasi-linearly with ω. In addition a shoulder develops centered around 700 cm^{-1} (90 meV). How these properties differ from an ordinary metallic film is the subject of the rest of this paper.

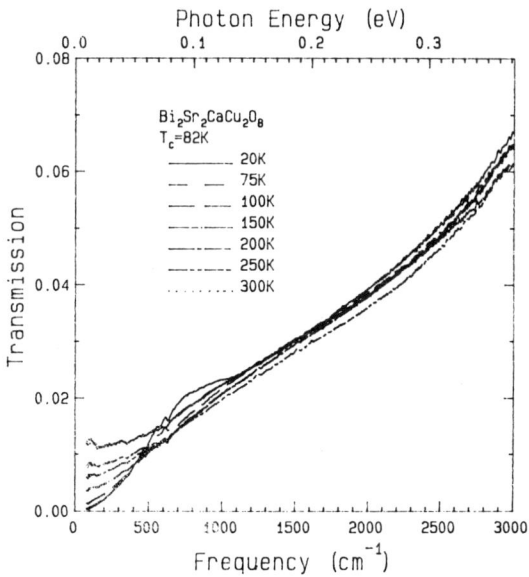

Fig. 2 Transmittance of a 1340 Å crystal of $Bi_2Sr_2CaCu_2O_8$.

B. Polarized transmittance

First, however, we show in Fig. 3 the polarized transmittance of the single-domain $Bi_2Sr_2CaCu_2O_8$ crystal at four temperatures. There is substantial anisotropy, despite the pseudo-tetragonal crystal structure of this material. At all temperatures the a axis is more transparent (less conducting) than the b axis; however, most qualitative features (such as the shoulder at 700 cm^{-1} and the quasi-linear increase in $\mathscr{T}$ with ω) are seen in both polarizations. The anisotropy in the conductivity is quite comparable to that in $YBa_2Cu_3O_{7-\delta}$ in this frequency region. That this anisotropy exists is surprising, because in $YBa_2Cu_3O_{7-\delta}$ the anisotropy is attributed[13,33] to the presence of chains along b and there are no chains in $Bi_2Sr_2CaCu_2O_8$.

C. Conductivity determination

We determined the frequency dependent conductivity of our samples by Kramers-Kronig analysis of the transmittance. The complex transmission amplitude is

$$t(\omega) = \sqrt{\mathscr{T}(\omega)}\, e^{i\theta(\omega)} \tag{6}$$

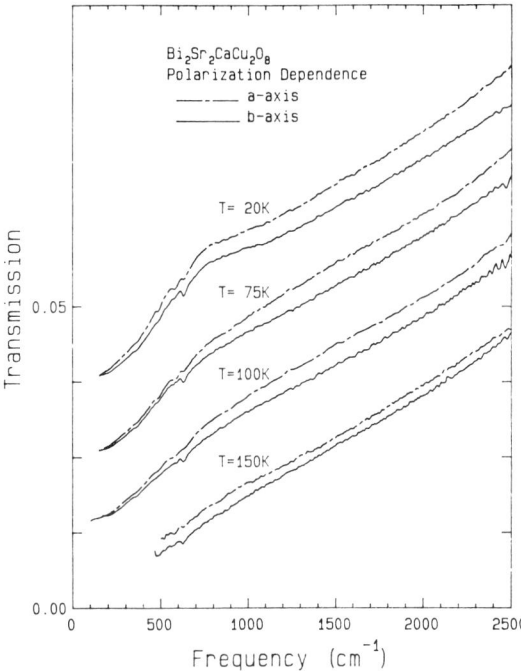

Fig. 3. The polarized transmittance of a single-domain Bi$_2$Sr$_2$CaCu$_2$O$_8$ crystal at four temperatures.

where $\mathcal{T}(\omega)$ is the measured transmittance and $\theta(\omega)$ the phase shift upon transmission. The latter quantity is related to $\mathcal{T}$ by a Kramers-Kronig integral just like the reflectance.[34]

Once the phase shift θ is known, the complex refractive index can be obtained by numerical solution of

$$\sqrt{\mathcal{T}}\, e^{i\theta} = \frac{4N}{(N+1)^2 e^{-i\delta} - (N-1)^2 e^{i\delta}} \quad (7)$$

where $N = \sqrt{\epsilon}$ is the complex refractive index—$\epsilon = \epsilon_1(\omega) + 4\pi\sigma_1(\omega)/\omega$ is the complex dielectric function—and $\delta = \omega N d/c$ the complex phase through the film. The only added complication is that the phase shift the radiation would have travelling through a thickness d of vacuum ($\omega d/c$) has to be added to θ prior to solving Eq. 7 for N. For extrapolations we used a power-law approach to unity at high frequencies and

$$\mathcal{T} = \begin{cases} a + b\omega^2 & T > T_c \\ b\omega^2 & T < T_c \end{cases}$$

at low frequencies.

Results of Kramers-Kronig analysis of the transmittance data are shown in Fig. 4, which shows $\sigma_1(\omega)$ over 150–3000 cm^{-1}. In the normal state the low-frequency conductivity approaches the dc conductivity and falls with increasing frequency as expected for the Drude response of free carriers. However, above ca 300 cm^{-1} the decrease in $\sigma_1(\omega)$ is closer to ω^{-1} than the ω^{-2} behavior expected for Drude free carriers. Furthermore, the T-dependence of $\sigma_1(\omega)$ at high frequencies is much smaller than at dc or

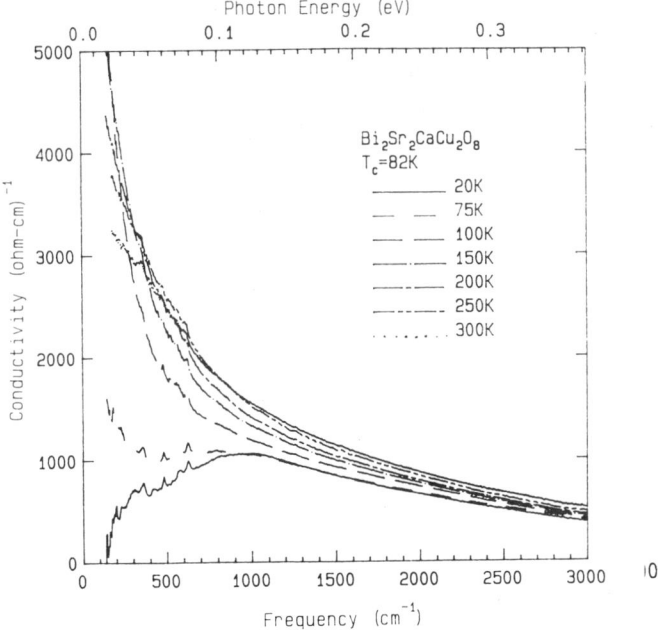

Fig. 4. Frequency-dependent conductivity of $Bi_2Sr_2CaCu_2O_8$ between 20 and 300 K.

low-frequencies. This is the "non-Drude" conductivity that is a well-known property of the high-T_c superconductors.[1]

IV. DISCUSSION

A. One-component model

An advantage of the Kramers-Kronig technique is that any of the optical constants can be calculated once the real and imaginary parts of the refractive index or dielectric function are known. In particular, the self-energy function $\Sigma(\omega)$ can be calculated by inverting Eq. 2:

$$\Sigma(\omega) = \frac{\omega_p^2/\omega}{\epsilon(\omega) - \epsilon_\infty} + \omega. \tag{8}$$

Note that $\epsilon(\omega)$ is negative, so that the first term on the right (which dominates at low frequencies) is negative. In order to make the calculation both ω_p and ϵ_∞ must be estimated. We can get ω_p from the oscillator strength sum rule and ϵ_∞ from the high-frequency behavior of the Kramers-Kronig dielectric function. In our analysis we used $\omega_p = 14,000$ cm^{-1} and $\epsilon_\infty = 4$.

From Σ one can calculate also the frequency-dependent effective mass and relaxation rate. In particular, the effective mass is

$$\frac{m_{eff}}{m_b} = 1 - \frac{\text{Re}\,\Sigma}{\omega}, \tag{9}$$

where m_{eff} is the effective mass and m_b the band mass. The results for m_{eff}/m_b are shown in Fig. 5. As expected for the MFL or NFL picture, the effective mass is enhanced at low frequencies, and then drops off to equal the band mass at high frequencies. However, the temperature dependence of the effective mass is stronger than predicted by these one component models. These give, in general, something like

$$\frac{m_{eff}}{m_b} = 1 + \lambda \ln \frac{\omega_c}{\pi T} \tag{10}$$

where λ is a dimensionless coupling constant and ω_c a cutoff frequency, typically taken to be 1200–8000 cm^{-1} (0.15–1 eV). Instead of this logarithmic T-dependence, the data in Fig. 5 is linear in T.

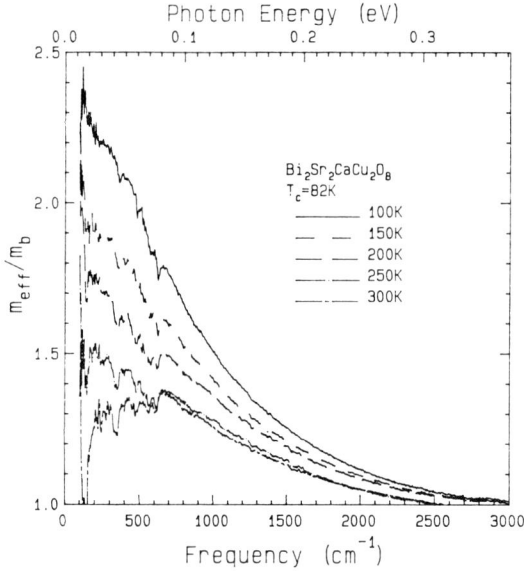

Fig. 5. Effective mass for $Bi_2Sr_2CaCu_2O_8$ at temperatures between 100 and 300 K, calculated from the Kramers-Kronig determined dielectric function.

Fig. 6 shows the imaginary part of the self-energy, calculated from Eq. 8. The one-component models predict that this function should be linear in temperature at low frequencies and then linear in frequency, at least up to some cutoff frequency ω_c. At low temperatures, $-\text{Im}\,\Sigma$ exhibits this behavior. Note that the cutoff frequency appears to be around 1200 cm^{-1} (0.15 eV) and at high temperatures the condition for true marginal Fermi liquid behavior, $\pi T \ll \omega_c$, is not met.

Another difference between the data shown in Fig. 6 and the MFL and NFL theories[16,17] is that the *slope* of $-\text{Im}\,\Sigma$ is not the same at all temperatures. According to the theory, this slope is essentially equal to λ, which in principle is not dependent on temperature. However, the data in Fig. 6 would suggest that $\lambda \propto T^{-1}$. This is also the implication of the T-dependence of the effective mass.

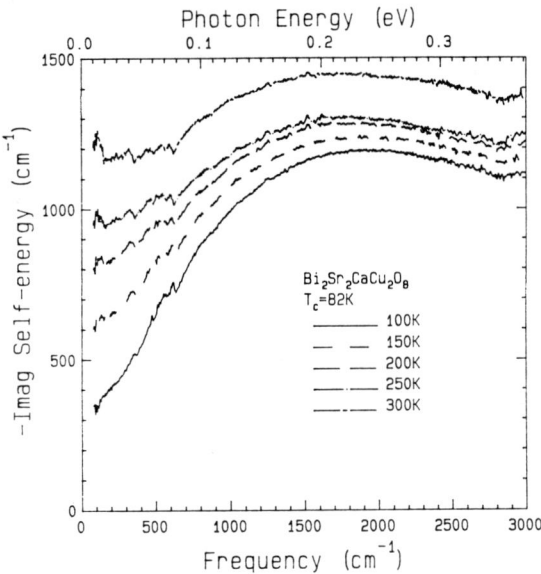

Fig. 6. Imaginary part of the electronic self energy as a function of frequency.

B. Two component picture

In the two-component approach, the dielectric function at each temperature is decomposed according to Eq. 3. This may be done through fits to model dielectric functions[2-4,9] but it can be done as well graphically.[9] We will use the latter approach here. To accomplish this, we make the *ansatz* that the 20 K conductivity in Fig. 4 is a good approximation to $\epsilon_{MIR} + \epsilon_\infty$. (As shown by Kamarás *et al.*[9] and illustrated in Fig. 1, the 20 K conductivity in $YBa_2Cu_3O_{7-\delta}$ does resembles quite closely the difference between the above-T_c total conductivity and a Drude contribution.) Thus, by subtracting the 20 K data from data at higher temperatures, we can obtain an estimate of the free carrier contribution ϵ_D. We can then find the Drude parameters relatively easily, since $\sigma_{1D} = \omega_{pD}^2 \tau/(1 + \omega^2 \tau^2)$, so a plot of σ_{1D}^{-1} vs. ω^2 gives a straight line with slope τ/ω_{pD}^2 and intercept $1/\omega_{pD}^2 \tau$.

Once we find the Drude parameters for the free carriers, we can then subtract a *calculated* ϵ_D from the data and take a look at the difference, which is an estimate of the midinfrared response. When we do so, we find that the above-T_c estimate of the mid-IR conductivity is quite similar to the below-T_c total conductivity, but that there are definite differences. Therefore, we repeated the above analysis, subtracting the average of the *above-T_c* mid-IR conductivity found in the previous step from the total conductivity. This iterative procedure converged after one iteration. The results are shown in Figs. 7 and 8.

Fig. 7 shows the free carrier contribution to the *ab*-plane conductivity of $Bi_2Sr_2CaCu_2O_8$ along with the calculated Drude conductivity. Above T_c, the Drude contribution has a nearly T-independent plasma frequency, $\omega_{pD} = 10,250 \pm 100$ cm^{-1} and a T-linear scattering rate, shown in the inset. Note the intercept at almost zero scattering rate at zero temperature.

This T-linear scattering rate combined with a T-independent plasma frequency is of

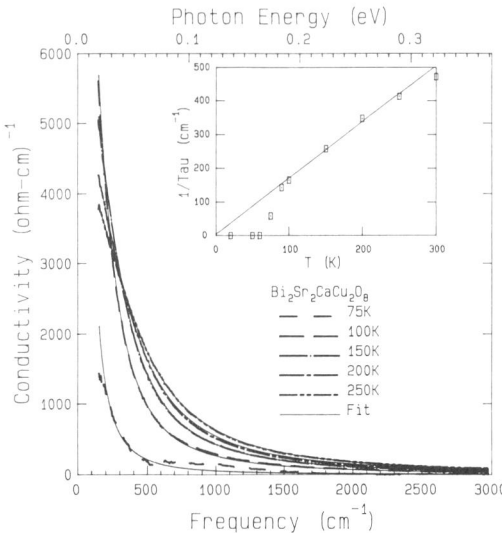

Fig. 7. The free carrier contribution to the ab-plane conductivity of $Bi_2Sr_2CaCu_2O_8$ at temperatures between 75 and 300 K. The thin lines are the calculated Drude conductivity at each temperature. The inset shows the scattering rate vs. temperature found from the analysis.

course completely consistent with the dc conductivity. From our parameters, we would estimate a 100 K resistivity of 100 $\mu\Omega$-cm, in good agreement with the measured dc value.

The scattering rate may be expressed as

$$\hbar/\tau = 2\pi\lambda k_B T \qquad (11)$$

where λ is the dimensionless coupling constant between the charge carriers and whatever T-dependent excitation is responsible for their scattering. From the inset to Fig. 7 we obtain $\lambda = 0.37$, a value which is a little larger than the 0.2–0.3 values estimated[15,9] for $YBa_2Cu_3O_{7-\delta}$.

The midinfrared contribution in Fig. 8 has a typical bound-carrier shape, with a absorption edge rising from below 200 cm^{-1}, some structure in the phonon region, and a broad peak around 1000 cm^{-1}. It is interesting that above T_c there is no T-dependence to the low-energy edge, some weak T-dependence between 700 and 1600 cm^{-1}, and no temperature dependence at higher energies. Below T_c the edge appears to shift to slightly *lower* frequencies and the T-dependence in the 700–1600 cm^{-1} region appears to be stronger. The antiresonance seen in $YBa_2Cu_3O_{7-\delta}$ around 400–500 cm^{-1} is seen here as well, but not so distinctly. Compare Figs. 1 and 8.

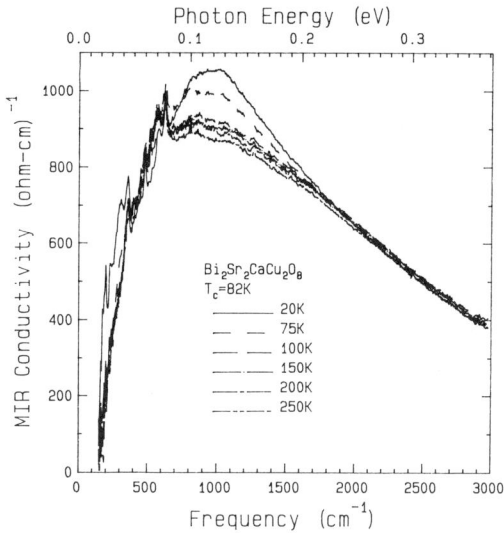

Fig. 8. The midinfrared contribution to the frequency-dependent conductivity of $Bi_2Sr_2CaCu_2O_8$ at temperatures between 20 and 300 K. The curves at and above 75 K were found by subtracting a Drude-model conductivity from the total conductivity of Fig. 4. The 20 K curve *is* the total conductivity.

C. The superconducting gap

The earliest far-infrared studies of ordinary metallic superconductors were carried out as transmittance measurements on thin films.[20,21] The combination of $\sigma_1 = 0$ for $\omega < 2\Delta$ and $\sigma_2 \propto 1/\omega$ gives a maximum in $\mathcal{T}_s/\mathcal{T}_n$ at $\omega = 2\Delta$. This is seen clearly in the work of Palmer and Tinkham[23] on Pb, where the onset in $\sigma_1(\omega)$ and the maximum in $\mathcal{T}_s/\mathcal{T}_n$ occur at identical frequencies. More recent measurements by Van Bentum and Wyder[35] on aluminum also show a maximum in $\mathcal{T}_s/\mathcal{T}_n$ very close to the energy gap.

The ratio $\mathcal{T}(T)/\mathcal{T}_n$ for the $Bi_2Sr_2CaCu_2O_8$ crystal data in Fig. 2 is shown in Fig. 9. The 250 K data is used for the "normal" state; the transmittance ratio to 100 and 200 K also is shown. For $T > T_c$, the ratio turns down at low frequencies because $\sigma_1(\omega)$ is greater at lower temperatures. For $T < T_c$, The low frequency transmittance falls to zero on account of the $1/\omega$ behavior of σ_2. At higher frequencies, a significant maximum develops in $\mathcal{T}_s/\mathcal{T}_n$, centered at 700 cm^{-1}.

The energy of the maximum in $\mathcal{T}_s/\mathcal{T}_n$ corresponds to $12k_BT_c$. Thus, a naive interpretation of the data in Fig. 9 would lead to this unreasonably large value for the gap. The actual situation is almost certainly more complicated. The frequency-dependent conductivity has been calculated for a MFL by Nicol *et al.*[36] The main result is that in the clean limit (which in this context means that the zero-frequency, normal-state scattering rate is smaller than the energy gap frequency) the absorption begins at 4Δ

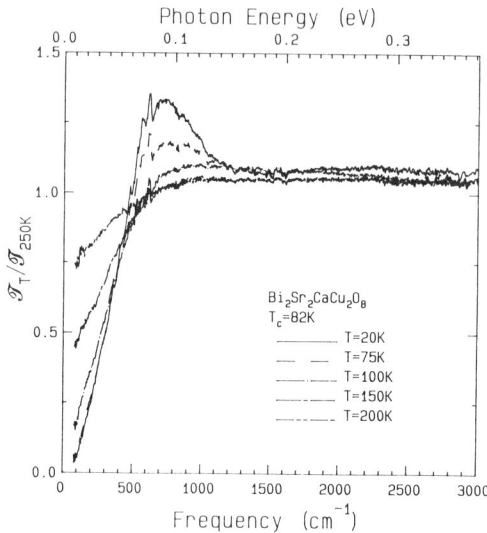

Fig. 9. Ratio of transmittance at temperature T to that at 250 K for a 1400 Å $Bi_2Sr_2CaCu_2O_8$ film.

not at 2Δ.[37] The reason is that for absorption to occur, the photon energy must be sufficient (1) to break a Cooper pair (2Δ) and (2) to create a charge or spin fluctuation (also 2Δ), for a total of 4Δ. Thus, if the 700 cm^{-1} ($12k_BT_c$) peak is associated with the gap in an MFL, then $2\Delta = 350$ cm$^{-1}= 6k_BT_c$. In addition, the calculations of Nicol et al.[36] imply that the $8k_BT_c$ feature of Schlesinger et al.[13] in $YBa_2Cu_3O_{7-\delta}$—if it is indeed related to the gap—would have to be re-interpreted as a 4Δ absorption.

In the two-component picture, the clean limit leads to nearly all the free-carrier oscillator strength appearing under the zero-frequency delta function.[9] Finite frequency absorption is due to the midinfrared band. In this picture, the shoulder which develops in $\mathscr{T}_s$ around 700 cm^{-1} is entirely due to the midinfrared oscillator and the collapse of the Drude peak to a delta function. The peak in $\mathscr{T}_s$ represents a passband between absorbing regions at zero and midinfrared frequencies. No spectroscopic gap is discernible in the infrared transmittance.

The value of the scattering rate is in accord with this interpretation. From the inset to Fig. 7 the scattering rate well below T_c may be estimated at ~ 50 cm^{-1}. Thus, both normal-state and superconducting-state conductivities are falling like $1/\omega^2$ above ~ 100 cm^{-1}. If the gap is near the 200 cm^{-1} BCS value, then most (90%) of the oscillator strength would occur under the zero-frequency delta function of the superconductor, and only a small amount (10%) would be above the gap. If 2Δ were bigger, even less oscillator strength would be available for the gap transition. The transmission is dominated by the dispersive $1/\omega$ behavior of σ_2 and the absorption due to the midinfrared band.

V. CONCLUSION

The normal-state infrared properties of the high-T_c materials is extremely unusual, with a strong non-Drude contribution which dominates over much of the far-infrared region. This non-Drude contribution may be understood either as a strong frequency-dependent scattering of carriers within a single-component approach or as the absorption by a second, strongly interacting component that exists in parallel to the free carriers. Analysis of the data within a single-component picture has qualitative features expected for a marginal or nested Fermi liquid, but differs in many details. Most notable of these differences is that the coefficient of the ω-linear term in the imaginary part of the self energy of the charge carriers is temperature dependent. Analysis within a two-component picture leads to a T-dependent, weak-coupling, free-carrier component and a nearly T-independent, strong-coupling midinfrared component.

In the superconducting state, a condensate forms which carries much the same oscillator strength as the free-carrier component above T_c. The remaining infrared absorption has an onset in the 150 cm^{-1} region, with structure in the phonon region. The total conductivity below T_c resembles closely the midinfrared conductivity above T_c. Even though these are single-domain, untwinned crystals, no absorption onset at 400–500 cm^{-1} ($8k_BT_c$) such as has been reported in $YBa_2Cu_3O_{7-\delta}$ is observed. A transmission maximum near 700 cm^{-1} ($12k_BT_c$) is seen below T_c. This feature can be interpreted either as a consequence of a two-component dielectric function or as due to 4Δ excitations from the superconducting condensate. For the latter interpretation, any gap in the optical conductivity has to be relatively smooth and independent of temperature, contrary to simple BCS behavior. In that sense, the two models are very similar: some kind of bound (either optically active or optically inactive) excitations, with binding energy of the order of 1000 K, exists well above T_c.

Acknowledgement: Research at Florida is supported by DARPA grant MDA-972-88-J-1006. Research at Stony Brook is supported by NSF grant DMR9016456. NSLS is supported by DOE through contract DE-AC02-76CH00016. We thank J. Carbotte, P. Littlewood and T. Timusk for useful discussions.

References

1. K. Kamarás, C.D. Porter, M.G. Doss, S.L. Herr, D.B. Tanner, D.A. Bonn, J.E. Greedan, A.H. O'Reilly, C.V. Stager, and T. Timusk, *Phys. Rev. Lett.* **60**, 969 (1988).

2. D.A. Bonn, A.H. O'Reilly, J.E. Greedan, C.V. Stager, T. Timusk, K. Kamarás, and D.B. Tanner, *Phys. Rev. B* **37**, 1574 (1988).

3. T. Timusk, S.L. Herr, K. Kamarás, C.D. Porter, D.B. Tanner, D.A. Bonn, J.D. Garrett, C.V. Stager, J.E. Greedan, and M. Reedyk, *Phys. Rev. B* **38**, 6683 (1988).

4. G.A. Thomas, J. Orenstein, D.H. Rapkine, M. Capizzi, A.J. Millis, L.F. Schneemeyer, and J.V. Waszczak, *Phys. Rev. Lett.* **61**, 1313 (1988).

5. M. Reedyk, D.A. Bonn, J.D. Garrett, J.E. Greedan, C.V. Stager, T. Timusk, K. Kamarás, and D.B. Tanner, *Phys. Rev. B* **38**, 11981 (1988).

6. R.T. Collins, Z. Schlesinger, F. Holtzberg, P. Chaudhari, and C. Feild, *Phys. Rev. B* **39**, 6571 (1989).

7. J. Schützmann, W. Ose, J. Keller, K.F. Renk, B. Roas, L. Schultz, and G. Saemann-Ischenko, *Europhys. Lett.* **8**, 679 (1989); U. Hoffmann et al., *Solid State Comm.* **70**, 325 (1989).

8. S.L. Cooper, G.A. Thomas, J. Orenstein, D.H. Rapkine, M. Capizzi, T. Timusk, A.J. Millis, L.F. Schneemeyer, and J.V. Waszczak, *Phys. Rev. B* **40**, 11358 (1989).

9. K. Kamarás, S.L. Herr, C.D. Porter, N. Tache, D.B. Tanner, S. Etemad, T. Venkatesan, E. Chase A. Inam, X.D. Wu, M.S. Hegde, and B. Dutta, *Phys. Rev. Lett.* **64**, 84 (1990).

10. K.F. Renk, H. Eschrig, U. Hoffman, J. Keller, J. Schützmann, and W. Ose, *Physica C* **165**, 1 (1990).

11. Joseph Orenstein, G.A. Thomas, A.J. Millis, S.L. Cooper, D. Rapkine, T. Timusk, L.F. Schneemeyer, and J.V. Waszszak, *Phys. Rev. B* **42**, 6342 (1990).

12. Z. Schlesinger, R.T. Collins, F. Holtzberg, C. Feild, G. Koren, and A. Gupta, *Phys. Rev. B* **41**, 11237 (1990).

13. Z. Schlesinger, R.T. Collins, F. Holtzberg, C. Feild, U. Welp, G.W. Crabtree, Y. Fang and J.Z. Liu, *Phys. Rev. Lett.* **65**, 801 (1990).

14. T. Holstein, *Phys. Rev* **96**, 539, (1954); *Ann. Phys. (N.Y.)* **29**, 410, (1964).

15. M. Gurvitch and A.T. Fiory, *Phys. Rev. Lett.* **59**, 1337 (1987).

16. C.M. Varma, P.B. Littlewood, S. Schmitt-Rink, E. Abrahams, and A. Ruckenstein, *Phys. Rev. Lett.* **63**, 1996 (1989); ibid, **64**, 497, (1990) (E).

17. A. Virosztek and J. Ruvalds, *Phys. Rev. B* **42**, 4064 (1990); *Physica B* **165&166**, 1267 (1990).

18. T. Timusk, C.D. Porter, and D.B. Tanner, *Phys. Rev. Lett.* **66**, 663 (1991).

19. P.B. Allen, *Phys. Rev. B* **3**, 305 (1971).

20. R.E. Glover and M. Tinkham, *Phys. Rev. B* **107**, 844 (1956); **108**, 243, (1957).

21. D.M. Ginsberg and M. Tinkham, *Phys. Rev.* **118**, 990 (1960).

22. H.D. Drew and A.J. Sievers, *Phys. Rev. Lett.* **19**, 697 (1967).

23. L.H. Palmer and M. Tinkham, *Phys. Rev.* **165**, 588 (1968).

24. Z. Schlesinger, R.T. Collins, D.L. Kaiser, and F. Holtzberg, *Phys. Rev. Lett.* **59**, 1958 (1987).

25. R.T. Collins, Z. Schlesinger, F. Holtzberg, and C. Field, *Phys. Rev. Lett.* **63**, 422 (1989).

26. T. Pham, H.D. Drew, S.H. Mosley, and J.Z. Liu, *Phys. Rev. B* **41**, 11681 (1990).

27. D.A. Bonn, J.E. Greedan, C.V. Stager, T. Timusk, M.G. Doss, S.L. Herr, K. Kamarás, C.D. Porter, D.B. Tanner, J.M. Tarascon, W.R. McKinnon, and L.H. Greene, *Phys. Rev. B* **35**, 8843 (1987).

28. M.S. Sherwin, P.L. Richards, and A. Zettl, *Phys. Rev. B* **37**, 1587 (1988).
29. T. Timusk et al, *Physica C* **162–164**, 841 (1990).
30. N.E. Bickers, D.J Scalapino, R.T Collins, and Z. Schlesinger, *Phys. Rev. B* **42**, 67 (1990).
31. T. Timusk and D.B. Tanner, *Physica C* **169**, 425 (1990).
32. L. Forro, D. Mandrus, R. Reeder, B. Keszei, and L. Mihaly, *J. Appl. Phys.* **68**, 4876 (1990).
33. B. Koch, H.P. Geserich, and T. Wolf, *Solid State Comm.* **71**, 495 (1989).
34. Frederick Wooten, *Optical Properties of Solids* (Academic Press, New York, 1972).
35. P.J.M van Bentum and P. Wyder, *Physica* **138B**, 23 (1986).
36. E.J. Nicol, J.P. Carbotte, and T. Timusk, *Phys. Rev. B,* submitted.
37. A "4Δ gap" has been suggested by P.B. Littlewood, private communication, and by J. Orenstein, S. Schmitt-Rink, and A.E. Ruckenstein (preprint).

FERMI SURFACE OF YBCO BY DHVA

J. L. Smith, C. M. Fowler, B. L. Freeman,
W. L. Hults, J. C. King, and F. M. Mueller

Los Alamos National Laboratory
Los Alamos, NM 87545, USA

INTRODUCTION

These proceedings demonstrate how far we have come in the last four years of high temperature superconductivity. Knowledge of the energy bands and Fermi surfaces from experiment has come rather late. Photoemission, which was presented at this workshop by Tobin and by Shen, first showed proof of the validity of the energy band calculations. Positron annihilation, presented by West, after a rough start, is now giving evidence of the Fermi surface. Both of these techniques involve electronic excitations and hence, although they show the Fermi surface, do not put as severe a constraint on various models for superconductivity as does the de Haas-van Alphen (dHvA) effect. This is a true measurement of the electronic ground state in an applied magnetic field where the frequency of oscillatory magnetization yields extremal cross-sectional areas of the Fermi surface.[1] We have already reported some of our Fermi surface work at two conferences[2,3] but present here discussion of several more important aspects of the work.

MEASUREMENT

We prepared the samples of $YBa_2Cu_3O_{6.97}$ by mixing dried, stoichiometric quantities of Y_2O_3, $BaCO_3$, and CuO in a high energy impact ball-mill for 24 hours. The powder was pressed into pellets and calcined in air at 940°C for 6 days with a brief excursion to 890°C every 4 hours.[4] The material was then ball-milled for 24 hours, pressed, and returned to the furnace for 2 days. This was done twice. After a final ball-milling the pressed powder was sintered in flowing oxygen at 960°C, slow cooled to 450°C, reheated and held at 960°C for 10 hours, slow cooled to 450°C and oxygenated there for 24 hours, and removed from the furnace at 200°C. This material showed an onset temperature at 93 K and a superconducting midpoint at 85 K in a squid

Fig. 1. Scanning electron microscope picture of the aligned sample powder in epoxy.

magnetometer in a field of 0.01 T. In a susceptometer running above 10 MHz the sample showed a transition midpoint at 91.7 K in zero field. Iodometric titration yielded an oxygen concentration of 6.97. Most of this sample was then ball-milled, mixed with epoxy, and c-axis aligned in a 4.2 T magnet.[5] Aligned powder allows us to minimize the eddy-current heating during the experiment and to have the equivalent of a single crystal when fields are parallel to the c-axes. Figure 1 shows an SEM photo of a lightly polished internal surface. Consistent with a particle size measurement, most of the grains can be seen to be within a 5-10 μm range. X-ray diffraction of the sample showed alignment of the c-axes to better than 2 degrees. No second phases were seen. As our experiment destroys the sample, many identical castings of the powder were made at the same time. These have been used for the dHvA measurements and for further characterization over the last two years.

One year ago a slice of the aligned material showed a midpoint of the transition at 82 K in a 14 MHz susceptibility apparatus and showed as narrow an NMR line as had ever been measured at Los Alamos. Recently, an aligned sample showed an onset at 93 K and a midpoint at 84 K in a squid magnetometer in a field of 0.01 T. A recent x-ray diffraction pattern of a freshly powdered chunk of the original sample pellet gave lattice parameters of a = 3.8238(2), b = 3.8935(6), and c = 11.6966(5) Å, which is consistent with an oxygen concentration of 6.90. Some second phases were seen. It appears that the sintered sample pellets have deteriorated a bit in two years, that the powder in epoxy is possibly a bit worse than freshly ground pure powder, and that YBCO powder in epoxy is remarkably stable over two years.

Next an epoxy YBCO sample and a piece of pure epoxy are machined into a pair of 1 mm diameter rods, about 5 mm long. Great care is taken to insure that these two rods are identical geometrically. Both rods are wound with about 90 turns of 30 μm insulated copper wire, which is lightly tacked down with thinned GE 7031 varnish. The coils are then matched in inductance, capacitance, and resistance to a few parts in 10^4 at an excitation frequency of 10 MHz by adjusting the coil wire. The coils are then connected in opposition electrically and in parallel geometrically and covered with epoxy. This process is what would normally be called balancing but, because of the MHz frequency range, is clearly a process of making an identical pair. This part of the experiment is critical as our signal is 0.05-0.2 V in the presence of about 1 kV of induced voltage across each coil. A 220 Ω, 0.1 W precalibrated carbon resistor is mounted near the sample to monitor temperature before applying the magnetic field. The signal from the coils is fed through a 550 kHz high pass filter into a bank of Tektronix equipment to record the data[2] and then to a single 50 Ω terminator. The filter removes more than 95% of the direct, pulsed-field dB/dT signal.

We cool the sample with liquid helium, from a 2 liter glass "liquid nitrogen" dewar, that passes through vacuum-jacketed, clear pyrex tubing. For 4 K the helium is pressurized and blown past the sample. For temperatures down to 2.3 K, a phenolic orifice is vacuum-greased into the pyrex just upstream from the sample. Several small vacuum pumps, within a few minutes, can slowly raise the column of helium up out of the dewar and finally through the orifice where it evaporates. It is not at all obvious that such a "soda straw" can raise a column of liquid when it is at its boiling point. Clearly, as the helium is rising in the tube it is already cooling a bit. The size of the orifice must be matched to the speed of the pumps. If the orifice is as much as 25 μm too large or too small in diameter (our orifice is 510 μm diameter, 7 mm long) the temperature at the sample can oscillate by as much as 0.5 K.

Magnetic fields to above 100 T with a time scale of tens of microseconds are produced with a destructive, flux-compression system.[6] A coil with a calibrated area of about 30 mm^2 is placed in a groove, outside of the pyrex tube but within the one-turn brass magnet coil, and its voltage is recorded along with the dHvA signal to yield the field, after integration. The oscillatory dHvA signal (or magnetization) is then made a function of the inverse of the applied magnetic field (for fields after the maximum field) and a Fourier transform is made of this function. The spectral content of this transform (in units of Tesla) directly yields extremal cross-sectional areas of the Fermi surface. When temperature dependences of the spectral content and the structure of various harmonics are known such things as electron effective masses, electron scattering rates, and signs of the charge carriers can be extracted and associated with different pieces of the Fermi surface.[1,2,3] Indeed, with the large, homogeneous magnetic fields and the high quality samples needed to perform dHvA experiments, the ground state of a metal becomes well-known.

RESULTS AND DISCUSSION

We have observed two areas of the Fermi surface (in various units):

A(kT)	A($Å^{-2}$)	A(at.units)	m^* (m_e)
0.56±5%	0.054	0.015	2.8
0.78±5%	0.075	0.021	4.5

The small area is in excellent agreement with the dHvA measurements of Kido et al.[7] They report an area of 0.54±0.03 kT with an effective mass of 2.1±0.5 m_e using a field modulation technique in the Tohoku hybrid magnets in fields up to 27 T. For our purposes, the electronic band structure calculations of Massida et al.[8] and Pickett et al.[9] are in agreement with each other, and the R and S centered areas are the obvious choices for our orbits. Our orbits are shown on the calculated Fermi surface in Fig. 2 in which we have assumed that the orbits are circular for purposes of illustration. Our areas are much closer to the values given by ref. 8.

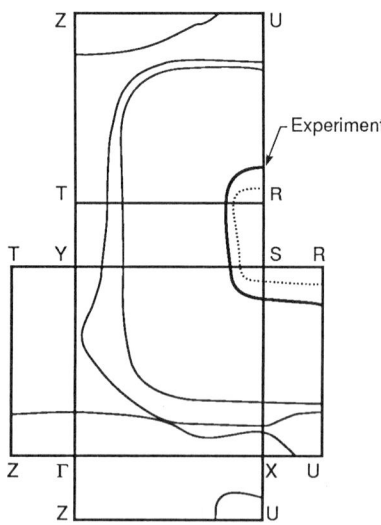

Fig. 2 Our Fermi surface areas drawn on the calculations of Pickett et al.[9] Their dotted line denotes strong sensitivity to oxygen doping.

It is usually assumed that to observe the dHvA effect an electron must complete one orbit without being scattered ($\omega_{cyc}\tau > 1$). An estimate for the shortest reasonable electron mean free path for YBCO at 2 K is about 400 Å.[10] The circumference for our larger area orbit at 100 T is 360 Å. However, the work of Kido et al.[7] would require a mean free path more than three times longer as their applied field is lower. This may also be a reasonable value, but an interesting argument was put forth at this workshop by Ong et al. They argue that their Hall data suggest a longer scattering time for cyclotron motion than for transport. This is a particularly appealing reason to make it possible to do dHvA measurements at lower fields than might have been estimated.

The major point of discussion of our results at this conference was the question of observing dHvA oscillations below H_{c2}. Clearly Kido et al. are working below H_{c2}. While we know of no published experimental results for H_{c2} that would suggest that our 100 T is also below H_{c2}, most people now believe that H_{c2} (c||H, T < 4 K for YBCO) is between 100 and 200 T. Furthermore, our continued failure to observe clear dHvA oscillations in increasing fields makes us believe that we are still fighting a flux lattice at 100 T. Markiewicz et al.[11] (and references therein) and informal discussions at this workshop (with Lowndes, Markiewicz, and Mueller about unpublished results) make it clear that dHvA works below H_{c2}, but more complex signals may be seen. Traditional Lifshitz-Kosevich theory of dHvA[1] is now a well-patched quilt laid upon a semi-classical model. It is clearly time to consider the full problem of the free energy of a superconductor in a high field. Nonetheless, our results can be placed under the old quilt by noting that, in high fields below H_{c2}, normal electrons exist in an inhomogeneous internal field, and this leads to dephasing or severe loss of amplitude in our signal. We expect, as the informal discussions at this conference point out, that when we excede H_{c2}, we will see a significant increase in signal amplitude as the internal field of the sample becomes homogeneous. Further work on this topic is underway, but there are no clear answers yet.

We are continuing these measurements with the goal of finding larger Fermi surface areas, presumably with higher masses by using higher fields (>H_{c2}) and lower temperatures.

ACKNOWLEDGEMENTS

We thank B.L. Bennett, P.C. Hammel, D.H. Herrera, M.F. Hundley, E.J. Peterson, A.P. Reyes, D.G. Rickel, J.F. Smith, K.D. Sowder, J.D. Thompson, and D.T. Torres for assistance and thank K.S. Bedell, R.E. Cohen, F. de la Cruz, I.E. Dzyaloshinskii, D.H. Lowndes, M.P. Maley, and R.S. Markiewicz for discussions. Work performed under the auspices of the U.S. Dept. of Energy, Office of Basic Energy Sciences, Division of Materials Sciences.

REFERENCES

1. D. Shoenberg, "Magnetic Oscillations in Metals," Cambridge University Press, Cambridge (1984).
2. F. M. Mueller, C. M. Fowler, B. L. Freeman, W. L. Hults, J. C. King, and J. L. Smith, to appear in Physica B.
3. J. L. Smith, C. M. Fowler, B. L. Freeman, W. L. Hults, J. C. King, and F. M. Mueller, to appear in Proc. 3rd Int. Symp. on Superconductivity, Springer-Verlag, Tokyo.
4. J. L. Smith, W. L. Hults, A. P. Clarke, and K. A. Johnson, Guidance and Control Information and Analysis Center, Chicago, GACIAC Report PR-89-02, p.11 (1989).
5. D. E. Farrel, B. S. Chandrasekhar, M. R. DeGuire, M. M. Fang, V. G. Kogan, J. R. Clem, and D. K. Finnemore, Phys. Rev. B36:4025 (1987).
6. C.M. Fowler, Science 180:261 (1973); C.M. Fowler, R.S. Caird, W.B. Garn, and D.J. Erikson, IEEE Trans. on Magnetics MAG-12:1018 (1976).
7. G. Kido, K. Komorita, H. Katayama-Yoshida, T. Takahashi, Y. Kitaoka, K. Ishida, and T. Yoshitomi, to appear in Proc. 3rd Int. Symp. on Superconductivity, Springer-Verlag, Tokyo; to appear in Proc. 2nd ISSP Int. Symp. on Physics and Chemistry of Oxide Superconductors, Springer-Verlag, Tokyo.
8. S. Massidda, J.J. Yu, A.J. Freeman, and D.D. Koelling, Phys. Lett. 122A:198 (1987).
9. W.E. Pickett, R.E. Cohen, and H. Krakauer, Phys. Rev. B42:8764 (1990).
10. M. Gurvitch, private communication.
11. R.S. Markiewicz, I.D. Vaquer, P. Wyder, and T. Maniv, Solid State Comm. 67:43 (1988).

DIRECT, EXPERIMENTAL EVIDENCE OF THE FERMI SURFACE IN $YBa_2Cu_3O_{7-x}$

H. Haghighi, J.H. Kaiser, S.L. Rayner, and R.N. West
University of Texas at Arlington, Arlington TX

J.Z. Liu and R. Shelton
University of California, Davis CA

R.H. Howell, P.A. Sterne, F. Solal, and M.J. Fluss
Lawrence Livermore National Laboratory, Livermore, CA

INTRODUCTION

We report new measurements of the electron-positron momentum spectra of $YBa_2Cu_3O_{7-x}$ performed with ultra-high statistical precision. These data differ from previous results in two significant respects: They show the D_2 symmetry appropriate for untwinned crystals and, more importantly, they show unmistakable, statistically significant, discontinuities that are evidence of a major Fermi surface section.

These results provide a partial answer to a question of special significance to the study of high temperature superconductors i.e. the distribution of the electrons in the material, the electronic structure. Special consideration has been given both experimentally and theoretically to the existence and shape of a Fermi surface in the materials and to the superconducting gap. There are only three experimental techniques that can provide details of the electronic structure at useful resolutions. They are angular correlation of positron annihilation radiation, ACAR, angle resolved photo emission, PE, and de Haas van Alphen measurements.

Positron ACAR measurements occupy a spot between angle resolved photo emission measurements and de Haas van Alphen measurements. The first, PE, can in one spectrum supply the energy spectrum for some selected momentum in the Brillion zone. This momentum selection is achieved by limiting the angular acceptance in the electron detector. In order to map the entire electron momentum distribution to see the full shape of the Fermi surface the measurement must be performed for each momentum bin. This is rarely done. Best photo emission angular acceptances are of the order of 2 degrees and the full momentum of the Brillion zone depends on photon energy but is typically 18 degrees[1,2]. This results in a momentum resolution of about 11 % of the reciprocal lattice vector and is the limit on which the momentum position of bands crossing the Fermi surface can be determined in photo emission. Photo emission is also performed on cleaved crystals at low temperatures and thus may not be determining normal state properties and is vulnerable to effects of surface contamination.

De Haas van Alphen measurements on the other hand measure the area enclosed by the Fermi surface in a plane defined by an external magnetic field. These are measurements of bulk magnetic properties in the normal state. However at present only orientated polycrystalline samples have been measured by this technique[3].

Positron ACAR measurements determine only the momentum of the electron in each annihilation event. Therefore there is no distinction of between electrons in bound and unbound states at the same momentum. Fortunately the Fermi surface is defined by limit in the momentum for occupied states set and can be easily seen as a sharp break or discontinuity in the measured distributions. More importantly the ACAR measurement maps out the entire momentum space even into higher order zones in a single experiment. Thus the continuous shape of a Fermi surface is determined in the same measurement and broadening of the Fermi surface can be detected by the sharpness of the discontinuity. These measurements have been very successful in studying Fermi surface driven order-disorder transitions in binary alloys where the shape and broadening of the Fermi surface are central issues in the physics of the transitions. Since the positron is a bulk probe there is no vulnerability to surface effects in the ACAR measurement. Also the sample temperature of the positron measurement can be set at any temperature so that the measurements can clearly be those of a normal state property.

Predictions of the existence of a continuously connected Fermi surface are not specific to all proposed theories for describing superconducting oxides. Consequently the analysis of the positron data for the existence of a Fermi surface must first be done independently of the predictions of any particular theory. This restriction presents a special challenge to the positron technique and requires levels of precision in the measurement beyond those of any previous experiments on elemental metals or alloys.

There have been several attempts to clearly demonstrate a Fermi surface using positrons[4-6]. Unfortunately the interpretation of earlier data on YBCO was complicated by low statistical precision in the earliest measurements and the use of highly twinned crystals throughout. Those significant features in earlier data sets that were statistically reproducible were in reasonable agreement with features expected from the effects of distortion of the positron and electron wave functions by the non-uniform charge densities in the oxide lattice[7].

EXPERIMENT

We have completed ACAR measurements on twin-free crystals of YBCO to ultra-high levels of statistical precision. The measurements were performed at the University of Texas at Arlington and the data were collected on their spectrometer. Untwinned YBCO samples were prepared by the LLNL program at the University of California at Davis.

The YBCO measurement was performed on a matrix of crystals that were made twin-free by annealing under compression along the a direction. This resulted in crystals that were twin-free over 90% of their surface. Measurements of the DC magnetic susceptibility showed that all of the crystals had an onset of superconductivity above 92 K. Two of the crystals had transitions significantly broader than the rest with mid points at 85 K. These data indicate that the level of oxygenation was high in all but isolated parts of two samples.

The samples were mounted in a matrix of six crystals, attached by epoxy to a web of 25 micron wires. The crystals were brought to a common alignment within 0.5 degrees by a process of repeated attachment, guided by Laue X-ray photography. This matrix was then mounted so that the c axis of the samples was along the line connecting the detector centers and the a and b sample axes were along the x and y data axes respectively. Data were collected in a 256 X 256 matrix with .143 mrad wide bins. The data were stored every 2×10^7 counts and after individual analysis to assure stability of the detection system were later summed into full distributions. Two distributions of 2.5×10^8 counts were obtained with the sample-detector alignment rotated by 90° between the two.

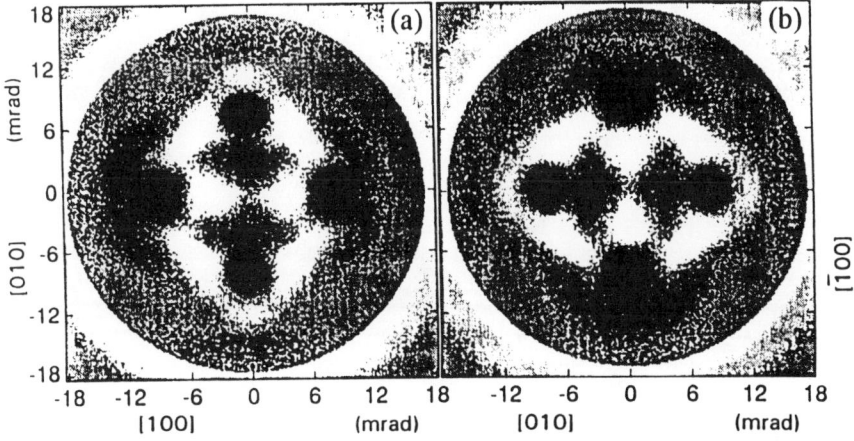

Figure 1. Residual anisotropy distributions for the two subsets of the data. The sample-spectrometer orientation has been rotated by 90° for data set b.

These data were then corrected for the efficiency function of the detector and the isotropic portion of the spectrum was subtracted producing anisotropic residuals as seen in figure 1 a and b. The data in figure 1 show a clear D2 symmetry in alignment with the crystalline axis as would be expected from contributions related to the Cu-O chains in untwinned samples and are identical within the statistics of the measurement. The data distributions were then rotated to the same orientation and summed to produce a single distribution containing 5×10^8 counts. Also since the underlying symmetry of the data are apparent it is permissible to bring the four equivalent quadrants into one. The anisotropic residual of this combined distribution is seen in figure 2 a and b in contour and isotropic views.

DISCUSSION

The data as displayed have several remarkable features. There are four-fold symmetric areas of positive or negative excursions that are consistent with the features observed in twinned samples and can be attributed to wave function effects. These effects are due in part to the non-uniform spatial distribution of the positron. Positron densities have been calculated to be highest in the region surrounding the CuO chains, high in the region of the CuO planes and low everywhere else. This has been verified in measurements on twinned crystals where the wavefunction effects were the statistically significant feature.

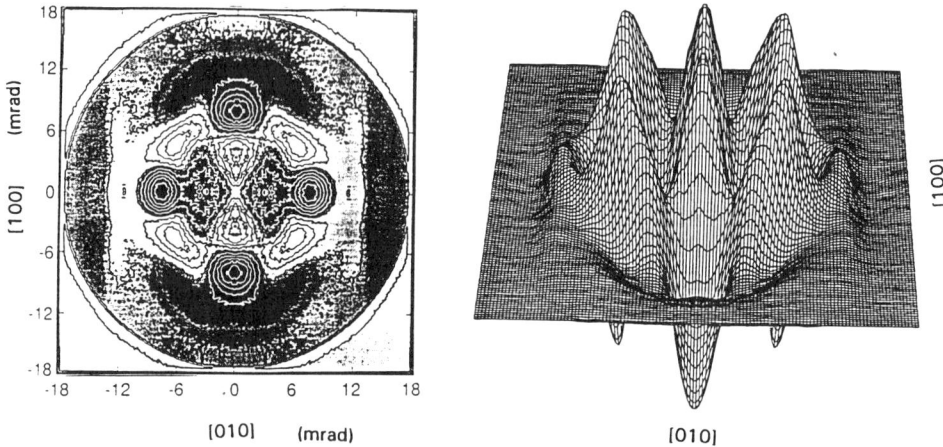

Figure 2. Residual anisotropy of the full data set shown in contour and isometric displays. The sharp changes in the data having two fold symmetry are continuous along the chains, vertical direction, for both low and high momentum values.

New features in the present data are the line discontinutites running parallel to the gamma-x direction seen both in the central, p=0, zone and at positions symmetrically placed at about +/- 4 pi/b, roughly 13 mrad from the distribution center. There is also evidence of a ridge at +/- 2 pi/b but it is obscured in both data displays by the peaks resulting from the wavefunction effects. The breaks at high momentum are significant at 10 standard deviations in the data in figure 2. They stand as convincing evidence of a Fermi surface at that momentum.

The ACAR distributions are measured in real momentum space and consequently are an extended pattern of the electronic distribution in the Brillion zone. Because of this any demonstration of the Fermi surface must be evident at momenta greater than the reciprocal lattice momentum. Observation of the breaks seen at high momenta in figure 2 is sufficient to justify further data manipulation. It is permissible in cases where there is a Fermi surface to sum all of the data back into a single zone using the LCW theorem.

Since we clearly see breaks in the data at high momentum we have applied the LCW operation to our data and the results are seen in figure 3. The Fermi surface sheets with D2 symmetry seen at momenta inside the central zone are now reinforced by the data from other zones and are clearly seen as breaks centered on gamma and running in the gamma-x direction.

The position of the experimentally observed Fermi surface is in good agreement with that expected from band theory[8-11]. Both real momentum calculations[10], single zone calculations[11] and our own LMTO results, shown in figure 4 predict the Fermi surface sheet that we see from the crossing of a band associated with the Cu-O chains. Fermi surface crossings in both the gamma and z plane are shown. There is even agreement between the data and the shape of the calculated Fermi surface through the zone.

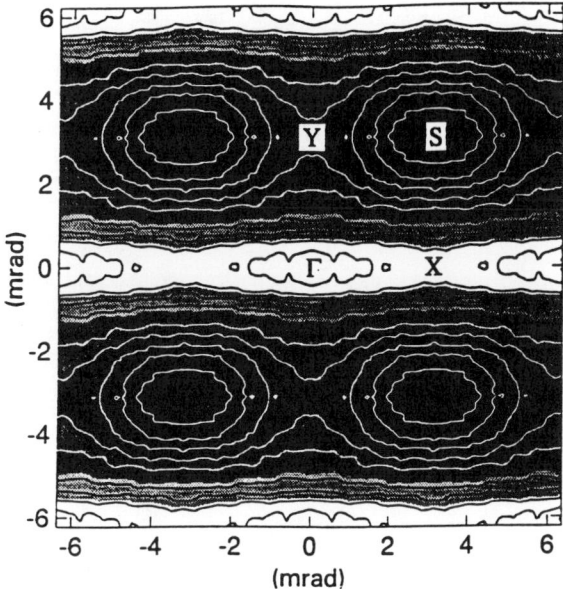

Figure 3. The full ACAR distribution reduced into a single zone using the LCW theorem.

Theory predicts that there are other Fermi surface features in the form of closed loops centered on the S point and associated with either CuO plane bands or hybrid states. We have no independent confirmation of these features in our data. Although there is a general depression at the S point, there are no sharp discontinutiies and it is known that a depression at S can be the result of wave function effects. Also these same wave function effects tend to diminish the size of any Fermi surface breaks associated with the CuO planes due to the propensity of the positron to overlap on the chains.

To obtain an unambiguous answer to the question of a Fermi surface on the CuO planes requires either much higher statistics or a different system. We are performing a similar experiment on single crystals of $La_{1.874}Sr_{.126}CuO_4$ with a sharp superconducting transition at 30 K using the LLNL spectrometer At the time of this report the data lack the statistical precision of the YBCO experiment just described.

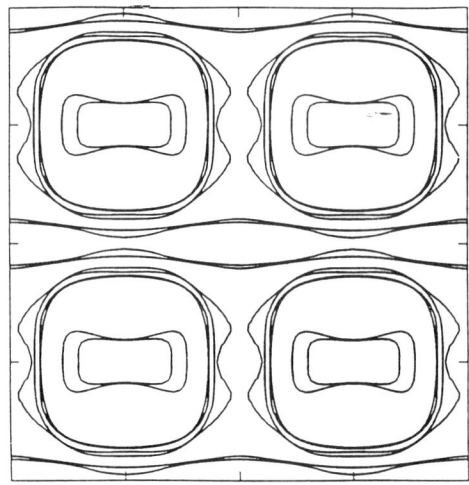

Figure 4. LMTO calculations of the Fermi surface crossings in both the gamma and z planes. The repeated zones are shown in the same format as the data of figure 3.

Work performed under the auspices of the U.S. Department of Energy by the Lawrence Livermore National Laboratory under contract N. W-7405-ENG-48, with the help of grants from the Robert A. Welch Foundation and the Texas Advanced Research Program.

REFERENCES

1. J. G. Tobin et al. unpublished.

2. C. G. Olsen et al., Phys. Rev. B42, 381 (1990).

3. F. M. Mueller et al., Physica B (in press).

4. L. C. Smedskjaer et al., Physica C156, 269 (1988).

5. M. Peter, L. Hoffman and A.A. Manuel, Physica C153-155, 1724 (1988).

6. H. Haghighi et al., J. Phys. Condens Matter 2, 1911 (1990).

7. P. E. A. Turchi et al., J. Phys. 2, 1635 (1990).

8. J. Yu, S. Massida, A. J. Freeman and D. D. Koelling, Phys. Lett. A122, 203 (1987).

9. W. E. Pickett, R. E. Cohen and H. Krakauer, Phys. Rev. B42, 8764 (1990).

10. A. Bansil, P. Mijinarends and L. C. Smedskjaer, Physica C172, 1975 (1990).

11. D. Singh, W. E. Pickett, E. C. von Stetten and S. Berko, Phys. Rev. B42, 2696 (1990).

THE ACUTE SPECTRAL STRUCTURE OF SINGLE-DOMAIN $YBa_2Cu_3O_{6.9}$

J. G. Tobin[A], C. G. Olson[B], F. R. Solal[A], C. Gu[B], J. Z. Liu[C] and M. J. Fluss[A]

[A]Lawrence Livermore National Laboratory, Livermore, CA 94550
[B]Ames Laboratory and Iowa State University, Ames, Iowa 50011
[C]Department of Physics, University of California, Davis, CA 95615

Abstract:

Extraordinarily sharp spectral features at binding energies near 1eV have been observed in the photoemission spectra of untwinned, single crystal $YBa_2Cu_3O_{6.9}$. This is the first observation of such distinctive electronic structure away from the near-Fermi Energy regime. It suggests that the entire valence band electronic structure, not merely the Fermiology, may provide insight into the nature of high temperature superconducting cuprates.

Discussion

High temperature superconductors, such as the cuprate $YBa_2Cu_3O_{6.9}$, have been the subject of intense interest and scientific investigation for the last several years. Of the tools used to investigate these materials, photoelectron spectroscopy has several significant advantages. Its desirable characteristics include the ability to probe the valence band electronic structure directly, with a capability to observe not just the superconducting gap region near the Fermi Energy but also the more tightly bound states as well. However, past investigations have often been hampered by the poor quality of the samples used. (This is merely another example of the truth of the axiom that materials research is only as good as the specimens utilized in the study.) In fact, it seemed that the only sharp, well defined spectral features were near the Fermi Energy and were often associated with the buildup of electronic density near the superconducting gap.[1] This empirical observation was consistent with the results of calculations of the band structure,[2,3] which suggested that the only resolvable structure would be in the simplified region near to the Fermi Energy.

Contrary to these notions, we have observed intense, sharp features in the angle-resolved photoelectron spectra of untwinned, single crystal $YBa_2Cu_3O_{6.9}$. An example of our data is shown in figure 1. The peak at a binding energy of 1eV is a factor of two taller than the other valence features and dwarfs the peak near the Fermi Energy. In this report, we will describe our study of this feature and suggest a possible origin of it.

The photoemission experiments were carried out at the University of Wisconsin Synchrotron Radiation Center on the 1 GeV ring, Aladdin. The work was performed using a high resolution photoemission spectrometer[1] on the Ames-Montana State Beamline.[4] This beamline has two monochromators, but in this study only the lower energy Seya-Namioka was used. The total instrumental resolution of this system has been demonstrated to

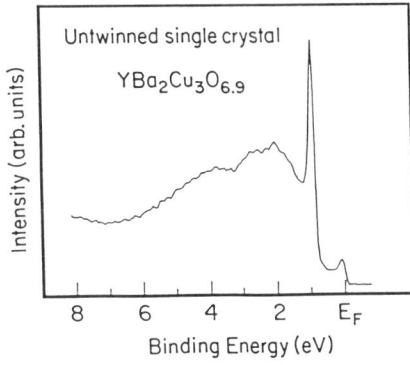

Figure 1. This photoelectron spectrum is of a cleaved YBa$_2$Cu$_3$O$_{6.9}$ sample, with the c axis normal to the surface. The broad scan was taken with hv = 24eV, an energy bandpass of 0.1eV, an angular resolution of ± 1° and an emission direction near $\bar{X}$. The two-dimensional Brillouin zone is defined in terms of the usual three-dimension zone, by collapsing the c-direction. Thus the line X-U becomes $\bar{X}$, Γ - Z becomes $\bar{Γ}$ and Y - T becomes $\bar{Y}$.

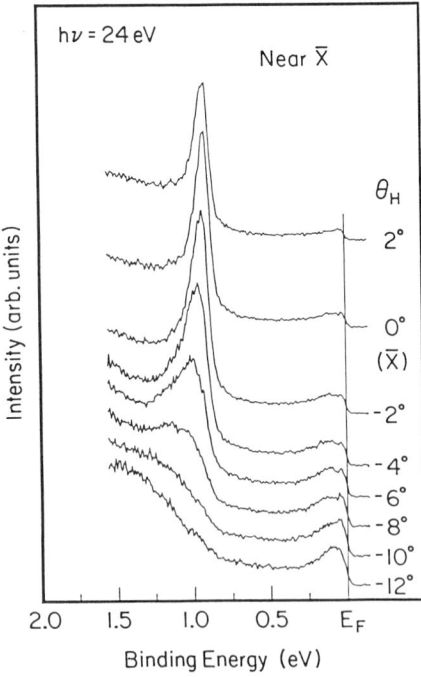

Figure 2. High resolution spectra of YBa$_2$Cu$_3$O$_{6.9}$ near the $\bar{X}$ point in the two-dimensional Brillouin zone. Θ$_H$ is defined in figure 3.

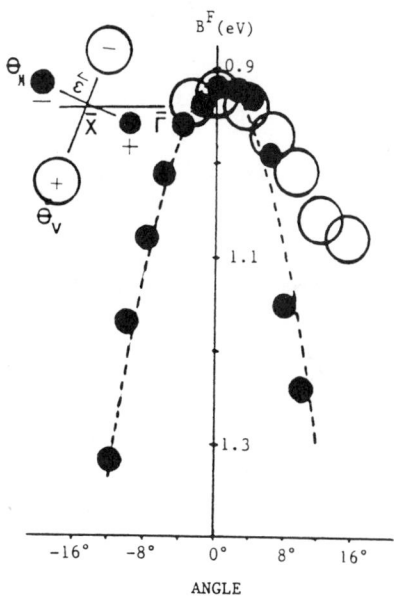

Figure 3. A mapping of energy versus angle (or momentum) is shown here. Two rotations of the analyzer were performed. One (Θ_H) was in the plane of the light polarization and the other (Θ_V) was perpendicular to it. The $\vec{a}$ vector and $\overline{\Gamma}\overline{X}$ axis were canted 20° relative to the plane of light polarization. This is shown schematically in the inset. Filled circles were used for the horizontal rotation (Θ_H) and hollow circles were used for the vertical rotation (Θ_V). Θ_H is greater than zero when rotating toward the sample normal. Θ_V is greater than zero when rotating below the plane of the light polarization. The angle of incidence of the light was about 40° relative to the sample normal. A parabola fit to the horizontal rotation data is also shown for comparison. Approximately 20° of polar rotation separates $\overline{\Gamma}$ from $\overline{X}$, at this photon $h\nu = 24$ eV.

be very high: The energy bandpass can be as low as 25 meV and the angular acceptance is ± 1°. The vacuum during these studies was quite good, with base pressures as low as 3 x 10^{-11} torr. Unless specified otherwise, samples were kept at 20K, including during the cleaving process.[5] All measurements were made upon samples cleaved under ultra high vacuum (10^{-11} torr) conditions. The untwinned, single crystals of YBa$_2$Cu$_3$O$_{6.9}$ were prepared as described elsewhere[6] and sample alignment was performed with both Laue backscattering and in-situ photoemission measurements of high symmetry points. The results presented here are part of a larger study including investigations of the fermiology of YBa$_2$Cu$_3$O$_{6.9}$. (Ref. 7.)

High resolution spectra indicate the feature near 1eV disperses rapidly and broadens as the binding energy increases. An example of the high resolution data is shown in figure 2. Here, the angle and thus momentum parallel to the surface are being varied. The data shown in figures 1 and 2 are essentially raw data: The number of counts normalized to photon flux, after a simple smoothing routine[8] has been performed. No subtractions or background fittings have been performed. Using this approach, it is easy to follow the systematic emergence of the peak at larger angles and the simultaneous effects of peak narrowing and dispersion to lower binding energies. At $\Theta_H = 0$, which is very near to the $\overline{X}$ point in the

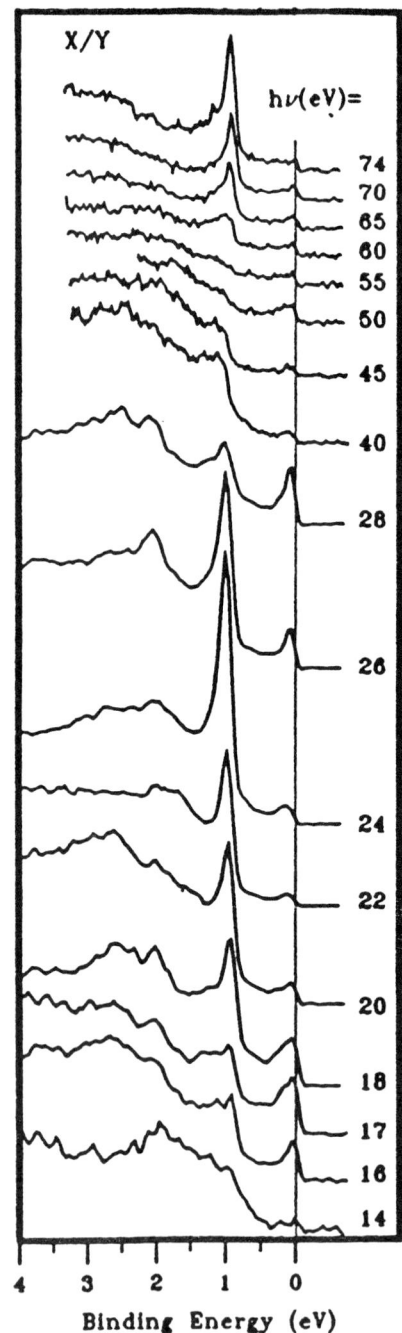

Figure 4. Angle resolved photoemission spectra of the $\bar{X}, \bar{Y}$ point of a twinned crystal, over a wide photon energy range.

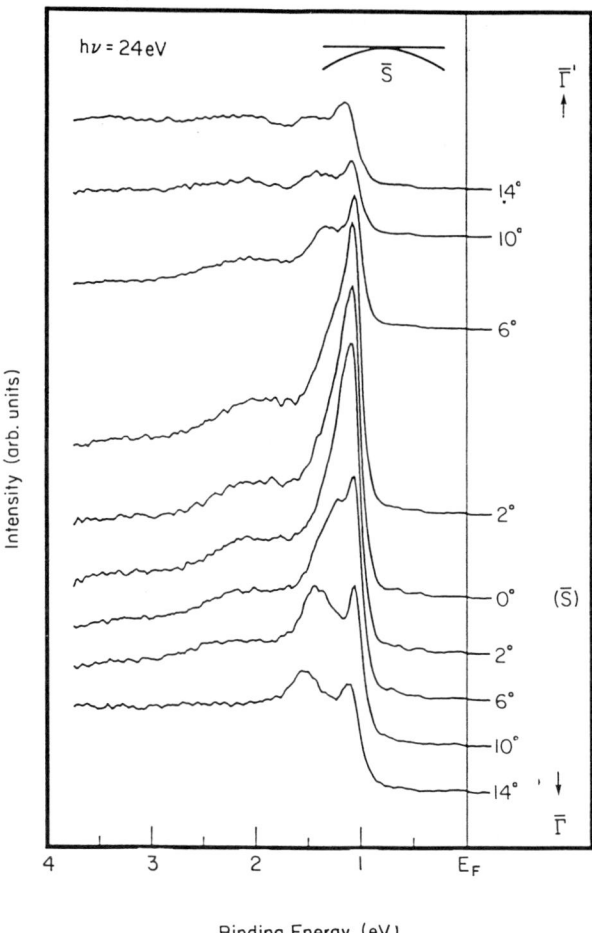

Figure 5. Shown here are angle resolved photoemission spectra along the $\overline{\Gamma}\,\overline{S}$ line of the two-dimensional Brillouin zone. These were taken at hv = 24eV.

two-dimensional Brillouin zone, the peak is at its minimum binding energy of 0.92eV and has a full-width-at-half-maximum of less than 0.15eV. For these spectra, the instrumental resolutions are 30 meV and ±1°

The state disperses to higher binding energy as the sampling position in reciprocal space is moved away from $\overline{X}$. This can be seen in figure 3, which contains a summary of the dispersion of the centroid of the structure for several different angular rotations. It is interesting that the horizontal rotation (approximately along $\overline{\Gamma}$ - $\overline{X}$ - $\overline{\Gamma}$) produces a more rapid dispersion than the vertical rotation (approximately $\overline{S}$ - $\overline{X}$ - $\overline{S}$). It is also of note that the front edge of the peak is always at a binding energy of about 0.9eV. The importance of this will be discussed below. Moreover, the two-dimensional nature of this peak has been established by the essentially identical dispersion curves for hv = 24eV and hv = 74eV. (Ref. 7.)

In general, this feature could be seen at many different photon energies, near the $\bar{X}$ and $\bar{Y}$ points of the Brilloiun zone. (See Figure 4.) The strong peak at $B^F \approx 1eV$ seems to be out of phase with significant emission near the Fermi Energy. The 1eV peak was consistently observed in several different samples with various orientations. However, rotation in a plane perpendicular to the light polarization did not produce the type of spectrum seen in figure 1. That sort of spectral dominance required a large projection of the photon electric vector in the plane of the rotation, suggesting that the parent state is of even symmetry.

The spectral structure discussed above was observed separately near the $\bar{X}$ and $\bar{Y}$ points of the Brillouin zone. Additionally, similar peaks were observed along the $\bar{\Gamma}$ - $\bar{S}$ line of the Brillouin zone. Shown in figure 5 are a series of spectra collected along $\bar{\Gamma}$ - $\bar{S}$. At $\bar{S}$, a large, sharp peak is seen at a binding energy of about 1.15eV. The development of this peak can be followed in the other spectra: Moving toward $\bar{S}$, two separate peaks at $B^F \approx 1.2eV$ and $B^F \approx 1.6eV$ coalesce into the single strong peak at $B^F \approx 1.15eV$ at $\bar{S}$. This general behavior is a confirmation of our sample alignment: increased degeneracy can be expected at high symmetry points such as $\bar{S}$ and the coalescence of different bands into a single peak is indicative of a high symmetry point. This condensation of peaks suggests a slightly different analysis of the data in figure 2: Two peaks are combining, one at a larger binding energy, moving down to $B^F \approx 0.92eV$ and another weaker feature (seen as a shoulder in the 4°, 6° and 8° spectra) at a lower binding energy. This lower binding energy feature appears to be non-dispersive, giving rise to the constant binding energy front edge at $B^F \approx 0.9eV$ of the peak.

It is important to note again that the binding energy of the structure at $\bar{X}$ and $\bar{Y}$ is not the same as at $\bar{\Gamma}$ and $\bar{S}$: Near $\bar{X}$ and $\bar{Y}$, the binding energy is 0.92eV, at $\bar{\Gamma}$ it is 1.3eV and at $\bar{S}$ it is 1.15eV. So while these peaks all appear to have a common origin, they are not identical and depend upon the position in the two-dimensional Brillouin zone.

The origin of the "1eV" peaks appears to be strongly connected to the CuO planes. First, they are observed at both the $\bar{X}$ and $\bar{Y}$ points as well as $\bar{\Gamma}$ and $\bar{S}$. This suggests that they are not related to the linear CuO chain states. Second, they appear at the top edge of the high electronic density regime for the CuO planes,[2] which exhibits what might be thought of as the remnant of an insulating gap beginning at $B^F \approx 1eV$. Thus, it is tempting to think of the "1eV" state as having split off from the usual CuO planes states due to some perturbation. This in many respects parallels the formation of Tamm surface states from bulk bands, but the relationship is less clear here. Because the entire photoemission experiment is relatively surface sensitive and the layered structure of these materials causes a two-dimension electronic structure, the assignment as some sort of surface state is less than completely convincing.[9] Additionally, it seems likely that the state will have O2p character, based upon angle-integrated photoemission study[10] and Cu 3p character, as indicated by its resonance behavior between $h\nu = 70eV$ and $h\nu = 74eV$. (Ref. 7.)

Summary

Remarkably sharp and intense electronic features away from the Fermi Energy have been observed in cleaved, untwinned, single-crystal $YBa_2Cu_3O_{6.9}$. They appear to be related to the electronic density-of-states of the CuO planes. The importance of utilizing the best possible samples and highest available resolution in such experiments is illustrated by the acute nature of these spectral peaks.

Acknowledgement

Work performed under the auspices of the U.S. Department of Energy by the Lawrence Livermore National Laboratory under contract No. W-7405-ENG-48 and Ames Laboratory under contract number W-7405-ENG-82. The synchrotron radiation investigations were performed at the University of Wisconsin Synchrotron Radiation Center.

References

1. C. G. Olson, R. Liu, A.-B. Yang, D. W. Lynch, A. J. Arko, R. S. List, B. W. Veal, Y. C. Chang, P. Z. Jiang and A. P. Paulikas, Science 245, 731 (1989).
2. S. Massida, J. Yu, A. J. Freeman and D. D. Koeling, Phys. Lett. A 122, 198 (1987); J. Yu, S. Massida, A. J. Freeman and D. D. Koeling, Phys. Lett. A 122, 203 (1987).
3. W. E. Pickett, R. E. Cohen and H. Krakauer, Phys. Rev. B 42, 8764 (1990).
4. C. G. Olson, Nucl. Instrum. Meth. A266, 205 (1988).
5. A. J. Arko, R. S. List, R. J. Bartlett, S. W. Cheong, Z. Fisk, J. O. Thompson, C. G. Olson, A.-B. Yang, R. Liu, C. Gu, B. W. Veal, J. Z. Liu, A. P. Paulikas, K. Vandervoort, H. Klaus, J. C. Campuzano, J. E. Schirber and N. C. Schinn, Phys. Rev. B 40, 2268, (1989).
6. J. Z. Liu, M. D. Lan, P. Klavins and R. N. Shelton, Phys. Lett. A144, 265 (1990).
7. J. G. Tobin, C. G. Olson, C. Gu, D. W. Lynch, J. Z. Liu, F. R. Solal, M. J. Fluss, R. H. Howell, J. C. O'Brien, H. B. Radousky and P. A. Sterne, submitted to Phys. Rev. B-15, 1991.
8. A. Savitsky and M. J. E. Golay, Anal. Chem. 36, 1627 (1964) and J. Steinier, Y. Termonia and J. Deltour, Anal. Chem. 44, 1906 (1972).
9. R. Claessan, G. Mante, A. Huss, R. Manske, M. Skibowski, T. Wolf and J. Fink, preprint, 1990.
10. B. W. Veal, J. Z. Liu, A. P. Paulikas, K. Vandervoort, H. Claus, J. C. Campuzano, C. G. Olson, A. -B. Yang, R. Liu, C. Gu, R. S. List, A. J. Arko, R. J. Bartlett, "Photoelectron Spectroscopy Study of $YBa_3Cu_3O_4$ with Varied Oxygen Stoichiometry: Possible Evidence for Strong-Coupling Superconductivity," Physica C 158, 276 (1989).

INFRARED STUDY OF THE ORDER PARAMETER IN $YBa_2Cu_3O_7$: ANALOGY TO SHORT RANGE ORDERING IN ITINERANT FERROMAGNETS

D. van der Marel[a,b], A. Wittlin[a,c,d], H.-U. Habermeier[a] and D. Heitmann[a]

[a]Max-Planck-Institut für Festkörperforschung, Heisenbergstrasse 1, D-7000 Stuttgart 80, Federal Republik of Germany

[b]Delft University of Technology, Faculty of Applied Physics Lorentzweg 1, 2628 CJ Delft, The Netherlands

[c]Institute of Physics PAN PL, 02-668 Warszawa, Poland research project RPBP 01.9

[d]Katholieke Universiteit Nijmegen, Toernooiveld 1, 6525 ED Nijmegen, The Netherlands

INTRODUCTION

We report on detailed measurements of the temperature dependence of the infrared reflectivity spectra of $YBa_2Cu_3O_7$. It is shown that the prominent edge[1] at $8k_BT_c$ shows practically no energy shift as a function of temperature, whereas the oscillator strength of this feature depends strongly on temperature with a marked critical behavior at the superconducting transition temperature. This behavior coincides with the temperature dependence of the superfluid fraction, which was determined from the spectra using two different methods. A precursor at the same energy position persists at temperatures above T_c. If the edge can be interpreted as the characteristic energy required to break up a pair, we arrive at the conclusion that the gap itself can not be considered as an order parameter in this case. The superfluid density on the other hand changes approximately as $1 - (T/T_c)^4$. We point out that a remarkable analogy exists to the situation in itinerant magnetism, where short range order persists above the Curie temperature resulting in an exchange splitting that does not vanish[2] at T_c.

EXPERIMENTAL RESULTS

C-axis oriented thin films with thicknesses between 2000 Å and 4000 Å were prepared using the pulsed laser deposition technique set up for *in situ* Y-Ba-Cu-O thin film growth. Experimental details have been published elsewhere.[3] Characterization of the films using scanning electron microscopy, optical microscopy, and X-ray diffraction revealed, that the films are single phase and c-axis oriented with a smooth surface. The substrates are single crystals of $SrTiO_3$ and twinned crystals of $LaAlO_3$ ([100] surface) with a wedged backside. Transmission electron microscopic measurements demonstrated a well defined substrate/film interface on an atomic scale. T_c of the samples was 89 K.

Reflectivities were measured in the range of 100 to 7000 cm^{-1}. Details on the measurement procedure, as well as on the data handling have been reported elsewhere[4]. The spectra were taken at 5 K intervals between 20 and 150 K, allowing a detailed study of the temperature dependence of the reflectivity. Some of the reflectivity curves are displayed in Fig. 1. A Kramers-Kronig analysis was used to determine the complex dielectric function. In Fig. 2a we show the resulting optical conductivity for a number of temperatures. The spectra are dominated by an absorption edge which rises steeply between 400 and 800cm^{-1}. The nature of this edge has been extensively discussed[5]. However, as we will see below, the temperature dependence of the oscillator strength of this edge strongly suggests an intimate relationship to the behavior of the order parameter. On the other hand the phenomenology, especially the temperature dependence, deviates in an essential way from BCS theory. We have to add here, that a small precursor of the same edge structure persists already above T_c, as one can see from the reflectivity plots. This is more difficult to recognize in the conductivity plots due to the steep Drude background. In Ref. 4 this is presented in more detail.

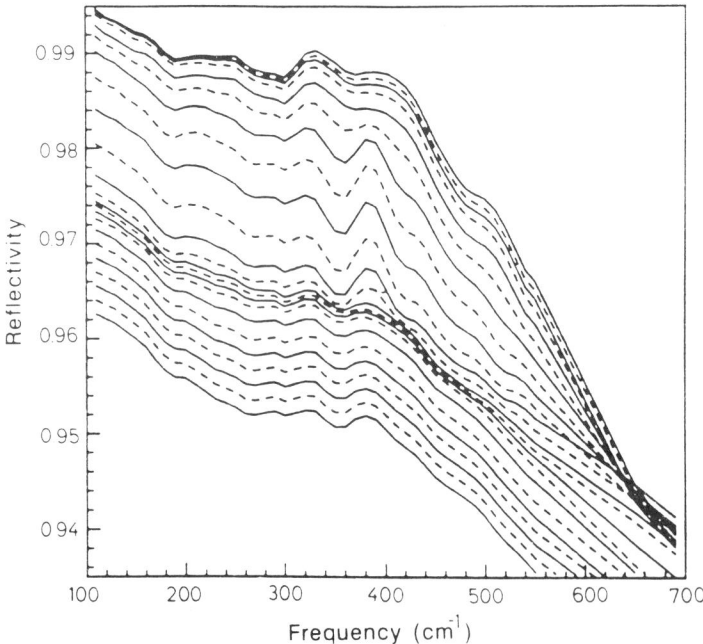

Fig. 1. Experimental reflectivity data of a thin film of $YBa_2Cu_3O_7$ on $SrTiO_3$ for various temperatures. From top to bottom: $T = 40K$ to $150K$ with steps of $5K$.

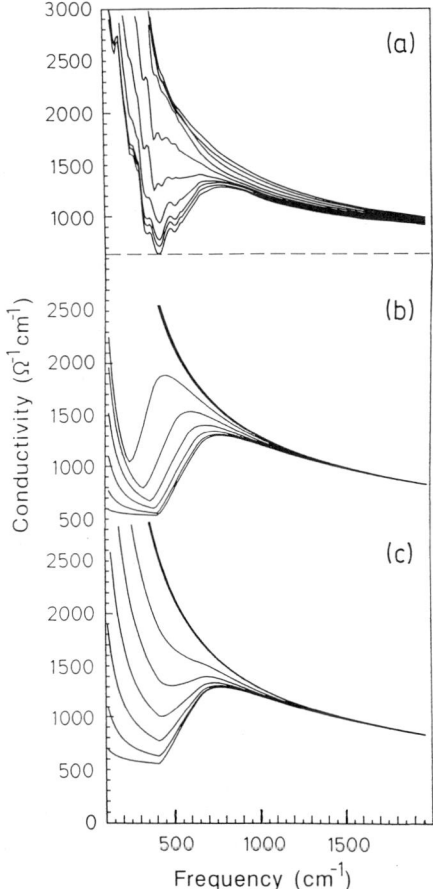

Fig. 2. Conductivity versus frequency :
(a) Determined from the reflectivity curves of Fig. 1 using Kramers-Kronig analysis,
(b) Theoretical plot using BCS theory,
(c) *Ibid.* using the two-fluid model.
From top to bottom: $T = 120$ to $30K$ with increments of $10K$.

The superfluid fraction was determined from the spectra in two different ways: The first method is to fit the following expression for the reflectivity at foton energies hν *below* 2Δ to the experimental data:

$$R = 1 - \sqrt{\frac{\nu}{\nu_{HR}}}\sqrt{\frac{\nu(\sqrt{\nu^2+\nu_X^2}-\nu_X)}{\nu^2+\nu_X^2}}$$

with

$$\nu_{HR} \equiv \frac{\nu_p^2(1-n_s)}{8\Gamma} \text{ and } \nu_X \equiv \frac{n_s\Gamma}{1-n_s}$$

where n_s is the superfluid fraction, Γ is the decay-rate of the Drude response of the non-condensed quasiparticles, and ν_p is the bare plasmon frequency. The latter parameter was obtained independently and corresponds to about 11000 cm^{-1}. The former two parameters both depend on temperature and were obtained by fitting the reflectivity curves below 300 cm^{-1} at each temperature. The above expressions are derived in Ref. 4 and follow from generalizing the Hagen-Rubens formula to the superconducting state. The second method is to apply the Glover-Ferrel-Tinkham rule[6], which relates the superfluid fraction to the missing area in the optical conductivity of the superconducting state. Both methods give the same result, which is displayed in Fig. 3a.

DISCUSSION

Also displayed in Fig. 3a are the temperature dependencies of the height of the $8k_BT_c$ edge as well as of its energy position, both relative to the normal state at 95 K. The solid curve is the empirical relation $n_s = 1 - (T/T_c)^4$, which is known to be a good approximation both for the description of the superfluid fraction of superfluid helium, as well as for the London penetration depth of YBa$_2$Cu$_3$O$_7$. The fact that the superfluid density and the oscillator strength of the $8k_BT_c$ edge behave similarly, strongly suggests that this edge is intimately related to the superconducting order parameter. In weak and strong coupling BCS theory the absorption threshhold corresponds to twice the superconducting gap parameter Δ in the former case [7], or to a Holstein process tied to 2Δ in the latter case[8,9]. In either case one expects the edge to shift considerably when T_c is approached, and close to the transition temperature Δ is proportional to $\sqrt{(n_s)}$. So in this respect there is no one-to-one correspondence between the $8k_BT_c$ edge structure and a BCS gap. To illustrate this point we display in Fig. 2b a calculation using weak coupling BCS theory with finite scattering, assuming that there is an even distribution of Δ's between 200 and 325 cm^{-1}, and that the gap scales with temperature in the same way as in weak coupling BCS theory. This absence of a temperature dependent shift becomes even more intriguing in view of the further observation that even *above* T_c a small but significant signature persists at the same energy. This has been used as an argument against the interpretation that this feature is related to the order parameter. However, as we see in Fig. 3a, the fact that the height of the edge follows almost exactly the same behavior as the superfluid density supports such an interpretation. Moreover, such a behavior, as was recently pointed out by Collins et al.[10], is also observed in the temperature dependence

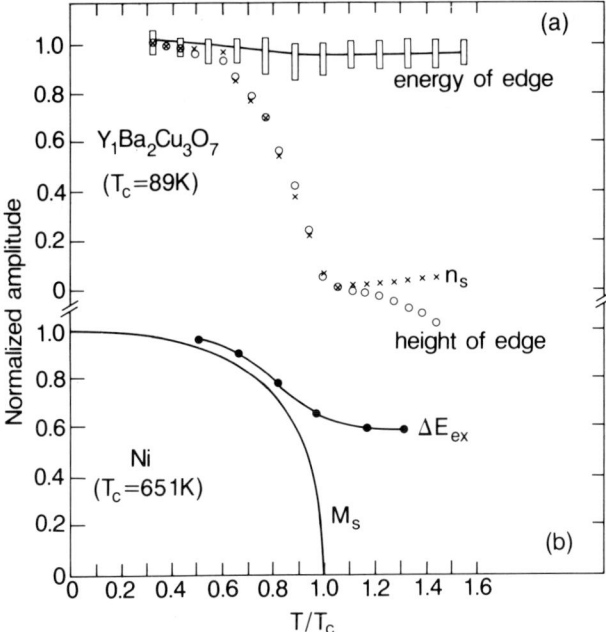

Fig. 3. Normalized amplitudes of the order parameter and energy splittings of a superconductor and an itinerant ferromagnet.
(a) Crosses: Superfluid fraction of $YBa_2Cu_3O_7$. Open circles: Height of the threshold at 400 cm^{-1}. The height was normalized to the conductivity at 95 K. The negative value at $T > T_c$ reflects a weak edge structure above the superconducting transition. Bars: Energy of the edge. The length of the bars reflects the estimated accuracy.
(b) Closed circles: The experimentally determined exchange splitting of the Ni 3d bands, using photo-electron spectroscopy. The results were taken from Ref. 2. The solid curve is a guide to the eye. Lower solid curve: Saturation magnetization of Ni.

of the NMR relaxation rate and in the penetration depth. Hence, the temperature dependence of the height of the edge reflects a universal behaviour observed in various other properties related to the superfluid fraction.

Although a microscopic understanding of this behavior is lacking, and a fit to either strong coupling theory or weak coupling theory doesn't seem to work at all temperatures, a fairly good fit can be made with the following empirical expression for the dielectric function[4]

$$\epsilon(\nu) = f_s(T)\epsilon_s + (1 - f_s(T))\epsilon_n$$

where $f_s(T)$ is the temperature dependent superconducting component, which we take proportional to $1 - (T/T_c)^4$, ϵ_s is the dielectric function of the superconducting component, and ϵ_n is the dielectric function of the normal component. In Ref. 4 it is shown that these dielectric functions of the two components can be expressed using simple assumptions of weak coupling theory with a large and temperature independent gap for the superconducting component and Drude theory for the normal component. The resulting curves are displayed in Fig. 2c. The number of parameters used is small, and the procedure is in practice very similar to the procedure of Collins et al.[10], who used in their calculations the experimental dielectric functions at T=30 and 100 K for the superconducting and the normal components respectively. The additional information that one obtains from modelling the dielectric functions in the way we did, is that at T=0 the observed conductivity corresponds reasonably well to what one would expect from weak coupling BCS theory with finite scattering. The only surprise is, that one would expect a T_c of about 160 K if the weak coupling result $2\Delta/k_b T_c = 3.5$ were still to hold[11]. In other words: The edge that we observe below T_c has the shape that one in fact expects to find given the normal state decay rate and plasma frequency. This also means that *if* there is a gap in the range between 200 and 500 cm^{-1} where one reasonably can expect it to occur, it should look very much like the edge that we and others actually see at about 400 cm^{-1}. The phenomenology underlying the properties of the superconducting condensate calls for an analogy with other strongly correlated systems near the phase transition. In that context we like to discuss an interesting analogy of the present system to itinerant ferromagnetism of materials with a strong on-site exchange interaction, where again an interplay between short range interactions and long range coherence takes place.

One of the simplest mean field treatments of a ferromagnet is the Stoner model. In this model an overall magnetization is created via an exchange splitting, resulting in a shift in energy of the spin-up bands relative to the spin-down bands[12]. The magnetization then follows from a different filling of the spin-split bands. Clearly in such a model the magnetization and the exchange splitting are proportional to each other. This is not unlike the situation in an ordinary BCS superconductor, where the superconducting gap parameter Δ is proportional to $\sqrt{(n_s)}$. Interestingly the transition metals on the right side of the 3d series do not behave like this, as has been beatifully demonstrated by Eastman et al. who used photo-emission to determine the exchange splitting of the Ni 3d bands as a function of temperature[2]. What is observed in those cases is, that the exchange splitting persists *above* the Curie temperature. This has been understood as a consequence of the fact, that there is still short range magnetic order above T_c, although the long range order is lost. The results of Collins et al. were replotted in Fig. 3b to facilitate comparison to our data. The close similarity of both plots seems to indicate, that the cuprate high T_c superconductors are characterized by a phase transition where some short range superconducting order persists above T_c, although the long range order, as reflected in the superfluid density, disappears more or less in the usual way at T_c. So far no bulk superconductors are known that exhibit a phase transition between states with long range order and short range order. On the other hand, this is a well studied phenomenon in 2-dimensional arrays of Josephson junctions[13]. In these systems a Kosterlitz-Thouless transition takes place *below* the transition temperature of the superconducting islands.

There are at least two factors which distinguish the high T_c cuprates from ordinary superconductors, and which may favor the rather unusual phenomenology outlined above. The first is the scale of the coherence length, which is of comparable magnitude as the average distance between charge carriers. The second is the layered two-dimensional nature of these materials. Due to the small coherence lenght Cooper-pairs have much smaller overlap than in classical superconductors, where the coherence length is of the order of thousands of lattice spacings. In the extreme limit of point-like Cooper-pairs with a vanishing coherence length the system consists of a system of Bose-condensed pairs. The phase-transition in such a situation would be a Bose-Einstein condensation of pre-existing pairs, which means that above T_c all charge carriers still exist in pairs, but the phase relationships are lost. This is the situation envisaged in theories of local pair condensation[14,15,16]. The high T_c superconductors are neither in the BCS limit of large Cooper-pair overlap, nor in the Bose-Einstein condensation limit of vanishing overlap between pairs. Instead estimates of the coherence length at zero temperature correspond to a situation where the Cooper-pairs are still partly overlapping. This is a situation which has not been fully understood at present. A much stronger pairing would still be required to remove all spatial overlap between pairs. It is interesting to note however, that theories based on the negative U Hubbard model place the highest T_c's in the region of intermediate U[16]. The layered two-dimensionality probably plays an important role here, as it would facilitate a phase transition of the Kosterlitz-Thouless type[17].

It seems to be interesting to further investigate the possible existence of local superconducting order, both experimentally and theoretically. This could help to understand the phenomenological similarities between itinarant ferromagnetism of the late 3d transition metals and superconductivity in materials with a small or intermediate ratio of the coherence length and the Fermi wavelength. In a recent theoretical study of the problem of a mixture of negative U centers and a fermionic wide conduction band Bar-Yam demonstrated that indeed a situation can exist where non-condensed local pairs exist above T_c in addition to wide band electrons[18]. Below the transition temperature both Bose condensation of local pairs occurs accompagnied by the formation of a BCS-like gap in the wide band part of the material.

CONCLUSIONS

The temperature depedency of the superfluid component and the position of the gap as observed in $YBa_2Cu_3O_7$ as observed with infrared spectroscopy is demonstrated to show non-BCS like behavior. The gap position does not shift with temperature, while the superfluid component is well approximated with a $1 - (T/T_c)^4$ behavior. A small signature at the gap position below T_c persists above T_c. It is discussed that this behavior may indicate that we have a situation analogous to what is observed in certain itinirant ferromagnets, where the long range order vanishes at T_c, whereas the short range order persists above the transistion temperature resulting in a non-vanishing exchange splitting.

REFERENCES

1. Z. Schlesinger, R. T. Collins, F. Holtzberg, C. Feild, U. Welp, Y. Fang and J. Z. Liu, Phys. Rev. Lett. 65: 801 (1990).
2. D. E. Eastman, F. J. Himpsel and J. A. Knapp, Phys. Rev. Lett. 40: 1514 (1978).
3. H.-U. Habermeier and G. Mertens, Physica C 162-164: 601 (1989).
4. D. van der Marel, M. Bauer, E. H. Brandt, H.-U. Habermeier, D. Heitmann, W. König, and A. Wittlin, Phys. Rev. B, in print; D. van der Marel, H.-U. Habermeier, D. Heitmann, W. König, and A. Wittlin, Physica C, in print.

5. R. T. Collins, Z. Schlesinger, F. Holtzberg, and C. Feild, Phys. Rev. Lett. 63: 422 (1989); G. A. Thomas, J. Orenstein, D. H. Rapkine, M. Capizzi, A.J. Mills, R. N. Bhatt, L. F. Schneemeyer, and J. V. Waszczak, Phys. Rev. Lett. 61: 1313 (1988); A. Wittlin, L. Genzel, M. Cardona, M. Bauer, W. König, E. Garcia, M. Barahona, and M. V. Cabanas, Phys. Rev. B 37: 652 (1988); A. D. Wieck, E. Batke, U. Merkt, D. Walker, and A. Gold, Solid State Commun. 69: 553 (1989); J. Schützmann, W. Ose, J. Keller, K. F. Renk, B. Roas, L. Schultz, and G. Saemann-Ischenko, Europhys. Lett. 8: 679 (1989); Z. Schlesinger, R. T. Collins, F. Holtzberg, C. Feild, and A. Gupta, Phys. Rev. B 41: 11237 (1990); T. Timusk, S. L. Herr, K. Kamaras, C. D. Porter, D. B. Tanner, D. A. Bonn, J. D. Garrett, C. V. Stager, J. E. Greedan, and M. Reedyk, Phys. Rev. B 38: 6683 (1988); K. Kamaras, S. L. Herr, C. D. Porter, N. Tache, D. B. Tanner, S. Etemad, T. Venketasan, E. Chase, A. Inam, X. D. Wu, M. S. Hegde, and B. Dutta, Phys. Rev. Lett. 64: 84 (1990).
6. R. E. Glover and M. Tinkham, Phys. Rev. 104: 844 (1956); M. Tinkham and R. E. Ferell, Phys. Rev. Lett. 2: 331 (1959).
7. M. Tinkham, "Superconductivity", McGraw-Hill, New York, 1965.
8. N.E. Bickers, D.J. Scalapino, R.T. Collins, and Z. Schlesinger, Phys. Rev. B 42: 67(1990).
9. W. Lee, D. Rainer, and W. Zimmermann, Physica C 159: 535 (1989); P. B. Allen, and D. Rainer, Nature 349: 396 (1991).
10. R. T. Collins, Z. Schlesinger, F. Holtzberg, and C. Field, Phys. Rev. B, in print; Z. Schlesinger, these proceedings.
11. P.B. Littlewood and C. M. Varma, "Phenomenology of the normal and superconducting states of a marginal Fermi liquid.", presented at "Theory of high T_c syperconductivity", Dubna, U.S.S.R., july 1990. These authors propose that in high T_c materials the reflectivity feature corresponds to 4Δ, using clean limit/strong coupling arguments with an electronic excitation spectrum providing the 'glue' instead of phonons. However, the complete absence of any structure at 2Δ only results from treating the problem in the extreme limit, which is not the case according to our results.
12. An overview of the physics of long range order, both in superconductivity and in magnetism is given in "Long range order in solids" by R. M. White and T. H. Geballe, Edited by H. Ehrenreich, F. Seitz and D. Turnbull, Supplement 15, Academic Press, New York (1979).
13. H. J. van der Zant, H. A. Rijken, and J. E. Mooij, Journal of Low Temperature Physics 79: 289 (1990).
14. D. Emin, and M.S. Hillery, Phys. Rev. B 39: 6575 (1989).
15. L. J. de Jongh, Physica C 152: 171 (1988).
16. R. Micnas, J. Ranninger, and S. Robaszkiewicz, Rev. of Modern Phys. 62: 113 (1990).
17. V. Cataudella, and P. Minnhagen, Physica C 166: 442 (1990).
18. Y. Bar-Yam, Phys. Rev. B 43: 359 (1991); *ibid.* 2601 (1991).

THERMOELECTRIC POWER AND RESISTIVITY MEASUREMENTS OF $Pr_xY_{1-x}Ba_2Cu_3O_{7-\delta}$ UP TO 1200 K AND A THEORETICAL ANALYSIS

B. Fisher,[a] J. Genossar,[a,b] L. Patlagan,[a] G. Koren,[a] J. Ashkenazi[c] and C. G. Kuper[a]

[a]Physics Department and Crown Center for Superconductivity,
Technion — Israel Institute of Technology, 32000 Haifa, Israel

[b]C.N.R.S.-S.N.C.I.-M.P.I., 166X, 38042 Grenoble Cedex, France

[c]Physics Department, University of Miami, Coral Gables, FL 33124, U.S.A.

INTRODUCTION

Much recent research has been devoted to the $Pr_xY_{1-x}Ba_2Cu_3O_{7-\delta}$ system ("PYBCO"), for a wide range of values of x and δ, in both its normal ("n-PYBCO") and superconducting ("s-PYBCO") states. The reason for the interest in this system is the exceptional role of Pr in the "123" structure. With increasing x, the critical temperature T_c for superconductivity (SC) falls, until for $x \gtrsim 0.6$, its SC is quenched. The experimental results and two proposed mechanisms for the quenching of SC in PYBCO have been discussed in a number of recent papers (see Refs. 1-6 and references quoted therein).

In this paper, we study the changes in the electronic structure of n-PYBCO with changing Pr content x. Some electronic properties,[1] are consistent with the view that the valency of Pr in PYBCO is close to 4+. Moreover, these properties change with increasing x in much the same way as they do with falling oxygen content in $YBa_2Cu_3O_{7-\delta}$ (YBCO), suggesting that the Pr "swallows" conducting holes, and thus apparently confirming the valency 4+. However, there is another class of properties,[7-9] which seem to indicate that the valency of Pr in PYBCO remains 3+.

The experimental results which we present here show that the effect of Pr on the resistivity ρ and on the thermopower S of n-PYBCO diminishes with increasing temperature T. At 1000 K, the transport properties of PYBCO are very similar to those of YBCO. This suggests that at high temperatures the effective valency of Pr is very close to 3+. We can thus understand why, in contrast to Ce or Tb, Pr forms the "123" structure: at the high temperature needed for preparation of the material, Pr, like Y, is trivalent. The observed weak dependence of the transport properties on x at high temperature is in marked contrast with the strong x dependence at low temperatures, where 4+ valency seems to prevail.

EXPERIMENT

Ceramic samples of PYBCO were prepared by the standard procedure[2,6], but with special care to ensure their homogeneity. Three types of ceramic samples were prepared for each composition: (a) pellets for gravimetric measurements, (b) pellets with embedded gold wire contacts for resistivity measurements at low temperatures, and (c) long bars with embedded gold wires for simultaneous high-temperature measurements of ρ and S. The pellets were formed by uniaxial compression, and the bars in an isostatic press.

Thin films of $PrBa_2Cu_3O_{7-\delta}$ (PBCO) were grown by KrF excimer laser ablation-deposition on to (100) substrates of $SrTiO_3$. The conditions for deposition were the standard ones for YBCO films.[10] All the films were prepared starting from a single ceramic PBCO pellet taken from the batch of sample CN0 (see below). X-ray diffraction was used to check phase purity in both the ceramic samples and the films, as well as the epitaxy in the films. Electrical contacts on the films were formed by evaporating gold or silver.

Above room temperature, two kinds of measurement were made on ceramic samples :
1. "Slow measurements" at temperatures $T \gtrsim 550° - 600°C$, under equilibrium conditions [oxygen pressure $P(O_2) = 1$ or 0.01 bar]. The oxygen content of the samples was estimated from T and $P(O_2)$, using the available phase diagrams for YBCO[11] and PBCO.[12]
2. "Fast measurements" below $\sim 550°C$. In this temperature range the oxygen equilibration time becomes very long; hence "slow" measurements were not practicable. Instead, the measurements were made as quickly as possible, in order to preserve the frozen-in oxygen content.

For low-temperature measurements ($T \lesssim 300$ K), sets of ceramic samples with and without contacts were equilibrated together and quenched to room temperature. Changes in oxygen stoichiometry of the samples were determined from the weight loss of the accompanying contactless samples. We estimate that gravimetric measurements measure changes of δ to better than ± 0.01. Thin-film samples were also studied over this same temperature range. The films were oxygenated at $\sim 400°C$, and $P(O_2) = 1$ bar, and then cooled very slowly to room temperature. From the phase diagram,[12] we estimate that their oxygen deficit is $\delta \lesssim 0.04$.

High temperatures

In Figs. 1 and 2 we plot the resistivity ρ and the absolute thermopower S against temperature T for ceramic samples with $x = 0.2, 0.4, 1.0$, under $P(O_2) = 1$ and 0.01 bar respectively. Note the different temperature ranges for the "slow" (equilibrium) and the "fast" (frozen-in oxygen composition) measurements. We also include data obtained by other laboratories.[6,13] A remarkable result is that variation of the Pr content x hardly affects the high-temperature transport properties of PYBCO, in contrast to the dramatic x-dependence at low temperatures. In Fig. 1, above 1150 K, the curves for all samples with different Pr content converge and while the graph for $\rho(x=1)$ rises slower than for $\rho(x=0)$, the spread of the values of ρ for all samples is within a factor of 1.5. For $S(x=1)$ there is an indication of a shallow maximum at high temperatures, around 1100 K.

"Fast" measurements at $P(O_2) = 0.01$ bar show trends similar to those obtained under $P(O_2) = 1$ bar. The differences in the absolute values of ρ and S are due to changes in oxygen content.[11,12] For $x < 1$, $d\rho/dT > 0$ and $S(T)$ exhibit plateaux in the vicinity of room temperature. For $x = 1$, the plateau regime of S shrinks to a dip between 500 and 800 K with $S \cong 110$ μV/K. The absolute magnitude of $d\rho/dT$ decreases with increasing temperature but its sign remains negative. From the phase diagram,[12] and from S, we estimate that for PBCO, under these conditions, $\delta \sim 0.50$. The curves of S vs. T for low temperatures, based on data from other groups, join our curves reasonably well around 300 K.

Low temperatures

Below room temperature we have measured only $\rho(T)$. In this regime we found very varied behavior. For ceramic samples, we present separately the cases of (a) low resistivity and (b) high resistivity. (a) Where the maximum value of ρ did not exceed $10^{-2} \Omega$cm, ρ was plotted vs. T on a linear scale (Fig. 3). (b) For $\rho \gtrsim 10^{-2} \Omega$cm plots of $\log \rho$ vs. $1/T$ seemed more suitable (Fig. 4). For ceramic and thin-film measurements with $x = 1$, plots of $\log \rho$ vs. $T^{-1/3}$ are found appropriate (Fig. 5).

Curves for $x=0.2$, $\delta \sim 0$ and for $x=0.4$, $\delta \sim 0$, shown in Fig. 3, are similar to the corresponding plots in Ref. 13. Here T_c falls with increasing x, while the normal-state resistivity increases with T and x. In this limited range of T, x and δ, the resistivity ρ appears to depend only on T and the single parameter $q \equiv (x + \delta)$. For $x=0.2$ and $\delta=0.4$, $d\rho/dT$ becomes negative below ~ 150 K. In this temperature range, both ρ and $|d\rho/dT|$ increase appreciably for small increases in q (see the upper two curves in Fig. 3).

Table I summarizes the results for ceramic samples with large ρ (and $|d\rho/dT|$). In Fig. 4, $\log \rho$ is plotted vs. $1/T$ for some of the samples. For many of the samples, the observed behavior is consistent with a temperature-activated transport interpretation: for large δ, the material is a semiconductor with a single activation energy – see for example, graphs b, c and f in Fig. 4. In other cases two distinct slopes, below and above T_b could be distinguished in the $1/T$ graph [e.g. curve e and run g (not shown)]. More complicated is case d, which shows

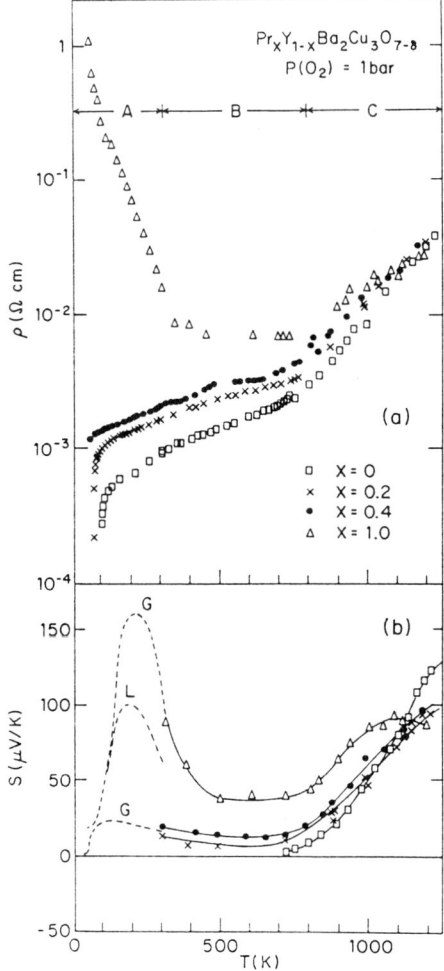

FIG. 1. Thermopower S and resistivity ρ of $Pr_xY_{1-x}Ba_2Cu_3O_{1-\delta}$ as function of temperature under $P(O_2) = 1$ bar. In region C – "slow" (equilibrium) measurements (see text). In region B – "fast" measurements. In region A – data taken from literature: G – Ref. 13; L – Ref. 6.

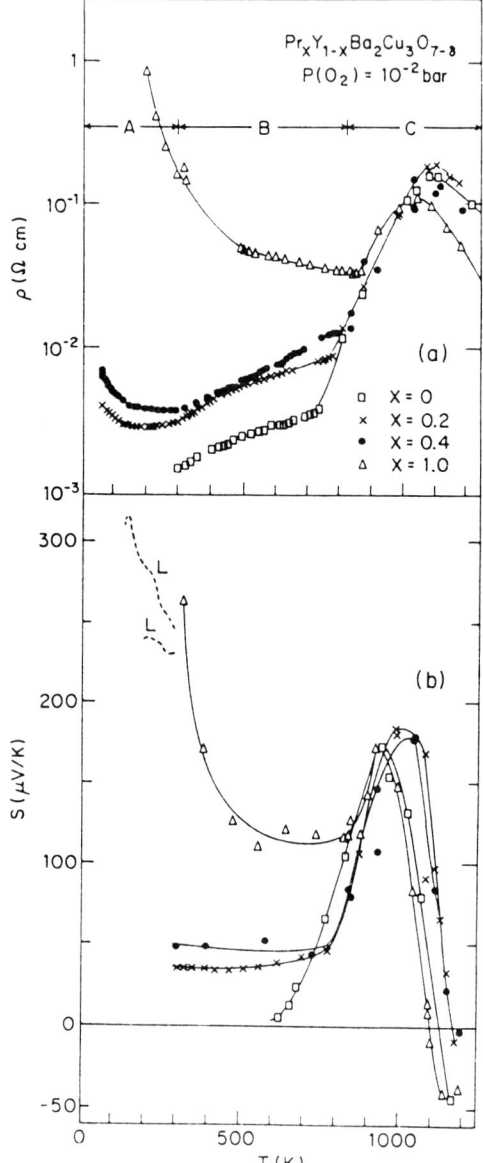

FIG. 2. S and ρ of $Pr_xY_{1-x}Ba_2Cu_3O_{1-\delta}$ as function of temperature under $P(O_2) = 0.01$ bar. In region C – "slow" measurements. In region B – "fast" measurements. In region A: L – data taken from Ref. 6.

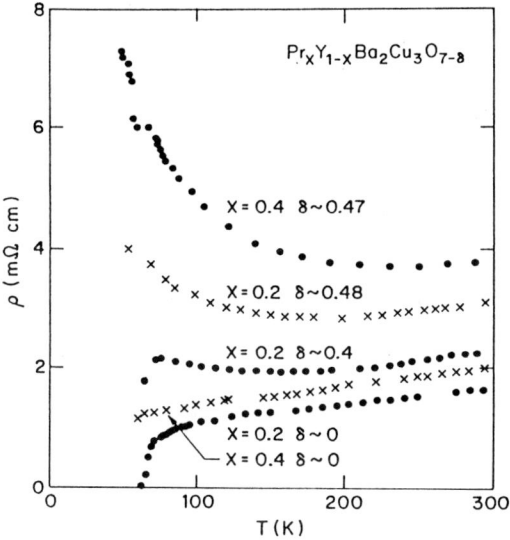

FIG. 3. Resistivity of $Pr_xY_{1-x}Ba_2Cu_3O_{1-\delta}$ as function of temperature.

TABLE I. Characterization of the PYBCO samples used in "low temperature" resistivity measurements. x and δ indicate the Pr and O contents respectively (see text). E_H and E_L are activation energies taken from the $\log \rho$ vs. $1/T$ graphs above and below T_b respectively. The results of runs a–f are shown in Fig. 4.

run	x	δ	E_H/k	E_L/k	T_b
a	1.0	~0	VRH		
b	1.0	0.40	~1000 K	—	—
c	1.0	0.92	1000 K	—	—
d	0.0	0.98	VRH (?)		
e	0.2	0.82	2400 K	1600 K	170 K
f	1.0	1.02	~4000 K	—	—
g	0.0	0.62	1000 K	700 K	130 K
h	0.4	0.86	VRH (?)		

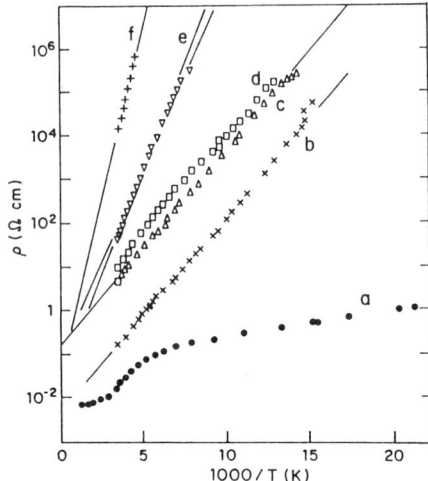

FIG. 4. Semilog plot of ρ vs. $1/T$ for $Pr_xY_{1-x}Ba_2Cu_3O_{1-\delta}$. The Pr and the (nominal) oxygen content for each of the graphs a–f is given in Table I.

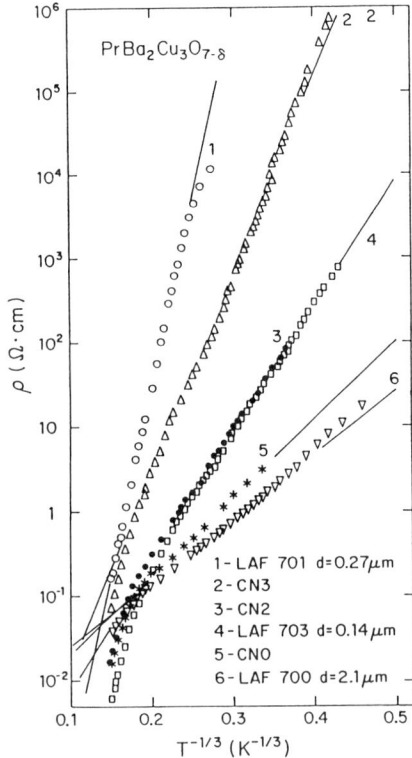

FIG. 5. Semilog plot of ρ versus $T^{-1/3}$ for three epitaxial films (LAF 700, 701 and 703) and three ceramic pellets (CN0, CN2 and CN3).

slight curvature in the plot, but is very close to the fairly straight line of sample c. The latter is straight over 4 orders of magnitude of ρ. (It should be noted that case d is pure YBCO, while c is pure PBCO).

Some samples exhibit more complicated behavior (see *e.g.* graph a and sample h of Table I). Here the curves deviate strongly from straight lines. Such behavior in PBCO ($x=1$, $\delta=0$) is not compatible with a single, temperature-activated mechanism. In order to study this behavior further, we made a number of low-temperature resistivity measurements on fully-oxygenated ceramic and thin-film samples of PBCO. In Fig. 5 we plot $\log \rho$ vs. $T^{-1/3}$ for three ceramic and three thin-film PBCO samples. The films are denoted by LAF, followed by a serial number; the ceramic samples are labeled CN followed by a serial number. The thickness of each film is given below the figure. The extended linear regions in Fig. 5 can be interpreted as evidence for two-dimensional variable-range hopping (VRH)[14] in the PBCO system. However, plotting $\log \rho$ against $T^{-1/4}$ — characteristic of VRH in three dimensions — gives about as good a linear behavior. Our conclusion is that electrical transport in these PBCO samples is characteristic of VRH, but on the basis of our data we cannot decide the dimensionality.

INTERPRETATION

The normal-state transport properties of orthorhombic YBCO have been interpreted in terms of the "Narrow Band" model.[15] This model assumed a narrow conduction band (CB) of width W, with two states per formula unit containing n effectively noninteracting electrons. It was shown[15] that in pure YBCO $n = 1 + \delta$, that is, the $(1 - \delta)$ chain oxygen atoms create $(1 - \delta)$ holes in the CB. The thermopower S is given by:

$$S = (k/e)\ln[n/(2-n)]. \qquad (1)$$

The narrow-band model predicts[16] that S will be independent of temperature, for constant band filling and at sufficiently high temperatures ($T \gtrsim W/4k$). Here k is Boltzmann's constant ($k/e = 86\ \mu\text{V/K}$), and n is the number of CB electrons per unit cell.

In PYBCO we find that at high temperature ($T \gtrsim 1000$ K), the graphs of ρ and of S overlap for samples of different Pr content; *i.e.* ρ and S become independent of the Pr content x, and are functions only of δ and T. Thus at high temperature Pr behaves like the trivalent ion Y^{3+}. However with falling temperature, the presence of Pr manifests itself more and more strongly. At constant oxygen content (regime B in Figs. 1 and 2), a plateau appears. This tendency towards saturation of $S(T)$ is seen for all the curves in Figs. 1 and 2 in the region of constant composition; this justifies the use of Eq. (1) for $0 \leq x \leq 1$. At $P(O_2) = 1$ bar, around 600 K, one obtains $S \sim 0$ for $x = 0$, and $S \sim 40\ \mu\text{V/K}$ for $x = 1$. In the latter case, Eq. (1) gives $n = 1$ for YBCO and $n = 1.23$ for PBCO. Thus, in PBCO ($x=1$) at $T \sim 600$ K and $\delta \sim 0$, only 0.23 additional holes in the CB have been filled, over and above the band filling in YBCO. Hence, at least at these high temperatures, the effective valency of Pr is closer to 3 than to 4. The graphs for $x = 0.2$ and $x = 0.4$ in Fig. 1 point to the same conclusion. The insensitivity of transport properties to changes of Pr content underlines the equivalence of Pr and Y at elevated temperatures.

As the temperature falls, S starts to increase. This suggests that electrons are transferred into the conduction band, presumably from the Pr ions, *i.e.* formation of Pr^{4+} ions. Below ~ 220 K, S shows a maximum; both the temperature where this maximum occurs and its magnitude increase with x. We propose that this effect is connected with transfer of electrons from the $4f$ states of the Pr atom to the conduction band at lower temperatures. In our proposed model for PYBCO we assume that the energy levels of the Pr f-electron overlap the CB, but that the CB states and the f states retain their identities. Consequently, the chemical potential (μ) and the temperature T determine the relative electron populations of the CB and the Pr($4f$) states. (We assume that *only* the above two types of states enter into the determination of μ.)

An atom of Pr in PYBCO has two possible ionic states: 4+ (with one f electron) and 3+ (with two f electrons). We shall consider here only a single Pr^{4+} level, E_1, of multiplicity g_1, and a single Pr^{3+} level, E_2, of multiplicity g_2. The standard statistics of localized levels in thermal equilibrium (see e.g. Eq. 28.30 in Ref. 17), gives the condition for neutrality (for small δ):

$$n = 1 + \delta + x/[1 + a\exp(\beta\mu)] , \quad (2)$$

where n is the number of CB electrons per formula unit, $\beta = 1/kT$, $a = g\exp(-\beta E_0)$, $g = g_2/g_1$ and $E_0 = E_2 - E_1$. The conductivity σ and the absolute thermopower S are calculated from standard relations[14]:

$$\sigma = -\int \sigma(E,T)(df/dE)dE , \quad (3a)$$

$$S = \sigma^{-1}\int (E-\mu)\sigma(E,T)(df/dE)dE . \quad (3b)$$

For simplicity we assume a box-shaped function for $\sigma(E)$, independent of temperature (see inset in Fig. 6). More realistic choices for $\sigma(E)$ affect only the details, but not the general shape of the resulting curves. In Fig. 6 we plot the thermopower S, for several values of x and δ.

The adjustable parameters are E_0 and g. As can be seen in Figs. 1 and 2, at high temperatures ($T \gtrsim 400$ K) and at constant oxygen content (region B), S becomes independent of temperature. It can be shown that in the model proposed, at $kT/W \gtrsim 0.5$, the exponential term in Eqs. (3a) and (3b) becomes almost constant. To fit the experimental high-temperature saturation values of S we need the parameter $g \simeq 6$; this implies that in the relevant energy range there are many more two-electron states than single-electron states. We choose E_0 at the top edge of the conduction band (see Fig. 6) in order to fit S_{max} for $x = 0.4$. This value of g gives a good fit for all the $S(T)$ data — both ours and from Ref. 13 — over the range $x \lesssim 0.6$.

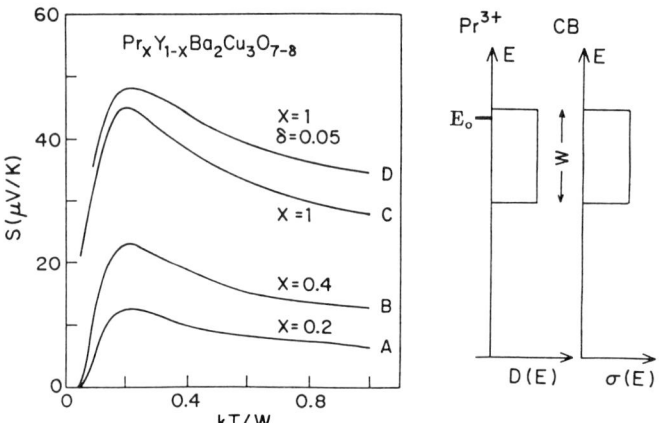

FIG. 6. Calculated thermopower S, as function of kT/W, for our narrow-band model, assuming a constant density of states $D(E)$, a constant conductivity function $\sigma(E)$ and a constant Pr^{3+} ionic level (see right half of the figure). Parameters used are : $g = 6$, $\delta = 0$ except for curve D, where $\delta = 0.05$, E_0 – at the top edge of the conduction band.

According to this model, the resistivity will remain "metallic" with increasing x. The substitution of Pr will affect the density of carriers, but this has only a relatively minor effect on ρ. Scattering by Pr ions, and disorder due to coexistence of Pr^{3+} and Pr^{4+} ions, may affect the transport mechanisms, and can lead to Anderson localization for larger x. The narrow CB in our model is probably a coherent many-body effect, which is disrupted by the Pr level. This will result in a decrease in mobility (and hence in $\sigma(E,T)$). Since, in Eq. (3b), $\sigma(E,T)$ appears both in the numerator and in the denominator, S is not affected greatly by such decrease, while ρ is.

The high multiplicity ratio ($g \simeq 6$) of the relevant f level causes electron transfer to the Pr levels at the expense of the conduction band. Note, however, that this depletion of the CB electron population is not necessarily a proof of high multiplicity of energy levels (as assumed above); it can also come about if the energy of the ionic states decreases linearly with temperature.

For PBCO ($x = 1$), we find that no choice of temperature-independent parameters can reproduce the low-temperature experimental results. The data for ρ show that the low-temperature conduction mechanism for $x = 1$ is very different from that for $x \lesssim 0.5$. The appearance of VRH (see Fig. 5) shows that (at least at low temperatures[18]) the electron states around the Fermi energy are localized for $x \simeq 1$ and $\delta \simeq 0$. A more elaborate model should be worked out for this regime. It will be interesting to study S in more detail at low temperatures in the VRH regime. Emin[19] suggests that for three-dimensional VRH transport, S will tend to zero as $T^{1/2}$. A low temperature drop in S has indeed been observed in Refs. 6, 13 (see Fig. 1); however, further work (both theoretical and experimental) will be needed.

CONCLUSIONS

1. In $Pr_xY_{1-x}Ba_2Cu_3O_{7-\delta}$, at high temperatures, ρ and S are independent of the Pr content x, and are a function of the oxygen content *only*. (*I.e.* at high temperatures Pr behaves like the trivalent ion Y).

2. For small x and δ the low-temperature resistivity is determined by a single variable $q = (x+\delta)$. Our interpretation is that q simply reflects the number of conducting holes; Pr doping and oxygen deficit reduce this number additively.

3. For $x \simeq 1$ and $\delta \simeq 0$, the low-temperature conduction is probably by VRH, but our data do not determine the dimensionality (2D or 3D).

4. The narrow-band model[15] can be applied to PYBCO. The dependence of S on T for $0 \leq x \lesssim 0.4$ fits a simple model consisting of a narrow conduction band and praseodymium atomic levels, independent of each other, with the Pr^{3+} levels above the center of the CB.

More details on this work have been published elsewhere.[20]

ACKNOWLEDGMENTS

We are grateful to Dr. M. Reisner for X-ray analysis. The support of the Israel Academy of Sciences (grant # 138-0021) and of the Israel—U.S.A. Binational Science Foundation (grant # 86-00421) are gratefully acknowledged.

REFERENCES

1. Y. Gao, A. Kebede, P. Pernambuco-Wise, M. Kuric, J.E. Crow, R.P. Guertin, T. Mihalisin, N.D. Spencer, and D.W. Cooke, these proceedings.

2. J.L. Peng, P. Klavins, R.N. Shelton, H.B. Radousky, P.A. Hahn and L.Bernandez, Phys. Rev. B **40**, 4517 (1989).

3. X.X. Tang, A. Manthiram and J.B. Goodenough, Physica C **161**, 574 (1989).

4. I. Felner, U. Yaron, I. Nowik, E.R. Bauminger, Y. Wolfus, E.R. Yacoby, G. Hilscher and N. Pillmayr, Phys. Rev. B **40**, 6739 (1989).

5. J.J. Neumeier T. Bjørnholm, M.B. Maple and Ivan K. Schuller, Phys. Rev. Lett. **63**, 2516 (1989).

6. M.E. López-Morales, D. Ríos-Jara, J. Tagüeña, R. Escudero, S. La Placa, A. Bezinge. V.Y. Lee, E.M. Engler and P.M.Grant Phys. Rev. B **41**, 6655 (1990).

7. J.S. Kand, J.W. Allen, Z.X. Shen, W.P. Ellis, J.J. Yeh, B.W. Lee, M.B. Maple, W.E. Spicer and I. Lindau, J. Less-Common Met. **148**, 121 (1989).

8. S. Horn,, J. Cai, S.A. Shaheen, Y. Jeon, M. Croft, C.L. Chang and M.L. den Boer, Phys. Rev. B **36**, 3895 (1987).

9. U. Neukirch, C.T. Simmons, D. Sladeczek, C. Laubschaft, O. Strebel G. Keindl and D.D. Sarma, Europhys. Lett. **5**, 567 (1988).

10. G. Koren, A. Gupta, R.J. Baseman, M. Lutwyche and R.B. Laibowitz, Appl. Phys. Lett. **55** 2450 (1989).

11. T.B. Lindemer, J.F. Huntley, J.E. Gates, A.L. Sutton, J. Brynestad, C.R. Hubbard and P.K. Gallagher, Oak Ridge report ORNL/TM – 10899 (1989).

12. K. Kishio, T. Hasegawa, K. Suzuki, K. Kitazawa and K. Kueki, Mat. Res. Soc. Symp. Proc. **156**, p. 91; (San Diego Conf.- April 1989); (M.R.S.- Pittsburgh, 1989).

13. A.P. Gonçalves, I.C. Santos, E.B. Lopez, R.T. Henriques, M. Almeida and M.O. Fiqueiredo, Phys. Rev. B **37**, 7476 (1988).

14. N.F. Mott and E.A. Davis, *Electronic Processes in Non-crystalline Materials*, (Clarendon Press, Oxford 1971).

15. B. Fisher, J. Genossar, I.O. Lelong, A. Kessel and J. Ashkenazi, J. Supercond. **1**, 53 (1988); J. Genossar B. Fisher, I.O. Lelong, J. Ashkenazi and L. Patlagan, Physica C **157**, 320 (1989); J. Genossar B. Fisher, and J. Ashkenazi, Physica C **162-164**, 1015 (1989); B. Fisher, J. Genossar, L. Patlagan, I.O. Lelong and J. Ashkenazi, *ibid*, p. 1207.

16. S. Bar-Ad, B. Fisher, J. Ashkenazi and J. Genossar, Physica C **156**, 741 (1988).

17. N.W. Ashcroft and N.D. Mermin, *Solid State Physics* (Holt, Rinehart and Winston, New York, 1976).

18. Note that this is also the regime where antiferromagnetism occurs — see Refs. 1, 4.

19. David Emin, Phys. Rev. Lett. **35**, 882 (1975).

20. B. Fisher J. Genossar, L. Patlagan and J. Ashkenazi, Phys. Rev. B **43**, 2821 (1991); B. Fisher, G. Koren, J. Genossar, L. Patlagan and E.L Garstein, Physica C, **176** (1991).

DETERMINATION OF THE COULOMB PARAMETERS OF

$La_{1.85}(Sr,Ba)_{0.15}CuO_4$ FROM AUGER SPECTROSCOPY

AND ELECTRONIC BAND STRUCTURE

R. Bar-Deroma,[a] J. Felsteiner,[a] R. Brener,[b] J. Ashkenazi[c]
and D. van der Marel[d]

[a]Physics Department, Technion, 32000 Haifa, Israel

[b]Solid State Institute, Technion, 32000 Haifa, Israel

[c]Physics Department, University of Miami, Coral Gables, FL 33124, U.S.A.

[d]Faculty of Applied Physics, Delft University of Technology, Delft
The Netherlands

INTRODUCTION

Numerous experimental and theoretical studies have been made in order to understand the mechanism responsible for high-T_c superconductivity. In particular, electron correlations are expected to be significant. Their strength depends roughly upon the ratio between the relevant intra-atomic Coulomb integral U and the electronic bandwidth. Auger electron spectroscopy (AES) is particularly suitable for determining the Coulomb interaction parameter U since the experimental Auger line shape is determined by a final two-hole density of states (DOS).[1] Recently, AES measurements for $La_{2-x}(Sr,Ba)_xCuO_4$ (LaCuO) have been reported[2] but no information about the Cu($3d$) and O($2p$) Coulomb parameters was given. Such information has, however, been given for $YBa_2Cu_3O_7$ (YBaCuO).[3,4] In this work, Auger electron high-resolution oxygen KLL and copper $L_{2,3}VV$ line shapes of LaCuO have been measured. Details have been published elsewhere.[5]. An analysis based on the comparison of these line shapes and those generated from local density approximation (LDA) band structure calculations is used to determine the values of the Cu($3d$) and the O($2p$) Coulomb parameters. In this analysis we used an extended version of Cini and Sawatzky's expression for the theoretical Auger line shape,[6] taking into account the multiplet structure of the Auger final states and the point symmetry of the oxygen sites. As a result we find a difference of about 2 eV between the O($2p$) Coulomb parameters corresponding to the planar and apical oxygen atoms in the compound.

EXPERIMENT

The measurements of the Auger electron high-resolution spectra were performed in a computerized Perkin-Elmer scanning Auger spectrometer (PHI model 590A). Single phase polycrystalline pellets of $La_{1.85}Sr_{0.15}CuO_4$, $La_{1.85}Sr_{0.15}CuO_4$ and $YBa_2Cu_3O_7$ were prepared by a standard procedure of melting high purity binary oxide powders. The structure and the superconductivity properties of the samples were checked by x-ray powder diffraction and T_c measurements. Prior to each measurement the samples were cleaned *in situ* in the UHV chamber (typical base pressure 5×10^{-10} Torr) by scraping with a diamond file. Auger spectra taken after the cleaning procedure showed negligible contamination level.

The Auger line shapes were recorded in a single-pass cylindrical mirror analyzer operated at 0.3% resolution with a 3 keV, 0.5 μA rastered primary electron beam in the first derivative mode, using 1 eV modulation voltage. The "true" Auger line shapes were obtained after integration and appropriate background subtraction.[7] The inelastic scattering contribution was removed by deconvoluting[8] an electron backscatter spectrum taken at the same kinetic energy as the Auger transition under study. The calibration of the kinetic energy scale relative to the vacuum level was performed by measuring the positions of the Auger electron peaks of sputter-cleaned silver and copper standards.

Figure 1 presents the experimental Auger $L_{2,3}VV$ spectra of Cu in $La_{1.85}Sr_{0.15}CuO_4$, $La_{1.85}Ba_{0.15}CuO_4$, and $YBa_2Cu_3O_7$. These spectra exhibit two components which correspond to the L_3VV and L_2VV transitions with no significant differences for the three compounds. At a distance of 5.5–7 eV from the main L_3VV line an asymmetry on the low energy side can be seen. This is due to multiplet splitting in the ^1G, ^3P, ^1S and ^3F d^8 final states and additional multiplets corresponding to the d^7 final states satellite lines.[3]

Figure 2 presents the experimental Auger spectra of oxygen for the same three compounds. We distinguish four lines which correspond to the KL_1L_1, $KL_1L_{2,3}$ (^1P and ^3P) and $KL_{2,3}L_{2,3}$ transitions in oxygen. There is an extra line at the high kinetic energy side which corresponds to the MNN transition in lanthanum and which does not appear in the $YBa_2Cu_3O_7$ spectrum. This peak was subtracted.

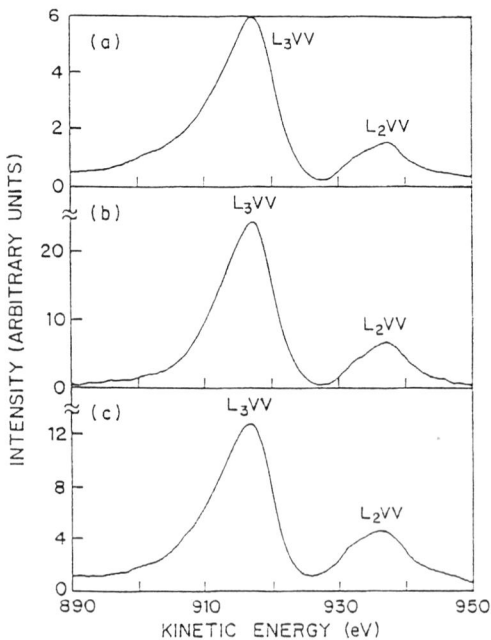

FIG. 1. The experimental Auger spectrum for copper in: (a) $La_{1.85}Sr_{0.15}CuO_4$; (b) $La_{1.85}Ba_{0.15}CuO_4$; (c) $YBa_2Cu_3O_7$. The kinetic energy is relative to the Fermi level.

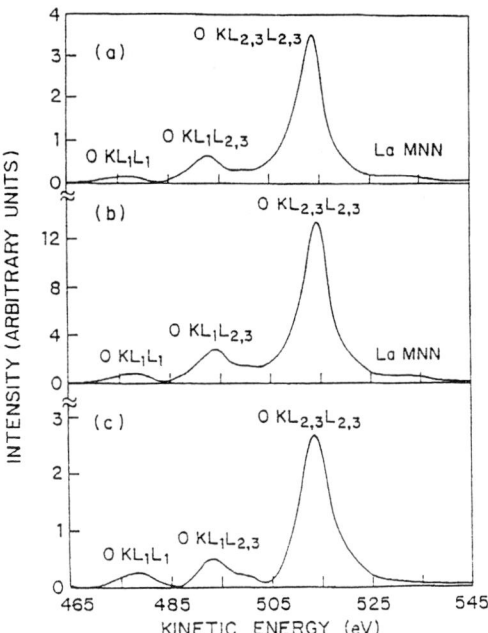

FIG. 2. The experimental Auger spectrum for oxygen in: (a) $La_{1.85}Sr_{0.15}CuO_4$; (b) $La_{1.85}Ba_{0.15}CuO_4$; (c) $YBa_2Cu_3O_7$. The kinetic energy is relative to the Fermi level.

TABLE I. The experimental Auger energy peak positions (E_k is in the kinetic energy scale relative to the Fermi energy, and E_A is in the two-hole binding energy scale) and XPS core level binding energies for $La_{1.85}Sr_{0.15}CuO_4$ (LaSrCuO), $La_{1.85}Ba_{0.15}CuO_4$ (LaBaCuO) and $YBa_2Cu_3O_7$ (YBaCuO). The energy is in eV units.

	LaSrCuO		LaBaCuO		YBaCuO	
	E_k	E_A	E_k	E_A	E_k	E_A
$Cu(L_3VV)$	917.2	16.3	917.3	16.0	916.7	16.1
$Cu(L_2VV)$	937.2	15.8	937.0	16.0	936.3	16.7
$O(KL_{23}L_{23})$	513.5	15.0	514.4	14.1	513.6	15.3
$O(KL_1L_1)$	477.2	51.7	477.6	51.3	478.0	50.9
	Binding Energies					
$Cu\ E(L_2)$	$953.0^{(9)}$		$953.0^{(9)}$		$953.0^{(4)}$	
$Cu\ E(L_3)$	$933.5^{(10)}$		$933.3^{(9)}$		$932.8^{(4)}$	
$O\ E(K)$	$528.5^{(10)}$		$528.5^{(10)}$		$528.9^{(4)}$	

In Table I we present the energy position of the different Auger transitions in the kinetic energy (E_k) and the two-hole binding energy (E_A) scales. The relevant core level binding energies, as measured by X-ray photoelectron spectroscopy (XPS), are also shown in Table I. They define the threshold energy of the Auger transition according to the rule E_k(Auger) $= E_c$(XPS) $- E$, where E is the excitation energy of two correlated holes, and E_c(XPS) is the initial state energy of the Auger process. Comparison between published Auger results[4,11] of the YBaCuO and LaCuO class of compounds and our experimental results show a good agreement.

THEORY

Our LDA calculations of the partial density of states (PDOS) used the linear muffin-tin orbitals (LMTO) band-structure method. The virtual-lattice approximation enabled us to take into account the Sr and Ba for non-integer doping parameters x. Within the framework of the LMTO method this was done by taking a weighted average of the LMTO potential parameters of the alloy constituents according to their atomic percentage.

Figure 3 presents the total DOS and the PDOS for each atom in $La_{1.85}Sr_{0.15}CuO_4$, with O1 and O2 denoting the in-plane and out-of-plane oxygen sites, respectively. In general, similar results are obtained for $La_{1.85}Ba_{0.15}CuO_4$. We must note that these PDOS's represent to a good approximation the $3d$ band for the copper and the $2p$ band for the oxygen. Our present results are in general agreement with those obtained in other calculations.

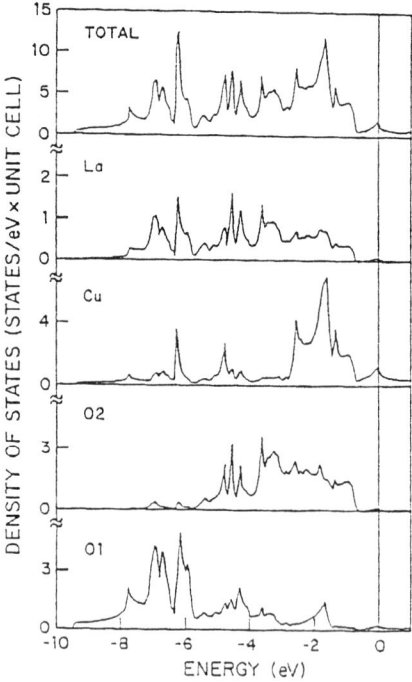

FIG. 3. The total and partial density of states for $La_{1.85}Sr_{0.15}CuO_4$. O1 denotes the oxygen atoms in the CuO_2 planes and O2 denotes the apical oxygen atoms.

The Cu $L_{2,3}VV$ Auger line shape was calculated using the Cini-Sawatzky expression[6] for the two-hole local DOS in a completely filled nondegenerate tight-binding band:

$$A(E) = \frac{N(E)}{[1 - UH(E)]^2 + [\pi U N(E)]^2} , \qquad (1)$$

where $N(E)$ is the self-convolution of the one-electron PDOS and $H(E)$ is the Hilbert transform of $N(E)$. For unfilled bands, the full-band expression can still be used as an approximation,[12] provided the number of holes per valence state is much smaller than one and U is not too large.

We will assume that this is still correct for the oxygen atoms, as the O^{2-} ions have a closed shell configuration. The deviations from the p^6 configuration in the ground state due to hybridization and hole doping are then treated by truncating the density of states obtained from the LMTO calculations at E_F. The situation is different for the Cu(3d) states, as the Cu^{2+} ion has an open d-shell with 9 d-electrons. Nevertheless, in order to extract an order of magnitude estimate of U on the copper sites, we will make a comparison between calculated spectra based on the Cini-Sawatzky expression and the measured spectra.

A generalized version of the Cini-Sawatzky expression, taking into account the point-symmetry of the O(2p) levels and the full multiplet structure of the p^4 final states, was used to calculate the Auger line shape of the oxygen $KL_{2,3}L_{2,3}$ Auger line (see Appendix of Ref. 5). This more complete analysis of the line shape is required in order to determine the two different values of U on the oxygen atoms O1 and O2. A similar approach has been used to study the LVV Auger spectra of transition metal impurities.[13]

A comparison between the calculated and experimental Auger line shapes is used to get the value of U that appears as a parameter in Eq. (1). The calculated line shape was convoluted with an appropriate Gaussian line in order to take into account the instrumental broadening and the broadening due to the life time of the holes during the Auger transition.

INTERPRETATION

Figure 4 presents the calculated and experimental copper $L_{2,3}VV$ Auger line shapes in $La_{1.85}Sr_{0.15}CuO_4$. From comparison between calculation and experiment we find $U_d = 7 \pm 2$ eV. Similar results, with $U_d = 8 \pm 2$ eV, were obtained for Ba-based LaCuO compound. By a similar method, Balzarotti et al.[4] have obtained $U_d \sim 7$ eV in YBaCuO using the calculated PDOS results of Temmerman et al.[14]. We cannot make a detailed comparison of Auger line shapes because the multiplet structure and the satellite structure due to the mixed valent (d^9-d^{10}) state of the copper ion were not considered.

In the LaCuO compounds the oxygen atoms occupy two inequivalent crystallographic sites: the planar O1 and the apical O2. Consequently, these two oxygen atoms differ in (i) the local density of states of the 2p valence bands, (ii) the energy positions of the core states (chemical shifts), and (iii) the Coulomb interaction parameter U. The band calculation resulted in two different PDOS's for the O1 and O2 atoms. These PDOS's are to be related to the experimental Auger line shape, composed of the contributions of both types of oxygen atoms.

In the first step of the numerical calculation of the oxygen KLL Auger line shape, the generalized version[5,13] of Eq. (1) for O1 and O2 was used which included the convolution of the p_x, p_y and p_z PDOS's. In the second step, the interaction terms were included. The oxygen KL_3L_3 spectra consist of three components. The ^1G and the ^1S peaks have U-parameters of $F^0+0.04F^2$ and $F^0+0.4F^2$, respectively.[15] The ^3P forbidden transition can be neglected due to its low weight. It is well established that the value of F^2 which determines the multiplet splitting is almost unchanged upon inserting atoms into a solid.[16] Therefore, we use here the value $F^2 = 6.2$ eV, which is obtained from the ^1G-^1S splitting in optical transitions of atomic oxygen.[17] On the other hand, F^0 depends strongly on screening effects in the solid and

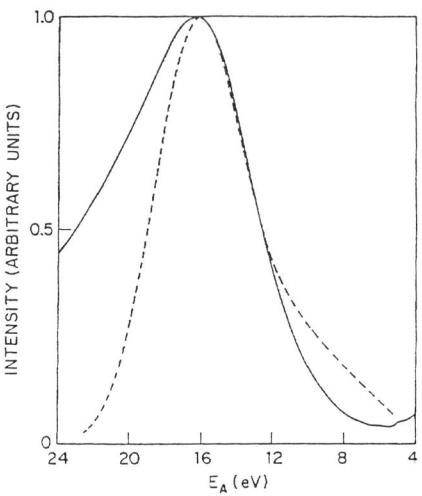
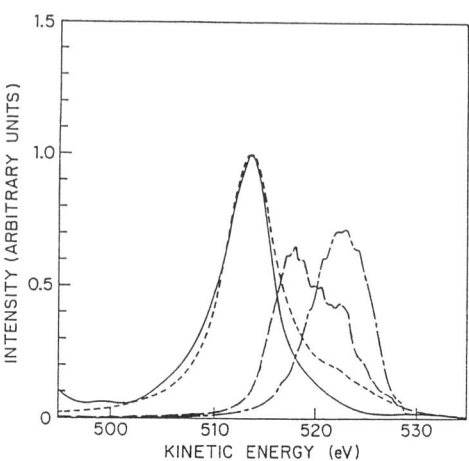

FIG. 4. Comparison between the experimental (solid) and calculated (dashed) Auger line shapes of the L_3VV transition of copper in $La_{1.85}Sr_{0.15}CuO_4$. E_A is the two-hole binding energy.

FIG. 5. Comparison between the experimental (solid) and calculated (short dashed) Auger line shapes of the $KL_{2,3}L_{2,3}$ transition of oxygen in $La_{1.85}Sr_{0.15}CuO_4$. The long dashed and dotted-dashed lines represent the two-hole densities of states of occupied $O(2p)$ orbitals on the O1 and O2 sites respectively.

TABLE II. $O(2p)$ centroids, and Coulomb parameters, in units of eV, used for fitting the oxygen KLL Auger spectra. Four different fits are presented here using: 1. the present LMTO results for $La_{1.85}Sr_{0.15}CuO_4$ (LaSrCuO); 2. the present LMTO results for $La_{1.85}Ba_{0.15}CuO_4$ (LaBaCuO); 3. the FLMTO results of Ref. 18 for La_2CuO_4 (LaCuO); 4. the centroids of the FLAPW results of Ref. 20 for La_2CuO_4. The binding energy of the $O(1s)$ core level is either taken as 528 eV for both O1 and O2 (abbreviated by UC), or shifted for the O1 atom to a value higher by the splitting between the O1 and O2 centroids (abbreviated by SC), as explained in the text. $U(^1G)$ and $U(^1S)$ are Slater integrals for the two multiplet components in the Auger spectra. The relative intensity of the ^1G peak is six times the ^1S intensity. We used the atomic value $F^2 = 6.2$ eV to calculate $U(^1G)$ and $U(^1S)$. Life time broadening of 2.5 eV and Gaussian broadening of 2.0 eV are used.

		Centroid	F^0	$U(^1G)$	$U(^1S)$
Present LMTO LaSrCuO	O1(UC)	-5.7	3.0	3.2	5.5
	O1(SC)	-5.7	5.0	5.2	7.5
	O2(UC)	-3.1	7.0	7.2	9.5
Present LMTO LaBaCuO	O1(UC)	-5.3	2.5	2.7	5.0
	O1(SC)	-5.3	4.5	4.7	7.0
	O2(UC)	-3.2	6.5	6.7	9.0
Ref. 18 FLMTO LaCuO	O1(UC)	-3.8	6.2	6.4	8.7
	O1(SC)	-3.8	8.2	8.4	10.7
	O2(UC)	-1.9	10.2	10.4	12.7
Ref. 20 FLAPW LaCuO	O1(UC)	-3.9	5.1	5.3	7.6
	O1(SC)	-3.9	6.2	6.4	8.7
	O2(UC)	-2.8	7.3	7.5	9.8

may even be site dependent due to differences in local screening properties. In the fit to the experimental data, two different values of F^0 were used for the two inequivalent oxygen sites O1 and O2.

Furthermore, we still have to incorporate the $O(1s)$ core level positions of the two oxygen types. On the one hand, observation seems to show that difference in core level binding energy, for the two types of oxygen, is small or non-existant, this for samples with clean sample surfaces[3]. On the other hand, a full-potential LMTO (FLMTO) band structure calculation for La_2CuO_4,[18] yields a rather large chemical shift of approximately 2 eV towards higher binding energy of the planar $O(1s)$ relative to the apical oxygen level. This value is very close to the centroidal relative shift between the occupied $O(2p)$ levels (see Table II). In order to account for this ambiguity in the relative positions of the core levels, we have repeated the calculations under two different assumptions. In one set of calculations we have put the $O(1s)$ threshold of both the apical and planar oxygens at 528 eV as obtained experimentally. In the other set of calculations we have put the $O(1s)$ threshold of the apical oxygen O2 at 528 eV (as before), but shifted the threshold of the planar oxygen O1 to an energy which is higher by the centroidal relative shift between the occupied $O(2p)$ levels (as pointed above, the values calculated in Ref. 18 for these shifts are very close to each other). In the last step, after proper normalization, we add the spectra of the two inequivalent oxygens. Finally, a convolution of the spectrum is made with a Gaussian, as explained before. An intrinsic lifetime broadening was taken into account by including an imaginary part in the denominators of the generalized version of Eq. (1). Table II presents the parameters used in our fits.

Figure 5 presents the calculated oxygen $KL_{2,3}L_{2,3}$ Auger line together with the experimental data for the Sr-doped LaCuO compound. No difference in the fit is observed when the two above assumptions for the position of the O1 core level are used. Similar results were obtained for the Ba-doped LaCuO compound. Although, close to the main peak, the calculated line shape is in reasonable agreement with the experimental one, there is a significant discrepancy on the high kinetic energy side of the spectrum, resulting in a much higher intensity of the calculated spectrum relative to the measured one. The high intensity of the calculated Auger line results from mixing of the quasi atomic two-hole bound state with the two-hole continuum on the high kinetic energy side. The latter continuum corresponds to having the holes located on two different sites, where they do not interact, as expressed by the on-site interaction Hamiltonian leading to the Cini-Sawatzky expression. In the present LMTO calculations the single-hole continuum extends up to 9.5 eV and therefore the two-hole continuum extends up to 19.0 eV. Consequently, the quasi atomic state is located near the edge of the two-hole continuum, leading to a strong mixing. It should be noted that if we use as our input the FLMTO band structure results,[18] which have narrower oxygen p bands, the calculated Auger spectrum would have considerably less weight on the high kinetic energy side than the results shown in Fig. 5.[19]

In order to check to what extent our fit for the oxygen Coulomb parameters depends on the band structure results used as an input, we have repeated the calculations using the results of two other band structure calculations. One fit[19] was carried out using the FLMTO results,[18] mentioned above. The other fit was done by replacing our centroidal relative shift between the occupied $O(2p)$ levels by those obtained in the FLAPW calculation of Pickett et al.[20] for La_2CuO_4 (here the centroids were obtained by digitizing the PDOS plots of Ref. 20, with uncertainties of 0.5 eV due to the digitizing procedure). The results of these two fits are presented together with our fit in Table II. An important conclusion from all the fits presented here is that one needs two different values of F^0, with the lower value corresponding to the planar oxygen O1. This result is quite insensitive to the details of the fit, being qualitatively the same for the three different band structure calculations used. It basically reflects the fact that the Auger spectrum has a single strong peak, whereas the centroids of the occupied PDOS's of the two oxygen sites differ by 1.1–2.6 eV, for the different band structure results used for the fits. The splitting in F^0 becomes smaller by about a factor of 2 when core level shifts identical to the centroidal shifts are assumed. This is due to the fact that the PDOS represents a one-hole spectrum while the Auger kinetic energy represents a two-hole spectrum. As seen in Fig. 5, in order to get the peak positions of the O1 and O2 contributions to coincide, the shift in the centroids of the two-hole spectrum has to be compensated by a difference in U_p. A smaller value of U_p for the planar oxygen could be due to the stronger screening in the CuO_2 planes, which is consistent with the fact that the conduction takes place in the planes.

The present conclusion, that there are two different U_p values, might be criticized on the basis that it is difficult to believe the occurrence of a cancellation of a splitting in the Auger peak by a splitting in the Coulomb parameter, and that the obtained splitting is due to inaccuracies in the band structure results used for the fits. Nevertheless, we note that three different band structure calculations give qualitatively the same results. It should also be pointed out that the cancellation of a splitting in the Auger peak by a splitting in the Coulomb parameter is not completely accidental. The position of the $2p$ centroid in an oxygen atom is partly determined by the average interaction of a $2p$ electron with the other $2p$ electrons in the atom. Though these interactions are not affected by exactly the same screening processes as in the Coulomb interaction between the two holes in the Auger process, there should be some similarity between them. Therefore a larger F^0 parameter for O2 than for O1, as determined from the Auger analysis, is consistent with a higher energy (*i.e.* an energy closer to E_F) for the O2 than for the O1 $2p$ centroid. Thus, there is a theoretical consistency between splittings in the centroids and in F^0 in the direction that diminishes the splitting in the Auger peak.

On the basis of the different values of Coulomb parameters appearing in Table II, we can conclude that U_p is approximately 5 and 7 eV on the O1 and O2 atoms respectively, with uncertainties of about 2 eV. The direction of the splitting between them, is however beyond these uncertainties, if our analysis is valid. AES evaluations of the intra-site oxygen U_p in Cu_2O have been made by Tjeng *et al.*,[21] who determined a value of $F^0 = 5.4 \pm 0.5$ eV. McMahan *et al.*[22] find by constrained-occupation LDA calculations on La_2CuO_4 that F^0 is 3.6 eV on O1 and 3.0 eV on O2. Thus they find smaller U_p values with an opposite anisotropy than ours.

In summary, we have determined the on-site Cu($3d$) and O($2p$) Coulomb interaction parameters in $La_{1.85}Sr_{0.15}CuO_4$ and $La_{1.85}Ba_{0.15}CuO_4$ by comparing experimental Auger electron spectra with those derived from band-structure calculations. Values of 7-8 eV are found for the Cu($3d$) Coulomb parameter. It is found that the O($2p$) Coulomb parameter have two different values of approximately 5 and 7 eV for the planar and apical oxygen sites, respectively.

ACKNOWLEDGMENTS

We are grateful to I. Felner for supplying the samples and to O. K. Andersen and O. Jepsen for allowing to use their band structure results prior to publication. This research was supported in part by the U.S.–Israel Binational Science Foundation.

REFERENCES

1. D. E. Ramaker, Appl. Surface Sci. **21**, 2443 (1985).

2. H. Nakayama, H. Fujita, T. Nogami, and Y. Shirota, Physica C **155**, 149 (1988).

3. D. van der Marel, J. van Elp, G. A. Sawatzky, and D. Heitmann, Phys. Rev. B **37**, 5136 (1988).

4. A. Balzarotti, M. De Crescenzi, N. Motta, F. Patella, and A. Sgarlata, Phys. Rev. B **38**, 6461 (1988); A. Balzarotti, in *Studies of High Temperature Superconductivity*, edited by A. Narlikar (Nova Science, Commack, NY, 1989), Vol. 4.

5. R. Bar-Deroma, J. Felsteiner, R. Brener, J. Ashkenazi and D. van der Marel, submitted to Phys. Rev. B.

6. M. Cini, Solid State Commun. **20**, 681 (1976); G. A. Sawatzky, Phys. Rev. Lett. **39**, 504 (1977).

7. J. A. D. Matthew, M. Prutton, M. M. El Gomati, and D. C. Peacock, Surf. Interf. Anal. **11**, 173 (1988).

8. M. C. Burrell and N. R. Armstrong, Appl. Surface Sci. **17**, 53 (1977).

9. N. T. Liang, K. H. Li, Y. C. Chou, M. F. Tai, and T. T. Chen, Solid State Commun. **64**, 761 (1987).

10. N. Nücker, J. Fink, B. Renker, D. Ewert, and C. Politis, Z. Phys. B **67**, 9 (1987).

11. D. D. Sarma and C. N. Rao, J. Phys. C **20**, L659 (1987); H. H. Madden, D. M. Zehner, and J. R. Noonan, Phys. Rev. B **17**, 3074 (1978); *ibid* Phys. Rev. B **18**, 2023 (1978); B. R. Chakraverty, D. D. Sarma, and C. N. Rao, Physica C **156**, 413 (1988); D. E. Ramaker, N. H. Turner, J. S. Murday, L. E. Toth, M. Osofsky, and F. L. Huston, Phys. Rev. B **56**, 5672 (1987); D. E. Ramaker, N. H. Turner, and F. L. Huston, Phys. Rev. B **38**, 11368 (1988).

12. M. Cini and C. Verdozzi, Nuovo Cimento **9** ,1 (1987).

13. M. Vos, D. van der Marel, and G. A. Sawatzky, Phys. Rev. B **29**, 3073 (1984); M. Vos, G. A. Sawatzky, M. Davies, P. Weightman, and P. T. Andrews, Solid State Commun. **52**, 159 (1984).

14. W. M. Temmerman, Z. Szotec, P. J. Durham, G. M. Stocks, and P. A. Sterne, J. Phys. F **17**, L319 (1987).

15. J. C. Slater, *Quantum Theory of Atomic Structure* (McGraw-Hill, New York, 1960).

16. D. van der Marel and G. A. Sawatzky, Phys. Rev. B **37**, 10674 (1988).

17. C. E. Moore, Atomic Energy levels, NBS Circular **467** (U.X. GPO, Washington, DC, 1958) Part I.

18. O. Jepsen and O. K. Andersen, unpublished; J. Zaanen, O. Jepsen, O. Gunnarsson, A. T. Paxton, O. K. Andersen, and A. Svane, Physica C **153-155**, 1636 (1988).

19. G. E. Rietveld, D. van der Marel, O. K. Andersen, and O. Jepsen, unpublished.

20. W. E. Pickett, H. Krakauer, D. A. Papaconstantopoulos, and L. L. Boyer, Phys. Rev. B **35**, 7252 (1987); W. E. Pickett, Rev. Mod. Phys. **61**, 433 (1989).

21. L. H. Tjeng, J. van Elp, P. Kuiper, and G. A. Sawatzky, unpublished.

22. A. K. McMahan, J. F. Annett, and R. M. Martin, Phys. Rev. B **42**, 6268 (1990).

SOME TRANSPORT PROPERTIES IN YBa$_2$Cu$_3$O$_7$

A.T. Seshadri and B. Subrahmanyam

Department of Physics
Indian Institute of Technology, Madras-36, INDIA

INTRODUCTION

Ever since the discovery of High Tc Superconductor[1] YBa2 Cu3 O7, many researchers all over the world have started determining/studying the various physical parameters in this material at Room Temperature, Low Temperatures and particularly around its transition temperature (Tc). HALL EFFECT is one such study as it leads to the determination of some important Transport Properties like, type of charge carriers (electrons (e) or holes (p), no. of charge carriers (n)). Mobility (μ) of the charge carrier etc. in the material.

Sample preparation

High purity powders of Y2 O3, BaCO3 and CuO were mixed in the stoichiometric ratio and they were ground into finer grains and this homogeneous mixture was heated at 900°C for 24 hours in an air ove. The sintered powder was reground again into a fine powder and pelletised using a rectangular die of dimensions 2 cm x 1 x 1 mm at a pressure of 5 tonnes/m3 and sintered at 900 °C for 24 hours and later at 600°C for 24 hours, both being done in a constant flow of oxygen. The sample was then allowed to cool naturally.

The samples prepared were found to be superconducting as evidenced by various test techniques like coil technique, microwave absorption technique, resistivity measurements and T_c was found to be 91 K in all the cases. The X-ray diffraction of the sample was taken and compared with that of the standard X-ray diffraction chart of YBa2 Cu3O7 from literature [2] and these two were agreeing.

MEASUREMENTS, RESULTS, DISCUSSIONS

HALL EFFECT MEASUREMENTS

When a current flows in a semiconductor located in a magnetic field such that a component of the field is perpendicular to the current, then a voltage is developed across the specimen mutually at right angles to both the current and the field. This is known as Hall Effect. The schematic diagram is shown in Fig.1.

The circuit diagrams used to determine the resistivity (ρ) and Hall Voltage (V_H) of the material YBa2Cu3O7 are shown in Fig.2 and Fig.3 respectively. The measurements were carried out over a wide range of temperatures (300 K to 77 K) using a bath cryostat (Fig.4) fabricated by the authors.

The specimen (YBa2Cu3O7) of dimension 2 cm x 1 cm x 0.8 mm was mounted vertically down on the cold finger of the cryostat [Fig.4] which is kept between the pole-pieces of an electromagnet. Four probe technique was used to measure both Hall Voltage and resistivity of the sample. The maximum magnetic field that could be obtained was 0.37 T.

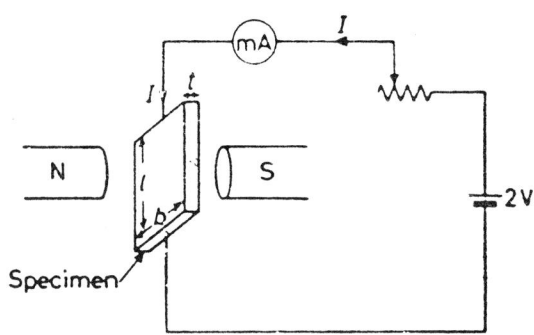

Fig.1. Schematic Diagram of Hall Effect

DETERMINATION OF TRANSPORT PROPERTIES

$$R_H = \frac{V_H \cdot t}{I_x \cdot B_z} \quad \ldots \quad (1)$$

$$R_H = \frac{1}{ne} \quad \ldots \quad (2a)$$

or

$$= \frac{1}{(np)} \quad \ldots \quad (2b)$$

$$\mu = \frac{R_H}{\rho}$$

R_H is the Hall coefficient of the material
V_H is the measured Hall Voltage
t is the thickness of the sample
I_x is the current through the samples
B_z is the magnetic field applied
n is the charge carrier concentration (/unit vol.)
ρ is the resistivity of the material
μ is the mobility defined as the drift velocity of the charge carriers per unit Electric field.

The experimentally determined values of R_H, n and for the material YBa2 Cu3O7 at Room Temperature are presented in Table 1, along with the literature values are comparison.

TABLE 1. COMPARISON OF RESULTS

	$\rho \times 10^{-3}$ (ohm-cm)	$R_H \times 10^{-9}$ (m^3/c)	n $\times 10^{21}$ (/cm^3)	μ (cm^2/V/S)
Author's values	1.74	3.62	1.72	2.08
Zhao Young et al [2]	2.50	5.50	1.10	2.20
Rickettes et al [3]	2.20	5.68	1.50	2.84

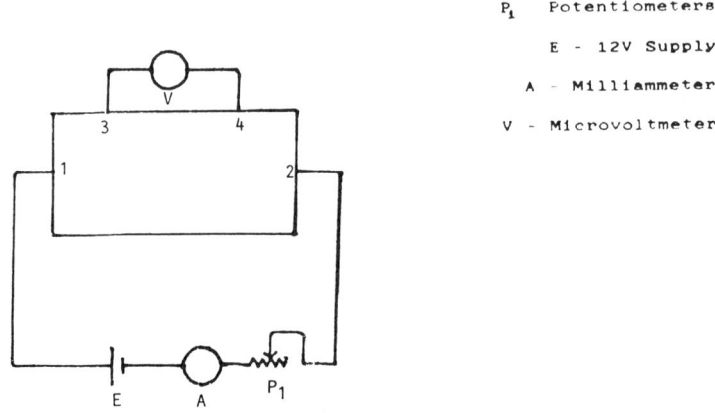

P₁ Potentiometers
E - 12V Supply
A - Milliammeter
V - Microvoltmeter

Fig.2. Circuit Diagram for Resistivity Measurement

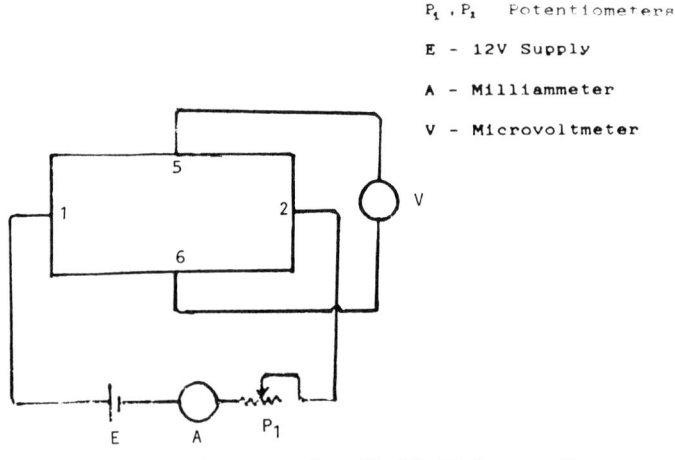

P₁ , P₂ Potentiometers
E - 12V Supply
A - Milliammeter
V - Microvoltmeter

Fig.3. Circuit Diagram for Hall Voltage Measurement

The mobility of the charge carrier is defined as the drift velocity acquired when unit electric field is applied. μ has been determined using the relation $\mu = R_H/\rho$, at various temperatures (300 K to 80 K) and the graph drawn between and T is shown in Fig.5. The variation of R_H with T shown in Fig.6.

Even though the variation of mobility with temperature is more or less the same as that of R_H with T, still, at and below T_c (= 91), the value of has shot up to ∞ and continues

228

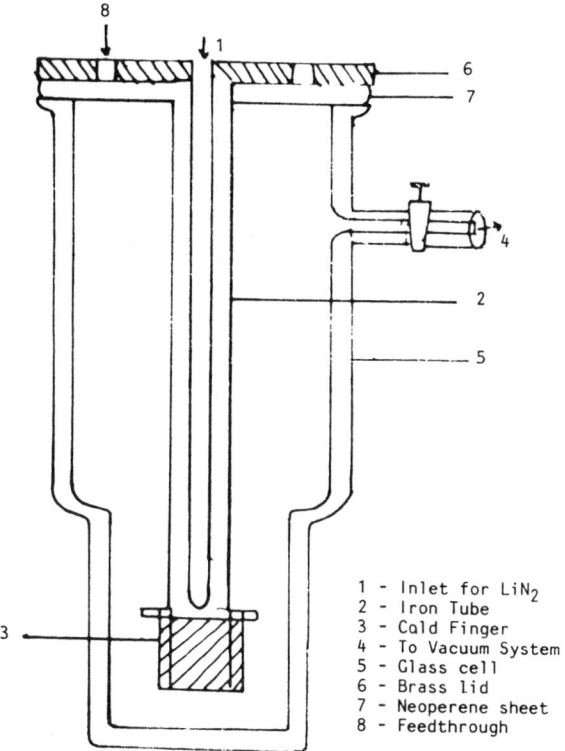

Fig.4. Cryostat

1 – Inlet for LiN$_2$
2 – Iron Tube
3 – Cold Finger
4 – To Vacuum System
5 – Glass cell
6 – Brass lid
7 – Neoperene sheet
8 – Feedthrough

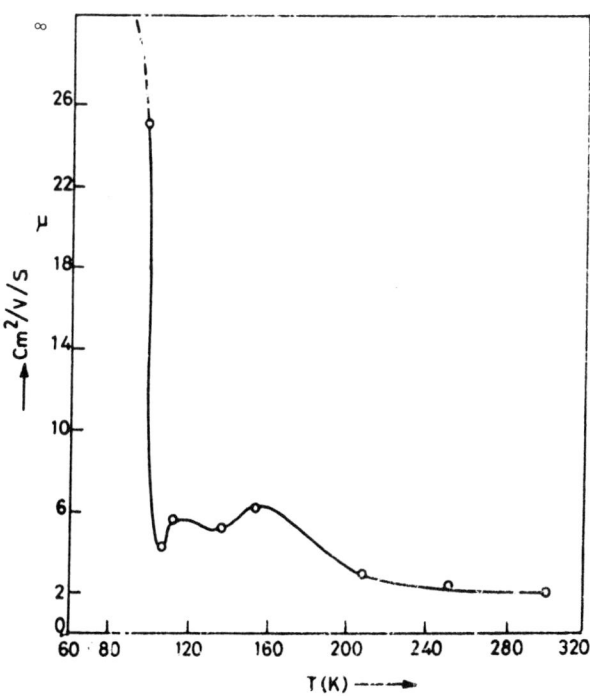

Fig.5. Variation of mobiliy (μ) with temperature (TK)

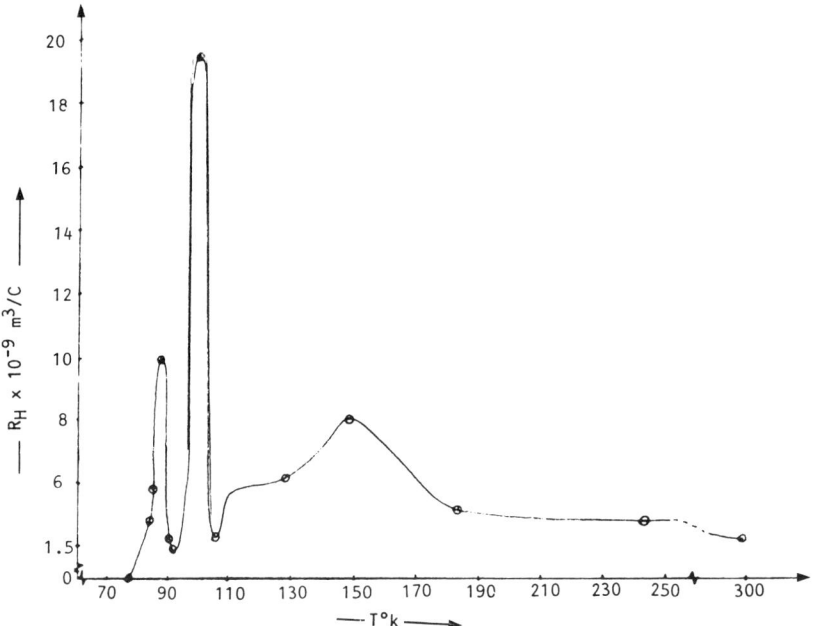

Fig.6. Variation of Hall Coefficient (RH) with Temperature (TK)

to be so. i.e., the charge are highly mobile at and below T_c, even though the material is under the influence of magnetic field, giving a finite Hall Voltage at T_c.

ACKNOWLEDGEMENT

The authors thank the Material Science Research Centre of our Institute for helping us in the sample preparation.

References

1. J.G. Bednorz, K.A. Muller, Z.Phys. P189, 1986.

2. Zhao Yang et al, Solid State Communications, Vol.64, p.885, 1987.

3. B.W. Ricketts et al, Solid State Communications, Vol.64, No.10, p 1287, 1987.

ON TRANSPORT PROPERTIES OF HTSC OXIDES

Marco Affronte and Davor Pavuna

Department of Physics
Swiss Federal Institute of Technology (EPFL)
CH-1015 Lausanne, Switzerland

We discuss the dominant features of electron transport properties of some HTSC oxides that were reported in recent experimental studies. We argue that there seem to exist some 'reference compositions', like $Y_1Ba_2Cu_3O_{6+x}$ with x=0.93±0.04, or YBCO '124' phase. These materials exhibit T_c^{MAX} as a function of the oxygen content and the former has metallic normal state resistivity in all crystallographic directions with $\rho_c < \rho_{Mott} \sim 10 m\Omega cm$. One can interpret transport properties of these 'metallic reference materials' within conventional band picture and, as a first approximation, account for transport properties of other, more complex HTSC oxides (usually with $T_c < T_c^{MAX}$, $\rho_c > \rho_{Mott}$), by taking into account the strong anisotropy, disorder and correlation effects.

INTRODUCTION

In order to establish a set of reliable experimental facts on transport properties of newly discovered family of materials, it is useful to introduce the concept of the 'reference' material. In 'old metal physics' one usually starts with the concept of the normal metal to develop a simple band picture. Subsequently, as resistivity rises due to various 'perturbations' one gradually introduces concepts like Mathiessen's rule, Friedel-Anderson s-d model, quantum interference effects etc. It is perhaps useful to check whether some of the concepts of the contemporary metal physics can provide the starting picture for transport properties of at least some HTSC oxides, before 'ab-initio' introducing new concepts, like holon/spinon contributions to transport.

In order to do that we shall first concentrate on the 'best metals' among these oxides, that we shall call 'reference materials'. We will argue that in these materials one can use fairly conventional band approach analysis and only subsequently introduce more evolved concepts in more complex situations in these solids. Crudely speaking, it's like establishing a subgroup of HTSC oxides that plays a role equivalent to the 'normal metal' in the 'old metal physics'. The anisotropy and correlations are subsequently integrated as corrections to an initially simple picture. While such an approach might appear rather artificial and non-appealing to theoreticians it may be useful to experimentalists, particularly within a current consensus (partly due to P.W.Anderson [1]) that one should introduce minimum interpretation into data analysis and rather present well organized experimental facts on any physical property.

EXPERIMENTAL EVIDENCE

We start with the analysis of the out-of-plane conductivity, σ_c, or its resistive counterpart, ρ_c. If ρ_c is above the Mott limit, one can argue that there is a coherent band transport in the Cu-O plane but non-metallic transport mechanism normal to the planes. This was indeed found by Ong et. al. who carefully analyzed resistivity anisotropy in seven single crystals of $Y_1Ba_2Cu_3O_7$ and concluded that $\sigma_c \sim 100(\Omega cm)^{-1}$ is comparable to the Mott minimum metallic conductivity, $\sigma_{min} \sim 0.3 e^2/(ha^3) \sim 97(\Omega cm)^{-1}$ $\{\rho_{Mott} \sim 10 m\Omega cm\}$[2]. Hence they argued that one should 'ab initio' introduce novel concepts like hole boson conduction in the Cu-O plane and electron tunneling between planes, as proposed by Anderson and Zou [3]. These arguments ('the Princeton approach'), together with the discovery of linear temperature dependence of resistivity up to several hundred Kelvin [4], were often invoked in discussions in favour of new (complex) concepts in transport and theory of HTSC oxides, in general.

We argue that recent experimental results on transport properties of YBCO family do not necessarily require 'the Princeton approach' [2,3]. Several groups have, in fact, independently measured 'metallic alike', positive temperature dependence (TCR) of c-axis resistivity of $Y_1Ba_2Cu_3O_{7-\delta}$ [5,6] schematically presented in Figure 1. The group from Tokyo reported that only fully oxidized $Y_1Ba_2Cu_3O_{7-\delta}$ ($T_c=90K$) shows positive TCR (with $\rho_c<10 m\Omega cm$) possibly due to inserted CuO chains between CuO_2 planes which create a Cu-O-Cu-O sequence along c-axis [8]. All other $Y_1Ba_2Cu_3O_{7-\delta}$ single crystals show lower T_c-s and negative TCR. As schematically shown in Figure 2, we conjecture that metallic c-axis resistivity, $\rho_c<\rho_{Mott}$ is measured in $Y_1Ba_2Cu_3O_{6+x}$ within the compositional range $x=0.93\pm0.04$ where these compounds exhibit the highest $T_c=94K$ [7]. This conjecture is most likely true for all fully doped systems (with $T_c=T_c^{MAX}$) of the 'YBCO family': $Y_1Ba_2Cu_{2+m}O_{6+m-\delta}$, where $m=1,1.5,2$ corresponds to the number of chains. However, it has to be verified experimentally.

To further clarify our argument we take the YBCO '124' phase. The '124' material is probably <u>the</u> 'reference' ionic metal for discussions of the normal state transport properties of HTSC oxides. It is a stable phase with $T_c=82K$ [7]. It has low resistivity, $\rho_{300}=100\mu\Omega cm$, and a small Hall coefficient, $R_H=1.4 \times 10^{-10} m^3/C$, almost temperature independent above 140K [9]. The temperature dependence of resistivity can be successfully fitted by the Bloch-Grüneisen model [10], which implies the phonon scattering mechanism. Band calculations estimates of R_H are in good agreement with the experimental data [11]. Hence transport in the '124' phase seems to be fairly well described in terms of the conventional band model. To our knowledge no one has measured c-axis resistivity, ρ_c, in '124' phase but according to our conjecture it should be metallic and lower than Mott limit of $\sim 10 m\Omega cm$.

In contrast, the transport in $Nd_{1.85}Ce_{0.15}CuO_{4-\delta}$ seems to be more complicated. Chemical analysis and some Hall effect measurements [12-14] indicate that transport is due to electrons, contrary to the predictions based on the band calculations [15]. However, there is no consensus on the sign of the Hall coefficient in $Nd_{1.85}Ce_{0.15}CuO_4$ crystals, as there are also reports of the positive Hall coefficient [16]. This discrepancy is probably due to the difficulty in controlling the oxygen stoichiometry in this material. Nevertheless, a very interesting feature of this compound is the temperature dependence of resistivity. It was found that it follows a $(T/T_F)^2 \ln(T_F/T)$ dependence, consistent with a two-dimensional Fermi liquid model [17]. This indicates that in $Nd_{1.85}Ce_{0.15}CuO_4$ the correlation effects have to be taken into account to explain the transport properties.

Finally, we note some of the most striking new results in YBCO '123' phase. The temperature and oxygen doping dependences of ab-thermopower of $Y_1Ba_2Cu_3O_{7-\delta}$ have

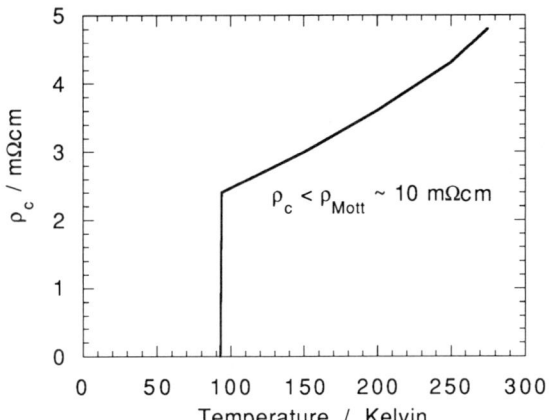

Figure 1. Schematic presentation of $\rho_c(T)$ (adopted from [5,6,8]) in best metallic 'reference samples' as defined in Figure 2 and discussed in the text.

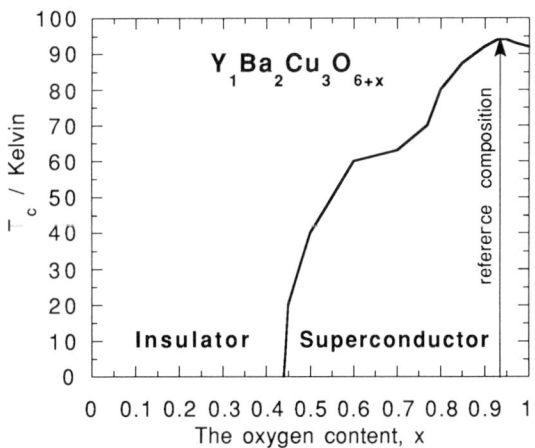

Figure 2. Schematic phase diagram of $Y_1Ba_2Cu_3O_{6+x}$. Compositions with highest T_c ($x\sim0.93\pm0.04$ [7]) within a given HTSC family we consider as 'reference metallic samples' and conjecture that these exhibit metallic resistivity in all crystallographic directions.

recently been consistently accounted by conventional band theory [18]. Furthermore, measurements of resistivity tensor on the un-twinned crystals [6] show strong anisotropy along all 3 crystallographic directions, while transport measurements on YBCO/PrBCO multi-layers [19] show features of an essentially two dimensional system. Although these results are not entirely surprising, they indicate that some of the observed 'anomalies' of transport properties in HTSC oxides may be caused by their pronounced anisotropy.

SUMMARY

1. While physical properties of HTSC oxides can be fairly complex, there seem to exist some 'reference metallic compositions', like $Y_1Ba_2Cu_3O_{6.93}$ or YBCO '124', that exhibit maximum T_c for a precise oxygen content [7]. The former has metallic normal state resistivity in all crystallographic directions with $\rho_c < \rho_{Mott} \sim 10 m\Omega cm$. Our conjecture implies that in fully oxidized '124' phase $\rho_c < \rho_{Mott} \sim 10 m\Omega cm$; this has to be verified experimentally. One can fairly successfully interpret the transport properties of these metallic 'reference materials' within conventional band picture.

2. As a first approximation one can then interpret the transport properties of other, more complex HTSC oxides (usually with $T_c < T_c^{MAX}$, $\rho_c > \rho_{Mott}$), as 'departures' from these 'reference' metallic systems by taking into account the strong anisotropy, disorder and correlation effects.

While such simplified, 'experimentalists' scheme does not rule out more evolved theoretical approach to electronic properties, which may well be necessary, it facilitates classification of reliable experimental facts and consequently helps theorists in their gradual advancements in microscopic theory of HTSC oxides.

ACKNOWLEDGEMENTS: We gratefully acknowledge financial support by PTT (Bern), constructive comments of L. Forro (EPFL), W. Sadowski and T. Graf (Geneva University) and many fruitful discussions with participants of the Miami workshop.

REFERENCES

[1] P.W. Anderson, Physics Today, p.9, September 1990
[2] S.J. Hagen et al., Phys Rev B **37**, (1988)
[3] P.W. Anderson and Z.Zou, Phys. Rev. Lett. **60**, 132 (1988)
[4] M. Gurvitch and A.T. Fiory, Phys. Rev. Lett. **59**, 1337 (1987)
[5] L. Forro et. al., Physica Scripta **41**, 365 (1990)
[6] T.A. Fridmann et al., Phys. Rev. B **42**, 6217 (1990)
[7] T. Graf, G. Triscone, J. Müller, J. Less-comm. Met. **159**, 349 (1990)
[8] T.Ito, H.Takagi, S.Ishibashi, T.Ido, S.Uchida, Nature **350**, 596 (1991)
[9] M. Affronte et al., Physica C **172**, 131 (1990)
[10] S. Martin et al. Phys. Rev B **39**, p.9611 (1989)
[11] S. Massidda et al., submitted to Physica C
[12] W. Sadowski et al., J. of Less Comm. Metals, **164-165**, 824 (1990)
[13] S.J. Hagen et al., submitted to Phys. Rev. B (1991)
[14] Y. Hidaka and M. Suzuki, Nature, **338**, 635 (1989)
[15] Z.Z. Wang et al., Phys. Rev.B, **43**, 3020 (1991)
[16] C.C. Tsui et al., Physica C **161**, 415 (1989)
[17] N. Hamada et al., Phys. Rev. B **42**, 6238 (1990)
[18] J.L. Cohen et al. Phys. Rev. Lett. **66**, 1098 (1991)
[19] M. Affronte et al., Phys. Rev.B, in print (1991)

NORMAL-STATE TRANSPORT PROPERTIES OF $YBa_2Cu_3O_{7-\delta}$:

CONVENTIONAL METALLIC PICTURE

J. L. Cohn[a], S. A. Wolf[a], V. Selvamanickam[b], and K. Salama[b]

a) Materials Physics Branch, Naval Research Laboratory, Washington, DC 20375

b) Texas Center for Superconductivity at University of Houston, Houston, TX 77004

INTRODUCTION

The normal-state and superconducting properties of the high-T_c cuprate superconductors offer a great challenge for theoretical explanation. Because many of the normal-state properties (electrical resistivity, optical conductivity, Hall coefficient) are unlike those observed in typical metals, some theories have stressed the non-Fermi-liquid[1] or marginal Fermi-liquid[2] nature of these materials. Others[3] have emphasized a conventional Fermi-liquid picture, taking into account a small value of the Fermi energy and anisotropic crystal structure.

We focus our attention in this paper on the transport properties of crystalline $YBa_2Cu_3O_{7-\delta}$ (YBCO) with particular emphasis on the thermoelectric power.[4] Investigating intrinsic properties requires the study of high-quality crystals. Although YBCO is the most widely studied and well-characterized cuprate material, difficulties associated with oxygen instability persist. The oxygen content of single crystals is difficult to control and to quantify. For these reasons the effect of small oxygen deficiencies ($\delta \leq 0.2$) on the normal-state transport in crystalline YBCO remains poorly understood. Yet the electronic structure of "92K" YBCO is quite sensitive to small changes in oxygen content in this regime near full oxygenation. Arguably the most dramatic example of this is the change in sign of the thermoelectric power (S) in polycrystalline specimens[5] which occurs for $\delta \approx 0.1$. In view of this result a multicarrier picture appears necessary for $YBa_2Cu_3O_{7-\delta}$ The temperature dependence of S, and specifically the occurence of a peak (or minimum for S < 0) in the range T_c < T < 150K, has been attributed to a variety of different mechanisms appropriate for metals.[4] A consistent correlation between the oxygen-doping and temperature dependence of S is a crucial, but missing, ingredient to a complete description of the cuprate normal state.

In discussing the thermopower we draw on measurements of the electrical conductivity to develop a conventional metallic description of the normal state. Our picture of the thermopower incorporates strong phonon-drag effects and two carrier species: holes on the planes and electrons on the chains. The most striking observation is the appearance of a sharply temperature dependent, positive contribution to S for T ≤ 160K, which is observed irrespective of the sign of S (i.e. independent of δ). A related feature is manifested in the electrical resistivity. We demonstrate that these observations can be explained by a freeze-out of carrier-phonon Umklapp processes involving holes in the CuO_2 planes. The data are consistent with the calculated[6] and measured[7] Fermi surfaces for $YBa_2Cu_3O_{7-\delta}$, and support a picture of strong coupling between the planar holes and optical-mode phonons.

EXPERIMENTAL

Preparation of the liquid-phase processed (LPP) specimens is described elsewhere.[8] This material is composed of large, twinned crystalline grains (up to 4mm dimensions in the ab-plane and 10-20μm thick) which are highly oriented along the c-axis during growth. X-ray analysis indicates that the misorientation of the c axes is less than five degrees. Small amounts of Y_2BaCuO_5 are present as distinct entities between the grains. The high degree of crystalline integrity of this material is evidenced by low ab-plane electrical resistivities [ρ(300K)=250-500 ($\mu\Omega$-cm)] and transport critical current densities[9] comparable to those measured in single crystals. The two LPP specimens studied (8×2×1 mm^3) had sharp resistive transitions ($\Delta T_c \leq 0.5K$), with R=0 at 92.5K. The LPP material is especially useful in this study because full oxygenation ($\delta \approx 0$) is more readily achieved by oxygen anneal than it is in crystals. The oxygen content of the LPP specimens is inferred by comparing the measured thermopower curves with the data of Ouseph and Ray O'Bryan,[5] as discussed below. The two $YBa_2Cu_3O_{7-\delta}$ crystals were grown by a flux method[10] and are designated CRYSTAL1 and CRYSTAL2. CRYSTAL1 had $\Delta T_c \approx 1K$ (R=0 at 90K) and CRYSTAL2 had $\Delta T_c \approx 0.3K$ (R=0 at 92.5K). Following reports discussing anomalies in the magnetization of single crystals[11] we have conducted x-ray and magnetization studies[12] of these and similarly-prepared crystals which indicate a correlation between c-axis lattice parameter (a measure of oxygen content) and features of the magnetic hysteresis. From these results we estimate oxygen deficiencies of $\delta \approx 0.16$ and $\delta \approx 0.08$ for CRYSTAL1 and CRYSTAL2, respectively.

RESULTS AND DISCUSSION

The thermopower was measured using a steady-state technique, employing a differential chromel-constantan thermocouple, glued to the specimen with varnish, and Cu Seebeck probes silver painted onto fired-on silver pads. Temperature gradients were typically 0.4K-1.0K. Linearity in the V(ΔT) response was confirmed throughout the temperature range by varying the heater power. The precision of the measurements is ±0.1μV/K.

Figure 1 shows the thermopower versus temperature for each of the four samples. Also shown is data for specimen LPP1 following an oxygen anneal (150H at 450°C in 1 atmosphere of flowing oxygen), and designated LPP1A. The change in sign and magnitude of the thermopower supports a picture of multiband conduction,[5] with the dominant carrier species changing from electron-like to hole-like with increasing δ. We will return to the issue of the sign change below. Comparing the curves in Fig. 1 with those of ref. 5 we estimate δ values for our specimens as: 0.15 (CRYSTAL1), 0.12 (LPP1), 0.06 (CRYSTAL2), 0.02 (LPP1A), 0.01 (LPP2). Note that these values for the single crystals are in accord with our estimates by independent means, as discussed above.

We begin our analysis of the thermopower within the context of a conventional metallic picture, in which contributions to S are due to

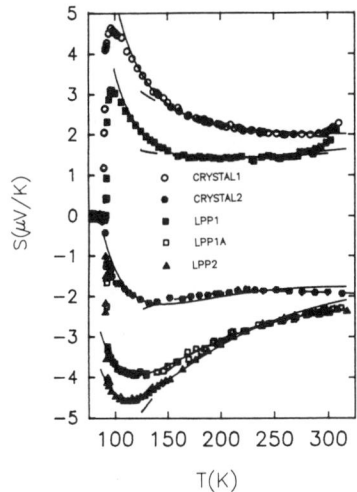

Fig. 1. Thermopower versus temperature. The dashed lines are fits to the simple metallic expression and the solid lines include the umklapp term.

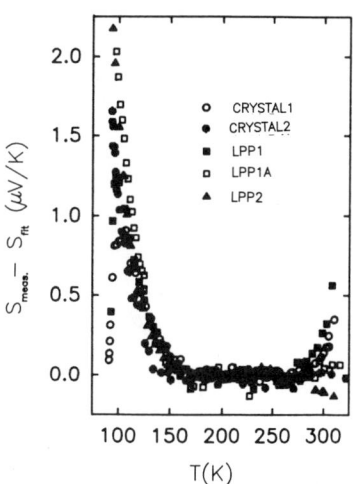

Fig. 2. Difference between measured thermopower and the dashed lines in Fig. 1.

carrier diffusion (S_d) and phonon drag (S_g). At high temperatures (typically $T \geq \Theta_D/4$), simple metals can be described by the form,[13] $S=S_d+S_g=AT+B/T$. This is representative of the thermopower in, for example, the noble metals for the range 77K $\leq$ T $\leq$ 250K. We see that this expression also fits the data in Figure 1 (dashed curves) for 160K $\leq$ T $\leq$ 275K. We find A,B > 0 for the crystal and LPP1, and A,B < 0 for LPP1A and LPP2.

A sharp, positive upturn away from the simple metallic expression is observed in all specimens for T < 160K. Interestingly this additional contribution to S appears to be nearly the same in magnitude and temperature dependence for all samples, as shown in Fig. 2 where we plot the difference $S_{meas.} - S_{fit}$ versus T. Note that it is this component of S which gives rise to the very sharp peak when S > 0 and to the broader minimum when S < 0. Here we see that these features appear to arise from a common mechanism.

Figure 3 shows the electrical resisitivity data $\rho(T)$ and its derivative, $d\rho/dT$, for crystal 2. There is considerable deviation from linearity throughout the temperature range. This behavior is now widely observed in high-quality $YBa_2Cu_3O_{7-\delta}$ crystals.[14] Of particular interest is the apparent increase in negative curvature (i.e. increase in $d\rho/dT$) that occurs at T $\leq$ 160K, coinciding with the increase in the positive thermopower component (Fig. 2). We suggest that the features in both transport coefficients arise from the same mechanism.

Several authors[4] have suggested that the peak might be caused by an enhancement of the phonon-drag thermopower due to superconducting fluctuations (conventional superconducting fluctuations decrease S toward S=0 and thus cannot explain the data). We have investigated this possibility by measuring the thermopower of sample LPP2 in a 6-T magnetic field applied along the c axis (Fig. 4). The magnetothermopower of the thermocouple and Cu leads (which amount to < 2%) have been accounted for. We find that, within the measurement uncertainty, there is no field

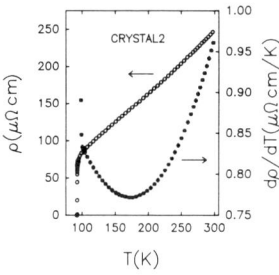

Fig. 3. Electrical resisitivity and its derivative for CRYSTAL2.

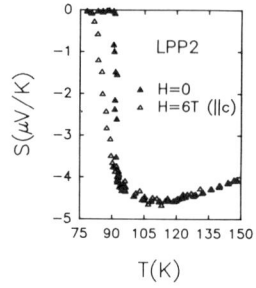

Fig. 4. Magnetothermopower for specimen LPP2.

dependence of S for $T \geq 95K$ (the magnetothermopower for $T < T_c$ is due to flux motion). The magnetoresistance of this specimen is also negligible for $T > 95K$. Hikita and Suzuki[15] have demonstrated that the magnetoresistance for single-crystal YBCO in a field of 15 Tesla is less than 1% for $T > 95K$, insufficient to account for all of the curvature in the zero-field $\rho(T)$ data. We conclude that superconducting fluctuations play a negligible role in determining the nature of the in-plane thermopower peaks (or minima) and the resistivity curvature at $T > 95K$ in $YBa_2Cu_3O_{7-\delta}$.

We propose a new explanation of these features which involves a freeze-out of carrier-phonon umklapp (large-angle) scattering. The assumptions implicit in the result $S_g \propto 1/T$ do not take into account the shape of the Fermi surface (FS) and its effect on the phonon population. For example, when the FS does not touch the Brillouin Zone boundary, the phonon wavevectors which may induce umklapp events are restricted to $q \geq q_{min}$ (see Fig. 5). This condition translates, through the population factor and the dispersion relation, into a temperature criterion for the "freeze-out" of U-processes. In the situation depicted in Fig. 5, for $T \ll \Theta^* = \hbar\omega(q_{min})/k_B$, we should anticipate a U-process phonon-drag thermopower given by $S_g^U \sim (1/T) \exp(-\Theta^*/T)$, reflecting the freeze-out of the Bose-Einstein distribution. A similar exponential decay would be expected in the electrical resistivity where umklapp events play an important role. This picture has been applied successfully in studies of the alkali metals,[14] where FS's are nearly spherical.

Band structure calculations[6] and photoemission experiments[7] for $YBa_2Cu_3O_{7-\delta}$ support a picture in which quasi-two-dimensional portions of the FS, arising from the CuO_2 planes, do not touch the BZ boundary throughout a substantial portion of the zone. Thus there is precedent for precisely the situation described above. For a simply-connected FS (the plane-derived surfaces are "rounded squares"), the scattering vector $K = k' - k$ tends to be directed opposite to q in an umklapp event. This leads to a phonon drag that is opposite in sign to that of the charge of the carriers. Thus for holes in the CuO_2 planes, S_g^U should be negative. A freeze-out of U-processes involving these planar holes might then be expected to yield a rapidly increasing, positive contribution to S that could account for the behavior observed in the experiments. With regard to the phonon spectrum we note that the measured acoustic phonon branches in $YBa_2Cu_3O_{7-\delta}$ do not extend in energy beyond ~10 meV (115K).[16] Since the thermopower data require $\Theta^* \geq 160K$, we conclude that if a U-process freeze-out is to

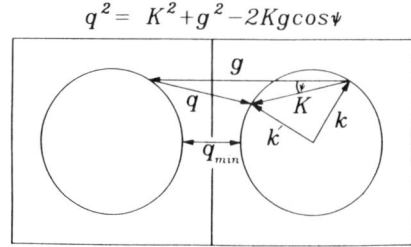
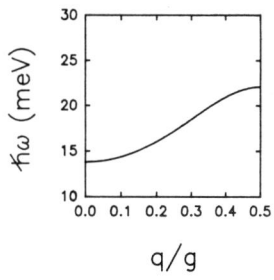

Fig. 5. Schematic diagram of a simple Fermi surface on a square lattice, showing the geometry of the umklapp process.

Fig. 6. Phonon branch used to calculate the Umklapp term.

explain our results, the relevant phonons must be optical modes.

We have recently presented[17] a calculation of S_g^U following the phonon-drag thermopower expression due to Bailyn.[18] The CuO_2 plane-derived FS are approximated by a single Fermi circle (Fig. 5). For simplicity we have considered a single phonon branch, and employed an expression for $\omega(q)$ which is quadratic in q for small q and flattens at the zone boundary, approximating the form of the measured[16] optical branches in the range 10-30 meV. The resulting Umklapp term, $S_{g,pl}^U$, is added to the simple metallic expression, and the data in Figure 1 refitted with the constraint that the magnitude of the umklapp term, represented by an adjustable prefactor, be the same for all specimens. Note that the term B/T accounts for the normal-process (small-angle scattering) phonon drag from the planes, $S_{g,pl}^N$, as well as $S_{g,ch}$. Excellent agreement with the data (solid lines in Fig. 1) is found using the phonon mode shown in Fig. 6. The values of the other parameters and a more detailed discussion may be found in ref. 17.

Based on our results we arrive at the following picture of transport in $YBa_2Cu_3O_{7-\delta}$. For $\delta \approx 0$ we postulate two conduction bands: holes in the planes and electrons on the chains.[19] This supposition regarding the chains is supported by positron-lifetime experiments.[20] The overall diffusion component of the thermopower (in a twinned crystal) is negative, presumably reflecting a chain mobility higher than that of the planes.[21] Oxygen vacancies at this level of deficiency ($\delta \leq 0.16$) are on the chain site. The removal of oxygen should have two principal effects, both of which will contribute to the observed sign change of S: a decrease in S_{ch} due to a reduction in the average charge on the chains and a decrease in the effective chain mobility due to oxygen disorder. Measurements of the planar Cu bond-valence sums[22] indicate that the hole concentration in the planes is nearly constant for $\delta \leq 0.2$. This tends to support our observation that $S_{g,pl}^U$ is the same in the specimens studied.

ACKNOWLEDGEMENTS

We are grateful for the efforts of T. Vanderah in making the $YBa_2Cu_3O_{7-\delta}$ crystal, to E. Skelton for conducting x-ray measurements, and to R. J. Soulen, A. Erlich, M. Miller, M. Osofsky, M. Reeves, for many fruitful discussions. One of us (J. L. C.) acknowledges fellowship support from the Office of Naval Technology. The work at the Texas Center for Superconductivity at University of Houston was supported under prime grant MDA972-88-G-0002 from the Defense Advanced Projects Agency and the state of Texas.

REFERENCES

1. P. W. Anderson, Science **235**, 1196 (1987); R. B. Laughlin, Phys. Rev. Lett. **60**, 2677 (1988).
2. C. M. Varma, P. B. Littlewood, S. Schmitt-Rink, E. Abrahams, and A. E. Ruckenstein, Phys. Rev. Lett. 63, 1996 (1989).
3. V. Z. Kresin and S. A. Wolf, S. S. Commun. 63, 1141 (1987); Phys. Rev B41, 4278 (1990).
4. For a recent review of thermopower in the cuprates see, A. B. Kaiser and C. Uher, in <u>Studies of High Temperature Superconductors</u>, Vol. 7, edited by A. V. Narlikar (Nova Science Publishers, New York, 1990).
5. P. J. Ouseph and M. Ray O'Bryan, Phys. Rev. B41, 4123 (1990).
6. J. Yu, S. Massida, A. J. Freeman, and D. D. Koelling, Phys. Letts. A122, 203 (1987); H. Krakauer, W. E. Pickett, and R. E. Cohen, J. Supercond. 1, 111 (1988).
7. J. C. Campuzano, G. Jennings, M. Faiz, L. Beaulaigue, B. W. Veal, J. Z. Liu, A. P. Paulikas, K. Vandervoort, H. Claus, R. S. List, A. J. Arko, and R. J. Bartlett, Phys Rev. Lett. **64**, 2308 (1989).
8. K. Salama, V. Selvamanickam, L. Gao, and K. Sun, Appl. Phys. Lett 54, 2352 (1989).
9. T. L. Francavilla, V. Selvamanickam, K. Salama, and D. H. Liebenberg, Cryogenics **30**, 606 (1990).
10. T. A. Vanderah, M. S. Osofsky, C. K. Lowe-Ma, E. A. Skelton, D. E. Bliss, and M. W. Decker, to be published, J. Crystal Growth.
11. M. Daeumling, J. M. Seuntjens, and D. C. Larbalestier, Nature 346, 332 (1990).
12. M. S. Osofsky, J. L. Cohn, S. Qadri, E. A. Skelton, R. J. Soulen Jr., and S. A. Wolf, unpublished.
13. F. J. Blatt, <u>The Thermoelectric Power of Metals</u>, (Plenum, New York, 1976).
14. T. A. Friedmann J. P. Rice, J. Giapintzakis, and D. M. Ginsberg, Phys. Rev B39, 4258 (1990).
15. M. Hikita and M. Suzuki, Phys. Rev. B39, 4756 (1989).
16. W. Reichardt, D. Ewert, E. Gering, F. Gompf, L. Pintschovius, B. Renker, G. Collin, A. J. Dianoux, and H. Mutka, Physica **156 & 157B**, 897 (1989).
17. J. L. Cohn, S. A. Wolf, V. Selvamanickam, and K. Salama, Phys. Rev. Lett. **66**, no. 8 (1991).
18. M. Bailyn, Phys. Rev. 157, 480 (1967).
19. V. Z. Kresin and S. A. Wolf, Physica **169C**, 476 (1990).
20. Y. C. Jean, J. Kyle, H. Nakanishi, P. E. A. Turchi, R. H. Howell, A. L. Wachs, M. J. Fluss, R. C. Meng, H. P. Hor, J. Z. Huang, and C. W. Chu, Phys. Rev. Lett. **60**, 1069 (1988); V. Z. Kresin and H. Morawitz, J. Superconductivity 3, 227 (1990).
21. T. A. Friedmann, M. W. Rabin, J. Giapintzakis, J. P. Rice, and D. M. Ginsberg, Phys. Rev. B42, 6217 (1990); U. Welp, S. Fleshler, W. K. Kwok, J. Downey, Y. Fang, G. W. Crabtree, and J. Z. Liu, Phys. Rev. B42, 10189 (1990).
22. R. J. Cava, A. W. Hewat, E. A. Hewat, B. Batlogg, M. Marezio, K. M. Rabe, J. J. Krajewski, W. F. Peck Jr., and L. W. Rupp, Physica **165C**, 419 (1990).

ELECTRON TUNNELING AND ACOUSTIC MODE IN HIGH-T_c SUPERCONDUCTORS

A. Kallio

Department of Theoretical Physics, University of Oulu
Linnanmaa, SF-90570 Oulu 57, FINLAND

ABSTRACT

We show that the electron-hole liquid model can explain some of the anomalous features found in the tunneling conductivity for NIS-junctions. The zero bias conductivity is shown to be due to the mobile fermion component present in the superconducting state. The deviation of the measured shapes from the BCS-curves is proposed to be due to the existence of the acoustic plasmon mode predicted by the model. The openings observed in the density of states function are proposed to be due to the crossing of the sound mode and pair breaking excitations. The asymptotic behaviour of the density of states is different depending whether the crossing takes place or not. In the present calculation we used BCS-pairs as the model for the bosons. If one could measure the zero bias conductivity as a function of temperature, also above T_c, it would give information about the pair breaking and hence offer the possibility to decide between various models for the bosons.

INTRODUCTION

Electron tunneling is a powerful method for measuring the temperature dependence of the BCS-energy gaps and the corresponding density of states near Fermi-surface[1]. For this reason tunneling experiments have been a major undertaking also in the case of high-T_c superconductors. For NIS-junctions the results generally obtained differ from the predictions of BCS-theory mainly on three points: (i) The tunneling conductance is non-zero[2,3] near $T = 0$. (ii) If the inner U-shaped curve for the differential conductance is fitted with the usual BCS-formula, the large-voltage region goes wrong[3]. (iii) The V shaped curves seem to exist[4] well above T_c.

The aim of this paper is to propose that these features can be understood within the electron-hole-liquid model (EHL)[5,6]. This model assumes the existence of mobile fermion component in the superconducting state, in addition to the mobile bosons. For hole-doped superconductors the charges are -e for the fermions and +2e for the bosons. The simplest interpretation is that the bosons are bound $1S_0$-states of two oxygen holes in CuO_2-planes in which case EHL-model is a generalization of the conventional BCS. For the electron-doped superconductors the charges would be of opposite signs.

The superfluid state for EHL would be analogous to a mixture of ^{3}He and ^{4}He. Such a system exhibits two types of collective excitations: a high-lying plasmon mode and an acoustic plasmon mode. The electron sound velocity u, associated with the acoustic mode, was shown[6] to be of the order of magnitude but larger than the fermi velocity v_F corresponding to the fermion density at T= 0. The mixture is superconducting between two boson concentration points x_1 and x_2 such that $x_1 < x < x_2 < 1$. Outside this range the sound mode is Landau-damped and $T_c = 0$.

In analogy with ^{3}He + ^{4}He -mixtures the bosons are superfluid (and superconducting) below $T_\lambda = T_c$, where the superfluid density vanishes $n_s (T_c) = 0$. The bosons may exist at temperatures higher than T_c up to the temperature T_{BCS}. The normal component of bosons consists of two parts: the pair breaking density $n_{BP}(T)$ and the density $n_{BA}(T)$ of exitations into the sound mode. At any temperature below T_c one has by the Landau two fluid model

$$n_S(T) + n_{PB}(T) + n_{BA}(T) = n_B, \qquad (1)$$

where n_B is the constant boson density. The electrons belong to the normal component but are decoupled from n_{PB} and n_{BA} by the charge conservation[6].

If we assume the BCS-model for the bosons and $T_{BCS} >> T_c$, the effect of $n_{PB}(T)$ should be moderatelly small in Eq. (1) below T_c, but continues to exist above T_c. In majority of experiments it is found that $2\Delta/k_B T_c$=6-11. If in the EHL-case the bosons are of BCS structure with $2\Delta/k_B T_{BCS}$=2-3, their binding would then be within conventional BCS-range.

On the basis of this 3D-theory for a uniform mixture, one can understand the doping, the pressure dependence of T_c for both electron doped and hole-doped superconductors, the linear term and the jump at T_c of the specific heat[6]. These results were obtained without specifying what the internal structure of the bosons is. In what follows we will assume the BCS-structure for them in order to see how far we can understand the tunneling results still without pinpointing any specific binding mechanism.

ELECTRON TUNNELING

Within the EHL-model, with BCS-bosons, the tunneling current will get contribution from the electrons and the bosons broken into quasi holes[7]:

$$\begin{aligned}I_{NS}(T,V) &= A\mid T\mid^2 N_e(0) N_2(0) eV \\ &+ A\mid T\mid^2 N_2(0) \int_{-\infty}^{\infty} N_{hs}(E)[f_F(E) - f_F(E+eV)]dE \end{aligned} \qquad (2)$$

Here $N_e(0)$ is the density of states for the fermion component and $N_2(0)$ is the corresponding quantity for the other metal. The density of states for the quasi holes may be defined by

$$N_{hs}(E) = \begin{cases} \dfrac{E N_h(0)}{\sqrt{E^2 - \Delta^2}} g_h(E) \\ 0, E < \Delta \end{cases} \qquad (3)$$

The function $g_h(E)$ is a measure of deviation from BCS and also contains the restriction for E to be below the Fermi-surface, since the limits of integration are infinite, in Eq. (2), and possible band structure effects.

This then gives for the differential conductivity the obvious expression

$$\frac{dI_{NS}}{dV} = \sigma_e + \sigma_e \int_{-\infty}^{\infty} \frac{N_{hs}(E)}{N_e(0)} \left[-\frac{\partial f_F(E+eV)}{\partial(eV)} \right] dE \qquad (4)$$

One observes immediately that for small T and $eV \gg \Delta(0)$ one gets the limits

$$\frac{dI_{NS}}{dV} = \sigma_e + \sigma_e \frac{N_{hs}(e|V|)}{N_e(0)} \to \sigma_e + \sigma_e \frac{N_h(0)}{N_e(0)} g_s(E) \qquad (5)$$

If we take the pure BCS-case with $g_s(E) = 1$, then for moderately large E one should get a constant density of states ratio $N_h(0)/N_e(0)$ from Eq. (5). The value of this ratio can show large sample dependence due to the disorder which can effect $N_e(0)$. For moderate voltages one measures the density of states of the quasi holes near the (common) Fermi-surface in terms of the density of states $N_e(0)$ for electrons. This of course is different from the quantity measured for ordinary BCS-superconductors. Another difference is the appearance of zero bias conductivity σ_e which here is due to the electrons present in the superconducting state as a part of the normal component. We call the density of states behaviour of this kind, the non-crossing case.

This result may be compared with the experimental data from ref. 3 for $Bi_2Sr_2Ca Cu_2O_{8+\sigma}$. Fig. 1a shows the existence of a zero bias conductance σ_e for V=0. Hence the electron component in the EHL-model offers a natural explanation for the "leakage current". Fig. 1b gives the conductivity also for very large V which shows that in this limit very small density of states ratio is obtained. The BCS-theory fit in ref 3 seems to describe very well the inner part of the U-shaped curve, but the outer part goes wrong as we predict. The model used there for the density of states

$$D(E,\Gamma) = Re[(E - i\Gamma)/\sqrt{(E - i\Gamma)^2 - \Delta^2}\,] \qquad (6)$$

has the advantage to also fit the zero bias conductivity but this has no theoretical basis. Even in the present case one would have theoretical[8] basis for using complex Δ with

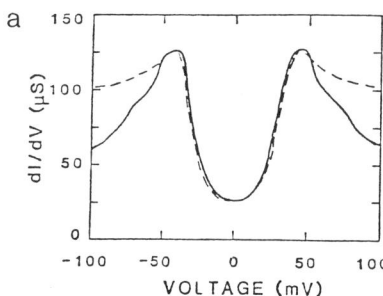
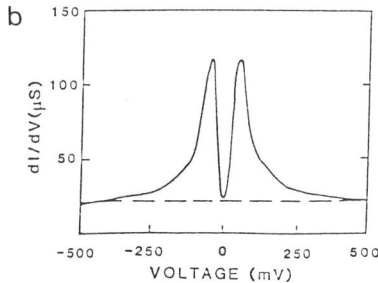

FIG. 1. Behaviour of differential conductivity in the non-crossing case. (a) Data and the fit with BCS-formula from Eq. (6) of ref. 3. (b) Large voltage behaviour of the conductivity from ref. 3.

$$D(E, \Delta(E)) = E \, Re[(E^2 - \Delta^2)^{-1/2}], \tag{7}$$

since also here the damping of quasiparticles due to the creation of real phonons can occur. Further on we discuss the possibility of creating quasiparticles by the acoustic plasmon mode in the crossing case. Even without crossing the existence of sound mode below the quasiparticle exitation can give contribution to the imaginary part. Needless to say one can obtain a good fit for the U-shapes with this conventional density of states model. However, to obtain good fit for the large-V region also modification of Eq. (7) by $g_h(E)$ of Eq. (3) is needed. We believe that the fit by Eq. (7) gives much smaller width but it has to be energy dependent.

The clearest effect of the EHL-sound mode may have been observed in the case the electron-doped superconductor NdCeCuO. In complete analogy with Eq. (5) the small-T large-V behaviour would now be

$$\frac{dI_{NS}}{dV} \approx \sigma_h + \sigma_h \frac{N_{es}(e\,|\,V\,|)}{N_h(0)} \to \sigma_h + \sigma_h \frac{N_e(0)}{N_h(0)} \tag{8}$$

predicting again a constant differential conductivity. But, on the contrary, for many junctions the density of states is experimentally found to rise[9,10]. Can one explain the V-shape of curves observed[9] for electron superconductor $Nd_{2-x}Ce_xCuO_{4-y}$?

If we accept the above theory for the density of states function there is a further consequence from the fact that the sound mode and the pair breaking excitation curves can cross at two energies, the lower one near 2Δ and the higher one at an energy larger than 2Δ. The existence of these thresholds would also produce two openings in the density of states function because of the repulsion of the crossing excitations. The position of the crossings is determined by the equation

$$E = \hbar u k = \sqrt{\xi_{\vec{k}_1}^2 + \Delta^2} + \sqrt{\xi_{\vec{k}_2}^2 + \Delta^2}, \tag{9}$$

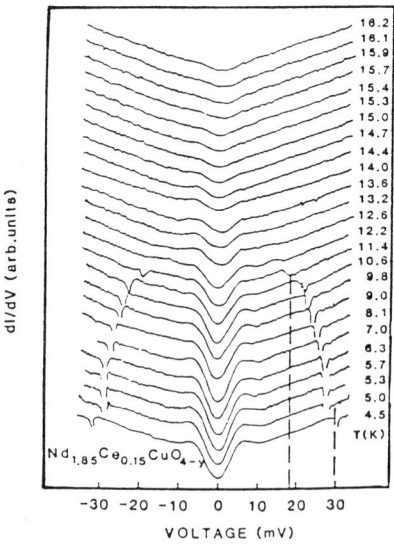

FIG. 2. Experimental results from ref. 9 for $Nd_{2-x}Ce_xCuO_{4-y}$ exhibiting the effect of crossing: the opening anomalies (marked with vertical lines).

where $\vec{k}_1 = \vec{q}+\vec{k}/2$, $\vec{k}_2 = -\vec{q}+\vec{k}/2$ and $\vec{q}$ is a vector on the Fermi-surface: $|\vec{q}| = k_F$. Eq (9) contains the energy and momentum conservation. In the most simplified model of spherical Fermi-surface we have approximately

$$E_2 = 4mu^2 \gg 2\Delta \tag{10}$$

for the sound mode energy at the second crossing. In between the two energies E_1 and E_2 the sound mode is damped by breaking up bosons into two quasi particles. Since the two excitation curves lie close to one other they repel and get mixed. In particular, near E_1 and E_2, the energy states have openings which should show up in the density of states function. Such openings may have been seen in ref. 9 (see Fig. 2). In fact the opening anomaly found there would correspond to quasiparticle energy $E_2/2 = 2mu^2$. Since the sound velocity should diminish at higher temperatures as it does in the case of He-liquids we anticipate that this opening moves to smaller energy at higher temperatures.

If this idea is correct, one can use electron tunneling to measure also the temperature dependence of u. Why then only one opening shows up? The reason may be the smallness of the opening near $E_1/2$ whereas at $E_2/2$ the curves are nearly parallel and correspondingly the opening is larger. A sketch of the situation is shown in Fig. 3. With the values $E_2/2 \equiv 30 meV$ for the opening anomaly from ref. 9 we obtain for the electron sound velocity the value $5.2 \cdot 10^4$ m/s with effective mass $m = m_e$ for the electron doped NdCeCuO-superconductor. For hole-superconductors BiSrCuO one obtains in the same way sound velocities of the order $u \sim 10^5$ m/s using the data of ref. 10 (sample S1) at T=4.2 K.

This however is not the whole story yet. Since the sound mode and pair breaking excitations are heavily mixed, also in between the crossing points E_1 and E_2, one will have quasiparticles with the sound mode dispersion in the range $E'_1 < E_q < E'_2$ where $E'_1 \lesssim E_1/2$ and $E'_2 \gtrsim E_2/2$ and one obtains an additional contribution $\Delta\sigma$ to the tunneling conductivity from them. In the electron doped case this amounts to

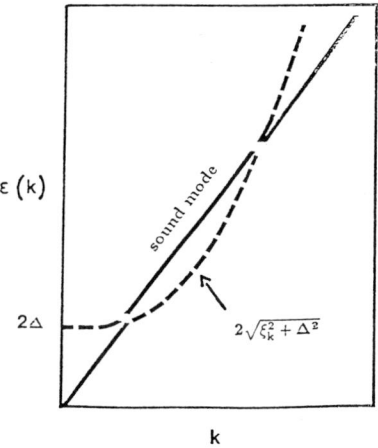

FIG. 3. Qualitative picture of the crossing by Eq. (9) with $\vec{q} \perp \vec{k}$, giving minimum energy for the two quasiparticles.

$$\Delta\sigma_e = \sigma_h \int_{-\infty}^{\infty} \frac{N_{es}(E)}{N_h(0)} g_s(E) \left(-\frac{\partial f(E+eV)}{\partial(E+eV)} \right) dE, \qquad (11)$$

where $g_s = 0$ inside the openings near crossings and outside the intergrations range $E'_1 < E < E'_2$. Inside the range one obtains

$$g_s(E) = \frac{1}{2}(E/mu^2)^{3/2}. \qquad (12)$$

To derive this expression one can start the k-integration with z-axes in the direction of vector $\vec{q}$ and with the quasiparticle energy

$$E_k = \frac{\hbar u k}{2} \simeq \sqrt{\xi^2_{\vec{q}+\vec{k}/2} + \Delta^2} \qquad (13)$$

In the case of crossing one clearly obtains a mechanism for an increasing density of states. Collecting all terms we obtain for the crossing case the differential conductivity for $T \approx 0$

$$\frac{dI}{dV} = \sigma_h + \sigma_h \frac{N_e(0)}{N_h(0)} [Re\left(\frac{|E|}{\sqrt{E^2 - \Delta^2}} \right) + \beta (E/mu^2)^{3/2}]. \qquad (14)$$

The coefficient β takes care of some other effects not included in the present simplified treatment. With parameters $\Delta = 4meV$, $\Gamma = 1meV$ one can obtain very good fit for the inner part since the zero bias conductivity σ_h in the present approach is separated from the fit. The outer part is fitted better with a linear energy dependence, but with $\beta = 1/2$ Eq. (14) gives very satisfactory fit for the whole energy range as is shown in Fig. 4. From the experiments referred to here, there seems to be evidence that in the non-crossing case the density of states goes to a constant but for the crossing situation we obtain an $E^{3/2}$-behaviour. Interestingly the same arguments applied to a two-dimensional case would give instead of Eq. (12) $g_s(E) = E/mu^2$, a linear behaviour which is experimentally observed in ref. 9. Unfortunately the parabolic effects due to the barrier structure[11] makes the analysis of the experimental situation more complicated. On the other hand the theory behind the extra term in Eq. (12) is only an estimate. Since we would need an imaginary part of Δ for $|eV| < 2\Delta$ then at least near the lower crossing this can have an effect. In any case it is premature to draw conclusions about the dimensionality since other experiments show also a non-linear behaviour.

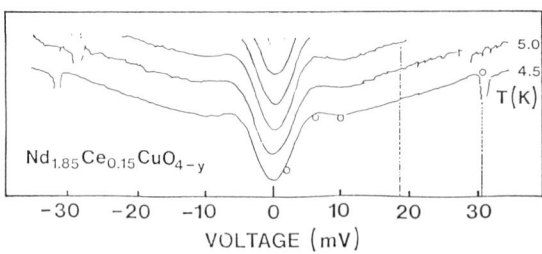

FIG. 4. Theoretical points calculated from Eq. (14) with $\Delta = 4meV$, $\Gamma = 1meV$ and $\sigma_h N_e(0)/N_h(0) = 3.11 \cdot \sigma_h$ as compared with experiment from ref. 9.

For hole-doped superconductor YBCO we have estimated from Raman back ground continuum the electron sound velocity to be roughly 10^6 m/s. In the case of YBCO we have been able to deduce[12] the temperature dependence of u near T_c using the cubic term of the electronic specific heat which has now been measured[13] with sufficient accurancy. The value obtained at T_c is smaller and from the temperature T=40 K downwards it climbs up. If our findings there are correct the temperature dependence obtained would resemble that of the second sound in ^{4}He. Since most probably one can create acoustic phonons in the range $E_1' < \hbar\omega < E_2'$ by shining the junction with suitable light pulse one can externally modultate the tunneling current. Such phenomanon may have been observed in SIS-junction recently[14]. The occurence of analogous non-equilibrium processis via optical phonons in conventional BCS-superconductors has been extensively studied[15]. Since the crossing with optical phonons also creates similar openings the identification becomes difficult and therefore the values for u reported here are preliminary.

Our final point concerns the temperature dependence of the zero bias points. The components of normal liquid density $n_{PB}(T)$ and $n_{BA}(T)$ of Eq. (1) can be calculated with the boson and fermion distribution functions f_B and f_F for the case of electron doping as follows:

$$n_{PB}(T) = \frac{1}{3}\frac{\hbar^2}{m_e^*} \int \frac{k^2 d^3k}{(2\pi)^3}\left(-\frac{\partial f_F}{\partial E}\right) \cong \frac{1}{3}\frac{\hbar^2}{m_e^*}\frac{k_F^2}{2\pi^2}N_e(0) \int N_{es}(E)dE\left(-\frac{\partial f}{\partial E}\right) \quad (15)$$

$$n_{PA}(T) = \frac{1}{3}\frac{\hbar^2}{m_B} \int \frac{k^2 d^3k}{(2\pi)^3}\left(-\frac{\partial f_B}{\partial E}\right) \sim (T/T_c)^4 \quad (16)$$

By Eq. (4), if only the BCS pair breaking is included, we obtain approximately

$$\left.\frac{dI_{NS}}{dV}\right|_{V=0} \cong \sigma_h + \sigma_h \frac{N_e(0)}{N_h(0)}\frac{3m_e^* 2\pi^2}{\hbar^2 k_F^2} n_{PB}(T) \quad (17)$$

If the BCS-theory were valid up to T_{BCS}, the low temperature and the high temperature behaviours are known to be[16]

$$\frac{n_{PB}}{n_B} \cong \frac{1}{2}\sqrt{\frac{2\pi\Delta(0)}{k_B T}} exp\left(-\frac{\Delta(0)}{k_B T}\right), \quad for \ T \to 0 \quad (18)$$

$$\frac{n_{PB}}{n_B} \cong \frac{T}{T_{BCS}}, \quad for \ T \to T_{BCS} \quad (19)$$

In the case of crossing in Eq. (15) one will get a contribution $\Delta\sigma$ also from the quasiparticles broken by the sound wave. This can be taken into account by calculating the corresponding correction to $n_{PB}(T)$. This would modify the BCS-behaviours in Eqs. (17)-(19).

If the zero bias curve i.e. the left hand side of Eq. (17) is measured, it would be another way of checking the present theory. In particular, it would furnish information about the applicability of the BCS-thory for the bosons both below and above T_c.

The main ingredients still missing in the EHL-model are effects of the plane-structure and the band structure effects other than the effective mass. At the present we are confident that the fermion component has to be present in the high-T_c superconductors. Not only is it required to understand the specific heat linear term but now also to understand the surface resistance[18], where the normal componet is found to be non-zero at low temperatures. Therefore the existence of normal component due to the electrons, even at zero temperature and hence finite surface resistance, is readily understood by EHL-model. Because the normal component $n_{PA}(T)$ due to the acoustic excitation in Eq. (16) gives $(T/T_c)^4$-behaviour it becomes understandable why the two fluid model gives such a good fit to penetration depth as measured by muon-spin-rotation[19]. In the case of high-T_c superconductors there is no reason to call it a phenomenological model anymore. Besides the NIS-junction also Josephson junctions SIS and hence SQUIDs should show a non-BCS-behaviour again due to the normal component electrons and the crossing.

As a conclusion we have indicated that the electron-hole liquid model can at least partly explain the three anomalous features of the tunneling conductivity stated at the beginning. The treatment of having an imaginary part in Δ is not entirely groundless since the acoustic mode excitation lies below the quasi particle excitations and hence can contribute to the imaginary part of Δ for small voltages. Besides the Raman scattering back ground[17] now also the electron tunneling exhibits the effects of the EHL-sound mode. For the first time we have an experimental estimate of the electron sound velocity $u_e \approx 5 \cdot 10^4$ m/s for an electron doped superconductor $Nd_{2-x}Ce_xCuO_{4-y}$. Therefore further confirmation of the existence the anomalous opening structure would be most useful. If the opening anomalies are real, the tunneling conductivity offers a convenient method to measure also the temperature dependence of the electron sound velocity.

The present theory predicts the V-shaped curves up to $T=T_{BCS}$. Hence the "normal" state conductance should be measured above T_{BCS}. Since $T_{BCS} > T_c$ the weak coupling BCS-theory may be sufficient in describing the intrinsic structure of the bosons.

It is most likely that the sound mode predicted by the EHL-model also partly contibutes in the binding of the bosons. We may do this reasoning by including also the plane structure within the EHL-model. The zeroth order situation could be taken to be electrons and holes on separate adjacent sheets, orthogonal to c-direction. Such a 3D-stack of planes would show also an acoustic mode which helps in binding the heavier effective mass component of the two into bosons. From these we obtain the following scenario: We now have an infinite system of planes stacked in c-direction where every second plane contains electrons and the others hole-bosons, all interacting via the Coulomb force, so that whole system becomes 3-dimensional. Again one obtains an electron sound mode. However the sound velocity in c-direction would be zero and hence by Landau tangent criterion the bosons are not superfluid. If, however, the electrons and/or the bosons are allowed to hop in the c-direction the situation may be changed in this respect and the system becomes superfluid. This would then explain the decreasing resistivity with increasing temperature in c-direction for $T<T_c$, since hopping becomes "easier" at higher temperature. The resulting 3D EHL-model may be more "correct" than the present one, because the sound mode there can have directional dependence etc., but qualitatively it will not be very different from the present, much simpler uniform EHL-model.

ACKNOWLEDGEMENTS

This work was supported by the Finnish Academy of Sciences, and, in part, by the office of Naval Research, through Department of the Navy Grant N00014-91-J-1224, and by the University of Miami.

REFERENCES

1. I. Giaever, Phys. Rev. Lett. **5**, 147 (1960); I. Giaever and K. Megerle, Phys. Rev. **122** 1101 (1961).

2. Q. Huang, J.F. Zasadzinski, K.E. Greg, J.Z. Liu and H. Claus, Phys. Rev. **B40**, 9366 (1989).

3. R. Escudero, E. Guarner and F. Morales, Physica **C166**, 15 (1990).

4. P.W. Anderson, Physics Today, September 1990, P 9.

5. A. Kallio and X. Xiong, Phys. Rev. **B41**, 2530 (1990); A. Kallio, X. Xiong and M. Alatalo, in Proceedings of "Recent Progress In Many-Body Theories", Arad, Israel, 1989, Vol. 2 (Plenum, 1990).

6. A. Kallio and X. Xiong, Phys. Rev. **B 43** (1991) (in press).

7. M. Tinkham, Introduction to Superconductivity, McGraw-Hill, 1975 p. 47.

8. D.J. Scalapino, J.R. Schrieffer and J. Wilkins, Phys. Rev. **148**, 263, (1966); G.M. Eliashberg, Zh. Experim. i Teor. Fiz. **38**, 966, (1960).

9. T.Ekino and J. Akimitsu, Phys. Rev. **B40**, 7364 (1989).

10. T.Ekino and J. Akimitsu, Phys. Rev. **B40**, 6902 (1989).

11. E.L. Wolf, Principles of Electron tunneling (Clarendon) (1989), P 38.

12. A. Kallio and X. Xiong, Report 56, University of Oulu, Department of Theoretical Physics (1990).

13. J.W. Loram, K.A. Mirza and P.F. Freeman, Physica **C171**, 243 (1990).

14. R. Laiho, E. Lähderanta, L. Säisä, Gy. Kovacs and G. Zsolt, I. Kirschner and I. Halasz, Phys. Rev. **B42**, 347, (1990).

15. L.R. Testardi, Phys. Rev. **B4**, 2189, (1971); C.S. Owen and D.J. Scalapino, Phys. Rev. Lett. **28**, 1559, (1972); W.H. Parker and W.D. Williams, Phys. Rev. Lett. **29**, 924, (1972).

16. E.M. Lifshitz and L.P. Pitaevskii, Statistical Physics, Part 2 (Pergamon) 1980 p. 91 and 163.

17. A. Kallio and V. Apaja, Report n:o 55, University of Oulu, Department of Theoretical Physics (1990).

18. G. Müller, N. Klein, A. Brust, H. Chaloupka, M. Hein, S. Orbach, H. Piel and D. Reochke, J. Supercond. **3**, 235 (1990).

19. B. Pümpin et al., Phys. Rev. **B42**, 8019, (1990).

PLASMONS IN HIGH-T_c CUPRATE SUPERCONDUCTORS

Jae H. Kim[1], Ivan Bozovic[2], James S. Harris, Jr.[1],
Wen Y. Lee[3], Chang-Beom Eom[1], Theodore. H. Geballe[1]

Stanford University[1], Varian Research Center[2]
IBM Almaden Research Center[3]

Although vast amounts of experimental data have been accumulated and extensive theoretical efforts have been made, the mechanism of high-T_c superconductivity has yet to be elucidated. Many theoretical models depart from the conventional electron-phonon interaction mechanism and consider alternative or additional interactions caused by other excitations, such as; plasmons, excitons, or magnons. Reliable information on the relevant low-energy electronic excitations is essential to clarify the pairing mechanism. Optical spectroscopy is generally a useful source of such information.[1-3] In particular, study of plasmons not only helps understanding their possible role in the pairing interaction,[4-11] but also provides valuable information on charge carriers via the plasma frequency, defined by

$$\omega_p \equiv (4\pi N e^2/m^*)^{1/2}$$

where N is the volume carrier density and m^* is the optical effective mass.

In this paper, we will concentrate exclusively on the metallic planes at room temperature and discuss the plasmon characteristics of the free carriers in $La_{1.85}Sr_{0.15}CuO_4$ (c-axis oriented film),[12] $YBa_2Cu_3O_7$ (c-axis oriented film),[13] $Bi_2Sr_2CaCu_2O_8$ (single-crystal),[13] and $Tl_2Ba_2Ca_2Cu_3O_{10}$ (c-axis oriented film).[14] We employ a novel technique which combines normal-incidence reflectance and spectroellipsometric measurements.[13] The technique removes the uncertainty due to high-frequency extrapolations in the customary Kramers-Kronig transformation procedure and makes it possible to determine the key spectral functions quite accurately of each high-T_c cuprate.[13] Two important features are clearly shared by all four high-T_c cuprate superconductors. First, the plasmon spectrum of each high-T_c cuprate contains a single optic plasmon.[12-14] Second, the dielectric loss function exhibits a characteristic quadratic frequency-dependence:[12-14]

$$\mathrm{Im}(-1/\epsilon) = \beta\omega^2$$

up to some limit frequency ω_q, which is smaller than, but quite close to, the screened plasma frequency $\tilde{\omega}_p \approx \omega_p/\sqrt{\epsilon_\infty}$.

High-Temperature Superconductivity
Edited by J. Ashkenazi et al., Plenum Press, New York, 1991

The preparation of the superconducting samples was described in detail elsewhere.[12-14] Our near-normal incidence reflectance measurements were made with a Bio-Rad FTS-40V Fourier-transform infrared (FTIR) spectrometer (far infrared), a Bio-Rad FTS-40 FTIR spectrometer (midinfrared), and a Perkin-Elmer Lambda 9 spectrophotometer (near infrared, visible, and near ultraviolet). For measurements in the midinfrared range, a Spectra-Tech IR-PLAN Microscope was coupled to the Bio-Rad FTS-40 FTIR spectrometer. The microscope was equipped with a dedicated HgCdTe detector (cooled with liquid nitrogen) and provided spatial resolution better than 100 μm. No significant changes in reflectance were observed when the spot was moved across the film surface. For measurements in the visible range, a Perkin-Elmer 60 mm Integrating Sphere was coupled to the Perkin-Elmer Lambda 9 Spectrophotometer in order to record the diffusive reflectance spectra. Spectra taken on different spectrometers agreed quite well with one another in the spectral regions of overlap. Spectroellipsometric measurements were performed on a Model 445A15 Rudolph Instruments Automatic Spectroellipsometer.

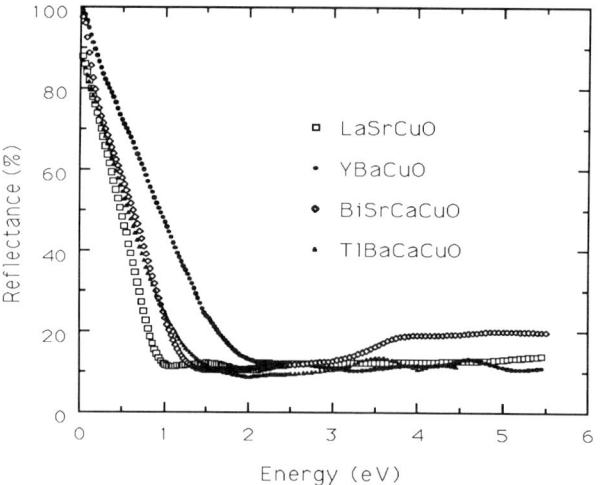

Figure 1. Normal-incidence reflectance of the four cuprates.

Figure 1 shows the normal-incidence reflectance spectra of the four high-T_c cuprate samples. In order to determine the key spectral functions, we utilized the novel technique introduced in Ref. 13. First, we determine the real and imaginary parts of the complex refractive index, n and k as follows: (I) assuming that R_H is constant for $\hbar\omega \geq \hbar\omega_H = 4.5$ eV, we employ the Kramers-Kronig transformation to determine the uncorrected phase $\theta_{\text{uncorr}}(\omega)$; (II) from the pseudodielectric function, we calculate $\theta_{\text{ellips}}(\omega)$ for $1.5 \text{ eV} \leq \hbar\omega \leq 4.5 \text{ eV}$; (III) we calculate

$$\Delta\theta(\omega) = \theta_{\text{ellips}}(\omega) - \theta_{\text{uncorr}}(\omega)$$

and fit $\Delta\theta(\omega)$ by a polynomial

$$p(\omega) = a_1\omega + a_3\omega^3 + \cdots + a_{2r+1}\omega^{2r+1};$$

and (IV), utilizing the corrected phase

$$\theta_{\text{corr}}(\omega) = \theta_{\text{uncorr}}(\omega) + p(\omega),$$

we calculate

$$n = (1-R)/(1+R-2\sqrt{R}\cos\theta_{\text{corr}})$$

and

$$k = 2\sqrt{R}\sin\theta_{\text{corr}}/(1+R-2\sqrt{R}\cos\theta_{\text{corr}}).$$

From n and k, we calculate the complex dielectric function (Figure 2), the dielectric loss function (Figure 3), and other spectral functions of interest.

There is clearly a strong similarity in the overall shape of the key spectral functions of all samples. Indeed the low-energy optical response in the high-T_c cuprates is dominated by the infrared absorption of free carriers (holes), which are believed to be confined within the Cu-O layers. The corresponding electronic states do not differ much from one cuprate to another. In all high-T_c cuprates under study here, we observe a single optic plasmon located near 1 eV. The value of the screened plasma frequency for each system, determined by the position of the zeros of ϵ_2, is given by: 0.8 eV for $La_{1.85}Sr_{0.15}CuO_4$, 1.4 eV for $YBa_2Cu_3O_7$, 1.0 eV for $Bi_2Sr_2CaCu_2O_8$, and 1.2 eV for $Tl_2Ba_2Ca_2Cu_3O_{10}$. Apparently, T_c does *not* scale with $\tilde{\omega}_p^2$.

Quite strikingly, the low energy part of the dielectric loss function in *every* system is *quadratic* with frequency, i.e.,

$$\text{Im}(-1/\epsilon) = \beta\omega^2$$

up to a certain limit frequency ω_q, which is smaller than, but quite close to, the screened plasma frequency.

The value of the limit frequency is: ~0.7 eV for $La_{1.85}Sr_{0.15}CuO_4$ and ~1.0 eV for $YBa_2Cu_3O_7$, $Bi_2Sr_2CaCu_2O_8$, and $Tl_2Ba_2Ca_2Cu_3O_{10}$. Different β values are found for different cuprates. A possible connection of this feature with the LEG behavior has already been pointed out.[12–15] The LEG model[16–19] applies to charge carriers more or less confined to planes (relatively free to move within individual metallic slabs but relatively limited in motion in the direction perpendicular to the slabs). The quadratic law is consistent with a detailed calculation of the dielectric loss function[15] within the LEG model. We believe that the quadratic law is univer-

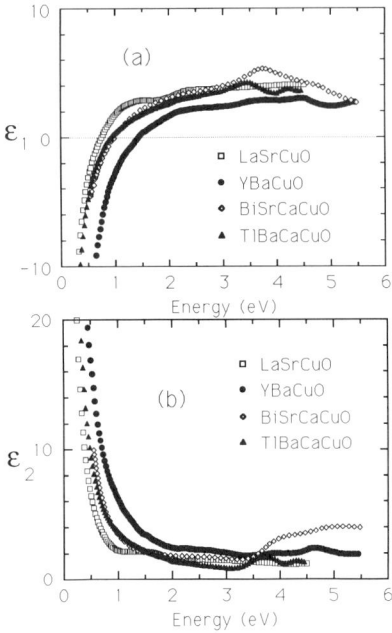

Figure 2. The real (a) and the imaginary (b) part of the dielectric function of the four cuprates.

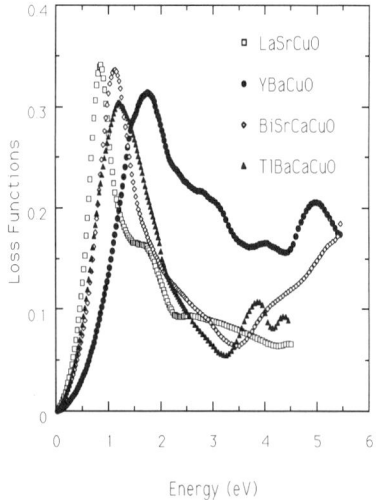

Figure 3. The dielectric loss function $\mathrm{Im}(-1/\epsilon)$ of the four cuprates.

sally obeyed in all cuprate superconductors. This peculiar feature of high-T_c cuprates, which is absent in conventional BCS superconductors, deserves further investigation.

It was shown recently[20] that $\beta \propto 1/N$. Since we can determine β quite precisely, this also provides an accurate, contactless method of monitoring the carrier density. Applying that method to the four compounds under study, it was found[20] that T_c does *not* scale with the volume density N – in accordance with the above statement that it does not scale with the plasmon frequencies. However, T_c does scale[20] with the slab carrier density (where the closely spaced pairs or triplets of Cu-O planes are assumed to be coupled together to act as distinct metallic slabs). That was interpreted in Ref. 20 as evidence for nearly-2D superconductivity in these cuprates.

In conclusion, from the key spectral functions of the four representative high-T_c cuprates, we have identified two features shared by them all: (I) there is a single optic plasmon around 1 eV and (II) the dielectric loss function depends quadratically on frequency nearly up to the screened plasma frequency. We have also pointed out some important implications of these observations.

This research was supported by the U. S. Defense Advanced Research Projects Agency-U. S. Office of Naval Research under contract No. N00014-88-0760, by the U. S. Air Force Office of Scientific Research under Contract No. F49620-88-C-004, and by the Stanford Center for Materials Research under NSFMRL program.

References

1. D. E. Aspnes and M. K. Kelly, IEEE J. Quantum Electron. **25** (1990) 2378.
2. T. Timusk and D. B. Tanner, *in*: "Physical Properties of High Temperature Superconductors," D. M. Ginsberg, ed., World Scientific, Singapore (1989).
3. G. Wendin, Physica Scripta T **27** (1989) 31.
4. V. Z. Kresin and H. Morawitz, Phys. Rev. B **37** (1988) 7854.
5. V. Z. Kresin and H. Morawitz, J. Supercond. **1** (1988) 89.
6. J. Ruvalds, Phys. Rev. B **35** (1987) 8869.
7. A. Griffin, Phys. Rev. B **37** (1988) 5943.
8. J. I. Gersten, Phys. Rev. B **37** (1988) 1616.
9. J. Ashkenazy, C. G. Kuper, and R. Tyk, Physica B **148** (1987) 366.
10. J. Ashkenazy, C. G. Kuper, and R. Tyk, Int. J. Mod. Phys. B **1** (1987) 965.
11. J. Ashkenazy, C. G. Kuper, and R. Tyk, Sol. St. Comm. **63** (1987) 1145.
12. J. H. Kim, I. Bozovic, C. B. Eom, T. H. Geballe, and J. S. Harris, Jr. to be published in Physica C.
13. I. Bozovic Phys. Rev. B **42** (1990) 1969.
14. I. Bozovic, J. H. Kim, J. S. Harris, Jr. and W. Y. Lee Phys. Rev. B **43** (1991) 1169.
15. H. Morawitz, I. Bozovic, and V. Z. Kresin, submitted to Phys. Rev. B.
16. P. B. Visscher and L. M. Falicov, Phys. Rev. B **3** (1971) 2541.
17. A. L. Fetter, Ann. Phys. **88** (1974) 1.
18. S. Das Sarma and J. J. Quinn, Phys. Rev. B **25** (1982) 7603.
19. A. C. Tselis and J. J. Quinn, Phys. Rev. B **29** (1984) 3318.
20. Ivan Bozovic, J. Supercond., in press.

THE ROLE OF MAGNETISM IN HIGH-T_C SUPERCONDUCTIVITY

G.C. Vezzoli[1], B.M. Moon[2], M.F. Chen[1],
T. Burke[3,4] and F. Craver[1]

1. U.S. Army Materials Technology Laboratory
   Emerging Materials Division
   Ceramics Research Branch
   Watertown, MA 02172

2. University of Illinois
   Department of Materials Science
   Champaign - Urbanna
   Illinois

3. U.S. Army Electronics and Devices
   Laboratory
   Pulsed Power Center
   Fort Monmouth, New Jersey 07703

4. Rutgers University
   Department of Electrical Engineering
   BUSCH Campus
   Piscataway, New Jersey 08854

ABSTRACT

A magnetic contribution toward the transition to the high-T_c superconductivity state is described from our recent experimental data on effects of paramagnetic moment and magnetic field sweep rate. Also previous related information taken from our studies and the work of others is reviewed and interpreted. The magnetic contributor is interpreted to be related to spin fluctuations from antiferromagnetism and to a spin density wave. Ultimately, this spin density wave couples with a charge density wave due to the system of holes. In terms of quasiparticles this interpretation relates to spinons and holons.

1.0 INTRODUCTION

Since the discovery of high-Tc superconducting materials[1,2] there has been a concerted effort to determine whether the phenomenon of high-temperature superconductivity was related in any major way to magnetic interactions. To address this question an earlier workshop was held in Oct 1988 at the National Institute of Standards and Technology (formerly the National Bureau of Standards) on the subject of magnetic interactions.[3] Papers were presented on the spin bag theory, the role of the substituted rare earth, antiferromagnetism, magnetic coupling resulting from hybridization of f and p electrons, and muon spin rotation. From these studies it was concluded that magnetic interactions existed in high-Tc superconductors, however, it was not explicitly established whether these interactions related to the high-Tc mechanism itself. The purpose of the present work is to report new experimental research findings (section 3) which suggest that magnetic interactions are indeed related to the high-Tc transition mechanism, and to review related work conducted in our laboratory and elsewhere. In so doing we shall concentrate primarily on spin fluctuations from antiferromagnetism and how this magnetic property is coupled to electronic properties also believed to be contributing to the transition.

2.0 Review and Interpretation of Previous Work

2.1 Antiferromagnetism in the YBaCuO System

It has been known that broadly speaking, the coexistence of ferromagnetic ordering and low-T_c superconductivity is confined to <u>narrow</u> temperature intervals, but that antiferromagnetism (AFM) and low T_c superconductivity can coexist over an <u>extended</u> temperature range. With the discovery of the new high-T_c materials, however, AFM and superconductivity were <u>believed</u> to be thermodynamically separated by a cusp in a phase diagram as reproduced in Fig. 1, and therefore were thought not to coexist.[4] However, very recent and elegant work by Budnick and co-workers[5], using muon spin rotation techniques has reported that there is indeed a subtle form of overlap of these two characteristic properties of high-Tc generic materials as shown in Fig. 2. However, this reported overlap is still not without some controversy.

Because of the correlation[6] between $Y_1Ba_2Cu_3O_{7-\delta}$ (in the idealized A_2BX_4 form $(Ba_2^{2+}Cu_2^{2+})(Y_1^{3+}Cu_1^{3+})O_{8-x}$) and the antiferromagnetic K_2NiF_4 or K_2MnF_4 structural archetypes[7] we examined in the summer of 1987 the possibility of static AFM at the Cu(2) sites in $Y_1Ba_2Cu_3O_{6.9}$. We were advised[8] to conduct this study at very low magnetic fields (<0.1T) using the MIT SQUID. The results are shown in Fig. 3, and although reflecting considerable scatter/fluctuation, this figure shows the structure of a static Néel condition at 400K$\pm \Delta$.[9] This observation may admittedly be associated with non-uniformities in the material, however, was not detected at higher magnetic fields of 0.5 and 1.0T. Shortly afterward, the detailed neutron diffraction work of Tranquada et al[10A] was published showing for <u>non-superconducting</u> stoichiometries of $Y_1Ba_2Cu_3O_x$ where $6.0 \leq x \leq 6.15$, a dynamic Néel temperature existed for the Cu(2) spins at 390$\pm$10K. The very important Tranquada et al work was

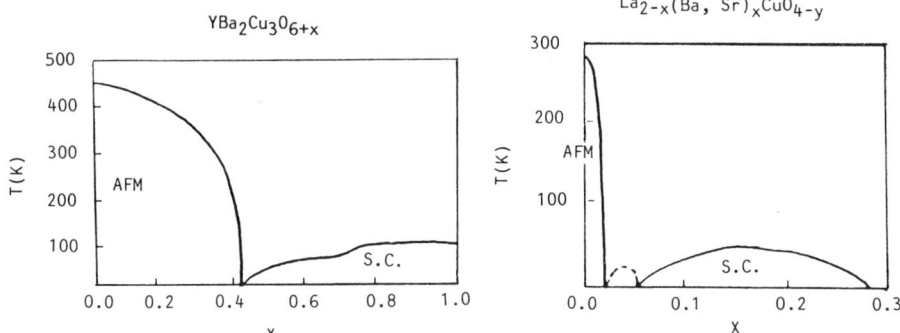

Fig. 1. Experimental phase diagrams for $YBa_2Cu_3O_{6+x}$ and $La_{2-x}(Ba, Sr)_xCuO_{4-y}$.

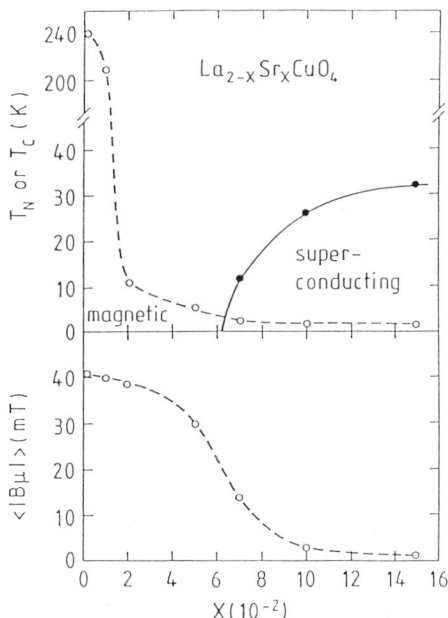

Fig. 2. Phase diagram for $La_{2-x}Sr_xCuO_4$ showing small overlap of antiferromagnetic and superconducting phases using muon spin rotation.

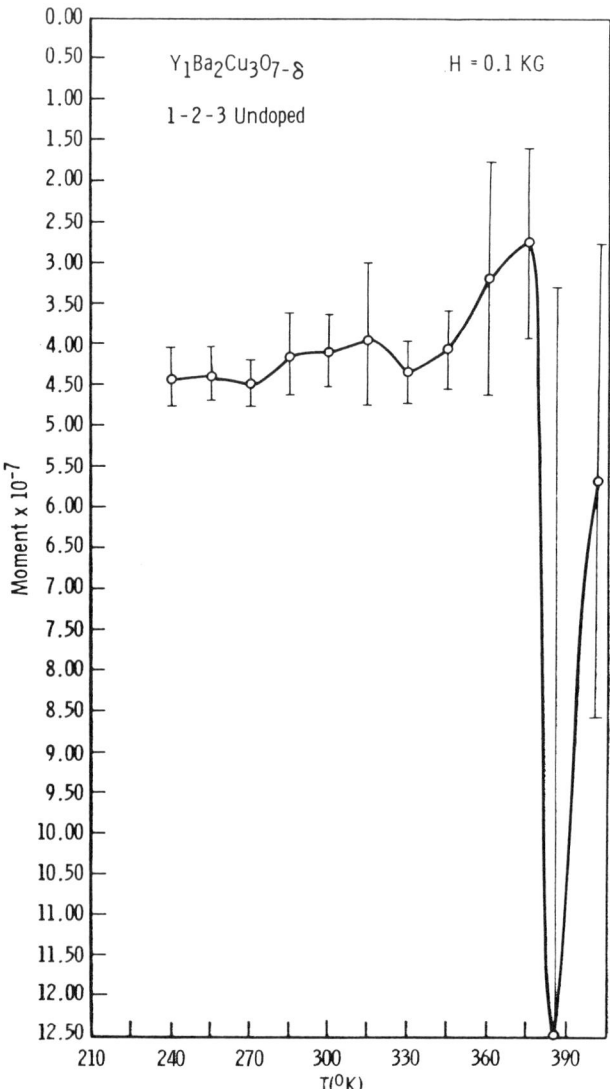

Fig. 3. Squid magnetometer study of $Y_1 Ba_2 Cu_3 O_{7-\delta}$ (Sn-doped) suggesting possible Neel condition at low field (150G) at 400K.

discussed intensely at the American Physical Society Meeting in New Orleans in March 1988, and AFM in these stoichiometries was confirmed by other laboratories.[10B] Even earlier, AFM was detected in $La_{2-x}Sr_xCuO_{4-y}$.[11] The molecular orbital hybridization associated with high-T_c materials (such as the d-p Π overlap in the chain region of $Y_1Ba_2Cu_3O_{7-\delta}$) will only modify the interpretations of the material to be presented herein, but will not alter the conclusions. This is because the idealized ionic A_2BX_4 configuration is essentially a bona fide extrema state at some stage of the change-transfer process which establishes bound holes on the oxygens, and allows for spin flucutuations at Cu(1) sites.

The significance of antiferromagnetism relative to superconductivity is understood actually in terms of fluctuations or deviations from the AFM state which are referred to in general as spin fluctuations. The spin fluctuations refer to the reversals of the direction of spin or the sign of m_s at a very high frequency. However, herein we address a specific form of the spin fluctuation which is localized at the Cu(1) site and out of phase with the AFM system at the planar sites in the $Y_1Ba_2Cu_3O_{7-\delta}$ material. These deviant unstable states which are localized or bound, upset the delicate balance of perfect spin-compensation, characteristic of AFM, and can themselves act as mediators inducing Cooper-pairing via spin-spin correlations with conduction electrons. The AFM related to the rare earth (RE) spin in $RE_1Ba_2Cu_3O_{7-\delta}$ had been also identified such as for $Gd_1Ba_2Cu_3O_{7-\delta}$ showing a $T_N \sim 0.3K$.[12a] Since Gd is strongly paramagnetic and its moment is likely to attack Cooper pairs in its vicinity, it was believed that the central ion (Gd or Er or Ho, etc) was magnetically isolated from the superconducting region of the unit cell.[12a] This unfavorable spatial domain for superconductivity was then proven theoretically,[12b,c] and for Gd shown to extend to about 0.55A from the valence shell of the central ion.

2.2 Crystal Chemistry Description of Stoichiometric Dependence of Cu(2) AFM in $Y_1Ba_2Cu_3O_{7-\delta}$

The antiferromagnetic stoichiometries of $O_{6.0}$ and $O_{6.15}$ yield in the idealized ionic states the compounds $Y_1Ba_2Cu_2^{2+}Cu_1^+O_6^{2-}$ and $Y_1Ba_2Cu_3^{2+}O_{6.15}^{2-}$. In the former compound the Cu(1) chain ions are in a $d^{10}s^0$ state which has no effect in upsetting the antiferromagnetism in the Cu(2) planes since it contains no unpaired spins. In the latter compound all the Cu ions are in 2+ and hence in d^9, allowing for complete antiferromagnetism at planar Cu and chain Cu sites, thus also not causing a deviation from antiferromagnetism. Therefore, it is clear that neutron diffraction measurements, capable of detecting long range dynamic antiferromagnetic ordering will verify AFM for these stoichiometries. In the $O_{6.6}$ stoichiometry, on the other hand, the resulting compound should have a cation composition idealized as $Y_1Ba_2Cu_{2.8}^{2+}Cu_{0.2}^{3+}$. The Cu(1) d^9 states, which then occupy only a portion of the non-planar Cu sites and with unpaired valence spin (constituting in this ionic stoichiometry 80% of the Cu(1) sites) can cause an upset or deviation of the balance of spin compensation that exists in the planar region. Thus in such a stoichiometry, long-range antiferromagnetic order will not be observed, nor will it be observed for O_x stoichiometries where 6.5<X<7.0. However, since the charge on these valence states is fluctuating at very high frequency (>10^{13}Hz), there will arise cases where short range time-dependent antiferromagnetic order may indeed exist and may be measurable at low fields (with a SQUID) to leave the spins relatively unperturbed (thus explaining the observations in Fig 3). This may lead to an overlap region of AFM and high-Tc superconductivity, as discussed earlier.

2.3 Effect of Rare Earth Substituted for Y in $Y_1Ba_2Cu_3O_{7-\delta}$

Fig. 4 gives a plot from Ref 2 of: the temperature T_o for the deviation from linearity in R vs T; Tc; and the temperature ($T_{R=0}$) for zero resistance, for many of the rare earth ions when substituted for Y^{3+} in $RE_1Ba_2Cu_3O_{7-\delta}$. It is shown that a large enhancing or elevation in T_o, and a small enhancing in $T_{R=0}$, directly correlate with the spin and the effective magnetic moment of the rare earth substituting element, with a peak occuring in the Eu-Gd range. This enhancement of a superconducting property by a rare earth 4f ion is in direct and almost mirrored contrast (Fig 5) to the diminishing effect in T_c that the rare earths have when they are added to low Tc elements such as lanthanum.[14] Fig. 6 gives another plot of the high-Tc enhancement effect using a different set of data[15] and shows again (as in Fig. 4) a direct correlation with m_s and μ_{eff}.

The negative or unfavorable effect given in Fig. 5 is normally explained by the tendency of paramagnetic (and ferromagnetic) moments to destroy or breakdown the weak Cooper-pairing bond. This breaking by paramagnetic moments occurs even in high-Tc superconductors. We have calculated (in Ref 12c) that for $Gd_1Ba_2Cu_3O_{7-\delta}$ the $Gd^{3+}(4f^7)$ moment destroys Cooper-pairs that are within approximately 0.5A of the ions' periphery. Therefore, in high-Tc materials the paramagnetic ion seems to have the dual role of destroying Cooper pairs near itself yet still favoring the triggering of the superconductivity transition itself as indicated in the Fig. 4 pre-onset (closed circle) data.

The cause of the enhancing effect, on T_o and $T_{R=0}$ in Figs 4 and 6, is not yet established. It is appealing, however, to search for an explanation of this effect in terms of kinetic-influenced magnetic phenomena that are believed to be a possible triggering mechanism of high-Tc superconductivity, such as the previously described spin fluctuations from antiferromagnetism. In our earlier study (Ref 12c) it was shown by a modified Rudderman-Kittel analysis that the magnetic moment of the rare earth ion which is substituted for Y in $Y_1Ba_2Cu_3O_{7-\delta}$ tends to polarize the localized unpaired and uncompensated d^9 valence spin existing at Cu(1) sites when a charge transfer excitation

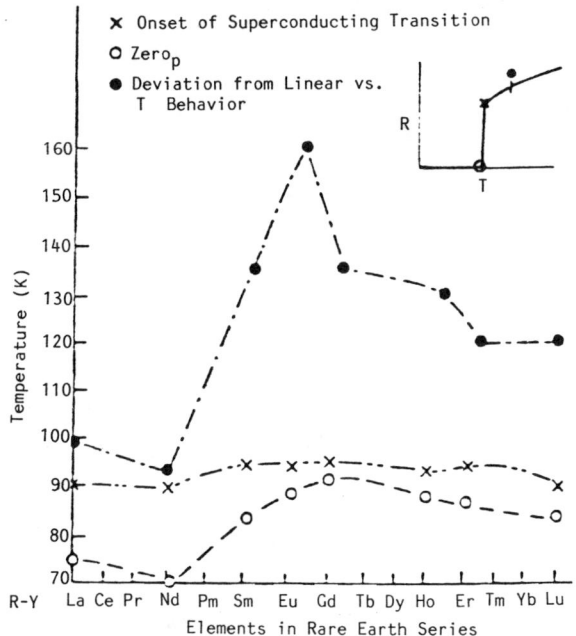

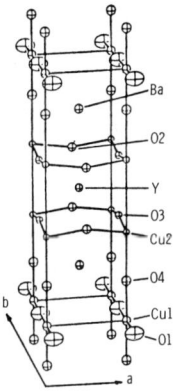

Fig. 4. Upper rare earth in R.E.Ba$_2$Cu$_3$O$_{7-\delta}$ enhancement effect of temperature (To) for deviation from linearity in R vs T and of temperature T$_{R=0}$ for zero electrical resistance. Tc remains relatively unaffected. Data from group of Ref.2. The effects on To and T$_{R=0}$ correlate with the spin and effective magnetic moments of the R.E. (see Fig. 5,6)

Lower: Unit cell of R.E. Ba$_2$Cu$_3$O$_{7-\delta}$

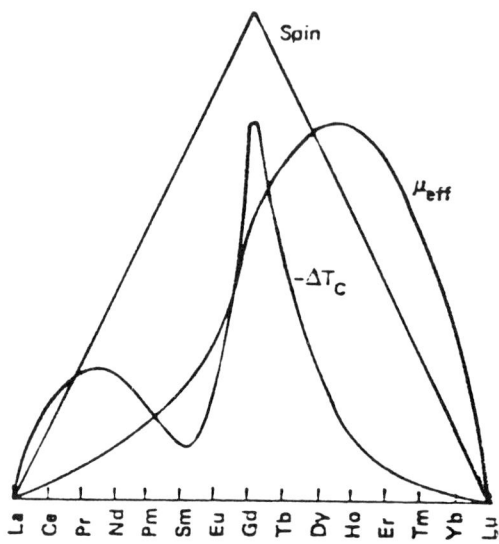

Fig. 5. The change in T_C for low-T_C lanthanum due to the addition of other rare earths as impurities in lanthanum. The effect is opposite to that shown in fig. 1 for high-T_C materials, suggesting opposite roles of internal magnetism regarding low-T_C and high-T_C materials.

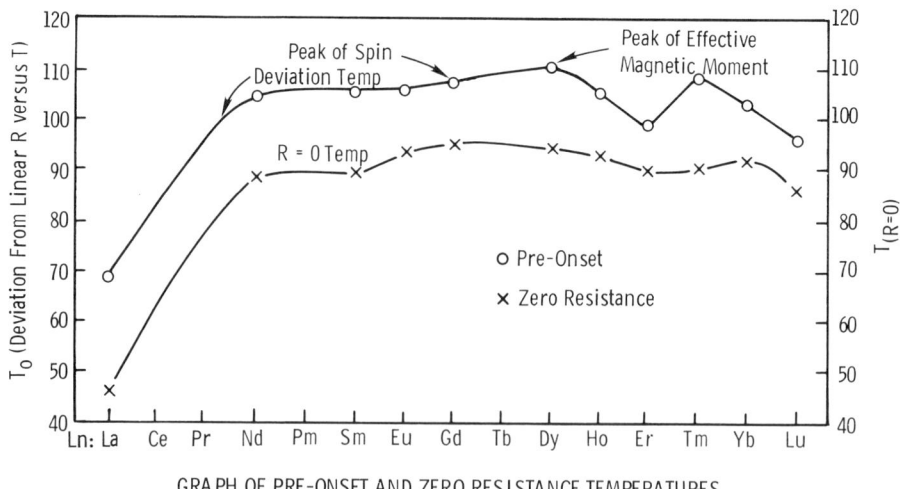

Fig. 6. Rare earth in RE $Ba_2Cu_3O_{7-\delta}$ enhancement effect of To and $T_{(R=0)}$ correlating with spin and effective magnetic moment (as in Fig. 4.) using the data of Ref. 15.

occurs creating an odd number of Cu^{2+} states at those lattice locations. (This charge transfer occurs to create neutralized unit cell conditions when the oxygen $O_{7-\delta}$ stoichiometry changes as δ increases from zero to a value <0.5). The calculation indicated that an indirect exchange correlation in the form of a spin density wave exists between the spin of Gd^{3+} (or other appropriate rare earth) and that of the d^9 $Cu(1)^{2+}$, and establishes a short range AFM coupling between the two. This AFM indirect correlation is enhanced as the effective magnetic moment and spin of the paramagnetic ion increases, the net result being an ordering of these local d^9 Cu(1) spin fluctuation moments. This ordering should promote ease of spin-spin coupling as a mediator for high-Tc. This is shown diagramatically in Fig. 7. The net result quantum-mechanically is a spin density wave and relates to a quasi-particle spinon.

2.4 Related and Anamalous Effects in $La_{1.6}Sr_{0.2}R.E._{0.2}CuO_4$

When a rare earth is substituted into the lanthanum strontium cuprate structure according to $La_{1.6}Sr_{0.2}R.E._{0.2}CuO_4$, the deviation from linearity in R vs T shows the opposite tendency (a second order change toward increasing resistance) with decreasing temperatures (for Gd, Eu, Sm) compared to $Y_1Ba_2Cu_3O_{7-\delta}$.[16] For Nd, Pr, and La the deviation is identical to the shoulder of the resistance drop (namely approximately Tc). Nonetheless this deviation temperature, T_o, still indicative of a phase transition, shows (see Fig. 8) the same peaking trend as does the T_o enhancing effect in Fig. 4. However, the almost linear decreasing trend of $T_{(R=0)}$ shown in Fig. 8 contrasts the soft crested behavior in $T_{(R=0)}$ in Figs 4 and 6. This indicates the magnetic factors influencing T_o are in some ways similar for $RE_1Ba_2Cu_3O_{7-\delta}$ and $La_{1.6}Sr_{0.2}R.E._{0.2}CuO_4$, but the magnetic properties affecting $T_{(R=0)}$ are different for the two materials.

2.5 Possible Explanation for the Deviation Region in R vs T Denoting T_o

It is believed that high-Tc superconductivity is fundamentally a manifestation of a Ginzburg-Landau phase transition, having the properties peculiar to and characteristic of a thermodynamic phase transition. We have observed [12c, 17] that the presence of low-frequency small oscillations (in resistance vs time at constant temperature) commence at the temperature(T_o) for deviation from linearity in R vs T in both $Y_1Ba_2Cu_3O_{7-\delta}$ and $Bi_2Sr_2Ca_2Cu_3O_{/O}$. Small oscillations have been observed earlier relative to low-Tc superconductors,[18] and ascribed to fluctuations in domain boundaries. We believe that the small oscillations are due to some type of instability that is developed when the initiation or triggering temperature for Cooper-pairing is attained. This initiating or pre-onset temperature we interpret to be related to a temperature where Cooper-pairing forces compete with pair-scission forces such that Cooper-pairing is not complete along a spatial path. In that sense, if given lengthy time or catalysis action at

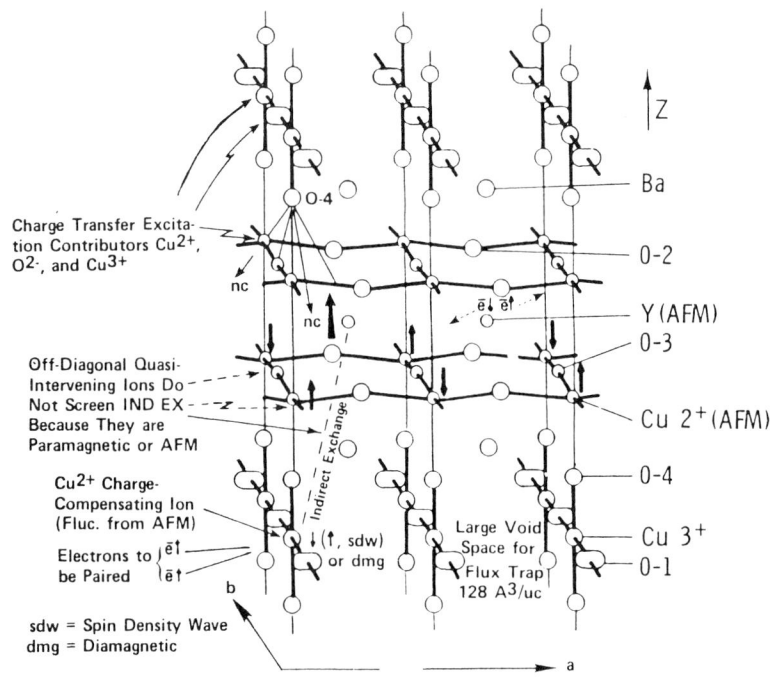

Fig. 7. Schematic of indirect exchange correlation between magnetic moment of central rare earth ion and unpaired valence spin of d^9 state when charge transfer excitation occurs causing a Cu(1) chain ion to change from $3+(d^8)$ to $2+(d^9)$

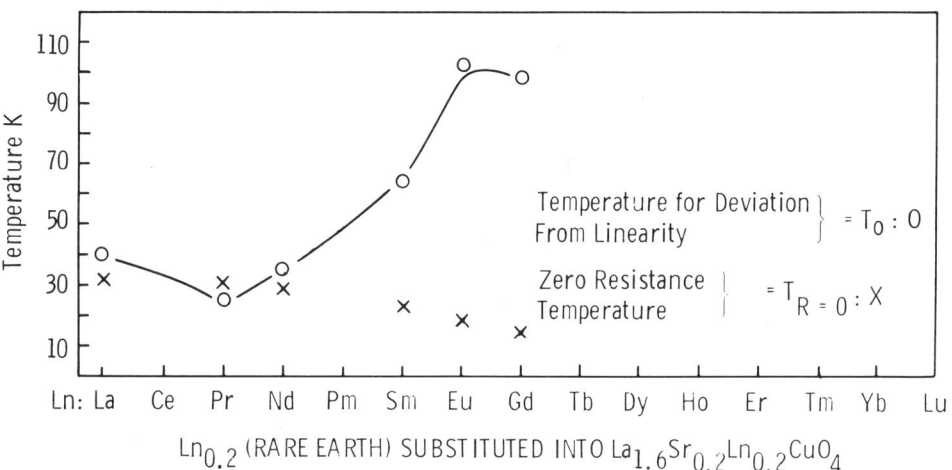

Fig. 8. Graph of To (upward deviation from linearity in R vs T or T_c shoulder without deviation in R vs T), and $T_{R=0}$ for rare earth substituted into $La_{1.6}Sr_{0.2}RE_{0.2}CuO_4$, showing similar behavior for To relative to To in $R.E.Ba_2Cu_3O_{7-\delta}$ but differing behavior in $T_{(R=0)}$.

such a temperature at least one domain for pairing should exist at that temperature. T_o takes on the appearance of identifying a <u>kinetic boundary</u> associated with a phase transition.

2.6 Importance of Diamagnetism

It is an empirical fact that unlike low-Tc superconductors, the high-Tc materials are composed of principal cations that are both <u>multivalent</u> and in their zero valence atomic state <u>diamagnetic</u> as well (i.e. Cu, Tl, Bi, Pb, Sb, and Ga). This observation further attests to the presence of charge transfer excitations which transiently ($\gamma \sim 10^{13}$Hz) create a zero valence state and in so doing establish a hole concentration bound to the oxygen sites or virtual excitons, this state apparently playing a significant role in estabishing a high-Tc material. This diamagnetic contribution may be related to facilitating the Meissner transition or to effectuating superconductivity current at the surface of the material that persists at fields greater than H_{c1}. The diamagnetism gives rise to orbital motion behaving like non-restrictive current loops in the d^{10} state. This orbital motion is altered by an externally-applied magnetic field so as to oppose the applied B-field.

3.0 Time Dependence of Flux Readmission at Constant Temperature and Constant Current

We have conducted a preliminary study of the recovery of electrical resistance in $Y_1Ba_2Cu_{3-x}Ga_xO_{7-\delta}$ (prepared by conventional mixing, calcining, sintering and annealing) as a function of magnetic field sweep rate using conventional 4-terminal electrical resistance measurements within a high-field Bitter magnet. (See Ref 12b for technique). In characterizing this polycrystalline material (2mm X 2mm X 4mm) EDAX measurements show x=0.2 to 0.8% and further characterization tentatively suggests that Ga^{3+} substitutes for Cu^{3+} at chain sites. The material shows strong levitation, high density, very low porosity, and very high electrical conductivity in the normal state (see Figs 9 and 10). In Figs 11 and 12 we plot new data on the recovery of electrical resistance at 83.7K as a function of the sweep rate of the magnetic field intensity. The data show that the effect of rate is most clearly observed at low field where the response of resistance recovery is lowest for the fastest sweep rate (18T in 30 sec). Thus for the most rapid sweep rate the resistive properties were not restored until a magnetic field of 1.5T was exceeded. The time-response for resistance recovery is about 2.5 sec. This kinetic parameter or dependence is in keeping with a phase transition phenomenon and is thought to be associated with fluxoid depinning time criteria as related to the pinning property of defects. Additional Ga substituted samples yielded the same or similar results, however substitution of In for Ga caused the loss of superconductivity. This may be due to the absence of multi-valence in indium.

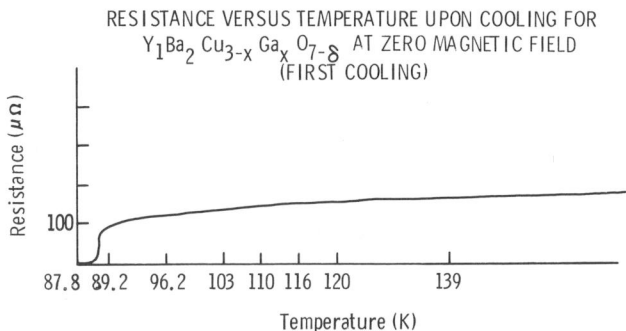

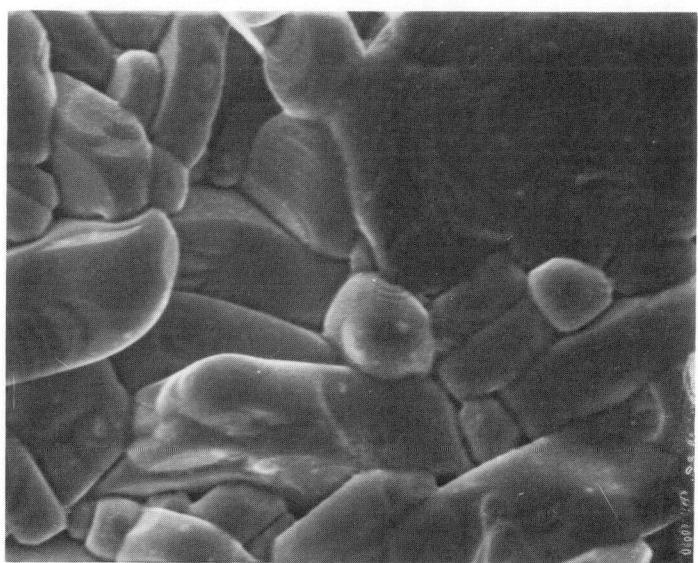

Fig. 9. Resistance vs temperature (trial 1) for superconducting $Y_1Ba_2Cu_{3-x}Ga_xO_{7-\delta}$. (Inset: Scanning Electron Micrograph of sample).

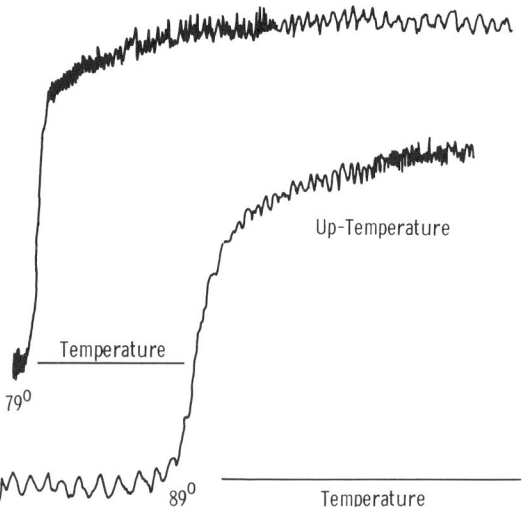

Fig. 10. Resistance vs temperature (trial 2: cooling and re-warming) for superconducting $Y_1Ba_2Cu_{3-x}Ga_xO_{7-\delta}$

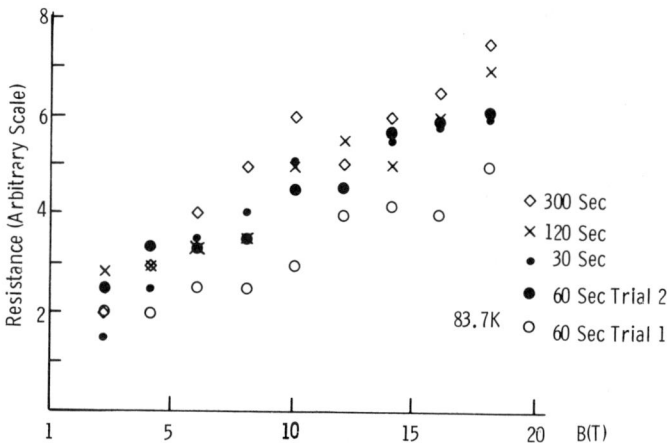

Fig. 11. Time-dependance data on resistance recovery of superconducting $Y_1Ba_2Cu_{3-x}Ga_xO_{7-\delta}$ as a function of magnetic field sweep rate.

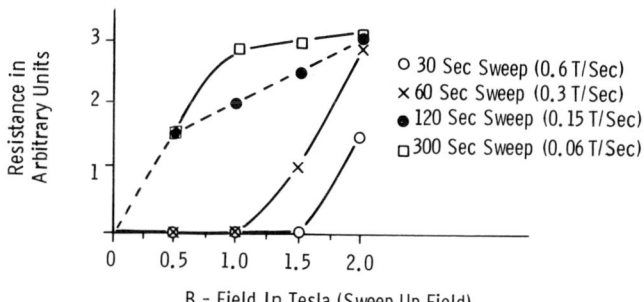

Fig. 12. Reduced data of recovering resistance of superconducting $Y_1Ba_2Cu_{3-x}Ga_xO_{7-\delta}$ as a function of magnetic field sweep rate.

4.0 Relationship of Magnetism to Mechanism Present in High-Tc Superconductivity

Since an isotope effect has been established in some high-T_c materials, with notable exceptions not being inexplicable, a phonon-related contribution to the mechanism is unsurprisingly believed to be present. However, all of our own earlier work (Refs 3,6,12,17) indicate that an electronic mechanism and a magnetic mechanism strongly supplement a phonon-electron correlation. Furthermore, a purely phonon basis for high-T_c offers no straightforward explanation of the anomalies observed in the Hall coefficient of ceramic high-T_c materials at temperatures slightly above T_c,[12,17] nor does it offer an explanation why only multivalence principal cations which are diamagnetic in their atomic states give rise to high-T_c compounds. We believe that the electronic mechanism is based on bound-holes and/or virtual-excitons (short lifetimes) which arise due to charge transfer excitations in a marginal ferroelectric material with very high internal electric perturbation fields that are consequences of crystal structure building block polyhedra. The holes in the $0°$ state and their associated electrons in the $O^{-\eta}$ state (-1.96 for O(2) and O(3)) can be considered in an excitonic relationship of very *short* lifetime and therefore a *virtual* exciton. This electronic effect based on holes and including a Jahn-Teller distortion, gives rise to a charge density wave which we believe couples to a spin density wave, thereby producing high-T_c superconductivity. The spin density wave has its origin in a magnetic phenomenon which relates to a spin-spin correlation between an unpaired valence spin (localized to a site) that upsets internal antiferromagnetism, and the spins of free electrons of opposite linear momenta, with the distance between the free electrons being no longer than a coherence length (where they themselves are correlated by net attraction). The presence of paramagnetic centers such as the 4f rare earths causing dipolar and higher magnetic fields would be expected to affect these spin-spin interactions and their phase, supporting the data of Figs 4,6, and 8. Calculations are now underway to evaluate the above interpretation quantitatively.

5.0 Conclusions

1. When a rare earth is substituted for Y in $Y_1Ba_2Cu_3O_{7-\delta}$, the paramagnetic moment of the rare earth enhances the superconductivity properties T_o and $T_{R=0}$ in a manner directly related to the spin and effective magnetic moment of the rare earth.[12b,2]

2. The above is also true for the La-Sr- Lanthanide-Cu-O system but only with respect to T_o.[15,16]

3. The substitution of Ga^{3+} for Cu(1) in $Y_1Ba_2Cu_3O_{7-\delta}$ causes enhanced normal state conductivity and increased grain size and density.

4. The recovery of electrical resistance of the above material is affected by the rate of application of external magnetic field such that the recovery of resistive properties is most sluggish for the fastest sweep rates (18T/30 sec), supporting a kinetically influenced phase transition interpretation for high-T_c

References

1. J.G. Bednorz and R.A. Mueller, Z. Phys. B64, (1986) 189

2. C.W. Chen, R.H. Hor, R.L. Meng, L. Gao, E. J. Huang and Y.Q. Wang, Phys. Rev. Lett. 58(1987) 405, C.W. Chu, Phys. Rev. Lett 58(1987) 1891, C.W. Chu, APS Meeting March 1987

3. Proceedings of High-Tc Superconductors: Magnetic Interactions, ed by L.H. Bennet, Y. Flom, and G.C. Vezzoli, Singapore, World Scientific 1989

4. C.S. Wang in Ref. 3, p. 64

5. J.I. Budnick, B. Chamberland, A. Weidinger, Ch. Niedermayer, A. Golnik, R. Simon, E. Recknagel, and C. Baines, in Ref. 3 p. 206

6. G. Vezzoli, R. Benfer, W. Spurgeon in Novel Superconductivity ed by S. Wolf and V. Kresin, Plenum, New York, p. 1017

7. R.J. Birgeneau, H.J. Guggenheim, and G. Sirane, Phys. Rev. B5 (1979) 2211

8. Richard Frenkel, private communication

9. A. Austin, R. Frankel and G. Vezzoli, unpublished, work conducted in Sept 1977 at Francis Bitter National Magnet Lab, MIT

10A. J. Tranquada, D. Coz, W. Klimman, H. Moulden, G. Shirane, M. Suenaga, P. Zaliker, D. Vaknin, S. Sinha, M. Alvarez, A. Jacobson, and D. Johnson, Phys. Rev. Lett. 60(1988) 156

10B. See Abstracts Bull Amer. Phys. Soc March 1988; G12(6) & H 6(8)

11. D. Moncton, Science 236(1987) 780. J.R. Thompson, S. Sekula, D. Christen, B. Sales, L. Boatner and Y. Kim, Phys. Rev B 36(1987) 718

12. a. J.R. Thompson, D.K. Christen, S.T. Sekula, B.C. Sales and L. Boatner Phys. Rev. B 36(1)(1987)836.
    b. G. Vezzoli, T. Burke, B. Moon, B. Lalevic, A. Safar, HGK Sundar, R. Bonometti, C. Alexandra, C. Rau and K. Waters, J. Magn & Magn Mat. 79, (1989) 146.
    c. G.C. Vezzoli, J. Magn & Magn Matt 82(1989) 335

13. G.C. Vezzoli in Ref. 3, p. 116

14. B. Matthies, H. Suhl, and E. Corenzwit, J. Phys. Chem. Solids 13 (1959) 156; Phys. Rev. Lett 1(1958) 92

15. M.B. Maple, Y. Dalichaouch, J. M. Ferreira, R.R. Hake, B.W. Lee, J.J. Neumeier, M.S. Torikachvili, K.N. Yang, H. Zhou, R.P. Guertin, and M.V. Kuric, Physica B 148, (1987) 155

16. J.M. Tarason, L.H. Greene, W.R. McKinnon, G.W. Hull, Solid State Commun. 63(1987) 499

17. G.C. Vezzoli, J. Magnetism and Magnetic Materials 88, 351 (1990)

18. R. Weber, Phys Rev 72, 1241 (1947); D. Baird, Can J. Phys 37, 20 (1959); B. Lalevic, J. Appl Phys 31, 234 (1960) and 35, 1785(1964)

A UNIFIED APPROACH TO A DESCRIPTION OF THE CUPRATE SUPERCONDUCTORS: MAJOR NORMAL AND SUPERCONDUCTING PARAMETERS AND THE MECHANISM RESPONSIBLE FOR THE HIGH T_C

V. Z. Kresin*, H. Morawitz** and S. A. Wolf***

*Lawrence Berkeley Laboratory, University of California Berkeley CA 94720
** IBM Almaden Research Center, San Jose CA 95120-6099
*** Naval Research Laboratory, Washington DC 20375-5000

ABSTRACT

We present a unified analysis which allows us to describe the normal and superconducting properties of the high T_C oxides as well as determine the underlying mechanism of high T_C. A number of crucial experiments have been performed recently and as a result, the basic physics of high T_C superconductivity has been clarified.

I. INTRODUCTION

During the last four years there has been significant progress in understanding the properties of the high T_C oxides [1,2]. The situation is now entirely different from the time immediately after the discovery of high T_C superconducting oxides [1], when the nature of the new materials was completely unknown. Of course, there are still many unresolved problems (also true with respect to conventional superconductivity). Nevertheless, we think that the major principles of the underlying physics, including the origin of high T_C, are clarified.

During the last several years we have been involved in the development of the theoretical framework describing the properties of the layered cuprates. Our main goal was to understand what makes these materials unique. In this paper we are going to describe the unified approach which has been developed by us [3-8] during these years.

II. Normal Properties

Let us start from the analysis of the normal properties of the new oxides. This problem is directly related to the origin of high T_C. It is very important to know the values of normal parameters such as effective mass, the Fermi energy, the Fermi velocity, electron-phonon coupling constant, etc. The important thing is that the values of all major superconducting parameters such as coherence length, critical temperature, etc., depend directly on the parameters describing the normal state. As we know, the high T_C oxides are doped anisotropic materials. Only Ba-(K,Pb)-Bi-O [9] is an exception; it has a cubic structure and we will mention its properties below. Our main focus is on cuprates which have a layered structure.

In addition, we are going to focus on the doped or metallic state. It is important to realize that it is only the doped or metallic materials that undergo transitions into the superconducting state; as a result, our analysis is directly related to the origin of high T_C. Note also that the La-Sr-Cu-O system deserves special attention because of the simplicity of its structure. At the same time it contains the major ingredient, namely the Cu-O plane. This system plays a role similar to the hydrogen atom in atomic physics; it is the best test system for understanding the basic principles of high temperature superconductivity.

With respect to the normal properties, one should, first of all, evaluate the major parameters such as the effective mass m^*, the Fermi energy E_f, and the Fermi momentum p_f. In other words, we are talking about the usual parameters whose values are well known for usual metals (see,e.g.[10]). The major parameters for La-Sr-Cu-O are presented in Table I (we will discuss the properties of Y-Ba-Cu-O below). The values of these parameters have been evaluated by us in [4,5]. The method used in the evaluation will be described below, but initially we want to stress the unique features of the material. In order to do this, we have also included typical parameters for conventional metals.

One can see directly from Table I that the cuprates are characterized by uniquely small values of the Fermi energy, E_f and the Fermi velocity, v_f. The Fermi energy is almost two orders of magnitude smaller in La-Sr-Cu-O than in usual metals.

We believe that the small values of the Fermi energy and the Fermi velocity along with large anisotropy are the key features of the cuprates.

TABLE I Normal state parameters of conventional metals and the cuprates

Parameters	Conventional	LaSrCuO	YBaCuO	
m^*	$1-15 m_e$	$5 m_e$	$5 m_e$	$25 m_e$
E_F(eV)	5-10 eV	0.1 eV	0.3 eV	
k_F(cm)$^{-1}$	$\sim 10^8$	3.5×10^7	--------	
v_F(cm-sec)	$1-2 \times 10^8$	8×10^6	10^7	2×10^6

We would like to point out that the effective mass m^* appears to be relatively large. It is important to emphasize that this value of the effective mass represents a so-called renormalized value, which is an experimentally observed quantity. It enters the density of states and can be determined, for example, by heat capacity or de-Haas-van- Alphen measurements. It is not equal to the band value of the effective mass (that is to say to the value for the "frozen" lattice) obtained from electronic structure calculations.

On the other hand, the Fermi momentum, p_f, is comparable with values found in conventional superconductors. This is important, because it means that there is sufficient phase space for pairing.

At this point we will briefly describe our approach (for a more detailed description see[4,5]). We employed the method which was widely used during the "Golden Age" of the physics of metals, namely "Fermiology" (see e.g.[11]). In doing this, we assumed that the conducting high T_C oxide is a metallic system, or, in other words, it is characterized by the presence of a Fermi surface (indeed, by definition, the metallic state is a solid with a Fermi surface). Anisotropy of the system is reflected in the shape of the Fermi surface, its deviation from a spherical shape. As in a typical metal, the parameters of the Fermi surface can be reconstructed using some experimental data.

Of course, the use of the Fermi liquid concept (see e.g.[12]) for unknown materials like the cuprates is not obvious. There is considerable controversy over this issue. Our conviction about the applicability of the concept of the Fermi surface to the doped cuprates has been based on our analysis which resulted in the set of parameters in Table I which then allowed us to develop an unified approach to the physics of high T_C. As we mentioned before, the Fermi momentum appears to be relatively large, and thus it allows us to define the Fermi surface. Recent experimental data confirmed such a viewpoint. First of all, photoemission data [13] show the presence of the Fermi edge. Moreover, very recent experiments [14] demonstrated the presence of de-Haas-van-Alphen oscillations. Thus, the existence of the Fermi surface is well established.

To reiterate what we said above, our approach is based on Fermiology, and it implies the use of some experimental data which are sensitive to the shape of the Fermi surface, in order to evaluate the major parameters (see Table I). First of all, we assumed the Fermi surface to be cylindrically shaped (Fig.1; this shape corresponds to a layered structure; to first approximation one can neglect the interlayer transitions).

Evaluation of the Sommerfeld constant which is proportional to the density of states leads to the following expression [4,5]

$$m^* = 3hd_c\gamma/\pi k_B^2 \qquad (1)$$

Here γ is the Sommerfeld constant, and d_c is the interlayer distance. It is important to stress that both Eq. (1) as well as Eq. (2)

(see below) are valid for a cylindrical Fermi surface. It is interesting to note that the density of states υ, and, therefore, the Sommerfeld constant ($\gamma \propto \upsilon$), for an isotropic 3D system is proportional to m^*p_f. As a result, υ_{3D} increases with carrier concentration, n ($p_f \propto n^{1/3}$). That's why in the BCS theory, a small carrier concentration is considered a negative factor (according to BCS theory T_c depends directly on the density of states); as a consequence, we observe low values of T_c for semiconducting superconductors. The situation is different for layered metals. There the density of states, $\upsilon_L \propto m^* d_c^{-1}$ does not depend directly on the carrier concentration. As a result, the relatively small carrier concentration in the cuprates does not contradict the large value of T_c in these new materials.

Eq.(1) express m^* in terms of the experimentally measured quantity γ. In addition, one can derive an expression for the Fermi energy which has the form for a cylindrical Fermi surface:

$$E_f = \pi^2 k_B^2 n / 3\gamma \qquad (2)$$

The values of $\gamma \equiv \gamma(0)$, the zero temperature limit, and n can be obtained from heat capacity data (see [5]) and Hall effect data (see e.g.[16]). Note that the determination of γ is not a simple task because the system is in the superconducting state and the critical field is large.This problem has been solved in [15] by analyzing the dependence of γ on magnetic field, H, using a model appropriate for a Type II superconductor. Indeed $\gamma(H_{C2}) = \gamma(0)$. Using a linear approximation for $\gamma(H)$ (this is analogous to the well known Bardeen-Stephen law for flow resistance), one can obtain $\gamma \approx H_{c2} (\delta\gamma/\delta H)$. The linear approximation is valid up to values of H approaching H_{c2}; it fails only in the small region near H_{c2} (because of the vortex-vortex interaction) and at very low fields near the Meissner region. Therefore, the approximation works very well for an estimation of $\gamma(0)$. The value of $H_{C2} \approx 90$ T has been used in order to evaluate $\gamma(0)$. Note that this value of H_{C2} can be evaluated in a self-consistent way as we will show in Sec.IIIa. Furthermore, the carrier concentration for the La-Sr-Cu-O compound can be determined from Hall effect data and is approximately equal to 3×10^{21} cm^{-3} [16] (a similar value has been obtained using a chemical method [17]). We want to point out that the ability to use the Hall effect is related to the simple band structure of this compound which leads to a weak temperature dependence of the Hall coefficient. The situation with the Y-Ba-Cu-O compound is more complicated, because this system contains two sets of carriers (two bands, see below, Sec.IIIc) and this leads to a strong temperature dependence of the Hall coefficient.

Using the expressions (1) and (2) derived in [4,5], we have evaluated the values of the major parameters presented in Table I.

Let us stress again the small values of E_f and v_f, which are much smaller than in conventional metals.

III. Superconducting Properties

In this section we are going to describe some unique features of the cuprates in the superconducting state. These particular features are not directly related to the pairing mechanism in the oxides, but are mainly due to the exotic normal properties discussed in the previous section.

a) <u>Coherence Length</u>: The coherence length is one of the key parameters describing the superconducting state. Its value can be estimated from the well-known expression $\xi_0 = hv_f/2\pi T_c$. Using the values of the Fermi velocity (see Table I) $v_f \cong 8 \times 10^6$ cm/sec and $T_c \cong$ 40K, we obtain $\xi_0 \cong 25$ Å. One can use a more precise definition of $\xi_0 = hv_f/\pi\Delta(0)$, where $\Delta(0)$ is the energy gap which is directly related to T_c, $\Delta(0) = aT_c$. In the BCS theory $a = 1.76$, but strong coupling effects lead to an increase in a; e.g. for La-Sr-Cu-O $a \cong 2.5$ (see below, Sec.IV and [5]). As a result, a better estimate of ξ_0 is 20Å. Note that estimates made by magnetic measurements are consistent with this calculation.

This value of the coherence length is very small relative to its value in the conventional superconductors ($\approx 10^3$ - 10^4 Å); this is a very important property of the new materials. The short coherence length in La-Sr-Cu-O is mainly due to a small value of the Fermi velocity (see Table I), although the higher T_c also contributes to the decrease in ξ_0 ($\xi_0 \sim v_f/T_c$).

Using this value of the coherence length and the expression $H_{C2} = \Phi_0/2\pi\xi^2$, where Φ_0 is the flux quantum, one can estimate H_{C2}. ξ is the Ginzburg-Landau coherence length which is directly related to ξ_0 (the BCS coherence length). Our evaluation (see [5a]) leads to a value of $H_{C2} \cong 90T$. It is important to note that this is the value that was used on order to evaluate $\gamma(0)$ and obtain values of the major parameters (see Table I), including the Fermi velocity v_f. The fact that we obtain now the same value of H_{C2} illustrates the self-consistency of our method.

A uniquely small value of the coherence length is a key feature of the cuprates. The small size of ξ_0 is directly related to the exotic values of the normal parameters (see sec.II). Let us now consider the ratio $\Delta(0)/E_f$. This ratio is an important parameter in the physics of superconductivity; it estimates what fraction of the carriers are directly involved in pairing. This parameter is small in conventional superconductors (10^{-4}). In the cuprates the situation is entirely different. A small value of the Fermi energy (along with a large value of the gap) leads to a large value of the ratio: $\Delta(0)/E_f \approx 10^{-1}$. This means that a significant fraction of the carriers are paired. Of course, this corresponds to a short average distance between paired carriers which further implies a short coherence length.

b) **Two-gap Superconductivity**: A short value of the coherence length presents a unique opportunity to observe multigap structure in the cuprates. This fact had been predicted in our papers [1b,4].
Recently, this effect has been observed in [18]; however, the most convincing observation was obtained by NMR [19,20].

The two-gap structure is caused by the presence of different overlapping energy bands; each band is characterized by its own energy gap. Since each band has its own properties, e.g. densities of states, this immediately leads to distinct values of the energy gap in each band.

At first sight, one might expect to find a manifestation of the multigap structure in conventional superconductors, because the presence of overlapping energy bands is a common feature of usual metals. In connection with this, it is worth noting that the two gap model was introduced shortly after the BCS theory [21a] and then developed in detail in [21b,c]. However, it is well known that despite the presence of overlapping bands in many conventional metals, their superconducting properties have been described accurately by a one-gap model. The reason for this is directly related to the value of the coherence length. The presence of overlapping bands is crutial, but, nevertheless is insufficient for observing a multigap structure. The essential criterion [22] is that the mean free path l, must be larger than ξ_o. Indeed, the presence of impurities leads to interband transitions. If, in fact, $l < < \xi$ (this case corresponds to most conventional superconductors because of their large value of coherence length), the strong scattering results in averaging and in the applicability of the one-gap picture. In other words, the large scale of the pair correlation makes the observation of the multigap picture in the usual superconductors unrealistic. The small value of the coherence length in the cuprates leads to an entirely different situation.

Therefore, two conditions have to be satisfied in order to observe a two-gap structure:
　　1. Presence of two different energy bands and
　　2. The mean free path must be larger than the coherence length.

The second condition can be met in some of the high T_c oxides. However, the La-Sr-Cu-O compound, has a relatively simple band structure and since it does not satisfy the first criterion, its properties can be described by the one gap model. A different situation exists in Y-Ba-Cu-O: it contains two different energy bands, a quasi-two-dimensional and a quasi-one-dimensional band. Therefore, both conditions are satisfied for this material, and we expect the presence of two different energy gaps [4,5].
Y-Ba-Cu-O contains two superconducting subsystems related to the planes and to the chains in the structure. The analysis of the properties of the superconductor with two gaps should be carried out

by two energy gaps, Δ_1 and Δ_2 and by three (!) independent coupling constants: $\lambda_{11}, \lambda_{22}$, and λ_{12} (for a more detailed description see the papers [5]).

The major parameters of the Y-Ba-Cu-O compound are presented in Table II. In terms of Fermiology, the two-band picture corresponds to a Fermi surface containing two different parts (see Fig. 1).

We will now describe several important features of two-gap superconductivity:

1. Because of the interband coupling, the system is characterized by a single value of T_C. However, the values of the ratios $\Delta_1(0)/T_C$ and $\Delta_2(0)/T_C$ can differ drastically from the value of $\Delta(0)/T_C$ obtained in one-gap theory, even in a weak coupling approximation (see Fig. 2).

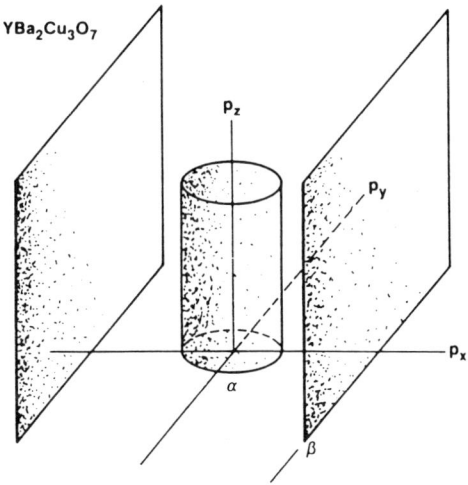

Fig. 1. Fermi Surface for Y-Ba-Cu-O.

TABLE II. Superconducting parameters of conventional metals and the cuprates

Parameters	Conventional	LaSrCuO	YBaCuO		
T_C	< 23K	40K	95K		
$2\Delta/k_B T_C$	< 4.4	~ 5	5-8	2-3.5	
Δ/E_F	10^{-4}	10^{-1}	2×10^{-1}	10^{-1}	
			λ_a	λ_b	λ_{ab}
λ_{el-ph}	< 2	2.5	>1	< 0.5	~ 0
$\xi_0(\text{Å})$	10^3-10^4	20	15	7	
H_{ez} (T)	< 60T	< 90T	140T	280T	

281

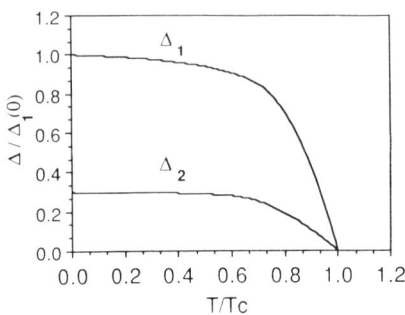

Fig. 2. A qualitative sketch of the temperature dependences for the energy gaps for a two-gap superconductor.

The same is true for $d\Delta/dT$ near T_c.

2. The presence of two bands is favorable for superconductivity. In other words, the value of T_c is larger relative to its value for the one-gap case.

3. Each energy band contains its own set of pairs with its own value of the coherence length. Different values of the energy gaps and the coherence lengths lead to peculiar transport and electromagnetic properties of the system.

Two distinct gap values for the Y-Ba-Cu-O compound have been observed in a tunneling experiment [18]. The most convincing evidence comes from NMR experiments[19,20]. In particular, we want to stress the importance of the analysis carried out in the paper [20]. The authors measured the temperature dependence of the Knight shift for two different Cu sites and clearly demonstrated the presence of two gaps.

Thus in summary, the short value of the coherence length has resulted in a unique opportunity to observe a two-gap structure.

c) <u>Positron Annihilation Lifetime, Impedance, Critical Behavior</u>: An interesting phenomenon with respect to positron annihilation in the cuprates has been described in [23]. It turns out that the transition into the superconducting state leads to an increase in a positron-annihilation lifetime in La-Sr-Cu-O and this shift, $\Delta\tau = \tau_s - \tau_n$, increases with decreasing temperature below T_c. This phenomenon is interesting because it has been observed only in the cuprates. A study of conventional superconductors did not reveal any shift in the lifetimes relative to the normal state, therefore, we are dealing with a phenomenon which is an unique feature of the new materials.

The positron entering the oxide annihilates mainly with localized electrons. However, its interaction with the localized electrons which results in the annihilation, is screened by the delocalized carriers, and the screening appears to be noticeably affected by the pairing (for a detailed analysis see [24a]). One can show that the shift in the lifetime is equal to: $\Delta\tau = \gamma(\Delta/E_f)^2 \ln(E_f/\Delta)$. Therefore, its value depends directly on the parameter Δ/E_f (see above). A dependence on $(\Delta/E_f)^2$ was also obtained in [24b]. For conventional materials the ratio Δ/E_f is very small; one can obtain, for example, for Al, $(\Delta(0)/E_f)^2 \approx 10^{-9}$. Such a small shift is unobservable. Converesely, for La-Sr-Cu-O, the situation is entirely different: a large value of $\Delta(0)$ along with small value of the Fermi energy leads to an observable shift of several percent. The increase in the shift with decreasing temperature below T_C is due to the temperature dependence of the energy gap. The same parameter, $\Delta(0)/E_f$ describes the pair fluctuations in the region near T_C. Therefore, one can expect unusual critical behavior of the cuprates. Based on the small value of E_f, G. Deutscher [25] predicted the possibility of observing such behavior in the cuprates, similar to that observed in liquid HeII; this effect, indeed, has been observed experimentally in the cuprates [26].

We also point out that the short coherence length leads to an inequality: $\delta >> \xi_o$, where δ is the magnetic penetration depth. This means that the behavior of the cuprates in an external electromagnetic field is described by London electrodynamics. The evaluation of the impedance [27] shows that this quantity is proportional to the parameter $(\delta/\xi_0)^2$. A short coherence length leads to a large value of this parameter and consequently a large value of the surface resistance vis-a-vis conventional superconductors, and this conclusion may explain recent experimental data.

As was mentioned above, the properties described in this section are not related directly to the mechanism of high T_C. They are consequences of the exotic normal properties described in Sec.II and all fit together in a unified and coherent manner.

In the next section we are going to describe our approach to the origin of high T_C. We will show that the peculiar electronic and lattice structure of the materials along with exotic values of the normal and superconducting parameters enable us to understand the origin of high T_C in the same unified way.

IV. ORIGIN OF HIGH T_C

This section is concerned with the origin of high T_C in the cuprates. The superconducting state in conventional superconductors is caused by the electron-phonon interaction (BCS theory). At the same time it is known that the pairing can be mediated by other excitations.

We think that the high T_C is due to strong coupling of the carriers with low energy excitations (generalized phonon or phonon-plasmon mechanism), namely with phonons and a peculiar phonon-like electronic acoustic branches ("electronic" sound). We also want to point out that the phonon spectrum in the cuprates has a number of unusual features.

Initially we are going to describe the major properties of the collective excitations; then we will discuss the mechanism of high T superconductivity.

Phonons

We think that phonons play a key role in high T_c superconductivity. Such statements may sound old-fashioned, because it means a similarity with conventional superconductors. However, we want to stress that we are dealing with an exotic phonon system in the cuprates. These materials contain soft optical modes, which are anharmonic. As a result, the lattice dynamics is very peculiar and differs drastically from that in conventional metals.

Ferroelectricity of the perovskites is a definite manifestation of unusual lattice dynamics. That's why the initial motivation by Bednorz and Muller was quite logical.

It was interesting to find that in the doped conducting La-Sr-Cu-O compound neutron scattering [28] showed the tetragonal-orthorhombic transition to be driven by a zone boundary soft tilt mode of the CuO units. A parallel may be drawn between these rigid-body like motions of sub-units of the high T_c cuprates and organic superconductors, in which the very nature of molecular structure leads to new orientational lattice modes-librons, which hybridize with the center-of-mass displacement longitudinal and transverse phonons [29].

Additional indications for unusual phonon dynamics were found in pulsed neutron experiments [30a] sensitive to short range order and in an extended X-ray absorption fine structure (EXAFS) [30b]. Neutron scattering experiments performed by the Karlsruhe group [31] probing the phonon dispersion relation over the entire brillouin zone found strong indication of multi-phonon contributions.

It has been proposed [32a,b] that a double-well potential be used a starting point for some of the oxygen dynamics. Theoretical support for a potential comes from LDA calculations [29]. Such a potential can affect the lattice dynamics and the carrier-phonon interaction.

An anharmonicity leads to an appearance of non-linear terms in the carrier-phonon interaction. A very general study of such high order effects was made in [33]. We have started [34] to investigate the effect of a higher-order electron-optical phonon interaction in the framework of the generalized Eliashberg formalism. By calculations of the effect of such a specific interaction (a two-phonon exchange) on the transition temperature, one can show that it leads to an additional increase in T_c.

Plasmons in layered cuprates. "Electronic" sound

The concept of plasmons is not a new phenomenon in solid state physics. Plasmons describe the collective motion of the carriers relative to the lattice. Usual metals are characterized by the following dispersion relation for the plasmons: $\omega = \omega_0 + aq^2$; therefore, the plasmon spectrum has a finite value ω_0 at $q=0$. The situation in layered materials is quite different. In fact, in the early seventies there was a surge of interest in layered metals such as TaX

(X=S,Se,Te) and intercalated graphite, which became superconducting. In contrast to isotropic metals, these materials have considerable anisotropy in their conductivities along and perpendicular to the layer planes. As a consequence, the screening properties of such a system are quiet different from the isotropic case and were analysed in some detailed by various theorists (see e.g.[35]. The term layered electron gas (LEG) describes the basic physical concept, namely an infinite set of two-dimensional layers of carriers, described as a 2D electron gas, separated by electronically inactive (insulating) spacer layers.

It was shown for a single layer [36] that the collective (plasmon) mode frequency goes to zero for a small in-plane wave-vector vector κ. In the LEG model the Coulomb interaction between charge carriers on different layers is included exactly, while the polarization of the system is treated by including the electron-hole pair responce calculated for a single layer with 2D plane-wave wave-functions for the carriers. In addition, the dielectric responce of the intervening, non-metallic region is included by using the low frequency limit of its dielectric constant, $\varepsilon \cong 3\text{-}6$ to screen the interaction between conducting layers.

A direct consequence of the layering of the carriers is that the spatial periodicity along the normal to the planes allows us to define a corresponding wave-vector $q_z = \pi/d_c$, which labels the phase relations between charge oscillations on different planes. The calculation utilizes a 2D Fourier transform describing the in-plane Coulomb interaction, while the direction perpendicular to the layers is described by a discrete index. The fully screened interaction can be studied algebraically in the random phase approximation by calculating the polarizability in a single layer and solving the Dyson equation for the screened interaction [35,36,6c,d]. The eigenvalues of the equation $\varepsilon(\kappa,\omega) = 1 - V(\kappa,q_z)\Pi(\kappa) = 0$ then define layer plasmon dispersion relations. Most strikingly, the completely in-phase motion of all layers $q_z = 0$ has a frequency $\omega = \omega_p$, which corresponds closely to the isotropic (bulk) plasma frequency. All other branches with $q_z = 0$ go to zero frequency as their transverse wave vector κ (which is proportional to the inverse of the wavelength of the density oscillation in the plane) goes to zero. This breaking of the degeneracy of the standard three-dimensional plasmon excitation into different branches of charge oscillation modes filling in the low frequency and transverse wave-vector space of the spectrum is, in our view, a very important aspect of layered conductors and superconductors, in particular, the copper oxide high T_c superconductors.

It is also possible to include a small hopping term t in the c-axis direction; one now find a small gap at $\kappa = 0$ of order $4t$.

There are several novel consequences of the splitting of the collective excitation spectrum into layer plasmon branches. These arise specifically because of the existence of a low frequency electronic response. The first of these is the possibility of pairing the carriers in the copper oxides by exchange of acoustic plasmons [6]; we have proposed such a contribution to the high T_c's in the layered cuprates some time ago. Our view is that such pairing occurs in addition to strong electron-phonon coupling (see below).

A second consequence of the low frequency plasmon spectrum is the possibility of a temperature dependent term in the electron loss function because of the thermal population of the acoustic plasmon branches [37].

Finally, it is also possible to relate the layer plasmon spectrum to $Im(\varepsilon^{-1})$ where $\varepsilon=\varepsilon_1+i\varepsilon_2$ is the dielectric function of the layered metal with ε_1 and ε_2, the real and imaginary parts, respectively. Calculation of this function in the LEG reveal that it is proportional to ω^2 [8,37], and it has been observed experimentally (see below,Sec.V).

Therefore, the plasmon spectrum in layered conductors such as the cuprates represent the plasmon band (Fig. 3 a).

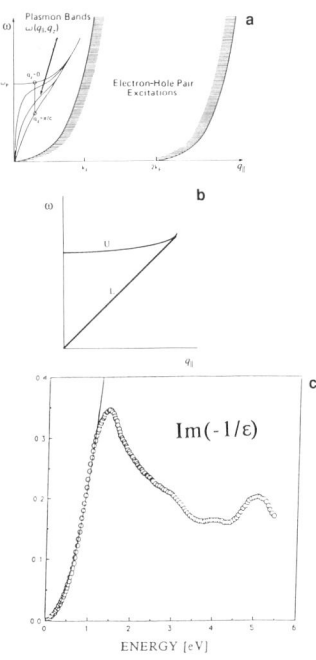

Fig. 3. a) Plasmon band, b) Upper and lower branches, c) Loss function

The density of states is peaked near the upper ($q_z=0$) and low ($q_z = \pi/d_c$) boundaries. Qualitatively, one may visualize the plasmon band as a set of two branches (Fig. 3), the upper (U) branch is similar to that in usual metals. A very important feature of the layered metals is the appearance of the lower (L) branch, which has an acoustic dispersion law and can be called "electronic" sound. One should stress also that the slope of this acoustic branch is of order of v_f (not the sound velocity as for usual phonons), however, the value of v_f

is small (see Table I), and therefore, we have an additional phonon-like branch with large phase space.

Determination of carrier-phonon coupling parameter

As was mentioned above, we think that high T_C is due to a generalized phonon mechanism, that is to strong coupling $\lambda_{e\text{-}ph}$ to lattice vibrations (phonons); additional coupling $\lambda_{e\text{-}pl}$ comes from an electronic phonon-like branch (" electronic' sound). The total coupling is $\lambda = \lambda_{e\text{-}ph} + \lambda_{e\text{-}pl}$. The phonon energies are in the range $\Omega \cong 15$ meV for La-Sr-Cu-O compound and $\Omega \cong 25$ meV for Y-Ba-Cu-O. The plasmons are characterized by a larger energy scale (≈ 60 meV).

It is very important to determine the strength of the coupling between the carriers forming the paired state. Indeed, the value of T_C is determined by two parameters: the strength of the coupling and the energy scale W, so that $T_C = T_C(\lambda, W)$. For example, in the BCS model W is a scale of the phonon energies Ω, and $\lambda_{e\text{-}ph}$ is the electron-phonon coupling constant.

If λ is small, then high T_C in the cuprates can be provided by a large energy scale, much greater than the phonon energies. On the other hand if λ is large, one can expect phonons to play an essential role.

The determination of electron-phonon coupling is not a trivial task. The use of the conventional technique, tunneling spectroscopy, has been frustrated by a short coherence length; only recently it has been performed (see below, Sec. V).

In the absence of high quality tunnelling spectroscopy data, which is the best way to determine λ, the electron phonon coupling parameter, we have presented an alternative method based on the analysis of a fundamental bulk property, heat capacity [4d,5a].

Electron phonon coupling (EPC), along with temperature dependence of the phonon distribution lead to the dependence of the Sommerfeld constant γ ($\gamma = C_{el}/T$) on temperature [38]. This manifests itself in a deviation of the electronic specific heat from a simple linear law. This dependence is described by the equation

$$\gamma(T) = \gamma(0)\{1 + \rho[[\kappa(T)/\kappa(0)] - 1]\} \tag{3}$$

where $\gamma(0) = \gamma^0(1+\lambda)$ and γ^0 is the band value of the Sommerfeld constant, $\lambda = 2\int d\Omega g(\Omega) \Omega^{-1}$; $\rho = \lambda[1+\lambda]^{-1}$ and $\kappa(T) = 2\int d\Omega g(\Omega) \Omega^{-1} Z(T/\Omega)$ where Z is a universal function that has been derived in [38] and Ω is a phonon frequency. It can be readily seen from Eq. X.1 that at low temperatures (T→0) the Sommerfeld constant approaches the band value renormalized by the EPC; i.e $\gamma(0) = \gamma^0(1+\lambda)$; it means that the carriers are "dressed" by phonons and other excitations.

The major contribution to the coupling comes from the low frequency region ($\Omega \approx 15$ meV for La-Sr-Cu-O). Let us denote the corresponding strength by λ_0, so that $\lambda = \lambda_0 + \lambda'$; λ' describes the coupling to the acoustic plasmon mode and to higher phonon branches. Therefore, $\gamma(0) = \gamma^0(1+\lambda_0+\lambda')$. One can show [5], that $\gamma(T_c) = \gamma^0(1+\lambda')$, so that the intensity of the thermal motion at T_c is sufficient to destroy the main part of the phonon "cloud". Then we obtain

$$\lambda^* = [\gamma(0)/\gamma(T_c)]-1 \qquad (4)$$

where $\lambda^* = \lambda_0 / (1+\lambda')$; therefore λ^* represents the minimum value of the electron-phonon coupling constant.

Using Eq.(4) one can evaluate the value of λ^*. The value of the electron-phonon coupling constant is even larger.

Thus, if we can use experimental data to extract the values of the low and high temperature Sommerfeld constant then we can get an estimate the minimum value of the electron-phonon coupling constant, λ_{e-ph}.

The value of $\gamma(0)$ can be obtained from the analysis of the dependence $\gamma(H)$, see above, Sec.II. The value of $\gamma(T_c)$ can be obtained from the measurements of the jump ΔC in heat capacity at T_c and the expression $\Delta C/\gamma(T_c)T = \beta$, where β describes the jump in the strong coupling theory (see a detailed analysis in [4,5]).

As a result, we obtained the value $\lambda_{e-ph} \approx 2.5$, and it means a strong carrier-phonon coupling (let us remember, that for Pb, which is a conventional superconductor with strong coupling, $\lambda_{e-ph} \approx 1.4$).

One can estimate the value $\lambda_{e-pl} \approx 0.3-0.4$. As a result, the total coupling constant in La-Sr-Cu-O is of order $\lambda \approx 2.8-3$.

The mechanism of high T_c

The question of how high can T_c be due to the exchange of some particular excitation may be broken in two distinct parts : 1) the question of the existence of a particular excitation , and 2) is the coupling of the carriers to this (or these) excitation sufficient?

The answer to the first question for phonons and acoustic plasmons is positive; we are talking about really existing excitations. By analyzing the phonon and plasmon spectra and the coupling between the carriers and these excitations, we can answer the question on the origin of high T_c.

The interaction with phonons is strong and plays a key role. In the presence of phonons and the electronic phonon-like branch T_c can be estimated from the expressio [3-6]

$$T_c = T_c^{ph}[\omega_{pl}/T_c^{ph}]^\upsilon \qquad (5)$$
with $\qquad T_c^{ph} = 0.25\Omega/(e^{2/\lambda_{eff}}-1)^{1/2}$

where $\upsilon = \lambda_{pl}(\lambda_{ph}+\lambda_{pl})$ and ω_{pl} is the average frequency for L branch.

Using the values obtained for La-Sr-Cu-O , namely $\Omega = 15$ meV,

$\lambda_{ph} = 2.5$, $\lambda_{pl} = 0.2$-0.3, ω_{pl} is of order 0.1 eV, we arrive at a value for T_c of 40 K. The exact evaluation for Y-Ba-Cu-O requires use of the two-gap model [5], but our estimate also gives good agreement with the experimental value. Note also that the non-monotonic dependence of T_c on the carrier concentration[17b] can be explained on the basis of strong electron-phonon coupling theory [6c].

It is important to stress that our analysis does not contain any additional assumptions and is based on the analysis of real excitations and their properties.

Note that the importance of the phonons is usually minimized because of the small value of the oxygen isotope effect. However (for a more detailed discussion see [6c]) the isotope effect, particularly for complicated compounds, is not at all a simple and straightforward phenomenon. It is well known that several conventional superconductors have a very small isotope effect even though the mechanism of superconductivity is definitely phononic. The presence of a polyatomic lattice, the effects of the isotope substitution on the elastic constants, plasmon contibution, etc. lead to a situation when it is inappropriate to correlate the value of the isotope coefficient with the contribution of the phonons to T_c. Another argument against phonons is connected with the evaluation of λ from normal transport data [39]. However, this analysis dealt with the so-called transport coupling constant λ_{tr} which is related but nevertheless different from the λ entering T_c. In addition, it is necessary take into account the anisotropy of the system and the two-band structure (for Y-Ba-Cu-O); and finally, but no less important, a small value of E_f (see Sec.II). This latter fact requires that the temperature dependence of the chemical potential $\mu(T)$ be considered in the high temperature region. All these factors affect the value of λ.

Our analysis leads to the conclusion that the unusual lattice dynamics as well as the layered structure in the cuprates leads to an exotic version of the BCS theory with a strong interaction between low energy excitations (phonons and phonon-like plasmon branch) and the carriers.

V. RECENT CRUCIAL EXPERIMENTS

During the last year there has been remarkable progress in the physics of high T_c superconductivity. This progress is mainly due to the improved quality of the samples, e.g. better thin film and single crystal preparation.

We started to develop our approach (see previous sections) immediately after the discovery of high T_c. According to our theory, cuprates are characterized by a Fermi surface with exotic parameters (see Sec. II). Their collective excitations contain anharmonic soft phonons and peculiar acoustic plasmons. In the superconducting state, there is a unique opportunity to observe a multigap structure. As for the origin of high T_c, we have developed our approach based on the generalized phonon (phonon-plasmon) mechanism (see Sec III).

Our method is based on a theoretical analysis of experimental data, but, nevertheless, it is a self-consistent theoretical scheme. There were many predictions. Below, we are going to describe some of the recent critical experimental results confirming our theory. At the present time, one can certainly form the foundation for the physics of the high T_c.

1. Fermiology, Photoemmission, De-Haas-van-Alphen Oscillations

A number of interesting papers on photoemission (see e.g.[13]) contain a wealth of information. Here we'd like to stress the observation of the Fermi edge. This is a clear manifestation of the presence of the Fermi surface. In addition, the value of the Fermi energy has been determined in [13]: $E_f = 0.2$ eV for Y-Ba-Cu-O. This value is in an excellent agreement with that obtained in [4,5], see also Table II.

The presence of the Fermi surface has been also established in [14]; the authors have observed de-Haas-van-Alphen oscillations. This technique is a conventional method in Fermiology (see e.g.[11]) The problem with such technique applied to the cuprates is that it is necessary to have a normal metal at a low temperature (all resonance signals are washed out at higher temperatures). However, the cuprates become normal only above a rather high T_c; therefore, a large value of T_c is an obstacle for such experiments. This problem has been overcome in [14] with the use of the explosive generation of superstrong magnetic fields, which both destroy the superconductivity and provide the field for the resonance.

The presence of the de-Haas-van-Alphen oscillations is direct evidence of the applicability of the concept of a Fermi surface to the cuprates. In addition, the authors have determined the values of the effective mass in different directions.We have evaluated the "cyclotron" mass, m^* [4,5], which we find equal to $5m_e$ for the planes in Y-Ba-Cu-O. The value $m^*=4.6m_e$, obtained in[14] is very close to our value, and it provides additional support for our theory.

2. Acoustic plasmons, Ellipsometric spectroscopy, Loss function

According to theoretical analysis [6,8], the excitation spectrum of the layered cuprates contains additional acoustic plasmon branches ("electronic" sound). Our evaluation of the electron energy loss function[8] has resulted in the dependence: $Im(\varepsilon^{-1}) \propto \omega^2$ in the region of small frequencies. Such a dependence is connected with the structure of the plasmon band (see above, Fig.3).

An experimental study [40], based on the use of ellipsometric spectroscopy, has revealed this quadratic dependence (see Fig.3c), in agreement with [8]. Therefore, the existence of the plasmon band and the low-lying acoustic branch has been confirmed experimentally.

3. S-wave pairing

The generalized phonon mechanism assumes S-wave pairing. There are currently three experiments that strongly indicate that the pairing in the cuprates is s-wave. The first is a detailed measurement of the temperature dependence of the magnetic penetration depth at very low temperatures. The absence of a finite slope is strongly indicative of S-wave symettry.

A very elegant set of proximity effect experiments is described in [41]. A composite ring containing sections of both Y-Ba-Cu-O and Pb has demonstrated a persistent current without any depression of the order parameter in the Pb. This means that the order parameter in the cuprates has the same S-symmetry as the conventional superconductor (Pb).

Finally, NMR Knight-shift data show that S-pairing is a the most favorable type of symmetry consistent with their results [20].

All these experiments give a strong indication that the exchange by lattice or electronic (non-magnetic) excitations is the mechanism for pairing.

4. Two-gap superconductivity, NMR experiments, Thermopower

The NMR paper [20] contains another important experimental observation. The detailed analysis of the temperature dependence of the Knight shift has shown the presence of two superconducting subsystems in Y-Ba-Cu-O and a clear two-gap structure.

A very sensitive analysis of the thermopower S [42] has resulted in observations of a change of sign of S as a function of the oxygen content in a very narrow range of oxygen contents between 6.9 and 6.98. Such a dramatic change is caused by the presence of two groups of carriers (holes(planes) and electrons(chains)).

5. Strong coupling to low energy modes, tunnelling spectroscopy

The method of tunneling spectroscopy based on the inversion of the Eliashberg equation (see the reviews[43]) is an unique technique which allows the determination of the strength of the coupling and the nature of the pairing interaction. The function $\alpha^2(F(\omega))$ (where $F(\omega)$ is a phonon density of states and α^2 describes the coupling) contains important information. The coupling constant which determines the value of T_c is equal to: $\lambda = 2\int d\Omega \alpha^2(\Omega) F(\Omega) \Omega^{-1}$

The structure of the function $\alpha^2(F(\Omega)$ allows one to determine which frequency region makes a major contribution to the pairing.

Tunneling spectroscopy measurements of Ba-K-Bi-O and Nd-Ce-Cu-O compounds have been made [44]. The function $\alpha^2 F$ for the both these materials were obtained by the Rowell-Mcmillan inversion of the Eliashberg equation. The Ba-K-Bi-O was shown to be phonon-mediated superconductor because the all the peaks coincide with measured neutron diffraction phonon peaks. As for the n-type layered cuprate it has been shown that the pairing is caused by low energy excitations. A definite assignment of the structure seen to the phonon and/or plasmon branches requires detailed neutron spectroscopy and electron energy loss measurements.

Recently R. Dynes et.al. [45] have reported tunneling spectroscopy measurements of Y-Ba-Cu-O. This investigation is the further development of the analysis reported in [18]. The function obtained by the inversion of the Eliashberg equation contains a peak near $\Omega \approx 25$ meV. According to neutron data [31], this region corresponds to the maximum of the phonon density of states. Therefore, the result [45] represents direct experimental evidence of the importance of the phonons for the appearance of high T_c. The coupling constant λ, obtained in [45], appears to be greater than 2; this means that we are dealing with strong electron-phonon coupling. The data presented in [45] should be refined; one should take into account the presence of two bands in Y-Ba-Cu-O and therefore, several functions $\alpha_i F$, the plasmon branch, etc, but, nevertheless, this reconstruction of $\alpha^2 F$, the peak at the phonon frequency, a large calculated value of T_c (~ 60 K) are key indications of the importance of the phonons.

There are still many unresolved problems in the physics of high T_c (dynamics of doping, Fermiology of Bi-and Tl-based cuprates, vortex dynamics, etc.), but, nevertheless, one can formulate the basic principles of the physics describing properties of these exotic materials.

ACKNOWLEGEMENTS

We thank the ONR for their support of this work. One of us (VZK) was supported under ONR Contract #N00014-90-F0005. We also thank J. Bardeen, I. Bozovic, G. Deutscher, D. Gubser and D. Scalapino for stimulating discussions and specially thank R. Soulen for a critical reading of the manuscript.

REFERENCES

1. A.Bednorz and K.A.Mueller, Z.Phys. B$\underline{64}$, 189 (1986)
2. M.Wu et al., Phys.Rev.Lett. $\underline{58}$, 908 (1987) ; H.Maeda et al., Jpn. J. Appl. Phys. Lett. $\underline{27}$, 209 (1988) ; Z. Sheng and A. Hermann, Nature 33 $\underline{2}$, 55 (1988)
3. a) V. Kresin, Phys.Rev. B$\underline{35}$, 8716 (1987) ; b) Solid State Comm. $\underline{51}$, 339 (1987); c) Phys.Lett. A $\underline{122}$, 434 (1987)
4. a) V. Kresin and S. Wolf, in _Novel Superconductivity,_ S.Wolf and V.. Kresin, eds., p.287 Plenum, NY (1987) ; b) Solid State Comm. $\underline{63}$,1141 (1987); c) J.of Superconductivity $\underline{1}$,143 (1988). d) V. Kresin, G.Deutcher, and S.Wolf, ibid.,1, 327 (1988)
5. V. Kresin and S. Wolf ,a) Phys. Rev. B $\underline{41}$, 4278 (1990); b) Phys.C $\underline{169}$, 476 (1990); c) _Fundamentals of Superconductivity,_ Plenum, NY(1990)

6. V. Kresin and H. Morawitz, a) Ref.4a), p.445 ; b) Phys.Rev.B 37, 7854 (1988); c) J. of Superconductivity 1,108 (1988); d) Phys. Lett. 145, 368 (1990). e) Solid State Comm. 74, 1203 (1990)
7. a) S. Wolf and V. Kresin, Phys.C 162-164, 217 (1989); b) in Advances in Superconductivity II, T. Ishiguro, K. Kajimura, Eds., p.447, Springer-Verlag, Tokyo (1989)
8. a) H. Morawitz and V. Kresin, Phys.C 162-164, 1471(1989); b) Bull. APS 29, 2234 (1989); c) H. Morawitz, I. Bozovik, V. Kresin (preprint)
9. R. Cava et al., Nature 333, 814 (1988)
10. N. Ashcroft and N. D. Mermin, Solid State Physics, Holt, Rinehart and Winston, New York, (1976)
11. A. Cracknell and K. Wong, The Fermi Surface, Clarendon, Oxford (1973)
12. E. Lifshitz and L. Pitaevskii, Statistical Physics, p.II, Pergamon Press
13. T. Takahashi et al. Nature 334, 691 (1988); A. Arko et al., Phys. Rev. B40, 2268 (1989); J. Imer et al., Phys. Rev. Lett. 62, 336 (1989)
14. F. Mueller et al., Bull. APS 35, 550 (1990); (preprint)
15. N. Phillips et al., Ref.4a, p.739; R. Fisher et al., J. of Superconductivity, 1, 231 (1988)
16. N. Ong et al. Phys. Rev. B 35, 8807 (1987); M. Suzuki, ibid, B 39, 2312 (1989)
17. a). J. Torrance et. al., Phys. Rev. Lett. 61,1127 (1988); b) J. Torrance et. al., ibid 61, 1127 (1988)
18. M. Gurvitch et al., Phys. Rev. Lett. 63,1008 (1989)
19 W. Warren et al. Phys. Rev. Lett. 59, 1860 (1987)
20. S. Barrett et al., Phys. Rev. B41, 6283 (1990)
21. a) H. Suhl et al., Phys. Rev. Lett. 3, 552 (1959) ; b) B. Geilikman et al., Sov. Phys.-Solid State 9, 642 (1967); c) V. Kresin, J. Low Temp. Phys. 11, 519,(1973)
22. P. Anderson, J. Phys. Chem. Sol. 11, 26 (1959)
23. Y. Jean et al., Phys. Rev. Lett. 60,1069 (1988)
24. a). V. Kresin and H. Morawitz, J. of Super. 3, 227 (1990); b) R. Benedek and H. Schuttler, Phys. Rev. B 41, (1990)
25. G. Deutcher, Ref. .4a), p.293
26. a) R. Fisher et al ., Phys. Rev. B 38,11942 (1988) ; b) D. Ginsberg et al., Physica C153-155,1082 (1988)
27. V. Kresin, J. of Super., 3, 177 (1990)
28. P. Boeni et al., Phys. Rev. B38, 185 (1988)
29. H. Morawitz, Phys. Rev. Lett. 34, 1567 (1975)
30. a) W Dmowski et. al., Phys. Rev. Lett.,61, 2608 (1988), b) S. Conradson and I. Raistrick, Science 243,1340 (1990)
31. a) W. Reichartdt et al. Phys.C 162-164, 464 (1989).; b) H. Rietschel et al., Phys.C 162-164,1705 (1989)
32. a) J. Hardy and J. Flocken, Phys. Rev. Lett. 60, 219 (1988); b) N. Plakida et al. Europh. Lett. 4,1309 (1987); c) H. Morawitz and V. Kresin, Bull. APS 29, 2234 (1989)
33. B. Geilikman, J. Low Temp. Phys. 4, 189 (1971)

34. H. Morawitz and V. Kresin, in Proc. of the Int. Conf.on the Electronic Structure of High T_c Superconductors, J. Fink, H.Kurzmany, G.Mehring, Eds., Kirchberg, Austria, Springer-Verlag (in press)
35. P.Visscher and L.Falicov, Phys. Rev. B$\underline{3}$, 2541 (1971); A.Fetter, Ann.Phys., N.Y. $\underline{88}$,1 (1974); P.Hawrylak et al., Phys. Rev. B$\underline{32}$, 4272 (1985)
36. F. Stern, Phys. Rev.Lett. $\underline{18}$, 5646 (1967)
37. V. Kresin and H. Morawitz, J. Opt. Soc. Am. B $\underline{6}$ (3),,490 (1989)
38. V. Kresin and G. Zaitsev, Sov. Phys.-JETP $\underline{47}$, 983 (1978); G.Grimvall, The Electron-Phonon Interaction in Metals North-Holland, Amsterdam (1984)
39. M. Gurvitch and A. Fiory, Phys. Rev. Lett., $\underline{59}$,1337 (1987); Ref. 4a, p.663
40. I. Bozovik, Phys. C $\underline{162\text{-}164,}$ 1239 (1989); Phys.Rev.B$\underline{42}$,1969 (1990)
41. G. Yee et al., Phys. C, $\underline{161}$,195 (1989); J. of Super., $\underline{3}$, 197 (1990)
42. J. Cohn et al., preprint
43. W. McMillan and J.Rowell, in Superconductivity, ed. by R. D. Parks, Dekker, New York (1969) Vol 1, p.561; E.Wolf, Electron Tunnelling Spectroscopy, Oxford, NY (1985)
44. Q. Huang et al., Nature (in press)
45. R.Dynes, ASC , Snowmass,1990 (invited talk)

ELECTRON-HOLE ASYMMETRY: THE KEY TO SUPERCONDUCTIVITY

J.E. Hirsch

Department of Physics
University of California, San Diego, La Jolla, CA 92093

The observation that superconductors have predominantly positive Hall coefficients was made already in 1962 by Chapnik. The discovery of superconductivity in hole-doped oxides in 1986, and the measurement of a positive Hall coefficient in electron-doped oxides in 1989, further underscore the existence of a distinct asymmetry between electrons and holes with respect to superconductivity, which calls for an explanation. No such explanation is provided by most proposed superconductivity mechanisms including the electron-phonon interaction. Here we discuss the simplest possible local interaction that breaks electron-hole symmetry in solids. This interaction causes holes to be heavier than electrons, and is repulsive for electrons and attractive for holes. Thus, it provides an explanation for the above observations. It leads to superconductivity in the absence of any other attraction mechanism, and the resulting theory exhibits many characteristic features observed in the high T_c oxides. We review the predictions of the theory and their relationship with existing and future experiments in high T_c oxides. A brief discussion of the implications of this mechanism for the understanding of superconductivity in "conventional materials" is also given.

I. CHAPNIK'S RULE

Empirical rules play an important role in the development of science. Whether or not they directly guide towards the correct theory, when a consistent theory finally emerges its agreement with empirical rules can immediately lend it strong support, especially if the theory was not specifically devised to account for those rules. To draw an example from the area of molecular biology (following a leading theorist in our field), the fact that Watson and Crick's base pairing model naturally explained Chargaff's rules on the proportion of the different bases in DNA strands immediately lent strong support to that model. In the field of superconductivity, in contrast, empirical rules (such as those used by B. Matthias and others to find new superconducting compounds) and theory have usually been strangely at odds. One may wonder why.

Here I would like to focus on one such rule, that was formulated almost 30 years ago by I.M. Chapnik,[1] hence I will call Chapnik's rule: "the existence of holes is important for the occurrence of superconductivity". Chapnik observed that among the elements essentially all the superconductors have positive Hall coefficient, and that non-superconductors like the simple and noble metals have negative Hall coefficient (note that even Al has a positive Hall coefficient at high fields).[2] Chapnik later extended his observation to various compounds and suggested that it would be profitable to systematically study the Hall coefficient in a variety of superconducting compounds where it has not been measured. In essentially all known cases the low temperature, high field Hall coefficient being negative in all directions implies absence of superconductivity.

Although some experimentalists were probably aware of this rule in the past and guided by it in the search for new superconducting compounds,[3] it did not come to the attention of theorists, and there has been to my knowledge not a single attempt to provide theoretical justification for it. Perhaps for this reason Chapnik's rule was unknown to the vast majority of the current generation of physicists (including myself), and when the high T_c oxides were discovered in 1986 it did not become immediately obvious that they were a prime example of this rule: they became superconducting when doped with holes, exhibiting a positive Hall coefficient, and superconductivity disappeared on further doping when the Hall coefficient changed sign.[4]

The situation became less clear-cut again in January 1989, when the discovery of "a superconducting copper-oxide compound with electrons as the charge carriers" was announced,[5] in apparent violation of Chapnik's rule. Fortunately, by that time the theory to be discussed here was already sufficiently developed that we were able to immediately predict that in this case "theory would prove experiments wrong" and that hole carriers would be found in these materials when more careful transport and spectroscopic measurements were performed.[6] It was also suggested how O holes would naturally be induced by electron doping of these materials due to the absence of apex oxygens in the structures.[6,7] The evidence that hole carriers indeed exist in these materials is mounting[8,9] and in particular Hall coefficient measurements on single crystals are finding positive values for magnetic field perpendicular to the planes, as predicted.

The conventional electron-phonon mechanism of superconductivity is manifestly particle-hole symmetric, and there is no element in the theory that would suggest that the sign of the curvature of the Fermi surface could make a difference in the magnitude of the coupling constants that enter the theory. Nor do most other mechanisms for superconductivity proposed in the past as well as for the new materials, such as excitonic-, plasmon-, magnon- or anharmonic phonon-mediated pairing, make a distinction between electrons and holes. In view of the vast empirical evidence in favor of Chapnik's rule for the old as well as the new materials we believe it is imperative for these theories to address the issue and suggest an explanation within their framework. Instead, electron-hole asymmetry is the cornerstone of the theory to be discussed in this paper, which in fact leads to what I would term a "strong form" of Chapnik's rule: *all superconductors are hole superconductors*. By that I mean that it is the pairing of hole carriers that drives superconductivity, in all cases.

II. THE BASIC PRINCIPLE: HOLES ARE HEAVIER THAN ELECTRONS

The essence of high T_c is, I believe, clearly displayed in the plot by Torrance and coworkers,[10] reproduced from their paper in Fig. 1. The qualitative behavior of T_c versus hole concentration shown here has since been found by many other workers in a wide range of samples and materials. The Figure also indicates that states are "localized" for small hole doping and the system is a "normal metal" for large hole doping. This refers to the rapid decrease in the normal state resistivity that occurs upon hole doping, that is observed in all the oxide superconductors.

The way to read Fig. 1 that leads directly to the theory under discussion here is as follows: a few holes added to the insulator have difficulty propagating as single objects, hence they are almost localized and the normal-state resistivity is very high. In this regime the holes pair as the temperature is lowered and the system becomes a superconductor (the initial rise in T_c versus hole doping is simply related to the increasing number of carriers.) When enough holes have been added to the system it becomes a "normal metal" with a large mobility for holes in the normal state, and in this regime holes don't pair and don't become superconducting. The conclusion that holes pair *because* they have difficulty propagating individually, *in order* to increase their ability to propagate, and cease to pair when their ability to propagate individually has become large, is almost inescapable. In fact, it can be derived in a phenomenological way from the single assumption that the hole mobility is an increasing function of hole concentration, which can be "read off" from measurements of resistivity versus doping.

We assume that the increase in hole mobility with doping arises from a decrease in the hole's effective mass, or equivalently an increase in the hopping amplitude t between nearest neighbor sites. To lowest order in the hole density n_h we have then for the hopping amplitude t as function of hole concentration:

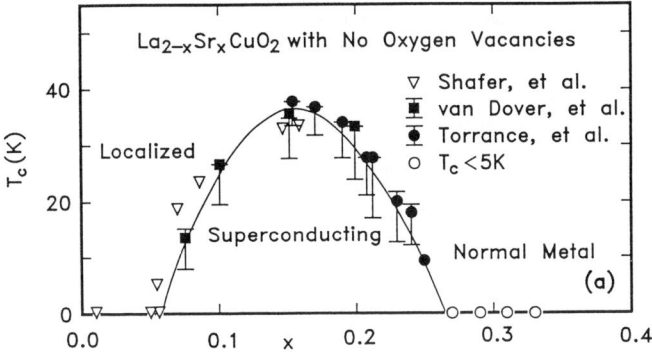

Fig. 1. T_c versus Sr content in $La_{2-x}Sr_xCuO_4$, from Ref. 10. $x/2$ is expected to correspond to number of holes added per O atom in the planes.

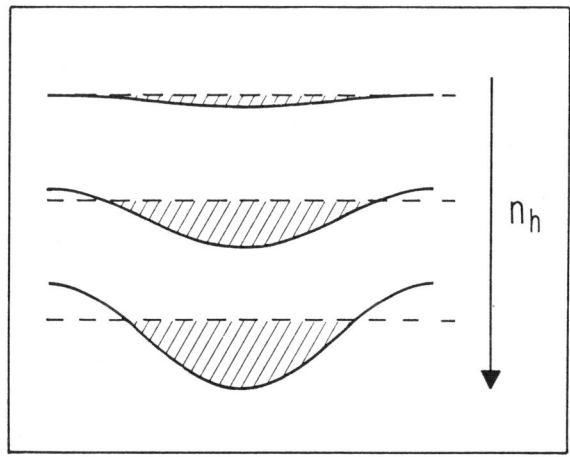

Fig. 2. The relevant electronic band for various hole contents (schematic). We expect this band to arise from overlap of O $p\pi$ orbitals in the planes.

$$t(n_h) = t_h + \Delta t\, n_h \quad , \tag{1}$$

with

$$\Delta t = \frac{\partial t}{\partial n_h} > 0 \quad . \tag{2}$$

The fact that for a few holes in the band the states are localized indicates that t_h in the oxides is close to zero, and the observed bandwidth at finite doping originates mainly in the Δt term in Eq. (1). Fig. 2 shows schematically the single-particle band for different values of hole doping.

Now it is reasonable to assume that the density dependance of the hopping amplitude arises from the local hole density dominantly, i.e. the hole density in the two sites involved in the hopping process. For a hole of spin σ hopping between sites i and j the exclusion principle prevents another hole of the same spin from being at those sites so that the hopping amplitude t_{ij} depends on the occupation of those sites by holes of opposite spin:

$$t_{ij} = t_h + \Delta t(n_{i,-\sigma} + n_{j,-\sigma}) \quad . \tag{3}$$

The kinetic energy for holes is then given by

$$H_{kin} = - \sum_{\langle ij \rangle, \sigma} t_{ij} (c^+_{i\sigma} c_{j\sigma} + h.c.) = - \sum_{\langle ij \rangle, \sigma} (t_h + \Delta t (n_{i,-\sigma} + n_{j,-\sigma}))(c^+_{i\sigma} c_{j\sigma} + h.c.) \tag{4}$$

and thus the density dependance of the hopping naturally gives rise to an interaction term in the Hamiltonian from the last term in Eq. (4):

$$V_h = -\Delta t \sum_{\langle ij \rangle, \sigma} (n_{i,-\sigma} + n_{j,-\sigma})(c^+_{i\sigma} c_{j\sigma} + h.c.) \quad . \tag{5}$$

This "hopping interaction" is most attractive for a few holes in a filled band, it becomes smaller as more holes are added and is repulsive when the band is more than half-full with holes (less than half-full with electrons). It gives rise to superconductivity with a T_c versus concentration dependance of the form shown in Fig. 1, as discussed in later sections. We have proposed that this interaction is responsible for superconductivity in all solids.[11]

In principle one could imagine various different origins for the physics discussed above, leading to the same phenomenological Hamiltonian. For example, it has been pointed out that a single hole in an antiferromagnetic background cannot propagate easily because it leaves behind a "string" of overturned spins, while two holes can propagate together without disrupting the background.[12] This point of view has been independently emphasized in a recent series of papers by Izuyama[13], and it is also a simple way to understand pairing by fractional statistics as advocated by Laughlin.[14] It leads naturally to a picture of heavy single holes that pair to increase their mobility, that could be described "phenomenologically" by a kinetic energy of the form Eq. (4). In such a picture pairing would occur in doping a half-filled band antiferromagnetic insulator in either direction (i.e. with electrons or holes), which would be in apparent agreement with some observations. However, we emphasize that our point of view here is very different.

The Hamiltonian Eq. (4) leads to pairing with arbitrarily short coherence length depending on the relative sizes of t_h and Δt. In fact, for $t_h=0$ two holes will propagate only if they are on the same or nearest neighbor sites, so that the coherence length is less than a single lattice spacing. It is interesting to contrast the physics described by Eq. (4) with the case of the attractive Hubbard model:

$$H = -t \sum_{\langle ij \rangle, \sigma} (c^+_{i\sigma} c_{j\sigma} + h.c.) + U \sum_i n_{i\uparrow} n_{i\downarrow} \tag{6}$$

with $U<0$. This model is widely used as a paradigm for short coherence length superconductors, and can be thought of also as an "effective model" describing more general forms of density-density attractive interactions. Here, short coherence length also occurs for small t (or large U), but in this regime the mobility of the pairs becomes very small, since the pair effective hopping is

$$t_{eff} = t^2/|U| \quad . \tag{7}$$

In other words, the pair interaction limits the mobility, in contrast to the interaction in Eq. (4) that enhances the mobility. The fact that the ultimate result of pairing is to superconduct, that is an enhanced ability for the carriers to propagate, favors a Hamiltonian of the form Eq. (4) rather than Eq. (6).

Beyond these "philosophical" remarks, there is a concrete difference between the physics described by Eq. (4) and Eq. (6) in the regime of short coherence length (relevant to the high T_c oxides). In the attractive Hubbard model as the coherence length becomes shorter the pair binding energy E_b (U) increases relative to the effective hopping Eq. (7):

$$E_b >> t_{eff} \qquad (8)$$

and because of this the transition temperature for Bose condensation is much lower than for pair formation. Thus, the superconducting state is destroyed by center of mass excitations of the pairs rather than pair dissociation, and pairs still exist above T_c. In contrast, in the case described by Eq. (4) the effective hopping for a pair remains finite as the coherence length is reduced and in the presence of Coulomb repulsion the pair binding energy is usually much smaller than the pair hopping amplitude:[15]

$$E_b << t_{eff} \, . \qquad (9)$$

As a consequence, the finite temperature transition is a BCS pair-unbinding transition even in the regime of very short coherence length. The absence of pairs above T_c in the oxide superconductors inferred from NMR experiments[16] thus supports the physics described by Eq. (4) rather than by Eq. (6). In fact we believe that the combined observations of short coherence length and absence of pairs above T_c *cannot be described* by a Hamiltonian involving only density-density interactions (as in Eq. (6) and generalizations) and imply a kinematic pairing mechanism as described by Eq. (4).

III. MICROSCOPIC ORIGIN OF Δt INTERACTION

Within the framework of a single-band tight binding model the interaction Δt is simply the "hybrid" matrix element of the interaction between quasiparticles:[11]

$$\Delta t = \int d^3r d^3r' |\varphi_i(r)|^2 V_{ee}(r-r') \varphi_i^*(r') \varphi_j(r') \qquad (10a)$$

with $V_{ee}(r-r')$ the effective (repulsive) interaction between electrons at positions r, r'. The wavefunctions φ_i, φ_j represent atomic orbitals or Wannier functions at sites i and j. The single particle hopping matrix element is given by

$$-t_0 = \int d^3r \, V_{ei}(r - R_j) \varphi_i^*(r) \varphi_j(r) \qquad (10b)$$

with V_{ei} the (attractive) potential for an electron due to the ion at lattice site R_j. Because of the opposite signs of V_{ee} and V_{ei} the signs of $(-t_0)$ and Δt are always opposite. This implies that the interaction generated by Δt is always attractive (repulsive) for quasiparticle states at the top (bottom) of the electronic energy bands. In other words, the constraint that states at the top of the band have to be orthogonal to those below them ensures that the interaction Eq. (10a) is most attractive for the band states with highest quasiparticle energies. With the convention that the phase of the wavefunctions φ_i, φ_j is the same in the region between sites i and j, Δt and t_0 are positive.

However, this is only part of the story. An important contribution to Δt arises from the modification of the electronic wavefunction in an atom by the presence of other electrons. This effect was described by a phenomenological model with a pseudospin degree of freedom,[17] and more recently with a tight binding model with two orbitals per site.[18] These models lead to a reduced hopping rate for holes as compared to electrons due to the larger modification of the "background" to the hole when it hops from site to site. In the tight binding model the physics is depicted in Fig. 3. We assume two atomic orbitals, s and s', and Coulomb repulsions such that two electrons go into the higher single-particle level. The hopping amplitude is smaller when there are more electrons (fewer holes) due to the overlap matrix element between initial and final states.

Both of the effects discussed above are of the same sign and we believe it is a combination of these that leads to values of Δt of the magnitude necessary to give rise to superconductivity. In fact, without taking into account the modification of atomic states by the presence of other electrons the value of Δt obtained from Eq. (10a) is close to zero when orbitals that are orthogonal at nearest neighbor sites are used.[19] Thus it is fundamentally the disruption that a hole causes in its environment when it propagates that limits its ability to do so, while a single electron outside closed shells can propagate essentially unimpeded. In this respect the charge state of the ions also plays an important role: the largest Δt will occur when in the filled band the

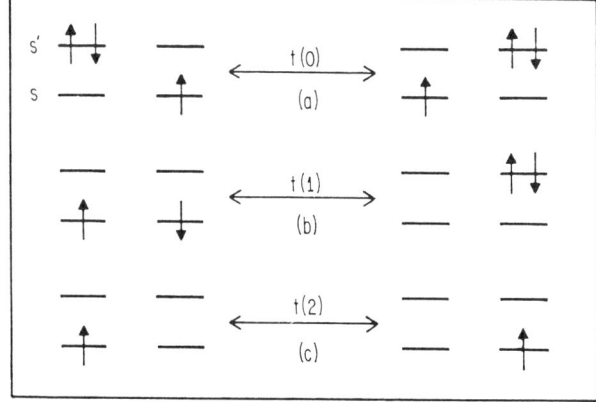

Fig. 3. Electrons (arrows) hopping between neighboring sites in the tight binding model of Ref. 18, with 2 orbitals per site. As the number of electrons (holes) increases (decreases) the hopping amplitude decreases: $t(2) \gg t(1) \gg t(0)$. $\Delta t = t(1) - t(0)$.

ion is most unstable, i.e. for holes propagating through negatively charged anions. In such a situation a hole will cause a large change in the state of the anion.[17] This is the case in the oxide superconductors where our picture suggests that the carriers that give rise to superconductivity are pπ holes conducting by direct hopping through O$^=$ anions.

There is a direct "empirical" way to determine the magnitude of Δt in a diatomic molecule, by the difference in excitation energies of the molecule with different number of electrons. As the simplest possible example, discussed in Ref. 20, the difference in excitation energies of H_2^+ and H_2^- gives a direct measure of the interaction Δt in that case. More generally through spectroscopic analysis one can obtain information on the magnitude of Δt by fitting optical spectra, and in fact such a term was "discovered" experimentally by Zaanen et al. in attempting to interpret spectroscopic data in Ni fluorides.[21] The configuration dependance of hopping matrix elements in tight binding models has also been recently emphasized by Gunnarson et al.[22] Zawadowski [23] has also recently considered the effect of such terms on superconductivity in a two-band model.

IV. THE THEORY

The theory of hole superconductivity based on the interaction Eq. (5) was developed by F. Marsiglio and the author. I will summarize some of the main aspects here and in the next section, and the reader is referred to the references (24-27,11,15) for additional information.

We start from the single band Hamiltonian:

$$H = -\sum_{<ij>,\sigma} t_{ij} (c_{i\sigma}^+ c_{j\sigma} + h.c.) + \sum_{<ij>,\sigma} (\Delta t)_{ij} (c_{i\sigma}^+ c_{j\sigma} + h.c.)(n_{i,-\sigma}+n_{j,-\sigma}) + \sum_{<ij>} V_{ij} n_i n_j \quad (11)$$

and assume that the quantities t_{ij} and $(\Delta t)_{ij}$ are proportional:

$$(\Delta t)_{ij}/t_{ij} = \alpha \quad (12)$$

which is a reasonable assumption since both originate in the same overlap matrix element, as seen in Eq. (10). Here, $c_{i\sigma}^+$ creates an electron of spin σ at site i (in contrast to Eq. (4) where $c_{i\sigma}^+$ is a *hole* creation operator). In Fourier space, Eq. (11) is:

$$H = \sum_{k,\sigma} \varepsilon_k c_{k\sigma}^+ c_{k\sigma} + \sum_{k,k',\sigma,\sigma'} \left[(-\alpha)(\varepsilon_k + \varepsilon_{k'} + \varepsilon_{k+q} + \varepsilon_{k'-q}) + V(q)\right] c_{k+q\sigma}^+ c_{k'-q\sigma'}^+ c_{k'\sigma'} c_{k\sigma} \quad (13)$$

with

$$\varepsilon_k = \sum_j e^{-ikR_j} t_{0j} \quad (14a)$$

$$V(q) = \sum_j e^{-iqR_j} V_{0j} \quad (14b)$$

The essential feature of the second term in Eq. (13) is that the interaction depends not only on momentum transfer q as is usually the case but also on the incoming momenta k and k'. Such an interaction cannot exist in a free electron system where the wave functions are plane waves.[11] Thus, the discreteness imposed by the lattice of ions is absolutely essential to the phenomenon of superconductivity.

The pair interaction for scattering from (k, -k) to (k',-k') is given by

$$V_{kk'} = -2\alpha (\varepsilon_k + \varepsilon_{k'}) + V(k-k') \quad (15)$$

and it is clearly least repulsive for largest $\varepsilon_k, \varepsilon_{k'}$, i.e. at the top of the band. To conform to our notation in previous papers we do a particle-hole transformation and obtain

$$V_{kk'} = 2\alpha (\varepsilon_k + \varepsilon_{k'}) + V(k-k') \quad (16)$$

describing now the scattering of hole quasiparticles of energy $\varepsilon_k, \varepsilon_{k'}$. Although in principle the problem could be solved for arbitrary $V(k-k')$ it becomes enormously simpler if we have to deal only with energy variables rather than momenta. This is the case if we only keep on-site and nearest-neighbor Coulomb repulsions, and also assume that the hopping involves only nearest neighbor sites. The pair interaction becomes:

$$V_{kk'} = U + 2\alpha (\varepsilon_k + \varepsilon_{k'}) + \alpha' \varepsilon_k \varepsilon_{k'} \quad (17)$$

where we have assumed

$$\alpha' = V_{ij}/t_{ij} \tag{18}$$

independent of direction, and $U=V_{ij}$ for $i=j=0$ is the on-site repulsion. Here we have left out some terms arising from $V(k-k')$ that are odd under $k \to -k$ or $k' \to -k'$ that drop out in the subsequent development. We should also point out that the origin of band energies is chosen such that

$$\sum_k \varepsilon_k = 0 \tag{19}$$

as follows from Eq. (14a) taking $t_{ij}=0$ for $i=j$.

The pair interaction Eq. (17) has a very simple form and in fact can be considered as having wider applicability than the conditions invoked above to derive it. It follows simply from the assumption that the dominant variation in the interaction arises from quasiparticle energy rather than momentum:

$$V_{kk'} = V_{eff}(\varepsilon_k, \varepsilon_{k'}) \tag{20}$$

assuming symmetry in the energy indices. Then,

$$U = V_{eff}(0,0) \tag{21a}$$

$$\alpha = \frac{1}{2} \frac{\partial V_{eff}}{\partial \varepsilon_k}\bigg)_{0,0} \tag{21b}$$

$$\alpha' = \frac{\partial^2 V_{eff}}{\partial \varepsilon_k \partial \varepsilon_{k'}}\bigg)_{0,0} \tag{21c}$$

and higher order contributions are neglected. As will be seen below, the assumption Eq. (20) has the fundamental consequence that the gap is constant over the Fermi surface.

We use the usual BCS equation

$$\Delta(\varepsilon_k) = -\frac{1}{N} \sum_{k'} V(\varepsilon_k, \varepsilon_{k'}) \Delta(\varepsilon_{k'}) \frac{\tanh(\beta \frac{E_{k'}}{2})}{2E_{k'}} \tag{22a}$$

and the constraint condition for the density of holes

$$n = 1 - \frac{1}{N} \sum_k \frac{\varepsilon_k - \mu}{E_k} \tanh(\beta \frac{E_k}{2}) \tag{22b}$$

with the quasiparticle energy given by

$$E_k = \sqrt{(\varepsilon_k - \mu)^2 + \Delta^2(\varepsilon_k)} \tag{23}$$

and μ the chemical potential. The sums in Eq. (22) can be transformed to integrals over energy using the single particle density of states. Eq. (22a) and the form of the potential Eq. (17) imply that the gap function $\Delta(\varepsilon_k)$ has a linear dependance on band energy, which we parametrize as:

$$\Delta(\varepsilon_k) = \Delta_m \left(-\frac{\varepsilon_k}{D/2} + c\right) \tag{24}$$

where D is the single-particle bandwidth and the parameters Δ_m and c are determined by solution of Eq. (22), as discussed in the references. The quasiparticle gap is obtained from minimization of Eq. (23) as:

$$\Delta_0 = \frac{\Delta(\mu)}{a} \quad (25a)$$

$$a = \left(1 + \left(\frac{\Delta_m}{D/2}\right)^2\right)^{1/2} \quad (25b)$$

and in terms of it the quasiparticle energy is

$$E_k = \sqrt{a^2 (\varepsilon_k - \varepsilon_0)^2 + \Delta_0^2} \quad (26)$$

with

$$\varepsilon_0 = \mu + \frac{1}{a} \frac{\Delta_m}{D/2} \Delta_0 \quad (27)$$

Note that a shift in the minimum of E_k occurs due to the energy dependance of the gap.

A fundamental feature of the interaction Eq. (17) is that because it depends strongly on the position of the states in the band it can give rise to superconductivity even if it is repulsive throughout the band.[24] This is analogous to the "pseudopotential effect" in electron-phonon theory.[28] The gap function Eq. (24) will change sign in the region where $V_{kk'}$ is most repulsive, leading to a solution of Eq. (22) even for $V_{kk'}>0$ throughout the band. This is an appealing feature of the mechanism discussed here and is absent in other mechanisms for superconductivity proposed originating in non-retarded interactions.[29]

A second important feature of this model is that its solution is given accurately by BCS theory, contrary to most other models[15,26]. This is because superconductivity only occurs for a dilute concentration of (hole) carriers. It is well-known that BCS theory accurately describes the superconducting ground state in dilute systems even in the regime of short coherence length,[30] and in the model discussed here the finite temperature transition is also described by BCS theory due to the fact that the pair binding energy is much smaller than its hopping rate, as discussed in the previous section.

Quantitative solution of the Eqs. discussed here is given in the references. It is found that very "plausible" values of the interactions in the model give rise to superconductivity with critical temperatures of the magnitude observed. The condition for occurrence of superconductivity in the limit of small hole concentration is

$$\frac{\Delta t}{t_h} > \sqrt{\left(1 + \frac{U}{2zt_h}\right)\left(1 + \frac{V}{2t_h}\right) - 1} \quad (28)$$

with U and V the on-site and nearest-neighbor Coulomb repulsion, Δt the hopping interaction Eq. (5), z the number of nearest neighbors and t_h the single hole hopping amplitude at the top of the band

$$t_h = t_0 - 2\Delta t \quad (29)$$

with t_0 the hopping amplitude at the bottom of the electron band, Eq. (10b). In the limit $t_h \to 0$ (for $\Delta t = t_0/2$) the criterion Eq. (28) reduces to the simple form:

$$\Delta t > \sqrt{\frac{UV}{4z}} \quad (30)$$

Analytic forms for the critical temperature in weak and strong coupling regimes are given in the references. The weak coupling T_c equation has been independently derived by Micnas et al.[31] In the next section we summarize the main results of the theory for a variety of observables.

V. CONSEQUENCES AND COMPARISON WITH EXPERIMENT

We summarize here the main consequences of the theory and its relationship to existing and future experiments:

(1) <u>Doping dependance of T_c</u>

Because of the energy dependance of the interaction Eq. (17) superconductivity is restricted to low hole density. Physically, the hopping interaction becomes increasingly less attractive as holes are added to a filled band. T_c versus n_h exhibits the typical behavior shown in Fig. 4. We have included the effect of the single particle hopping renormalization Eq. (1). As discussed in Ref. 25 within a single band model there is not much freedom in adjusting the range of hole concentration where T_c is non-zero which is typically in the range of 0.15 to 0.2 holes per 0 atom, in agreement with observations (note that the hole concentration given in Fig. 1 is per Cu ion in the plane). The behavior shown in Fig. 1 has been reproduced in a variety of samples and little doubt remains that it is intrinsic.

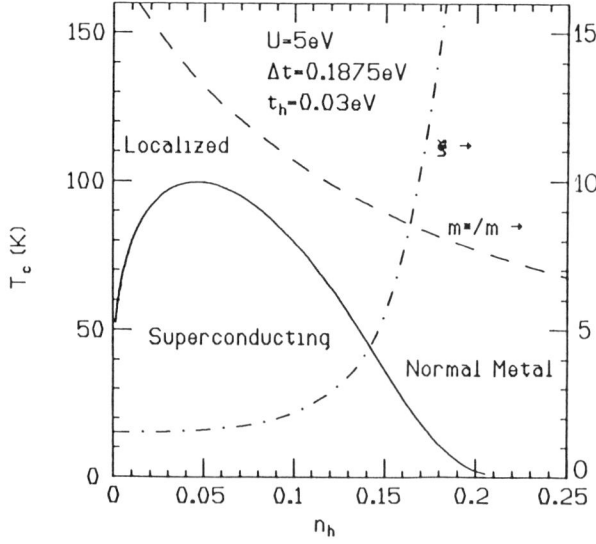

Fig. 4. T_c versus hole concentration n_h in the single band model for a representative set of parameters. The behavior of the coherence length ξ (as defined in Ref. 26) and effective mass versus n_h is also shown.

(2) <u>Gap isotropy</u>

The gap Eq. (24) is constant over constant energy surfaces even for anisotropic band structures. This followed from the fact that the hopping interaction has the same distance dependance as the single particle hopping (Eq. (12)) together with the assumption Eq. (18). In fact, some deviations from this can occur in an anisotropic material due to deviations from Eq. (18), i.e. the Coulomb repulsion will in general have a different distance dependance. However, the anisotropy introduced by this effect turns out to be only a few percent even in very anisotropic structures, as shown in Ref. 25. There is evidence from photoemission experiments that the gap is isotropic in the plane,[32] but no information yet on this question in directions off the plane. Observation of a large anisotropy would be inconsistent with the theory.

(3) Pressure dependance of T_c

Because the hopping interaction Δt increases when the atoms are closer together, pressure will in general increase T_c in this model. More specifically however, in an anisotropic material T_c will increase rapidly with pressure in the planes and is only weakly dependent on pressure perpendicular to the planes. Quantitative results are given in Ref. 25. With increasing doping the model predicts the pressure derivative to increase.

Initial reports suggested that pressure perpendicular to the planes has a similar effect to hydrostatic pressure,[33] in contradiction to the theory. However, recent measurements of thermal expansion coefficients indicate an anisotropic pressure dependance[34] as predicted by the theory, as do measurements of critical temperature in strained multilayer films.[35] More extensive experiments on this question are highly desirable, as well as measurements of the change in pressure dependance of T_c with doping. The theory also predicts a larger in-plane stiffness for electron-doped compared to hole-doped materials[36] which should be experimentally verifiable.

(4) Tunneling asymmetry

Because of the energy dependance of the gap the tunneling density of states exhibits an asymmetry of universal sign: the tunneling current in an N-I-S junction should be larger for a negatively-biased sample. Quantitative estimates are given in reference 27. Although many tunneling experiments appear to show an asymmetry of the sign predicted the evidence is not yet conclusive. It would be desirable that the polarity convention be clearly stated in reports of tunneling experiments.

(5) Strong to weak coupling cross-over

Because of the single particle hopping density dependance Eq. (1) as well as the energy dependance of the interaction Eq. (17) a cross-over from strong to weak coupling occurs as the hole concentration increases. In particular the normal state effective mass and the superconducting coherence length are predicted to *monotonically* decrease and increase respectively with hole density, as shown in Fig. 4. Because the band becomes very narrow as the hole density decreases the chemical potential can fall below the bottom of the band giving rise to behavior very different to "conventional BCS", in particular the gap will not go to zero as $T \rightarrow T_c$.[26]

We are aware of a single report on measurements of H_{c2} versus hole concentration,[37] which shows monotonically decreasing behavior in agreement with the theory. More experiments are highly desirable. The monotonic decrease in normal state effective mass predicted should be evident in a variety of normal state properties but we have not perfomed a detailed analysis. Unusual behavior of the gap versus temperature has been reported in various experiments,[38] in particular tunneling.

(6) Other properties

Unusual behavior in normal state properties can result from a very narrow band, as we expect in this theory for low hole concentration. This is discussed for example by Scalapino et al.[39] and Ashkenazi et al.[40] Various other observables in the superconducting state like spin susceptibility, ultrasonic attenuation, NMR relaxation rate and frequency dependent conductivity follow behavior very similar to "conventional BCS", except for effects arising in the low density regime from the extreme narrowness of the band.[25,41] The observed absence of NMR relaxation peak appears to contradict the theory but could be related to a variety of effects, and it should be kept in mind that such a peak has often not been observed in "conventional" type II superconductors.[42]

(7)   Effect of disorder

The effect of impurity scattering on the critical temperature in this theory has recently been studied by Marsiglio.[43] He finds that non-magnetic impurity scattering can have an important pair-breaking effect due to the energy dependance of the gap function Eq. (24). In the low-doping strong-coupling regime non-magnetic and magnetic impurities cause a reduction in T_c of comparable magnitude. This finding is in qualitative agreement with observations on the effect of various experiments where Cu ions in the plane are replaced by other transition-metal ions.[44]

(8)   Multi-band effects

In the oxide superconductors we envisage superconductivity as arising from pairing of hole carriers that conduct through direct hopping between 0 anions, i.e. in a band formed by 0 $p\pi$ orbitals in the plane. This is in contradiction with the more commonly held view that doped holes go into 0 $p\sigma$ orbitals, that appears to be supported by certain experiments.[45] We believe that further analysis of these experiments will prove this view wrong.

Nevertheless, in the metallic regime conduction will occur also through the band formed by $0p\sigma$ - $Cud_{x^2-y^2}$ orbitals, and while these carriers do not play a central role in the superconductivity they can give rise to observable effects. To describe this situation we have recently studied a simple two-band model with hole-like and electron-like carriers.[46] One of the effects of the second band is to give rise to a second, smaller, superconducting gap associated with the (weaker) binding of the electron carriers, and as a consequence additional structure in tunneling characteristics, that resembles experimental observations.[47] Another more fundamental effect that is specific to the model discussed here is the expected behavior of the Hall resistivity near the critical temperature: as the magnetic field is increased at temperatures somewhat below T_c the model predicts the Hall resistivity to turn negative first as the weakly bound electron pairs are driven normal, and to turn positive at larger fields as the hole carrier pairs also break up. Such behavior is observed experimentally in a variety of samples.[48]

VI. DISCUSSION

The theory discussed above rests on the premise that electron-hole asymmetry is a fundamental feature of solids. This premise is in contradiction with commonly held views[49] but is clearly not demonstrably wrong. Its fundamental origin is to be sought in the physics of the atoms that form the solids and is clearly demonstrated by the simple fact that the periodic table manifestly does not exhibit left-right symmetry across its center.

Starting from this premise, and with the single assumption that holes are heavier than electrons, a theory of superconductivity results that in its simplest form exhibits a remarkable number of features observed in the high T_c oxides. The theory makes in addition a number of detailed predictions that should be experimentally verifiable in the near future. Furthermore, while these materials display most clearly the physics that leads to superconductivity, they are by no means qualitatively different from other superconductors; similarly the theory discussed here naturally extends to all superconductors and thus offers the possibility of a unified explanation for superconductivity in all solids.

The electron-phonon mechanism is the accepted origin of superconductivity in "conventional" materials[50]. However, for many observations supposedly understood within that mechanism, the mechanism discussed here provides alternative natural explanations.[11] Isotope effect and phonon structure in tunneling, conventionally understood from phonon modulation of the hopping t, would be expected in our model from a (weaker) phonon modulation of Δt. The propensity of superconductors to lattice instabilities, and their large normal state resistivity, are conventionally understood as arising from large electron-phonon coupling; in our theory they are naturally expected from the fact that <u>antibonding</u> states are occupied in superconductors and from the large hole effective mass respectively. For many other observations like Matthias's rules[51], Chapnik's rule and the existance of high T_c oxides the mechanism discussed here has a natural explanation and the electron-phonon mechanism does not.

So does the electron-phonon interaction or the hole nature of the carriers lead to superconductivity in "conventional" materials? Given any superconductor, the theory discussed here suggests that if its ionic mass is increased to infinity without altering other properties the superconducting T_c would remain finite, while if instead its Fermi surface is altered to eliminate all regions of negative curvature T_c would go to zero. The electron-phonon mechanism would suggest the opposite behavior. Until experimentalists succeed in performing these "thought experiments" the question may not be strictly speaking settled; however, we suggest that it is already answered by the examples of Pb and Li respectively, and more generally by the systematics displayed by superconducting elements and compounds that show absolutely no correlation with the magnitude of the ionic mass but a strong correlation with the electron/hole-like nature of the carriers.

In my view the fact that from a very simple principle such a large number of consequences emerge that are in agreement with observations represents compelling evidence that the principle is correct and the resulting theory describes reality. It would be indeed remarkable if nature had failed to take advantage of such a simple way to achieve superconductivity in solids while at the same time exhibiting so many features that point towards this explanation.

<u>Acknowledgements</u>: I am grateful to F. Marsiglio for collaboration in this work, to the National Science Foundation for support under NSF-DMR-8918306, to J. Torrance for permission to reproduce Fig. 1, and to the organizers of this meeting and particularly to J. Ashkenazi for their kindness and hospitality.

<u>References</u>

1. I.M. Chapnik, Sov. Phys. Doklady <u>6</u>, 988 (1962); Phys. Lett. <u>72A</u>, 255 (1979); J. Phys. F <u>13</u>, 975 (1983).
2. C.M. Hurd, "The Hall Effect in Metals and Alloys", Plenum, New York, 1972, Chpt. 3.
3. I am told for example that B. Matthias was well aware of this rule (D. Wohlleben, private communication).
4. S. Uchida et al., Jpn. J. Appl. Phys. <u>26</u>, 440 (1987); H. Takagi et al., in "Mechanisms of High Temperature Superconductivity," ed. by H. Kamimura and A. Oshiyama, Springer, Berlin, 1989, p. 238 and references therein.
5. Y. Tokura, H. Takagi and S. Uchida, Nature <u>337</u>, 345 (1989). A large number of theoretical and experimental papers followed emphasizing that electron-hole symmetry in oxides was now an established fact. See for example the commentary by R. Pool in Science <u>243</u>, 1436 (1989).
6. J. E. Hirsch, "Electron Superconductivity in Oxides?", UCSD Report, March 2, 1989. (Submitted to Nature and Science correspondance sections, journals declined publication.)
7. J. E. Hirsch and F. Marsiglio, Phys. Lett. A <u>140</u>, 122 (1989).
8. N. Nucker et al., Z. Phys. B <u>75</u>, 421 (1989).
9. J. M. Tarascon, private communication, July 1989; Y. Tokura et al., reported at the APS March meeting, 1990; Y. Iye et al., reported at APS March meeting, 1990.
10. J. B. Torrance et al., Phys. Rev. Lett. <u>61</u>, 1127 (1988); Phys. Rev. B<u>40</u>, 8872 (1989).
11. J. E. Hirsch, Physica C <u>158</u>, 326 (1989); Phys. Lett. A<u>138</u>, 83 (1989); Mat. Res. Soc. Symp. Proc. Vol. <u>156</u>, 349 (1989).
12. J E. Hirsch, Phys. Rev. Lett. <u>59</u>, 228 (1987); Y. Takashashi, Z. Phys. B<u>67</u>, 503 (1987).

13. T. Izuyama, Physica B <u>169</u>, 995 (1990), and references therein.
14. R. B. Laughlin, in "Mechanisms of High Temperature Superconductivity," ed. by H. Kamimura and A. Oshiyama, Springer, Berlin, 1989, P. 76.
15. J. E. Hirsch, Physica C <u>161</u>, 185 (1989).
16. M. Takigawa et al., Phys. Rev. B <u>39</u>, 7371 (1989).
17. J. E. Hirsch, Phys. Lett. A<u>134</u>, 451 (1989); J. E. Hirsch and S. Tang, Phys. Rev. B <u>40</u>, 2179 (1989); Sol. St. Comm. <u>69</u>, 987 (1989); J. E. Hirsch and F. Marsiglio, Phys. Rev. B <u>41</u>, 2049 (1990).
18. J. E. Hirsch, Phys. Rev. B (in press).
19. M. S. Hybertsen et al., Phys Rev. B <u>41</u>, 11068 (1990).
20. J. E. Hirsch, Chem. Phys. Lett. <u>171</u>, 161 (1990).
21. J. Zaanen, C. Westra and G. Sawatzky, Phys. Rev. B<u>33</u>, 8060 (1986).
22. O. Gunnarson and N. E. Christensen, Phys. Rev. <u>B42</u>, 2363 (1990) and references therein.
23. A. Zawadowski, Phys. Scripta <u>VT27</u>, 66 (1989) and references therein.
24. J. E. Hirsch and F. Marsiglio, Phys. Rev. B<u>39</u>, 11515 (1989); Physica C <u>162-164</u>, 591 (1989).
25. F. Marsiglio and J. E. Hirsch, Phys. Rev. B<u>41</u>, 6435 (1990); Physica C <u>162-164</u>, 1451 (1989).
26. F. Marsiglio and J. E. Hirsch, Physica C <u>165</u>, 71 (1990); Physica C <u>171</u>, 554 (1990).
27. F. Marsiglio and J. E. Hirsch, Physica C <u>159</u>, 157 (1989).
28. N. N. Bogoliubov, Nuovo Cimento <u>7</u>, 794 (1958); P. Morel and P. W. Anderson, Phys. Rev. <u>125</u>, 1263 (162).
29. R. Micnas, J. Ranninger and S. Robaskiewicz, Rev. Mod. Phys. <u>62</u>, 113 (1990).
30. A. J. Legget, J. Phys. (Paris) <u>41</u>, C7 (1980); P. Nozieres and S. Schmitt-Rink, J. Low Temp. Phys. <u>59</u>, 195 (1985).
31. R. Micnas, J. Ranninger and S. Robaskiewicz, Phys. Rev. B<u>39</u>, 11653 (1989).
32. C. G. Olson et al., Sol. St. Comm. <u>76</u>, 411 (1990).
33. M. F. Crommie et al., Phys. Rev. B <u>39</u>, 4231 (1989).
34. C. Meingast et al., Phys. Rev. B <u>41</u>, 11299 (1990).
35. A. Gupta et al. Phys. Rev. Lett. <u>64</u>, 3191 (1990).
36. J. E. Hirsch and F. Marsiglio, Physica C <u>172</u>, 265 (1990).
37. M. Suzuki and M. Hikita, Jap. Jour. Appl. Phys. <u>28</u>, L1368 (1989).
38. See various contributions in these proceedings.
39. D. J. Scalapino, R. T. Scalettar and N. E. Bickers, in "Novel Superconductivity", ed. by S. E. Wolf and V. Z. Kresin, Plenum, New York, 1987, p.475.
40. J. Ashkenazi et al., these proceedings.
41. F. Marsiglio and J. E. Hirsch, unpublished.
42. B. G. Silbernagel et al., Phys. Rev. <u>153</u>, 535 (1967); W. Fite, II, and A. G. Redfield, Phys. Rev. Lett. <u>17</u>, 381 (1966)).
43. F. Marsiglio, unpublished.
44. J. M. Tarascon et al., Phys. Rev. <u>337</u>, 7458 (1988).
45. F. J. Adrian, Physica C <u>171</u>, 505 (1990) and references therein.
46. J. E. Hirsch and F. Marsiglio, Phys. Rev. B<u>43</u>, 424 (1991).
47. M. Gurvitch et al, Phys. Rev. Lett. <u>63</u>, 1008 (1989).
48. Y. Iye, S. Nakamura and T. Tamegai, Physica C <u>159</u>, 616 (1989); S. J. Hagen et al., Phys. Rev. B <u>41</u>, 11630 (1990).
49. W. Heisenberg, Ann. der Phys. <u>10</u>, 883 (1931); N. W. Ashcroft and N. D. Mermin, "Solid State Physics", Holt, Reinhart and Winston, New York, 1976, Chpt. 12.
50. See for example various chapters in "the last nail in the coffin of superconductivity": "Superconductivity", ed. by R. D. Parks, Marcel Dekker, New York, 1969.
51. B. T. Matthias, Prog. in Low Temp. Phys. Vol. 2, ed. by J. C. Gorter, North Holland, Amsterdam, 1957, p. 138.

NON-BCS FEATURES OF TUNNELING AND POINT-CONTACT SPECTROSCOPY DATA OF

ORGANIC SUPERCONDUCTORS AND HIGH-T_c CUPRATES

M. Weger

Racah Institute of Physics, The Hebrew University
Jerusalem, Israel

R.F. Wehrhahn

Theoretical Nuclear Physics, University of Hamburg
Hamburg, Germany

ABSTRACT

Point-contact spectra of some two-dimensional organic superconductors show very sharp structure in three energy regions: Around zero, at the phonon frequency, and at twice the phonon frequency. No structure is observed near the BCS gap. The zero-bias anomaly is similar to that observed in some high-T_c superconductors. The structure at the phonon frequency seems to be related to structure observed in the cuprates by IR spectroscopy. We present a phenomenological picture describing these non-BCS-like data in terms of a strongly energy-dependent gap-function $\Delta(E)$. We present theoretical arguments accounting for this phenomenological gap function in terms of the breakdown of some common approximations made in the strong-coupling theory of superconductivity: Specifically, the replacement of $|\Delta(E)|^2$ by $\Delta^2(E)$ in the zero-temperature Dyson equation, and the neglect of possible strong energy-dependences of the Coulomb pseudopotential.

1. POINT CONTACT SPECTROSCOPY MEASUREMENTS ON α_t-$(BEDT-TTF)_2I_3$

Point-contact spectroscopy and tunneling measurements were carried out on several phases of the organic superconductor $(BEDT-TTF)_2X$ ($X=I_3$, IAuI) (shortly denoted ET) for several years by several groups [1-5]. The data depend very much on the quality of the samples, and we obtained the best results on the α_t phase, which is obtained from the α phase of ET by annealing at about 70 C for several hours (or days) [6]. In these experiments, two crystals were pressed against each other, with the current flowing in the a-b plane. The dV/dI vs. V curves are shown in Fig. 1 for T=1.5 K, as well as higher temperatures. It is seen that there is a sharp minimum in the resistivity at zero bias, a sharp maximum at about 10 meV, and another maximum at about 20 meV. These structures are more pronounced in the second-derivative curve d^2V/dI^2 vs. V (Fig. 1b).

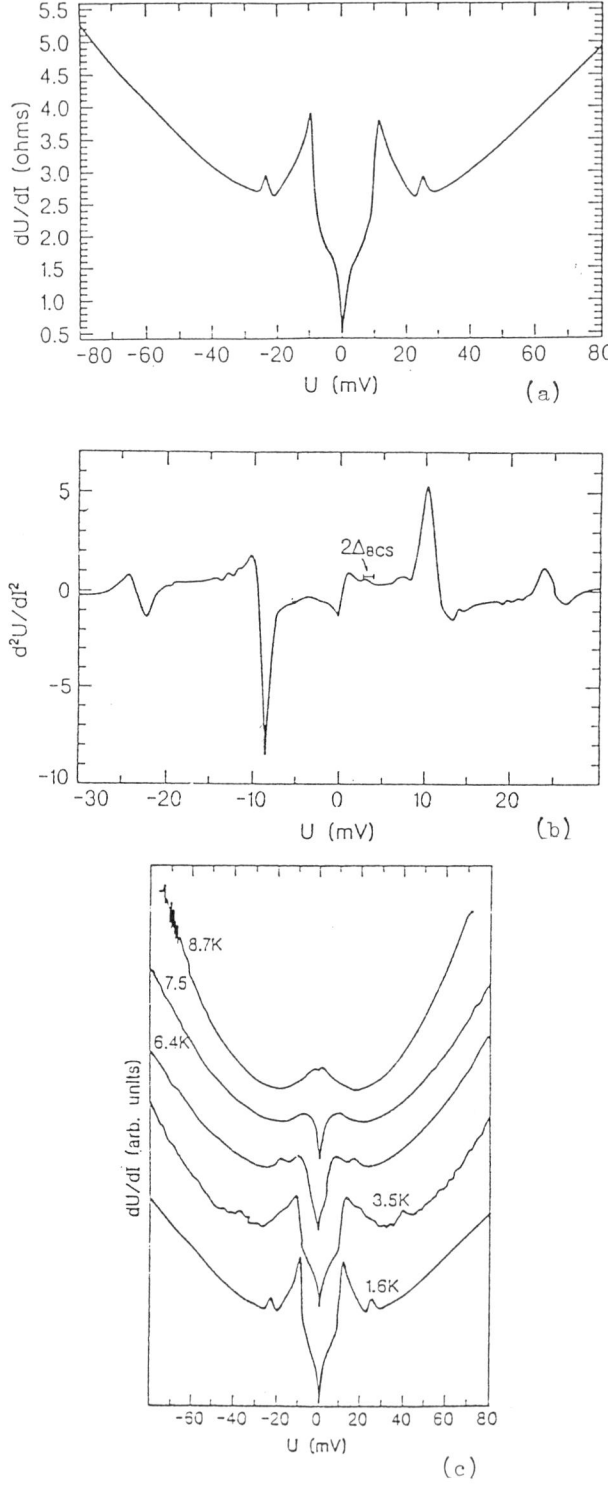

FIGURE 1

The point-contact resistance of $\alpha_t(\text{BEDT-TTF})_2 I_3$ (a) and its derivative (b) as function of voltage at T=1.6 K; (c) The resistance at several temperatures

Table 1. Explanation see text.

	T [K]	$\delta_M^{(0)}$ [mV]	$\delta_{1/2}^{(0)}$ [mV]	$\delta_M^{(1)}$ [mV]	$\delta_{1/2}^{(1)}$ [mV]	$\delta_M^{(2)}$ [mV]	$\delta_{1/2}^{(2)}$ [mV]	T_c [K]
β-(BEDT-TTF)$_2$I$_3$ (point contact) [3]	0.08	0.5	0.21					1.35
	1.4	0.5	0.3	5.5	3.75			(?)
β-(BEDT-TTF)$_2$IAuI (vacuum tunneling) [3]	0.47	0.7	0.35					4.1
Hawley et al [1]	2.4			4	2.5			4.1
α_t-(BEDT-TTF)$_2$I$_3$ (point contact) [5]	1.4		0.25	5.2	4.5	12.5	11.5	8.0

The observation of structure in three distinct energy ranges; namely near zero bias (denoted by a superscript 0), near a voltage of 10 meV (for s-n-s or s-i-s structures) or half this value (for s-n or s-i-n structures) (denoted by a superscript 1), and at double this value (denoted by a superscript 2) is common in various ET salts. In Table I we present the voltages (per individual s-n or s-i junction) of the maxima of the resistance (or conductance, for the tunneling data), denoted by a subscript M, as well as the width at half-maximum, (roughly, the maximum of the second derivative) denoted by a subscript ½.

Striking in its absence is the structure at Δ_{BCS}, the BCS gap, being about 1.2-1.3 meV (i.e. 1.76 to 2 times T_c) for the α_t phase. This "Giaever" structure is usually the strongest observed structure in tunneling [7] and point-contact spectroscopy [8] measurements.

The zero-bias anomaly was also observed in YBCO [9] and Ba$_{1-x}$K$_x$BiO$_3$ [10], at about 0.2 - 0.25 of Δ_{BCS}, and is reported in this conference by Greene [11]. Its origin has not yet been accounted for.

The structure at $2\delta^{(1)} \approx 10$ meV is very close to the phonon frequency $\omega_{ph} = 4$ meV [12]. If, following the McMillan-Rowell theory, we set: $\delta_M^{(1)} = \omega_{ph} + \Delta_{BCS}$, then this structure coincides with the phonon frequency to within 5%. We can attain this high accuracy because the data are extremely sharp (Fig. 1b), and because the phonon spectrum is very close to an Einstein spectrum (Fig. 2, [12]). In this respect, ET possesses an advantage over the cuprates, since for them there are several phonon branches and we do not have a single-branch Einstein spectrum.

The structure at $\delta^{(2)}$ seems to be the second harmonic of $\delta^{(1)}$, although sometimes it moves somewhat with increasing temperature above or below $2\delta^{(1)}$.

A structure at $\omega_{ph} + \Delta_{BCS}$ is seen also in tunneling, and point-contact spectroscopy, in normal superconductors [7], but there it is much weaker than the "Giaever" structure at Δ_{BCS}. Thus, the present structure is not at all BCS-like.

In early work, there was an attempt to associate $\delta^{(1)}$ with the BCS gap, thus very large values of $2\Delta_{BCS}/k_B T_c$ were suggested. Since $\delta^{(1)}$ coincides so closely with the phonon frequency, we do not attempt to adopt this approach. As the temperature is raised, the structure broadens rapidly, thus we cannot establish definitely a temperature-dependence of $\delta^{(1)}$.

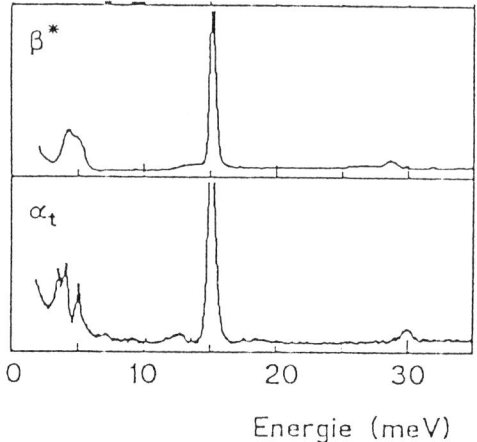

FIGURE 2

The Raman spectra of ET_2I_3; There is a sharp phonon line at 4 meV. The line at 15 meV is due to I_3 stretching vibrations, and is not strongly coupled with the conduction electrons.

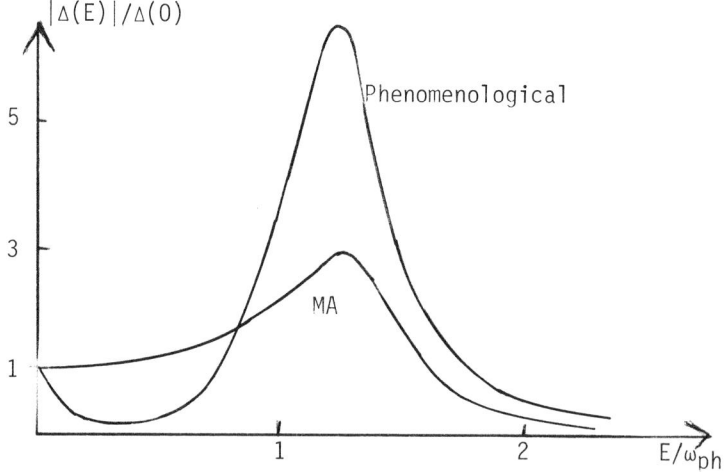

FIGURE 3

$|\Delta(E)|$ as function of E for the Morel-Anderson calculation (MA) and for our phenomenological model.

Thus, we are faced with the three questions:
(1)   What is the origin of the zero-bias anomaly ?
(2)   Why is the Giaever structure at Δ_{BCS} absent ?
(3)   Why is the phonon structure $\delta^{(1)}$ so enormously strong ?

Point-Contact vs. Tunneling Spectroscopy

Point-contact spectroscopy involves s-n-s and s-n structures, while tunneling spectroscopy involves s-i-s, s-i-n structures. In point-contact spectroscopy, the gap causes the resistance to attain a <u>minimum</u> at zero bias, due to Andreev reflection, and it rises sharply at Δ ; while in tunneling spectroscopy, the resistance at zero bias is maximum, and it has a sharp minimum at Δ (as found by Giaever). Actually, the transition between these two limits is continuous, as the reflectivity of the barrier is increased. This continuous transition is described by the Blonder-Tinkham-Klapwijk theory [8]. While most of our work involves point-contact spectroscopy, we did some vacuum tunneling as well. In Table I, the data on ET-IAuI are tunneling measurements by Hawley et al[1] as well as Poppe [3]. The resistivity at zero bias is indeed a maximum. A transition between these two limits as the <u>temperature</u> is changed was observed in ET-I_3 [3]; at 1.5K we observe a resistance minimum at zero bias, while as the temperature is lowered to 0.08 K, the maximum changes continously into a minimum.

The fact that the $\delta^{(0)}$ structure is seen in point-contact spectroscopy as a minimum in R, and in tunneling as a maximum, proves that this structure is <u>not</u> due to a Josephson effect.

2. A POSSIBLE PHENOMENOLOGICAL EXPLANATION

A plot of $|\Delta(E)|$ vs. E for a superconductor with an Einstein spectrum is shown in Fig. 3a, following Morel & Anderson [13], who calculated $\Delta(E)$ analytically for the case of weak-coupling. At $E = \omega_{ph}$, $|\Delta(E)|$ has a peak of about $3\Delta(0)$. When dispersion is present, this peak is smeared out [14].

Our data can be described by a curve as shown in Fig. 3b, in which $\Delta(E)$ <u>falls</u> sharply for E≠0, i.e. <u>away from</u> the Fermi surface. Thus, the lowest excitation energy: $E(\xi) = \sqrt{\xi^2+\Delta^2(\xi,E)}$ is obtained not at the Fermi level ($\xi=0$), but away from it, where $\Delta(\xi,E)$ has fallen well below $\Delta(0)$, and this minimum excitation energy is smaller than $\Delta(0)$. We associate this minimum excitation energy with the zero bias anomaly. This interpretation also accounts for the absence of the Giaever structure; it is pushed down and becomes the zero-bias anomaly. We have obtained this kind of behavior from a theoretical model [15]. A minimum excitation energy well below $\Delta(0)$ ($=\Delta_{BCS}$) is in accord with measurements of the London penetration depth by Kanoda et al [16].

A harmonic at $2\omega_{ph}$ is obtained by conventional theory when strong-coupling effects are considered. Morel & Anderson [13] point out a singularity at this energy, however they show that its residue vanishes in the weak-coupling limit. It is not surprising that this singularity appears under strong-coupling conditions.

A particularly dramatic effect of strong-coupling is a large increase in $\Delta(E)$ at: $E = \omega_{ph}+\Delta(0)$. This occurs <u>because</u> $\Delta(E)$ approaches E in this region; consequently, the term $1/\sqrt{E^2-|\Delta(E)|^2}$ tends to diverge. Obviously, this cannot happen for weak-coupling conditions, but under strong coupling conditions, $\Delta(0)$ is of order ω_{ph} (say, a factor of 3 smaller, for $\lambda \simeq 1$), thus $|\Delta(\omega_{ph})|$ is close to ω_{ph}. This increase accounts for the huge $\delta^{(1)}$ structure described in section 1.

3. THEORETICAL INTERPRETATION

In BCS theory, the excitation energy E is given by: $E = \sqrt{\xi^2+|\Delta(\xi)|^2}$, where $\xi = v_F(k-k_F)$ is the one-electron energy. Note that the absolute value $|\Delta(\xi)|$ has to be taken. The phase of $\Delta(\xi)$ is arbitrary and the excitation energy E does not depend on it. In contrast, in the Eliashberg equation [17], the expression: $\sqrt{E^2-\Delta^2(E)}$ appears in the denominator, without the absolute value. The phase of $\Delta(E)$ is defined so that $\Delta(0)$ (or, rather, $\Delta(E)$ at a value of E satisfying $E = \Delta(E)$) is real. Therefore, the phase of $\Delta(E)$ is no longer arbitrary. The replacement of $|\Delta(E)|$ by $\Delta(E)$ is essential for the deformation of the path of integration in the complex-E plane, since $|\Delta(E)|$ is not analytic and therefore we cannot use Cauchy's theorem [18]. This deformation of the path of integration plays an essential role in Eliashberg theory, and the elegant algebraic formulation of this theory by Bergmann & Rainer [19]. This replacement is justified by Eliashberg for the case of moderate coupling by showing that for small E, $\text{Im}[\Delta(E)] \ll \text{Re}[\Delta(E)]$ (p. 702 of ref. 17). From Eliashberg's argument, it is not obvious that this replacement is justified for the case of <u>very strong</u> coupling.

Morel & Nozieres [20] derived the Dyson equations for the case of strong coupling, and show that indeed, the absolute value of $\Delta(E)$ must be used. In Fig. 4 we reproduce the Feynmann diagrams from their paper; their Dyson equations are:

$$G(k,\omega)=G_o(k,\omega)+G_o(k,\omega)\Sigma_1(k,\omega)G(k,\omega) - G_o(k,\omega)\overline{\Sigma}_2(k,\omega)F(k,\omega)$$

$$F(k,\omega)= G_o(k,-\omega) \Sigma_2(k,\omega) G(k,\omega) + G_o(k,-\omega) \Sigma_1(k,-\omega) F(k,\omega)$$

It is clear that since the arrows of Σ_2 point inwards in the first equation, Σ_2^* (i.e. $\overline{\Sigma}_2$) has to be used, while in the second equation, they point outwards and therefore Σ_2 has to be used. The absolute value appears

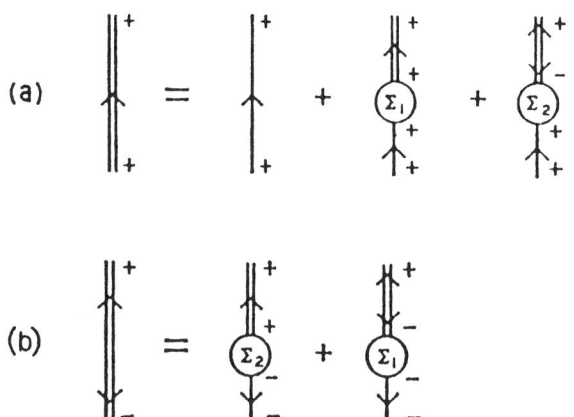

FIGURE 4

Feynman Diagrams for the Dyson equation for the normal (G, (a)) and anomalous (F, (b)) Green's functions, (from Morel & Nozieres).

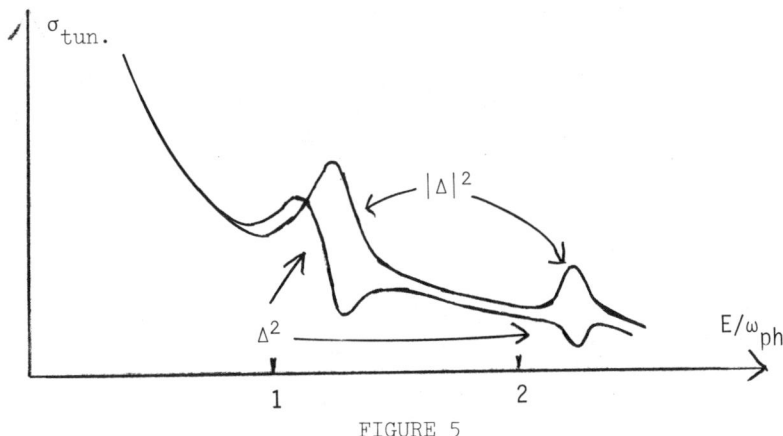

FIGURE 5

The tunneling conductivity as function of voltage for the conventional theory ($\text{Re}[E/\sqrt{E^2-\Delta^2(E)}]$) and our proposed theory ($E/\sqrt{E^2-|\Delta(E)|^2}$). The difference near $E \simeq \omega_{ph}$ is rather small, but at $E \simeq 2\omega_{ph}$, the signs are opposite.

explicitely in their formulas (10), (11). To our surprise, we found that this paper is not cited in most textbooks that deal with Eliashberg theory. The absolute value gives rise to a bootstrap effect; $1/\sqrt{E^2-|\Delta(E)|^2}$ becomes extremely large for $E \simeq \omega_{ph}+\Delta(0)$, and this further enhances $\Delta(E)$. This accounts for the large peak in Fig. 3.

In the experimental verification of strong-coupling theory, $\Delta^2(E)$ (rather than $|\Delta(E)|^2$) was employed [7],[14]. We calculated numerically the tunneling conductivity for both cases, i.e. $\text{Re }E/\sqrt{E^2-\Delta^2(E)}$ and $E/\sqrt{E^2-|\Delta(E)|^2}$, for the case of moderate coupling. For an Einstein spectrum, this requires a value of λ not larger than about 0.4, while for a spectrum with dispersion, λ can be about 1. We show the curves in Fig. 5. The difference is seen to be small, and probably less than the difference between the calculated and measured values for Pb. Thus, accord with experiment cannot tell us whether to use $\Delta^2(E)$ or $|\Delta(E)|^2$ for "conventional" superconductors.

The difference between the Δ^2 and $|\Delta|^2$ expressions is particularly striking for $E \simeq 2\omega_{ph}$. Here, the two expressions differ in sign. The Δ^2 expression predicts a minimum in the tunneling conductivity, while the $|\Delta|^2$ expression a maximum. This is because the anomaly in $\text{Re}[\Delta(E)]$ is small, and the anomaly in $\text{Im}[\Delta(E)]$ gives a contribution of opposite sign in both cases. For Andreev reflection, as we discussed in section 1, a minimum in the tunneling conductivity (pseudogap) gives rise to a maximum in Andreev reflection, i.e. a minimum in the resistivity, and a maximum in the tunneling conductivity gives rise to a maximum in the (point-contact) resistivity. Thus, the experimental curve (Fig. 1a) is in accord with the $|\Delta|^2$ expression, rather than with the Δ^2 expression.

In Fig. 6 we plot the one-electron energy ξ as function of E, as well as the density of states: $N(E) = n(\xi) |d\xi/dE| \simeq n(E_F) |d\xi/dE|$. We see that ξ is not a monotonic function of E, as it is for weak coupling. This non-monotonic behavior is a necessary consequence of strong coupling. As a result, there are values of ξ for which 3 values of E are possible. This means that the weight-function $A(\xi,E)$ for these values of ξ, has peaks at three values of E, i.e. it is not at all Lorentzian (quasi-particle

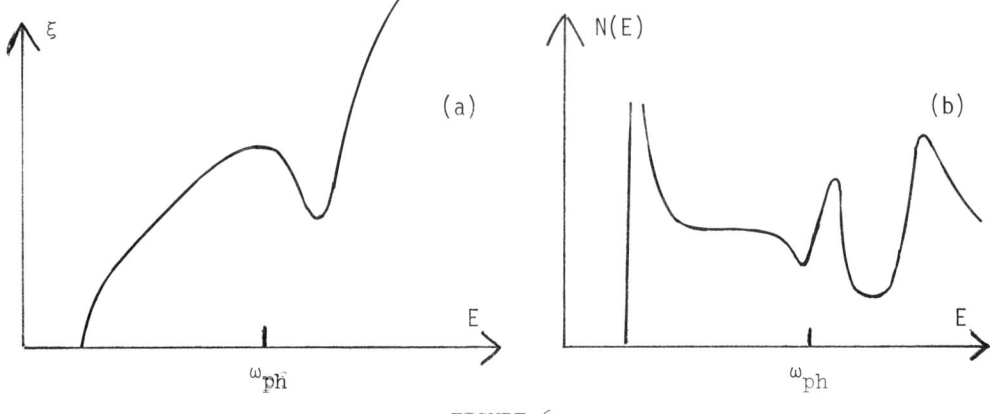

FIGURE 6

(a) The one-electron energy ξ as function of the excitation energy E for strong coupling. (b) The density-of-states $N(E) = n(\xi) |d\xi/dE|$ (multiplied by a weight factor where several values of E exist for one value of ξ).

like). The fact that for strong coupling the weight function is not Lorentzian was emphasized by Engelsberg & Schrieffer already for the normal state [21]. Therefore such a behavior for the superconducting state is not so totally unexpected. (Because of the non-monotonic behavior, the expression for the density of states $N(E)$ must be properly modified by the weight function).

The <u>minimum</u> in the density-of-states at $E \simeq \omega_{ph}$ is in excellent accord with the IR transmissionmeasurements of Tanner [22], which indicate the presence of a minimum in the conductivity at this frequency, below T_c. It is also in accord with the non-BCS like behavior of the IR reflectivity, observed by Wieck et al, Van der Marel et al, Schlesinger et al and others [23], which is one of the main features discussed at this conference.

The anomalous behavi<u>or appears</u> when $|\Delta(E)|$ increases and becomes of order E ; then, the term $1/\sqrt{E^2-|\Delta(E)|^2}$ becomes very large. This also causes the binding energy to increase very much. In the BCS picture, most of the contribution to the binding energy W comes from the region where $E \simeq \Delta(0)$ therefore $W \simeq W_o \simeq \frac{1}{2}n(E_F)\Delta^2(0)$. Here, most of the binding energy comes from the region where $E \simeq \omega_{ph}$, and W is much larger than W_o. We call this "Deep Pairing". We discuss some experimental consequences of Deep Pairing elsewhere [24]. In particular, Deep Pairing causes a deviation from BCS-like behavior to a two-fluid-like behavior, since the binding energy of an individual boson (Cooper pair) is characterized by W, while the interaction between the bosons is characterized by W_o. When $W_o \ll W$, we have nearly non-interacting bosons, which is the characteristic feature of a two-fluid model.

We also considered the case of a strong Coulomb potential with a cutoff at a rather low energy ω_1. If the cutoff is sharp, the solution of the Dyson Ladder equation deviates from the solution of the Eliashberg equation [25]. Within the BCS formulation, the sharp cutoff gives rise to a solution in which the gap function $\Delta(E)$ oscillates as function of E with a period of $(4/3)\omega_1$. These solutions are stable in the presence of a very strong Coulomb interaction. They also account for many features observed in the cuprates and in organic superconductors. We discuss these oscillatory solutions in some detail elsewhere [24], and therefore shall not dwell on them here.

We should note that the difference between Δ^2 and $|\Delta|^2$ becomes irrelevant at $T = T_c$, since in the linearized equation, this term drops out altogether. Most calculations nowadays deal with the linearized equation (following the elgant representation of Bergmann & Rainer [19]), rather than with the $T=0$ equation. As we see from Fig. 1, the peculiar features of the I-V curve are indeed seen only below about 2 K, i.e. below $T_c/4$, and not close to T_c. In contrast, the effects caused by the sharp cutoff of the Coulomb interaction, do not depend on this difference and persist at $T=T_c$.

4. CONCLUSION

We present experimental data on point-contact-spectroscopy of $(BEDT-TTF)_2 I_3$, that show a strong zero-bias anomaly, and a very strong signal at: $E = \omega_{ph} + \Delta(0)$, and a second harmonic at twice this value. The normal Giaever structure is absent.

Theoretically, we consider the Dyson Ladder equation, which was written down by Eliashberg, Nambu, Morel & Anderson, Morel & Nozieres, Schrieffer, Scalapino, & Wilkins and others in the early 1960's. We solve this equation numerically for strong-coupling conditions, for the case of an Einstein spectrum, and find deviations from the solutions of the Eliashberg equations. These deviations are due to a strongly energy-dependent Coulomb potential, and to the retention of the absolute value $|\Delta(k,E)|^2$ (instead of $\Delta^2(k,E)$) in the zero-temperature gap equation. We suggest that these deviations account for the non-BCS like features of high-T_c superconductivity.

REFERENCES

1) M.E. Hawley, K.E. Gray, B.D. Terris, H.H. Wang, K.D. Carlson, J.M. Williams, Phys. Rev. Lett. 57(1986)629.
2) A. Nowack, M. Weger, D. Schweitzer, H.J. Keller, Solid State Commun. 60(1986)199.
3) A. Nowack, U. Poppe, M. Weger, D. Schweitzer, Z. Physik B 68(1987)41.
4) G.V. Kamarchuk, K.I. Pokhodnya, A.V. Khotkevich, I.K. Yanson, Sov. J. Low Temp Physics, 16(1990)419.
5) M. Weger, A. Nowack, D. Schweitzer, Synthetic Metals 41(1991)1885.
6) D. Schweitzer, P. Bele, H. Brunner, E. Gogu, U. Haeberlen, I. Hennig, T. Klutz R. Swietlik, H.J. Keller, Z. Physik B 67(1987)489.
7) W.L. McMillan, J.M. Rowell, in "Superconductivity", R.D. Parks Ed. Marcel Dekker 1969.
8) G.E. Blonder, M. Tinkham, T.M. Klapwijk, Phys. Rev. B 25(1982)4515.
9) M. Gurvitch, J.M. Valles, A.M. Cucolo, R.C. Dynes, J.P. Garno, L.F. Schneemeyer, J.V. Wasczak, Phys. Rev. Lett. 63(1989)1008.
10) D.G. Hinks, B. Dabrowski, P.R. Richards, J.P. Jorgensen, Shiyou Pei, J.F. Zasadzinski, Physica C 162-164(1989)1405.
11) L.H. Greene, this conference.
12) R. Swietlik, D. Schweitzer, H.J. Keller, Phys. Rev. B 36(1987)6881.
13) P. Morel & P.W. Anderson, Phys. Rev. 125(1962)1263.
14) J.R. Schrieffer, D.J. Scalapino, J.W. Wilkins, Phys. Rev. Lett. 10 (1963)336.
15) R. Englman, M. Weger, B. Halperin, Physica C 169(1990)314; in "Physics & Material Science of High T_c Superconductors", R. Kossowsky, S. Methfessel, D. Wohlleben Edts. Kluwer, Dordrecht, 1990. (p. 53).

16) K. Kanoda, K. Akiba, K. Suzuki, T. Takahashi, G. Saito, Phys. Rev. Lett. 65(1990)1271.
17) G.M. Eliashberg, Sov. Phys. JETP 11(1960)696.
18) R. Englman, M. Weger, B. Halperin, Physica C 162-164(1989)1339; M. Weger & R. Englman, Physica A 168(1990)324.
19) G. Bergmann & D. Rainer, Z. Phys. 263(1973)59.
20) P. Morel & P. Nozieres, Phys. Rev. 126(1962)1909.
21) S. Englsberg & J. R. Schrieffer, Phys. Rev. 131(1963)993.
22) D.B. Tanner, this conference.
23) A.D. Wieck, E. Batke, U. Merkt, D. Walker, A. Gold, Solid State Comm. 69(1989)553. Z. Schlesinger, R.T. Collins, F. Holtzberg, C. Field, U. Welp, G.W. Crabtree, Y. Fang, J.Z. Liu, Phys. Rev. Lett. 65(1990)801. D. van der Marel, A. Wittlin, H.U. Habermeier, M. Bauer, D. Heitmann, this conference.
24) M. Weger & Y. Pereg, Solid State Commun. 76(1990)39; M. Weger, Physica B 171(1991)
25) R. Wehrhahn & M. Weger, to be published.

PROPERTIES AND SUPERCONDUCTIVITY OF LARGE BIPOLARONS

David Emin

Solid State Physics Division, 1152
Sandia National Laboratories
Albuquerque, NM 87185

INTRODUCTION

The proposal of this article may be summarized simply. Specifically, ferroelectric-like materials (defined here as materials for which the ratio of the static and optical dielectric constants, $\varepsilon_0/\varepsilon_\infty$, is exceptionally large, >> 2) provide mediums within which individual electronic carriers can pair to form large bipolarons, bipolarons that extend over several atomic sites. Large bipolarons, like large polarons, are presumably mobile. As such, large bipolarons act as mobile charged bosons. As a result, in sufficient (yet moderate) densities large bipolarons can condense into a superconducting state. Thus, the unusual superconductivity of modest densities of charge carriers in ferroelectric-like materials, e.g., WO_3, $SrTiO_3$, $BaBiO_3$ and CuO_2-based materials, may be bipolaronic superconductivity.

This article will summarize three sets of studies. First, the conditions for the formation of large bipolarons are considered. Large-bipolaron formation can occur in ferroelectric-like materials. Second, the nature and distinctive features of bipolaronic superconductivity are addressed. Bipolaronic superconductivity (type II) is possible only within a limited range of carrier densities. Third, distinctive features of the electronic, optical, vibrational and magnetic properties of large bipolarons and of bipolaronic superconductors are described. These features appear consistent with properties of the "ferroelectric"-based superconductors.

Why consider bipolarons?

The notion of polaronic charge carriers in the novel superconductors (e.g., $La_{1.85}Sr_{.15}CuO_4$) arises naturally because of the exceptionally strong electron-phonon interactions of their insulating parents (e.g., La_2CuO_4). In particular, charge carriers typically form polarons or bipolarons in four classes of solids. First, strong electron-phonon interactions and polaron formation are generally associated with ionic solids. In these instances, a large electron-phonon interaction arises from the strong long-range Coulombic interactions between charge carriers and the anions and cations of an ionic solid.[1,2] In fact, polarons received their name from their formation in ionic (polar) solids. Second, carriers associated with oxygen anions generally possess a very strong short-range electron-lattice interaction. This strong interaction arises because the binding of the outermost electrons of an O^{2-} ion requires the presence of the positive charge of surrounding cations. As a result, the highest occupied electronic states of an oxygen ion are exceptionally sensitive to the positions of the adjacent cations. That is, the oxygen-valence orbitals possess a very strong electron-lattice interaction. Third, the relative ease of producing atomic displacements in ferroelectric-like materials promotes formation of polaronic carriers. Finally, polaron and bipolaron formation is fostered in low-dimensional electronic systems because of the relative ease of the electronic localization through which charge carriers self-trap to form polarons.[3,4] The insulating parents of the high-temperature superconductors are obvious candidates for the formation of polaronic charge carriers.

High-Temperature Superconductivity
Edited by J. Ashkenazi *et al.*, Plenum Press, New York, 1991

Large Versus Small Bipolarons

Self-trapping occurs with the binding of carriers in a potential well produced by atoms displaced from their carrier-free equilibrium positions. Self-trapping is energetically stable when the binding energy of the carriers exceeds the strain energy of producing the atomic displacements. A polaron is a quasiparticle composed of a self-trapped carrier and the atomic displacement pattern associated with the self-trapping.

The strain-related lowering of the energy of two noninteracting self-trapped carriers is enhanced if the carriers reside at the same location rather than being separated. As a result, the atomic strain associated with a polaron provides a driving force for the coalescence of two carriers into a pair. If this driving force is sufficient to overcome the Coulomb repulsion of the carriers, two carriers will find it energetically favorable to form a bipolaron. Then, with a nondegenerate ground state, it is energetically favorable for the two carriers to form a singlet. Thus, the formation of a bipolaron is analogous to the formation of a covalent bond. Furthermore, as in bond formation, the requirement that no more than two electrons occupy a nondegenerate state impedes forming multipolarons with more than a pair of self-trapped carriers centered about a site.

The nonlinearity of the self-trapping phenomenon results in there being two distinct types of self-trapped electronic states.[3-5] The self-trapped carrier may be confined to the smallest unit compatible with the atomicity of the solid (e.g., an atom or a bond). Then, the polaron or bipolaron is referred to as small. Alternatively, the self-trapped state may extend over several units. Then, the polaron or bipolaron is referred to as large.

A self-trapped carrier only moves when the atoms associated with the self-trapping alter their positions. Because of the compact nature of a small polaron, its intersite motion requires some atoms to move distances much greater than their zero-point amplitudes. As a result, small polarons are so massive that they readily localize (e.g., self-trapped holes in alkali halides). Therefore, small bipolarons do not move coherently and cannot give rise to superconductivity. By contrast, since large polarons extend over several sites they can adjust continuously to atomic motion. Therefore, they move in a coherent manner (e.g., self-trapped electrons in the alkali halides). Since large bipolarons are (in analogy with large polarons) mobile, they may form a superconducting ground state.

FORMATION OF A LARGE BIPOLARON

The conditions under which charge carriers will form stable large bipolarons have been studied.[4,6] It is found that large bipolarons cannot form in covalent solids with short-range (deformation-potential-like) electron-lattice interactions. Furthermore, even the strong long-range electron-lattice interaction of a typical ionic solid (e.g., alkali halide) is not strong enough to support the formation of a large bipolaron.

Large bipolarons can only form in multidimensional ionic materials with exceptionally displaceable ions, i.e., with especially large values of $\varepsilon_0/\varepsilon_\infty$ (>> 2). These ferroelectric-like materials have long-range electron-lattice interactions, proportional to $(1/\varepsilon_\infty - 1/\varepsilon_0)$, roughly twice as strong as typical polaron materials (alkali halides) for which $\varepsilon_0 \approx 2\varepsilon_\infty$. In addition, the stable formation of a large bipolaron is favored in systems with 1) strong short-range components of the electron-lattice interaction (e.g., oxides) and 2) less than three-dimensional electronic dimensionality. Thus, the CuO_2-based materials are prime candidates for the formation of singlet large bipolarons. In these materials the large values of ε_0, and hence the strong long-range electron-lattice coupling, arises from easily displaced ions lying off the CuO_2 sheets.

BIPOLARONIC SUPERCONDUCTIVITY

For bipolaronic superconductivity the ground state of a collection of large bipolarons must support resistanceless flow. In particular, the absence or presence of superconductivity depends on the properties of the ground state and low-lying excitations of large bipolarons. These collective properties depend upon interactions between large bipolarons.

Large bipolarons in a solid are first modeled as a collection of mobile bosons of charge q ($|q|$ = 2e) that interact only via their Coulomb repulsion.[7] For overall charge neutrality, the bosons are presumed to be in a medium with a uniform charge density of opposite sign and equal net magnitude to that of the bosons. Each charged boson is assigned a mass equal to that of an independent bipolaron, m. The Coulomb interaction between bosons is reduced by the static dielectric constant of the carrier-free medium, ε_0. Altering the size and shape of large bipolarons from their minimum-energy value is represented as exciting internal states of the large bipolarons.[7]

Quantum mechanics ($\hbar > 0$) permits two distinct ground states for a collection of charged particles.[8] At low carrier densities, $n^{1/3} < (2mq^2/\varepsilon_0\hbar^2)$, the Coulomb interactions dominate the kinetic energy causing the particles to condense into a regular array, a Wigner crystal. In the complementary regime, $n^{1/3} > (2mq^2/\varepsilon_0\hbar^2)$, the ground state forms a fluid. Since a superconducting ground state must be fluid, the density of particles must exceed a minimum value. As a result, bipolaronic superconductivity requires a sufficiently large density of large bipolarons.

For the ground state to flow without resistance, the excitation spectrum of the ground state must be such that the flow will not be impeded by the creation of excitations. Landau's condition for resistanceless flow is that the minimum value of $E(p)/p$ be nonzero.[9] The excitation spectrum of interacting charged bosons satisfies this condition as $p \to 0$ because $E(p) \to \hbar\omega_p$ as $p \to 0$, where $\omega_p = 4\pi q^2 n/\varepsilon_0 m$, as a result of the long-range Coulomb interactions.[7,10] Therefore, a fluid ground state of charged bosons is resistanceless. The energy gap, $\hbar\omega_p$, also ensures a Meissner effect.[7]

Since self-trapped carriers adiabatically follow the atomic motion, the bipolaronic ground state is characterized by long-range coherence of the atomic motion. The coherence length is the length scale below which excitations manifest single-particle-like behavior. With the excitation spectrum of the charged Bose fluid first obtained by Foldy,[10] the coherence length, ξ, is defined by $\hbar^2/2m\xi^2 = \hbar\omega_p$: $\xi = (\hbar/2m\omega_p)^{1/2}$.[7] For low carrier densities ($\simeq 10^{21}$ cm^{-3}), large effective masses (m $\gg m_e$, where m_e is a free electron's mass) and large static dielectric constants ($\varepsilon_0 > 20$), $\xi \ll \lambda_L = c/\omega_p$, the London penetration depth, where c is the speed of light. Therefore, bipolaronic superconductors are expected to be type II.

A distinctive feature of bipolaronic superconductivity is the limited range of carrier concentrations over which it can occur. With too low a density the bipolaronic ground state will be a Wigner solid rather than a superconducting fluid. This minimum density falls as delocalization becomes easier: as ε_0 rises or m falls. This feature may explain why doped SrTiO$_3$ (with $\varepsilon_0 \approx 20,000$ near 1 K) becomes superconducting with a carrier density of only $\simeq 10^{19}$ cm^{-3}.[11] Furthermore, there is an upper limit to the carrier density for bipolaronic superconductivity. This limit arises because bipolarons can only form in sufficiently low densities so that they do not overlap one another. The existence of a finite range of bipolaron densities for superconductivity requires that m be small enough and ε_0 be large enough.[7] These conditions favor bipolaronic superconductivity arising from large (rather than small) bipolarons in ferroelectric-like materials.

Bipolaronic superconductivity is described in terms of a two-fluid picture. The resistanceless superconducting fluid is associated with occupancy of the ground state. The ground state is viewed as a uniform density of identical bipolarons. The normal fluid arises from occupancy of excited states. These excited states involve dynamic atomic displacements that produce a distribution of local densities, sizes, shapes and motions of the bipolarons. For simplicity in what follows, the excited states are modelled as having a distribution of bipolaronic radii. Then, in the ground state all bipolarons have one common radius. As the temperature is raised from absolute zero, occupancy of the ground state of the boson system is decreased. The density of the superconducting fluid then falls as the temperature is raised.

Mass of an Isolated Large Bipolaron

The mass of an independent bipolaron is a fundamental property that affects the collective properties of large bipolarons. To determine the mass of a large bipolaron, note that a self-trapped carrier only moves when atoms alter their positions. In the adiabatic (strong-coupling) approach to polaron motion, the effective mass of a bipolaron is a weighted sum of the masses of the atoms whose motion enables the self-trapped electronic carrier to move.[4] For example, the diagonal elements of a bipolaron's effective mass in an isotropic medium are:

$$m = \Sigma_i M_i (\Delta d_i/a)^2, \tag{1}$$

where Δd_i is the shift of the equilibrium position of the i-th atom, with mass M_i, when the bipolaron moves a lattice constant, a.

Displacements of atoms nearest a self-trapped carrier are affected most by its motion. That is, the bipolaron's effective mass is dominated by displacements of atoms close to the self-trapped carrier. Thus, the summation in Eq. (1) is dominated by contributions from a finite number of atoms, N_{bp}. However, the typical magnitude of a displacement of one of these N_{bp} atoms, $<\Delta d_i>$, varies inversely with N_{bp}. As a result, the bipolaron's mass varies inversely with its size, N_{bp}, and is proportional to the mass of the atoms involved in the polaronic displacements.

Since the mass of a bipolaron depends on its size and shape, it is also sensitive to the structure of a solid. For example, in the CuO_2-based layered materials one may envision a large bipolaron being confined within contiguous CuO_2 layers.[6] Then, the mass of the large bipolaron associated with in-plane motion will fall as the thickness of the bipolaron, determined by the number of contiguous layers, increases. The in-plane mass will saturate at the bipolaron's three-dimensional value when the net thickness of the bipolaron is no longer determined by the number of contiguous sheets.

Motion perpendicular to the CuO_2 layers is qualitatively different from that within CuO_2 layers. Such motion requires a large bipolaron to transfer between groups of contiguous CuO_2 layers. Presuming little overlap between the bipolaron at successive groups of layers, the interplanar motion is analogous to that of a small bipolaron. However, since a large bipolaron extends among multiple sites within each group of CuO_2 layers, the magnitude of the band-narrowing effect will be reduced from that characterizing a true small bipolaron. Specifically, the adiabatic bandwidth of a small bipolaron is $\hbar\omega \sqrt{S} \exp(-S)$, where ω is the characteristic vibrational frequency and $S \gg 1$.[12] However, for the interplanar tunneling of a large bipolaron, the increased size of the bipolaron reduces S of a small bipolaron to $S_L \sim S/N_{bp}$. Nonetheless, the mass of a large bipolaron for motion perpendicular to the CuO_2 planes with lattice constant b will be very large, $> \hbar/2\omega b^2$. However, it is uncertain if the scattering rate for motion in the interplane direction, $1/\tau$, is always sufficient to suppress coherent transport perpendicular to the CuO_2 planes: $1/\tau > \omega \sqrt{S_L} \exp(-S_L)$.

Transition Temperature

Bipolaronic superconductivity occurs when the fluid ground state of a system of interacting bipolarons is occupied. The superconducting transition temperature, T_c, is the highest temperature for which the ground state has finite occupation, the Bose condensation temperature. Below T_c the total density of carriers, n, exceeds the density of carriers associated with excitations, n_{normal}. Following Landau's treatment,[9] one finds[7]

$$n_{normal} = (2\pi\hbar)^d \int d\underline{p} \ \{p^{1-d}\partial[p^{d+1}/dmv(p)]/\partial p\} \ B[E(p), T], \tag{2}$$

where d is the system's dimensionality; $v(p) = \partial E(p)/\partial p$ and $B[E(p), T]$ are the "velocity" and Bose factor for the bipolarons' collective excitations at temperature T, respectively. At T_c, $n = n_{normal}$ and the bosons' chemical potential vanishes. For free particles, $E(p) = p^2/2m$, Eq. (2) reduces to the Bose condensation formula for free bosons as the curly bracket term is unity.

The transition temperature has been determined for interacting charged bosons in a layered structure comprising groups of h contiguous layers that contain large bipolarons separated by planar carrier-free regions of width b.[7] Several general features are noteworthy. T_c increases as a bipolaron's mass falls or as the density of bipolarons is increased. However, as noted earlier, bipolaronic superconductivity vanishes when the carrier density is too large to support bipolaronic carriers. Furthermore, T_c falls as the static dielectric constant rises. This feature may explain why the superconductivity in $SrTiO_3$ (with $\varepsilon_0 \approx 20{,}000$ at 1 K) doped to a carrier density $\approx 10^{21}$ cm^{-3} is restricted to about 1 K.[11] Finally, T_c is found to fall as the width of the carrier-free regions, b, is increased.

Since bipolaron motion depends on the motion of atoms, isotopic substitutions will affect the mass of a large bipolaron, and, through m, T_c. As illustrated in Eq. (1), the bipolaron's mass is proportional to a sum of the masses of the solid's atoms weighted by a factor measuring the change of their equilibrium positions with the translation of the bipolaron by a lattice constant. As noted above, the effective mass is dominated by atomic displacements close to the bipolaron. If these short-range atomic displacements do not involve atoms of specie s, then $\alpha_s = \partial nT_c/\partial nM_s = 0$. Thus, a small isotope effect with oxygen can be understood if the carrier is associated with occupation of oxygen orbitals and the nearest-neighbor atomic displacements are of the adjacent metal cations. Furthermore, altering the structure to provide some readily displaceable oxygen atoms can introduce a significant oxygen-isotope effect. Thus, both small and large isotope effects can be understood or, at least, rationalized.

Finally, the superconductivity of large bipolarons provides an explanation of the rise of T_c in the CuO_2 materials with the number of contiguous CuO_2 layers until a saturation value of T_c is reached.[4,6] In particular, it is recalled that T_c in a layered material varies inversely with a bipolaron's in-plane mass, m. In addition, a large bipolaron's (in-plane and out-of-plane) masses fall as a disklike bipolaron confined to CuO_2 planes enlarges by spreading out over contiguous CuO_2 sheets. Taken together, these two results imply that T_c for bipolaronic superconductivity rises in the CuO_2-based materials as the number of contiguous CuO_2 planes is increased. This increase will cease once the net thickness of the contiguous CuO_2 layers is great enough so that it no longer determines the thickness of the bipolaron.

DISTINCTIVE PROPERTIES OF LARGE BIPOLARONS

Only observation of distinguishing characteristics of large-bipolarons can establish their existence in a particular material. Some unusual features may be used to identify large bipolarons in lightly doped semiconductors. Other distinctive properties describe large bipolarons in electrical conductors. Yet other characteristics may indicate large bipolarons in a superconducting state. Here, electronic, optical, vibrational and magnetic properties indicative of large bipolarons are succinctly described.

DC Conductivity

The nature of a charge carrier affects how it is scattered. Here the scattering of a large bipolaron (or polaron) by acoustic phonons is considered.[4,13] A large bipolaron, like a multi-site impurity state, only interacts effectively with phonons of wavelengths much greater than the diameter of the self-trapped state, 2R. By virtue of its large size, only long wavelength acoustic phonons scatter a large bipolaron. Since these phonons are of low momentum, $\hbar k \ll \hbar/R$, they are ineffective in scattering a large bipolaron.

In these scattering events acoustic phonons may be regarded as bouncing off the large bipolaron with momentum transfer $2\hbar k$. The net scattering rate of a bipolaron by acoustic phonons is

$$1/\tau \propto \int_0^{1/R} dk\, (2\hbar k/m)\, [\exp(\hbar\omega_k/\kappa T) - 1]^{-1} \approx \kappa T/msR, \tag{3}$$

when $\kappa T > \hbar s/2R$, where the acoustic phonon dispersion relation at low frequencies is ω_k

= s|k| and s is the sound velocity. Then, the mobility, $\mu = q\tau/m \approx qsL/\kappa T$ and the resistivity of a temperature-independent density of large bipolarons is proportional to T. A similar result should emerge for quasiparticles associated with collective motion of large-bipolarons.

Infrared Conductivity

The ac conductivity of large bipolarons (or polarons) possess two distinctive features associated with two different frequency ranges. At frequencies below the relevant phonon frequencies, $\Omega < \omega$, the motion of a large bipolaron can be observed. At higher frequencies, observations probe only the self-trapped electronic carrier since the atoms cannot respond.

First, since large bipolarons move itinerantly, $\text{Re}[\sigma(\Omega)]$ manifests a Drude-like fall off of the conductivity with increasing applied frequency, Ω: $\text{Re}[\sigma(\Omega)] \propto \sigma(0)/(1 + \Omega^2\tau^2)$. The temperature dependence of this fall off is distinctive since, as noted in the preceeding section, $1/\tau \propto \kappa T$. In addition, because of the difficulty of scattering a heavy particle (the bipolaron) with particles of low momentum (low-energy acoustic phonons), τ will be longer than experience with nonpolaronic carriers suggests: $\tau \propto m$. Thus large-polaronic carriers can have high mobilities despite possessing large masses (e.g, as in alkali halides).

Second, at higher frequencies $\text{Re}[\sigma(\Omega)]$ will display an absorption band associated with exciting self-trapped electronic carriers. For large polarons or bipolarons, as with multisite impurity states, absorbing an infrared photon can excite a self-trapped carrier to a higher level of its self-trapping potential well or to a state that is resonant with the continuum. The density of states associated with the continuum rises with excitation energy while the matrix element for photon absorption to a resonant-continuum state falls for final-state wavelengths exceeding the diameter of the self-trapped state, 2R. As with impurity states, these effects produce an absorption band with a nearly temperature-independent asymmetric shape (if the phonon broadening is of subsidiary importance). As a large polaron or bipolaron is increased in size, its electronic levels become shallower and its absorption band narrows and shifts toward lower photon energies. In particular, increasing the number of contiguous CuO_2-sheets of a CuO_2-based material to increase a bipolaron's thickness should shift the absorption band toward lower energies. In the superconducting state of a bipolaronic superconductor the Drude-like tail should disappear (since $\tau \to \infty$) while the absorption band, caused by the existence of large bipolarons, should remain. For large bipolarons in a semiconductor, the intensity of the polaronic absorption, proportional to the density of large bipolarons, should increase with increasing temperature with the carrier density.

The infrared absorption of large polarons (or bipolarons) differs significantly from that of small polarons (or bipolarons). The small-polaron absorption band arises from the photon-induced transfer of a self-trapped carrier from the site of its localization to a neighboring site.[14] Since the absorption energy, the difference of the electronic energies of these two states, is strongly dependent on the atomic configuration at the time of excitation, the small-polaron absorption band broadens significantly with increasing temperature. A marked temperature dependence occurs even when minimized by the carrier only interacting with optical phonons.[14]

Thermoelectricity

A fundamental difficulty in establishing the presence of bipolarons is showing that a carrier consists of a pair of self-trapped electrons. For example, one must be able to distinguish between large polarons and large singlet bipolarons. In particular, one must show that the carrier's electronic charge is of magnitude 2e or that the carrier has no spin. Measurements of the Seebeck coefficient can provide means for doing this.

In a semiconductor with itinerant charge carriers the dc conductivity and Seebeck coefficient are approximated by $\ln(\sigma/\sigma_0) = -E/\kappa T$ and $S = (\kappa/q)(E/\kappa T + A)$, respectively, where E is the energy for generating carriers and A is near temperature-independent constant of order unity. Eliminating $E/\kappa T$ between these two relations yields an expression for the carriers' charge:

$$q = [A + \ln(\sigma_0/\sigma)]\kappa/S. \qquad (4)$$

This observation suggests a method of determining the carriers' charge. Specifically, after ascertaining the semiconducting behavior of the dc conductivity and Seebeck coefficient, one determines q from the slope of a plot of $A + \ln(\sigma_0/\sigma)$ against S/κ. With available data, I find values of q close to 2e in La_2CuO_{4+x}[15] and near 1.7e in $YBa_2Cu_3O_{6.25}$.[16]

The Seebeck coefficient, the entropy per charge of a carrier, can detect the change of a carrier's entropy arising from aligning its spin in an applied magnetic field. The entropy associated with a spin-1/2 is

$$S_{spin} = \kappa\{\ln[2\cosh(\beta g\mu_B H)] - (\beta g\mu_B H)\tanh(\beta g\mu_B H)\}$$

$$\approx \kappa[\ln 2 - (\beta g\mu_B H)^2/2], \qquad (5)$$

for $\beta g\mu_B H \ll 1$. Equation (5) illustrates the reduction of the carrier's entropy with the aligning effect of a magnetic field. Because the Seebeck coefficient of a conductor is small, the magnetic-field dependence arising from aligning the spin of a carrier can be detected.[17] Thus, the absence of any observable magnetic-field dependence of the Seebeck coefficient in the normal state of superconducting $YBa_2Cu_3O_7$, despite a 300 % change predicted from Eq. (5), has been interpreted as indicating that the normal-state charge carriers are singlets.[18]

Photoemission from Large Bipolarons

Photoemission occurs when an electron absorbs a high-energy photon and escapes from a solid as an energetic free electron. Since this absorption is much faster (10^{-17} sec) than atomic motion (10^{-13} sec), the atoms remain fixed during the excitation process. In probing polaronic carriers, photoemission excites electrons from nearly static self-trapped states. In this sense, photoemission of polarons is similar to that of defect states. Nonetheless, some properties of the photoemission of polaronic carriers distinguish it from that of defects or impurities.

With bipolaron formation, the self-trapped carriers themselves induce the formation of the potential wells in which they are bound. Therefore, the density of self-trapping states and the intensity of photoemission from these states are proportional to the carrier density. Even establishing equality of the electronic chemical potential of a sample with that of a metallic contact (by transferring electrons between them) alters the number of self-trapped states. The dependence of the intensity of photoemission on carrier density distinguishes polaronic carriers from (free or trapped) nonpolaronic carriers. Changing the density of self-trapped carriers alters the density of carrier-related states and the photoemission intensity of these states. By contrast, changing the density of nonpolaronic carriers in a typical metal primarily alters the occupation of existing electronic states (observed with photoemission and inverse photoemission) but has little effect on the densities of these states.

The size and shape of multisite self-trapped states change as atoms move. Photoemission is essentially an instantaneous probe of the distribution of self-trapping states that exist at the time of the photons' absorption. At any moment the radii of self-trapped states are distributed about the self-trapping radius of a bipolaron of lowest energy. The bipolaron's energy is the sum of the self-trapped state's electronic energy and the solid's strain energy. Therefore, the radius that minimizes a bipolaron's energy is not the radius that minimizes the self-trapped state's electronic energy. In particular, at the radius of the lowest-energy bipolaron, the electronic binding energy is generally increasing rapidly with decreasing radius. As a result, the distribution of electronic binding energies of large bipolarons will be asymmetric about that of the lowest-energy bipolaron. That is, the distribution will be broader for states with exceptionally large binding energies than for states with especially small electronic binding energies. This asymmetric distribution of electronic binding energies should be a general feature of polaronic states.

As the temperature is lowered below T_c, bipolarons progressively condense into a single collective groundstate. Concomitantly, the asymmetric broadening of the photoemission spectra resulting from the distribution of bipolaronic sizes and shapes is suppressed. As a result, the photoemission spectra for a bipolaronic superconductor should become progressively more peaked (about the electronic binding energy of the minimum-energy bipolaron) as T is lowered below T_c.

A localized state with a spatial extent in direction $\underline{\ell}$ of R_ℓ may be expressed as a superposition of Bloch states with wavevectors $k_\ell \sim \leq \pi/R_\ell$.[2] Therefore, photoexcited electrons from a localized state will have a distribution of momenta that describes the shape of the localized state. Moreover, for self-trapped carriers the shape of the localized state itself depends on the electronic structure. In particular, the size of a self-trapped state in a given direction of a simple structure is proportional to the electronic bandwidth in that direction.[2-4] In other words, the electronic structure affects the size and shape of a self-trapped state and therefore the momentum distribution of emitted electrons.

Positron Annihilation

Positron annihilation studies probe the electronic states of a solid. In these studies a positron injected into a sample tends to be attracted to negatively charged regions. Ultimately a positron and an electron annihilate one another emitting a pair of gamma rays. In the CuO_2-based "doped" insulators, the CuO_2 planes are nominally negatively charged, composed of two O^{2-} ions for every Cu^{2+} ion. Thus, positrons are presumably attracted to the CuO_2 layers, where hole-like bipolarons may exist.

Holes are self-trapped when atomic displacements locally reduce the density of positive nuclear charge thereby attracting fewer electrons. The presence of these localized centers may also create traps for positrons. For example, the self-trapping centers that bind holes may themselves also trap positrons. Alternatively, interbipolaron regions might act as traps. In either case, if adding hole-like bipolarons to CuO_2 sheets creates positron annihilation centers, the positron lifetime may fall in spite of the layers becoming less negative overall. Such an ocurrence would explain the observation that the addition of holes to CuO_2 sheets (e.g., by replacing La with Sr in La_2CuO_4) decreases the positron lifetime.[19]

The positron annihilation rate will fall as density variations are reduced, if, for example, the positron annihilation rate varies inversely with the size (radius) of the carrier-related trapping center. Then, since the bipolaronic ground state is homogeneous, the positron lifetime will increase as the ground state becomes occupied upon cooling below T_c. This argument is consistent with observations of a progressive increase of the positron lifetime as the temperature is lowered below T_c.[19]

These ideas, while not constituting a developed theory, provide a rationalization of unusual features observed in positron-annihilation studies of the novel superconductors. In particular, the abrupt rise of the positron lifetime as T is lowered below T_c is neither predicted nor observed for standard metallic superconductors. The key ideas of this discussion are that, unlike the situation in metals, carriers introduce charge inhomogeneities that provide localized annihilation centers for the positrons. These inhomogeneities are reduced in the superconducting ground state.

Internal Atomic Vibrations of a Large Polaron or Bipolaron

Self-trapping is associated with carrier-stabilized displacements of atomic equilibrium positions. Moreover, in some circumstances, self-trapping also affects the vibratory motions of the atoms associated with a self-trapped state. In particular, within the adiabatic approach these effects occur because the total potential energy experienced by the atoms, $V_T(\cdots\underline{R}_g\cdots)$, is the sum of that in the absence of charge carriers, $V_L(\cdots\underline{R}_g\cdots)$, and the electronic energy of the self-trapped carrier as a function of atomic positions, $E(\cdots\underline{R}_g\cdots)$.

The atomic motions associated with self-trapped states of radius R have been studied.[20] Two distinct situations arise. If $\lambda/2 > 2R$, the phonon half-wavelength exceeds the diameter of the self-trapped state. Then, the phonon-related atomic displacements are uniform across the extent of the self-trapped state. In these instances, the atomic displacements associated with the phonon do not redistribute the self-trapped charge amongst the sites occupied by the self-trapped carrier. As a result, the vibrational frequency is not shifted. Rather, the sole effect of self-trapping is to displace equilibrium atomic positions. This is the situation for a small (single-site) polaron since then the small-radius criterion is always satisfied. Alternatively, if $\lambda/2 < 2R$, a phonon will produce a nonuniform deformation across the length of a self-trapped carrier. The concomitant sloshing of the self-trapped charge locally shifts the vibrational frequency by an amount proportional to the square of the electron-lattice coupling strength.[19] That is, large polarons or bipolarons produce a local shifting of the frequencies for short-wavelength vibrations, $\lambda/2 < 2R$. With a sufficiently strong electron-lattice coupling, this effect yields different vibrational frequencies in the vicinity of the self-trapped carriers than in carrier-free regions. Thus, large polarons and bipolarons can result in "extra" vibrational modes albeit of only short wavelength.

Orbital Magnetism of Large Bipolarons

A singlet large bipolaron has no net spin and presumably quenched orbital angular momentum. Nonetheless, the application of a magnetic field can induce significant (Van Vleck) paramagnetism.[21] If the thermal energy is much less than the energy separation between ground and excited states, the Van Vleck susceptibility is nearly temperature independent. Specifically, a bipolaron's paramagnetic susceptibility per unit volume is $\chi = (e^2/2m_e^2c^2) \Sigma_n |<n|L_H|0>|^2/\Delta E_n$, where the ground and excited self-trapped electronic states are denoted by $|0>$ and $|n>$, ΔE_n is the excitation energy of the n-th state and L_H is the component of the orbital angular momentum aligned with the magnetic field. With rotational symmetry about the direction of the applied field, conservation of angular momentum ensures the vanishing of the matrix element. However, the matrix element exists in asymmetric situations.

Consider a bipolaron of a disklike morphology (as envisioned in the CuO_2-based materials) having the radius R and the thickness 2a. The paramagnetic susceptibility vanishes with the magnetic field applied perpendicular to the disk (the CuO_2 planes). However, with the magnetic field aligned parallel to the disk (considering only the first excited state with odd parity in both directions parallel to the disk), $\chi \approx (2e^2/m_ec^2)(R^2 - a^2)^2/(R^2 + a^2)$. This paramagnetism rises with the bipolaron's asymmetry, R/a, to dominate the bipolaron's diamagnetism, $\approx e^2R^2/8m_ec^2$. Thus, an asymmetric large bipolaron possesses a large ($\propto R^2$) asymmetric induced orbital paramagnetism.

In the normal state, bipolarons are distributed in size but assume a common radius in their collective (homogeneous) ground state. Because of the nonlinear dependence of a bipolaron's susceptibility on R, the net susceptibility must fall as the bipolarons condense into their groundstate.[21] NMR can probe the carriers' magnetic susceptibility. The sensitivity of different nuclei to the bipolarons' magnetic state depends on the location of the nuclei relative to the bipolarons. The carriers' contribution to NMR frequency ("Knight") shifts may be indicated by their changing on cooling through T_c. Then, the absence of a change in the NMR frequency shifts in the CuO_2-based materials when the magnetic field is aligned perpendicular to the CuO_2-layers[22] may indicate the carriers' asymmetric susceptibility. Distinctively, the paramagnetic susceptibility for disklike large bipolarons vanishes for magnetic fields aligned perpendicular to the disks' circular surface.

With bipolaronic carriers, the net spin-lattice relaxation rate should fall monotonically upon decreasing temperature below T_c. In particular, relaxation of nuclei can occur via low energy excitation of the bipolarons' state that alter the bipolarons' susceptibility. However, this relaxation is precluded by the presence of the gap, $\hbar\omega_p$, when the bipolarons are in their collective ground state.

SUMMARY

Bipolaronic superconductivity occurs with the condensation of a sufficient density of mobile bipolarons. While small (single-site) polarons readily localize, large (multisite) polarons are generally mobile. In analogy with this situation, bipolaronic superconductivity is presumed to only occur with large bipolarons. Pairing of carriers to form large bipolarons can occur in ionic solids with especially displaceable ions: materials for which the ratio $\varepsilon_0/\varepsilon_\infty$ is $>> 2$. Superconductors produced by adding carriers to such materials (e.g., $SrTiO_3$, WO_3, $BaBiO_3$ and the CuO_2-based materials) may be bipolaronic superconductors. Bipolaronic superconductivity is possible if the density of large bipolarons is sufficient to provide a fluid ground state while not being so large that the bipolarons overlap one another.

Unusual properties may be used to identify large bipolarons. The resistivity of a temperature-independent density of large polarons or bipolarons (arising from their scattering by acoustic phonons) is proportional to temperature. Concomitantly, the carriers' scattering time is unusually long and is inversely proportional to temperature. In addition, with large-polaronic carriers a temperature-independent asymmetric absorption band arises from photoexciting electrons from the potential well that self-traps them. Measurements of the Seebeck coefficient provide a means of establishing the pairing of charge carriers. Singlet carriers have Seebeck coefficients that are independent of magnetic field. Furthermore, measurements of the semiconducting conductivity and Seebeck coefficient can be used in tandem to determine if the carrier has a charge of $|e|$ or $2|e|$. The photoemission of polaronic carriers is distinguished from that typical of metals. With polarons or bipolarons the densities of carrier-related states (near the Fermi level) depend on the carrier density. Observations of positron-annihilation measurements differ qualitatively from those of conventional superconductors. These differences can be rationalized with the novel superconductors being bipolaronic superconductors. With large polarons or bipolarons the vibrational spectra can possess "extra" vibrational modes at short wavelength caused by the internal intersite motion of the self-trapped charge. The orbital paramagnetism of a large bipolaron is distinguished from the spin paramagnetism of an unpaired carrier by its asymmetry. Disklike large bipolarons in layered compounds can have large induced orbital moments aligned parallel to the layers. A discussion of the experimental situation exceeds the scope of this article. Nonetheless, it should be noted that evidence of all these distinctive effects has been reported in the high-T_c superconductors.

ACKNOWLEDGEMENT

This work was performed under auspices of the U.S. Department of Energy and funded in part by its Office of Basic Energy Sciences, Division of Materials Sciences under contract #DE-AC04-76DP00789.

REFERENCES

1. L. D. Landau, Phys. Z. Sovietunion 3, 644 (1933).
2. H. Frohlich in Polarons and Excitons (Plenum, New York, 1963) p. 1.
3. D. Emin and T. Holstein, Phys. Rev. Lett. 36, 323 (1976).
4. D. Emin and M. S. Hillery, Phys. Rev. B 39, 6575 (1989).
5. Y. Toyozawa, Prog. Theor. Phys. 26, 29 (1961).
6. D. Emin, Phys. Rev. Lett. 62, 1544 (1989).
7. D. Emin and M. S. Hillery, in preparation.
8. E. P. Wigner, Phys. Rev. 46, 1002 (1934).
9. L. D. Landau, Zh. Eskp. Teor. Fiz. Nauk, 59, 592 (1941).
10. L. L. Foldy, Phys. Rev. 124, 649 (1961).
11. J. F. Schooley, W. R. Hosler and M. L. Cohen, Phys. Rev. Lett. 12, 474 (1964).
12. T. Holstein, Ann. Phys. (N.Y.), 8, 343 (1959).
13. H.-B. Schuttler and T. Holstein, Phys. Rev. Lett., 51, 2337 (1983); Ann. Phys. (N.Y.), 166, 93 (1986).
14. D. Emin, Adv. Phys. 24, 305 (1975).

15. M. F. Hundley, private communication.
16. B. Fisher, J. Genossar, L. Patlagan, I. O. Lelong and J. Ashkenazi, Physica C 162-164, 1207 (1989) Fig. 2.
17. P. M. Chaikin, J. F. Kwak and A. J. Epstein, Phys. Rev. Lett. 42, 1178 (1979).
18. R. C. Yu, M. J. Naughton, X. Yan, P. M. Chaikin, F. Holtzberg, R. L. Greene, J. Stuart and P. Davies, Phys. Rev. 37, 7963 (1988).
19. Y. C. Jean, J. Kyle, H. Nakanishi, P. E. A. Turchi, R. H. Howell, A. L. Wachs, M. J. Fluss, R. L. Meng, H. P. Hor, J. Z. Huang and C. W. Chu, Phys. Rev. Lett. 60, 1069 (1988).
20. D. Emin, Phys. Rev. B 43, 8610 (1991).
21. D. Emin, Phys. Rev. B 43, 2633 (1991).
22. M. Takigawa, P. C. Hammel, R. Heffner and Z. Fisk, Phys. Rev. B 39, 7371 (1989).

ANOMALOUS DYNAMICS AND HIGH T_c

J.W. Flocken

Department of Physics
University of Nebraska at Omaha
Omaha, Nebraska 68182

J.R. Hardy and H.M. Lu

Department of Physics and Center for Electro-Optics
University of Nebraska at Lincoln
Lincoln, Nebraska 68588

ABSTRACT

A discussion is presented of the role of anharmonic lattice motion in enhancing the superconducting transition temperature in high T_c systems. In particular, the role of double-well motion and the manner in which it can enhance the Rowell-Dynes coupling constant λ are discussed. Evidence for such motion is presented in the general context of the unusual dynamics of ionic molecular solids. Finally, we discuss recent experimental and theoretical findings which increasingly indicate that high T_c can be understood within an extended form of the BCS theory.

INTRODUCTION

After almost five years since its discovery[1,2], the basic mechanism responsible for high temperature superconductivity remains elusive. Immediately after the report of a T_c value $>$ 90K for yttrium barium copper oxide (YBCO)[2], a strong consensus developed that such a large T_c value effectively ruled out the standard BCS mechanism, even in its strong-coupled Eliashberg form. The further discovery that the oxygen isotope dependence of T_c was very weak[3], essentially crystallized this belief and led to a vast amount of theoretical work on purely electronic mechanisms. While this produced no clear consensus, it did tend to reveal one drawback: while such mechanisms generally produce high T_c values for the cuprate-based materials, they also can predict high T_c's for materials that, in reality, show no superconducting transition whatsoever. However, these efforts were bolstered by the apparent lack of any phonon/lattice involvement, manifested by neutron scattering studies[4,5] which, while revealing interesting structural transformations in the $T_c \sim$ 40K systems lanthanum (barium/strontium) copper oxides (LBCO, LSCO), failed to show any clear anomalies associated with the onset of superconductivity. At the same time, calculations of T_c by the standard BCS techniques[6] apparently ruled out the possibility of producing values $>$ 30-40K, and these only if suspiciously low values were used for the appropriate average lattice frequencies (suspicious, in that they were much below those of typical copper-oxygen "stretch" vibrations, to which the strongest coupling is to be expected).

As a consequence, the last three years have seen a vast amount of additional work on exotic mechanisms with no conclusive results. A major factor in this ambiguity has been the lack of clear experimental evidence for the operation of such mechanisms. The one candidate that can be said to have been seen experimentally is antiferromagnetic spin fluctuations in the cuprates.[7] However the situation is very unclear. While it is certainly true that non-superconducting cuprates do possess antiferromagnetic order, as they are doped/oxygenated into the superconducting phase, the static antiferromagnetism disappears and either becomes fluctuating or vanishes completely. Indeed, it has been argued that the two phases are separated by a first order transition and consequently do not coexist except as a two-phase mixture.[8]

Over the same period of time, particularly recently, there has accumulated a body of serious evidence that indicates with increasing strength that lattice-related BCS mechanisms were too hastily dismissed. Moreover, some of this new evidence, in addition to supporting a BCS-like mechanism, directly erodes the support for an exotic mechanism, to the point that it is now argued that such a mechanism is required because it explains anomalies in the *normal* state properties of high T_c cuprates. However, this immediately raises the problem of explaining how it is that very high quality and very low quality specimens of YBCO can have virtually identical T_c values, while their normal state properties are likely to be very different.

ANOMALOUS DYNAMICS

Approximately three years ago, others and ourselves[9-11] first suggested that high T_c could be reconciled with strong-coupled (Eliashberg) BCS theory if the *dynamics* of these novel materials were anomalous. Our suggestion was based on a body of work on apparently unrelated systems, in particular halide-based perovskites such as RbCaF$_3$.[12] For these systems, we found that structural instabilities were associated with double wells in the lattice potential energy surface; and, most significantly, that these wells were *always* present, even in the high temperature, undistorted phase. Furthermore, it was found by molecular dynamics simulations that these wells had as their signature anomalously large thermal ellipsoids.[13]

To illustrate this Figs. 1(a) and (b) compare the results for KCaF$_3$, (which is unstable owing to the double-well potential shown in Fig. 1(c)) and CsCaF$_3$, which is stable. The temperature is $\sim$ 800K, 200K above the theoretical phase transition temperature of 600K (c.f. the 550-560K experimental value).

It was at this point that a possible connection with high T_c materials emerged, since these too are characterized by anomalously wide thermal ellipsoids which, in the cases of LBCO and LSCO, are also associated with structural instabilities. We were therefore led to ask if such anomalous dynamical behavior could affect predicted values of T_c. Fortunately, we discovered that in 1971 Hui and Allen[14] had performed a simple model calculation for an anharmonic Einstein oscillator, and we were thus able to extend their formalism to the case of a double-well oscillator at finite temperatures.[11]

In order to predict T_c, the relevant functions are the Rowell-Dynes coupling constant λ, its second moment $<\omega^2>$, and the expression for the maximum possible T_c, T_c^m (which becomes the actual value if $\lambda \sim$ 5-10).

These are given by

$$\lambda = N(0) \sum_{kk'}^{FS} \sum_{n'>n} \frac{|\langle n|M_{kk'}x|n'\rangle|^2}{E_{n'} - E_n}(f_n - f_{n'}), \qquad (1)$$

[where N(0) is the density of electron states at the Fermi surface (*FS*), M$_{kk'}$x is the electron-phonon matrix element between the electronic states $|k\rangle$ and $|k'\rangle$ on that surface, $|n\rangle$ and $|n'\rangle$ are oscillator states corresponding to energies E$_n$ and E$_{n'}$, respectively and f$_n$ and f$_{n'}$ are the thermal weighting factors; f$_n$ = exp($-\beta E_n$)/$\sum_{n'} e^{-\beta E_{n'}}$) where $\beta = 1/kT$, T being the temperature and k Boltzmann's constant.]

$$<\omega^2> = \frac{N(0)}{\hbar^2 \lambda} \sum_{kk'}^{FS} \sum_{n'>n} |\langle n|M_{kk'}x|n'\rangle|^2, \qquad (2)$$

$$\times (E_{n'} - E_n)(f_n - f_{n'}),$$

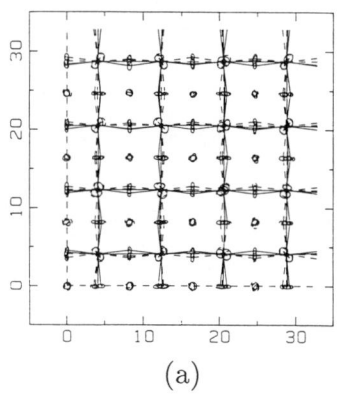

(a)

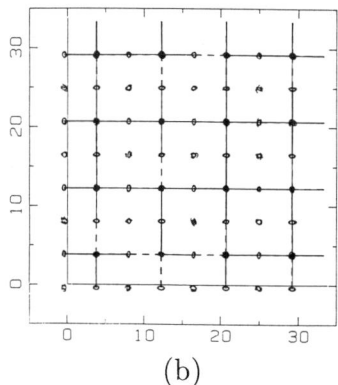

(b)

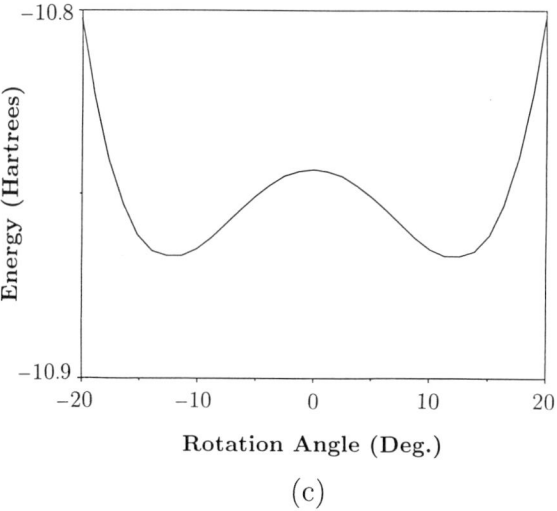

(c)

Fig. 1(a) Structure and thermal ellipsoids for KCaF$_3$ at 800K: the apparent slight static distortion is in fact a very slow fluctuation.

(b) Structure and thermal ellipsoids for CsCaF$_3$ at 800K.

(c) Double well for the [100] rotations in Fig. 1(a). Energy is for a 40 ion unit cell.

and

$$T_c^m = 0.18 \left[\sum_{i,\alpha} \eta_{i\alpha}^2 \frac{1}{M_i} \right]^{\frac{1}{2}} \tag{3}$$

where $\eta_{i\alpha}$ is the total carrier lattice coupling constant for the α displacement component of the i'th ion, and M_i is the mass of that ion (evidently the O^{2-} ions are likely to be the dominant contributors).

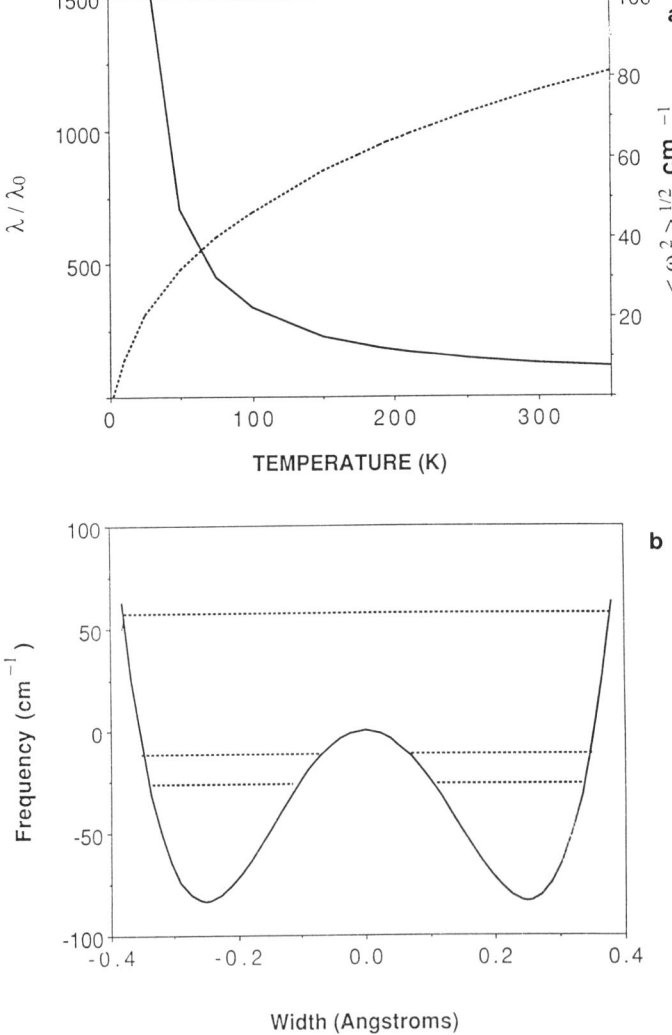

Fig. 2(a) Plots of the functions $<\omega^2>$ and λ as a function of temperature for the double well in Fig. 2(b). λ is normalized with respect to λ_0 which is unity for a 20 amu mass in a 600 cm^{-1} harmonic well.

(b) Double well for the chain oxygen motion in YBCO after Ref. 16. Energy is per chain O^{2-} ion.

In Fig. 2(a), λ and $<\omega^2>^{1/2}$ are shown for the double-well in Fig. 2(b): the reason for using this particular double well will emerge shortly. λ_0 was taken as unity for an atom of mass 20 amu moving in a harmonic well with a frequency of $600cm^{-1}$ (70-80 mev) typical of oxygen "stretch" motion in ionic oxides, e.g. cuprates. As can be seen, the enhancement of λ is dramatic – so much so that if the "grad V" coupling terms (the McMillan η's in Eq. 3) which give $\lambda = 1$ for a $600cm^{-1}$ harmonic well, have even a 10% component in the direction of the present well ($\eta^2 \sim 10^{-2}$) $\lambda \simeq 3$-4 at 100K! Or alternately if there are, as is the case for the $T_c \simeq 100$K materials, many oxygens per unit cell, a few percent coupling into relatively few of their degrees of freedom can easily give $\lambda > 5$ for the entire complex.

The origin of these vast enhancements is not far to seek; it lies in the existence of one or two close "tunnel split" levels at the bottom of the double-well (see Fig. 2(b)). These in turn derive from the fact that the classical action $\int p dq$ is zero for a trajectory which just touches the central maximum leading, classically, to a zero frequency oscillation. Quantum mechanically this is disallowed since $\int p dq \geq h/2$. Hence one has two closely spaced levels, since the action is varying very rapidly with energy in this region. However, this argument does not depend on the one dimensionality. So long as there is some region of negative curvature (a "hump") at the one, two, or three dimensional well center, there will be at least one closely spaced pair of levels. The actual magnitude of the effects they produce naturally depends on how pronounced is the central maximum. Specifically, for these effects to be large, its height above the minima must be comparable with that of the ground state (as is clearly the case for our specific well). We thus arrive at the quite remarkable conclusion that for *any* system having a *multidimensional* double-well motion with at least one component comparable with our model well, even what might normally be regarded as an insignificant carrier-lattice coupling can drive λ into the 5-10 regime and thus ensure the $T_c = T_c^m$. Moreover, one can see from Eq. 3 that if the η's are largest for the lightest atoms T_c^m can become particularly large. For a purely harmonic lattice, this would be of no practical interest, because the effective frequency for the light atoms would be very high. As a consequence λ would be very small, and the actual T_c would be $\ll T_c^m$.

With this in mind, it is instructive to examine the hierarchy of structures from perovskite to YBCO shown in Fig. 3. Two features emerge, first that it is a hierarchy of decreasing symmetry, and second that it is one of increasing openness. Specifically, while in the perovskite structure all the octahedral vertices are shared and the symmetry is cubic, in the La_2CuO_4 structure the apices are unshared and the symmetry is tetragonal, while in the YBCO structure the basic unit is derived from three octahedra, two of which have been converted into pyramids by the removal of one vertex, and the third has been converted into a plane by the removal of two opposed vertices.

That this is indeed a hierarchy of instability has now been explicitly demonstrated by Cohen et al[15] who have shown, by full LAPW total energy calculations, that the octahedral tilt modes in La_2CuO_4 have a double-well character, and that the zone boundary bending mode of the chain oxygens in YBCO is also of an even more pronounced double-well nature[16], the double well in question being that shown in Fig. 2(b). The magnitude of this latter instability is such that it alone can probably render bistable the motion of the entire complex of ions. In addition there are almost certainly other double-well motions – indeed it is very likely that all transverse motions of the chain oxygens are similarly unstable, including the zone center mode, which is polar. Given these facts, it would appear that λ values > 5 at 100K should be easy to produce, at least for YBCO; and thus a $T_c = T_c^m$.

In previous work we made this point[17,18], however, at that time, the only value of T_c^m we could cite was that of Allen et al[6], calculated using the "rigid muffin tin" approximation (RMTA), which for YBCO, was $\simeq 40$K; i.e. less than half the measured value. Fortunately, recent calculations by Cohen et al[16,19], and Zeyher[20], have now revealed that the RMTA is qualitatively incorrect for the high T_c materials, which behave as ionic metals in which the free carrier density is so low that the Coulomb part of the electron-ion interaction is very ineffectively screened, in particular it is likely that the local field contribution[21] (which can be very large in perovskites and perovskite-derived systems) is completely unscreened. At present, revised values of T_c^m have not been reported, but from Zehyer's work[20], we estimate a value ~ 100K ($\pm \sim 20\%$) which appears encouraging.

ISOTOPE EFFECT ON T_c: NEW FINDINGS

If we now return to the basic question of BCS versus non-BCS mechanisms, the principal argument for the latter has, during the past year, undergone a major setback. This has resulted

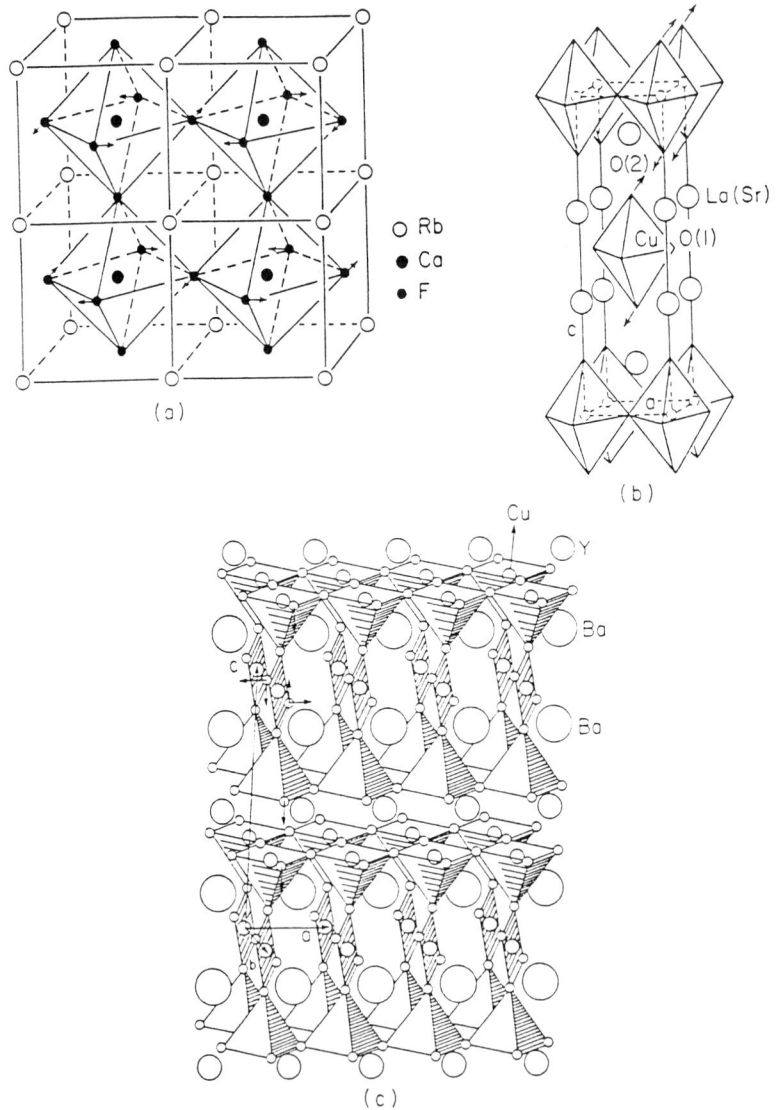

Fig. 3 (a) Perovskite structure for RbCaF$_3$ (also for $T_c = 30K$ BKBO).
(b) Structure of La$_2$CuO$_4$ in the high temperature tetragonal (HTT) phase.
(c) Structure of YBCO: YBa$_2$Cu$_3$O$_7$

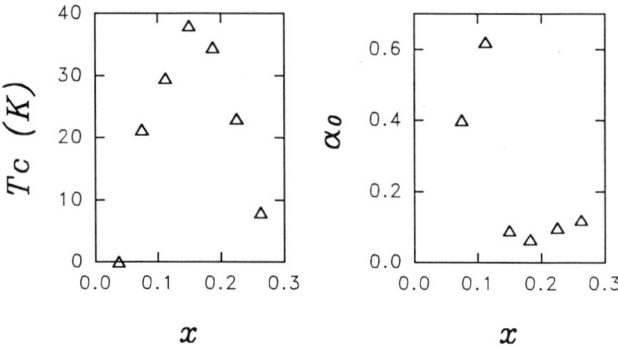

Fig. 4 Variation of T_c and α_0: $[T_c \propto (\text{atomic mass})^{-\alpha_0}]$ for LSCO as a function of Sr^{2+} concentration x after Ref. 22.

from experimental work by Crawford et al[22] on the variation of the exponent α_0 in the $T_c \propto$ (atomic mass)$^{-\alpha_0}$ expression as a function of Sr^{2+} concentration (x) in LSCO. The results are shown in Fig 4: They are remarkable. Most striking is the *fall* in α_0 by $\sim$ 500-600% between $x = 0.12$ and $x = 0.15$ from a value *larger* than the "ideal" BCS value of 0.5 when $T_c = 30K$ to a value ~ 0.1 when $T_c = 40K$: the latter value previously being argued as clear proof of a large non-BCS mechanism. Clearly this argument is invalid since $T_c = 30K$ ($\alpha_0 \geq 0.5$) is evidently of BCS origin: hence if there is any non-BCS contribution to $T_c = 40K$ it constitutes only $\sim 30\%$ of the total, and it is thus difficult to see how its presence has reduced α_0 by *500-600%*.

The argument is further weakened by the fact that Crawford et al[22] also examined the dependence of the frequencies of two infrared absorption lines on x. These showed clear anomalies over the $x = 0.12$-0.15 range, pointing to anomalous dynamical behavior over this concentration range. Moreover, Axe et al[23] had already reported independently an even more dramatic anomaly in LBCO: T_c drops to zero (or ~ 0) at $x \simeq 0.12$ before recovering to 35K (the maximum value for LBCO) at $x \simeq 0.15$, behavior correlated with a structural transformation that appears at $x \simeq 0.12$ and disappears at $x \simeq 0.15$.

Most recently work by Franck et al[24] on Pr^{3+} and Ca^{2+} doped YBCO has shown, that as T_c is reduced by doping, α_0 increases from ~ 0.02 to BCS-like values in a monotonic fashion, which is most consistent with a single mechanism that is being continuously varied.

Certainly the LSCO results, and possibly the YBCO data, point to anomalous dynamics as, in some way, the cause of small isotope shifts and this is clearly corroborated by the very recent work of Cohen et al[20] who have now shown that, in addition to the double well associated with the single tilt of the CuO_4 about octahedra [010] they found earlier, there is an even deeper well associated with a tilt about [110] to produce the low temperature tetragonal (LTT) phase observed[23] for LBCO.

What may occur in LSCO is that the [010] component of the [110] rotation, which for small values of x is harmonic because of the large static [100] tilt, as x increases and the static tilt decreases assumes a symmetric double-well character: hence λ increases and with it T_c. However since the motion is less constrained, it is possible for the structure to adjust to maximize T_c (e.g., a very slight change in the [100] static tilt could do this) and thus minimize the free energy. Thus if T_c is reduced by isotopic replacement such freedom will be employed to minimize that reduction. Moreover, since a static [110] tilt never develops in LSCO, this implies that the associated double well is too shallow to drive a structural instability. Thus, this well remains symmetric, and T_c is maximal for those x values (0.12-0.15) where the structure is closest to this second instability. In LBCO the static [110] rotation does develop over this range of x, and the fact that T_c drops to zero when this occurs is clear evidence that high T_c is strongly dependent on coupling to *this motion* while it remains *dynamic*.

The difference between LSCO and LBCO has a simple explanation: Ba^{2+} is significantly larger than La^{3+} while Sr^{2+} is essentially the same size. Thus since each Ba^{2+} is surrounded by four apical oxygen ions lying along $<110>$ directions they can most effectively accommodate the mismatch by relaxing along those directions – hence driving the rotations about $<110>$ axes. However, as more La^{3+} ions are replaced by Ba^{2+} ions, it becomes increasingly favorable energetically to assume the high temperature tetragonal phase (which would probably be the stable form for 100% replacement).

In the case of YBCO, the effect of substitution on T_c is probably partially due to a reduction in the number of free carriers, but could also reflect a reduction in bistability by aliovalent ion substitution. Any loss of symmetry in the unit cell could favor one or other component of any double-well motions, thus rendering the wells asymmetric and strongly reducing the local contribution to λ. In addition, the ability of the structure to adjust to minimize the effect of isotope substitution will be progressively reduced: hence the α_0 values will increase.

UNIFICATION OF CUPRATE AND NONCUPRATE HIGH T_c MATERIALS

One can perhaps sum up the situation for the cuprates by saying that there is increasing evidence for strong-coupled BCS-like superconductivity in these systems, both from theory and experiment. However, there is still controversy over explaining much of the data, e.g., the temperature dependence of the normal state resistivity. In that specific context, we should draw attention to the fact that double-well λ's are strongly *temperature dependent* and fall *drastically* [see Fig. 2(a)] at higher temperatures: consequently, the argument that rules out large λ's on the basis of resistivity saturation[25] is *invalid* since it *assumes* that λ is temperature independent. Indeed the entire Eliashberg theory probably needs re-examination, since it makes the same assumption [earlier we assumed $\lambda(T) \gg 5$ when $T \leq T_c^m$].

Previously it has been necessary to accommodate these ambiguities because the "ideal" behavior of an actual high T_c system was unknown. However, the very recent work of Hinks et al[26,27] on potassium doped barium bismuthate (BKBO) has now provided a paradigm. In particular, they have obtained tunneling data that are "clean" and unambiguous with classic phonon structure and clear BCS behavior. When inverted these give a $\lambda \sim 1$ for a $T_c \sim 20\text{-}25\text{K}$ junction. Also the isotope effect $\alpha \simeq 0.4$ is at the upper limit of that allowed by realistic phonon calculations. Moreover this system can give a T_c value of 34K[28] and recent work on mixed doping with K^+ and Rb^+ has produced a $T_c = 37\text{K}$ system.[29]

Previously it has been argued that BKBO is an oddity: it is a BCS system for which T_c just happens to be high, and thus to overlap slightly the high T_c's of the cuprates which have a basically exotic origin. However in Ref. 26, the Argonne group also report tunneling data on cerium doped neodymium copper oxide (NCCO) which are qualitatively identical to their BKBO data, with clear phonon structure, and the capability of being inverted to give $T_c \simeq 20\text{K}$ (c.f. 20-25K, observed) and $\lambda \sim 1$. Since NCCO is a cuprate, and indeed has magnetic properties qualitatively similar to the other cuprates[30], this finding effectively proves that superconductivity and magnetism are distinct and unrelated in at least one cuprate. This would appear to rule out any connection between superconductivity and unusual electronic properties of copper, at least for NCCO.

Furthermore, one can see that there is now no need to invoke any exotic mechanism to obtain high T_c, since if $\lambda \sim 1$ for NCCO gives a $T_c \cong 20\text{K}$, then if λ could be enhanced to ~ 5, T_c would be equal to T_c^m.

An estimate of $<\omega^2>$ for both NCCO and BKBO by numerical integration of the $\alpha^2 F(\omega)$ data given in Ref. 26 gives $T_c^m \cong 50\text{-}60\text{K}$ for *both* materials. Clearly these are "high T_c's" ($\sim$ three times T_c for Nb$_3$Ge).

It thus can be most plausibly argued that the $T_c \sim 100\text{K}$ cuprates are simply materials for which $\lambda \geq 5$ and T_c^m is higher by a factor of 1.5-2 due to radically different geometry and/or a larger density of states at the Fermi surface (since the number of free carriers is so low relative to A15 materials, it is much easier to manipulate the Fermi energy by doping, etc.).

An additional benefit is that there could now be a *unified* mechanism for high T_c in both cuprates *and* BKBO if we can reconcile the double-well picture with the apparent standard BCS behavior of the latter material. This can easily be achieved when one realizes that the maximum of T_c versus K^+ doping occurs at the stability limit of the perovskite structure.[31] At this point the dynamics will qualitatively resemble those of our KCaF$_3$ system presented earlier, except that, even at 0K, the double-well which drives the transition at lower K^+ concentrations is just too weak to do so. Thus we have the maximum width symmetric well possible in this system: this maximizes λ and thus T_c. Furthermore, as was the case for LSCO, we can argue that the very fact that T_c drops to zero in the *rotationally* distorted phase automatically implies that a major component of λ was due to this rotational motion. Indeed the low frequency peaks in the $\alpha^2 F(\omega)$ curves of Ref. 26 appear to show some evidence of anomalies that may reflect the double-well motion. Thus this motion varies smoothly from the highly anomalous behavior in YBCO to quasi-harmonic behavior in BKBO as the ground state rises above the central hump.

There is one final and most intriguing implication of this unity between BKBO and the cuprates to which we have previously drawn attention.[32] Exactly as we have just argued to unify NCCO and the "high T_c" cuprates, we can now argue that for BKBO, if λ could be enhanced by structural manipulation, specifically by making it more "open" and less symmetric (c.f. perovskite and YBCO), T_c could be enhanced at least to 50-60K. One possibility we suggested[32] is to split the octahedra into two pyramids (again c.f. YBCO). This was actually done quite recently, but for LSCO, by Cava et al[33], and T_c did in fact jump from 40 to 60K.

If this were in fact possible for bismuthates, then not merely would the association of cuprates (and by implication exotic mechanisms peculiar to copper) with high T_c have been broken, but a new class of high T_c materials would be created which, being based on bismuth, could prove more tractable for applications. This is exemplified by the BKBO tunneling work which clearly shows that fabrication of excellent tunnel junctions is considerably easier for that system, than it is for cuprates.

ACKNOWLEDGEMENTS

Work at the University of Nebraska was supported by the Office of Naval Research, by the Army Research Office, and by the Research Initiative of the State of Nebraska. We wish to acknowledge helpful discussions with Professors S. K. Kurtz and H. Krakauer and Drs. L. L. Boyer, W. Pickett, R. E. Cohen, J. Serene, and D. A. Papaconstantopolous, together with access to preprints of their work, and supporting computations by Professor R. A. Guenther and Drs. P. J. Edwardson, and D. P. Billesbach.

REFERENCES

1. J.G. Bednorz and K.A. Muller, "Possible High T_c Superconductivity in the Ba-La-Cu-O System," Z. Phys. B, 64:189 (1986).

2. M.K. Wu, J.R. Ashburn, C.J. Torng, P.H. Hor, R.L. Meng, L. Gao, Z.J. Huang, Y.Q. Wang and C.W. Chu, "Superconductivity at 93 K in a New Mixed-phase Y-Ba-Cu-O Compound System at Ambient Pressure," Phys. Rev. Lett., 58:908 (1987).

3. D.E. Morris, R.M. Kuroda, A.G. Markelz, J.H. Nickel and J.Y.T. Wei, "Small Oxygen Isotope Shift in $YBa_2Cu_3O_7$," Phys. Rev. B, 37:5936 (1987).

4. P. Day, M. Rosseinsky, K. Prassides, W.I.F. David, O. Moze and A. Soper, "Temperature Dependence of the Crystal Structure of the Ceramic Superconductor $La_{1.85}Sr_{0.15}CuO_4$: A Powder Neutron Diffraction Study," J. Phys. C:, 20:429 (1987).

5. P. Böni, J.D. Axe, G. Shirane, R.J. Birgeneau, D.R. Gabbe, H.P. Jenssen, M.A. Kastner, C.J. Peters, P.J. Picone and T.R. Thurston, "Lattice Instability and Soft Phonons in Single-Crystal $La_{2-x}Sr_xCuO_4$," Phys. Rev. B, 38:185 (1988).

6. P.B. Allen, W. E. Pickett and H. Krakauer, "Anisotropic Normal-State Transport Properties Predicted and Analyzed for High-T_c Oxide Superconductors," Phys. Rev. B, 37:7482 (1988).

7. J.M. Tranquada, A.H. Moudden, A.L. Goldman, P. Zolliker, D.E. Cox, G. Shirane, S.K. Sinha, D. Vaknin, D.C. Johnston, M.S. Alverez, A.J. Jacobson, J.T. Lewandowski and J.M. Newsam, "Antiferromagnetism in $YBa_2Cu_3O_{6+x}$," Phys. Rev. B, 38:2477 (1988).

8. B.H. Brandow (this volume).

9. R. Jagadish and K.P. Sinha, "Electron-Fluctuating Distortion Interaction in the New High T_c Superconductors," Pramana-J. Phys. (India), 28:L317 (1987).

10. N.M. Plakida, V.L. Aksenov and S.L. Drechsler, "Anharmonic Model for High-T_c Superconductors," Europhys. Lett., 4:1309 (1987).

11. J.R. Hardy and J.W. Flocken, "Possible Origins of High-T_c Superconductivity," Phys. Rev. Lett., 60:2191 (1988).

12. L.L. Boyer and J.R. Hardy, "Theoretical Study of the Structural Phase Transition in $RbCaF_3$," Phys. Rev. B, 24:2577 (1981).

13. P.J. Edwardson, L.L. Boyer, R.L. Newman, D.H. Fox, J.R. Hardy, J.W. Flocken, R.A. Guenther, and W. Mei, "Ferroelectricity in Perovskitelike NaCaF$_3$ Predicted *Ab Initio*," Phys. Rev. B, 39:9738 (1989).

14. J.C. Hui and P.B. Allen, "Effect of Lattice Anharmonicity on Superconductivity," J. Phys. F, 4:L42 (1974).

15. R.E. Cohen, W.E. Pickett and H. Krakauer, "First Principles Phonon Calculations for La$_2$CuO$_4$," Phys. Rev. Lett., 62:831 (1989).

16. R.E. Cohen, W.E. Pickett and H. Krakauer, "Theoretical Determination of Strong Electron-Phonon Coupling ni YBa$_2$Cu$_3$O$_7$," Phys. Rev. Lett., 64:2575 (1990).

17. J.R. Hardy and J.W. Flocken, "Possible Origin of High T_c Superconductivity," Ferroelectrics, 92:175 (1989).

18. J.R. Hardy and J.W. Flocken, "Vibronic Origins of High T_c," Phase Trans., 22:121 (1990).

19. R.E. Cohen, W.E. Pickett and H. Krakauer (this volume).

20. R. Zeyher, "Importance of Long-range Electron-phonon Coupling in High-T_c Superconductors," Z Phys. B–Condensed Matter, 80:187 (1990).

21. P.W. Anderson and E.I. Blount, "Symmetry Considerations on Martensitic Transformations: 'Ferroelectric' Metals?" Phys. Rev. Lett., 14:217 (1965).

22. M.K. Crawford, M.N. Kunchur, W.E. Farneth, E.M. McCarron III and S.J. Poon, "Anomalous Oxygen Isotope Effect in La$_{2-x}$Sr$_x$CuO$_4$," Phys. Rev. B, 41:282 (1990).

23. J.D. Axe, A.H. Moudden, D. Hohlwein, D.E. Cox, K.M. Mohanty, A.R. Moodenbaugh and Y. Xu, "Structural Phase Transformation and Superconductivity in La$_{2-x}$Ba$_x$CuO$_4$," Phys. Rev. Lett., 62:2751 (1989).

24. P. Franck, J. Jung, M.A.K. Mohamed, S. Gygax and I.G. Sproule, (this volume).

25. M. Gurvitch and A.T. Fiory, "Resistivity of La$_{1.825}$Sr$_{0.175}$CuO$_4$ and YBa$_2$Cu$_3$O$_7$ to 1100 K: Absence of Saturation and Its Implications," Phys. Rev. Lett., 59:1337 (1987).

26. Q. Huang, J.F. Zasadzinski, N. Tralshawala, K.E. Gray, D.G. Hinks, J.L. Peng and R.L. Greene, "Tunnelling Evidence for Predominantly Electro-phonon Coupling in Superconducting Ba$_{1-x}$K$_x$BiO$_3$ and Nd$_{2-x}$Ce$_x$CuO$_{4-y}$," Nature, 347:369 (1990).

27. D.G. Hinks et al (this volume).

28. A.W. Sleight, "Chemistry of High Temperature Superconductors," Science, 242:1519 (1988).

29. D. Tseng and E. Ruckenstein, "Structure and Superconductivity of BaBiO$_3$ doped with Alkali Ions," J. Mater. Res., 5:742 (1990).

30. J. Akimitsu, H. Sawa, T. Kobayashi, H. Fujiki and Y. Yamada, "Successive Magnetic Phase Transitions in Nd$_2$CuO$_4$," J. Phys. Soc. Japan, 58:1646 (1989).

31. D.G. Hinks, B. Dabrowski, J.D. Jourgensen, A.W. Mitchell, D.R. Richards, S. Pei and D. Shi, "Synthesis, Structure and Superconductivity in the Ba$_{1-x}$K$_x$BiO$_{3-y}$," Nature, 333:836 (1988).

32. J.R. Hardy and J.W. Flocken, "Possible Vibronic Origins of High T_c Superconductivity: Non-Cuprate High T_c's?", Ferroelectrics, 105:3 (1990).

33. R.J. Cava, B. Batlogg, R.B. van Dover, J.J. Krajewski, J.V. Waszczak, R.M. Fleming, W.F. Peck Jr., L.W. Rupp Jr., P. Marsh, A.C.W.P. James and L.F. Schneemeyer, "Superconductivity at 60 K in La$_{2-x}$Sr$_x$CaCu$_2$O$_6$: The Simplest Double-Layer Cuprate," Nature, 345:602 (1990).

DYNAMIC JAHN-TELLER THEORY OF HIGH-T_c SUPERCONDUCTIVITY

K.H. Johnson

Department of Materials Science and Engineering
Massachusetts Institute of Technology
Cambridge, MA 02139

D.P. Clougherty

Department of Physics
University of California
Santa Barbara, CA 93106

M.E. McHenry

Department of Materials Science
Carnegie-Mellon University
Pittsburgh, PA 15213

INTRODUCTION

Several experiments reported at this Workshop offer compelling evidence for the presence of *anharmonic* oxygen vibrations ("nonlinear phonons") and *double-well* potentials in high-T_c superconductors. Also reported were extensive experimental and theoretical evidence for marginal *Fermi-liquid* character of the *hybrid* O(2p)-Cu(3d) hole/electron states responsible for both the normal-state metallic and superconducting properties of high-T_c oxides.

The most direct route from Fermi-liquid character to lattice anharmonicity *via* double-well potentials is the cooperative *dynamic Jahn-Teller effect*.[1-4] Here the real-space orbital topology and degeneracy of the Fermi liquid generate double-well potentials between which the oxygen ions rapidly vibrate (or tunnel), dynamically breaking and restoring the orbital degeneracy.[4] The *dynamic* Jahn-Teller effect is closely related to, but should be distinguished from, the *static* Jahn-Teller effect, where there is a permanent symmetry-breaking lattice distortion or instability to one of the double-well minima, removing all or part of the orbital degeneracy at the Fermi energy (E_F). Dynamic and static *pseudo* Jahn-Teller effects,[1] due to "mixing" of *nearly-degenerate* orbitals at E_F, are virtually indistinguishable from the normal Jahn-Teller effect.

Here we review a dynamic Jahn-Teller mechanism for Cooper pairing, first proposed in 1983[2] and applied to high-T_c oxides since 1987,[3] originally *predicting* oxygen double wells, anharmonicity, and their electronic origin in the degenerate Fermi-liquid orbital topology.[4]

FERMI-LIQUID ORBITAL TOPOLOGY IN HIGH-T_c CUPRATES

Common to all the high-T_c cuprates are the approximately square-planar CuO_4 molecular complexes shown in Fig. 1(b), forming the CuO_2 planes. In $YBa_2Cu_3O_7$ CuO_4 complexes are found in both the "puckered" CuO_2 planes (Fig. 3(d)) and chains (Fig. 3(b)). Key to understanding both the normal metallic and superconducting states of cuprates is the fact that the relative ordering and orbital characters of the $dp\pi^*$ and $dp\sigma^*$ levels in a CuO_4 complex (Fig. 1(b)) are significantly different from those of the ligand-field levels for a typical ionic transition-metal complex ML_4 (e.g. FeO_4) (Fig. 1(a)). This is due to the strong $Cu(d\pi^*)$-$O(p\pi)$ *antibonding hybridization* which promotes substantial $O(p\pi)$-$O(p\pi)$ *bond overlaps* ψ_+ and ψ_- within and parallel to the CuO_4 plane, as revealed in the respective orbital contour maps of Fig. 2. Single occupancy of the *localized* b_{1g} $dp\sigma^*$ spin-orbital (Fig. 1(b)) is responsible for the antiferromagnetic nonmetallic state of *undoped* cuprates. *Doping* empties the b_{1g} orbital, producing *delocalized* holes (or electrons) in a narrow band corresponding to the *degenerate* e_g $pd\pi^*$ and e_u $p\pi$ orbitals, appearing as new states above the mainly oxygen-like valence band.[5,6] Although there has been much controversy about σ- vs. π-symmetry of hole/electron states responsible for high-T_c superconductivity, these *hybrid* $Cu(d\pi^*)$-$O(p\pi)/O(p\pi)$-$O(p\pi)$ orbitals are an ideal basis for, and indeed are symmetry compatible with, the x,y- (a,b-) polarization of the Δ_5-symmetry hybrid Fermi-liquid states of the CuO_2 planes observed in photoemission measurements for $Bi_2(Ca,Sr,La)_3Cu_2O_8$ and $Nd_{2-x}Ce_xCuO_4$.[7,8] Along with $O2(p\pi)$-$O3(p\pi)$ bond overlaps ψ_+ and ψ_- *parallel* to the "puckered" CuO_2 planes in $YBa_2Cu_3O_7$ (Fig. 3(d)), there is also significant bond overlap between the *apical* $O4(p_y\pi)$ and *chain* $O1(p_z\pi)$ orbitals promoted by $Cu1(d_{yz}\pi^*)$ antibonding hybridization (Fig. 3(b)), in agreement with the band structure[9-11] and photoemission[12] around E_F.

ANHARMONICITY AND THE DYNAMIC JAHN-TELLER EFFECT

Recent experimental evidence for *axial oxygen double wells and strong oxygen anharmonicity* in $YBa_2Cu_3O_7$[13] is also consistent with these coupled *degenerate* $O4(p\pi)$-$O1(p\pi)$ and $O2(p\pi)$-$O3(p\pi)$ *bonding* orbitals at E_F (Fig. 3(b-d)), which are ideal precursors to the *dynamic-Jahn-Teller* effect.[14] Here dynamical interconversion or tunneling between the double-well potentials (Fig. 3(a)) of these degenerate $O(p\pi)$-$O(p\pi)$ "bonds" induces rapid anharmonic oxygen vibrations δQ ("nonlinear optical phonons") of maximum frequency[2]

$$\omega_c \approx h(m/M)^\beta / 2md^2 \qquad (1)$$

and amplitude $\delta \approx (m/M)^\beta d$, where h = Planck's constant, m = electron mass, M = oxygen atomic mass, d = distance between $O(p\pi)$-$O(p\pi)$ "bonds" of opposite phase ψ_+ and ψ_- (Fig. 3(d)), and $0 < \beta \leq 1/2$ is the dynamic-Jahn-Teller *anharmonicity exponent*,[1,2] dependent on the *bond overlap population* at E_F (a function of doping) according to Fig. 4.[3] $O(p\pi)$-$O(p\pi)$ overlap at E_F decreases from a maximum of *17%* in $YBa_2Cu_3O_7$ to *5%* in $La_{1.85}Sr_{0.15}CuO_4$ to only *3%* in $Nd_{1.85}Ce_{0.15}CuO_4$ and $Ba_{0.6}K_{0.4}BiO_3$, *increasing* β (Fig. 4) while causing ω_c and δ to *decrease* toward the *harmonic limits*

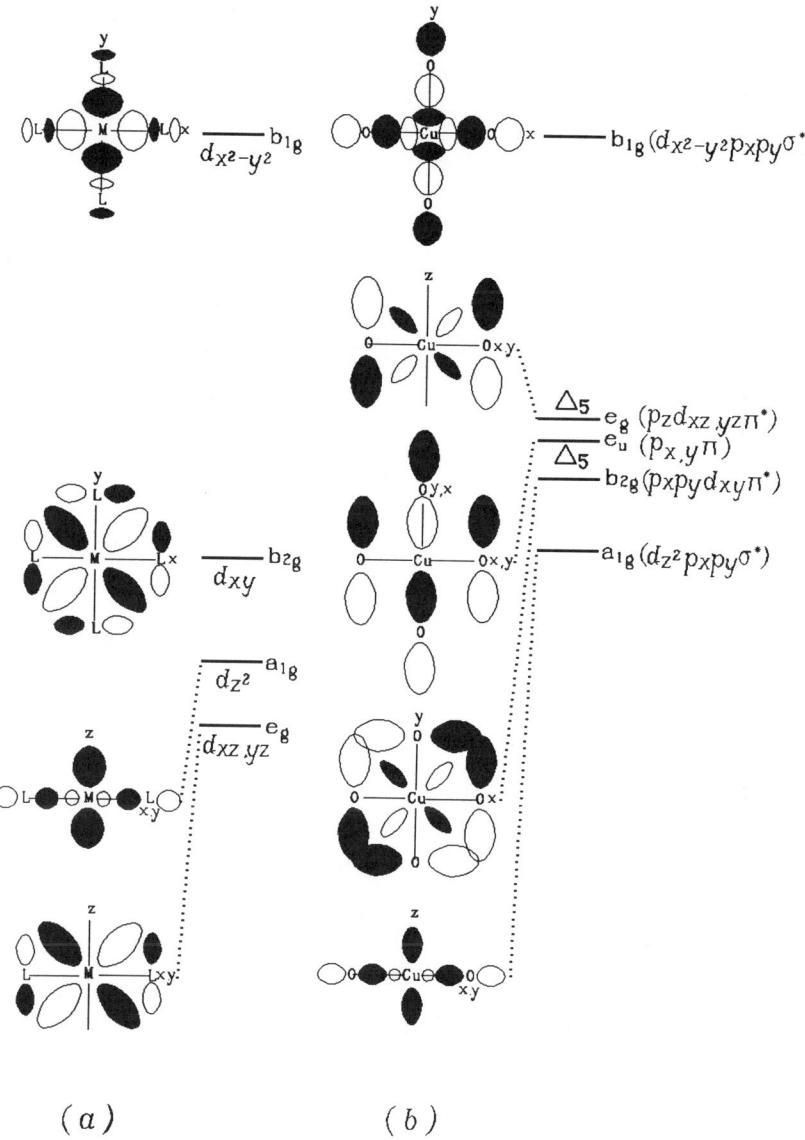

Fig. 1. (a) Ligand-field orbitals of a typical ionic square-planar transition-metal complex ML4; (b) molecular orbitals of a square-planar CuO4 complex, showing the combined effects of Cu(3d)-O(2p) σ- and π^*-antibonding and O($p\pi$)-O($p\pi$) bonding.

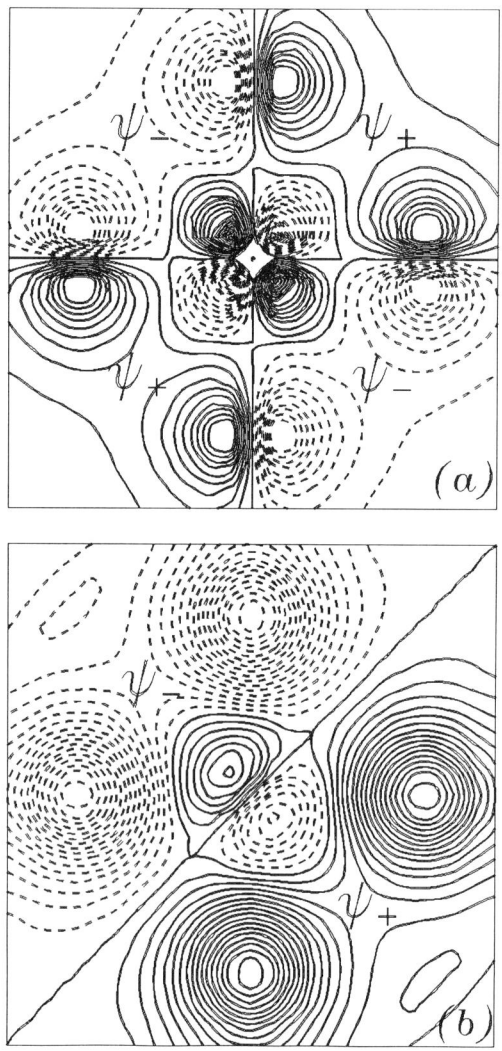

Fig. 2. Wavefunction contour maps for the (a) b_{2g} and (b) e_g orbitals of Fig. 1(b), plotted respectively in the (a) CuO_4 xy-plane and (b) the xy-plane 1Å above and parallel to the CuO_4 plane.

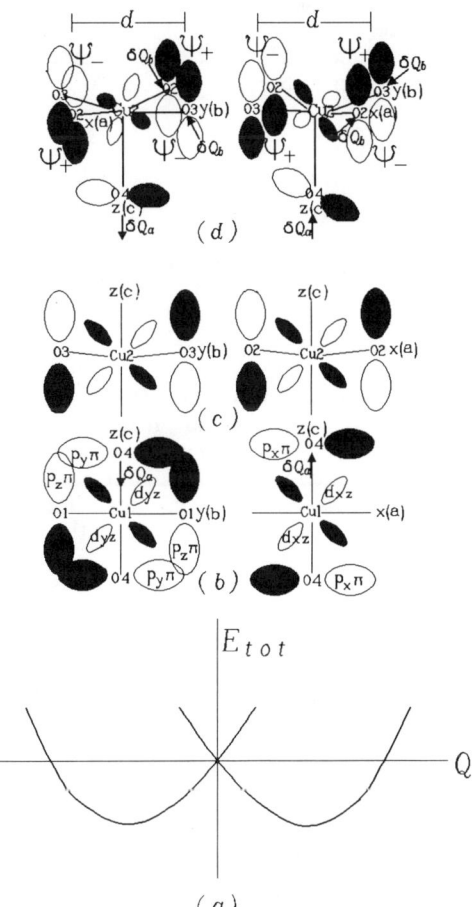

Fig. 3. (a) Double-well potentials and (b-d) corresponding doubly degenerate hybrid O($p\pi$)-Cu($d\pi^*$) Fermi-liquid orbital topologies of YBa$_2$Cu$_3$O$_7$ for (b) the chains and (c,d) the "puckered" CuO$_2$ planes, coupled via axial O4 atoms.

$$\omega_D \approx \lim_{\beta \to 1/2} \omega_c = h/2(mM)^{1/2}d^2 \qquad (2)$$

and $\delta_D \approx (m/M)^{1/2}d$. Approximation (2) for the *Debye frequency*, first suggested by Weisskopf,[14] yields remarkably accurate Debye temperatures.[3] In $Ba_{0.6}K_{0.4}BiO_3$ and $Nd_{1.85}Ce_{0.15}CuO_4$, where $O(p\pi)$-$O(p\pi)$ overlap at E_F is only *3%*, $\beta \approx 0.35$ (Fig. 4) implies $\hbar\omega_c \approx 0.07eV$ and $\delta \approx 0.08$Å. This cutoff energy $\hbar\omega_c$ for anharmonic oxygen vibrations, originally predicted on the basis of the dynamic Jahn-Teller mechanism,[3,4] is confirmed by the experimental values of high-frequency ("optical") phonons in $Ba_{0.6}K_{0.4}BiO_3$ and $Nd_{1.85}Ce_{0.15}CuO_4$ reported at this Workshop and elsewhere.[15] In $La_{1.85}Sr_{0.15}CuO_4$ and $YBa_2Cu_3O_7$, where $O(p\pi)$-$O(p\pi)$ overlaps at E_F are *5%* and *17%*, respectively, $\beta \approx 0.3$ and 0.17 (Fig. 4) lead to much larger $\hbar\omega_c \approx 0.1$-$0.5$eV and $\delta \approx 0.1$-0.5Å.

DYNAMIC JAHN-TELLER COUPLING AND HIGH-T_c COOPER PAIRING

It is widely believed and has been reaffirmed at this Workshop that high-T_c superconductivity in both the cuprates and "copper-free" $Ba_{1-x}K_xBiO_3$, involves the formation of *s-wave Cooper pairs*, as in conventional superconductors. A complete solution of the Cooper pairing problem must include both the *screened Coulomb repulsion* between the electrons and an *attractive pair potential*, the latter being mediated by virtual harmonic phonons in conventional BCS superconductors. In the standard *"two-square-well"* model[16] for Cooper pairing, these two contributions are assumed to add up to an effective pair potential that is sharply cut off in *momentum space*, resulting in long coherence lengths. In high-T_c oxides, however, where the *coherence length* is relatively short, a *real-space* representation of pairing is more appropriate. The partial delocalization of holes/electrons parallel to the CuO_2 planes along $O(p\pi)$-$O(p\pi)$ *bonding channels* of opposite phase ψ_+ and ψ_- (Fig. 3(d)) suggests a Coulomb repulsion that is largely screened out beyond the ψ_+-ψ_- distance d by the intervening positive lattice ions (Cu in cuprates and Bi in $Ba_{1-x}K_xBiO_3$). This screened Coulomb repulsion is the first term in the effective pair potential (3) below and is shown in Fig. 5 as a steep repulsive potential for $r \leq d$. The *dynamic-Jahn-Teller*-induced oxygen ion displacements $\delta \approx (m/M)^\beta d$ produce an *attractive "dipolar"* component $-e^2\delta/d^2$ of the effective pair potential that acts over a distance approximately equal to $l \approx (M/m)^\beta d$, the effective distance an electron/hole travels along the $O(p\pi)$-$O(p\pi)$ bonding channels before the displaced oxygen ions return to their original positions. The combined short-range Coulomb repulsion and dynamic Jahn-Teller-induced attraction lead to the effective *real-space "two-square-well"* pair potential

$$V(r) \approx \begin{cases} e^2/d & \text{for } 0 < r \leq d \\ -e^2/l & \text{for } d < r \leq l = (M/m)^\beta d. \end{cases} \qquad (3)$$

This expression and Fig. 5 imply that *the distance l over which the dynamic Jahn-Teller-induced attractive potential acts decreases sharply with increasing anharmonicity, i.e. with decreasing $0 < \beta \leq 1/2$*. For this potential, Schrödinger's equation

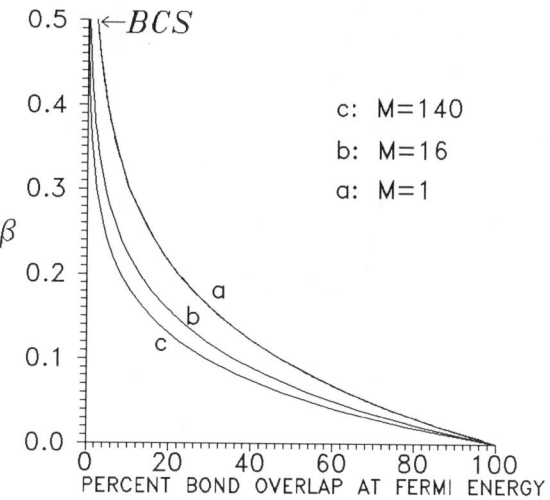

Fig. 4. Dynamic Jahn-Teller anharmonicity exponent β plotted as a function of bond overlap at the Fermi energy.

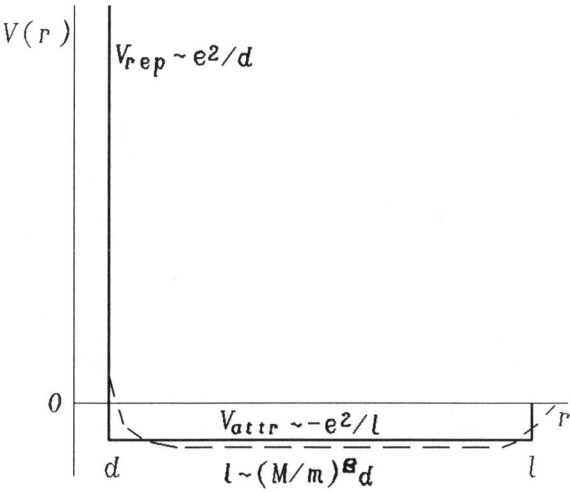

Fig. 5. Dynamic-Jahn-Teller-induced real-space Cooper-pair potential.

$$[p^2/2\mu + V(r)]\Psi(r) = E\Psi(r) \tag{4}$$

for an electron/hole pair of *reduced* mass μ, *relative* distance r, *relative* linear momentum p, *total* linear momentum $P = 0$, *total* energy E, and *opposite* spins can be solved analytically with the pair wavefunction[2,14]

$$\Psi(r) = (2/R)^{1/2} \sum_{p' > p_F} a(p') \sin(p'r/\hbar), \tag{5}$$

subject to the boundary condition $\Psi(d) = 0$ where the the pair potential (3) is strongly repulsive (Fig. 5), and normalized to a large sphere of radius $R \geq l$. This gives the pair binding energy $2E_F - E$ and leads to the following equation for the transition temperature T_c:[2]

$$1 = (2me^2d/\hbar^2)[1-(m/M)^\beta] \int_0^{\hbar\omega_c} E^{-1} \tanh(E/2k_B T_c) dE, \tag{6}$$

where k_B is Boltzmann's constant and $\hbar \equiv h/2\pi$. Since $k_B T_c << \hbar\omega_c$ for all physical values of β and d, integration of equation (6) can be carried out *exactly*, leading to the following *"weak-coupling"* type formulae for T_c and the *isotope shift* α:[2,3]

$$k_B T_c = 1.13\hbar\omega_c \exp\{-\hbar^2/2me^2d[1-(m/M)^\beta]\}$$
$$= 1.13[\hbar^2(m/M)^\beta/4\pi md^2]\exp\{-\hbar^2/2me^2d[1-(m/M)^\beta]\} \tag{7}$$

$$\alpha \equiv -\partial \ln T_c/\partial \ln M = \beta\{1-\hbar^2(m/M)^\beta/2me^2d[1-(m/M)^\beta]^2\}. \tag{8}$$

Plots of T_c and α vs. β and d from formulae (7) and (8) are shown in Fig. 6 for $M = 16$. Dependence on atomic mass M is revealed by plots of T_c vs. β and d for $M = 1$ and $M = 140$ in Fig. 7. Curves "a" in Figs. 6 and 7 ($T_c \leq 25°K$) exemplify most *ordinary and organic superconductors*, where *nearest-neighbor* $p\pi$ (or in transition metals, $d\delta$) bonding at E_F implies $d \leq 2Å$.[2,3] Because of the larger distances d between *second-neighbor* $O(p\pi)-O(p\pi)$ "bonds" (Fig. 3), high-T_c oxides fall around curve b of Fig. 6 for $\beta \cong 0.17-0.35$, or $\sim 3-17\%$ $O(p\pi)-O(p\pi)$ overlap at E_F (Fig. 4). YBa$_2$Cu$_3$O$_7$ and Bi$_2$CaSr$_2$Cu$_2$O$_8$ are near peak b of Fig. 6(a), where $T_c \cong 90°K$ and $\alpha \cong 0$ (Fig. 6(b)), while La$_{1.85}$Sr$_{0.15}$CuO$_4$ is near curve b ($d \cong 2.7Å$) of Fig. 6 for $\beta \cong 0.28$ for $T_c \cong 40°K$ and $\alpha \cong 0.22$. In "electron-doped" Nd$_{1.85}$Ce$_{0.15}$CuO$_4$, $O(M=16)$, Nd($M=142$), and Ce($M=140$) masses are *coupled via* significant Nd,Ce($4f$)-O($2p$) hybridization, yielding an average of Figs. 6 and 7(b) for $d \cong 2.7Å$ and $\beta \cong 0.31$, a smaller $T_c \cong 24°K$, and very small oxygen isotope effect $\alpha \cong 0.08$. Ba$_{0.6}$K$_{0.4}$BiO$_3$ is on curve b of Fig. 6 for $\beta \cong 0.34$, or $\sim 3\%$ $O(p\pi)-O(p\pi)$ overlap at E_F (Fig. 4), giving $T_c \cong 30°K$ and $\alpha \cong 0.3$. $(T_c)_{max} \cong 230°K$ for $d \cong 8Å$ and $\beta \cong 0.1$, or $\sim 33\%$ $O(p\pi)-O(p\pi)$ overlap at E_F (Fig. 4), far enough into the *extreme anharmonic region* $\beta \leq 1/4$ for *static Jahn-Teller lattice instabilities*[1] to interfere with Cooper pairing. Note in Figs. 6 and 7 that $(T_c)_{max} \cong 230°K$ is *independent* of the atomic mass M.

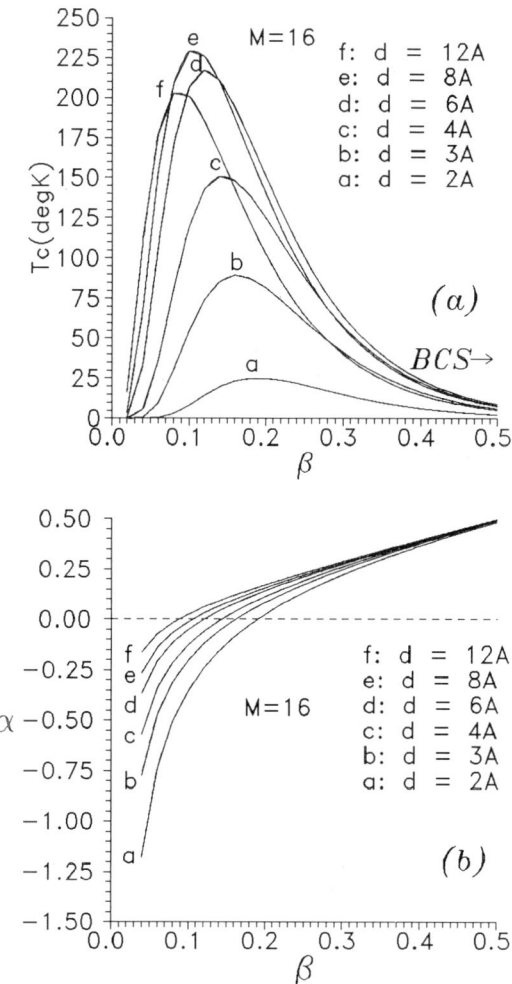

Fig. 6. (a) T_c as a function of anharmonicity β and bond distance d, obtained from formula (7) for $M = 16$; (b) isotope effect α as a function of anharmonicity β and bond distance d, obtained from formula (8) for $M = 16$.

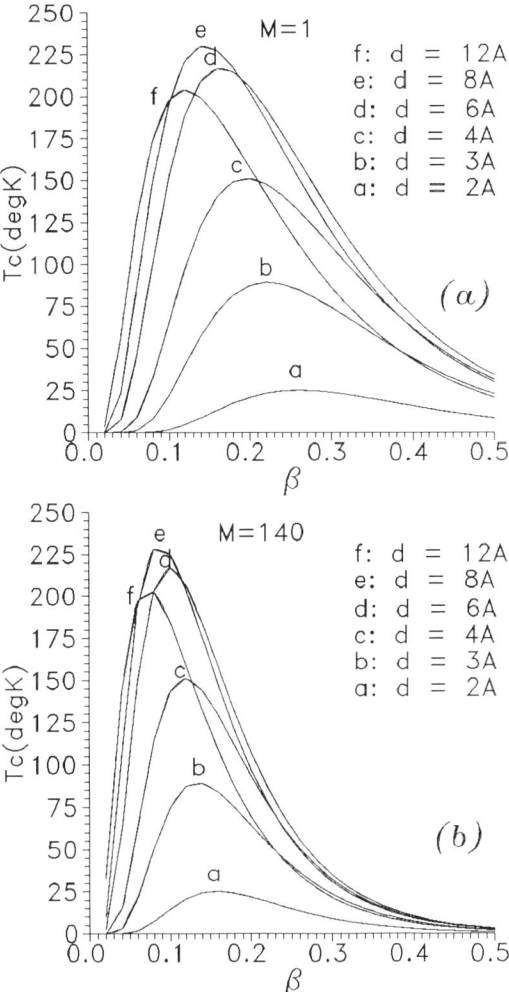

Fig. 7. T_c as a function of anharmonicity β and bond distance d, obtained from formula (7) for (a) $M = 1$ and (b) $M = 140$.

THE HARMONIC (BCS) LIMIT OF DYNAMIC JAHN-TELLER COUPLING

For bond overlaps at E_F less than 1%, $\beta \cong 1/2$ according to Fig. 4, and formulae (7) and (8) reduce respectively to

$$\lim_{\beta \to 1/2} k_B T_c = 1.13 \hbar \omega_D \exp(-h^2/2me^2 d) \tag{9}$$

$$= 1.13[h^2/4\pi(mM)^{1/2}d^2]\exp(-h^2/2me^2d)$$

$$\lim_{\beta \to 1/2} \alpha = \tfrac{1}{2}, \tag{10}$$

where expression (2) has been used. Formula (9) is similar to one originally proposed by Weisskopf[14] in his reformulation of BCS theory. In this *harmonic* $\beta = 1/2$ limit, the distance l for the attractive potential (Fig. 5) greatly increases and dynamic Jahn-Teller coupling reduces to conventional virtual phonon-electron coupling.[2]

As $O(p\pi)$-$O(p\pi)$ bond overlap at E_F increases to approximately 10% through the dynamic Jahn-Teller *anharmonic* coupling region $1/4 < \beta < 1/2$ (Fig. 4), pairing still appears to have some "phonon" contribution but, according to formulae (1), (7), and (8), with significantly increasing frequency ω_c and decreasing isotope effect α shown in Fig. 6(b). For $\beta \cong 1/4$, the short range of the dynamic-Jahn-Teller-induced attractive component of the pair potential in Fig. 5 is beyond the scope of BCS theory, which cuts off the potential sharply in momentum space.[16] For bond overlap at E_F greater than 10%, $\beta < 1/4$ (Fig. 4), and dynamic Jahn-Teller coupling *mimics* an "electronic" pairing mechanism with small-to-vanishing isotope effect α, while ω_c is in the "optical" range with $\hbar\omega_c \to 0.5 E_F$ as $\beta \to 0$. In this strongly *anharmonic* region of large bond overlap, *static* Jahn-Teller oxygen atomic displacements[1] to the double-well potential minima in Fig. 3 can compete with the dynamic Jahn-Teller coupling responsible for pairing, promoting *lattice instabilities*. Fig. 6(a) implies that high T_c-values in oxides are possible only for relatively large d (and therefore large lattice constant) and *strong anharmonicity* or $\beta < 1/2$, i.e significant $O(p\pi)$-$O(p\pi)$ overlap at E_F (Fig. 4).

CONCLUSIONS

Cooper pairing and trends of T_c and α in all the known high-T_c oxides can be explained by dynamic-Jahn-Teller-induced rapid anharmonic oxygen vibrations between double-well potentials associated with the orbital degeneracy of the Fermi liquid. This mechanism also provides a quantum-chemical basis for understanding organic and conventional superconductors,[2] where dynamic Jahn-Teller coupling reduces to virtual harmonic phonon coupling (BCS theory) as $\beta \to 1/2$ and $\hbar\omega_c \to \hbar\omega_D$ for bond overlaps at E_F less than 1% (Fig. 4). Negative α's for small β (Fig. 6(b)), i.e. for large bond overlap at E_F (Fig. 4), explain the *inverse* isotope shifts and anharmonicity observed in unusual superconductors such as Pd(H,D)$_x$.[2,17] Finally, the upper limit $(T_c)_{max} \approx 230°K$ $(-45°F)$ in Fig. 6 is *not* restricted to oxides, as is evident in the T_c-vs.-(β,d) plots for $M = 1$ and $M = 140$ in Fig. 7. Thus "room-temperature" superconductivity is ruled out in *any* material, if the dynamic Jahn-Teller theory is truly universal, as we first pointed out in 1987.[3]

ACKNOWLEDGEMENTS

We are grateful to the AMP Corporation for sponsoring this research, and to Dr. George Cvijanovich for helpul discussions and encouragement.

REFERENCES

1. I. B. Bersuker and V. Z. Polinger, "Vibronic Interactions in Molecules and Crystals," Springer-Verlag, New York (1990).
2. K. H. Johnson and R. P. Messmer, Synth. Metals 5:193 (1983).
3. K. H. Johnson, M. E. McHenry, C. Counterman, A. Collins, M. M. Donovan, R. C. O'Handley, and G. Kalonji, in "Novel Superconductivity," S. A. Wolf and V. Z. Kresin, eds., Plenum, New York (1987), p. 563; Physica C 153-155:1165 (1988); K. H. Johnson, D. P. Clougherty, and M. E. McHenry, Mod. Phys. Lett. B 3:1367 (1989); K. H. Johnson, Phys. Rev. B 42:4783 (1990).
4. D. P. Clougherty, K. H. Johnson, and M. E. McHenry, Physica C 162-164:1475 (1989).
5. T. Takahashi, H. Matsuyama, H. Katayama-Yoshida, K. Seki, K. Kamiya, and H. Inokuchi, Physica C 170:416 (1990).
6. H. Romberg, M. Alexander, N. Nücker, P. Adelmann, and J. Fink, Phys. Rev. B 42:8768 (1990).
7. B. O. Wells, P. A. P. Lindberg, Z. -X. Shen, D. S. Dessau, W. E. Spicer, I. Lindau, D. B. Mitzi, and A. Kapitulnik, Phys. Rev. B 40:5259 (1989).
8. Y. Sakisaka, T. Maruyama, Y. Morikawa, H. Kato, K. Edamoto, M. Okusawa, Y. Aiura, H. Yanashima, T. Terashima, Y. Bando, K. Iijima, K. Yamamoto, and K. Hirata, Phys. Rev. B 42:4189 (1990).
9. J. Yu, A. J. Freeman, and S. Massida, in "Novel Superconductivity," S. A. Wolf and V. Z. Kresin, eds., Plenum, New York (1987), p. 367.
10. W. E. Pickett, Rev. Mod. Phys. 61:433 (1989).
11. M. Hirao, T. Uda, and Y. Murayama, in "Proceedings of the CIMTEC Symposium on High-Tc Superconductors, Trieste, Italy," Elsevier, Amsterdam (1990).
12. A. J. Arko, R. S. List, J. Bartlett, S. -W. Cheong, Z. Fisk, J. D. Thompson, C. G. Olson, A. -B. Yang, R. Liu, C. Gu, B, W, Veal, J. Z. Liu, A. P. Paulikas, K. Vandervoort, H. Claus, J. C. Campuzano, J. E. Schirber, and N. D. Shinn, Phys. Rev. B 40:2268 (1989).
13. S. D. Condradson, I. D. Raistrick, and A. R. Bishop, Science 248:1394 (1990); J. Mustre de Leon, S. D. Conradson, I Batistic, and A. R. Bishop, Phys. Rev. Lett. 65:1675 (1990); S. D. Conradson, J. D. Mustre de Leon, I. Batistic, and A. R. Bishop, in this Workshop Proceedings.
14. V. F. Weisskopf, Contemp. Phys. 22:375 (1981).
15. Q. Huang, J. F. Zasadzinski, N. Traishawaia, K. E. Gray, D. G. Hinks, J. L. Peng, and R. L. Greene, Nature 347:369 (1990); D. G. Hinks, B. Dabrowski, J. D. Jorgensen, and J. Zasadzinski, in this Workshop Proceedongs.
16. L. L. Daemen and A. W. Overhauser, Phys. Rev. B 38:81 (1988).
17. K. H. Johnson and D. P. Clougherty, Mod. Phys. Lett. B 3:795 (1989).

μSR MEASUREMENTS OF PENETRATION DEPTH IN EXOTIC SUPERCONDUCTORS

Y.J. Uemura

Department of Physics, Columbia University, New York City
New York 10027, USA

The magnetic field penetration depth λ has been measured in high-T_c cuprate, bismuthate (BKBO), organic, Chevrel phase and heavy-fermion superconductors, using muon spin relaxation (μSR) and/or some other bulk methods. Fermi energies ϵ_F of these systems are estimated from λ, based on the relation $1/\lambda^2 \propto n_s/m^*$ (superconducting carrier density / effective mass). A plot of transition temepratue T_c versus $T_F = \epsilon_F/k_B$ demonstrates that all of these "exotic" systems possibly belong to a unique group of supercopnductors characterized by high transition temperature relative to n_s/m^* and T_F. This feature clearly distinguishes these exotic superconductors from ordinary BCS superconductors.

I. Introduction

The magnetic field penetration depth λ is one of the most fundamental parameters of superconductivity. Together with many collaborators, the author has performed[1-7] measurements of λ in high-T_c cuprate, bismuthate (BKBO), organic, Chevrel phase and heavy-fermion (HF) superconductors using the technique of muon spin relaxation (μSR). The penetration depth λ is related to the superconducting carrier density n_s and the effective mass m^* as $1/\lambda^2 \propto n_s/m^*$. In an approximation for non-interacting electron gas, the Fermi energy ϵ_F is directly proportional to n_s/m^* in 2-dimensional systems, while $\epsilon_F \propto n_s^{2/3}/m^*$ in 3-d systems can also be calculated using n_s/m^*.

In this paper, we present a plot of T_c versus Fermi temperature $T_F = \epsilon_F/k_B$, including the above mentioned exotic superconductos as well as ordinary BCS superconductors. The plot demonstrates theat these exotic systems have much higher T_c/T_F ratios than ordinary BCS systems. We provide thermodynamic arguments to duscuss implications of this plot for the condensation mechanisms of these superconductors.

II. μSR method

Positive muon spin relaxation (μ^+SR) measurements have been extensively performed in the study of high-T_c cuprate and other superconductors[1,2,8]. When an external field H_{ext} ($H_{c1} \ll H_{ext} \ll H_{c2}$) is applied to a type-II superconductor, it penetrates the specimen by forming a lattice of flux vortices[9]. The local magnetic field B within the specimen then has a distribution with a width ΔB. The field B is close to H_{ext} near the vortex core, while somewhat smaller at the center of the adjacent vortices. The width ΔB is determined by λ. When λ is sufficiently longer than the distnance between the adjacent vortices, ΔB is proportional to $1/\lambda^2$ and nearly indenpendent of H_{ext} (ref. 10).

In μ^+SR experiments[11], a beam of spin polarized positive muons is stopped in the specimen: the μ^+ is thermalized within $10^{-10}s$ after implantation (without losing its spin polarization), and occupies an interstitial site in the crystal until it emits a positron after a time t (mean life time $\tau_\mu = 2.2\mu s$). These positrons are emitted preferentially along the muon spin direction. When H_{ext} is applied perpendicular to the initial muon spin direction, the precession of the muon spin around H_{ext} produces a sinusoidal modulation of the exponential life time histogram:

$$N(t) = N_o exp(-t/\tau_\mu)[1 + AG_x(t)cos(\omega_\mu t + \phi)],$$

where $A \sim 0.2$ is the initial asymmetry, $\omega_\mu \sim \gamma_\mu H_{ext}$ is the precession frequency ($\gamma_\mu = 2\pi \times 13.5 \times 10^3 rad/G$ is the gyromagnetic ratio of μ^+), and the relaxation function $G_x(t)$ describes the depolarization of muon spins.

The local field distribution in the vortex state causes the depolarization. Usually, $G_x(t)$ is approximated by a Gaussian function $exp(-\sigma^2 t^2/2)$ which defines the muon spin relaxation rate σ. Since σ is proportional to ΔB, one can directly determine the penetration depth from the relation $\sigma \propto \Delta B \propto 1/\lambda^2$.

In general, $1/\lambda^2$ is given as a function of n_s and m^* as

$$\frac{1}{\lambda^2} = \frac{4\pi n_s e^2}{mc^2} \times \frac{1}{1 + \xi/l}$$

with the correction factor related to the coherence length ξ and the mean free path l. For superconductors close to the clean limit ($\xi \ll l$), the correction factor essentially becomes unity. Fortunately, this is the case[12] for high-T_c cuprate, organic (BEDT), and Chevrel-phase systems, as well as the HF superconductor UPt$_3$. Thus, we can use the muon spin relaxation rate σ as a measure of n_s/m^* as $\sigma \propto 1/\lambda^2 \propto n_s/m^*$.

The temperature dependence of σ and λ reflects the reduction of n_s at finite temperatures caused by the thermal pair-breaking excitation beyond the energy gap. This feature allows the study of symmetry of the energy gap. The orientational dependence of σ reflects the anisotropy of m^* in quasi-2d systems. These aspects have been described elsewhere[2,8,13,14]. In this paper, we focus on the absolute values of $\sigma(T \to 0)$ which represent values of n_s/m^* in the ground state for each of the different systems.

III. Results on various superconductors

Cuprates

We have performed μSR measurements of $\sigma(T)$ in more than 30 different hole-doped cuprate compounds[4,5] belonging to the families of the so-called 214, 123, 2212, and 2223 superconductors. We used ceramic specimens, because they have more spatially homogeneous concentration of doped carriers compared to the single crystal specimens. As shown in Fig. 1, we found[4] remarkable correlations between T_c and $\sigma(T \to 0) \propto n_s/m^*$ in these systems. Upon doping hole carriers into compounds with the same crystal structure (such as the 214 series), T_c initially increases with increasing n_s/m^*, then shows a saturation followed by a suppression in the so-called over-doped region. The initial increase of T_c follows a straight line $T_c \propto n_s/m^*$ which is common to all the different systems. The results on oxygen depleted YBa$_2$Cu$_3$O$_y$ systems[4] agree well with those on the Pr doped (Y$_{1-x}$Pr$_x$)Ba$_2$Cu$_3$O$_7$ (ref. 5), indicating that T_c is determined by n_s/m^* regardless of the chemical method used to control the hole concentration on the CuO$_2$ planes.

The relaxation rate σ in ceramic specimens reflects an angular average $<\lambda>$ of λ. In highly 2-dimensional (2-d) systems, such as the cuprates, σ is determined predominantly by the in-plane penetration depth λ_{in} defined for $H_{ext} \perp$ c-axis: it is known[15,14,3] that $<\lambda> = 1.2 \sim 1.3\lambda_{in}$. The relaxation rate σ is determined by the 3-d density n_s of the carriers which produce the screening current. One can, however, convert the 3-d density to the 2-d areal density n_{s2d} by using the average interplane distance c_{int}. Thus, we can deduce n_{s2d} in a series of conversion $[\sigma, \lambda] \to [\sigma_{in}, \lambda_{in}] \to [n_s/m^*$ for $H_{ext} \perp$ conducting plane$] \to [n_{s2d}/m^*]$. Since all the cuprate systems shown in Fig. 1 have $c_{int} = 6 \pm 1$ Å, the horizontal axis of this figure is essentially proportional to n_{s2d}/m^*. Thus, Fig. 1 can be enjoyed as a plot of T_c versus $\epsilon_F \propto n_{s2d}/m^*$ in the case of the cuprate systems.

Organics

The 2-d organic superconductor (BEDT-TTF)$_2$Cu(NCS)$_2$ has been studied by Uemura et al.[1,6,16] and by Harshman et al.[17], in the μSR measurements using single crystal specimens with H_{ext} applied perpendicular to the conducting b-c planes. Except for some difference in the results of the temeprature dependence $\sigma_{in}(T)$, the absolute values $\sigma_{in}(T \to 0) = 0.14 \sim 0.16 \mu s^{-1}$ agree well in the two independent measurements. As shown in Fig. 1, $\sigma_{in}(T \to 0)$ of this system gives a point lying close to the straight line obtained for the cuparates.

Bismuthates

We have also performed μSR studies on the 3-d perovskite superconductor (Ba$_{1-x}$K$_x$)BiO$_3$ (ref. 7). The points from two specimens of this series also lie rather close to the straight line from the cuprates. This result suggests that, with respect to the relation between T_c and n_s/m^*, the dimensionality of these oxide systems does not play major role. This motivates us to compare the results of these oxide high-T_c superconductors to some other 3-d superconductors.

Chevrel phase compounds

μSR studies on Chevrel phase superconductors have been carried out by Birrer et al.[18] at PSI and by our group[1,7] at TRIUMF. The plot of σ versus T_c of several non-magnetic Chevrel systems exhibits features remarkably similar to those observed in the cuprates. Namely, T_c initially increases following the same straight line as found for cuprates, then shows saturation and suppression with increasing $\sigma \propto n_s/m^*$.

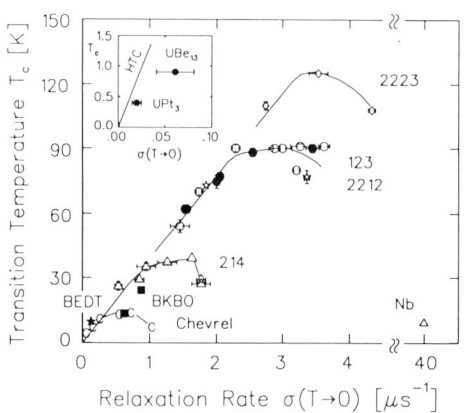

Fig. 1. Plot of T_c versus $\sigma(T \to 0) \propto n_s/m^*$ of cuprates (ref. 4, 5), BKBO (closed square, ref. 7), Chevrel phase systems (C: ref. 7, 18), BEDT (closed star, ref. 7, 1), based on the μSR measurements, and Nb based on the bulk measurement of λ (ref. 23). The inset figure shows the same plot for the HF systems with the values of σ estimated from the bulk penetration depth measurements of Gross et al. (ref. 21). The straight line in the inset corresponds to the linear relation found for the cuprates.

Heavy − fermion systems

μSR measurements of λ in UPt$_3$ have been carried out by Broholm et al.[19] and by our group[20,1]. The relaxation rate σ measured in low external fields $H_{ext} \sim 150 - 200G$ shows some increase with decreasing temperature below T_c in both studies. The results in high external field $H_{ext} \sim 4kG$ by Luke et al.[20,1], however, show essentially no change of $\sigma(T)$ below T_c. In general, measurements in higher field are less sensitive to the field broadening due to extrinsic effects (e.g., demagnetization field, closeness to H_{c1}, etc.), and thus more reliable for estimating λ. Here we refer to the high field μSR results which set the limit $\lambda \geq 10,000$Å. The high field results[20,1] in UBe$_{13}$ also show no change, setting a similar limit for λ.

Due to background relaxation caused by nuclear dipolar fields, it is difficult to determine $\sigma \leq 0.05\mu s^{-1}$ in μSR measurements. The small σ and large λ in these HF systems are direct consequences of heavy effective mass m^*. For more precise estimates of λ, here we refer to the bulk flux confinement study of Gross et al.[21] who reported $\lambda = 11,000$Å in UBe$_{13}$ and 19,000Å in UPt$_3$. Converting these values to σ expected in hypothetical μSR studies free from background, we obtain the inset figure of Fig. 1. This figure demonstrates that even the points from these HF superconductors lie rather close to the straight line found for the cuprates.

Niobium

It is interesting to see the case for more usual type-II superconductors. μSR studies in Nb was reported by Herlach et al.[22]. The results are given only for the temeprature-field region in the superconducting state colse to the boundary with the normal state. This is presumably due to large relaxation rate σ. More accurate estimate of λ can be obtained from the bulk masurements of Maxfield and McLean[23] who obtained $\lambda(T \to 0) = 470$Å. Converting this value into σ, we obtain a point in Fig. 1 for Nb which lies very far away from the points of the other systems. This reflects the large carrier density and yet modest T_c of Nb.

IV. Plot of T_c versus T_F

Using the values of $\sigma \propto 1/\lambda^2 \propto n_s/m^*$, we can estimate the Fermi energy ϵ_F. This is straight-forward in 2-d systems, since $\epsilon_F = (\hbar^2\pi)n_{s2d}/m^*$. In 3-d systems, we need additional independent information on n_s or m^* or their combination. We use the Sommerfeld constant $\gamma \propto n_s^{1/3}m^*$ to calculate $\epsilon_F = (\hbar^2/2)(3\pi^2)^{2/3}n_s^{2/3}/m^* \propto \sigma^{3/4}\gamma^{-1/4}$. Due to the low power (1/4) for γ, ϵ_F is determined predominantly by σ in 3-d systems as well. In this way, we can estimate ϵ_F within the approximation of a non-interacting electron gas[7]. The estimated values of ϵ_F would, however, serve as a good yardstick for representative energy scales of each system.

Figure 2 shows a plot of T_c versus $T_F = \epsilon_F/k_B$ thus obtained using the values in Fig. 1. We also included points for Al, Sn and Zn using known values of ϵ_F and T_c. We find that the transition temperatures of the cuprates, BKBO, BEDT, Chevrel phase, and HF superconductors are ranging between 1/10 and 1/100 of their Fermi tempratures T_F. For convenience, we shall call these systems as "exotic" superconductors henceforth. This is in a clear constrast to the case of ordinary BCS superconductors where $T_c/T_F = 1/1000 \sim 1/10,000$. The points from these exotic superconductors roughly follow a universal linear trend $T_c \propto T_F$, extended from the linear correlation found in the cuprate systems.

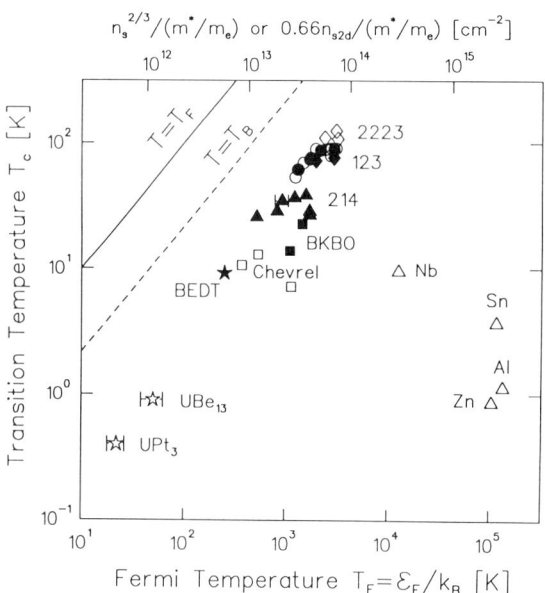

Fig. 2. A log-log plot of T_c versus the Fermi temprature $T_F = \epsilon_F/k_B$ estimated from the results of σ in Fig. 1 (combined with the interplanar distance c_{int} for 2-d and the Sommerfeld constant γ for 3-d systems). The values of T_F for Sn, Al, and Zn are based on knowledge from other estimates. The broken line represents the B-E condensation temperature T_B of the ideal boson gas with corresponding $n_s^{2/3}/m^*$, the boson mass $2m^*$ and density $n_s/2$.

Such a relation between T_c and T_F is not expected in the weak-coupling BCS theory where the pairing interaction with energy scale $\hbar\omega_B$ is retarded (i.e. $\epsilon_F \gg \hbar\omega_B$). Since these exotic systems have relatively small Fermi temperatures, we can consider the case with $\epsilon_F \leq \hbar\omega_B$. For such a non-retarded interaction, the linear relation $T_c \propto T_F$ can be expected, as discussed by Emery and Reiter[24]. This argument provides a possible explanation for the linear trend.

In the 2-d cuprate and organic systems, the effective mass m^* has been determined in quantum oscillation studies to be about 2-4 times the bare electron mass m_e. Then, one can calculate the areal carrier density n_{s2d} from σ, and estimate the number of superconducting pairs within the area of ξ^2 on the conducting plane. In the 2-d $YBa_2Cu_3O_7$ and the BEDT systems, we find only several pairs overlapping with each other, as discussed in refs. 1 and 6. This is in clear contrast to ordinary BCS superconductors which have more than 1000 Cooper pairs overlapping within the "coherence area" ξ^2. This "pair counting" argument provides another way to present the large ratio T_c/T_F in the exotic 2-d systems.

The other extremum to be compared with these systems is the case for the ideal Bose gas where each pair of fermions exist without overlapping with other pairs (i.e., one pair per ξ^2). Thus, we find that the situation for the exotic superconductors interpolates the cases for BCS Cooper pairs and ideal independent local bosons. To further illustrate this situation, we calculate the Bose-Einstein condensation temperature T_B for ideal Bose gas. Since the B-E condensation occurs when the thermal wave length of each boson becomes comparable to the interboson separation, T_B is given as a function of the 3-d boson density n_B and the boson mass m_B as $T_B \propto n_B^{2/3}/m_B$. Assuming $n_B = 0.5 n_s$ and $m_B = 2m^*$, we calculated T_B for given $n_s^{2/3}/m^*$, as plotted by the broken line in Fig. 2.

We note that T_c of the exotic systems is $1/5 \sim 1/50$ of T_B. This again suggests closeness of these systems to local boson systems. This feature encourages formation of theories which interpolate between BCS and BE condensations. The boson-fermion theory of Lee and Friedberg[25] represents an effort along such direction. In the case of B-E condensation, the transition temprature T_B is not related to $\hbar\omega_B$ as long as $T_B \ll \hbar\omega_B$. This feature might be related to the universal linear trend of T_c versus T_B or T_F which holds for various different systems with presumably a wide variety of pairing interaction energies $\hbar\omega_B$.

Phenomenologically, the exotic superconductors share some important features, i.e., high H_{c2} and short ξ, highly correlated electronic structures, intrinsic type-II features near the clean limit situation, etc. The universal linear trend in Fig. 2, together with these features, suggests that some or even all these exotic superconductors may share a new condensation mechanism, different from ordinary BCS condensation. Theoretically, this results may be related to a non-retarded interaction among fermions[24] and/or local pairing[26] and/or some tendency towards B-E condensation[25]. These three concepts are somewhat different, yet sharing a common tendency towards a greater role of n_s and m^* in determining T_c compared to that of $\hbar\omega_B$.

Figure 2 represents the first attempt (to our knowledge) to plot T_c of various superconductors versus Fermi energy/temperature. Traditionally, it has been common to produce plots of T_c versus the Sommerfeld constant γ which is proportional to the density of states at the Fermi energy (see, for example, ref. 27). Correlations between T_c and γ can be expected when the system is described by the BCS theory. In contrast, the plot of T_c versus T_F helps elucidate fundamental features of superconductors which may be different from ordinary BCS systems.

Acknowledgement

The author would like to acknowledges illuminating discussions with V.J. Emery and T.D. Lee, and collaboration on μSR experiments with J.H. Brewer, R.F. Kiefl, G.M. Luke, A.W. Sleight, M.B. Maple, G. Saito, S. Uchida, D. Hinks, W.N. Hardy, and many other scientists. This work is financially supported by the Packard Foundation (YJU-Columbia) and NSF DMR-89-13784 (at Columbia).

References

1. For most recent review of results from our Columbia-TRIUMF collaboration, see Y.J. Uemura, Proc. (invited papers) Int. Conf. of Low Temperature Physics LT-19, Brighton, England, 1990, Physica B, in press.
2. Y.J. Uemura et al., J. Phys. (Paris) 49, C8-2087 (1988).
3. Y.J. Uemura et al., Phys. Rev. B38, 909 (1988).
4. Y.J. Uemura et al., Phys. Rev. Lett. 62, 2317 (1989).
5. C.L. Seaman et al., Phys. Rev. B42, 6801 (1990).
6. Y.J. Uemura et al., Proc. of Int. Conf. on Organic Superconductors, Lake Tahoe, 1990, ed. by V. Kresin and W.A. Little, Plenum, New York (1990), pp. 23-29.
7. Y.J. Uemura et al., submitted for publication.
8. For μSR studies in high-T_c systems, see, for example, H. Keller, IBM J. Res. Develop 33, 314 (1989); Y.J. Uemura et al., Physica C162-164, 857 (1989); and references therein.
9. See, for example, A.G. Redfield, Phys. Rev. 162, 387 (1967).
10. P. Pincus et al., Phys. Lett. 13, 31 (1964); E.H. Brant, Phys. Rev. B37, 2349 (1989). These two papers give about 30% different values of the constant α for $\sigma = \alpha \times 1/\lambda^2$. For consistency with our earlier work, here we use the constant α derived by Pincus et al. after modifying it for the triangular lattice. The choice of this constant does not affect the overall argument of the present paper.
11. For general aspects of the μSR technique, see proceedings of previous international conferences in this field, Hyperfine Interactions 6 (1979); 8 (1981); 17-19 (1984); 31 (1986); 63-65 (in press).
12. For the cuprates, see K. Kamaras et al., Phys. Rev. Lett. 64, 84 (1990), and recent resistivity studies. For BEDT, see G. Saito, Physica C162-164, 577 (1989); N. Toyota et al. J. Phys. Soc. Japan 57, 2616 (1988). For Chevrel systems, see Ø. Fischer, Appl. Phys. 16, 1 (1978); M. Decroux et al., J. Low Temp. Phys. 73, 283 (1988).
13. G. Aeppli et al., Phys. Rev. B35, 7129 (1987); D.R. Harshman et al., Phys. Rev. B36, 2386 (1987); Phys. Rev. B39, 851 (1989); B. Pümpin et al., Phys. Rev. B42, 8019 (1990).
14. E.M. Fogan et al., presented at the μSR'90 Conference, Oxford, 1990, Hyperfine Interactions 63-65, in press.
15. W. Barford and J.M.F. Gunn, Physica C156, 515 (1988). A c-axis oriented specimen of YBa$_2$Cu$_3$O$_7$ (ref. 2) shows that $\sigma_{in} \propto 1/\lambda_{in}^2$ for single crystals is about 1.4 times larger than the angular averaged value.
16. Y.J. Uemura, L.P. Le, G. Saito et al., unpublished.

17. D.R. Harshman *et al.*, Phys. Rev. Lett. 64, 1293 (1990).
18. P. Birrer *et al.*, presented at the μSR'90 Conference, Oxford, 1990, Hyperfine Interactions 63-65, in press.
19. C. Broholm *et al.*, Phys. Rev. Lett. 65, 2062 (1990).
20. G.M. Luke *et al.*, submitted for publication.
21. F. Gross *et al.*, Physica C162-164, 419 (1989).
22. D. Herlach *et al.*, presented at μSR'90, Hyperfine Interactions 63-65, in press.
23. B.W. Maxfield and W.L. McLean, Phys. Rev. 139, A1515 (1965).
24. V.J. Emery and G. Reiter, Phys. Rev. B38, 4547 (1988).
25. R. Friedberg and T.D. Lee, Phys. Rev. B40, 6745 (1989).
26. R. Mincas, J. Ranninger and S. Rabaszkiewicz, Rev. Mod. Phys. 62, 113 (1990)
27. R. Tournier *et al.*, J. Magn. Magn. Matrs. 76-77, 552 (1988) studied relations between T_c and γ of some HF and cuprate systems.

ON THE ORBITAL ANGULAR MOMENTUM OF Cu 3d HOLES IN HIGH T_c SUPERCONDUCTORS:

IS A Cu $3d_{3z^2-r^2}$ BIPOLARON THE SUPERCONDUCTING PAIR ?

A. Bianconi, A.M. Flank*, P. Lagarde* C. Li, I. Pettiti, M. Pompa, and D. Udron

*INFM research unit of University of Rome, Department of Physics
P. A. Moro 2, 00185 Roma, Italy; and * LURE, bat. 209D, 91405 Orsay, France*

The orbital angular momentum m_ℓ of the Cu 3d holes in high T_c superconductors has been measured by Cu L_3 x-ray absorption spectroscopy. The symmetry of the itinerant states $3d^9\underline{L}$ induced by doping at the Fermi level is deduced to be a mixture of 3d holes with $m_\ell=\pm 2$ and $m_\ell=0$ orbital angular momentum and oxygen O $2p_{x,y}$ holes in the molecular orbital combination of local b_1 and a_1 symmetry, $\underline{L}(b_1)$ and $\underline{L}(a_1)$. The coupling of the itinerant charges with lattice dynamics and doping induced structural changes, point toward a polaron associated with the $3d_{3z^2-r^2}$ electronic states.

1. INTRODUCTION

The polarized x-ray absorption spectroscopy provides a direct method to measure the orbital angular momentum of the unoccupied electronic states in the conduction band of high T_c superconductors. Therefore it gives complementary information on the itinerant states at the Fermi level observed by angular resolved photoemission which measures the band dispersion. We have measured the polarized Cu L_3 x-ray absorption of several high T_c superconductors [1] and we have obtained the symmetry of the $3d^9\underline{L}$ states by joint analysis of the O K-edge and Cu L_3 spectra. The dynamics of the lattice and the structural changes associated with doping have been studied by the analysis of the extended x-ray absorption fine structure EXAFS and x-ray absorption near edge structure XANES at the Cu K-edge.

The changes of the Cu L_3 spectra from the insulating to the metallic phase provides direct experimental evidence for the formation of the $3d^9\underline{L}$ states. We show that from the integral of the oscillator strength of the Cu 2p->3d transitions in the **E**//**c** Cu L_3 edge x-ray absorption spectra it is possible to extract the number n_{dz} of the $3d_{3z^2-r^2}$ holes on the total

---

Invited talk at Miami Workshop on *Electronic Structure and Mechanisms for High Temperature Superconductivity* Miami January 3-9 1991

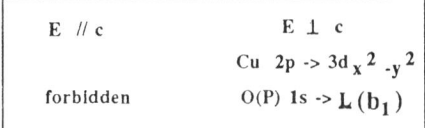
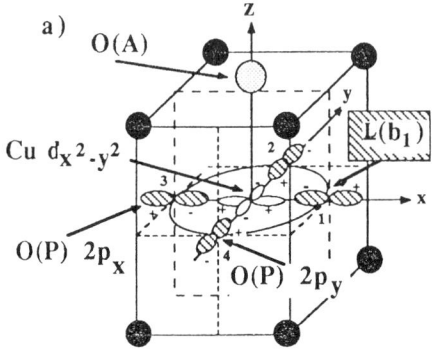
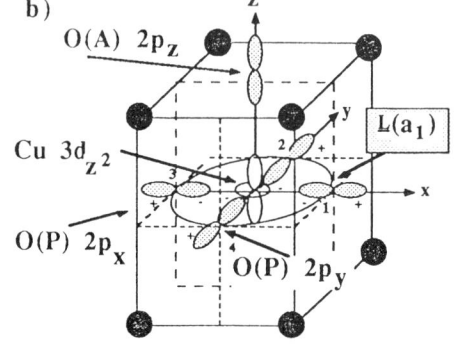

Fig.1 Pictorial view of the valence states of symmetry b_1, (panel a), a_1, (panel b) that can be reached with polarized Cu L_3 edge and O K edge absorption.

number of 3d holes, n_{dz}~15-20% in $La_{1.85}Sr_{0.15}CuO_4$ and $Bi_2Sr_2CaCu_2O_8$ which is as large as the hole doping δ, i.e. the number of holes per Cu site induced by doping, has been found. The interpretation of the polarized O K-edge spectra provides additional information on the oxygen 2p holes. These results clearly show that the Cu 3d holes in the conduction band at the Fermi level have both $3d^9_{x^2-y^2}$, and $3d^9_{3z^2-r^2}$ symmetry and the ligand holes in the oxygen O $2p_{x,y}$ orbitals have local b_1 and a_1 symmetry, $\underline{L}(b_1)$ and $\underline{L}(a_1)$, see Fig.1.

The energy splitting Δ_z between the out-of-plane and the in-plane-polarized Cu 2p->3d transitions is interpreted as a measure of the energy splitting at the Γ point between two bands with mainly $3d_{3z^2-r^2}$ ($m_\ell=0$) and $3d_{x^2-y^2}$ ($m_\ell=2$) symmetry. The energy splitting Δ_z is modulated by doping, and its average value for each family of high T_c superconductors increases with the maximum T_c of the family.

The present findings support a picture of the electronic structure of high T_c superconductors where two sets of different electronic states are close to the Fermi level giving experimental support to several theories for the pairing mechanism which consider the presence of two bands with different symmetry around the Fermi level [2-12]. However the actual scenario can be more complex than that described by theories considering a single $[CuO_2]_\infty$ layer because the states with $3d^9_{3z^2-r^2}$ character in the $[CuO_2]_\infty$ plane are expected to be mixed with out of plane electronic states like the BiO derived band in BiCaSrCuO family or with the linear chain band in YBaCuO family.

The modulation of the orbital angular momentum m_ℓ of the Cu 3d holes by charge carriers, giving a large electron-phonon coupling by partial dynamic suppression of the Jahn Teller splitting, was at the basis of the driving idea of Bednorz and Müller [13] on the road that led to the discovery of the high T_c superconductivity. The present findings that the Cu 3d orbital angular momentum is not frozen in the $m_\ell=2$ state expected for the insulating system opens the problem of the coupling of the itinerant $3d^9\underline{L}$ states with lattice dynamics.

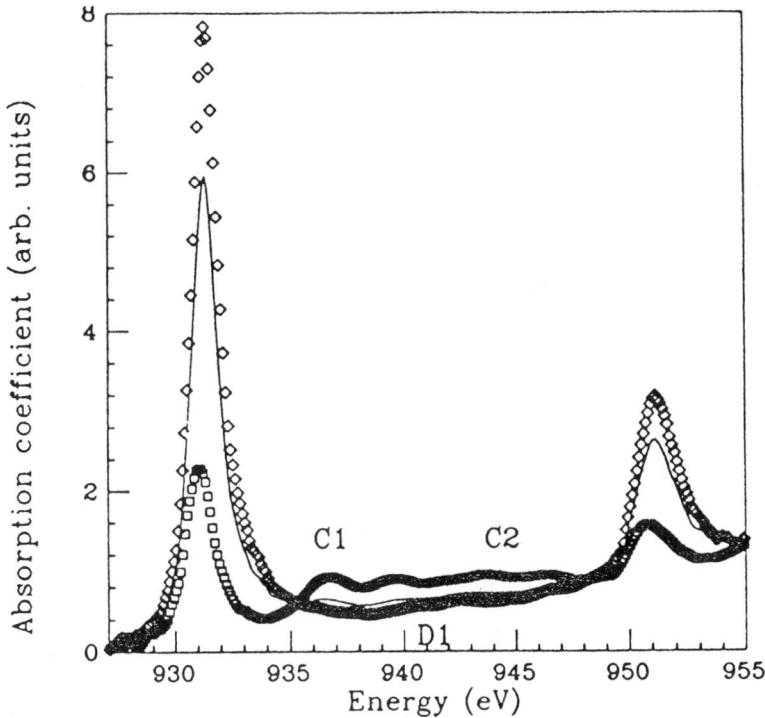

Fig.2 Polarized E⊥c (diamonds) and E∥c (squares) Cu $L_{2,3}$-edge XANES spectra of $Bi_2CaCu_2O_8$ measured in the total electron yield mode. The unpolarized spectrum (solid line) measured at the magic incidence angle (35⁰) is shown.

2. Cu L_3 AND O K EDGE ABSORPTION SPECTRA

The polarized Cu L_3 x-ray absorption spectra (see Fig.2) have been measured at the synchrotron radiation source of the Laboratoire pour l'Utilisation du Rayonnement Electromagnetique (LURE), Orsay [14-18]. The monochromator was a double beryl 1010 crystal and absorption spectra have been recorded by total electron yield method on single crystals. The linearly polarized Cu L_3-edge x-ray absorption near edge structure (XANES) of $Bi_2Sr_2CaCu_2O_8$ (Bi2212) is shown in Fig.2. The spectra have been interpreted by the full multiple scattering approach in the real space [19,20], see Fig.3. In this approach the transition rate from the Cu 2p core level to unoccupied valence states is calculated. The final state wavefunction of the excited electron is calculated in a potential having the form of non-overlapping muffin tins for a cluster of neighboring atoms around the central absorbing atom. The cluster of 5 shells of neighboring atoms is constructed by using the atomic coordinates given by crystallography investigation of $Bi_2Sr_2CaCu_2O_8$. The oxygen atoms in the $[CuO_2]_\infty$ plane are classified as planar oxygen O(P) and the oxygen atoms in the SrO planes are classified as apical oxygen O(A). The cluster includes four Cu ions of the neighbor CuO_4 square planes of the central Cu and one Cu ion from a different $[CuO_2]_\infty$ plane, the fifth shell includes one Bi ion belonging to the BiO layers.

The transitions giving origin to the XANES spectrum shown in Fig.3 can be classified as Cu 2p-> εd, i.e. the transitions to the $\ell=2$ final states at energies ε in the continuum, to be distinguished from the Cu 2p->3d transitions giving origin to the white line. The Fermi level is found at 11.6 eV above the average interstitial potential $\overline{V_0}$. The polarized spectra over a range of 20 eV can be predicted in term of one-electron dipole ($\Delta \ell=+1$) transition Cu 2p->εd, probing the unoccupied d-like ($\ell=2$) density of states projected on Cu site with orbital angular momentum $m_\ell=0, 1$ in the **E**∥**z** spectra, and the $m_\ell=2, 1$, and 0 in the

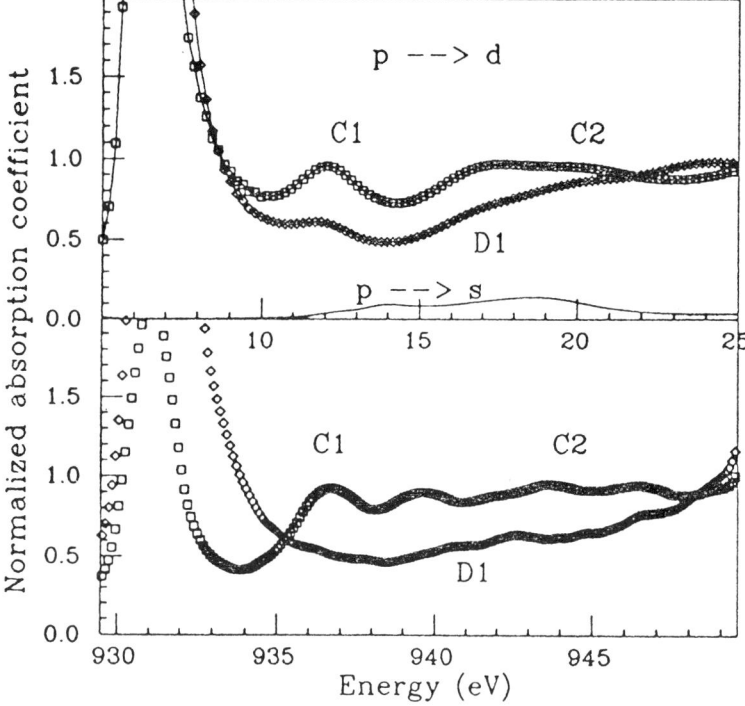

Fig.3 Calculated cross section for the Cu 2p-> d polarized spectra (**E**⊥**c** diamonds, **E**//**c** squares) and the Cu 2p-> s channels giving the Cu L3-edge XANES spectrum of Bi$_2$Sr$_2$CaCu$_2$O$_8$ in the fully relaxed potential and including the energy dependent broadening, the zero if the energy scale is the average interstitial potential $\overline{V_0}$. In the lower panel the blow up of the experimental polarized XANES spectra is shown.

E⊥**c** spectra. The oscillator strength for the dipole allowed transitions ($\Delta \ell=-1$) Cu 2p-> εs has been found to be a factor 100 weaker than the 2p-> 3d transitions. The Coulomb interaction in the final state between the Cu 2p core hole and the excited Cu 3d electron is found to be 5.5 eV forming a bound state below the continuum threshold, the well known Cu L3 white line. On the contrary the core hole induces a nearly rigid red shift of about 1 eV of the high energy conduction bands. The comparison between the experimental and the calculated Cu L3-edge XANES spectra is shown in Fig.3.

The $E \perp c$ polarized oxygen K-edge absorption spectrum of Bi2212 has been measured by Krol et al. [21] by means of the $K\alpha$ soft-x-ray fluorescence yield, that is a bulk sensitive detection method, by using the Dragon monochromator at the Brookhaven National Light Source. In Fig.4 we report the one-electron calculation of the oxygen K-edge $E \perp c$ x-ray absorption spectrum compared with the experimental spectrum. The $E \perp c$ oxygen K-edge absorption spectrum has been calculated in the real space by using the multiple scattering formalism for a cluster of 40 atoms with the planar O(P) oxygen at its centre. The transition

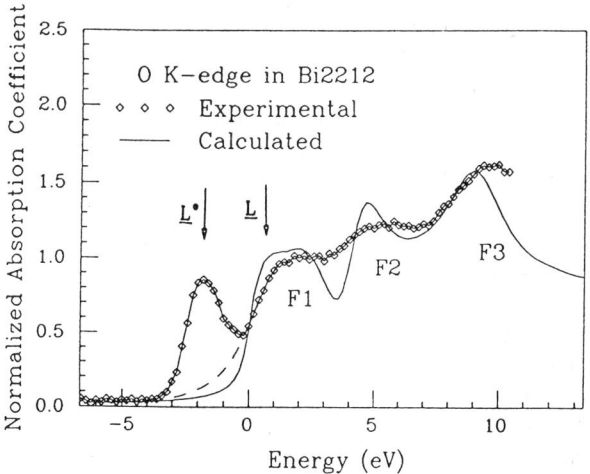

Fig.4 Calculated polarized $E \perp c$ K-edge absorption spectra of the planar oxygen in the $[CuO_2]_\infty$ plane in $Bi_2Sr_2CaCu_2O_8$ for the fully relaxed potential, in the presence of the oxygen 1s hole, including the energy dependent broadening (solid line) compared with the experimental oxygen 1s absorption spectra. The zero of the energy scale is at the Fermi energy position in the one-electron calculations in the unrelaxed potential.

rate from the atomic O 1s core level to the final states in the conduction bands with p-like character ($m_\ell = \pm 1$) is calculated. A good agreement between the one-electron calculations and the experimental spectrum is found in the energy range above the Fermi level. In the insulating material $Bi_2Sr_2YCu_2O_8$ the peak $\underline{L}^*$ (at 528 eV in the experimental spectrum) is missing [22] as in La_2CuO_4 [23] and the peak $\underline{L}$ at 530 eV (in the experimental spectrum) at the absorption threshold is well predicted by the one-electron calculations as the O 1s-> 2p transition where the excited electron is filling the hole in the Cu 3d - O 2p band. It can be described as the O $1s^1 2p^6$ Cu $3d^{10}$ final state.

3. SYMMETRY OF THE $3d^9\underline{L}$ STATES INDUCED BY DOPING

In the metallic phase the peak $\underline{L}^*$ (at about 528 eV photon energy) appears in the oxygen K-edge XANES. It can be described as the O $1s^1 2p^6$ Cu $3d^9$ final state. It is associated with transitions from the O 1s level to ligand holes $\underline{L}^*$ of the $3d^9\underline{L}^*$ states induced by doping with $2p_{x,y}$ character in the $[CuO_2]_\infty$ plane, because it is found only in the $\mathbf{E}\perp\mathbf{c}$ polarized spectrum. Therefore the symmetry of the $\underline{L}^*$ ligand holes can be b_1, a_1 and b_2. In fact the final states O $2p_{x,y}$ of the planar oxygens O(P) that can contribute to the absorption cross section from the O 1s level in the $\mathbf{E}\perp\mathbf{c}$ oxygen K-edge could form three set of molecular orbitals:

i) the b_1 molecular orbitals $\underline{L}(b_1)$, shown in Fig.1, formed by the combination of the O $2p_{x,y}$ orbitals of the 4 planar oxygens coordinated by Cu, in the $[CuO_2]_\infty$ layer, with local b_1 symmetry $(1/\sqrt{4})(+p_{x1} - p_{y2} - p_{x3} + p_{y4})$ that are mixed with the Cu $3d_{x^2-y^2}$ orbitals, where the phases of the orbitals have been chosen with the same convention used in ref.3.;

ii) the a_1 molecular orbitals $\underline{L}(a_1)$ formed by the combination of the O $2p_{x,y}$ orbitals of the 4 planar oxygens coordinated by Cu, in the $[CuO_2]_\infty$ layer, with the local a_1 symmetry $(1/\sqrt{4})(+p_{x1} + p_{y2} - p_{x3} - p_{y4})$ that are mixed with the Cu $3d_{3z^2-r^2}$ orbital, see Fig.1.

iii) the b_2 molecular orbitals $\underline{L}(b_2)$ formed by the combination of the O $2p_{x,y}$ orbitals of the 4 planar oxygens coordinated by Cu in the $[CuO_2]_\infty$ layer with the local b_2 symmetry $(1/\sqrt{4})(+p_{y1} + p_{x2} - p_{y3} - p_{x4})$ that are mixed with Cu $3d_{xy}$ orbitals.

The lack of O 1s -> $\underline{L}^*$ transitions in the $\mathbf{E}//\mathbf{c}$ oxygen K edge spectrum rules out both 1) ligand holes with e symmetry, $\underline{L}(e)$ formed by the O $2p_z$ components of the planar oxygens O(P) mixing with the Cu $3d_{xz}$ and Cu $3d_{yz}$ orbitals $(1/\sqrt{2})(+p_{z1} - p_{z3})$; $(1/\sqrt{2})(p_{z2} - p_{z4})$, and 2) ligand holes formed by the O $2p_z$ components of the apical oxygen O(A) contributing to the molecular orbitals with the a_1 symmetry, i.e. out of plane $\underline{L}(a_1)$ orbitals.

Further information on the symmetry of the unoccupied states at the Fermi level is obtained by the polarized Cu L_3 edge XANES spectra. In the polarized Cu L_3 edge XANES spectra the core excitations to the Cu 3d partial density of states of the Cu 3d - O 2p states around the Fermi level, with a total of $1+\delta$ holes per Cu site, appear over the range 930-936 eV. In the insulating system BiSrYCuO (2212) a symmetric white line appears at ~931.2 eV. With increasing doping a long tail on the high energy side of the Cu 2p-> 3d white line appears due to the Cu $2p^5 3d^{10}\underline{L}^*$ final states probing the $3d^9\underline{L}^*$ states in the initial state in all classes of high T_c superconductors studied so far [14-18,24-28]. Moreover the hole doping induces subtle changes on the low energy side of the unpolarized Cu 2p->3d white line [24]. The polarized spectra of the metallic systems show a white line in the $\mathbf{E}//\mathbf{c}$ polarization at lower energy than that in the $\mathbf{E}\perp\mathbf{c}$ polarization shifted by Δ_z as shown in Fig.2. This is a common feature of all families of doped cuprate perovskites showing superconductivity where Δ_z is in the range of 200-600 meV [14-17] as reported in Fig.5.

Following the dipole selection rule for polarized x-ray absorption spectra the $\mathbf{E}\perp\mathbf{c}$ Cu L_3 edge absorption $\alpha_\perp(\omega)$ is given by the sum of partial contributions $I_{\ell,m}$ to the absorption cross section due to final states with orbital angular momentum ℓ, m: $\alpha_\perp(\omega) = I_{2,2}(\omega) + 1/6\, I_{2,0}(\omega) + 1/2\, I_{2,1}(\omega)$, and the $\mathbf{E}//\mathbf{c}$ Cu L_3 edge absorption spectrum $\alpha_{//}(\omega)$ is given by $\alpha_{//}(\omega) = 2/3\, I_{2,0}(\omega) + I_{2,1}(\omega)$. By assuming the lack of Cu 3d holes with e symmetry ($m_\ell = \pm 1$) as derived by the analysis of polarized oxygen K-edge spectra, we neglect the $I_{2,1}(\omega)$ contribution in the Cu L_3 edge absorption spectra. Therefore the relative number of the Cu $3d_{3z^2-r^2}$ holes n_{dz} on the total number of the Cu 3d holes has been extracted from the Cu L_3 edge absorption spectra by the integral of the absorption spectra over the energy range of the bound states below the continuum threshold

$n_{dz} = 1/2 \left(\int I_{//}(\omega)d\omega / \int I(\omega)d\omega \right) = \int I_{2,0}(\omega)d\omega / \int [2 I_{2,2}(\omega) + I_{2,0}(\omega)]d\omega$, i.e. one half of the ratio between the integral of the **E** ∥ **c** polarized spectrum and the integral of the unpolarized spectrum. The integrals are calculated in the range 928-935.5 eV of the bound states below the continuum threshold where all unoccupied Cu 3d states are pulled down by the core hole final state effect. We have found a value of n_{dz}=21±2% from the spectrum in Fig.2 of a Bi2212 crystal, but n_{dz} depends on the doping level as it has been already reported [16-17]. The values of n_{dz} that we have measured by an extended investigation of several families of high T_c superconductors $La_{2-x}Sr_xCuO_4$, $La_2CuO_{4.05}$, and $Bi_2Sr_2Ca_{1-x}Y_xCu_2O_8$ have ben found to be as large as the hole doping level δ.

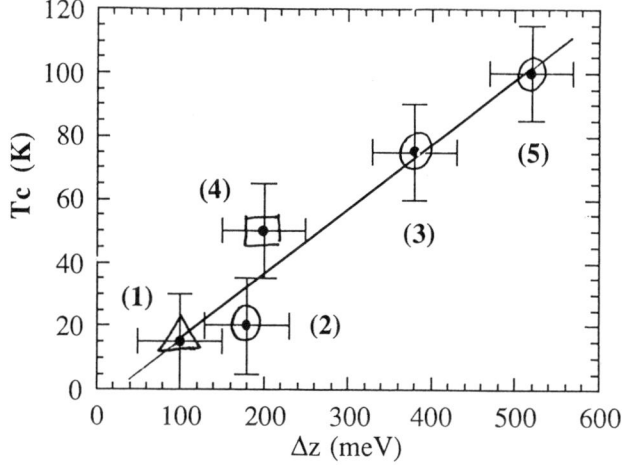

Fig.5 Average energy splitting Δ_z between in-plane and out-of-plane polarized white lines in each family of high T_c superconductors as function of the maximum T_c for each family: 1) LSCO, 2) Bi2201, 3) Bi2212 4)Y123 5)Bi2223

The value of the n_{dz} component in the range of 15-20% can be explained only if a second band with mainly $3d_{3z^2-r^2}$ character is very close (the energy splitting should be few hundreds meV) from the $3d_{x^2-y^2}$ band as clearly shown by theoretical calculations including correlation effects [3]. Two possible scenarios are possible: 1) the $3d_{3z^2-r^2}$ band is very close but always below the Fermi level or 2) it crosses the Fermi level. The presence of a second band crossing the Fermi level has been observed by angular resolved photoemission data reported by Takahashi et al. [29] for $Bi_2Sr_2CaCu_2O_8$. The energy separation between the two bands in Bi2212 at the Γ point was found to be about 300 meV. The presence of a second band crossing the Fermi level in the 110 direction and with BiO character out of the plane has recently been found at high hole doping level [30]. The presence of a second band in the YBa$_2$Cu$_3$O$_7$ crossing the Fermi level with the linear chain character out-of-plane is well established. This band has a Cu $3d_{3z^2-r^2}$ component in the [CuO$_2$]$_\infty$ plane for the **k** vector in the direction of the nearly collinear Cu-O-Cu bonds [31].

The energy splitting Δ_z between the in-plane and the out-of-plane polarized white line in the Cu L$_3$ x-ray absorption spectrum probes the energy splitting at the Γ point between the two bands because the core hole forming a bound atomic-like final state suppresses the dispersion of the valence states in the initial state. The atomic like electron-hole Coulomb interaction is assumed to be the same in the two polarizations. In the insulating material La$_2$CuO$_4$ no energy splitting Δ_z has been found in agreement with a single $3d^{10}$ final state. The value of $\Delta_z \sim 300$ meV in Bi2212 is in good agreement with angular resolved photoemission data [29] and the value $\Delta_z \sim 150$ meV found in YBa$_2$Cu$_3$O$_7$ [14] will be soon verified by undergoing photoemission experiments.

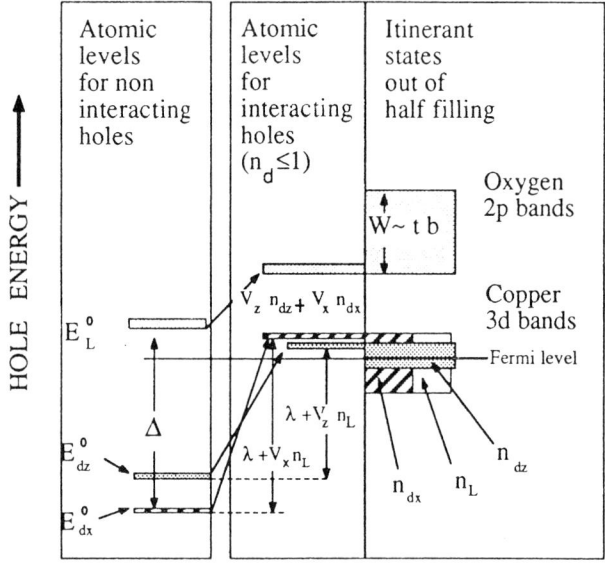

Fig.6 Pictorial view of the formation of two bands at the Fermi level in the frame of the slave boson formalism (ref. 3).

The energy separation Δ_{JT} (the Jahn Teller energy splitting) between the $3d_{x^2-y^2}$ and the $3d_{3z^2-r^2}$ levels in the CuO$_5$ cluster neglecting correlation effects has been calculated by the angular overlap method [32,33] and by the molecular calculations for a cluster of 40 atoms in Bi2212 and we have obtained $\Delta_{JT} \sim 1.3$ eV. The reduction of energy splitting Δ_z in the hole doped systems is assigned to the ligand holes. The interatomic Coulomb repulsion $V_x \sim 1-1.5$ eV between the O $2p_{x,y}$ hole and the Cu $3d_{x^2-y^2}$ hole pushes the $3d_{x^2-y^2}$ level toward higher energy and the actual energy splitting becomes $\Delta_z = \Delta_{JT} - (V_x - V_z)$ where V_z is the interatomic Coulomb repulsion between the O $2p_{x,y}$ hole and the Cu $3d_{3z^2-r^2}$ hole) if the ligand hole is assumed to be localized over a CuO$_4$ square plane unit. In the Fermi liquid description of the Cu 3d bands in the frame of the slave boson formalism [3] the energy splitting is given by $\Delta_{JT} - (n_L \cdot V_x - n_L \cdot V_z)$ where n_L is the number of the O $2p_{x,y}$ holes (see Fig.6). By increasing the ligand holes n_L the energy splitting Δ_z is expected to decrease in agreement with experimental findings [1,16,17].

It is important to remark the key role of the interatomic Coulomb repulsion V_x in the electronic structure of cuprate perovskite systems. In fact an increasing value of V_x determines an increase of the correlation energy gap in the insulating phase, $\Delta=E_L-E_d+V_x$ in the ionic limit (where E_L and E_d are the energy to create a ligand $\underline{L}(b_1)$ and a $3d^9{}_{x^2-y^2}$ hole respectively) and at the same time it suppresses the energy splitting Δ_z between the states with mainly $3d_{3z^2-r^2}$ and $3d_{x^2-y^2}$ symmetry. Going from one family to another the energy splitting Δ_z has been found to increase [1,16,17] with the maximum critical temperature that can be reached in the corresponding family as shown in Fig.5. This can be associated with different Δ_{JT} and different interatomic Coulomb repulsion $\Delta V=(V_x-V_z)$. The value of Δ_{JT} is controlled by the crystal structure and in particular by the Cu-O(A) distance while $\Delta V = (V_x-V_z)$ is controlled by the actual electronic charge density, the Cu-O bond and the Cu-O(P)-Cu bond angle, being maximum for the collinear geometry.

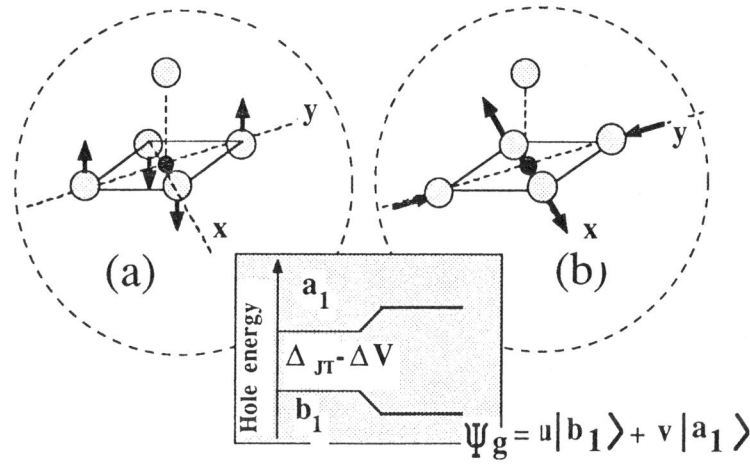

Fig.7 Dynamic distortions of a CuO_5 square pyramid mixing the $3d_{x^2-y^2}$ hole states with $3d_{3z^2-r^2}$ states: $S_{\beta_1}{}^{(b)}$ (panel a) and $S_{\beta_1}{}^{(s)}$ (panel b)

4. COUPLING OF CHARGE CARRIERS WITH THE LATTICE DYNAMICS

The parent insulating Cu(II) compounds of the high T_c superconductors exhibit the characteristic structure of Cu(II) Jahn-Teller ions: elongated CuO_6 octahedra in La_2CuO_4, square pyramids in $Bi_2YSr_2Cu_2O_8$ and $YBa_2Cu_3O_6$, and square planes in Nd_2CuO_4. The Jahn Teller effect characteristic of the Cu(II) ions, removes the degeneracy of the upper E_g states $3d_{x^2-y^2}$, ($m\ell=2$) and $3d_{3z^2-r^2}$, ($m\ell=0$) in the octahedral O_h symmetry by pushing up the energy of the $3d_{3z^2-r^2}$, $m\ell=0$, and lowering the energy of the $3d_{x^2-y^2}$, $m\ell=2$, by reducing the O_h symmetry with an elongation of the Cu-O bond in the **c** direction, or in the extreme cases by pushing away one or two oxygen ions forming Cu sites with a square pyramid or a square plane coordination. Therefore in the divalent cuprate perovskite the single hole per unit cell is stabilized in the Cu 3d derived states with the component of orbital momentum $3d_{x^2-y^2}$, $m\ell=2$. The pseudo Jahn Teller (JT) effect [33] predicts an instability of a Cu(II) cluster (elongated octahedra or a square pyramids) for distortions, shown in Fig.7, that mixes the non degenerate high energy level $3d_{3z^2-r^2}$, (a_1 molecular symmetry) with the $3d_{x^2-y^2}$ levels. The divalent Cu(II) systems are well known to show long range ordering of the pseudo Jahn Teller type distortions giving ordering of the angular momentum like in K_2CuF_4 [34].

The increasing Cu $3d_{3z^2-r^2}$ character of the Cu 3d holes with doping has suggested us to investigate the coupling between the electronic states and the structural dynamics of the Cu site local structure [16]. The relevance for superconductivity of the vibration modes mixing the $3d_{x^2-y^2}$ with $3d_{3z^2-r^2}$ hole states is demonstrated by the Raman results in the YBaCuO system [35,36]. The mode at 335 cm^{-1} corresponding to out-of-phase vertical vibration of two oxygen atoms in the $[CuO_2]_\infty$ plane, that becomes soft at T_c has the symmetry $S_{\beta 1}^{(b)}$, shown in Fig.7, mixing the $3d_{x^2-y^2}$ with $3d_{3z^2-r^2}$ states. The asymmetric Fano lineshape of this Raman line indicates the interaction of this mode with electronic transitions.

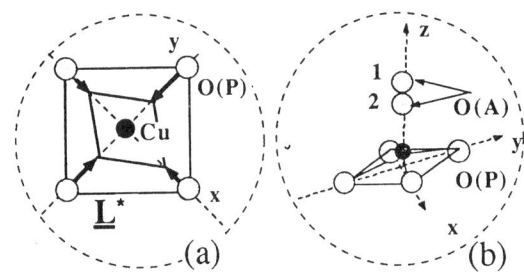

Fig.8 Pictorial view of the averaged Cu site distortions observed in $Bi_2Sr_2Ca_{1-x}Y_xCu_2O_8$ system with increasing the hole doping.

We have investigated the Cu site structural variations and its dynamics in the transition from the insulating to the metallic phase in the $Bi_2Sr_2Ca_{1-x}Y_xCu_2O_8$ system by Cu K-edge XANES and EXAFS [37,38]. We have found a contraction of the Cu-O(P) distance with doping, ~0.03 Å from x=1 to x=0, in agreement with the formation of O 2p holes in the $[CuO_2]_\infty$ plane, second a difference of about 0.12 Å between two short and two long Cu-O(P) distances has been found indicating the distortion of the CuO_4 square plane as shown in Fig.8. The EXAFS results are consistent with a Cu-O(A) distance of 2.45 Å nearly independent on the doping level and with an alternative second position for the apex oxygen at a 0.12 Å shorter Cu-O(A) distance, as it was found in $YBa_2Cu_3O_7$ [40]. The probability of the short Cu-O(A) distance becomes relevant for the over doped system $Bi_2Sr_2CaCu_2O_{8+\delta}$ where the critical temperature decreases.

All structural variations observed as function of doping are consistent with increasing mixing of the $3d_{x^2-y^2}$ with $3d_{3z^2-r^2}$ states induced by doping. The decrease of the energy splitting Δ_z with doping is in agreement with the contraction of the Cu-O(P) distance increasing ΔV, and with the increasing probability for the apex oxygen to occupy the short Cu-O(A) distance which will suppress Δ_{JT}. The difference of 0.12 Å between in plane Cu-O(P) distance gives a mixing between the $3d_{x^2-y^2}$ with $3d_{3z^2-r^2}$ states that can be large for small Δz as discussed in ref.39. The coupling term between the electronic states and the lattice vibrations is controlled by the Cu-Cu-O(P) angle θ shown in Fig.9 [33]. EXAFS analysis has given an average angle θ=9° nearly independent on doping but its standard deviation due to O(P) vibration amplitude has been found to be $\sigma_\theta=1.4°$ in the insulating phase, x=1, and $\sigma_\theta=6°$ for the sample x=.25 that exhibits the highest T_c. The large amplitude of the Cu-Cu-O(P) angle modulation associated with the presence of itinerant carriers shows that the carriers are associated with lattice vibrations of the in plane oxygens O(P). The vibration of O(P) oxygens with symmetry $S_{\beta 1}^{(b)}$ mixing the $3d_{x^2-y^2}$ with

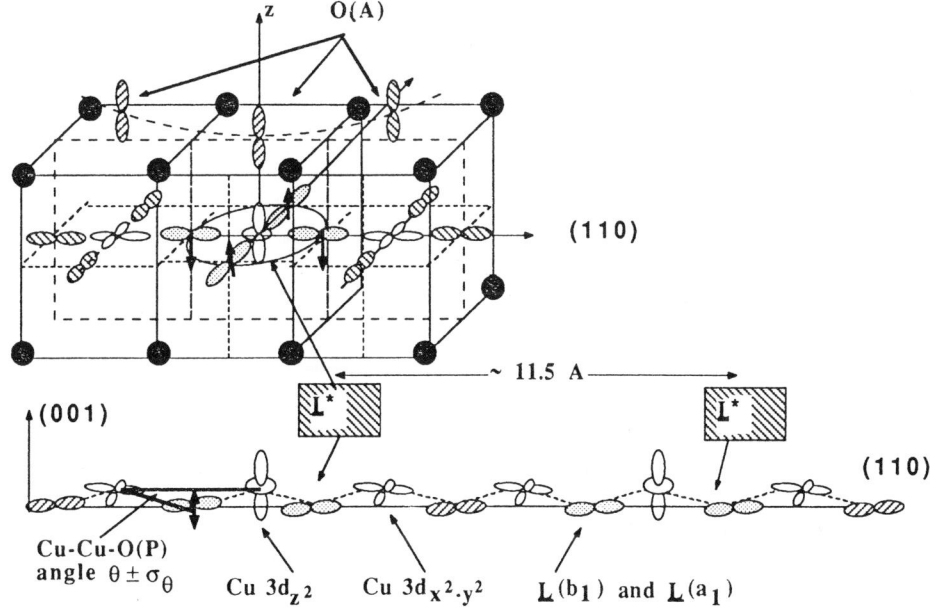

Fig.9 Pictorial view of a $3d_{3z^2-r^2}$ polaron with radius ~5.7 Å where the out-of-plane vibration of the O(P) oxygens is coupled with the ligand hole delocalized over four oxygens and coupled with the central Cu $3d_{3z^2-r^2}$ and the four Cu $3d_{x^2-y^2}$ holes in the neighbor corner sharing CuO4 units (only two neighbor corner sharing CuO4 units are shown in the figure). The instability of the apex oxygen between a long and short Cu-O(A) distance is indicated. It is interesting to remark that the distance along the (110) direction in Bi2212 between two polarons forming a bipolaron will be ~11.5 Å, close to the coherence length.

$3d_{3z^2-r^2}$ states is a possible candidate for the lattice dynamics associated with the itinerant charges forming a polaron. The presence a polaron in the high Tc superconductors is in agreement with infrared photoinduced absorption experiments [41].

The polaron shown in Fig.9 can be classified as a $3d_{3z^2-r^2}$ polaron. The fact that n_{dz}~δ indicates that for each ligand hole $\underline{L}^*$ one Cu 3d hole has the $3d_{3z^2-r^2}$ symmetry. The ligand holes $\underline{L}^*$ should be delocalized over at least five CuO4 units formed by a central CuO4 and four neighbor corner sharing CuO4 square plane units, in agreement with the optimum doping level of 20%. In the $3d_{3z^2-r^2}$ polaron the modulation of the orbital angular momentum of the 3d holes by the oxygen holes, via the interatomic Coulomb repulsion V_x, increasing the probability of the $m_\ell=0$ states, is associated with the increasing amplitude of the vibrations of the in plane oxygens O(P) in the **c** direction mixing the $3d_{3z^2-r^2}$ and the $3d_{x^2-y^2}$ states and with the instability of the apex oxygen in a double well [40]. We are looking forward further experimental investigation to verify the hypothesis of the $3d_{3z^2-r^2}$ bipolaron as a candidate for the pairing mechanism in the high T_c superconductors.

5. CONCLUSIONS

By polarized Cu L_3 x-ray absorption spectroscopy of a large set of crystals we have found that the orbital angular momentum of the Cu 3d holes is not frozen in the $m_\ell=2$ state. The number of the $3d_{3z^2-r^2}$ ($m_\ell=0$) holes on the total number of Cu 3d holes is close to the value of the doping level $n_{dz} \sim \delta$. The energy splitting between two different bands at the Γ point is of the order of few hundreds meV. The fact that the two bands have different symmetry implies the coupling between the electronic excitations and the lattice vibration with the appropriate symmetry that allow the mixing between the $3d_{3z^2-r^2}$ and the $3d_{x^2-y^2}$ states. The coupling between the itinerant carriers and the lattice dynamics found by EXAFS data points toward the hypothesis for the formation of a $3d_{3z^2-r^2}$ polaron.

REFERENCES

1. A. Bianconi in *International Conference on Superconductivity -ICSC*, S.K. Joshi, C.N.R. Rao, and S.V. Subranyam editors, (Proc. Int. Conf. on Superconductivity, Bangalore, Jan. 10-14, 1990) World Scientific, Singapore, pag. 448 (1990)
2. C. Castellani, C. Di Castro and M. Grilli *Physica C* **153-155**, 1659 (1988); C. Castellani, C. Di Castro and M. Grilli *Int. Journal of Modern Physics* **1**, 659 (1988)
3. M. Grilli, C. Castellani, and C. Di Castro *Phys. Rev. B* **42**, 6233 (1990)
4. H. Kamimura *Int. Journal of Modern Physics* **1**, 699 (1988)
5. S. Kurihara *Physica C* **153-155**, 1247 (1988)
6. T. Nishino, M. Kikuchi, and J. Kanamori *Solid State Commun.* **68**, 455 (1988)
7. V.V. Flambaum and O.P. Sushkov *Physica C* **159**, 595 (1989)
8. R. Englman, B. Halperin, and M. Weger *Solid State Commun.* **70**, 57 (1989)
9. J. Ruvalds *Phys. Rev. B* **35** 8869 (1987)
10. J. Ashkenazi and C.G. Kuper *Physica C* **153-155**, 1315 (1988); J. Ashkenazi, and C.G. Kuper in *High T_c Superconductors: Electronic Sructure,* ref.6, pag.43 (1989);
11. W. Weber, *Z. Phys. B - Condensed Matter* **70**, 323 (1988); W. Weber, *Advances in Solid State Physics* **28**, 141 (1988); A.L. Shelankov, X. Zotos, and W. Weber *Physica C* **153-155**, 1307 (1988); Yu B. Gaididei and V.M. Loktev *Phys. Stat. Sol.* (b) **147**, 307 (1988)
12. M. Jarrell, H.R. Krishnamurthy, and D.L. Cox *Phys. Rev. B* **38**, 4584 (1988); D.L. Cox, M. Jarrell, C. Jayaprakash, H.R. Krishna-murthy, and J. Diez *Phys. Rev. Lett.* **62**, 2188 (1989)
13. J.G. Bednorz and K.A. Müller *Rev. Mod. Phys.* **60**, 565 (1988)
14. A. Bianconi, M. De Santis, A.M. Flank, A. Fontaine, P. Lagarde, A. Marcelli, H. Katayama-Yoshida and A. Kotani.*Physica C* **153-155,** 1760, (1988); and *Phys. Rev. B* **38,** 7196 (1988)
15. A. Bianconi, P. Castrucci, A. M. Flank, P. Lagarde, S. Della Longa, A. Marcelli, H. Katayama-Yoshida, and Z.X. Zhao *Modern Physics Lett. B* **2**, 1313 (1988)
16. A. Bianconi, P. Castrucci, A. Fabrizi, M. Pompa, A.M. Flank, P. Lagarde, H. Katayama-Yoshida, and G. Calestani *Physica C* **162-164**, 209 (1989)

17. A. Bianconi, P. Castrucci, A. Fabrizi, M. Pompa, A.M. Flank, P. Lagarde, H. Katayama-Yoshida, and G. Calestani *Earlier and Recent Aspects of Superconductivity*, Springer Series in Solid State Sciences, Springer Verlag edited by K.A. Müller and J.G. Bednorz, pag.407 (1990)
18. A. Bianconi, C. Li, M. Pompa, A. Congiu-Castellano, S. Della Longa, D. Udron, A.M. Flank, and P. Lagarde *to be published*
19. A. Bianconi in *"X-ray Absorption: Principles, Applications, Techniques of EXAFS, SEXAFS, and XANES"*, ed.by D.C. Koningsberger, and R. Prinz, J. Wiley &Sons, New York, pag. 573 (1988)
20. A. Bianconi, J. Garcia, and M.Benfatto in*"Synchrotron Radiation in Chemistry and Biology I"*, edited by E.Mandelkow, Spriger Verlag, Berlin, Topics in Current Chemistry **145**, 29-67 (1988)
21. A. Krol, C.S. Lin, Z.H. Ming, C.J. Sher, Y.H. Kao, C.T. Chen, F. Sette, Y. Ma, G.C. Smith, Y.Z. Zhu, and D.T. Shaw *Phys. Rev. B* **42**, 2635 (1990)
22. H. Matsuyama, T. Takahashi, H. Katayama-Yoshida, T. Kashiwakura, Y. Okabe, S. Sato, N. Kosugi, A. Yagishita, K. Tanaka, H. Fujmoto and H. Inokuchi *Physica C* **160**, 567 (1989)
23. C.T. Chen, F. Sette, Y. Ma, M.S. Hybertsen, E.B. Stechel, W. M. C. Foulkes, M. Schluter, S.-W. Cheong, L.W. Rupp, Jr., B. Batlogg, Y.L. Soo, Z.H. Ming, A. Krol, and Y.H. Kao *Phys. Rev. Lett.* **66**, 104 (1991)
24. A. Bianconi, A. Congiu Castellano, M. De Santis, P. Rudolf, P. Lagarde, and A.M. Flank, and A. Marcelli *Solid State Commun.* **63**, 1009 (1987)
25. A. Bianconi, J. Budnick, A.M. Flank, A. Fontaine, P. Lagarde, A. Marcelli, H. Tolentino, B. Chamberland, C. Michel, B. Raveau, and G. Demazeau *Phys. Lett.* **127**, 285 (1988)
26. D.D. Sarma O. Strebel, C.T. Simmons, U. Neukirch, G. Kaindl, R. Hoppe, and H.P. Müller, *Phys. Rev. B* **37**, 9784 (1988)
27. M. Grioni, J.B. Goedkoop, R. Schoorl, F.M.F. de Groof, J.C. Fuggle, F. Schäfers, E.E.Koch, G. Rossi, J.-M.Esteva, and R.C. Karnatak *Phys. Rev. B* **39**, 1541 (1989)
28. M. Ronay, A. Santoni, A.G. Schrott, L.J. Terminello, S.P. Kowalczyk, and F.J. Himpsel *Sol. State Commun.* **77**, 699 (1991)
29. T. Takahashi, H. Matsuyama, H. Katayama-Yoshida, Y. Okabe, S. Hosoya, K. Seki, H. Fujimoto, M. Sato, and H. Inokuchi *Nature* **334**, 691 (1988); and *Phys. Rev. B* **39**, 6636 (1989)
30. B.O. Wells, Z.-X.Shen, D.S. Dessau, W.E. Spicer, C.G. Olson, D.B. Mitzi, A. Kapitulnik, R.S. List, and A. Arko, *Phys. Rev. Lett.* **65**, 3056 (1990)
31. O. Jepsen et al. (1991) to be published
32. M. Pompa *Thesis, University of Rome La Sapienza* (1990) to be published
33. M. Bacci *Chemical Physics* **104**, 191 (1986); *Jpn. J. Appl. Phys.* **27**, L1699 (1988)
34. K.I. Kugel and D.I. Khomskii *Sov. Phys. Usp.* **25**, 321 (1982) and *Sol. State Commun.* **13**, 763 (1973)
35. R.M. Macfarlane, H. Rosen, and H. Seki *Sol. State Commun.* **63**, 369 (1987)
36. M. Cardona, C.Thompsen *ref.2*, pag.79 (1989); T. Ruf, C. Thomsen, R. Liu, and M. Cardona *Phys. Rev. B* **38**, 11985 (1988)
37. I. Pettiti *Thesis, University of Rome La Sapienza* (1990) to be published
38. A. Di Cicco, S. Stizza, I. Pettiti, D. Udron, and A. Bianconi, to be published.
39. Y. Seino, A. Kotani and A. Bianconi *Jour. Phys. Soc. of Japan* **59**, 815 (1990)
40. J. Munstre de Leon, S.D. Conradson, I. Batistic, and A.R. Bishop, *Phys. Rev. Lett.* **65**, 1675 (1990)
41. C. Taliani, R. Zamboni, G. Ruani, F.C. Matacotta, and K.I. Pokhodnya *Sol. State Commun.* **66**, 487 (1988)

KOSTERLITZ-THOULESS-LIKE BEHAVIOR OVER EXTENDED RANGES OF TEMPERATURE AND LAYER THICKNESS IN CRYSTALLINE YBa$_2$Cu$_3$O$_{7-x}$/PrBa$_2$Cu$_3$O$_{7-x}$ SUPERLATTICES

Douglas H. Lowndes and David P. Norton

Solid State Division, Bldg. 2000
Oak Ridge National Laboratory
P. O. Box 2008, Oak Ridge, TN 37831-6056

INTRODUCTION

High-temperature superconductivity is associated with layered, quasi-two-dimensional (2D) crystal structures and with carriers moving in CuO$_2$ planes. Because the c-axis superconducting coherence length is very short (e.g., ξ_c ~ 3–6 Å, vs. a lattice constant c ~ 11.7 Å in YBa$_2$Cu$_3$O$_{7-x}$), questions arise whether isolated single-cell-thick layers of these materials are superconducting and, if so, how their superconductivity is affected, first, by their extreme anisotropy (including possible reduced dimensionality) and, second, by residual interlayer coupling or other interactions. We[1,2] and several other groups[3-5] recently fabricated epitaxial M×N "123"-family superlattices in which c-axis perpendicular (c$_\perp$) YBa$_2$Cu$_3$O$_{7-x}$ (YBCO) layers M unit cells in thickness are separated by N unit-cell-thick layers of semiconducting PrBa$_2$Cu$_3$O$_{7-x}$ (PBCO). The superlattice T$_{co}$(R = 0) was found to decrease rapidly with increasing PBCO layer thickness, but then to saturate at T$_{co}$ ~ 19 K, 54 K, 71 K, or 80 K, for structures containing 1-, 2-, 3-, or 4-cell-thick YBCO layers, respectively.[1,2] In contrast, recent measurements for Bi$_2$Sr$_2$(Ca$_{1-x}$Y$_x$)$_1$Cu$_2$O$_8$-based superlattices, in which superconducting (x=0.15) and semiconducting (x=0.5) layers also alternate, show almost no change in T$_{c(mid)}$ ~65 K with decreasing superlattice period.[6] These results imply that high-T$_c$ is an intrinsic property of CuO$_2$ bilayers, but that 3-dimensional interactions are present, at least for the "123"-family materials.

Consequently, it is necessary to distinguish the effect that reduced dimensionality alone has on the resistive transition shape and on T$_{co}$, as the thickness of a thin YBCO layer is reduced, from purely electronic effects within the three-dimensional YBCO/PBCO superlattice structure. Some of the electronic mechanisms that have been suggested include (1) electron transfer from PBCO to YBCO, resulting in hole filling in the YBCO layers and depression of T$_c$;[7,8] (2) leakage of the pair amplitude from YBCO into PBCO, with hole filling playing only a secondary role;[9] or, (3) an intercell pairing interaction that couples adjacent YBCO cells and raises T_c.[10] On the other hand, Rasolt, Edis, and Tesanovic recently suggested that YBCO/PBCO superlattices provide a 3D system in which the interlayer (c-axis) coupling can be quasi-continuously weakened to zero, so that a crossover to 2D behavior must occur with increasing PBCO thickness, accompanied by characteristic 2D dissipation (due to thermally generated free vortices) and a Kosterlitz-Thouless transition.[8] The experimental test suggested by Rasolt et al. for the 2D regime is that the superconducting properties should become constant with increasing PBCO layer thickness, as the YBCO layers become isolated.[8] Rasolt et al. also predict that charge transfer and hole filling have only a secondary effect on T_c. Ariosa and Beck have suggested that each conducting CuO_2 bilayer in a superlattice should be modeled as a 2D array of in-plane Josephson-coupled junctions, with no Josephson coupling along the c-axis.[11] However, c-axis coupling is included in their model, through electrostatic charging between the conducting planes in the YBCO layers. If the electrostatic coupling is sufficiently large, relative to the in-plane Josephson coupling, then characteristic 2D quantum phase fluctuations (using an XY model Hamiltonian) are expected, together with depression and broadening of the superlattice superconducting transition. Although Ariosa and Beck do not consider the shape of the resistive transition, it seems to us that the consequences of their physically different but mathematically related model would be similar to those of Rasolt et al..

II. BACKGROUND

1. Kosterlitz-Thouless (K-T) Behavior in Superconducting Films

In an uncharged 2D superfluid film, only bound vortex-antivortex (v-av) pairs exist at temperatures below T_c, but free vortices are generated by thermal fluctuations for $T > T_c$.[12-15] (Here we use T_c to denote the true temperature for the onset of dissipation, corrected for the screening and renormalizing effects of the vortices, whereas T_{KT} denotes the temperature above which the mean squared separation of an *isolated* v-av pair diverges.[14]) Pearl showed that in a superconducting film (*charged* superfluid) the potential (interaction) energy of vortices is logarithmic only out to a distance $\lambda_\perp \sim \lambda^2 / d_{eff}$, where λ is the (bulk) magnetic penetration depth, d_{eff} is the effective film thickness for superconducting correlations, and $\lambda_\perp$ is the magnetic penetration depth for an H field perpendicular to the film.[16] The significance of large $\lambda_\perp$ is that diamagnetic currents in the superconductor are ineffective in shielding the vortices and their interaction remains logarithmic out to large separations. Beasley, Mooij, and Orlando showed that, in practice, $\lambda_\perp$ can be comparable to the lateral dimensions of typical thin-film specimens (~mm), so that characteristic K-T behavior should be expected in thin superconducting films that have high sheet resistance, $R_\square$.[17] For the highly anisotropic HTSc materials, d_{eff} may be as short as ξ_c, the c-axis superconducting coherence length. For YBCO, with $\lambda_{ab}(0) \sim 0.15$ µm and $\xi_c(0) \sim 3\text{-}6$ Å, this implies that $\lambda_\perp(0) \sim 50$ µm, at least approaching sample dimensions, so that K-T behavior is a realistic possibility.

2. Characteristic "Signatures" of K-T Behavior

Recent experiments[18-22] (see below) have focused on four characteristic "signatures" of a K-T transition. Because each of these may be obscured over part of the accessible temperature or magnetic field range by other competing effects, it is necessary in general to observe more than one signature in order to unambiguously identify the K-T transition.

The first signature of a K-T transition is that for $T_c < T \ll T_{cB}$, where T_{cB} is the the mean-field (bulk, or BCS) superconducting transition temperature,[14,15] thermally generated free vortices[12,13] produce a nonzero resistivity with temperature dependence[23,24]

$$R(T)/R_N \sim \xi_+(T)^{-2} = \xi_{ab}(T_c)^{-2} \exp[-2b'(\tau_c/\tau)^{1/2}]. \tag{1}$$

In Eq. (1) $\xi_+(T)$ is the phase correlation length for the superconducting order parameter,[23,24] $\xi_{ab}(T)$ is the Ginzburg-Landau correlation length within the a-b plane, $\tau_c = (T_{cB} - T_c)/T_c$, $\tau = (T - T_c)/T_c$, and b' is a constant of order unity. From Eq. (1) it follows that a plot of log R(T) vs. $\tau^{-1/2}$ should be linear. In practice, the condition $T_c < T \ll T_{cB}$ may be difficult to achieve (T_c and T_{cB} may be close together) and paraconductivity,[25-28] due to superconducting amplitude fluctuations above T_{cB}, also may interfere with observation of the characteristic inverse-square-root singularity in log R(T).

The second signature of a K-T transition is the power-law I–V characteristic[23, 24, 29] (nonlinear resistivity) that is induced by a current for $T < T_c$. This is due to the opposite effects that a current has on vortices and antivortices, resulting in unbinding of v-av pairs. In theory it is expected that[23, 24, 29]

$$V \sim I^{N(T)}, \tag{2}$$

so that log V vs. log I has a non-ohmic and T-dependent slope, $N(T) = [1 + \pi K(T)]$. Here K(T) is the K-T "stiffness" parameter for phase fluctuations,[12,13] and is related to the 2D superfluid density, $n_s^{2D}(T) = n_s^{3D}(T) d_{eff}$, by[23, 24]

$$K(T) = n_s^{2D}(\hbar^2/m^*k_BT). \tag{3}$$

The third signature of a K-T transition is a "universal" jump[29,30] in n_s^{2D}, hence in N(T), as T_c is approached from below. This occurs because K(T) approaches the constant value $2/\pi$ at T_c, while $n_s^{2D} = 0$ for $T > T_c$ but $n_s^{2D} \neq 0$ for $T \to T_c^-$.

The first three signatures all are observed in zero magnetic field. However, a weak magnetic field also can be used to unbind v-av pairs,[19,23] and it is found that

$$d[\ln \rho]/d[\ln H] = (\pi/2) K(T), \tag{4}$$

where ρ is the resistivity. Thus, K(T) can be obtained from magnetoresistance data. However, both the T and H ranges for this behavior are limited, because H-field-driven unbinding will merge with thermal unbinding for $T > T_c$, and as H is increased the dissipation due to free vortex motion also will begin to dominate that due to H-driven v-av unbinding.[19]

3. K-T Behavior in Single-Crystal and Thick-Film HTSc Specimens

The large anisotropy of the coherence lengths ($\xi_{ab} >> \xi_c$) in HTSc materials, together with their small absolute values of ξ_c, strongly suggest that quasi-2D behavior should be present in the superconducting order parameter. In fact, several groups[18-22] recently obtained direct experimental evidence from electrical transport measurements that the anticipated quasi-2D behavior does occur in very narrow temperature ranges, ΔT, just below T_{cB}, even in nominally 3D single-crystal (platelet) or thick-film specimens of $YBa_2Cu_3O_{7-x}$ (for which $\Delta T \sim$ 0.2-1.0 K[18, 20, 21]), $Bi_2Sr_2CaCu_2O_8$ (Bi-2212, $\Delta T \sim$ 2 K[19]), and $Tl_2Ba_2CaCu_2O_8$ (Tl-2212, $\Delta T \sim$ 1 K[22]). Because $R_\square$ is small for such near-bulk specimens, the earlier work of Beasley et al. suggests that the K-T transition temperature, T_c, should be very close to T_{cB}, consistent with the narrow ΔT ranges that have been found. At such temperatures extrinsic flux pinning is expected to be negligible so that any vortex motion observed should be intrinsic. All four of the characteristic signatures of a K-T transition described above have been observed in these experiments.

Fiory et al. apparently were the first to observe K-T behavior in a clean type-II superconductor, using ~500-Å-thick YBCO films.[18] From measurements of the complex sheet impedance they determined the kinetic inductance, from which the temperature dependence of the a-b plane magnetic penetration depth, $\lambda_{ab}^2(T)$, was determined. Two types of K-T behavior were observed, both an algebraic decay-law behavior of λ^{-2} with a $[(T-T_c)/T_c]^{1/2}$ cusp in the exponent, and the "universal" jump in superfluid density at T_c. Their observations are consistent with phase fluctuations of the entire 500-Å-thick film, driven by quantized vortices, with a dissipation peak in the narrow range between $T_c \sim$ 88.4 K and $T_{cB} \sim$ 89.4 K ($\Delta T \sim$ 1 K).[18]

Martin et al. showed that 2D phase fluctuations due to vortex generation most likely cause the broadening of the superconducting transition in Bi-2212 single-crystal platelet specimens with dimensions $L_a \sim L_b \sim$ 1 mm and $L_c \sim$ 2 μm.[19] They observed the $\tau^{-1/2}$ singularity in log R(T) (due to thermal generation of free vortices) for $T \geq T_c$. This behavior suggested a narrow range of thermally activated v-av depairing, between T_c = 84.7 K and T_{cB} = 86.8 K.[19] They were not able to induce vortex pair-breaking using a current (in order to observe the power-law I-V characteristic) but instead used weak magnetic fields to unbind v-av pairs for $T < T_c$, and observed the power-law behavior in the magnetoresistance. Martin et al. estimated that the effective thickness of the superconducting sheets in Bi-2212 is ~15 Å, comparable to the c-axis CuO_2-bilayer spacing.[19] They also pointed out that the transverse (c-axis) vs. in-plane coupling anisotropy, $K_\perp/K$, (as estimated by the conductivity ratio) is ~10^{-2} for YBCO but is ~10^{-5} for Bi-2212, and suggested that the relatively larger c-axis coupling in YBCO might be too large for bulk YBCO crystals to display characteristic 2D behavior (e.g., dissipation due to 2D vortex excitation).

Nevertheless, Yeh and Tsuei recently made very precise and detailed measurements of the electrical transport properties of ~ 0.5 × 0.4 × 0.03 mm^3 YBCO single-crystal platelets and obtained direct evidence of 2D fluctuations, and of a K-T transition occurring only ~0.2 K below T_{cB} = 93.15 K.[20] They observed the $\tau^{-1/2}$ singularity in log R(T) for $T_c < T < T_{cB}$, as well as the power-law I-V characteristic and the universal jump in superfluid density at T_c.[20] The small separation of T_c and T_{cB} is consistent with the low sheet resistance of the CuO_2 bilayers in YBCO single crystals.[17] Ying and Kwok[21] also subsequently observed several signatures of 2D fluctuations and a K-T transition in laser-ablated epitaxial YBCO thin films, including the power-law I-V behavior, the universal jump, and scaling of log R(T) with $\tau^{-1/2}$. They also studied the paraconductivity (amplitude fluctuations) above T_{cB} and drew attention to its "1D-like" behavior at temperatures near T_{cB}. They suggested this might result from imperfections in laser-ablated films, which generally are polycrystalline (with symmetry-related 90° twin boundaries), even though epitaxial.[21] Ying and Kwok suggested that these boundaries might affect the I–V characteristics and possibly the apparent dimensionality of the transition.

Finally, Kim et al. have studied the electrical transport properties of ~0.7-μm-thick c-axis-perpendicular Tl-2212 films.[22] They observed nonlinear I–V and magnetoresistance behavior, together with log R(T) scaling as $\tau^{-1/2}$, consistent with 2D fluctuations and a K-T transition, with T_c = 99.0 K and T_{cB} = 100.2 K.[22]

4. Superlattice Growth Conditions

The YBCO/PBCO superlattices used in this work were grown by an in situ pulsed laser ablation process that has been described elsewhere.[1,2,31] Briefly, stoichiometric YBCO and PBCO targets were mounted in a multitarget "carousel" that permits rotation of the individual targets as well as their interchange. The pulsed KrF (248 nm) excimer laser beam was brought to a line focus on the target; this line was scanned over the target in a direction perpendicular to its length, to produce a region of only slowly varying film thickness on the substrate heater face; epitaxially polished (100) $SrTiO_3$ substrates were bonded to the heater using silver paint. Superlattices were grown at a heater (or estimated substrate) temperature of 730°C (~670°C) in 200 mTorr oxygen, at a deposition rate of 1.1 Hz and ~0.1 nm/laser pulse. Superlattice periods measured directly by x-ray diffraction agreed with those calculated from the measured film thicknesses and total number of laser shots to within a few percent.[1,2]

III. EXPERIMENTAL RESULTS

As was described in the Introduction, Rasolt, Edis, and Tesanovic recently suggested[8] that reduced dimensionality (resulting in a Kosterlitz-Thouless transition) is the primary factor governing the shape and T_{co} (R=0) values for YBCO/PBCO superlattices that contain well-isolated YBCO layers. In order to test this idea directly, we have used the Kosterlitz-Thouless theory to analyze the resistive transition shapes for superlattices of various layer thicknesses. We find that, for structures encompassing a remarkably wide range of superconducting (YBCO) and "barrier" (PBCO) layer thicknesses, log R(T) does

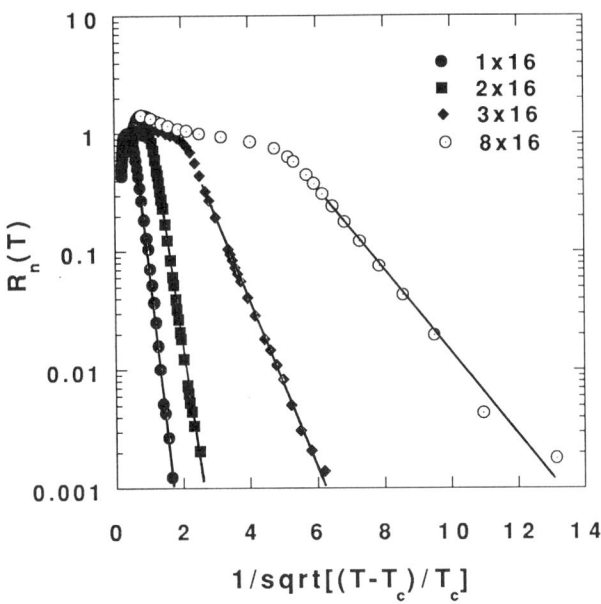

Fig. 1. Superconducting transitions for M(YBCO)×16(PBCO) superlattices with M = 1, 2, 3, and 8 c-axis unit cells, plotted as log $R_n(T)$ vs. $\tau^{-1/2}$, where $R_n(T)$ is the normalized resistance $R(T)/R(100\ K)$ and $\tau = (T - T_c)/T_c$. The T_c values determined by best least-squares fits to the data are given in the text.

scale as $\tau^{-1/2}$, as expected[23, 24] for a system in which dissipation is due to 2D fluctuations. Here $\tau = (T - T_c)/T_c$, and T_c may be interpreted as the temperature at which bound vortex-antivortex pairs first are broken by thermal excitation, resulting in free vortices and the onset of dissipation. The T_c values determined by fitting the resistive transition shape are close to, but slightly below, our measured zero-resistance values (T_{co}).

Figure 1 shows the results of plotting log $R_n(T)$ vs. $\tau^{-1/2}$, where $R_n(T) = R(T)/R(100\ K)$, for the M(YBCO) × 16(PBCO) superlattice structures with M = 1, 2, 3, and 8 unit cells per YBCO layer. These are the structures with the thickest PBCO layers, for which the superconducting properties no longer change with barrier layer thickness, the condition suggested by Rasolt et al.[8] as clearly requiring a crossover to 2D resistive behavior. As seen in Fig. 1, very good fits are obtained to the anticipated K-T resistive behavior, and from these plots we extract least-squares best-fit values for the K-T transition temperature of T_c = 14 K, 44 K, 70 K, and 86.5 K, for the structures with 1-, 2-, 3-, and 8-cell-thick YBCO layers, respectively.

We also have examined the resistive behavior of a wide range of other superlattices, for which the barrier layers are not so thick, and we find evidence that 2D fluctuations may be important quite generally in these structures. Figure 2 shows the effect on the resistive transitions of reducing the barrier layer thickness, for structures with 2-cell-thick (left) and 3-cell-thick (right) YBCO layers. Good straight-line behavior is obtained in plots of log $R_n(T)$ vs. $\tau^{-1/2}$, until the PBCO barriers become very thin (N = 1). The T_c values

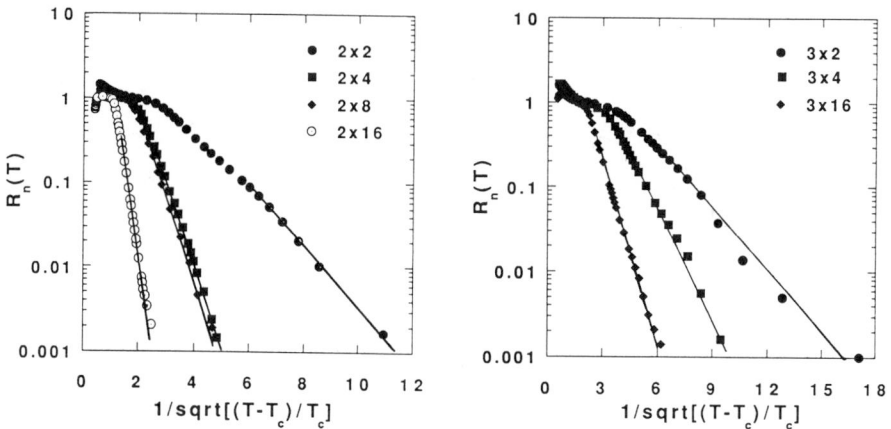

Fig. 2. Superconducting transitions for [left] 2(YBCO)×N(PBCO) and [right] 3(YBCO)×N(PBCO) superlattices, plotted as log $R_n(T)$ vs. $\tau^{-1/2}$.

determined by least-squares fitting to the data are 44, 62, 65, and 75.5 K for the 2×N structures (with N = 16, 8, 4, 2, respectively), and 70, 79, and 82.1 K for the 3×N structures (with N = 16, 4, and 2, respectively). It is seen that the best-fit T_c values rise rapidly toward T_{co} (~ 92 K) for a thick YBCO film, as the thickness of the PBCO barriers is reduced. T_c also rises rapidly as the YBCO superconducting sheets are made thicker.

The effect of 2D fluctuations in broadening the superlattice resistive transitions can be illustrated by plotting both T_c (K-T) and T_{cOnset} for a series of superlattices with fixed barrier layer thickness (e.g., N = 16 unit cells), as the YBCO layer thickness is increased. As an estimate[14] of T_{cOnset} we use the temperature at which the break (or "corner") occurs at the intersection of the extrapolated normal-state and superconducting-state behavior in the log $R_n(T)$ vs. $\tau^{-1/2}$ plots of Figs. 1 and 2. (On a linear plot of $R_n(T)$ vs. T these "corners" were found to correspond to approximately the 75%-points on the resistive transitions.) In Fig. 3 we show T_{cOnset}, T_c, and a measure of the transition width, $\Delta T_c = T_{cOnset} - T_c$, plotted as a function of the YBCO layer thickness, for a series of superlattices with fixed barrier layer thickness (N = 16 PBCO unit cells). Figure 3 shows that the T_c values rise rapidly toward a thick YBCO film's T_{co} ~ 92 K, and that T_c and T_{cOnset} nearly have merged for only an 8-cell-thick YBCO layer. This merging of T_c and T_{cOnset} values is consistent with the very small ranges of K-T behavior (small separation of T_c and T_{cB}) found by Fiory et al.[18] and by Yeh and Tsuei[20] for much thicker YBCO films and platelets, if we take T_{cOnset} as an estimate for T_{cB}. [Another estimate of T_{cB} can be obtained by fitting $R_n(T)$ to the the Aslamazov-Larkin functional form for the paraconductivity,[25] in the temperature range above T_{cB} where the superconducting transition is just beginning, $R_n(T) \leq 1$. However, near T_{cB}, $R_n(T)$ of a superlattice is decreasing from near its maximum value, not falling linearly with temperature as does $R(T)$ for a single-component HTSc film; this significantly complicates the Aslamazov-Larkin fitting and may make less reliable the T_{cB} values obtained. Consequently, we have simply used the T_{cOnset} values defined above to estimate[14] the corresponding behavior of T_{cB}. The least-squares fitting to determine T_c values was done in the resistance range $R_n(T) < 0.3$ or < 0.1 (depending on the superlattice structure), well away from $T_{cOnset} \sim T_{cB}$.]

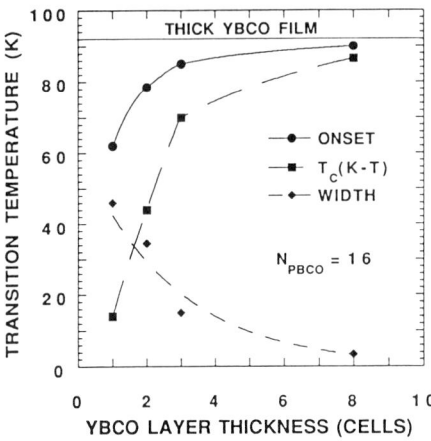
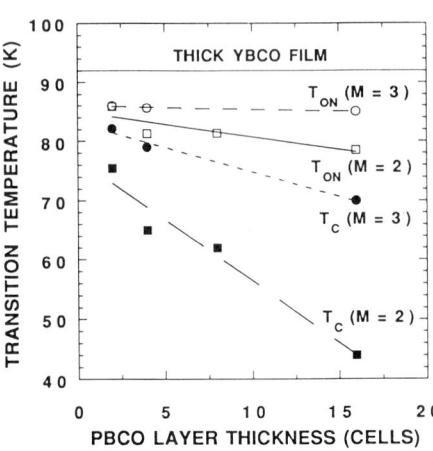

Fig. 3. Transition temperatures and widths as functions of YBCO layer thickness (in unit cells, for superlattices with well-isolatd YBCO layers ($N_{PBCO} = 16$).

Fig. 4. Onset and K-T transition temperatures as functions of PBCO barrier-layer thickness (in unit cells), for superlattices with 2- and 3-unit-cell-thick YBCO layers.

In Fig. 4 we show the dependence of T_{cOnset} and T_c on barrier-layer thickness, for the superlattices with 2- and 3-cell-thick YBCO layers. Note that only 3 unit cells of YBCO are needed for the onset temperature ($\sim T_{cB}$) to become essentially independent of the PBCO barrier-layer thickness, whereas T_c still varies significantly with barrier-layer thickness as either the 2- or 3-cell thick YBCO sheets become isolated from each other.

IV. DISCUSSION

Rasolt et al.[8] have interpreted the depression of $T_{cOnset} \sim T_{cB}$ that occurs for the thinnest (2- or 3-cell-thick) YBCO layers (see Fig. 3) as being due to electron transfer from PBCO to YBCO, which causes hole filling in YBCO and a decrease of T_{cB} (and, indirectly, also of T_c, as the superconducting layers become still more resistive). However, they find that the dominant effect on T_c is that due to reduced dimensionality and 2D fluctuations.[8]

The data presented in Figs. 1–4 are consistent with the log $R(T) \sim \tau^{-1/2}$ behavior expected[23, 24] for a Kosterlitz-Thouless transition; however, for the reasons discussed in section II, other experimental confirmation is needed. We emphasize that, unlike much earlier work[17,32], the evidence of K-T behavior presented in Figs. 1–4 occurs in multilayered *crystalline* (not amorphous or granular) films, for which the sheet resistance, $R_\square$, of the YBCO layers approaches the "limiting metallic resistance", $R_0 \sim \hbar/e^2 = 4.11$ kΩ/square, but may not greatly exceed it. If we assume that all of the conductance of our YBCO/PBCO superlattices is due to the YBCO layers, then from their measured total resistances at T = 100 K we can infer sheet resistances per YBCO layer (and per YBCO unit cell) of > 3,500 (> 3,500) $\Omega/\square$, > 1,300 (> 2,600) $\Omega/\square$, and > 700 (> 2,100) $\Omega/\square$, for the 1×16, 2×16, and 3×16 structures, respectively. In contrast,

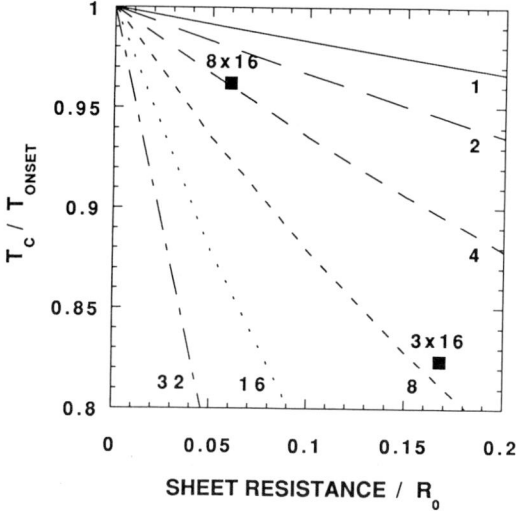

Fig. 5. Calculated and measured values of T_c/T_{cB}, as a function of dielectric constant ε and sheet resistance/YBCO layer, $R_\Box R_0$.

the $T = 100$ K sheet resistance of a single-cell-thick layer in a thick YBCO film ($\rho \sim 80\ \mu\Omega$-cm) is $R_\Box \sim 667\ \Omega/\Box$, if we take the effective sheet thickness as $d_{eff} \sim 12$ Å. Thus, single YBCO unit cells in superlattices are significantly more resistive than in a YBCO film. This additional resistance could be caused by hole filling,[7–9] especially for the thinnest YBCO layers, or by growth defects,[2,33] again particularly for the thinnest YBCO layers. If hole filling is present, the associated charge transfer should correspondingly *reduce* the resistance of the PBCO layers (by making their conductivity more hole-like), so that the true superlattice sheet resistances per YBCO unit cell may be even higher than those estimated above. Thus, hole filling, to the extent that it is present, may actually "push" the YBCO layers in YBCO/PBCO superlattices still deeper into the $R_\Box > 1$ k$\Omega/\Box$ regime, for which the work of Beasley et al.[17] indicates that 2D fluctuation effects should be significant, and particularly so if the effective dielectric constant due to the vortices is significantly greater than 1, as Fiory et al.[18] recently found for a YBCO film ($\varepsilon \sim 4.6$).

Fiory et al. have pointed out that their experiments on 500-Å-thick YBCO films apparently were the first in which K-T behavior was observed in a "clean" type-II superconductor.[18] However, if we use a free-electron-like formula for the normal-state conductivity of YBCO/PBCO superlattices, and assume one hole/unit cell in the YBCO layers, then the mean free path, in units of the a-axis lattice constant, is $\ell/a \sim 2.15\ R_0/R_\Box$, where $R_\Box$ in this equation (only) refers to a single-cell-thick layer. Hence, $R_\Box \geq 2$ k$\Omega/\Box$-unit cell corresponds to $\ell/a \sim 2$, and comparison with $\xi_{ab}(0) \sim 20$–30 Å shows that the superlattices with the thinnest YBCO layers may be closer to the "dirty" than the "clean" limit. This is interesting because Beasley et al. have shown that for a "dirty" thin-film superconductor with large κ ($= \lambda/\xi$, with λ and ξ being replaced by the a-b plane penetration depth and coherence length, respectively, for HTSc materials), and in the Ginzburg-Landau limit (i.e. for $\Delta T_c = T_{cB} - T_c \ll T_{cB}$), the depression of T_c below T_{cB} should be given by[17]

$$(T_c / T_{cB}) = 1 / [1 + 0.173\, \varepsilon\, (R_\square / R_o)], \qquad (5)$$

where ε is the dielectric constant due to the vortices and for a superlattice $R_\square$ is the sheet resistance/layer. Eq. (5) is plotted parametrically in Fig. 5 for a range of ε values, together with data for the 3×16 and 8×16 superlattices (the only ones for which the G-L condition $T_c \sim T_{cB}$ may be satisfied). Clearly, ε values ~4–8 are consistent with the data, and also in fairly good agreement with the value $\varepsilon = 4.6$ found by Fiory et al.[18] for YBCO.

The compositional integrity of superlattice structures is a critical issue in any interpretation of their properties. We recently used Z-contrast transmission electron microscopy (TEM) to examine several superlattices,[2,33] and were able to infer their growth mechanism.[33] For the moderate (average) supersaturations used to grow our superlattices, it was found that the $c_\perp$ structures are formed by a process of nucleation and growth of 3D islands.[33] The TEM studies suggest that the growing units are one Ba-Y-Ba structural unit (~11.7 Å) in height and are terminated at the Cu-chain plane. This island growth mechanism leads naturally to gradual roughening (gentle undulations) of the growing surface with increasing film thickness. The absence of structural defects upon island coalescence demonstrates that the height of the growing 3D islands must be an integral number of unit cells; direct TEM observations of a 1×8 structure show that the preferred island height is a single unit cell.[2,33] These observations agree with a recent study of YBCO film growth by Terashima, Bando, et al., in which RHEED oscillations and precision x-ray analysis were used to show that the unit cell is the minimum structural unit that can satisfy the requirements of stoichiometry and electrical neutrality.[34] However, the undulations do produce one-unit-cell "step" or "kink" discontinuities in the chemical continuity of a single-cell-thick layer, for a lateral movement of one island diameter. In our superlattices this could require the current in a YBCO layer to tunnel one cell along the c-direction for every 20–30 nm laterally, which will have no serious effect on the critical current density, J_c, in an M×N superlattice unless M = 1. For a 1×N superlattice, the average overlap between YBCO islands is zero, and the connectivity of the superconducting path through the single-cell YBCO layer is determined by "chance" areas of overlap at the edges of islands.[33] This should result in a significant reduction of J_c (which we observe)[35] and in clear weak-link behavior for 1×N structures. The relatively high $R_\square$ values found for superlattices with the thinnest YBCO layers (especially M = 1 and 2) may be due in part to current conduction along the c-axis at these one-cell step discontinuities.

Lobb, Abraham, and Tinkham have considered in detail how a vortex-unbinding phase transition would affect various properties of a square lattice of superconducting weak links.[36] They found that the interaction between junctions suppresses the Ambegaokar-Halperin temperature-dependent resistance[37] of an isolated junction, leaving the free vortex density as the dominant temperature-dependent factor in the expression for the weak-link array's resistance. Their resulting expressions for the temperature-dependent phase correlation length and resistance are very similar to Eq. (1), except for an additional temperature-dependent factor inside the exponential. This additional factor comes from the temperature-dependent coupling energy, $E_J(T)$, between adjacent superconducting islands,[36] which causes the resistance to be expressed in terms of a dimensionless temperature parameter, $k_BT/E_J(T)$. However, Lobb et al. also point out that if $T_c \ll T_{cB}$ then $E_J(T) \sim E_J(0) \sim$ constant, so that the expressions for R(T) become entirely equivalent for fitting purposes. Thus, the K-T behavior that we observe (Figs. 1 and 2) in YBCO/PBCO superlattices may be entirely consistent both with TEM studies and with the expected behavior of superconducting layers that are composed of an array of weak links.

V. CONCLUDING REMARKS

One possible interpretation of the K-T data analysis presented here is that reducing the thickness of well-isolated YBCO layers (to ≤ 100 Å) produces 2D fluctuations in an ever-widening temperature range ($T_{cB} - T_c$ increasing). Reducing the PBCO barrier-layer thickness apparently then increases the coupling between the conducting CuO_2 bilayers in the YBCO layers, toward the value ($K_\perp / K \sim 10^{-2}$) found in pure YBCO. Relative to thick-film and single-crystal HTSc specimens, YBCO/PBCO superlattice structures may provide a much larger temperature range for observation of 2D fluctuations, between T_{cB} and T_c values that appear to be separated by tens of degrees. Consequently, they may provide very interesting 3D systems for future systematic studies of how quasi-2D sheets couple together to produce the 3D superconducting state.

However, we believe that the nature of this coupling is not revealed by experiments to date. As we noted in the Introduction, the model of Ariosa and Beck[11] (involving an array of in-plane Josephson-coupled weak links, with electrostatic coupling along the c-axis) may lead to K-T resistive behavior very similar to that suggested by Rasolt et al.,[8] and in agreement with the earlier analysis by Lobb et al..[36] Furthermore, the apparently successful application of K-T theory to describe the resistive transition of YBCO/PBCO superlattices raises questions that are both troubling and intriguing. How can the large temperature range for apparent K-T resistive behavior be reconciled with the strong flux pinning in these materials (when T_c is not close to T_{cB}). How can we reconcile the sensitivity of T_c to both YBCO and PBCO layer thicknesses, with the apparent near-independence of $T_{c(midpoint)}$ from layer thickness in Bi-2212 superlattices? Finally, Fig. 5 suggests that, if free vortices are being generated in the YBCO layers of these superlattices, then they may provide very interesting systems in which to study vortex interactions.

This research is sponsored by the Division of Materials Sciences, U.S. Department of Energy under contract DE-AC05-84OR21400 with Martin Marietta Energy Systems, Inc.

REFERENCES

1. D. H. Lowndes, D. P. Norton, and J. D. Budai, *Phys. Rev. Lett.* **65**, 1060 (1990).
2. D. H. Lowndes, D. P. Norton, J. D. Budai, D. K. Christen, C. E. Klabunde, R. J. Warmack, and S. J. Pennycook, in *Progress in High-Temperature Superconducting Transistors and Other Devices*, ed. by R. Singh, J. Narayan, and D. T. Shaw, *SPIE Proceedings* **1394**, 150 (1990).

3. J.-M. Triscone, O. Fischer, O. Brunner, L. Antognazza, A. D. Kent, and M. G. Karkut, *Phys. Rev. Lett.* **64**, 804 (1990).
4. O. Brunner, J.-M. Triscone, L. Antognazza, and O. Fischer, Physica B **165-166**, 469 (1990).
5. Q. Li, X. X. Xi, X. D. Wu, A. Inam, S. Vadlamannati, W. L. McLean, T. Venkatesan, R. Ramesh, D. M. Hwang, J. A. Martinez, and L. Nazar, *Phys. Rev. Lett.* **64**, 3086 (1990).
6. M. Kanai, T. Kawai, and S. Kawai, *Appl. Phys. Lett.* **57**, 198 (1990).
7. R. F. Wood, *Phys. Rev. Lett.* **66**, 829 (1990).
8. M. Rasolt, T. Edis, and Z. Tesanovic, submitted to *Phys. Rev. Lett.*.
9. J. Z. Wu, C. S. Ting, W. K. Chu, and X. X. Yao, submitted to *Phys. Rev. Lett.*.
10. T. Schneider, Z. Gedik, and S. Ciraci, submitted to *Europhysics Letters*.
11. D. Ariosa and H. Beck, *Phys. Rev. B* **43**, 344 (1991).
12. J. M. Kosterlitz and D. J. Thouless, *J. Phys. C* **6**, 1181 (1972); J. M. Kosterlitz and D. J. Thouless, *Prog. Low Temp. Phys.* **7**, 373 (1978).
13. J. M. Kosterlitz, *J. Phys. C* **7**, 1046 (1974).
14. J. E. Mooij, in *Advances in Superconductivity*, ed. by B. Deaver and J. Ruvalds (Plenum Press, New York, 1983), p. 433.
15. J. E. Mooij, in *Percolation, Localization, and Superconductivity*, ed. by A. M. Goldman and S. A. Wolf (Plenum Press, New York, 1984), p. 325.
16. J. Pearl, *Appl. Phys. Lett.* **5**, 65 (1964).
17. M. R. Beasley, J. E. Mooij, and T. P. Orlando, *Phys. Rev. Lett.* **42**, 1165 (1979).
18. A. T. Fiory, A. F. Hebard, P. M. Mankiewich, and R. E. Howard, *Phys. Rev. Lett.* **61**, 1419 (1988).
19. S. Martin, A. T. Firoy, R. M. Fleming, G. P. Espinosa, and A. S. Cooper, *Phys. Rev. Lett.* **62**, 677 (1989).
20. N.-C. Yeh and C. C. Tsuei, *Phys. Rev. B* **39**, 9708 (1989).
21. Q. Y. Ying and H. S. Kwok, *Phys. Rev. B* **42**, 2242 (1990).
22. D. H. Kim, A. M. Goldman, J. H. Kang, and R. T. Kampwirth, *Phys. Rev. B* **40**, 8834 (1989).
23. B. I. Halperin and D. R. Nelson, J. Low Temp. Phys. 36, 599 (1979).
24. V. Ambegaokar, B. I. Halperin, D. R. Nelson, and E. D. Siggia, *Phys. Rev. B* **21**, 1806 (1980).
25. L. G. Aslamazov and A. I. Larkin, *Soviet Physics--Solid State* **10**, 875 (1968).
26. W. J. Skocpol and M. Tinkham, *Rep. Prog. Phys.* **38**, 1409 (1975).
27. P. P. Freitas, C. C. Tsuei, and T. S. Plaskett, *Phys. Rev. B* **26**, 833 (1987).
28. M. Tinkham, *Intro. to Superconductivity* (McGraw-Hill, New York, 1975).
29. D. R. Nelson and J. M. Kosterlitz, *Phys. Rev. Lett.* **39**, 1201 (1977).
30. P. Minnhagen, *Rev. Mod. Phys.* **59**, 1001 (1987).
31. D. H. Lowndes, D. P. Norton, J. W. McCamy, R. Feenstra, J. D. Budai, D. K. Christen, E. Jones, and D. Poker, *Mat. Res. Soc. Symp. Proc.* **169**, 431 (1990).
32. See Refs. 14, 15, and 30 for reviews of earlier work.
33. S. J. Pennycook, M. F. Chisholm, D. P. Norton, D. H. Lowndes, R. Feenstra, H. R. Kerchner, and J. O. Thomson, submitted to *Phys. Rev. Lett.*.
34. T. Terashima, Y. Bando, et al., *Phys. Rev. Lett.* **65**, 2684 (1990).
35. R. H. Kerchner, unpublished data.
36. C. J. Lobb, D. W. Abraham, and M. Tinkham, *Phys. Rev. B* **27**, 150 (1983).
37. V. Ambegaokar and B. I. Halperin, *Phys. Rev. Lett.* **22**, 1364 (1969).

LOCAL STRUCTURAL ANOMALY AT T_c OBSERVED BY NEUTRON SCATTERING

T. Egami[1], B.H. Toby[1], S.J.L. Billinge[1], Chr. Janot[2], J.D. Jorgensen[3], D.G. Hinks[3], M.A. Subramanian[4], M.K. Crawford[4], W.E. Farneth[4] and E.M. McCarron[4]

1. Department of Materials Science and Engineering, and Laboratory for Research on the Structure of Matter University of Pennsylvania, Philadelphia, PA 19104-6272
2. Institut-Laue-Langevin, 156X-38402 Grenoble Cedex, France
3. Division of Materials Science, Argonne National Laboratory, Argonne, IL 60439
4. Central Research and Development, E.I. du Pont de Nemours and Co., Experimental Station, Wilmington, DE 19898

INTRODUCTION

One would expect that the onset of superconductivity might leave a signature on the atomic structure of the high temperature superconducting oxides, if the electron-lattice interaction is strong and totally or partially accounts for the occurance of superconductivity. For this reason the structural change at T_c has been the subject of considerable interest. Although the initial report of a lattice distortion at T_c [1] was contradicted by later measurements [2,3] and it is now well established that long range order parameters such as the lattice constants do not show clearly discernible changes at or near T_c, there are strong experimental indications that some local, possibly dynamic, structural changes take place at the onset of superconductivity. For instance ion-channelling experiments [4-6] and EXAFS measurements [7,8] suggest such changes involving Cu and O atoms. Our pulsed neutron atomic pair-distribution analysis gave a more direct evidence of changes in the interatomic correlations in the vicinity of T_c [9]. In this paper we describe our recent neutron scattering data which strongly suggest that the local dynamic structural anomaly at T_c is a general phenomenon for high-T_c solids, and discuss the implication of these observations, particularly focussing on $Tl_2Ba_2CaCu_2O_8$.

REAL-SPACE ANALYSIS OF PULSED NEUTRON SCATTERING DATA

It may appear that the atomic structure of superconducting oxides has been precisely determined beyond any doubt by a large number of crystallographic studies [10]. A more careful look soon reveals, however, that this impression is seriously mistaken. For instance the thermal, or Debye-Waller, factors to describe the apparent atomic vibrations of some of the atoms, notably oxygen, are quite often anomalously large, with little dependence upon temperature. Such large "thermal factors" usually flag failure of the structural analysis, since they imply that there exist large uncertainties in the atomic position. Anomalously large thermal factors are in most cases not due to large amplitudes of lattice vibrations, but due to static or quasi-static atomic displacements that are not correctly resolved by the standard crystallographic structural determination.

The standard method of structural study, usually using x-rays from a sealed tube or neutrons from a thermal reactor, proceeding with indexing of the Bragg diffraction peaks, tends to fail in correctly describing these displacements for two principal reasons; the wavelength of the probe used is not sufficiently short, and the assumption of periodicity is not warranted. The spatial resolution of the diffraction experiment is determined by the wavelength of the probe, so that in order to resolve a displacement of an atom by δ, the wavelength of the probe must be shorter than 2δ. Only when the displacement has a long range periodicity which effectively renders a microscopic displacement into a macroscopic length change, can it be precisely determined by a probe with a long wavelength. The lattice constant can be determined very precisely, only because the lattice preiodicity in effect transforms the microscopic lattice constant, a, into a macroscopic length, Na, where N is the number of the unit cells in one dimension of the crystal. However, even when a high energy probe with a short wavelength is used, the crystallographic analysis fails to detect short range correlations among the displacements that often are important in undertanding the physical consequence of such displacements. The crystallographic study describes only the _average_ structure, since it assumes a-priori lattice periodicity and considers only the position and intensity of Bragg diffraction peaks. Non-periodic displacements can only be described in terms of partially occupied split atomic positions, without correlations of occupation, or by artificially enlarged thermal factors.

In order to observe these non-periodic atomic displacements accurately we have been using high energy probes such as pulsed neutrons and high energy x-rays from the synchrotron source, and the method of atomic pair distribution function (PDF) analysis. The PDF analysis has been widely but almost exclusively used for the structural studies of glasses and liquids. The PDF describes the distribution of the interatomic distances in the material averaged over all orientation and weighted by the neutron (or x-ray) scattering amplitude for each atom. It is obtained by Fourier-transforming the normalized diffraction intensity including the diffuse scattering intensity. The neutron coherent scattering intensity corrected for absorption and multiple-scattering and normalized is given by

$$I(\vec{Q},\omega) = N\langle b\rangle^2 S(\vec{Q},\omega)$$
$$= \sum_{i,j} b_i b_j e^{i(\vec{Q}\cdot\vec{r}_{ij} - \omega t_{ij})} , \qquad (1)$$

where $\vec{Q}$ and ω are the momentum and energy transfers during scattering, N is the number of atoms in the solid, $\langle...\rangle$ represents a compositional average, b_i is the neutron scattering length of the i-th atom, and $\vec{r}_{ij}$ and t_{ij} are the separations in space and time between the scattering events involving the i-th and j-th atoms. If the measurement is done on a powder sample, the structure factor should be averaged over all directions of $\vec{Q}$;

$$S(Q,\omega) = \frac{1}{4\pi} \int S(\vec{Q},\omega) d\Omega_Q . \qquad (2)$$

Furthermore when the neutron detector does not have an energy analyzer, as in the case of time-of-flight pulsed neutron measurements, the measured intensity is close to the total structure factor,

$$S_t(Q) = \int_{-\infty}^{\infty} S(Q,\omega) d\omega \quad , \tag{3}$$

except that Q depends slightly upon ω in the integration due to the Placzek shift [11]. If we neglect this shift for the time being, then from (1) and (3) we can see that only the scattering events with $t_{ij} = 0$ contribute to $S_t(Q)$. The PDF obtained as the inverse Fourier-transform of $S_t(Q)$,

$$\rho(r) = \rho_0 + \frac{1}{2\pi^2 r} \int [S_t(Q) - 1] \sin(Qr) Q dQ \quad , \tag{4}$$

where ρ_0 is the average density, then describes the <u>instantaneous</u> atomic correlations. Theoretically the range of integration in eq. (4) is from 0 to ∞. In practice, however, the maximum value of Q to which S(Q) can be determined is finite, and is less than $2k_i$ where k_i is the momentum of the incoming particle (probe). For the Fourier transformation to be accurate, the scattering intensities have to be determined up to high values of momentum transfer, Q. This requires a high energy, short wavelength probe to be used. The pulsed neutron source, rich in high-energy epithermal neutrons, satisfies this need quite nicely. As shown in Fig. 1, the PDF of polycrystalline Al at T = 50 K, obtained from $S_t(Q)$ determined with the SEPD spectrometer at the IPNS up to 30 Å^{-1}, is in excellent agreement with the calculated PDF up to 40 Å [12]. The PDF peak width in this case is determined by the zero-point atomic vibration.

STRUCTURAL ANOMALY IN THE VICINITY OF T_c

If the atomic motions are uncorrelated, the PDF peaks such as those in Fig. 1 are Gaussian with σ equal to $\sqrt{2\langle u^2 \rangle}$, where u is the atomic displacement along the direction of $\vec{r}_{ij}$. Thus the peak height, which is inversely proportional to σ, should monotonically decrease with increasing temperature. We found that some of the PDF peaks of high-T_c solids do not behave in such a way. Fig. 2 shows the height of the PDF peak at 3.4 Å for $Tl_2Ba_2CaCu_2O_8$ [Tl(2212)] (T_c = 110 K) as a function of temperature [9]. The solid line is the expected normal behavior, calculated from the phonon density of states measured by inelastic neutron scattering, and the vertical bars indicate statistical errors. Thus it is quite clear that there are statistically meaningful deviations from the normal behavior below T_c. Similar anomalies were found for $La_{2-x}Sr_xCuO_4$ (x = 0.12, 0.15), $YBa_2Cu_3O_7$, $Nd_{1.835}Ce_{.165}CuO_4$, and $(Ba_{.6}K_{.4})BiO_3$ at various distances as will be described in detail elsewhere. An example is shown in Fig. 3, for the Cu-O peak at 1.93 Å of $La_{1.88}Sr_{.12}CuO_4$.

DYNAMIC NATURE OF THE ANOMALY

By placing an energy analyzing crystal in front of the neutron detector it is possible to eliminate all, or most of, the inelastic scattering and obtain S(Q,0). Then the Fourier-transform,

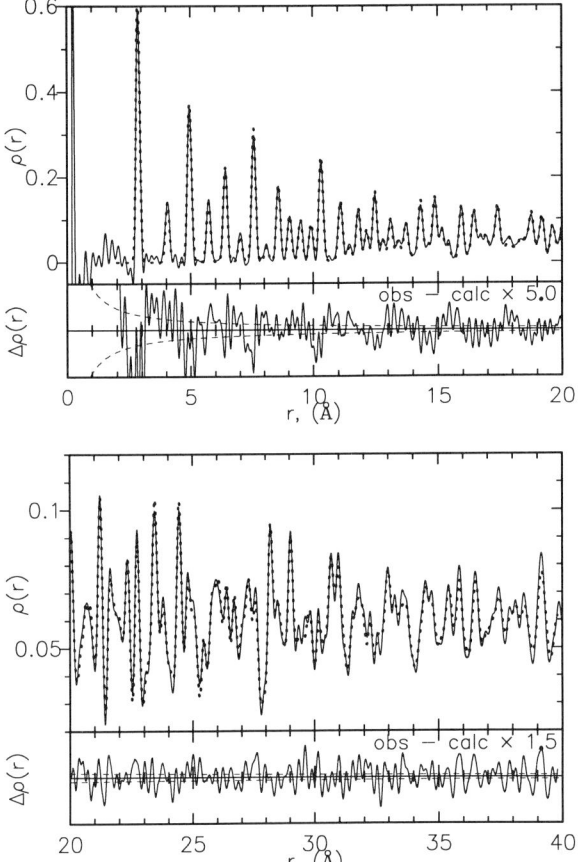

Fig. 1. The pair-distribution function of polycrystalline aluminium at T = 50 K, determined by the pulsed neutron scattering [12]. In the upper figure, the PDF is shown up to 20 Å, and in the lower figure up to 40 Å. In the top portion of each figure, the solid line is the experimental PDF, and the dotted line is the calculated PDF. The lower portion shows the difference between the experimental and calculated PDF's (solid line) and the estimated statistical error (dashed line), magnified by a factor of 5.0 (top), and 1.5 (bottom).

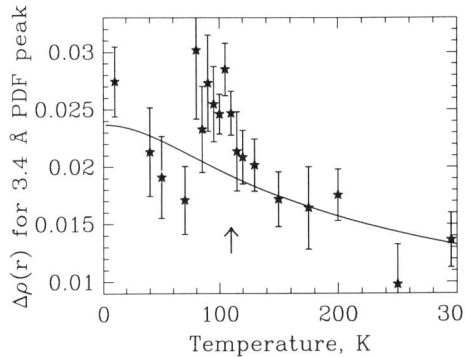

Fig. 2. Temperature dependence of the PDF peak height at 3.4 Å for Tl(2212). The solid line is the normal temperature dependence calculated from the phonon density of states, and the vertical bars indicate estimated statistical error. The arrow indicates T_c [9].

$$\rho_{el}(r) = \rho_0 + \frac{1}{\pi^2 r}\int [S(Q,0) - \exp(-2BQ^2)]\sin(Qr)QdQ \quad , \tag{5}$$

where $\exp(-2BQ^2)$ is the Debye-Waller factor, yields the elastic or static PDF which describes the auto-correlation of the time averaged single atomic density,

$$\rho_{el}(\vec{r}) = \int \rho_1(\vec{r})\rho_1(\vec{r}+\vec{R})d\vec{R} \quad , \tag{6}$$

where $\rho_1(r)$ is the single atomic density function, and averaged over the all directions of $\vec{r}$. The measurement of $S(Q,0)$ was made using the triple-axis-spectrometer at the IN-1 beamline of the High Flux Reactor of the Institut-Laue-Langevin (ILL) with the hot neutron source operating at 2400 K. The incident neutron energy was set at 265 meV, so that $S(Q,0)$ can be determined up to large values of Q (19.8 Å^{-1}). The energy resolution (HWHM) was 6.8 meV. The static or elastic PDF thus obtained for Tl(2212) is compared with the instantaneous PDF obtained with the identical termination condition in Fig. 4. Even though they were determined by two totally different techniques the two PDF's show remarkably good overall agreement. However, there are significant differences, for instance around 2.4 Å and around 3.4 Å. The temperature dependence of the 3.4 Å peak is quite different for these two as shown in Fig. 5, clearly indicating that the anomaly is dynamic in nature.

MONTE-CARLO STRUCTURE REFINEMENT

The PDF thus determined directly convey substantial amounts of real-space information, such as the temperature dependence discussed above. However, due to the orientational averaging and the self-convolution inherent to the auto-correlation functions, the PDF does not directly provide three-dimensional structural information. In order to derive more specific information, it is necessary to make comparisons to a PDF calculated for a model structure. By such comparison it sometimes becomes clear how the structure model needs to be modified to attain better agreement with the experimental PDF. For instance if the model has one PDF peak in a range of distances while the experimental PDF has a split peak, atomic positions must be bifurcated. Such an observation led us to conclude earlier that the apical oxygen in $La_{2-x}(Sr,Ba)_xCuO_4$ is shifted away from the c-axis, tilting the CuO_6 octahedra even in the tetragonal phase [13], and oxygen and thallium in the Tl-O plane of Tl(2212) are displaced to alter the local symmetry from tetragonal to orthrhombic [14]. In many cases, however, the real structure is complex, and computer refinement procedure becomes necessary. We define the agreement factor A by

$$A^2 = \int_{r_{min}}^{r_{max}}[\rho_{exp}(r) - \rho_{cal}(r)]^2 dr / \int_{r_{min}}^{r_{max}} \rho_0^2 dr \quad , \tag{7}$$

where $\rho_{exp}(r)$ is the experimentally determined PDF, $\rho_{cal}(r)$ is the calculated PDF, and r_{min} and r_{max} define the region in which the comparison is made. Our Monte-Carlo refinement process minimizes the A factor by the Metropolis algorithm of simulated annealing. For a random displacement $\Delta\vec{r}$ the change in the A factor, ΔA, is calculated. Then a random number R ($0 < R \le 1$) is generated and compared to

$$P = \frac{1}{e^{\Delta A/kT} + 1} \quad . \tag{8}$$

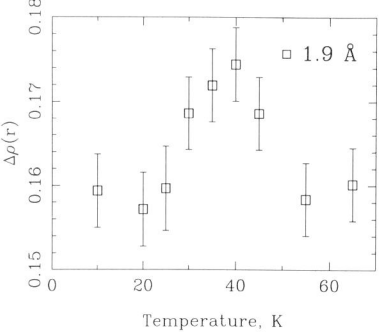

Fig. 3. Temperature dependence of the PDF peak height at 1.93 Å for $La_{1.88}Sr_{.12}CuO_4$. This distance corresponds to the Cu-O separation in the Cu-O plane. T_c of this solid is 32 K.

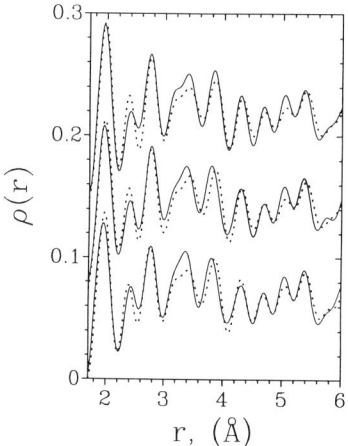

Fig. 4. Elastic, or time averaged PDF for Tl(2212) obtained at the ILL (solid line) and the instantaneous PDF obtained at the IPNS (dotted line), at T = 120 K (top), 90 K (middle) and 70 K (bottom).

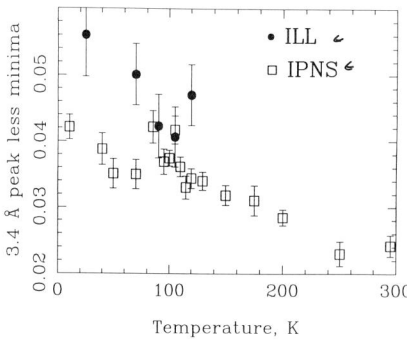

Fig. 5. Temperature dependence of the peak height at 3.4 Å, of static PDF (closed circle) and of instantaneous PDF (open square). The vertical bars indicate statistical error.

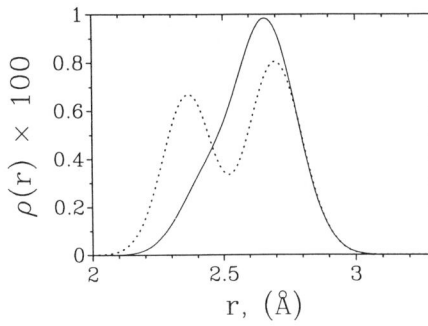

Fig. 6. Distribution of the Cu-O2 distance at T = 120 K, for the instantaneous correlation (dotted line) and time averaged correlation (solid line). The distances in each model were convoluted by a Gaussian peak with the width corresponding to the calculated vibrational amplitude.

If R is smaller than p, the displacement Δr is "accepted", that is, the structure is modified by Δr, and if R is larger than p, Δr is rejected. The role of the fictitous temperature T is to avoid the system from being trapped in a local minimum, and is adjusted to control the convergence.

The Monte-Carlo refinement applied to Tl(2212) showed that the distance between Cu and the apical oxygen (O2) is bifurcated for the IPNS (instantaneous) PDF, but not in the ILL (static) PDF, as shown in Fig. 6. This means that the O2 atom is in an apparent double-well potential, with the separation of bistable points by about 0.3 Å, and jumps between the wells with the frequency higher than the energy resolution of the ILL spectrometer, 6.8 meV. The bifurcation of the O2 atom, however, is hardly temperature dependent, persisting much above T_c. On the other hand the oxygen atoms in the Cu-O plane (O1) were found to become displaced along the c-axis perpendicular to the plane at low temperatures, with the average amplitude increasing with decreasing temperature over about 100 K above T_c and saturating to a value. Below T_c, about a half of O1 atoms are displaced from the plane by about $\pm$ 0.3 Å.

These O1 and O2 displacements are largely uncorrelated at high temperatures, but in the vicinity of T_c they become correlated, in such a way that when the O2 is closer to Cu, the O1 atoms nearby are diplaced from the Cu-O plane, and when the O2 is away from Cu, the O1 atoms nearby are not displaced from the plane, having normal vibrational amplitudes. Thus there are <u>two distinct environmnets of Cu</u>, one with a short O2-Cu distance and displaced O1, and the other with a long O2-Cu distance and undisplaced O1. The latter case is the typical CuO_5 half-octahedral environment found in many insulating cuprates, while the former is anomalous. Now each O1 atom has two O2 neighbors, so that if these two O2 were in different positions the O1 atom would be in frustration. The fact that the O1-O2 correlation is quite strong indicates that these two kinds of O configurations occur in clusters, so that a O1 atom has two O2 neighbors more often in the same configuration. Thus our picture is that there are regions or domains of normal and anomalous Cu environments below T_c.

For compounds other than Tl(2212) the Monte-Carlo simulation is still in progress and incomplete, but both in $La_{2-x}Sr_xCuO_4$ (x = .12, .15) and in $Nd_{1.835}Ce_{.165}CuO_4$ the anomaly is clearly associated with the unharmonic displacement of oxygen atoms around Cu just as in the case of Tl(2212). In $YBa_2Cu_3O_7$, however, the structural change is more subtle, and we have not obtained a clear evidence of bifurcation in the oxygen position so far, even though the evidence of some local structural change below T_c is unmistakable.

OXYGEN DISPLACEMENT AND ELECTRONIC STRUCTURE

In this section we speculate on the possible cause and consequence of the oxygen displacements. The main idea is that the displacement of oxygen atoms can cause non-linear effects on the electronic band structure near the Fermi surface, and can strongly influence the spatial distribution of charge carriers. Calculated band structure shows a number of bands close to the Fermi surface [15]. Oxygen displacements can shift the energy of some of the bands and make them touch and cut the Fermi level locally, creating extra carriers. Although the exact nature of such Fermi level crossing will vary from a system to system, and needs to be clarified by actual electronic structure calculations, it is possible to consider a hypothetical case and speculate on the phenomena.

As an example let us consider the case of d_{z^2} band. The displacement of the apical oxygen (O2) which reduces the Jahn-Teller distortion was considered from the beginning as the mechanism of strong electron-lattice interaction [16]. In the insulating Cu^{2+} (d^9) state, the d_{z^2} orbital (with the z-axis in the direction of O2) is occupied, while the hole resides in the σ state consisting of hybridized Cu-$d_{x^2-y^2}$ and O-p_x orbitals. If Cu is fully oxidized to Cu^{3+}, an additional hole goes to the d_{z^2} state, thus contracting the distance between Cu and O2. Therefore it is most natural to consider the O2 displacement in relation to the presence of holes in the d_{z^2} orbital.

In all the cuprates including Tl(2212), electronic band structure calculations based upon the crystallographic coordinates indicate that the d_{z^2} band is well below the Fermi level, and the σ band cuts across the Fermi level [15]. In addition, photo-emission studies suggest that in the doped samples the holes are mostly in the oxygen p (σ) band. These led to the conclusion held by many that the conducting holes are in the σ band, and the theory can be greatly simplified to consider only the hopping and the exchange between the Cu sites, described by the t-J Hamiltonian. However, the observed displacement of the apical oxygen suggests a strong possibility that the displaced apical oxygen may move the holes to the d_{z^2} band, by raising it above the Fermi level. Since the d_{z^2} band is completely below the Fermi level, as the apical oxygen is brought toward Cu the total energy would initially increase. But as the band starts to cut the Fermi level and holes start to occupy the d_{z^2} band, the attraction between the hole and the apical oxygen would reduce the energy, thus the total energy vs. the oxygen displacement would show double minima. This explains the origin of the bifurcated O2 positions. In other words the displacement of O2 atoms leads to formation of polarons in the d_{z^2} band.

This induction of holes to the d_{z^2} level is most likely done by collective displacement of adjacent apical oxygens, since it is likely to cost less energy to raise the top portion of the d_{z^2} band, which is already close to the Fermi level, above the Fermi level by collective oxygen displacement, than to bring an entire d_{z^2} state of one atom above the Fermi level by a displacement of a single apical oxygen. Thus the oxygen displacement is likely to result in a large polaron [17-19] than a small polaron. The size of the large polaron will be discussed later. The in-plane oxygen atom has a tendency of becoming displaced out of the plane, even in the insulating state [20]. The presence of holes in the d_{z^2} state would attract the in-plane oxygen (O1) toward the apical oxygen, and induce larger displacements of the in-plane oxygen atoms. This explains the correlation between O2 and O1.

OXYGEN INDUCED STATES AND SUPERCONDUCTIVITY

It is reasonable to assume that the area of the domain, in which the O2 atoms are collectively displaced, is proportional to the number of carriers in the domain. Then if this domain contains only one electron rather than two, its size has to be reduced by 30 %, costing more energy. The size of the domain is inversely related to the size of the hole pocket in the d_{z^2} band, that is above the Fermi level. Thus in order to reduce the domain diameter, ξ_B, by a factor of $1/\sqrt{2}$, the top of the band has to be raised above the Fermi level twice as much. Therefore it is likely to cost less energy to create a domain which contains two holes than to create two separate domains containing one hole each, if the Coulomb repulsion is not too strong. In addition if the top of the d_{z^2} band is at the X-point so that the wave function of the holes inside the domain has the X-point character consistent with the anti-ferromagnetic spin polarization, the energy of a hole pair is further reduced by forming an exchange coupled singlet state. Thus there is a realistic possibility that the domain thus created represents a bipolaron, or the negative-U center.

One way to estimate the size of the domain may be the following. The displacement of apical oxygens lifts the d_{z^2} band but at the same time lowers the Tl s-p band. This should result in the electron transfer from the Cu-O layer to the Tl-O layer, creating large dipolar moments. We may assume that these dipolar moments are screened by the moments due to the O2 displacement. If the collective displacement is limited to within one Cu-O plane, the number of O2 atoms involved in a collective displacement, N, may be estimated by

$$N = \frac{nR_c}{2\delta}, \quad (9)$$

where n is the number of holes in the domain, R_c is the distance between the Cu-O plane and the Tl-O plane (= 4.67 Å [10]), and δ is the magnitude of the O2 displacement. For n = 2 and δ = 0.3, we obtain N ≈ 16 (= 4 x 4). Thus the domain size, ξ_B, is about 4a (≈ 15 Å), comparable to the coherence length. Another possibility is that the domain size is related to the Fermi surface, by $\xi_B = \pi/k_F$, and the latent tendency of the CDW excitation produces the domain.

The bipolarons created by the oxygen displacement as described above would have a rather large mass, therefore their bose condensation temperature may be quite low. However, by coupling with the conduction electrons they may be able to produce superconductivity. At high temperatures the bipolarons would be Anderson-localized due to random motion of oxygen atoms, and they would move only by hopping. But as temperature is reduced and the oxygen motions become more coherent, they may start to form a narrow delocalized band, coupled through hybridization with the σ band. The localized negative-U centers forming a narrow band as the consequence of hybridization with the conduction band has a strong possibility as a mechanism of superconductivity [21]. In particular if the d_{z^2} bipolarons and the σ holes have different k's, since the former is likely to have the X- or Γ-point symmetry while the Fermi level is cutting the σ band in mid-zone, they will not directly hybridize, but interact only through the Coulomb repulsion and exchange, thus reducing the single particle scattering of the localized electron by the conduction electron which destroys pairing. As discussed by Micnas et al. [21] and more recently by Bar-Yam [22], the relevant Hamiltonian is,

$$H = \sum_{k,\sigma} \epsilon_k c^\dagger_{k,\sigma} c_{k,\sigma} + E_B \sum_q B^\dagger_q B_q + \sum_{q,k} [w_q c^\dagger_{k,\uparrow} c^\dagger_{-k+q,\downarrow} B_q + H.c.] \quad (10)$$

where B_q is the Fourier-transform of a boson anihilation operator for the local bipolarons. Note that here a single particle scattering term, $t_k [c^\dagger_k d_k + d^\dagger_k c_k]$, where d_k is an anihilation operator for the localized hole, is absent because the conduction holes and localized holes occupy different parts of the k-space. The mean-field solution resembles that of the BCS Hamiltonian.

The discussion above should apply without major changes to the case when the domain is formed on the band other than d_{z^2}. For instance in Tl(2212) the highest occupied bands (at the X-point) have O-p_π and O-p_z character [23]. As long as the domain develops at a different location in the k-space than the σ holes, the single particle scattering of the σ holes by the domain states is suppressed, and the pair interaction as in eq. (10) becomes dominant. The essential mechanism is that the virtual excitation of the bipolaronic pair in the domain pairs up the σ electrons. Therefore it may be possible to observe, below T_c, virtual electrons above the Fermi level at the X-point by photo-emission experiments.

The Monte-Carlo analysis of Tl(2212) indicates that about 40 % of Cu atoms are in the bipolaronic domains. From this we can estimate the concentration of the holes in the bipolaronic state to be 5 % with respect to Cu. Now if the scaling of the carrier concentration and T_c suggested by μ-SR [24] is correct, Tl(2212) should have the total carrier density of about 50 % with respect to the Cu density, similar to that in $YBa_2Cu_3O_7$. Thus the concentration of holes in the bipolaronic state is only about 10 % of the total carriers, and the majority of carriers remain in the σ band. This explains why the photo-emission measurements find the holes to be in the σ band. Furthermore, if the estimate of the total concentration of the carriers above is correct and all the carriers condense into the superconducting state, then there must be 4 ~ 5 Cooper pairs within the coherence length, suggesting that the system is in the weak-coupling limit. The local bipolaron mediated pairing mechanism described above indeed leads to the weak-coupling, even though T_c is high. When the doping concentration is increased the bipolarons start to overlap and become delocallized, thus the superconductivity will disappear, as the experimental results show.

CONCLUSIONS

Through the pair-distribution function analysis of neutron scattering data the structure of high temperature superconducting oxides were found to display anomalous behavior in the superconducting state. In particular the oxygen atoms around Cu which are displaced from the crystallographic atom sites, show more correlated collective unharmonic motion below T_c. We conjecture that these oxygen movements must originate from the presence of superconducting charge carriers, and propose a senario involving a bipolaronic band induced by the displaced oxygen atoms. While we are very far from proving this senario either experimentally or theoretically, our observation strongly indicates that the local atomic scale dynamic inhomogeneity must be playing a crucial role in producing superconductivity in oxides, and that experimental characterization of such inhomogeniety is of great importance. Since few traditional methods of structural study are suited to determining such local scale dynamics, more experiments specially designed to look for such effects need be carried out. The PDF analysis is just one of these attempts, and we hope that more varieties of other techniques will also be focussed in this direction.

ACKNOWLEDGMENTS

It is a particular pleasure to aknowledge very useful discussions with Drs. D. Emin, Y. Bar-Yam, E. Mele, L. F. Mattheiss, J. B. Goodenough and A. J. Arko. The work at the University of Pennsylvania was supported by the National Science Foundation through DMR90-01704 and DMR88-19885. The Intense Pulsed Neutron Source is operated as a user facility by the U.S. Department of Energy, Division of Materials Sciences, under contract No. W-31-109-Eng-38.

References

1. P.M. Horn, D.T. Keane, G.A. Held, J.L. Jordan-Sweet, D.L. Kaiser, F. Holtzberg and T.M. Rice, Phys. Rev. Lett., 59, 2772 (1987).
2. e.g., D.E. Cox, C.C. Torardi, M.A. Subramanian, J. Gopalakrishnan and A.W. Sleight, Phys. Rev., B38, 6624 (1988).
3. M.A. Rodriguez, D.P. Matheis, S.S. Bayya, J.J. Simmis, R.L. Snyder and D.E. Cox, J. Mater. Res., 5, 1799 (1989).
4. R.P. Sharma, L.E. Rehn, P.M. Baldo and J.Z. Liu, Phys. Rev. Lett., 62, 2869 (1989).
5. R.P. Sharma, L.E. Rehn, P.M. Baldo and J.Z. Liu, Phys. Rev., B40, 11396 (1989).
6. T. Haga, K. Yamaya, Y. Abe, Y. Tajima and Y. Hidaka, Phys. Rev., B41, 826 (1990).
7. S.D. Conradson and I.D. Raistrick, Science, 243, 1340 (1989).
8. J. Mustre de Leon, S.D. Conradson, I. Batistic and A.R. Bishop, Phys. Rev. Lett., 65, 1675 (1990).
9. B.H. Toby, T. Egami, J.D. Jorgensen and M.A. Subramanian, Phys. Rev. Lett., 64, 2414 (1990).
10. e.g., R.M. Hazen, in "Physical Properties of High Temperature Superconductors II", ed. D.M. Ginsberg (World Scientific, Singapore, 1990) p. 121.
11. G. Placzek, Phys. Rev., 86, 377 (1952).
12. B.H. Toby and T. Egami, to be published.
13. T. Egami, W. Dmowski, J.D. Jorgensen, D.G. Hinks, D.W. Capone, II, C.U. Segre, and K. Zhang, Rev. Solid State Sci., 1, 247 (1987), and in "High Temperature Superconductivity", eds. S.M. Bose and S.D. Tyagi (World Scientific, Singapore, 1987) p. 101.
14. W. Dmowski, B.H. Toby, T. Egami, M.A. Subramanian, J. Gopalakrishnan and A.W. Sleight, Phys. Rev. Lett., 61, 2608 (1988).
15. W.E. Pickett, Rev. Mod. Phys., 61, 433 (1989).
16. J.G. Bednorz and K.A. Muller, Z. Phys., B64, 189 (1986).
17. D. Emin, Phys. Rev. Lett., 62, 1544 (1989).
18. D. Emin and M.S. Hillery, Phys. Rev., B39, 6575 (1989).
19. J.B. Goodenough and J. Zhou, Phys. Rev., B42, 4276 (1990).
20. S.J.L. Billinge, P.K. Davies, T. Egami, and C.R.A. Catlow, Phys. Rev., B43, in press.
21. R. Micnas, J. Ranninger, and S. Robaszkiewicz, Rev. Mod. Phys., 62, 113 (1990).
22. Y. Bar-Yam, Phys. Rev., B43, 359 (1991), Phys. Rev., B43, 2601 (1991).
23. D.R. Hamann and L.F. Mattheiss, Phys. Rev., B38, 5138 (1988); L.F. Mattheiss, private communication.
24. Y.J. Uemura, et al., Phys. Rev. Lett., 62, 2317 (1989).

ANOMALOUS MAGNETIC PROPERTIES OF HIGH TEMPERATURE SUPERCONDUCTORS

R. E. Walstedt

AT&T Bell Laboratories
Murray Hill, NJ 07974

Introduction

In this paper we summarize the principal results from an NMR and magnetic susceptibility study of normal-state magnetism in $YBa_2Cu_3O_7$ (YBCO).[1] A more detailed account of this work is to be submitted for publication elsewhere.[2] High quality c-axis susceptibility data on oriented powder samples, published by Lee, Klemm and Johnston (LKJ),[3] show a downturn just above T_c which may reasonably be attributed to superconducting fluctuation diamagnetism (SFD). These authors sought to extract the latter contribution by subtracting $\chi_{ab}(T)$ from $\chi_c(T)$, arguing that the SFD contribution to the in-plane response $\chi_{ab}(T)$ is negligible and, further, that the temperature-dependent spin paramagnetism is essentially isotropic[4] and will therefore not contribute to the difference. Close to T_c, where the SFD susceptibility $\chi_\nu^{SF}(T)$ is of comparable magnitude with the spin paramagnetism $\chi_\nu^s(T)$, the foregoing approximation is not unreasonable. At higher temperatures, however, we argue that the anisotropy of $\chi_\nu^s(T)$ (ν = ab,c) must be taken into account for a quantitative treatment. In practice, this can be accomplished by using NMR shifts to calibrate $\chi_\nu^s(T)$ and thus isolate $\chi_\nu^{SF}(T)$, which produces only a negligible shift effect. In this paper, we describe experimental measurements and analysis which accomplish that goal and also present, for the first time, separate determinations of the chain ($\chi_{1\nu}^s(T)$) and plane ($\chi_{2\nu}^s(T)$) contributions to the spin paramagnetism.

Our basic viewpoint in this work is that of the Mila-Rice (MR) model of the hyperfine interactions in YBCO.[4] The paramagmetic susceptibilities are assumed to reside on the copper sites, with ligand (O^{2-}) contributions accounted for through orbital reduction factors. The copper sites are treated in terms of the spin Hamiltonian model for Cu^{2+} ions in a tetrahedrally distorted octahedral environment[5] with a single hole in a $d_{x^2-y^2}$ ground state. This localized picture with its attendant crystal field splitting gives a good account of the highly anisotropic Van Vleck paramagnetism and associated NMR shift. The spin hyperfine tensors, when augmented by a strong transferred coupling term between neighboring Cu sites, are also found to explain the unusual shift and relaxation behavior.[4,6] These and other high-frequency properties of this system are expected to carry over into a tight binding-like band when a modest density of mobile carriers is added to the CuO_2 planes of this structure.

We would also like to emphasize that the ionic spin-Hamiltonian model[5] has as one of its core features a spin-orbit induced electronic g-shift of as much as 10-15%. In the abovementioned narrow band environment, this will give rise to a significant anisotropy in the spin paramagnetism according to the relation

$$\chi^s_{1,2v} = g^2_{1,2v} \mu^2_B n(E_F)/2 \qquad (1)$$

for chain (Cu(1)) or plane (Cu(2)) sites, where $n(E_F)$ is the density of states for one spin direction. We note in passing that g-factor anisotropies of similar order are found in fcc copper metal.[7] For simplicity, MR chose to ignore the g-factor effect and assumed, further, that the spin susceptibility was the same on chain and plane sites. Here we relax the latter restriction and treat the g-factors with the ionic spin Hamiltonian.[5,8]

The principal objective, then, is to use the foregoing framework to obtain separately the contributions $\chi^s_{1v}(T)$, $\chi^s_{2v}(T)$ and $\chi^{SF}_v(T)$, using measured bulk susceptibility and NMR shift tensor data for $^{63}Cu(2)$ and $^{63}Cu(1)$ as input, as well as ancillary data from the literature. The fundamental equation for the measured susceptibility is

$$\chi^{expt}_v(T) = \chi^s_v(T) + \chi^{orb}_v + \chi^{dia} + \chi^{SF}_v(T), \qquad (2)$$

where

$$\chi^s_v(T) = \chi^s_{1v}(T) + 2\chi^s_{2v}(T). \qquad (3)$$

The shifts $K^s_{1,2v}$ are related to the corresponding χ's by $K^s_{1,2v} = \alpha^s_{1,2v} \chi^s_{1,2v}$, where the $\alpha^s_{1,2v}$ are spin hyperfine coefficients. Likewise, for the orbital (Van Vleck) susceptibility one has $\chi^{orb}_v = (K^{orb}_{1v} + 2K^{orb}_{2v})/\alpha_{orb}$, where $\alpha_{orb} = 22.41 <r^{-3}>$ is the orbital shift coefficient[9] assumed to have a common value for both sites. The orbital shift tensors $K^{orb}_{1,2v}$ for the two sites are available from the literature.[10] χ_{dia}, which represents the isotropic core diamagnetic contribution from all atoms in the crystal, can be derived from tabulated data.[11] Here, we adopt the value[12] $\chi_{dia} = -166 \cdot 10^{-6}$ emu/mole f.u.

From Eq.(2), then, one can extract values of $\chi^s_v(T)$ using data for $\chi^{expt}_v(T)$ and for the orbital shifts, but *only at temperatures high enough so that $\chi^{SF}_v(T)$ is negligibly small*. With this in mind, we recast Eq.(3) in terms of $\chi^s_v(T)$ and the spin components of shift:

$$K^s_{1v}(T)/\chi^s_v(T) = \alpha^s_{1v} - 2(\alpha^s_{1v}/\alpha^s_{2v})(K^s_{2v}(T)/\chi^s_v(T)). \qquad (4)$$

Eq.(4) predicts a linear relationship between the two experimentally derived quantities $(K^s_{1,2v}/\chi^s_v)$, where temperature is the implicit variable. This equation, if realized, yields experimental values for $\alpha^s_{1,2v}$ and thus partitions χ^s_v into its Cu(1) and Cu(2) constituents as desired.

There is, however, a practical difficulty with Eq.(4) which must be dealt with. For technical reasons, data for the shift components K^s_{1ab} and K^s_{2c} are prohibitively difficult to obtain,[13] and thus Eq.(4) must be formulated in terms of K^s_{1c} and K^s_{2ab} alone. We do this by introducing the spin paramagnetism anisotropy factors $\xi_{1,2} = \chi^s_{1,2c}/\chi^s_{1,2ab}$ and use these to reformulate Eq.(4) in terms of the measurable shift components. The ξ's will be discussed below using Eq.(1) and the ionic model.[8] Thus, we rewrite Eq.(4) for $v = ab$ and c, respectively, as

$$K^s_{1c}/\chi^s_{ab} = \alpha^s_{1c}\xi_1 - 2\xi_1(\alpha^s_{1c}/\alpha^s_{2ab})(K^s_{2ab}/\chi^s_{ab}) \qquad (5a)$$

and

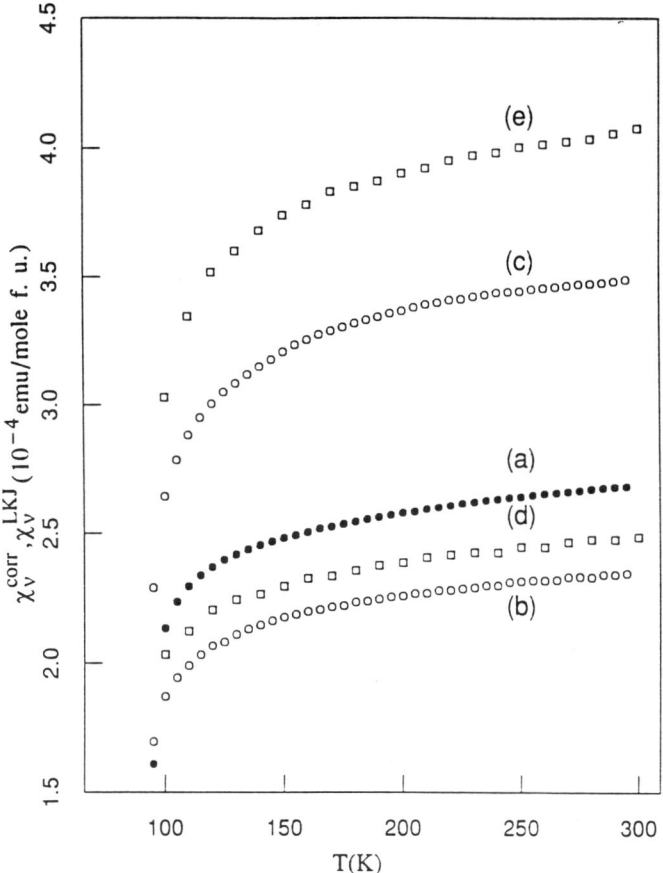

Fig. 1. Curve (a) shows susceptibility data for the original ceramic pellet from which the oriented powder sample used in this study was made. Curves (b) and (c) show ab plane and c axis susceptibility data, respectively, for the oriented powder sample prepared as described in the text. Curves (a) - (c) have all been corrected for small, spurious Curie terms by the method outlined in Ref. 14. Curves (d) and (e) are facsimile representations of the LKJ data with field orientation in the ab plane and along the c axis, respectively.

$$K_{1c}^s/\chi_c^s = \alpha_{1c}^s - 2\xi_2(\alpha_{1c}^s/\alpha_{2ab}^s)(K_{2ab}^s/\chi_c^s), \qquad (5b)$$

The "K/χ" quantities in Eqs.(5) are all obtainable from experiment by the methods described.

Experiment

We turn now to the experimental results. Shift and susceptibility data were taken on a high-quality ceramic sample of YBCO which had been ground to fine powder in an agate mortar and fixed in clear epoxy in a magnetic field of ~7 T. The original ceramic pellet exhibited a very small Curie-like term in the measured susceptibility, which became somewhat larger on grinding and fixing. A simple procedure has been used[2] to compare our data with those of LKJ and subtract off the Curie component,[14] which is

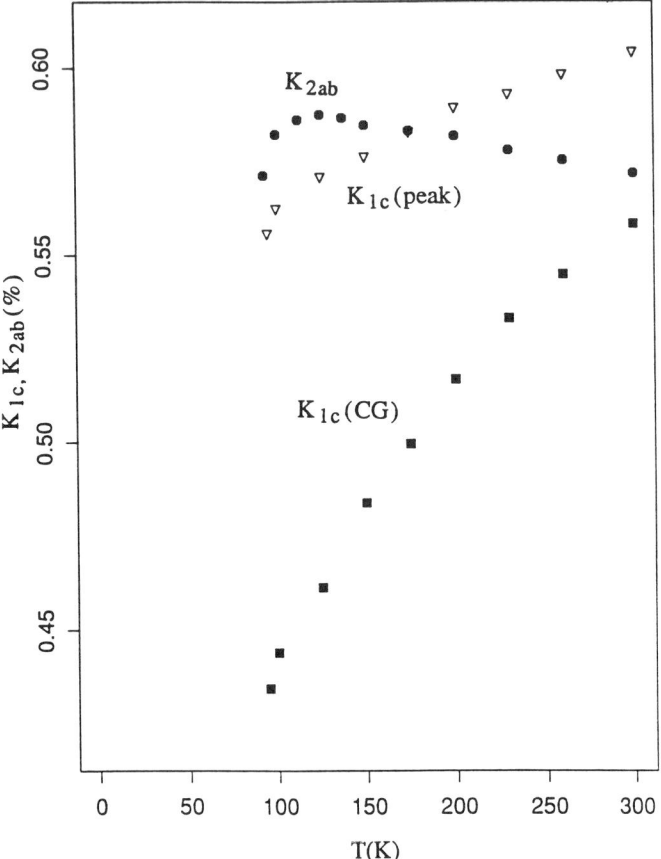

Fig. 2 Normal state ^{63}Cu NMR shift data for both chain and plane sites in YBCO are plotted as a function of temperature. The Cu(1)-site shifts are plotted for both the peak position (triangles) the center of gravity (CG - squares) of the lines.

spurious and not reflected in the NMR measurements. The corrected data are plotted in Fig. 1 along with a facsimile of the LKJ data. As is confirmed by detailed graphical comparison,[2,14] the present data are essentially equivalent to those from Ref.3 except for a minor difference in scale factor and magnitude of anisotropy $A_{expt} = \chi_c^{expt}/\chi_{ab}^{expt}$. The value of A_{expt} reflects the degree of orientation achieved for the powdered specimen, which is determined, in turn, by the morphology of the ceramic starting material. Quoting room temperature values throughout, the present work gives $A_{expt} = 1.49$, whereas the LKJ data give $A_{lkj} = 1.61$. In the discussions below, we regard A as a parameter to be varied, noting that under a wide range of conditions the susceptibility components χ_ν^{tot} for a perfectly ordered sample can be constructed from those measured on a partially oriented one *if the actual value of anisotropy A is known*. These corrected tensor components are given by[2]

$$\chi_c^{tot} = \frac{\chi_c^{expt}[A(A_{expt} + 1) - 2] - 2\chi_{ab}^{expt}(A - A_{expt})}{(A_{expt} - 1)(A + 2)} \quad (6a)$$

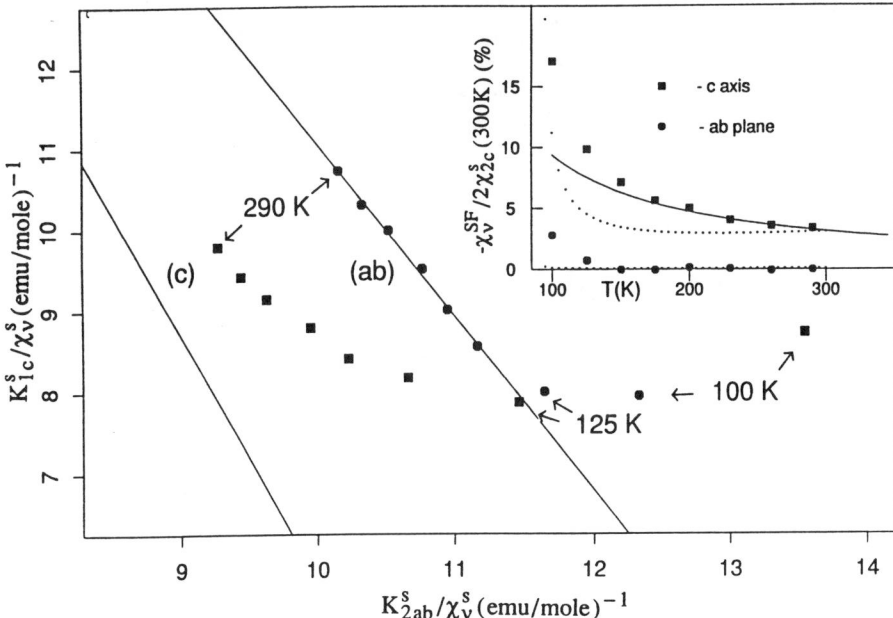

Fig. 3 Plot of (K^s_{1c}/χ^s_ν) vs. (K^s_{2ab}/χ^s_ν) for ν = ab and c. The solid line shown for ν = ab is a linear regression (Eq.(5a)) fit to the data for T > 125 K. The corresponding line for ν = c is a plot of Eq.(5b) with the coefficient parameters determined as described in the text. The inset shows the temperature-dependent diamagnetism which results from subtracting χ^s_ν values derived from the straight-line behavior of Eqs.(5a) and (5b), as well as the orbital and core diamagnetic contributions, from the experimental data. The solid curve passed is a fit to $\chi^{SF}_c(T) = B/T$ at the high-temperature end. The dotted lines show the model theory presented in Ref. 3.

$$\chi_{ab} = \frac{-\chi^{expt}_c(A - A_{expt}) + \chi^{expt}_{ab}(AA_{expt} + A_{expt} - 2)}{(A_{expt} - 1)(A + 2)} \quad (6b)$$

These relations are employed in the discussions to follow, where we conclude in fact that the true anisotropy is slightly larger than A_{lkj}.

NMR shift data were derived from spin echo spectra taken over the temperature range 95 K ≤ T ≤ 300 K. The data and analysis were refined to yield the temperature variation of the copper NMR shifts to a precision several times greater than the current values in the literature. This was required in order to resolve the rather weak temperature dependences of the normal state shifts. The results are shown in Fig. 2. For the Cu(2) site K_{2ab} rises gradually as T is lowered. Such an effect was reported some time ago for the ^{89}Y shift[15]; the present data are consistent with these earlier results.[2] Just above T_c the Cu(2) shift rolls over and descends abruptly. This small precursor develops into a major spin gap effect in oxygen-deficient YBCO.[16] Spin gap behavior has been identified in several other high-T_c systems as well.[16] The Cu(1) shift in Fig. 2 has a much stronger temperature dependence than that of the Cu(2) and opposite in sign. The associated NMR line is narrow, but asymmetrical at room temperature. It broadens asymmetrically as T is lowered.[2] Thus, the center of gravity (CG) of the line (Fig. 2, squares) shows a marked decline at lower temperatures, whereas the position of the peak (triangles) descends much more slowly. We identify the CG position as the shift $K_{1c}(T)$ to be correlated with $\chi^s_{1c}(T)$. The asymmetric broadening of the Cu(1) line is indicative of disorder, an effect which is surprising in a high-quality, fully oxygenated sample of this compound.

Analysis and Discussion

The contrasting temperature dependences of K_{1c} and K_{2ab} enable us to make plots of Eq.(5a) and (5b) with a reasonably wide range of variation for the plot variables K^s_{1c}/χ^s_v and K^s_{2ab}/χ^s_v. Note that the chain sites are clearly responsible for the downtrend observed for $\chi^{expt}_{ab}(T)$. Using the shift data of Fig. 2, the spin components are extracted by taking, e.g., $K^s_{1c}(T) = K_{1c}(T) - K^{orb}_{1c}$, where the values $K^{orb}_{1c} = 0.25\%$ and $K^{orb}_{2ab} = 0.28\%$ are taken from the literature.[10] The spin susceptibility components are derived from the data of Fig. 1 by subtracting off the orbital and core diamagnetic part as explained earlier, and are then adjusted to correspond with a chosen anisotropy value A using Eqs.(6). Plots of Eq.(5) for a chosen set of parameter values are shown in Fig. 3. Considering the in-plane (ab) susceptibility first, the data points are seen to form a reasonable straight line down to about 150 K, below which the points at 125 K and 100 K deviate noticeably upward. This upturn is interpreted as the onset of a small fluctuation diamagnetism in the ab plane, in what may be a dimensional crossover from two to three-dimensional fluctuations. The straight line drawn is a least-squares fit to the first six points, which from Eq.(5a) yields a value of α^s_{2ab} with which we can partition χ^s_{ab} into chain and plane components. To go further we need to estimate ξ_1 or ξ_2. Using parameter values derived from the shift and susceptibility data as explained below, we use Eq.(1)[8] to obtain ξ_2. This gives us χ^s_{2c}, with which we partition χ^s_c as well, and thus obtain all of the parameters in Eq.(5a) and (5b).

Before continuing our discussion of Fig. 3 we comment briefly on the parameter values used in our analysis. The hyperfine coefficients derived from Eq.(5a) as described above are dependent upon the value of α_{orb} (i.e., $<r^{-3}>$) used to determine the orbital susceptibilities as well as the crystal field and spin-orbit parameters which determine the g tensor. In Ref. 2, Eq.(1), (2), (3) and (5) are combined with expressions from the ionic model for the spin hyperfine coefficients and orbital effects to form a coupled, nonlinear set of equations which is straightforwardly solved. Here we shall only summarize a nearly optimal set of parameter values obtained in this way (Table I).

The parameter values in Table I are also dependent on our analysis of the c axis susceptibility in Fig. 3 to which we now turn. The discussion of Eq.(5a) using Fig. 3 yields the coefficient parameters for both Eq.(5a) and (5b). However, the values employed for $\chi^s_c(T)$ are critical for those results, and therein lies a problem. As we see from the plot of Eq.(5b) (Fig. 3, left side), *there is no obvious linear region*. From this we conclude that $\chi^{SF}_c(T)$ is appreciable at temperatures up to room temperature and possibly beyond. This is a very important, but unexpected result of this study. In order to proceed with the analysis it is therefore necessary to estimate the magnitude of $\chi^{SF}_c(T)$. We do this using the criterion that asymptotically $\chi^{SF}_c(T) \propto T^{-1}$, as given by the theory of Nagaosa and Lee[18]. The procedure is as follows. Adopting a trial value for $\chi^{SF}_c(300\ K)$, we then obtain $\chi^s_c(300\ K)$ by subtraction and, thus, the parameters for Eq.(5a) and (5b) as described above. Using the shift data, Eq.(5b) then yields the entire curve of $\chi^s_c(T)$, and by subtraction, $\chi^{SF}_c(T)$ as well. The magnitude of $\chi^{SF}_c(300\ K)$ is then adjusted so that the asymptotic variation of the SFD term is approximately T^{-1}. The results are shown in the inset to Fig. 3, where the estimated magnitude of χ_c^{SF}(300 K) is seen to be ~ 3% of the planar spin paramagnetism. The solid line passed illustrates T^{-1} behavior. The resulting linear behavior of Eq.(5b) is shown in Fig. 3, where the large discrepancy with the plotted data (squares) illustrates the sensitivity of this method. The same subtraction procedure is used to obtain values of χ_{ab}^{SF}(T), which onsets below 150 K as noted earlier (Fig. 3, inset).

There is one further point to discuss with regard to parameter selection, namely the value A of the room temperature anisotropy. The susceptibility values employed throughout the analysis are, of course, dependent on A via Eqs.(6). On careful examination, the only parameter critically dependent on A is ξ_1. Because the symmetry axis of the Cu(1) sites is in the a direction, a rough estimate from the value of ξ_2 might be $\xi_1 \sim 2/(1 + \xi_2) \sim 0.9$. The value of ξ_1 which results from $A_{lkj} = 1.61$ is 0.7, which we feel is unreasonably small. Taking A = 1.70 yields ξ_1 =0.91. This value is used to obtain the parameters in Table I. Taking A slightly larger than A_{lkj} is quite reasonable, since is it very unlikely that the LKJ powdered sample was perfectly oriented.

Table I

Room temperature parameter values for Cu-site hyperfine and susceptibility tensors, derived as decribed in the text from static magnetic and hyperfine measurements. Units are as follows: α's, (emu/mole)$^{-1}$; χ's, 10^{-4} emu/mole; $\Delta_{0(1)}$, eV; $<r^{-3}>$, a.u. These values of χ_ν^s correspond to local anisotropies $\xi_1 = 0.91$ and $\xi_2 = 1.28$.

ν	g_ν	α_ν^s	χ_ν^{orb}	χ_ν^s	Misc. parameters
1ab	-	-	54.1	95.7	$\Delta_0 = 2.04$
1c	-	35.6	20.1	86.9	$\Delta_1 = 2.33$
2ab	2.075	30.4	22.5	96.4	$<r^{-3}> = 5.57$
2c	2.345	-0.8	102.6	123.0	

Conclusions

In conclusion, we have succeeded in partitioning the spin paramagnetism in YBCO, finding comparable values for the Cu(1) and Cu(2) sites (Table I) as suggested earlier by Mila and Rice. The other parameter values yielded by the present analysis are discussed more fully in Ref. 2, but are not widely different from values derived in Ref. 4. We emphasize that this analysis is internally consistent and uses only static magnetic and hyperfine data. Reasonable parameter values require a slightly larger room-temperature susceptibility anisotropy than found with our own data or that of Lee, Klemm and Johnston.[3] This is not surprising, given the limitations of the magnetically oriented powder technique. A remarkably persistent temperature-dependent diamagnetic susceptibility term emerges from the analysis given, this term appearing to extend well beyond room temperature. This term is consistent in magnitude and temperature dependence with the gauge fluctuation theories of both Nagaosa and Lee[18] and of Ioffe and Kalmeyer[19]. An onset for in-plane diamagnetism near T ~ 125 K is also observed, possibly representing a crossover from two to three dimensional fluctuations as one approaches T_c.

The authors are indebted to P. A. Lee and L. B. Ioffe for informative discussions and for the communication of manuscripts before publication.

References

1. The other topic which was addressed during the workshop lecture is an NMR study of ^{63}Cu in the high-T_c compound $Bi_2Ca_2SrCu_2O_8$. A short account of that work has been submitted for publication: R. E. Walstedt, R. F. Bell and D. B. Mitzi, submitted to Phys. Rev. B, Rapid Communiations.

2. R. E. Walstedt, R. F. Bell, L. F. Schneemeyer, J. V. Waszczak and G. P. Espinosa, unpublished.

3. W. C. Lee, R. A. Klemm, and D. C. Johnston, Phys. Rev. Lett. **63**, 1012, (1989).

4. F. Mila and T. M. Rice, Physica C **157**, 561 (1989).

5. A. Abragam and B. Bleaney, *Electron Paramagnetic Resonance of Transition Ions*, (Oxford University Press, Oxford, 1970).

6. R. E. Walstedt, W. W. Warren,Jr., R. F. Bell and G. P. Espinosa, Phys. Rev. B**40**, 2572 (1989), and references therein.

7. D. L. Randles, Proc. R. Soc. London Ser. A**331**, 85 (1972).

8. These are given for, e.g., the Cu(2) site, by $g_{2ab} = 2 - 2\lambda/\Delta_1$ and $g_{2c} = 2 - 8\lambda/\Delta_0$. λ is the spin-orbit parameter (Ref. 5) and $\Delta_{0,1}$ are, respectively, the crystal-field energy splittings to the d_{xy} and (d_{yz}, d_{zx}) excited states. We use here the estimate $\lambda \approx -0.088$ eV from Ref. 5.

9. This expression assumes values of $<r^{-3}>$ in atomic units. The susceptibilities which result are in units of emu/mole f.u.

10. S. Barrett et al., Phys. Rev. B**41**, 6283 (1990).

11. A. Junod, A. Bezinge and J. Muller, Physica C **152**, 50 (1988).

12. B. Batlogg, private communication. This number results from a survey of core diamagnetism in related compounds, based on data given by Landolt-Börnstein, New Series, Vol. II/2, (Springer, Berlin/New York, 1966).

13. The reasons for this are as follows: K^s_{2c} is accidentally very small, rendering precise measurements nearly impossible. K^s_{1ab} is not a single value, but an average, because of large Cu(1) shift anisotropy in the ab plane. The resulting 2-dimensional powder pattern is very difficult to measure accurately.

14. If $\chi_{expt}(T)$ contains a small Curie term, C/T, and $\chi_{REF}(T)$ does not, C can be determined by plotting $T\chi_{expt}(T)$ vs. $T\chi_{REF}(T)$ and finding the intercept. Such plots in Ref. 2 show that $\chi_{expt}(T)$ -C/T (using reference data from LKJ) is then consistent with the temperature variation of the LKJ data to within the scatter of the measurements.

15. G. Balakrishnan, R. Dupree, I. Farnan, D. McK. Paul, M. E. Smith, J. Phys. C **21**, L847 (1988).

16. R. E. Walstedt, W. W. Warren, Jr., R. F. Bell, R. J. Cava, G. P. Espinosa, L. F. Schneemeyer, and J. V. Waszczak, Phys. Rev. B**41**, 9574 (1990); M. Takigawa, *et al.*, Phys. Rev. B **43**, 247 (1991).

17. T. M. Rice, Proceedings of the ISSP Symposium on the Physics and Chemistry of Oxide Superconductors (Tokyo, January, 1991).

18. N. Nagaosa and P. A. Lee, unpublished.

19. L. B. Ioffe and V. Kalmeyer, unpublished.

THE OXYGEN ISOTOPE EFFECT IN $(Y_{1-x-y} Pr_x Ca_y) Ba_2Cu_3O_{7-\delta}$

J.P. Franck[a], S. Gygax[b], J. Jung[a], M.A-K. Mohamed[a] and G.I. Sproule[c]

[a] Physics Department, University of Alberta
Edmonton, Alta., Canada, T6G 2J1

[b] Physics Department, Simon Fraser University
Burnaby, B.C., Canada, V5A 1S6

[c] Institute for Microstructural Sciences, National Research Council
Ottawa, Ont., Canada, K1A OR9

INTRODUCTION

Oxygen isotope effect measurements are presented in the system $(Y_{1-x-y} Pr_x Ca_y) Ba_2Cu_3O_{7-\delta}$. Substitution levels investigated are $x = 0.2, 0.3, 0.4$ and 0.5 with no Ca substitution; and $y = 0.5, 0.10, 0.15, 0.20, 0.25$ and 0.30 with a Pr substitution of 20% ($x = 0.2$). We find[1] that increasing Pr substitution leads to an increasing isotope effect, the isotope exponent α approaching 0.5. Ca substitution at constant Pr concentration reduces the isotope effect. The isotope exponent appears to be a function of mobile hole concentration, it decreases continuously from $\alpha \approx 0.5$ with increasing concentration of mobile holes.

EXPERIMENTAL

Samples of the required stoichiometry were prepared by the usual powder metallurgical process from high purity Y_2O_3, Pr_6O_{11}, $BaCO_3$, and CuO. Calcining in air was performed at temperatures between 900 and 930° C, with various intermediate grindings. For the Pr substituted samples we used a total calcining time of 125 hours, and for the Pr and Ca substituted samples a total of 620 hours.

At each composition we prepared three identical pellets from the calcined powders. The pellets were then sintered simultaneously in a quartz vessel, designed to ensure identical sintering conditions. Two of the pellets were heated in $^{16}O_2$ gas and one in $^{18}O_2$ (98.5% isotopic purity). The oxygen gas was constantly circulated, and purified by passing through traps at the temperature of liquid oxygen. The apparatus used has been previously described[2]. Sintering conditions were: for the Pr substituted samples: 7 hours at 935° C, 10 hours each at 450°, 350° and 250° C; for the Pr 20% and Ca substituted samples: 48 hours at 935° C, 18 hours at 500° C and 10 hours each at 400° and 300° C. The sintering conditions were designed to ensure full oxygenization, $\delta \leq 0.1$. X-ray diffraction showed a single superconducting phase isomorphous to the structure of YBC0. All samples are orthorhombic, with lattice constants slowly increasing with Pr concentration. A small amount (1.5 to 3%) of $BaCuO_2$ was present in most samples. Meissner fractions were determined at 20 Oe for the Pr substituted samples, they are 24% (x = 0.2), 21% (x = 0.3), 17% (x = 0.4) and 7% (x = 0.5).

Table 1. Transition Temperatures and Oxygen Isotope Exponents in the Systems $(Y_{1-x-y} Pr_x Ca_y) Ba_2Cu_3O_{7-\delta}$.

x	y	^{18}O conc. (at.-%)	$T_c^{R=0}$ (^{16}O) (K)	$\Delta T_{c,res}$ (K)	$\Delta T_{c,mag}$ (K)	α_{res}^b	α_{mag}^b	Δp^c
0.2	0	84.6	75.6	0.80	0.68	0.09	0.08	-0.08
0.3	0	79.0	60.4	1.07	1.25	0.15	0.17	-0.12
0.4	0	85.5	46.2	1.50	1.60	0.27	0.29	-0.16
0.5	0	85.9	30.6	1.65	1.70	0.45	0.49	-0.20
0.2[a]	0	--	72.5	0.90	1.35	0.10	0.16	-0.08
0.2[a]	0.05	84.0	74.2	0.76	0.94	0.08	0.11	-0.06
0.2[a]	0.10	78.0	74.5	0.47	0.49	0.05	0.06	-0.03
0.2[a]	0.15	77.0	73.3	0.44	0.60	0.05	0.07	-0.01
0.2[a]	0.20	76.0	71.5	0.48	0.47	0.06	0.05	+0.02
0.2[a]	0.25	75.0	69.8	0.28	0.68	0.03	0.08	+0.04
0.2[a]	0.30	83.0	68.2	0.40	0.45	0.05	0.06	+0.07

[a]The calcining and sintering conditions for this series are different from the first four comparison sets (see text).
[b]Oxygen isotope exponents for the measured temperature shifts.
[c]Estimated relative concentration of mobile holes per CuO plane.

The superconducting transition was observed resistively by the usual four-terminal method, and magnetically through low field dc magnetization (0.1 0e) and low field ac susceptibility (0.01 0e, 160 H$_z$). ^{18}O concentrations were obtained by SIMS. We found that the two samples treated in $^{16}O_2$ had coinciding resistive and magnetic transitions. In samples sintered in $^{18}O_2$ the transitions were shifted to lower temperatures. Shifts in the superconducting transition temperature were taken for the resistive transition at the R = 0 point, and for the dc magnetization at the extrapolation of the rapidly increasing shielding part to the point of no shielding. The transition temperatures obtained in this way closely coincide for both methods.

RESULTS

In Table 1 we give the observed shifts in transition temperature, and the oxygen isotope exponents α, where $T_c = \text{const.} \cdot M^{-\alpha}$. The isotope exponents in Table 1 are for the actually observed temperature shifts, no correction was made for incomplete ^{18}O exchange. The observed transition temperature shifts, and the isotope exponent, increase with increasing Pr concentration, approaching $\alpha = 0.5$ for 50% Pr substitution. The transition temperature continuously decreases with increasing Pr concentration. Ca substitution at constant Pr concentration shows a different effect. The transition temperature goes through a shallow maximum, as found already by Neumeier et al.[3] The isotope exponent decreases by about a factor 2 to 3 with increasing Ca concentration, reducing to the small values found for unsubstituted YBC0. These results show that the transition temperature is not the determining factor for the size of the observed isotope effect.

DISCUSSION

The effect of Pr and Ca substitution in YBC0 is somewhat controversial. A number of investigations show that Pr fills holes in the Cu0 planes, according to an effective valence near 4. Experimental evidence for this comes from Hall effect measurements[4], and the observation of increased penetration depth in μSR experiments[5,6]. Many other, particularly spectroscopic experiments, show evidence for a valence of 3 for Pr[7]. It may be possible to reconcile these findings by the assumption that Pr localizes holes, rather than filling them.[7] The substitution of Ca increases the hole concentration, as is generally accepted.

In order to compare our results with the mobile hole concentration, we used the analysis of Neumeier et al.[3] In this analysis, two effects of Pr substitution are recognized: Pr lowers the maximum transition temperature due to magnetic pair breaking, and it also influences the transition temperature due to hole immobilization. Neumeier et al. propose the following equation for the transition temperature T_c as function of Pr and Ca substitution:

$$T_c = (T_{co} - A(a - \beta x + y)^2 - Bx. \qquad (1)$$

The second term represents the influence of mobile hole concentration, with a the optimal hole concentration, y the creation of holes due to Ca substitution, and $-\beta x$ the localization of holes due to Pr substitution. The effective valance of Ca is then +2, and that of Pr is $+(3+\beta)$. The third term represents the influence of pair-breaking, it is independent of Ca concentration. We have fitted the expression (1) to our transition temperature data, and obtain the following constants: $T_{co} = 94$ K, $A = 240$ K, $B = 90$ K, $a = 0.06$, $\beta = 0.8$. These data compare well with those quoted by Neumeier et al., with the exception of A, where we find a smaller value (240 K vs. 425 K in ref. (3)). The effective valence of Pr for our data should then be +3.8 (vs. 3.95 in ref. (3)).

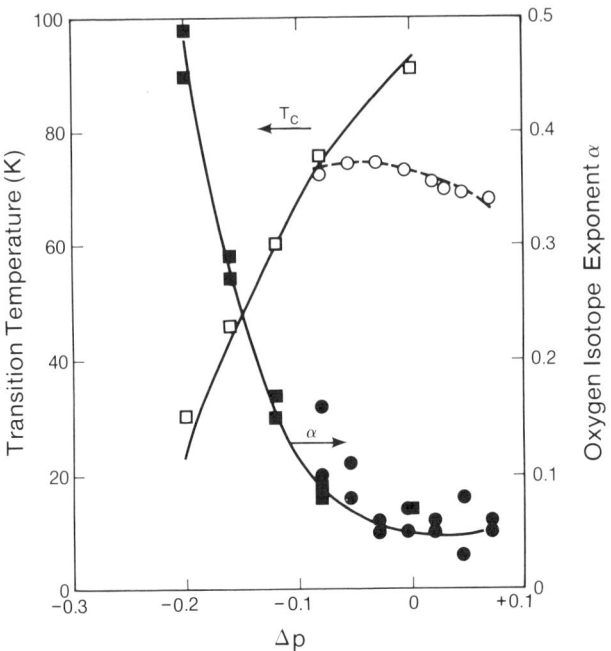

Fig. 1. The transition temperature T_c and the oxygen isotope exponent α as a function of estimated mobile hole concentration per CuO plane in the system $(Y_{1-x-y} Pr_x Ca_y) Ba_2Cu_3O_{7-\delta}$. Open symbols: transition temperatures, (left scale); solid symbols: oxygen isotope exponents, (right scale). The squares refer to the system $(Y_{1-x} Pr_x) Ba_2Cu_3O_{7-\delta}$, and the circles to the Ca substituted system $(Y_{0.8-y} Pr_{0.2} Ca_y) Ba_2Cu_3O_{7-\delta}$. The isotope exponent $\alpha = 0.07$ for unsubstituted $YBa_2Cu_3O_{7-\delta}$ with a T_c of 91.4 K represents an average of earlier determinations (Ref. (2). Fitting of the transition temperatures by equation (1) is shown. The line connecting the isotope exponents is only a guide to the eye.

We estimate the mobile hole concentration in our samples using this analysis. The changes in mobile hole concentration per CuO plane, relative to unsubstituted $YBa_2Cu_3O_7$, are included in Table 1. The total change in mobile hole concentration per plane for the data given here is 0.27, from a minimum for $x = 0.5$, $y = 0$, to a maximum for $x = 0.2$, $y = 0.3$.

The observation that the increased isotope effect in Pr substituted YBC0 can be reduced at constant Pr concentration by the substitution of Ca shows, that the magnetic pair-breaking effect is not responsible for the enhanced isotope effect. The isotope effect appears to depend primarily on the concentration of mobile holes.

In Fig. 1 we show the observed transition temperatures, as well as the observed isotope exponents, as a function of mobile hole concentration per CuO plane. The transition temperatures are not a unique function of mobile hole concentration. The oxygen isotope exponent α, however, appears to follow a common curve as a function of hole concentration. The exponent assumes its largest value of $\alpha \approx 0.5$ for the smallest hole concentration and continuously falls with increasing hole concentration towards the small values typical for pure YBC0.

A number of theoretical estimates have addressed the isotope effect in high T_c superconductors. It was found by Tsuei et al.[8], and also by Schachinger et al.[9] and by Friedel[10], that singularities, or at least pronounced maxima, in the carrier density of states, can considerably reduce the isotope effect expected in phonon mediated superconductors. Changes in mobile hole concentration move the Fermi surface relative to such an anomaly, and change therefore the observed isotope effect. The anomaly in the carrier density of holes has been speculated to arise from a van Hove singularity at slightly larger than half-filling[10,11,12], but it could in principle also arise through other mechanisms.

The results presented here show that large isotope effects, approaching those for a phonon-mediated BCS superconductor, can be observed in superconductors isomorphous with the 90 K superconductor $YBa_2Cu_3O_{7-\delta}$. It appears that the isotope effect in pure YBC0 is suppressed, possibly due to singularities in the carrier density of states. The results show that the interaction between carriers and phonons plays an important role in high T_c superconductors.

ACKNOWLEDGEMENT

This research was supported by various grants from the Natural Sciences and Engineering Research Council of Canada.

REFERENCES

1. J.P. Franck, J. Jung, M.A.-K. Mohamed, S. Gygax and G.I. Sproule, "Oxygen isotope effect in superconducting $(Y_{1-x} Pr_x) Ba_2Cu_3O_{7-\delta}$", Proceedings of the 19th International Conference on Low Temperature Physics, Brighton (1990), Part III, Physica B 169 (1991).
2. J.P. Franck, J. Jung, G. Salomons, W.A. Miner, M.A.-K. Mohamed, J. Chrzanowski, S. Gygax, J.C. Irwin, D.F. Mitchell and G.I. Sproule, "The oxygen isotope effect in superconducting $YBa_2Cu_3O_{7-\delta}$", Physica C 172:90 (1990).

3. J.J. Neumeier, T. Bjornholm, M.B. Maple and I.K. Schuller, "Hole filling and pair breaking by Pr ions in $YBa_2Cu_3O_{6.95 \pm 0.02}$", Phys. Rev. Letters 63:1516 (1990).
4. A. Matsuda, K. Kinoshita, T. Ishii, H. Shibata, T. Watanabe and T. Yamada, "Electronic properties of $Ba_2Y_{1-x}Pr_xCu_3O_{7-\delta}$", Phys. Rev. B 38: 2910 (1988).
5. D.W. Cooke, M.S. Jahan, R.S. Kwok, R.L. Lichti, T.R. Adams, C. Boekema, W.K. Dawson, A. Kebede, J. Schwegler, J.E. Crow and T. Mihalisin, "Transverse- and zero-field muon-spin-rotation investigation of magnetism and superconductivity in $(Y_{1-x} Pr_x) Ba_2Cu_3O_7$", Phys. Rev. B 41:4801 (1990).
6. C.L. Seaman, J.J. Neumeier, M.B. Maple, L.P. Le, G.M. Luke, B.J. Sternlieb, Y.J. Uemura, J.H. Brewer, R. Kadano, R.F. Kiefl, S.R. Krietzman and T.M. Riseman, "Magnetic Penetration Depth of $Y_{1-x}Pr_xBa_2Cu_3O_{6.97}$ Measured by Muon-spin Relaxation ", Phys. Rev. B 42:6801 (1990).
7. J. Fink, N. Nucker, H. Romberg, M. Alexander, M.B. Maple, J.J. Neumeier and J.W. Allen, "Evidence against hole filling by Pr in $YBa_2Cu_3O_7$", Phys. Rev. B 42:4823 (1990).
8. C.C. Tsuei, D.M. Newns, C.C. Chi and P.C. Pattnaik, "Anomalous isotope effect and Van Hove singularity in superconducting Cu oxides", Phys. Rev. Letters 65:2724 (1990).
9. E. Schachinger, M.G. Greeson and J.P. Carbotte, "Modification of the isotope effect due to energy-dependent electronic density of states", Phys. Rev. B 42:406 (1990).
10. J. Friedel, "The high-T_c superconductors: a conservative view", J. Physics, Condensed Matter 1:7757 (1989).
11. R.S. Markiewicz and B.G. Giessen, "Correlation of T_c with structure in the density of states in the new high-T_c superconductors", Physica C 160:497 (1989).
12. R.S. Markiewicz, "Van Hove singularities and high-T_c superconductivity: A Review", Intern. J. of Modern Physics B (to be published).

BOSONIC MECHANISM FOR HIGH TEMPERATURE SUPERCONDUCTORS

F.M. Mueller*, G.B. Arnold†, and J.C. Swihart‡

*Center for Materials Science, Los Alamos National Laboratory
Los Alamos, NM 87545

†Department of Physics, University of Notre Dame
Notre Dame, IN 46556

‡Department of Physics, Indiana University
Bloomington, IN 47405

INTRODUCTION

The temperature-dependent angle-resolved photoemission data of D.S. Dessau et al.[1] on Bi 2212 show strong modulations in the superconducting state when compared to the normal state. These are similar to those seen historically in standard tunneling experiments in lower temperature superconductors. We have analyzed the Dessau data using Nambu-Eliashberg theory assuming some (as yet unknown) boson exchange as the primary mechanism for the superconductivity.[2] The derived $\alpha^2 F$'s, λ's and μ^*'s show features which resemble those derived from inversions of other low-temperature superconductors, albeit that λ here for (110) data is about 8.67 and μ^* is approximately 0.15. Several Bosonic mechanisms are considered.

Temperature dependent angle-resolved photoemission experiments, although very different from tunneling in the laboratory environment in terms of techniques, probe similar physics close to the fermi level E_F. Both effects are proportional to the electron density of states. A major difference is that photoemission data only sense electrons from below E_F, whereas through biasing, tunneling experiments sense density of states structure both above as well as below E_F. We show in Fig. 1 a long count photoemission run of Dessau et al.[1] taken along the (110) direction, near the Fermi surface. It is tempting to apply Nambu-Eliashberg (NE) theory to these data.

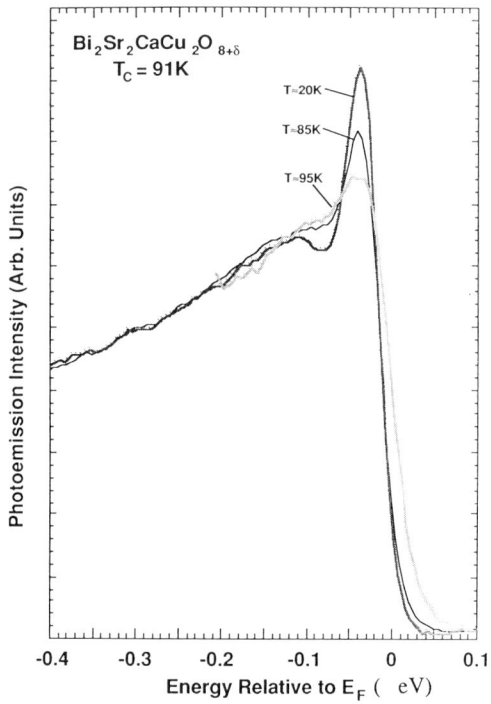

Fig. 1. Temperature Dependence of Angle Resolved Photoemission Data of Bi 2212 taken along the (110) direction (after Reference 1). As discussed in the text these features resemble Bosonic strong-coupling effects seen in lower temperature superconductors.

INTERPRETATION OF PHOTOEMISSION DATA

The typical energy resolution of photoemission data is about 10-15 meV, whereas that of tunneling data is better than perhaps 0.1 meV. Hence in applying NE theory to photoemission data we need to take into account a broad spectral resolution function in some detail. In Fig. 2 we show as the pluses a plot of

$$N'(E)^{exp} = \frac{\sigma_s(E)}{\sigma_n(E)} - 1 \quad , \tag{1}$$

where $\sigma_s(E)$ is the 20 Kelvin and $\sigma_n(E)$ is the 100 Kelvin data from Fig.1. One can see that $N'(E)^{exp}$ has strong-coupling modulations, similar to those $N'(E)$'s derived from tunneling data of conventional superconductors such as Hg or Nb.[3,4]

We have solved numerically the Eliashberg nonlinear coupled integral equations derived from the two components of Nambu's iso-spin equations[3,5] in which we have explicitly preserved their nonlinearity. Because the broadening of the experimental data is large in comparison to tunneling data, a direct inversion along the lines of

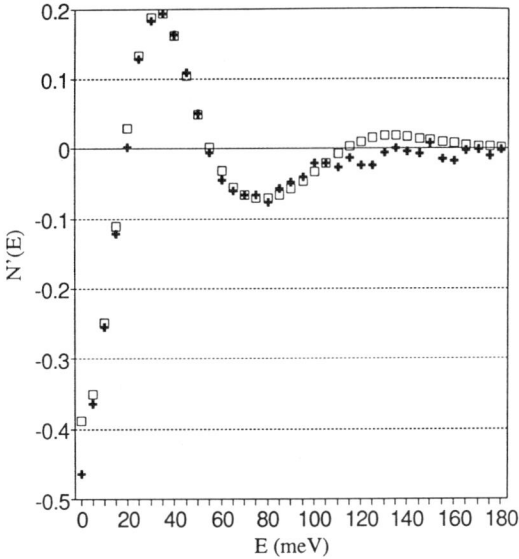

Fig. 2. Experimental and theoretical values of the normalized "tunneling function" $N'(E)$. The pluses are derived from the data of Fig. 1. These features resemble those seen from tunneling in lower temperature superconductors. The open squares denote the theoretical fit using the $\alpha^2 F$ of Fig. 3.

references (4) and (5) is not feasible. Instead, we adopt the approach described below.

The superconducting density of electron states N(E) is given by

$$N(E) = Re\left\{\frac{E}{\sqrt{E^2 - \Delta^2(E)}}\right\}, \qquad (2)$$

where $\Delta(E)$ is the τ_1 isospin component in NE theory. Equation 2 contains no photoemission detector resolution effects. This density of states broadened by the experimental spectral resolution function $P(E, E')$ is what we take to be the theoretical $N'(E)$ (plus one):

$$N'(E)^{th} = \int_{-\infty}^{\infty} P(E, E')N(E')dE' - 1, \qquad (3)$$

This is the quantity to be compared in a least squares sense with the experimental data $N'(E)^{exp}$ above. We have tried both Gaussian and Lorentzian forms for this resolution function.

For simplicity we approximate $\alpha^2 F$ by a sequence of analytic functions[6] rather than using a conventional evenly-spaced numerical function. The broadening of the data is such that fine features in $\alpha^2 F(E)$ are not directly reflected in it. Our goal is to determine the gross features in this function, leaving the calculation of finer features for the future. Initial results of this work are being reported in *Nature*.[7] Details of our procedure will be given in a later paper.[8]

In tunneling data on strong-coupling low T_c superconductors, it is known that the quantity we calculate is such that the location of peaks in $\alpha^2 F(E)$ corresponds to energies at which $|dN'/dE|$ is peaked. In the present case, the substantial smearing of structure by limited experimental resolution (below, we find this resolution to be about 16 or 17 meV) makes such identification unreliable. In addition, we find that the lowest energy peak in $\alpha^2 F$ (the only significant peak which we find) is at such a low energy that it falls quite close to the gap singularity. Indeed, the data are analogous to that for Hg,[3,4] where the gap is 0.83 meV and a large sharp peak in $\alpha^2 F$ occurs at $E_P = 2 meV$. The ratio $(E_P + \Delta_o)/\Delta_o$ is thus a little more than 3 for Hg. For our case, it is of order 1.5. In this sense, the data here are similar to an "extreme form" of Hg tunneling data.

One can see directly on looking at $N'(E)^{exp}$ that this function drops much more rapidly with energy than does the BCS $N'(E)$ just above the peak in $N'(E)$. This fast drop is an indication that there is a strong peak in $\alpha^2 F$ at about this energy (or lower) minus the gap energy.

RESULTS OF INVERSION

We carry out a "double" variational procedure (in both resolution function and in $\alpha^2 F$), and we find that the best fit is for Gaussian broadening (as opposed to Lorentzian broadening) with

$$\Delta_o = 18 meV, \quad \sigma = 15.9 meV, \quad \mu* = 0.15, \quad \lambda = 8.67, \quad \alpha^2 = 49.77 meV \quad,$$

and with the rms deviation devn = 0.135. Here σ is the standard deviation for the Gaussian resolution function, and α^2 is defined as the integral of $\alpha^2 F(E)$ over all E.

The $\alpha^2 F(E)$ function which produces the best fit to $N'(E)^{th}$ is shown in Fig. 3. The sharp, dominant peak at 10 meV is striking. If this peak is shifted upward to 15 meV, we find devn increases to 0.488; so, obviously, the fit is very sensitive to the position of this peak. We can say, with high confidence, that such a peak must occur very near this energy for this choice of gap value. Changing the width of the 10meV peak or the value of the cutoff energy for $\alpha^2 F(E)$ increases devn. The value of μ^* is less certain, but seems most likely to be between approximately 0.11 and 0.16. Changing the width σ of the resolution function or the gap Δ_o from the values given above also increases devn. We show in Fig. 2 as the open squares the calculated $N'(E)^{th}$ for the three-Lorentzian $\alpha^2 F$ of Fig. 3.

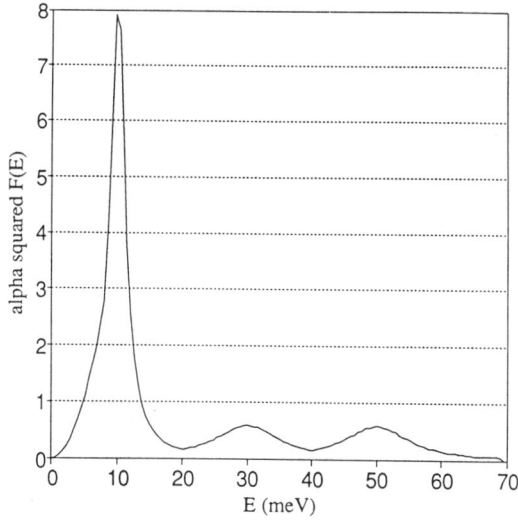

Fig. 3. Inverted $\alpha^2 F$ as a function of energy in meV. As discussed in the text, the sharp peak at 10 meV is the best resolved feature in this inversion.

We find that the gap and μ^* are somewhat sensitive to the nature of the resolution function, while the position of the dominant peak in $\alpha^2 F(E)$ is hardly changed at all. Because this peak position depends on the second zero-crossing of $N'(E)$, which is nearly invariant under the smearing operation, the invariance of the peak is as one would expect.

COMPARISON WITH OTHER EXPERIMENTS

Our inversion of the photoemission data gives an energy gap at $T = 0$ of $\Delta_o = 18$ meV. This results in a ratio $2\Delta_o/k_B T_c$ of 4.6 for the experimental $T_c = 91K$ or of 4.9 for our calculated $T_c = 85K$. How does this compare with energy gaps of $Bi_2Sr_2CaCu_2O_8$ measured by other means? The low-temperature gap of this material has been measured by a variety of techniques including far-infrared reflectance measurements of crystals of $Bi_2Sr_2CaCu_2O_8$ by Reedyk, et al.[9].

Applying a Kramers-Kronig analysis to their data, they obtained a real part of the conductivity that in the superconducting state has a sudden onset from a value at or near zero at a frequency of about 37.5 meV. Although these authors argue that "the sharp edges in the superconducting reflectance are not necessarily energy gaps," we are struck by the fact that this onset at 37.5 meV is identical (within the errors of our inversion process carried out on the photoemission data with its limited resolution) with our $2\Delta_o = 36$ meV.

Recent far-infrared transmission measurements on a single-crystal of $Bi_2Sr_2CaCu_2O_X$ with a $T_c = 76K$ show no energy gap,[10] but this can be understood by the fact that the high-temperature superconductors are near the clean limit.[10] However more recent tunneling and far-infrared transmission measurements by this same group [11] on 2212 give values of $2\Delta_o/k_BT_c$ of 7 to 12, "depending on the interpretation of the results."

A range of energy gaps have been reported from a number of tunneling measurements.[12,13] The ratio $2\Delta_o/k_BT_c$ runs from 3.4 to 8.7 (equivalent to from 13meV to 34meV for Δ_o), so that our gap of 18 meV falls in this range.

An energy gap $2\Delta_o$ of 58-60 meV was reported[14] using high resolution electron-energy-loss spectroscopy. Two earlier photoemission measurements[15,16] were interpreted to give energy gaps $2\Delta_o/k_BT_c = 7$ and 8 respectively. These values are considerably higher than our calculated value.

Results of recent femtosecond optical absorption studies[17] on high T_c superconductors also seem to indicate the existence of a sharp Bosonic peak at low energies in these materials that interacts strongly with the charge carriers.

The large peak in $\alpha^2 F$ at the low energy of about 10 meV leads to a large $\lambda = 8.7$. μ^* seems to be similar to that found for lower temperature superconductors, possibly indicating that Coulomb processes are similar in Bi 2212 to those of Nb_3Sn, but that λ is about four times larger. Direct numerical solution of the temperature dependent NE equations with these parameters produces a T_c of 85K. Because of the large λ, it is perhaps questionable to employ the Dynes-Allen formula to calculate T_c. Nevertheless, we find, on using this formula with parameters derived from Gaussian broadening, the range of 61.5K to 72.6K for T_c with 65K for the best fit. All of these values are close to the experimental value of 92K for Bi 2212.

The sharp low-energy peak in Fig. 3 has a profound influence on the temperature-dependent resistivity in the normal phase. As has been discussed by Allen et al.,[18] one needs $\alpha^2 F$ transport rather than our result in Fig. 3. We have assumed that the features of $\alpha^2 F$ transport are similar to $\alpha^2 F$, but the overall magnitude may differ.

From λ alone, one can easily calculate the normal-state scattering lifetime and hence the temperature dependence of the resistivity in arbitrary units. We find that the resistivity is very nearly linear in temperature for temperatures above T_c. This is what is seen experimentally,[19] but cannot be produced from models with Boson peaks at larger energies. To obtain the magnitude of the resistivity, we also need the Drude plasma frequency in addition to the electron lifetime.[18] Although the plasma frequency can be determined from infrared data, as pointed out in Ref. 18, "it is appropriate to be skeptical" of such determinations because various factors have to be subtracted from the data. We find it interesting that Allen et al.[18], after estimating λ to be less than one for both LSCO and YBCO, find that they cannot fit the magnitude of the experimental resistivity data. They conclude that one way they can fit this data is by "increasing λ by a factor of 5-10." Thus although they were considering different materials from BSCCO, they find that the resistivity data *may* be indicating that a large λ is required.

CONCLUSIONS

In summary, we have applied Nambu-Eliashberg theory to the Dessau et al. photoemission data and have assumed that these (110) data are typical. The results of inversion produce a μ^* similar to that found in other superconductors and with a gap Δ_o of 18 meV but with a λ near 8.7 and an $\alpha^2 F$ with a large narrow peak at about 10 meV. Several types of Boson mediated mechanisms have been proposed for high-temperature superconductors including[20] acoustic plasmons, bi-excitons, and various forms of phonons. Any of these models can be consistent with a dominant peak at about 10 meV (116 Kelvin). If the large peak is due to phonons, such phonons would be strongly coupled to the electrons. Hence phonon lifetimes would be short, and phonon widths would be broad. We are unaware of any measurements of the phonon density of states derived, for example, from neutron scattering experiments for Bi 2212. Such experiments would be most helpful for comparison with our $\alpha^2 F$, and would tend to rule out, or support, phononic mechanisms. Large λ's have been considered before[21,22] in YBCO.

In this work, we have considered photoemission data drawn from a small region of the Brillouin zone on the (110) line. We have assumed in performing our inversions that this point on the Fermi surface is typical. The electronic band structure[23] suggests that these states are Bi rich, but that other points on the Fermi surface have different parentage. Hence it is possible that the gap, $\alpha^2 F$, and λ are in fact dependent on direction. This would necessitate a more complex inversion scheme than the one we have applied, but we believe that the main results of our work would remain intact. This also points toward the utility of photoemission data such as that of Dessau et al.[1] to probe anisotropies.

The (110) data are taken close to a place of band crossings; so it may be possible that the spectral features that we have analyzed are due to this band effect and not to strong coupling. One of us (FMM) has examined unpublished photoemission data of C.G. Olson[24] on Bi2212 searching for the band-structure possibility. We have concluded that the view we present here (the strong-coupling $\alpha^2 F$ model) is the more likely explanation of the Desseau et al. data.

ACKNOWLEDGMENTS

Part of this work was performed under the auspices of the United States Department of Energy. We are particularly indebted to D.S. Dessau, B.O. Wells, Z.-X. Shen, W.E. Spicer, A.J. Arko, R.S. List, D.B. Mitzi, and A. Kapitulnik for making their experimental data available to us before publication. We would like to thank many colleagues for helpful discussions: P.B. Allen, K.S. Bedell, M.H. Cohen, S. Doniach, Z. Fisk, S.M. Girvin, C.G. Olson, D. Scalapino, J.D. Thompson, W. Tomasch, C. Uher, B.W. Veal, and J.F. Zasadzinski. We also thank P.B. Allen, J.P. Carbotte, L. Mihaly, C. Uher, and J.F. Zasadzinski for providing us with preprints of their papers. One of us (FMM) is especially grateful to C.G. Olson sharing his unpublished Bi2212 results.

REFERENCES

1. D.S. Dessau, B.O. Wells, Z.-X. Shen, W.E. Spicer, A.J. Arko, R.S. List, D.B. Mitzi, and A. Kapitulnik, *Phys. Rev. Lett.* **66**: 2160 (1991); and to be submitted.

2. See also J.P. Carbotte's *Properties of Boson Exchange Superconductors, Rev. Mod. Phys.* **62**: 1027 (1990).

3. D.J. Scalapino, *The electron-phonon interaction and strong-coupling superconductors*, in: "Superconductivity," R.D. Parks, ed., Marcel Dekker, New York (1969).

4. W.L. McMillian and J.M. Rowell, *Tunneling and strong-coupling superconductivity*, ibid.

5. E.L. Wolf and G.B. Arnold, *Physics Reports* **91**: 31 (1982).

6. J.R. Schrieffer, D.J. Scalapino, and J.W. Wilkins, *Phys. Rev.* **10**: 336 (1963).

7. G.B. Arnold, F.M. Mueller, and J.C. Swihart, *Phys. Rev. Lett.* **67**, 2569 (1991).

8. G.B. Arnold, F.M. Mueller, and J.C. Swihart, to be submitted to *Phys. Rev. B*.

9. M. Reedyk, D.A. Bonn, J.D. Garrett, J.E. Greedan, C.V. Stager, T. Timusk, K. Kamarás, and D.B. Tanner, *Phys. Rev. B* **38**: 11981 (1988).

10. L. Forro, G.L. Carr, G.P. Williams, D. Mandrus, and L. Mihaly, *Phys Rev. Lett.* **65**: 1941 (1990).

11. L. Forro, D. Mandrus, and L. Mihaly, preprint.

12. See references listed in Q. Huang, J.F. Zasadzinski, K.E. Gray, J.Z. Liu, and H. Claus, *Phys. Rev. B* **40**: 9366 (1989).

13. T. Walsh, J. Moreland, R.H. Ono, and T.S. Kalkur, *Phys. Rev. Lett.* **66**: 516 (1991).

14. J.E. Demuth, B.N.J. Persson, F. Holzberg, and C.V. Chandrasekhar, *Phys. Rev. Lett.* **64**: 603 (1990).

15. C.G. Olson, *et al.*, *Physica C* **162-164**: 1697 (1989).

16. Y. Petroff, *Physica C* **162-164**: 845 (1989).

17. J.M. Chwalek, C. Uher, J.F. Whitaker, G.A. Mourou, J. Agostinelli, and M. Lelental, *Appl. Phys. Lett.* **57**: 1696 (1990); and *ibid* **58**: 980 (1991).

18. P.B. Allen, W.E. Pickett, and H. Krakauer, *Phys. Rev. B* **37**: 7482 (1988).

19. D.B. Mitzi thesis, Stanford University (unpublished) (1989).

20. See for example, "High Temperature Superconductivity," K.S. Bedell, D. Coffey, D.E. Meltzer, D. Pines, and J.R. Schrieffer, ed., Addison-Wesley, New York (1989).

21. See for example, B.W. Veal *et al.*, *Physica C* **158**: 276 (1989).

22. V. Kresin (private communication).

23. S. Massidda, J. Yu, and A.J. Freeman, *Physica C* **152**: 251 (1988); H. Krakauer and W.E. Pickett, *Phys. Rev. Lett.* **60**: 1665 (1988); M.S. Hybertsen and L.F. Mattheiss, *Phys. Rev. Lett.* **60**: 1661 (1988).

24. C.G. Olson, (private communication).

THE ROLE OF THE AXIAL OXYGEN IN HIGH-T_C MATERIALS

J. Mustre de Leon, S.D. Conradson, P.G. Allen,
I. Batistić, and A.R. Bishop

Los Alamos National Laboratory
Los Alamos, NM 87545

I. INTRODUCTION

The axial oxygen, O(4), which serves as a bridge between the Cu(1)-O(1) chains and Cu(2)-O(2,3) planes in $YBa_2Cu_3O_7$, provides a natural way by which structural and electronic changes in the basal Cu(1)-O(1) planes (or equivalent planes in other materials) can influence the superconductivy occuring in the Cu(2)-O(2,3) planes. More specifically, O(4) vibrations are expected to couple to charge transfer between chains and planes.[1] Experimentally, several studies have shown that the O(4) site is involved in structural changes that also affect the superconductivity in $YBa_2Cu_3O_{7-\delta}$.[2,5] Infrared reflectivity studies indicate an anomalously large strength for the 155 cm^{-1} phonon mode associated with the [Cu(1)+O(4)]-Cu(2) vibration,[2] and excess intensity and frequency shift of the Cu(1)-O(4) 580 cm^{-1} mode that correlate with the superconducting order parameter.[3] Raman profiles also suggest that a structural instability related to the A_g ($\sim$ 500 cm^{-1}) Cu(1)-O(4) stretching vibration could be coupled to charge transfer.[4] Diffraction studies have shown a correlation between the Cu-O(4) bond lengths and T_c.[5]

Direct evidence for O(4) lattice anomalies across T_c, comes from ion channeling experiments.[6] We recently presented extended-x-ray absorption fine structure (EXAFS) from $YBa_2Cu_3O_7$ which indicate that the O(4) atom moves in a double-well potential that softens in a fluctuation region around T_c.[7] Theoretical support for the existence of double-well potentials associated with Cu-O(4) bonds, comes from charge transfer models (based on descriptions of the highly polarizable O(4) ions),[8,9] and dynamical Jahn-Teller coupling models.[10] These models lead to anharmonic potentials for the Cu-O(4) relative motion in $YBa_2Cu_3O_7$, predicting a more harmonic motion of the O(4) atom as the stoichiometry changes from $YBa_2Cu_3O_7$ to $YBa_2Cu_3O_6$.[8] Our analysis provides evidence of a dynamical coupling between the O(4) vibrations and in-plane electronic degrees of freedom involved in the superconductivity.

In this paper we present the major results from this analysis. We also present Cu–O(4) potentials and radial distribution functions, derived from EXAFS measurements in $YBa_2Cu_3O_{6.5}$ and $YBa_2Cu_{2.8}Co_{0.2}O_{7+\delta}$ samples, that indicate a correlation between O(4) anharmonicity and T_c. Finally, we present EXAFS from

TlBa$_2$Ca$_3$Cu$_4$O$_{11}$ which indicate an split postion for the axial oxygen and changes in the vicinity of T$_c$ similar to those observed in YBa$_2$Cu$_3$O$_7$.[7] The rest of this paper is organized as follows: In Sec. II we discuss the problem of EXAFS arising from an anharmonic system. In Sec. III we present Cu(1)–O(4) potentials and radial distribution functions (RDF's) as a function of temperature derived from Cu K-edge polarized EXAFS of an oriented sample of YBa$_2$Cu$_3$O$_7$ (T$_c$=92 K). In Sec. IV we present Cu–O(4) potentials and RDF's derived from Cu K-edge polarized EXAFS of oriented samples of YBa$_2$Cu$_3$O$_7$ (T$_c$=92 K), YBa$_2$Cu$_3$O$_{6.5}$ (T$_c$=52 K) and YBa$_2$Cu$_{2.8}$Co$_{0.2}$O$_{7+\delta}$ (T$_c$=25 K) at T=10 K. In Sec. V we present a qualitative analysis of axial oxygen contributions to the Cu-K edge EXAFS from a TlBa$_2$Ca$_3$Cu$_4$O$_{11}$ sample (T$_c$=118 K) as a function of temperature. Sec. VI contains the general conclusions from these three studies.

The samples, experimental methods, and EXAFS data reduction have been described elsewhere.[11] We only note that as the listed temperatures T_{nom} are those of the cold finger of the cryostat used in the experiment, the actual temperature of the sample may be as much as 5–10 K higher.

II. EXAFS FROM AN ANHARMONIC SYSTEM

In order to consider the effect of the relative motion of a given atomic pair in an arbitrary pair potential, we consider the statistical average of the EXAFS from a single bond. In the *single* particle approximation, this can be represented by

$$\langle \chi \rangle = \int dz g(z) \chi(k,r) \quad , \tag{1}$$

where, $\chi(k,r)$ denotes the single scattering EXAFS contribution arising from atoms located at a distance r from an absorbing atom, and the photoelectron momentum $k = [(2m/\hbar^2)(E - E_0)]^{1/2}$ is referenced to the arbitrary energy origin E_0. The radial distribution function (RDF), $g(z)$, is expressed in terms of single particle wave functions $\{\psi_i(z)\}$ and single particle energy levels $\{\epsilon_i\}$;

$$g(z) = \frac{\sum_i |\psi_i(z)|^2 e^{-\beta \epsilon_i}}{\sum_i e^{-\beta \epsilon_i}} \quad . \tag{2}$$

The temperature of the system T is introduced through $\beta = 1/k_B T$, and z denotes the relative displacement along the bond from the bond equilibrium position, i.e., $\vec{r} = \vec{R} + z\hat{R}$. The wavefunctions $\{\psi_i(z)\}$ are determined by solving the Schrödinger equation using the reduced mass for an isolated Cu-O pair, and a model potential, $V(z)$. In the systems discussed here we found that the functional form,

$$V(z) = \frac{a}{2}(z - z_1)^2, z \leq z_0; \quad \frac{b}{2}(z - z_2)^2, z \geq z_0 \quad , \tag{3}$$

where z_0 is determined by the continuity condition $V(z_0^+) = V(z_0^-)$, and a, b, z_1, z_2 are fitting parameters, lead to the best fits to experiment. In the standard harmonic analysis of XAFS data, $g(z)$ is assumed to have a Gaussian form that leads to the usual quadratic form for the Debye-Waller factor.[7] We note that the *single* particle approximation that leads to an effective pair potential is equivalent to the commonly used Einstein approximation.

III. THE Cu(1)–O(4) RDF IN YBa$_2$Cu$_3$O$_7$

Diffraction studies in YBa$_2$Cu$_3$O$_7$ have shown large anisotropic thermal ellipsoids for the O(4) ion, hinting at anharmonicity associated with this ion.[12] Also unpolarized EXAFS studies on unoriented samples have shown anomalies in the Debye-Waller factors across T_c.[13] Polarized EXAFS measurements separate the Cu–O(4) contributions from those arising from Cu–equatorial oxygens, allowing a more quantitative analysis of the Cu–O(4) signal.

The presence of a beat in the Cu–O(4) EXAFS contribution around $k = 12$Å^{-1}, signals two Cu(1)–O(4) distances, indicative of a double-well (Fig. 1). This beat dissapears for temperatures near T_c. A nonlinear squares fit, including *both* the Cu(1)–O(4) and Cu(2)–O(4) contributions was performed over the range $k = 3-14$ Å^{-1} on Fourier-filtered data (filtered over the ranges $1.0 \leq R \leq 2.0$ Å, $k = 2-15$ Å^{-1}) measured at T_{nom}=10, 83, 86, 88, 95, and 105 K, using as parameters to be determined: R, E_0, and the potential parameters a, b, z_1, z_2, for each bond. The temperature-independent EXAFS amplitude and phase functions in $\chi(k,r)$ were obtained from the EXAFS of Cu–O bonds in the $a-b$ plane, and the number of O(4) atoms at an average distance R from Cu(1) was fixed at 2. The comparison between experiment and the achieved fits is shown in Fig. 1, and the resulting Cu(1)–O(4) parameters are presented in Ref. 7. These results indicate that the effect of the Cu(2)–O(4) signal is negligible in the beat region, indicating the two positions for the O(4) ion.

From these fits a potential, $V(z)$, and a RDF, $g(z)$, for the Cu(1)–O(4) bond were extracted at the measured temperatures. These results indicate that the motion of the O(4) atom must be described quantum mechanically, since only the ground and first excited states are appreciably occupied even at $T_{nom}= 105$ K. The beat disappears in a fluctuation region around T_c because the separation between the minima of the potential $V(z)$ decreases by ~ 0.02 Å. This small change in distance is well within the sensitivity of these data because it moves the beat from $k = 12$ to beyond $k = 14$ Å^{-1}; the structural change lowers the potential barrier between the two wells (Fig. 2). As shown in Fig. 2 $g(z)$ exhibits 2 maxima located 0.13 Å apart for $T_{nom} \sim 10$ K and $T_{nom} > 86$ K. Within the fluctuation region, $T_{nom}= 83$ and 86 K, the separation between maxima decreases by ~ 0.02Å, leading to a decrease in the root mean square deviations of the Cu(1)–O(4) bond length, consistent with ion channeling results.[6] We do not yet have data to establish the lower limit of the fluctuation region. The changes in the RDF can be described as a result of an increased tunneling of the wave function through the potential barrier between the two wells (Fig. 2) (effective polarization of the O(4) may also contribute). The average of the two Cu(1)–O(4) distances is in good agreement with crystallographically determined values,[14] and it does not show appreciable variations as a function of temperature.

The motion of the O(4) atom in the deep double-well potential $V(z)$ can be described by a two-level Hamiltonian, $H = (\omega_T/2)\sigma_z$, such that $H|A\rangle = +(\omega_T/2)|A\rangle$, and $H|S\rangle = -(\omega_T/2)|S\rangle$, where σ_z is a Pauli spin matrix, and $|S\rangle, |A\rangle$ denote the (symmetric) ground state and (antisymmetric) first excited state, respectively, separated by an energy $\hbar\omega_T = \hbar\omega_1 - \hbar\omega_0$. Unlike a ferroelectric system at the order-disorder limit, where a double-well potential also occurs, for the fluctuation we observe, both sides of the well are nearly equally occupied over the entire temperature range, but the splitting between levels increases within the fluctuation region. Fig.

3 shows the tunneling frequency ω_T as a function of temperature. An increase of ~ 80 K in ω_T is observed for $T_{nom} = 83$ and 86 K. Since the local potential description of the Cu–O(4) motion does *not* correspond to the normal modes of the system, a precise identification of this increase in the tunneling frequency with frequency shifts in related Raman and infrared modes is *not* correct. Calculations, based on electron-phonon models with on-site anharmonicity derived from double-well potentials obtained from EXAFS and biphonon-electron coupling, predict the anomalously large strength and correct frequency shifts of the infrared active mode at 155 cm^{-1}.[15]

We note that the observed fluctuation is *not* directly driven by temperature, but rather is a result of the coupling of the electronic degrees of freedom involved in the superconducting transition and the elastic degrees of freedom leading to the potential describing the ionic motion. We assume a coupling between the proposed two level system and the superconducting order parameter Ψ, leading to a free energy F of the form:

$$F = \frac{\omega_T^0}{2} \langle \phi | \sigma_x | \phi \rangle - A|\Psi|^2 + \frac{B}{2}|\Psi|^4 + C|\Psi|^2 \langle \phi | \sigma_x | \phi \rangle, \quad (4)$$

where, $|\phi\rangle$ is a spinor denoting the state of the two level system, the Pauli matrix σ_x introduces hopping between $|S\rangle$ and $|A\rangle$, and ω_T^0 is the tunneling frequency in the absence of any coupling. We analyze the case in which the effect of the coupling on the order parameter is small. Taking into account the effect of fluctuations $\langle (\Delta \Psi)^2 \rangle$ (that diverge at T_c) about the mean field value of Ψ_0, we find a change in the tunneling frequency which agrees qualitatively with Fig. 3 however, a detailed quantitative comparison will require more experimental data in the fluctuation region. If the coupling between the superconducting order parameter and the nonlinear phonons is strong, it is insufficient to consider the effect of this coupling in the phonon system and the two equations that result from the minimization of the free energy F (*cf.* Eq. 4) must be solved self-consistently.

IV. EFFECT OF DOPING AND OXYGEN CONTENT ON THE Cu(1)–O(4) BOND

The main local structural changes achieved by the removal of the O(1) atoms from the YBa$_2$Cu$_3$O$_{7-\delta}$ orthorhombic phase (for $\delta \leq 0.55$) are a decrease of the Cu(1)–O(4) bond length (~ 0.03 Å) and a large increase of the Cu(2)–O(4) bond length (~ 0.08 Å).[16] At low doping levels, Co substitutes specifically for Cu(1) and induces similar structural changes in the Cu–axial O bonds, such as those obtained by oxygen removal.[17] The Cu–O contribution to the polarized ($\hat{\epsilon} \parallel c$) Cu K-edge EXAFS signal was isolated by Fourier filtering within the ranges $3 \leq k \leq 15$ Å^{-1}, and $1.0 \leq r \leq 2.0$ Å. Anharmonic fits, including *both* the Cu(1)–O(4) and Cu(2)–O(4), were performed in the range $4 \leq k \leq 14$ Å^{-1}, using the model potential in Eq. 3, and the same potential fitting parameters used in the study described in the previous section. The EXAFS phase and amplitude functions derived from the Cu–O signal from the $\hat{\epsilon} \perp c$ data were used in these fits. The coordination number for the Cu–O pairs was fixed at $N = 2$. The agreement between experimental data and the fits is similar to the one shown in Fig. 1. These fits and the resulting parameters for the Cu(1)–O(4) bond are presented in Ref. 18.

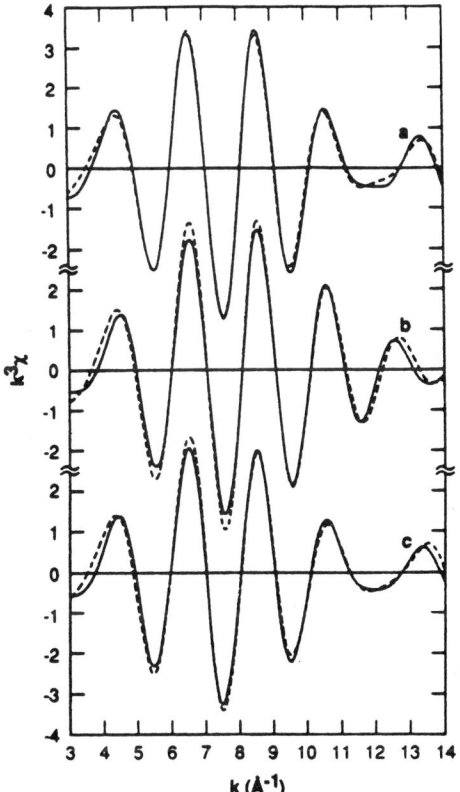

Fig. 1. Cu–O4 EXAFS, experimental data (solid line) and calculated EXAFS (solid line), at (a) $T_{nom} = 10K$, (b) $T_{nom} = 86$, and (c) $T_{nom} = 105K$.

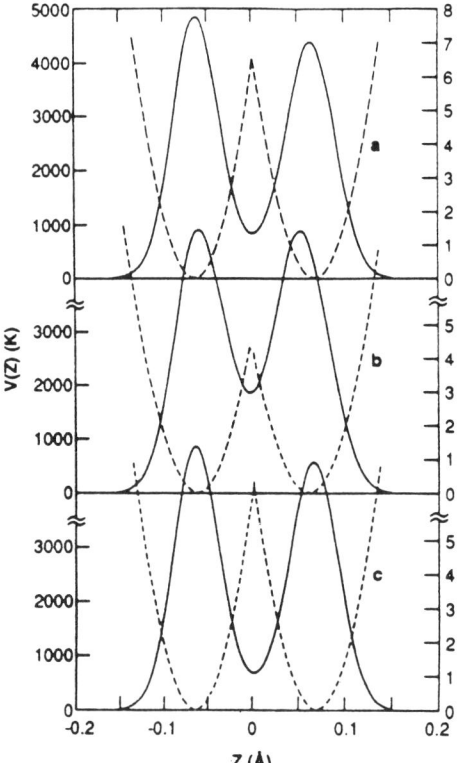

Fig. 2. RDF (solid line) and potential, $V(z)$, (dashed line) for: (a) $T_{nom} = 10$K, (b) $T_{nom} = 86$K, and (c) $T_{nom} = 105$K in $YBa_2Cu_3O_7$.

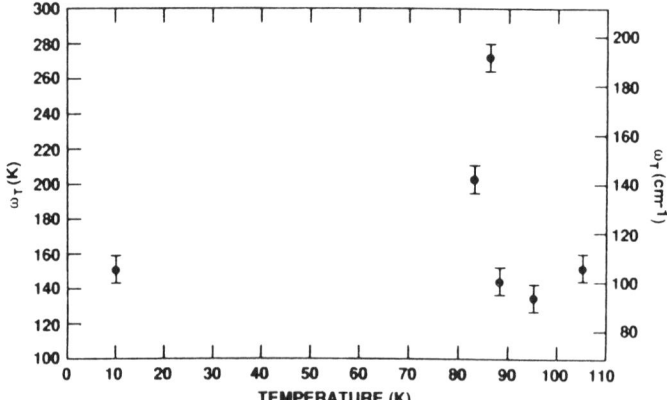

Fig. 3. Tunneling frequency, ω_T, between the two O4 sites, as a function of temperature in $YBa_2Cu_3O_7$.

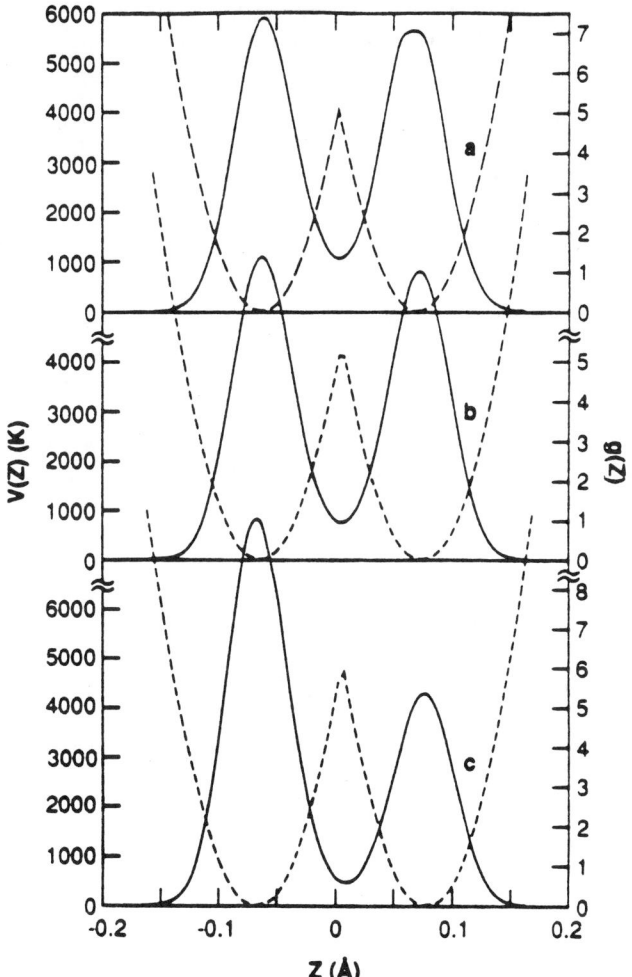

Fig. 4. RDF and potential, $V(z)$, for a) $YBa_2Cu_3O_7$, b) $YBa_2Cu_3O_{6.5}$, c) $YBa_2Cu_{2.8}Co_{0.2}O_{7+\delta}$.

In going from $YBa_2Cu_3O_7$ to $YBa_2Cu_3O_{6.5}$ and $YBa_2Cu_{2.8}Co_{0.2}O_{7+\delta}$, we observe the following changes: a) the minima of the double-well potentials move further apart – from $\Delta R = 0.13$ Å in $YBa_2Cu_3O_7$, $\Delta R = 0.14$ Å in $YBa_2Cu_3O_{6.5}$, to $\Delta R = 0.145$ Å in $YBa_2Cu_{2.8}Co_{0.2}O_{7+\delta}$ –; b) the RDF, $g(z)$, becomes more asymmetric in the same progression, going from near equal populations for both sites in $YBa_2Cu_3O_7$, to a ratio 2:1 for the population of the shorter site, compared with the longer site in $YBa_2Cu_{2.8}Co_{0.2}O_{7+\delta}$ (See Fig. 4). These structural changes produce an increase in the height and width of the potential barrier between the two wells leading to a decreased tunneling between the two sites. This change is quantified by the tunneling frequency as shown in Fig. 5. Although, there are not enough cases to establish quantitative trends, we can conclude that the tunneling frequency systematically decreases as the critical temperature decreases. We interpret these changes as an indication that the motion of the O(4) atom becomes more *harmonic* with decreasing T_c, i.e., the motion of the O(4) in each well becomes *decoupled*, and the occupation of only *one site* is favored. Indeed, in the case of $YBa_2Cu_{2.8}Co_{0.2}O_{7+\delta}$ (T_c=25K) it was possible to fit the $\epsilon \parallel c$ data using a purely harmonic treatment.[18]

In an attempt to understand the observed correlation between the tunneling frequency and T_c, we calculated T_c using the Eliashberg equation, modified to include the anharmonic phonons derived with potentials obtained from the EXAFS fit. Other parameters, i.e., the density of electronic states at the Fermi level and the (bare) electron-phonon interaction parameter were kept fixed in this calculation. We found that the predicted changes in T_c were much smaller than the actual variation in T_c in the studied materials. This suggests that the coupling between nonlinear phonons (derived from the double-well potential) and electrons is not the main factor determining T_c.[19] As mentioned in the previous section *not* only on-site anharmonicity, but *also* multiphonon-electron coupling terms seem to be important to describing spectroscopical data and possibly other properties.[15]

In going from anharmonic motion to the case of more harmonic motion, a decrease in the elastic-electronic coupling is expected.[8,9] Consequently, elastic anomalies across T_c involving the O(4) atom are expected to be smaller for the more harmonic systems, i.e., a smaller change in the tunneling frequency across T_c, should be observed in $YBa_2Cu_3O_{6.5}$ and $YBa_2Cu_{2.8}Co_{0.2}O_{7+\delta}$ compared with that observed in $YBa_2Cu_3O_7$, discussed in the previous section. This trend has been observed in the anomalous frequency shift and intensity peak of the 580 cm^{-1} infrared active mode in $YBa_2Cu_{3-x}Co_xO_{7+\delta}$.[3] As the Co content is increased and T_c decreases, the observed anomalies decrease in magnitude. Also, specific heat measurements in $YBa_2Cu_{3-x}Co_xO_{7+\delta}$ have reported the same trend for the anomalous contribution to the specific heat across T_c.[20]

V. THE Cu-O(4) BOND IN TlBa$_2$Ca$_3$Cu$_4$O$_{11}$

Atomic-pair-distribution analysis of diffraction data in $Tl_2Ba_2CaCu_2O_8$ have found a split position of the axial oxygen and changes in the local arrangement of O atoms in the Cu-O planes across T_c.[21] These results suggest that the axial oxygen in Tl-based superconductors exhibits a behavior similar to that observed in the axial oxygen in $YBa_2Cu_3O_7$, using EXAFS.

We have analyzed unpolarized Cu K-edge EXAFS data from a $TlBa_2Ca_3Cu_4O_{11}$ powder sample measured at T_{nom}=10, 100, 117, 127, 135, and 156 K, in order to study the structure and temperature dependence of the Cu-O(4) bond. In this case is not possible to isolate the Cu-O(4) signal from the Cu-equatorial oxygens, Cu-Ca and Cu-Cu signals, by direct Fourier filtering, and a multishell fit was used to isolate the Cu-O(4) signal.

The Cu-O(4) signal shows a beat in the region $k \sim 11$ Å^{-1} for T_{nom}= 10, 127, 135 and 156 K, while the beat is not present for T_{nom}= 100 and 117 K. Such a beat manifests itself as a a sudden jump of π radians in the EXAFS phase (Fig. 6). Consequently, the observed jump in the phase signals the presence of *two* distinct Cu-O(4) distances. Nonlinear least squares fits of these Cu-O(4) signal using Gaussian

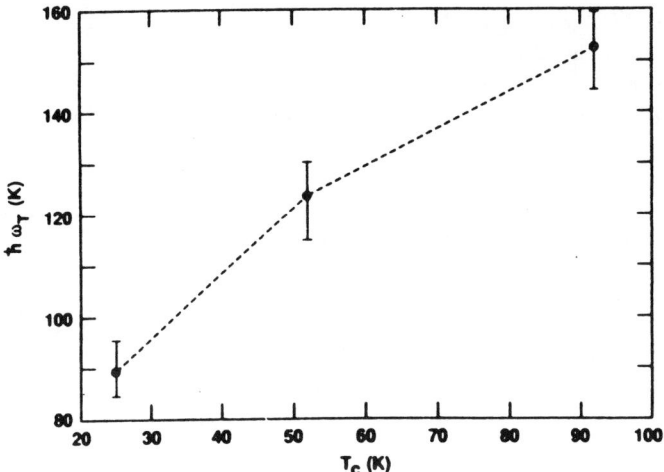

Fig. 5. Tunneling frequency, ω_T, as a function of T_c, in $YBa_2Cu_3O_7$, $YBa_2Cu_3O_{6.5}$, and $YBa_2Cu_{2.8}Co_{0.2}O_{7+\delta}$.

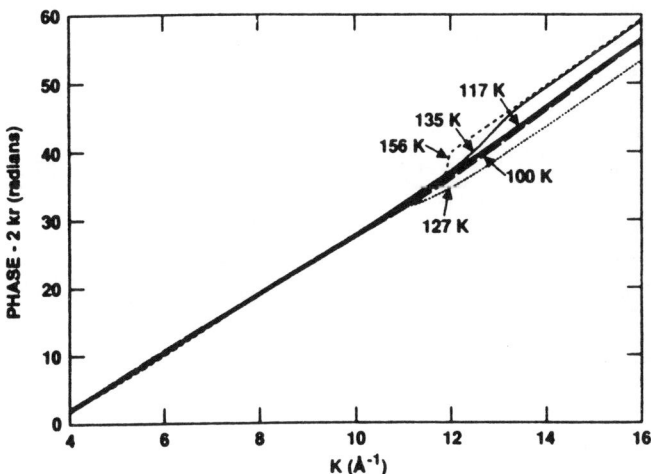

Fig. 6. Phase of the Cu–O4 EXAFS contribution in $TlBa_2Ca_3Cu_4O_{11}$.

RDF's for each oxygen peak indicate the presence of two Cu–O(4) distances at $R_1 = 2.57(4)$ Å and $R_2 = 2.74(4)$ Å for $T_{nom} \neq 100, 117$ K, while they indicate a separation $\Delta R = R_2 - R_1 < 0.1$ Å at these temperatures (possibly having a unique Cu–O(4) distance). These distances are in good agreement with Cu–O(4) distances observed in other Tl-based superconductors. They lead to Tl–O(4) distances consistent with the presence of Tl^{3+} as observed in photoemission studies.[22] In this case it is not possible to analyze the Cu–O(4) signal using an anharmonic local potential for the Cu–O(4) bond, as the introduction of a local single-pair potential is adequate only for nearest-neighbor bonds for which the the Einstein approximation for the relative pair motion is adequate.

The above analysis, together with the observation of two configurations of Tl and O atoms in the Tl-O planes of $Tl_2Ba_2CaCu_2O_8$[23], suggest the presence of two different charge states of the Tl ion, with evidence for charge transfer between the Tl-O planes and the Cu-O planes around the superconducting transition. This analysis points out the fact that the axial oxygen atom in Tl-based superconductors plays a similar role to that played by the O(4) atom in $YBa_2Cu_3O_7$, i.e., it serves as a vehicle for charge transfer between Tl-O and Cu-O planes, and its motion reflects electronic changes taking place in the Cu-O planes during the superconducting transition.

VI. CONCLUSIONS

The three studies discussed here present a unified view of the role of the axial oxygen in high-T_c materials. These studies show two different positions for the axial oxygen, which can be explained in terms of Cu–O(4) motion in an anharmonic double-well potential. These potentials arise naturally in electron-phonon models that allow charge tranfer between the (mixed valence) cations and the highly polarizable O^{2-} ions. In the oxygen deficient and Co-doped samples, the tendency of the cation to a full shell configuration supresses charge fluctuations leading to a more harmonic motion for the cation–oxygen pair. The observed temperature behavior of the Cu–O(4) RDF can be explained consistently as a result of the coupling between anharmonic phonons and the superconducting order parameter. These results show the influence of electronic degrees of freedom entering in the superconducting transition on the phononic degrees of freedom related with the axial oxygen. To address the relevance of the observed anharmonic phonons to the superconducting mechanism, it is necessary to formulate a microscopic model that includes the Cu-O in-plane electronic degrees of freedom and their interactions with electronic and phononic degrees of freedom related with the axial oxygen. We note, however that changes in the electronic properties of the Cu(2) planes are suggested by fluctuations in the XANES, which may support dynamic Jahn-Teller coupling.

This work was supported by the US DOE.

REFERENCES

1. P.B. Littlewood, C.M. Varma and E. Abrahams, *Phys. Rev. Lett.* **63**, 2602 (1989).
2. L. Genzel, *et al.*, *Phys. Rev.* B **40**, 2170 (1989).
3. H. S. Obhi and E. K. Salje, Physica (Amsterdam) **171C**, 547 (1990).
4. R. Zamboni, *et al.*, *Solid State Comm.* **70**, 813 (1989).

5. R. J. Cava *et al.*, Physica (Amsterdam) **165**C, 419 (1990).
6. R. P. Sharma *et al*, Phys. Rev. Lett. **62**, 2869 (1989).
7. J. Mustre de Leon *et al.*, Phys. Rev. Lett. **65**, 1678 (1990).
8. I. Batistic *et al.*, Phys. Rev. B **40**, 6869 (1989).
9. C. C. Yu and P. W. Anderson, *Phys. Rev. B* **29**, 2165 (1984).
10. K.H. Johnson *et al.*, Mod. Phys. Lett. **B3**, 1367 (1989).
11. S. D. Conradson *et al.*, Science **248**, 1394 (1990).
12. M. A. Subramanian *et al.*, J. Solid State Chem. **77**, 192 (1988).
13. H. Maruyama *et al.*, Physica (Amsterdam) **160**C, 524 (1989).
14. Compare, *e.g.*, J. J. Caponi *et al.*, Europhys. Lett. **3**, 1301 (1987); M. A. Beno *et al.*, Appl. Phys. Lett. **51**, 57 (1987); A. Williams *et al.*, Phys. Rev. B **37**, 7960 (1988).
15. A.R. Bishop *et al.*, in preparation.
16. S. Katano *et al.*, Jpn. J. Appl. Phys. **26**, L1049 (1987); R. J. Cava *et al.*, Phys. Rev. B **36**, 5719 (1987).
17. J. M. Tarascon *et al.*, Phys. Rev. B **37**, 7458 (1988); T. J. Kistenmacher, Phys. Rev. B **38**, 8862 (1988).
18. J. Mustre de Leon *et al.*, submitted to Phys. Rev. B.
19. N.M. Plakida *et al.*, Europhys. Lett. **4**, 1309 (1987); J.R. Hardy and J.W. Flocken, Phys. Rev. Lett. **60**, 2191 (1988).
20. J. W. Loram *et al.*, in Conference on High Temperature Superconductivity, Cambridge, U.K. (1990), to be published in Superconductor Science and Technology (1991).
21. W. Dmowski *et al.* Phys. Rev. Lett. **61**, 2608 (1988); T. Egami *et al.* in this volume.
22. T. Suzuki *et al.*, Physica (Amsterdam) **162-164**C, 1387 (1990).
23. B. H. Toby *et al.*, Phys. Rev. Lett. **64**, 2414 (1990).

THERMALLY ACTIVATED DEPINNING OF VORTICES IN HIGH-T_c SUPERCONDUCTORS

Ernst Helmut Brandt[1]

AT&T Bell Laboratories
Murray Hill, New Jersey 07974, USA

In high-T_c superconductors thermally activated depinning of the flux lines leads to the appearance of a finite resistivity when a magnetic field is applied. At low current densities this corresponds to a finite *diffusivity* of the magnetic flux, whose exponential temperature dependence explains a large variety of experiments without having to introduce concepts like flux-line lattice melting or phase transitions. Nonlocal elasticity and material anisotropy *soften* the flux-line lattice and strongly enhance its thermal fluctuations and pinning-caused distortions.

1. Thermally Activated Flux Motion

Soon after Abrikosov [1] had predicted the existence of quantized magnetic flux lines in type-II superconductors, Anderson [2,3] realized that these tiny current vortices should move under the action of an applied electric current density **J**. This vortex drift **v** dissipates energy and thus generates an electric field $\mathbf{E} = \mathbf{B} \times \mathbf{v}$ where **B** is the flux density or magnetic induction in the sample. The dissipation is caused by two effects: (a) By dipolar currents which surround each moving flux line (eddy currents) and which have to pass through the "normal conducting vortex core" in the model [4]. (b) By the retarded relaxation of the order parameter $\Psi(\mathbf{r})$ when the vortex core (a region of depressed Ψ) moves [5]. Since at low B the dissipation of the vortices is additive and since at the upper critical field $B_{c2}(T)$ the flux-flow resistivity ρ_{FF} has to reach the normal conductivity ρ_n, one approximately has $\rho_{FF} \approx \rho_n B/B_{c2}(T)$. A more quantitative treatment of this flux dissipation uses time dependent Ginzburg-Landau theory [6,7]. For reviews of flux motion see [8], and for extensions to layered superconductors [9]. A recent excellent overview on magnetic flux transport in superconductors is given by Clem [10].

In real superconductors, at small current densities $J < J_c$ the flux lines are pinned by inhomogeneities in the material, e.g., by dislocations, vacancies, interstitials, grain boundaries, precipitates, or by a rough surface. Only when J exceeds a critical value J_c do the vortices move and dissipate energy. As predicted by Anderson [1], thermally activated depinning of the flux lines may occur at finite temperatures T. In conventional superconductors this effect is not important; it is observed only close to the transition temperature T_c as *flux creep* [11]. Flux creep occurs when the superconductor is in a

---

[1] On leave from: Max-Planck-Institut für Metallforschung, Institut für Physik, Heisenbergstr. 1, D-7000 Stuttgart 80, FRG

critical state, which is established when the applied field B_a is increased or decreased: The magnetic flux then enters or exits until a *critical slope* is reached, i.e., a maximum and nearly constant gradient of $B = |\mathbf{B}|$. More precisely, the *current density* $\mathbf{J}$ reaches a maximum value J_c; one has $\mathbf{J} = (\partial H/\partial B)\nabla \times \mathbf{B} \approx \mu_0^{-1}\nabla \times \mathbf{B}$, where $H(B) \approx B/\mu_0$ is the (reversible) magnetic field which would be in equilibrium with the induction $\mathbf{B}$. In general, the current density in a flux-line lattice can have two different origins, a *gradient* of B or a *curvature* of the flux lines (or field lines) with direction $\hat{\mathbf{B}} = \mathbf{B}/B$. This is easily seen by writing $\nabla \times \mathbf{B} = \nabla B \times \hat{\mathbf{B}} + B\nabla \times \hat{\mathbf{B}}$. Thus, in simple geometries $J \approx \mu_0^{-1}\nabla B$. When thermally activated flux creep occurs, the field gradient, and the persistent currents and magnetization, slowly *decrease with a logarithmic time law*. Formally, this flux creep is equivalent to a highly nonlinear, current dependent flux-flow resistivity $\rho \propto \exp(J/J_1)$. Initially, the persistent currents, e.g. in a ring, cause a large ρ, but as the current decays, ρ decreases rapidly and so does the decay rate $-\dot{J} \propto \exp(J/J_1)$.

In high-T_c superconductors (HTSC), thermal depinning is observed in a much larger temperature interval below T_c. This "giant flux creep" [12,13] occurs mainly because the superconducting coherence length ξ ($\approx$ vortex core radius) is so small and the magnetic penetration depth λ is large, which means the elementary pinning *energy* of small pins (e.g. oxygen vacancies) is small, $U_p \approx (B_c^2/\mu_0)\xi^3 = (\Phi_0^2/8\pi^2\mu_0\lambda^2)\xi$. The elementary pinning *force* U_p/ξ, however, is *independent* of ξ in this estimate and is not necessarily small in HTSC.

A novel feature in HTSC is that thermally assisted flux flow (TAFF) [13,14] with a linear (ohmic) resistivity ρ is observed at small current densities $J \ll J_c$. Both effects, flux creep at $J \approx J_c$ and TAFF at $J \ll J_c$, are limiting cases of Anderson's general expressions for the electric field $E(B,T,J)$ caused by thermally activated flux jumps out of pinning centers [1,13-15], which may be written as [16]

$$E(J) = 2\rho_c J_c \exp(-U/k_B T) \sinh(JU/J_c k_B T) \quad (1)$$

where $J_c(B,T)$ (the critical current density at $T = 0$), $\rho_c(B,T)$ (the resistivity at $J = J_c$), and $U(B,T)$ (the activation energy for flux jumps) are *phenomenological parameters*. The physical idea behind Eqn.(1) is that the Lorentz force density $\mathbf{J} \times \mathbf{B}$ acting on the flux-line lattice (FLL) *increases* the rate of thermally activated jumps of flux lines or flux-line bundels along the force, $\nu_0 \exp[-(U-W)/k_B T]$, and *reduces* the jump rate for backward jumps, $\nu_0 \exp[-(U+W)/k_B T]$. Here $U(B,T)$ is an activation energy, $W = JBVl$ the energy gain during a jump, V the jumping volume, l the jump width, and ν_0 is an attempt frequency. All these quantities depend on the microscopic model, which is still controversial (cf. Sct. 2), but by defining a critical current density $J_c = JU/W = U/BVl$, only measurable quantities enter in (1). Eqn. (1) is obtained by subtracting the two jump rates to give an effective rate ν and then writing the drift velocity $v = \nu l$ and the electric field $E = vB = \rho J$.

For large currents $J \approx J_c$ one has $W \approx U \gg k_B T$ and thus $E \propto \exp(J/J_1)$ with $J_1 = J_c k_B T/U$. For small currents $J \ll J_1$ one may linearize the $\sinh(W/k_B T)$ in (1) and gets *ohmic* behavior with a thermally activated linear resistivity $\rho_{TAFF} \propto \exp(-U/k_B T)$. Combining (1) with the usual, not activated flux-flow resistivity ρ_{FF} valid at $J \gg J_c$, or with the square-root result for a model particle moving viscously across a sinusoidal potential [17,18], one gets for the resistivity $\rho = E/J$ the scenario:

$$\rho = (2\rho_c U/k_B T)\exp(-U/k_B T) = \rho_{TAFF} \quad \text{for } J \ll J_1 \quad \text{(TAFF)} \quad (2)$$
$$\rho = \rho_c \exp[(J/J_c - 1)U/k_B T] \propto \exp(J/J_1) \quad \text{for } J \approx J_c \quad \text{(flux creep)} \quad (3)$$
$$\rho = \rho_{FF}(1 - J_c^2/J^2)^{1/2} \approx \rho_{FF} \approx \rho_n B/B_{c2}(T) \quad \text{for } J \gg J_c \quad \text{(flux flow)}. \quad (4)$$

2. Vortex Glass State

Thermally activated depinning is the reason for the broadening of the resistivity transition at $T \approx T_c$ when a magnetic field is applied. More precisely, since in resistivity measurements J is small, one is in the TAFF region (2) [19]. Since this effect is more pronounced at large T, a room-temperature superconductor, if found, would probably exhibit a rather large resistivity at such high temperatures. This would thwart many applications.

At low temperatures the situations is different: At $T < 77$ K, e.g. in YBCO, ρ becomes unmeasurably small because of the factor $\exp(-U/k_B T)$ in (2), and at $T = 4$ K, Bi-based HTSC even outperform the conventional high-current wires of NbTi. Furthermore, recent "scaling theories" of flux creep [20-23] going beyond Anderson's idea (1), suggest that the ohmic TAFF regime (2) should *not* exist at very low J and low T. It is predicted that the jumping volume V and thus the activation energy U should diverge as $1/J^\alpha$ with $\alpha > 0$. Thus, $\rho \propto \exp[-(J_2/J)^\alpha]$ becomes truly zero at low J. Though this very low resistivity is hard to measure [24] and not relevant for application, the question for the existence of this "vortex glass state" with rigidly pinned FLL and $\rho = 0$ is of principal interest [25].

The advanced theories of flux creep [20-23] use the elasticity of the FLL to show that at $J \to 0$ larger and larger volumes (or kink-shaped nuclei [26]) have to be activated thermally to facilitate flux creep. It might be that an appropriate consideration of plastic deformation, of the always present dislocation loops in the FLL, modifies this prediction and yields a finite U as $J \to 0$. After all, it has been shown that such a vortex glass state does *not* occur in a "liquid" FLL (i.e. when thermal fluctuations are large, Sct. 5) [27] and also not in a two dimensional (2D) FLL [28]. In HTSC, flux pinning and thermal fluctuations become 2D when the "point vortices" (or pancake vortices, vortex dots) in the nearly isolated (weakly Josephson-coupled) superconducting CuO-layers interact mainly with the point vortices in the *same* layer, but only weakly (with interaction $< k_B T$) with those in *other* layers. This is the case in Bi-HTSC at sufficiently large B or T, where the vortex-dot lattices in different layers are decoupled [28].

Layered superconductors are described by the Lawrence-Doniach (LD) modification [29-30] of the Ginzburg-Landau (GL) theory. This ascribes to each layer its own order parameter $\Psi_n(\mathbf{r})$ and replaces the gradient perpendicular to the layers by a difference. The LD model, which can be derived from a tight-binding formulation of the BCS-theory [30], with decreasing coupling strength yields first the anisotropic GL theory, or the London theory when $B \ll B_{c2}$ and thus $|\Psi_n| \approx const. \cdot \exp[i\phi_n(\mathbf{r})]$, then weakly coupled layers, and finally completely separated layers with only *magnetic* interaction between the point vortices [31] and almost complete transparency for the in-layer component of B [32].

3. Flux Diffusion and the Irreversibility Line

The "irreversibility line" $T_{irr}(B)$ or $B_{irr}(T)$ which separates in the $B - T$-plane the regions of irreversible (low B, T) and reversible magnetic behavior, is easily explained by thermally activated depinning: Above this line, which may also be called "depinning line" $T_d(B)$ [33], the pins become ineffective and the vortices can move freely, giving rise to completely reversible magnetization curves, whereas below $T_d(B)$ hysteretic behavior is observed.

An interesting feature of the depinning line is that it depends on the size and shape of the specimen and on the time scale or frequency $\omega/2\pi$ of the experiment. This often overlooked effect, which means the $T_d(B)$ is *not* a genuine intrinsic property of the material, originates from the *diffusive character* of the flux motion [33-35]. Ohmic flux-flow resistivity ρ_{TAFF} (2) is equivalent to a linear diffusion of the flux lines, i.e., in simple geometries the diffusion equation

$$\dot{\mathbf{B}} = D\nabla^2\mathbf{B} \qquad (5)$$

holds with $D = \rho/\mu_0$ (in SI-units). Since in the TAFF regime $\rho = \rho_{TAFF} \propto \exp(-U/k_BT)$, one has for the thermally activated diffusion $D = \rho_{TAFF}/\mu_0 = D_0\exp(-U/k_BT)$. Thus, for sufficiently large times, small specimens, and large T, any change of the applied magnetic field or current completely *penetrates* the HTSC, which then is in the *resistive* state. At lower T, such a change influences the FLL only in a thin *surface layer* within the skin depth $\delta = (D/2\omega)$, and the superconductor behaves as if it were in the *Meissner* state with complete expulsion of the applied field. In between these two limiting cases, the surface current penetrates more or less deeply and causes a large dissipation peak when the skin depth coincides with a specimen dimension. This damping peak, occuring at a fixed value of $D(T, B)$, may be used to define an irreversibility line.

This simple "classical' picture explains a wide variety of experiments, which all define (slightly different) depinning lines [36]:

1) The broadening of the resistive transition $\rho(T, B)$ in a magnetic field.

2) The vanishing of irreversibility (hysteresis) in magnetization curves at $T_{irr}(B)$.

3) The maximum in the imaginary part of the a.c. susceptibility at $T_d(B)$ [34].

4) The divergence at $T_d(B)$ of the a.c. penetration depth measured by the screening of an a.c. field by a superconducting film between two coils [37].

5) The sharp maximum at $T_d(B)$ in the attenuation Γ of a vibrating HTSC glued on a silicon reed [38], or of a vibrating superconducting reed [33], and the corresponding reduction of the frequency enhancement. The applied field "pulls" at the flux lines in the periodically tilted superconductor and thereby enhances its resonance frequency as long as the vortices are pinned. If the pinning were ideal this would not cause any dissipation, but with the onset of thermal depinning the flux lines *move* relative to the atomic lattice in abrupt unpinning jumps. Therefore, the frequency enhancement decreases and the attenuation of the reed increases by several orders of magnitude, reaches a sharp maximum, and, when the vortices become nearly unpinned and move smoothly, the damping of the reed decreases again to a small value determined by the viscous motion of vortices, Eqn. (4). This behavior is similar to the depinning peak observed in classical superconductors even at zero temperature, where a similar step in the resonance frequency and peak in $\Gamma(B)$ is observed just below $B_{c2}(T)$ where pinning has to vanish [39].

6) The conduction noise in Bi-based HTSC films at constant current density [40]. A sharp peak in the noise power at a given frequency as a function of B and T yields a further depinning line. This noise is caused by depinning processes: each "plucking" of a vortex releases elastic energy of the VL, which then relaxes viscously with an exponential time law. Below $T_d(B)$ the noise is small since only few depinning processes occur; at $T = T_d(B)$ the noise is maximum; and above $T_d(B)$ it decreases again since the viscous motion of the thermally depinned vortices is smooth. The depinning line should shift to larger T or B when higher frequencies of the noise are selected.

7) The ultrasonic attenuation peak and sound velocity enhancement at $T_d(B)$ [41] were explained without any adjustable parameter just from the d.c. resistivity.

8) The magnetic irreversibility of the line width in Muon-Spin Rotation (μ^+SR) experiments [42]. This promising method probes the local magnetic field in a prescribable

depth. The frequency spectrum of the μ^+SR signal is proportional to the probability $n(B)$ to find a field B at the random position where the positive muon stops inside the sample and rotates its spin with circular frequency $\omega = \gamma_\mu B$ where $\gamma_\nu = 8.513 \times 10^8$ rad sec^{-1} T^{-1} is the gyromagnetic ratio of the muon. In a perfect FLL this field density $n(B)$ exhibits van-Hove singularities at the maximum, minimum, and saddle-point values of the periodic field $B(\mathbf{r})$, which are smeared by distortions of the FLL caused by pinning. This method is discussed in detail in [43]. Recently, high-resolution μ^+SR experiments in Niobium of high purity were obtained [44].

Sound attenuation 7) and μ^+SR 8) probe the VL also *far inside* a superconductor, whereas in the above methods 1) – 6) the FLL interacts with the outer world (the applied field or transport current) only at its *surface*, in a layer of thickness λ where shielding currents or transport surface currents exert forces on the vortex lines or on their end points (magnetic monopoles). The resulting compression or tilt of the VL then diffuses into the interior as discussed above and in [16].

For simple geometries the above experiments can be explained by solving the diffusion equation (5) or by just writing down the diffusion time $\tau = L^2/\pi D^2$ or skin depth $\delta = (D/2\omega)^{1/2}$; the damping peak is then expected when δ equals an appropriate specimen size L, or ω equals $1/\tau$. The depinning line can thus be shifted to lower B and T if very small specimens are used, e.g., HTSC grains embedded in epoxy [45].

In more complicated geometries one has to account for the anisotropy of flux-flow resistivity, not to be confused with the material anisotropy, which is very important in HTSC (see Scts. 4, 5 and [10]) but for brevity cannot be considered here. The correct equations for flux flow are $\mathbf{E} = \mathbf{B} \times \mathbf{v} = \mathbf{B} \times \mathbf{J} \times \mathbf{B}/\eta = (B^2/\eta)\mathbf{J}_\perp = \rho_\perp \mathbf{J}_\perp$ where $\eta = B^2/\rho_\perp$ is the flux-flow viscosity defined by $\mathbf{v} = \mathbf{J} \times \mathbf{B}/\eta$. This means: Due to the anisotropic nature of flux flow only the current density component $\mathbf{J}_\perp = \mathbf{J} - \mathbf{J}_\parallel$ *perpendicular* to $\mathbf{B}$ generates an electric field and dissipates energy. Dissipation by longitudinal currents, which may be caused by helical instabilities of the flux lines and by vortex-cutting processes [10], are considered in [35, 36]. The equation of motion for $\mathbf{B}(\mathbf{r}, t)$ now follows from the induction law $\dot{\mathbf{B}} = -\nabla \times \mathbf{E} = -\nabla \times \mathbf{B} \times \mathbf{J} \times \mathbf{B}/\eta$, which yields

$$\dot{\mathbf{B}} = \mu_0^{-1} \nabla \times \rho_\perp \hat{\mathbf{B}} \times \hat{\mathbf{B}} \times \nabla \times \mathbf{B} . \qquad (6)$$

Eq. (6) applies also when the resistivity $\rho_\perp$ is *non-linear*, e.g., given by (3). It may thus also be used to compute the history dependent penetration of flux into a superconductor with pinning, which is usually approximated by the static Bean model [46]. If $\rho_\perp$ is independent of J [cf. (2) and (4)] and if $\mathbf{B}$ varies little in space or time, (6) may be linearized to give the diffusion equation (5) plus a term $\rho_\perp \nabla \times \mathbf{J}_\parallel$ which vanishes when $\mathbf{J} \perp \mathbf{B}$ [35,36].

4. Non-Local Elasticity of the Flux-Line Lattice

The thermal fluctuations and the distortions of the FLL caused by pins crucially depend on its elasticity. The linear elastic energy U of a distorted lattice of atoms or flux lines may be obtained by integrating $u_\alpha(\mathbf{k})\Phi_{\alpha\beta}(\mathbf{k})u_\beta(\mathbf{k})$ over the Brillouin zone (BZ) in k-space. Here $\mathbf{u}(\mathbf{k})$ is the Fourier transformed displacement field with (for a FLL with B along z) *two* components u_x, u_y, and $\Phi_{\alpha\beta}(\mathbf{k})$ is the elastic matrix of the lattice. For a usual, local elastic (e.g. an atomic) lattice, one has $\Phi_{\alpha\beta} \propto k^2$ in the central part of the BZ, say for $k < 0.3 k_{BZ}$ where k_{BZ} is the radius of the circularized first Brillouin zone; for the FLL, $k_{BZ}^2 = 4\pi B/\Phi_0$. This means the elastic energy density is "local", i.e., proportional to the square of the *strains* at a given point. For the FLL, the relationship

$\Phi_{\alpha\beta} \propto k^2$ applies only for very small $k \ll 1/\lambda \ll k_{BZ}$. This means the elastic moduli $c_{11}(\mathbf{k})$ for compression and $c_{44}(\mathbf{k})$ for tilt of the FLL depend on $\mathbf{k}$ [47]. The physical reason for this unusual dispersion is that the interaction between flux lines (or flux-line segments if the vortices are curved) has a range λ which is typically *much larger* than the FLL spacing $a = (2\Phi_0/\sqrt{3}B)^{1/2}$. Since the interaction energy is contained in a sphere of radius $\approx \lambda$, the elastic energy becomes *nonlocal* [47] and the FLL is *much softer* for short-wavelength compression or tilt.

In general, $\Phi_{\alpha\beta}(\mathbf{k})$ is periodic in $\mathbf{k}$-space and given by a sum over all reciprocal lattice vectors of the FLL [47,48]. In continuum approximation, valid for $k \ll k_{BZ}$, one has

$$\phi_{\alpha\beta}(\mathbf{k}) = [\, c_{11}(\mathbf{k}) - c_{66} \,] k_\alpha k_\beta + \delta_{\alpha\beta} [\, c_{66} k_\perp^2 + c_{44}(\mathbf{k}) k_z^2 \,] \qquad (7)$$

where $k_\perp^2 = k_x^2 + k_y^2$ and $k^2 = k_\perp^2 + k_z^2$. Eqn. (7) applies to isotropic and uniaxially anisotropic superconductors with $B \| c$-axis. For uniaxially anisotropic superconductors with penetration depths λ_{ab} in the ab-plane and $\lambda_c = \Gamma \lambda_{ab}$ along the c-axis, the compression and tilt moduli for $B \| c \| z$, $\kappa \gg 1$, and $0 < b = B/B_{c2} < 1$ are [48,49]

$$c_{11}(\mathbf{k}) = (B^2/\mu_0)(1 + \lambda_c'^2 k^2)(1 + \lambda_{ab}'^2 k^2)^{-1}(1 + \lambda_c'^2 k_\perp^2 + \lambda_{ab}'^2 k_z^2)^{-1} \qquad (8)$$

$$c_{44}(\mathbf{k}) = (B^2/\mu_0)\Big[(1 + \lambda_c'^2 k_\perp^2 + \lambda_{ab}'^2 k_z^2)^{-1} + \lambda_c^{-2} k_{BZ}^{-2} \ln \tilde{\kappa}\Big] \qquad (9)$$

with $\tilde{\kappa} = [(1+\Gamma^2\kappa^2+\lambda_{ab}^2 k_z^2)/(1+b\Gamma^2\kappa^2+\lambda_{ab}^2 k_z^2)]^{1/2} + (1-b)/2$ and $\lambda_c' = \Gamma\lambda_{ab}' = \lambda_c/(1-b)^{1/2}$. The last term in (9) is the isolated-vortex contribution [48], which is important for $B \to 0$ or $k \approx k_{BZ}$ and which was constructed such as to yield the correct limits of c_{44} for $b \to 0$, $b \to 1$, $k \to 0$, $k_z \to \infty$. The shear modulus for $k \ll k_{BZ}$ is not dispersive, $c_{66} \approx B\Phi_0(1-b)^2/(16\pi\lambda_{ab}^2\mu_0)$.

5. Thermal Fluctuation of the Flux-Line Lattice

The thermal fluctuations $\langle u^2 \rangle$ of the flux-line positions were first calculated from local elasticity in [50] and in the extreme nonlocal limit in [51]. The correct calculation [49,52] yields a $\langle u^2 \rangle$ which practically coincides with [51] but is typically much larger than the local result [50]. This enhancement is an effect of the dispersion of $c_{44}(\mathbf{k})$ (9). One has

$$\langle u^2 \rangle = \langle u_x^2 + u_y^2 \rangle = k_B T \int_{BZ} \frac{d^3k}{8\pi^3} \Big[\Phi_{xx}^{-1}(\mathbf{k}) + \Phi_{yy}^{-1}(\mathbf{k}) \Big] \qquad (10)$$

$$\langle u^2 \rangle = \frac{k_B T}{2\pi^2} \int_0^{k_{BZ}} dk_\perp \, k_\perp \int_0^\infty dk_z \Big[\frac{1}{c_{66} k_\perp^2 + c_{44}(\mathbf{k})k_z^2} + \frac{1}{c_{11}(\mathbf{k})k_\perp^2 + c_{44}(\mathbf{k})k_z^2} \Big] \qquad (11)$$

$$\langle u^2 \rangle \approx k_B T \left(\frac{\mu_0}{4\pi B \Phi_0 c_{66}} \right)^{1/2} \times \left(\frac{B\kappa^2/2}{B_{c2} - B} \right)^{1/2} \times \frac{\lambda_c}{\lambda_{ab}} \qquad (12)$$

or for $B \ll B_{c2}$ with $c_{66} = B\Phi_0/16\pi\mu_0\lambda_{ab}^2$, $\langle u^2 \rangle \approx k_B T \mu_0 \lambda_{ab} \lambda_c k_{BZ}/\Phi_0 B = k_B T \mu_0 (4\pi/B\Phi_0^3)^{1/2} \lambda_{ab}\lambda_c$. The fluctuation (12) has the form: Local result [50], times a typically large nonlocal correction $(B\kappa^2/2B_{c2})^{1/2} \approx (B \ln \kappa / 4B_{c1})^{1/2} \gg 1$, times the large anisotropy ratio $\lambda_c/\lambda_{ab} = \Gamma$, where $\Gamma \approx 5$ for YBCO and $\Gamma \approx 60$ for Bi-HTSC.

In spite of its large thermal fluctuation, in my opinion it is *not* clear at present whether the 3D FLL "melts", what this melting means, or what the type of such a transition is. The Lindemann criterion $\langle u^2 \rangle^{1/2}/a = 0.1 \ldots 0.2$ for melting is satisfied at rather low T. As discussed in Sct. 3, many experiments which appear to indicate a phase transition, can be explained by the rather abrupt onset of thermally activated depinning. A melting of the 2D vortex-point lattice in the CuO-layers of Bi-HTSC probably does occur [28], and a Berezinsky-Kosterlitz-Thouless transition in HTSC films was indeed observed [53].

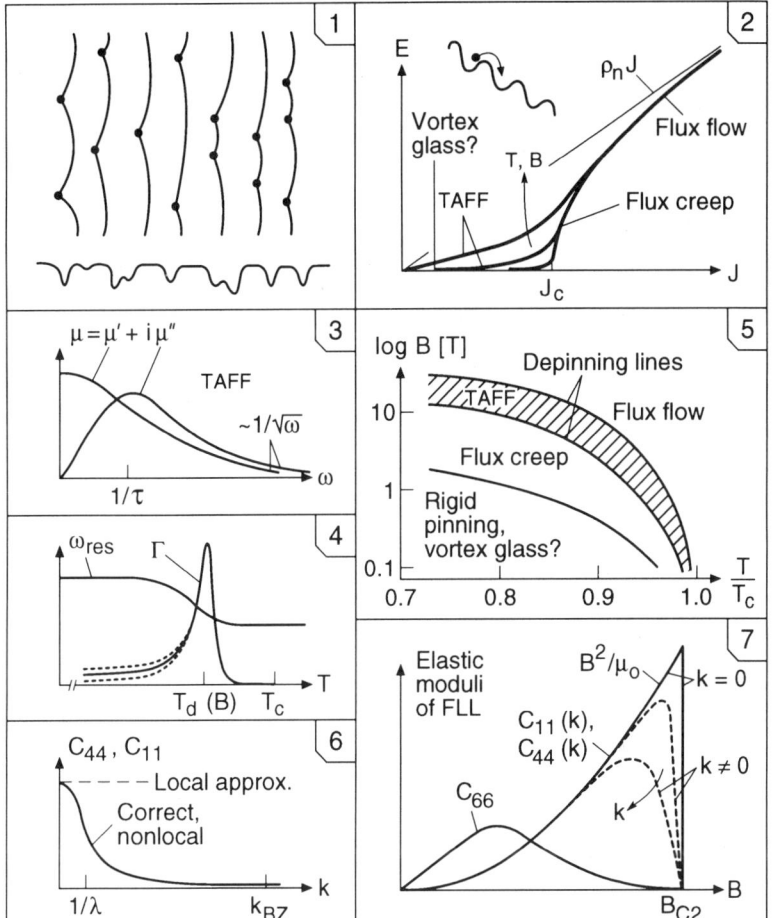

Figures 1–7. 1) Flux lines pinned by point pins, e.g., oxygene vacancy clusters. Bottom: Pinning potential. 2) Current–voltage curves, E = electric field, J = current density. $J \gg J_c$: flux flow, $E \approx J\rho_n$; $J \approx J_c$: flux creep, $E \propto \exp(J/J_1)$; $J \ll J_1 < J_c$: thermally activated flux flow (TAFF), $E \approx J\rho_n \exp(-U/k_BT)$, or, at low T: vortex glass, $E \propto \exp(-J_2^\alpha/J^\alpha)$. Top: Tilted periodic pinning potential with jumping flux-line bundle. 3) Complex a.c. susceptibility $\mu = \mu' + i\mu''$ in the TAFF state, described by a resistivity ρ_{TAFF} or flux diffusivity $D = \rho_{TAFF}/\mu_0 \propto \exp(-U/k_BT)$. 4) Resonance frequency ω_{res} and attenuation Γ of a superconductor performing tilt vibrations about the direction of an applied magnetic field. Maximum damping Γ and a step in ω_{res} occur when the time for flux-line diffusion into the specimen $\tau(B,T)$ coincides with $1/\omega_{res}$. Dashed: Hysteretic damping for various amplitudes. 5) The depinning lines in the field-temperature plane separate the region of flux flow (reversible) from the region of flux creep (slow logarithmic relaxation) and rigid pinning (completely irreversible). At the depinning lines TAFF occurs (flux diffusion). For YBCO, schematic from: P. Esquinazi et al., *Physica B* **165&166**, 1151 (1990). 6) Nonlocal tilt and compressional moduli c_{44} and c_{11} of the FLL versus the strain wavevector k. 7) The dispersive moduli $c_{11} \approx c_{44}$ and the shear modulus c_{66} of the FLL versus induction B for various strain wavevectors k.

References

[1] A. A. Abrikosov, *Zh. Eksper. Theor. Fiz.* **20**, 1064 (1950); *Soviet Phys. – J. Exp. Theor. Phys.* **5**, 1174 (1975).
[2] P. W. Anderson, *Phys. Rev. Lett.* **9**, 309 (1962).
[3] P. W. Anderson and Y. B. Kim, *Rev. Mod. Phys.* **36**, 39 (1964).
[4] J. Bardeen and M. J. Stephen, *Phys. Rev.* **140**, A1197 (1965).
[5] M. Tinkham, *Phys. Rev. Lett.* **13**, 804 (1964);
[6] C.-R. Hu and R. S. Thompson, *Phys. Rev.* **B 6**, 110 (1972); A. I. Larkin and Yu. N. Ovchinnikov, *Sov. Phys.–JETP* **37**, 557 (1973)(small B).
[7] R. S. Thompson and C.-R. Hu, *Phys. Rev. Lett.* **20**, 1352 (1971) (large B).
[8] M. Tinkham, *Introduction to Superconductivity* (McGraw-Hill, New York, 1975) p. 161 ff and p. 273 ff; L. P. Gor'kov and N. B. Kopnin, *Sov. Phys.–Uspechi* **18**, 496 (1976); V. V. Shmidt and G. S. Mkrtchyan, *Sov. Phys.–Uspechi* **17**, 170 (1974); A. I. Larkin and Yu. N. Ovchinnikov, in: *Nonequilibrium Superconductivity*, D. N. Langenberg and A. I. Larkin, eds. (Elsevier, Amsterdam, 1986), p. 493.
[9] B. I. Ivlev and N. B. Kopnin, *Phys. Rev. B* **42**, 10052 (1990); J. R. Clem and W. M. Coffey, *Phys. Rev. B* **42**, 6209 (1990).
[10] J. R. Clem, in: *Progress in Low Temperature Phyiscs*, R. Nikolski, ed. (World Scientific, Singapore, 1990), Vol. **25**, p. 64.
[11] M. R. Beasley, R. Labusch, and W. W. Webb, *Phys. Rev.* **181**, 682 (1969); C. Rossel et al., *Physica C* **165**, 233 (1990); P. Berghuis and P. H. Kes, *Physica B* **165 & 166**, 1169 (1990).
[12] Y. Yeshurun and A. P. Malozemoff, *Phys. Rev. Lett.* **60**, 2202 (1988).
[13] D. Dew-Hughes, *Cryogenics* **28**, 674 (1988).
[14] P. H. Kes, J. Aarts, J. van den Berg, C. J. van der Beek, and J. A. Mydosh, *Supercond. Sci. Technol.* **1**, 242 (1989).
[15] M. Tinkham, *Introduction to Superconductivity* (McGraw-Hill, New York, 1975) p. 175 ff.
[16] E. H. Brandt, *Z. Physik B* **80**, 167 (1990).
[17] V. Ambegaokar and B. I. Halperin, *Phys. Rev. Lett.* **22**, 1364 (1969); *Phys. Rev. Lett. Errata* **23**, 274 (1969); M. Inui et al., *Phys. Rev. Lett.* **63**, 2421 (1989).
[18] E. Conen and A. Schmid, *J. Low Temp. Phys.* **17**, 331 (1974).
[19] T. T. M. Palstra, B. Battlogg, R. B. van Dover, L. F. Schneemeyer, and J. V. Waszczak, *Phys. Rev.* **B 41**.
[20] M. P. A. Fisher, *Phys. Rev. Lett.* **62**, 1415 (1989).
[21] D. S. Fischer, M. P. A. Fischer, and D. A. Huse, *Phys. Rev. B* **43**, 130 (1991).
[22] M. V. Feigel'man, V. B. Geshkenbein, A. I. Larkin, and V. M. Vinokur, *Phys. Rev. Lett.* **63**, 2303 (1989).
[23] T. Nattermann, *Phys. Rev. Lett.* **64**, 2454 (1990).
[24] R. H. Koch et al. *Phys. Rev. Lett.* **63**, 1511 (1989); P. L. Gammel, L. F. Schneemeyer, and D. J. Bishop, *Phys. Rev. Lett.* (submitted).
[25] The vortex-glass state of a pinned FLL should not be confused with the "glass state" early proposed to explain the irreversibility line (Sct. 3), which was seen analogous to the De Almeida-Thouless line in spin glasses. This behavior was initially ascribed to the granularity of ceramic HTSC; later it was also observed in HTSC single crystals, indicating some "internal glassiness" with length scale $(\hbar/2eB)^{1/2}$ typically < 0.1 μm (the grain size is several μm); see: K. A. Müller, M. Takashige, and J. G. Bednorz, *Phys. Rev. Lett.* **58**, 1143 (1987); I. Morgenstern, K. A. Müller, and J. G. Bednorz, *Z. Phys. B* **69**, 33 (1987); C. Rossel, Y. Maeno, and I. Morgenstern, *Phys. Rev. Lett.* **62**, 681 (1989). For another type of glassiness caused by impurities in a

supercondcutor, see: R. Oppermann, *Z. Phys. B* **68**, 329 (1987); *Sol. St. Comm.* **65**, 1391 (1988); *Physica A* **167**, 301 (1990). I personally prefer to explain the irreversibility line in terms of pinning and depinning of the flux-line lattice, though possibly a connection between all these pictures might exist when the phase-slip of the superconducting order parameter at the core of a moving flux line is considered; see: M. Tinkham, *Phys. Rev. Lett.* **61**, 1658 (1988).

[26] S. Chakravarty, B. I. Ivlev, and Yu. N. Ovchinnikov, *Phys. Rev. Lett.* **64**, 3187 (1990); *Phys. Rev. B* **42**, 2143 (1990).

[27] M. V. Feigel'man and V. M. Vinokur, *Phys. Rev. B* **41**, 8986 (1990); V. M. Vinokur, M. V. Feigel'man, V. B. Geshkenbein, and A. I. Larkin, *Phys. Rev. Lett.* **65**, 259 (1990).

[28] M. V. Feigel'man, V. B. Geshkenbein, and A. I. Larkin, *Physica C* **167**, 177 (1990).

[29] W. E. Lawrence and S. Doniach, Proc. 12th Internatl. Conf. of Low Temperature Physics LT12 (E.Kanda ed., Academic Press of Japan, Kyoto, 1971) p. 361.

[30] R. A. Klemm, A. Luther, and M. R. Beasley, *Phys. Rev. B* **12**, 877 (1975); L. N. Bulayevskiĭ, *Int. J. Mod. Phys.* (a review) (in print).

[31] J. R. Clem, *Phys. Rev. B* (pancake vortices) (in print).

[32] V. M. Vinokur, P. H. Kes, and A. E. Koshelev, *Physica C* **168**, 29 (1990).

[33] P. Esquinazi, *Solid State Comm.* **74**, 75 (1990); A. Gupta et al., *Europhys. Lett.* **10**, 663 (1989); *Physica C* **162-164**, 667 (1989); A. Gupta, P. Esquinazi, and H. F. Braun, *Phys. Rev. Lett.* **63**, 1869 (1989); *Physica B* **165 & 166**, 1151 (1990).

[34] P. H. Kes et al., *Supercond. Sci. Technol.* **1**, 242 (1989).

[35] E. H. Brandt, *Z. Physik B* **80**, 167 (1990).

[36] E. H. Brandt, *Int. J. Mod. Phys. B* (a review) (in print).

[37] A. Hebard, P. Gammel, C. Rice, and A. Levi, *Phys. Rev. B* **40**, 5243 (1989).

[38] P. L. Gammel, L. F. Schneemeyer, J. V. Waszczak, and D. J. Bishop, *Phys. Rev. Lett.* **61**, 1666 (1988); comment: E. H. Brandt, P. Esquinazi, and G. Weiss, *Phys. Rev. Lett.* **62**, 2330 (1989).

[39] E. H. Brandt, P. Esquinazi, H. Neckel, and G. Weiss, *Phys. Rev. Lett.* **56**, 89 (1986); *J. Low Temp. Phys.* **63**, 187 (1986); P. Esquinazi, et al., *J. Low Temp. Phys.* **64**, 1 (1986); E. H. Brandt, *J. de Physique, Colloque* **C8** (No. 27, Vol. 48), 31 (1987).

[40] A. Maeda et al. *Physica B* **165 & 166**, 1363 (1990).

[41] J. Pankert, *Physica C* **168**, 335 (1990); J. Pankert, et al., *Phys. Rev. Lett.* **65**, 3052 (1990); P. Lemmens, et al., *Physica C* (in print).

[42] B. Pümpin, H. Keller et al., *Z. Phys. B* **72**, 175 (1988); P. Zimmermann et al., *Hyperfine Interactions* **63-65**, 25 (1990) V. G. Grebinnik et al., *dito*, p. 66.

[43] E. H. Brandt, *Phys. Rev. B* **37**, 2349 (1988); E. H. Brandt, *J. Low Temp. Phys.* **73**, 335 (1988); E. H. Brandt and A. Seeger, *Adv. Phys.* **35**, 189 (1986).

[44] D. Herlach, et al., *Hyperfine Interactions* **63-65**, 41 (1990), and to be published.

[45] J. Kober et al., *Phys. Rev. Lett.* (in print).

[46] C. P. Bean, *Rev. Mod. Phys.* **36**, 31 (1964); *J. Appl. Phys.* **41**, 2482 (1970).

[47] E. H. Brandt, *J. Low Temp. Phys.* **26**, 709; 735 (177); **28**, 263, 291 (1977); *Phys. Rev. B*, **34**, 6514 (1986).

[48] A. Sudbø and E. H. Brandt, *Phys. Rev. B* (in print); *Phys. Rev. Lett.* (submitted);

[49] A. Houghton, R. A. Pelcovits, and A. Sudbø, *Phys. Rev. B* **40**, 6763 (1989).

[50] D. R. Nelson and H. S. Seung, *Phys. Rev. B* **39**, 9174 (1989).

[51] M. A. Moore, *Phys. Rev.* **B 39**, 9174 (1989).

[52] E. H. Brandt, *Phys. Rev. Lett.* **63**, 1106 (1989).

[53] S. N. Artemenko, I. G. Gorlova, Yu. I. Latyshev, *Phys. Lett. A* **138**, 428 (1989).

RECENT μSR RESULTS ON MAGNETIC PROPERTIES OF CUPRATE MATERIALS[1]

E. J. ANSALDO

Department of Physics
University of Saskatchewan
Saskatoon, Canada S7N0K3

Presented at the Workshop on Electronic Structure and Mechanisms for High Temperature Superconductivity, University of Miami (3-9 January 1991)

ABSTRACT

An overview is given of recent muon spin rotation-relaxation (μSR) measurements of magnetic penetration depths, λ(T), on the $YBa_2Cu_4O_8$, $Y_2Ba_4Cu_7O_{15-\delta}$ and Bi-based cuprates, and of the interplay of magnetic ordering with superconductivity in the simpler double-layer cuprate $La_2MCu_2O_{6+\delta}$, M=(Sr,Ca). The dependence of T_c on carrier concentration in the $Y_2Ba_4Cu_7O_{15-\delta}$ family is similar to that of the $YBa_2Cu_3O_x$, almost linear, but monotonically increasing to the maximum T_c =90K, and with a gentler slope. Available data for Bi(Pb)SrCaCuO samples of different stoichiometries depart from such general trends and are rather iconsistent sample-to-sample, indicating that extrinsic factors (vortex lattice morphology and pinning effects, uncontrolled defect structures) can not be separated from the intrinsic (mechanisms, electronic and crystalline structure) behaviour of the mixed state for the Bi-based cuprate superconductors as yet. Static magnetic order was determined for the first time for $La_2MCu_2O_{6+\delta}$, M=(Sr,Ca). Oxygen excess results in a superconducting phase separating for M=Ca (similar to $La_2CuO_{4+\delta}$), but not for the M=Sr samples. In general, the results underscore the crucial rôle played by inhomogeneities and defect structures in the magnetic properties.

---

[1] New results presented here were obtained at TRIUMF (Canadian meson facility, Vancouver, B. C.) in collaboration with the TRIUMF-UBC μSR group, on samples prepared and characterized by the CSIC (Barcelona) and DSIR (New Zealand) groups led by X. Obradors and J. Tallon, respectively.

INTRODUCTION

The μSR technique provides unique insights on flux line lattice (FLL) properties and magnetic ordering in oxide superconductors and their precursors. The positive muon, bound to oxygen ions up to ca. 200K, is a truly microscopic probe of internal field distributions in the oxides. Such fields are due to the FLL in the mixed state and/or the effective internal fields due to electronic (or even anyonic) moments in ordered samples. In the latter case, measurements in zero or in longitudinal (ZF, LF) applied field help define the origin of the field distribution (e.g. antiferromagnetism vs. spin glass order), and its dynamics on a time-scale complementary to that of neutron diffraction. The technique is uniquely suited to the determination of coexistence with great sensitivity, but unfortunately without spatial discrimination, and to the study of new materials for which large crystals are not currently available, as illustrated by the $La_2MCu_2O_{6+\delta}$, M=(Sr,Ca) case below.

FLL IN THE MIXED STATE

As discussed in papers on $YBa_2Cu_3O_x$ and $La_{1.85}Sr_{0.15}CuO_4$,[1-6] and one- and two-layer Bi-oxides,[7-9] the penetration depth λ is determined (assuming that the muons are randomly distributed in the vortex lattice) from the relaxation of the muon polarization in an external field such that the separation between the vortices is smaller than λ. The μ-spin relaxation rate $\sigma = 1/T_2$ is then independent of applied field and given by the second moment $M_2 = <|\Delta H|^2>^{1/2}$ of the microscopic field distribution. The data for powder samples are fited well by Gaussian relaxation functions $e^{-(t/T_2)^2}$, but such approximation is inappropriate for single crystals, as discussed for the only single crystal experiment published so far.[5] The penetration depth is related to the relaxation time T_2, assuming a triangular vortex lattice, by[10] $\lambda = \sqrt{0.043\phi_o\gamma_\mu T_2}$, where ϕ_o is the magnetic flux quantum and γ_μ the muon's gyromagnetic ratio. In the simple London picture the relaxation rate is then $\sigma \propto n_s/m^*$, where n_s and m^* are the superconducting carrier density and effective mass, respectively, and the measurements provide an important database for testing attempts to understand the dependence of T_c on hole concentrations. For large anisotropy, the directionally averaged λ^{powder} so obtained is dominated by the in-plane (or "hard") penetration depth, given by $\lambda_{ab} = \lambda^{powder}/1.23$.[11] Consistency between values obtained for single crystals and sintered samples among themselves and with those from low field magnetization measurements in $YBa_2Cu_3O_x$ lends confidence to the technique.[12] The temperature dependence of λ may in principle be compared to theoretical predictions, i.e. strong vs weak coupling, dirty vs clean limit.[7,8] In practice, however, the comparison is obscured by extrinsic effects such as T-dependent pinning strength, FLL thermal motion, different defect structures for samples of similar composition, etc., i.e. heuristic factors affecting the topology of the FLL as a function of temperature, and which are of primary practical importance, especially for the bismuth-based family. A typical result is shown in Figure 1 for a ceramic sample of (BiPb)-2223.[13] The depinning temperatute, T_p, can be determined by means of zero field cooling μSR runs. For fully oxygenated $YBa_2Cu_3O_x$ it was found that T_p is very close to T_c, and the "irreversibility" line traced out in detail.[14] In contrast, for a variety of Bi-2212 and Bi(Pb)-2212 samples, the reversible effects were found to extend well below T_c, with T_p varying from 20 to ca. 50K depending on the sample preparation and applied magnetic field [7,15]. In addition, a variety of extrapolated relaxation rates $\sigma(0)$ obtain for different samples of similar T_c, while single crystals yield systematically lower, also variable, values. Thus a good deal of scepticism should be exercised when considering the consequences for mechanisms and electronic structures of current FLL experimental results on individual samples of Bi-based high-T_c superconductors. It may well be that results are dominated by largely uncontrolled and poorly characterised defect structures. Such effects have not been investigated in detail for the higher T_c BiPb-2223 or Tl-2223 systems as yet.

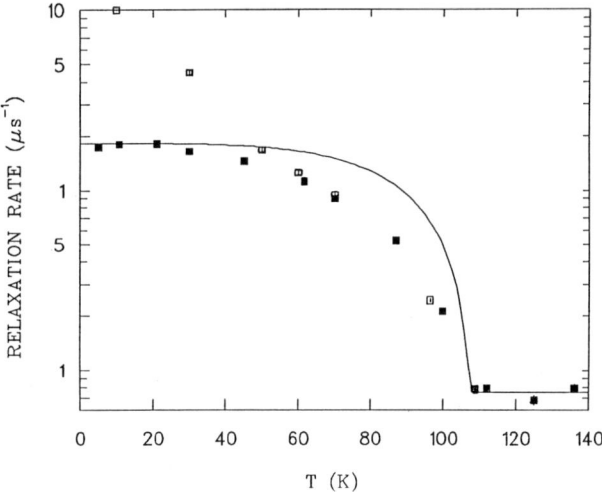

FIGURE 1. Temperature dependence of the relaxation rate for a sintered sample of $Bi_{1.8}Pb_{0.2}Sr_2Ca_2Cu_3O_{10}$, prepared as described in reference 13. Field cooled data (filled symbols) compared to the two-fluid model (solid line, to show the strong couplig limit) yield $\lambda_{ab}(0)=1720$Å. The open symbol data points were obtained after zero field cooling, and show that the onset of reversibility (depinning) is at $T_p = 70$K for this sample in the 0.4 T applied field, at higher temperature than for Bi-2212 samples in the same field (typically 20–40K).

Insofar as the μSR σ measures the penetration depth, results for $\lambda(T)$ below 20K always show only a small variation with temperature, compatible with conventional (nodeless order parameter) pairing in all cases for homogeneous samples. Anisotropic power law T-dependence of the μSR σ, indicating that the superconducting gap function has nodes on the Fermi surface, has been observed recently for crystals of the heavy electron case UPt_3,[16] incidentally demonstrating the ability of the technique to determine the intrinsic behaviour of the order parameter for less pathological cases. The μSR results for cuprates have thus been useful as a diagnostic tool, to study the low temperature behaviour and magnitude of the penetration depth, and to produce an empirical dependence of T_c on the ratio n_s/m^* (isotropic one band clean limit London picture), paralleling similar efforts where the carrier density is obtained by chemical or transport measurements. Figure 2 shows a compilation of (published) data[1-9] on $La_{2-x}Sr_xCuO_{4-y}$, $YBa_2Cu_3O_x$, Bi-based cuprates, and the data of Figure 1; plus points for $(BaPb)BiO_3$ and $(BaK)BiO_3$ samples to show that the general trends include the non–cuprate superconductors. In the $YBa_2Cu_3O_x$ system T_c increases directly proportional to $\sigma(0)$ as x increases from 6.65 (60K "plateau" region not studied in detail as yet), but then saturates for $x \geq 6.85$, even as $\sigma(0)$ continues to increase. The linear trend resumes with the 3-layer (BiPb)-2223 to saturate again for Tl-2223 samples. We have extended the systematics to the $YBa_2Cu_4O_8$ (Y-124)and $Y_2Ba_4Cu_7O_{15-\delta}$ (Y-247), with the results also shown in Figure 2, In this case T_c increases almost linearly with n_s/m^* without reaching a saturation at maximum doping. Structurally, Y-124 has saturated double-oxygen chains (thus a robust oxygen stoichiometry) and its unit cell may be considered as two Y-123 unit cells connected at the chains. Homogeneous samples of Y-247 consist of ordered alternate layers of Y-123 and Y-124 stacked in sequence, as a new phase and not just a multilayer construct.[16,17] The results show that the oxygen depletion, which occurs at the chains of the Y-123 blocks, affects the superconducting condensate in both the Y-123 and Y-124 layers, i.e. there is definite interlayer change transfer-coupling. This is in agreement with the fact that the normal state carrier density changes linearly with the oxygen content as measured by δ, and with the large increases in T_c under pressure that obtain in this system. It is interesting to speculate whether the difference with the $YBa_2Cu_3O_x$ is due to the $Y_2Ba_4Cu_7O_{15-\delta}$ system being more homogeneous and closer to a three-dimensional behaviour, or just a consequence of

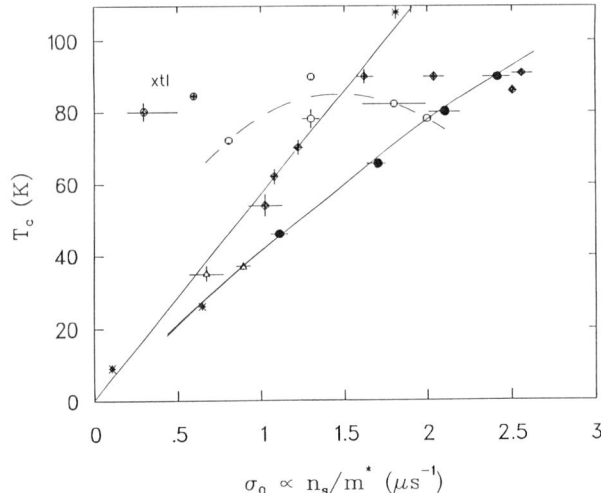

FIGURE 2. Plot of T_c vs extrapolated relaxation rate $\sigma(0)$ for $La_{1.85}Sr_{0.15}CuO_4$ (triangles), $YBa_2Cu_3O_x$ (diamonds), Bi-Sr-Ca-Cu-O (open circles),[1-9] sample of Figure 1 (asterisk), and new Y-124 and Y-247 results (filled circles, see text). The lines have been drawn to guide the eye for $YBa_2Cu_3O_x$ (straight line of reference 6) and $Y_2Ba_4Cu_7O_{15-\delta}$, and to indicate (dashed line) a tentative general trend for ceramic Bi-based 2212 samples. Only $x = 0.15$ $La_{2-x}Sr_xCuO_{4-y}$ (optimum doping) is shown, since other samples are clearly inhomogeneous. Notice the difference in results for two samples from different origins.[6,22] Data were obtained by the different authors in a variety of fields, from 0.1 to 1.8 T. Note the substantially lower value of $\sigma(0)$ for Bi-2212 crystals.[7,15] Unpublished data for ceramic samples of $(BaPb)BiO_3$ (T_c=10K) and $(BaK)BiO_3$ (T_c=26K) are also included for comparison.

electronic structure. This may be of relevance to the questions of two-dimensionality and the understanding of nanostructured epitaxial superlattices.

MAGNETIC ORDERING

The magnetic ordering of $La_{2-x}Sr_xCuO_{4-y}$ was elucidated early by µSR in conjunction with neutron diffraction[18,19]. The muon, as a local probe, responds to short range ordering (and is sensitive to small moments) thus complementing staggered magnetization measurements. The data showed that random freezing of fixed moments supplants the Nèel order as doping is increased. Evidence was also found for the frustration effects that produce ferromagnetically aligned in-plane pairs of copper spins, i.e. transition into a spin-glass phase.[19] Effects of the static ordering were also observed for higher values of the Sr concentration, in the superconducting range, which has been interpreted variously as due to an intrinsic connection between magnetic interactions and the superconductivity[20] or as evidence (reinforced by Meissner fraction reduction effects) for the important effect of electronic inhomogeneities.[21] The quasi-static magnetic freezing is totally absent above $x = 0.15$[22] down to 90 mK (although samples, especially crystals used in neutron diffraction, with nominal $x = 0.15$ have shown a sizable magnetically ordered component in the µSR signal, which are attributed to inhomogeneities). Similar ordering phenomena, i.e. Nèel ordering evolving into random magnetic order have also been detected in $YBa_2Cu_3O_x$ for x varying from 6.0 to 6.5, with definite spin-glass ZF and LF relaxation functions for $x = 6.45$.[23]

In view of the usual magnetic phase diagram, with long range magnetic ordering yielding to disorder (frustration) and eventually superconductivity (with possible magnetic fluctuations persisting), the $La_2MCu_2O_{6+\delta}$, M=(Sr,Ca) system was at first an enigma as a double-layer cuprate for which no evidence for superconductivity or magnetic order could be obtained in spite of its metal-insulator transition. Part of the enigma has been resolved recently for $La_2CaCu_2O_{6+\delta}$ by showing that Sr substitution for La (with careful control of the oxygen stoichiometry) results in bulk superconductivity with T_c up to 60K.[24] At about the same time, oxygen excess in $La_2CaCu_2O_{6+\delta}$ was shown to induce a phase separation with the superconducting phase having $T_c \sim 45K$.[25] We have performed initial μSR measurements in this system and observed clear magnetic signals in both cases, with notable differences. The $La_2SrCu_2O_{6+\delta}$ case is similar to $La_{2-x}Sr_xCuO_{4-y}$ and oxygen-defficient $YBa_2Cu_3O_x$, in that two signals are observed at 5 and 18 MHz (i.e. similar value of the ordered moment on copper). Such signals could be associated with muons bound to the apical oxygens and in the CuO_2 planes, respectively. The "freezing" temperature of over 300K is also similar to that of $YBa_2Cu_3O_x$, indicating similar exchange constants for both double-layer cases. The data in general indicate that the $La_2MCu_2O_{6+\delta}$,M=(Sr,Ca) structure is more disordered than that of $YBa_2Cu_3O_x$ or $La_{2-x}Sr_xCuO_{4-y}$, so current samples may appear spin-glassy to other techniques. Similar signals were obtained for $La_2CaCu_2O_{6+\delta}$, but with a frequency change at lower temperatures, indicating a second transition to a different ordered state. In addition, the superconducting fraction (about 10% of the sample) was definitely identified below 50K in an applied magnetic field, this phase separation similar to that observed for $La_2CuO_{4+\delta}$.[26]

CONCLUSIONS

The remarkable correlation between T_c and the ratio of carrier concentration to effective mass ($\sigma(0)$), first shown for a selected group of sintered copper oxide high-T_c samples, has been extended with the inclusion of $Y_2Ba_4Cu_7O_{15-\delta}$, $YBa_2Cu_4O_8$ and a variety of Bi-2212 samples.[7-9] The $Y_2Ba_4Cu_7O_{15-\delta}$ family also displays a dramatic increase of T_c with the carrier concentration, but in a monotonic fashion, to be compared to the saturated behaviour in the Y-123 family showing the sensitivity of results to the details of electronic structure. Even if the "universal correlation" of Reference 6 becomes more complex by inclusion of all relevant data, it still provides serious constraints for models of mechanisms. It should also be noted that such correlations are not the exclusive province of layered high-T_c materials, as shown by the inclusion of cubic Ba-oxide cases in Figure 2 and remarked at this Workshop and in the literature by others.[6,27] An example worth further study is the classical isotropic superconductor $PbMo_6OS_8$, with very small correlation length and sensitivity to local electronic defect structures.[28] By contrast, the Bi-oxide based samples have shown a lack of systematic behaviour, related to the softness and complex magnetic phase diagram of their FLL, shifting the interest in this case from mechanisms and electronic structure to the morphology and thermal stability of the FLL. The Bi-2111 of reference 9 is outside the systematics. For the 2212 in general, crystalline samples display considerable lower values of the relaxation rate, and lower depinning T_c's than ceramic samples. This rises the interesting possibility that the FLL in the former case is more "liquid" in structure, weakly pinned, while the FLL in ceramic samples is more inhomogeneous due perhaps to defects, effects enhanced by the extremely short coherence lenght in the Bi-oxides and their extreme anisotropy. Similar effects are found in mechanical studies of the FLL. Phase stabilizing atomic substitutions also affect T_p. Whether the results change at milliKelvin temperatures remains to be seen.

Given the current interest in recent theories (anyons, flux phases) that predict the existence of anomalous fields in two-dimensional structures, it shoud be remarked that from early on (for example Reference 1) the ZF μSR relaxation data showed no effects other than that due to nuclear dipole moments, confirmed recently in an accurate *ad-hoc* measurement.[29] In what regards magnetic ordering, the new results for the $La_2MCu_2O_{6+\delta}$,M=(Sr,Ca) family reafirm the striking similarity of freezing temperatures and electronic moments found for lamellar CuO_2 antiferromagnetic insulators, notwithstanding important differences in structures and coordination[30]. In this area the μSR data nicely complements the NMR and neutron diffraction[31] body of data.

In summary, this review attempts to present recent μSR results in an unbiased way, without attempting to extract far-reaching conclusions as to their significance for the mechanisms of high temperature superconductivity. It is the author's prejudice, however, that defects and electronic inhomogeneities play an essential rôle in the physical properties of the cuprates,[32] and it remains a challenge for experimentalists to determine their ultimate intrinsic level (as opposed to plain "poor" samples) and for theorists to incorporate them as part of their models *ab-initio*.

REFERENCES

1. G. Aeppli *et al.*, *Phys. Rev.* **B35**, 7129 (1987).
2. F. N. Gygax *et al.*, *Europhys. Lett.* **4**, 473 (1987).
3. D. R. Harshman *et al.*, *Phys. Rev.* **B36**, 2386 (1987).
4. W. J. Kossler *et al.*, *Phys. Rev.* **B35**, 7133 (1987).
5. D. R. Harshman *et al.*, *Phys. Rev.* **B39**, 851 (1989).
6. Y. J. Uemura *et al.*, *Phys. Rev. Lett.* **62**, 2317 (1989), and these Proceedings.
7. E. J. Ansaldo *et al.*, *Physica C* **162-164**, 259 (1989), and *NATO-ASI Series* **E181**, 411 (1989).
8. R. Lichti *et al.*, *J. Appl. Phys.* **54**, 2361 (1989), S.G. Barsov *et al. Hyp. Int.*, in press (1991).
9. P. Birrer *et al.*, *Physica C* **158**, 230 (1989).
10. E. H. Brandt, *Phys. Rev.* **B37**, 2349 (1988).
11. W. Barford and J. Gunn, *Physica C* **153-155**, 691 (1988).
12. L. Krussin Elbaum *et al.*, *Phys. Rev. Lett.* **62**, 217 (1989).
13. N. D. Spencer *et al.*, *J.J.A.P.* **28**, L1564, (1989).
14. B. Pümpin *et al.*, *Z. Phys. B***72**, 175 (1988),
15. B. Sternlieb *et al.*, *Physica C* **162-164**, 1679 (1989).
16. C. Broholm *et al.*, *Phys. Rev. Lett.* **65**, 2062 (1990).
16. J. Karpinski *et al.*, *Physica C* **160**, 449 (1989).
17. J. L. Tallon *et al.*, *Phys. Rev. B* **41**, 7220 (1990).
18. Y. J. Uemura *et al.*, *Physica C* **153-155**, 768 (1988).
19. D. R. Harshman *et al.*, *Phys. Rev.* **B39**, 851 (1989).
20. A. Weidinger *et al.*, *Phys. Rev. Lett.* **62**, 102 (1989).
21. D. R. Harshman *et al.*, *Phys. Rev. Lett.* **63**, 1187 (1989).
22. R. Kiefl *et al.*, *Phys. Rev. Lett.* **63**, 2136 (1989).
23. J. H. Brewer *et al.*, *Phys. Rev. Lett* **60**, 1073 (1989).
24. R. J. Cava *et al. Nature***332**, 814 (1990).
25. A. Fuertes *et al. Physica C***170**, 153 (1990).
26. E. J. Ansaldo *et al. Phys. Rev.* **B40**, 2555 (1989).
27. J. Hirsch, these Proceedings.
28. C. Rosser*et al. Physica C* **165**, 233 (1990).
29. R. Kiefl *et al.*, *Phys. Rev. Lett.* **64**, 2082 (1990).
30. E. J. Ansaldo *et al.*, to be published.
31. Contributions by H. Alloul and J. M. Tranquada, these Proceedings.
32. Contributions by R. S. Markiewicz and J. C. Phillips, these Proceedings.

TOPOLOGICAL EFFECTS IN DISORDERED PHASE OF TWO-DIMENSIONAL MAGNET

A.M. Tsvelik

Department of Physics
University of Florida
Gainesvile, Florida 32611

M.Yu. Reizer

Department of Physics
The Ohio State University
Columbus, Ohio 43210-1106

ABSTRACT

The CP^{N-1}-model for $N \gg 1$ is considered as a model describing a disordered spin liquid phase within a controlled approximation. We show that integration over configurations of fields with a nonvanishing topological charge give rise to the Chern-Simons term in the effective action for the RVB-gauge field. This term cannot be reproduced by standard $1/N$-expansion and the system has a phase transition at $N \to \infty$. The Chern-Simons term weakens confining ability of the gauge fields and removes monopoles. Therefore z-quanta composing the spin field behave almost like independent particles in agreement with the concept of the spin liquid.

INTRODUCTION

According to the concept of the spin liquid (or the RVB liquid) a spin system being disordered in the ground state has some gapless mode associated with a gauge field. This gauge field represents collective degrees of freedom of the spin system. Many authors claim that the spin liquid state describes a spin subsystem of the $t - J$ model, which is often used to describe copper oxide superconductors[1]. In this scenario the gapless gauge field serves to explain non-Fermi liquid electron properties of the normal state of these materials.

The gauge symmetry of the disordered magnetic state is very important and there is no reason for it to be somehow broken in this phase. A classical example of a gauge theory is Quantum Electrodynamics (QED). Elementary excitations in QED (photons) are gapless. It is not surprising that QED served as an effective theory for the disordered spin state[1]. However, existence of a massless mode in a disordered phase looks strange. Really, the exact solution of the spin-1/2 Heisenberg antiferromagnet on a honeycomb lattice with the first and the second nearest neighbors interaction at $J_2 = 1/2 J_1$ shows a spectrum of elementary excitation with a gap in the disordered phase[2].

In this paper we shall try to solve this contradiction. Note that the Chern-Simons term being added to the action of 2 + 1 QED makes photons massive but does not break the gauge invariance[3]. Thus we shall try to find a model of a quantum antiferromagnet which has a disordered ground state and show that the Chern-Simons term appears in the effective action when we treat the model nonperturbatively.

As a basis for description of the disordered phase we consider the $SU(N)$ invariant lattice Heisenberg antiferromagnet model with the Hamiltonian

$$H = \frac{1}{N} \sum_{(i,j),\alpha,\beta} S_\beta^\alpha(i) S_\beta^\alpha(j), \tag{1}$$

where S_β^α is the generator of $SU(N)$ group, which satisfy the following commutation relations

$$[S_\beta^\alpha(i), S_\sigma^\rho(j)] = \delta_{ij}\left(\delta_\beta^\rho S_\sigma^\alpha(i) - \delta_\sigma^\alpha S_\beta^\rho(i)\right), \tag{2}$$

and are subject to constraint $\sum_{\alpha=1}^{N} S_\alpha^\alpha = SN$. This model has been studied recently by the $1/N$ expansion (see Ref.4 and references therein). The continuous version of the model is called the CP^{N-1} model and has the action

$$S\{z, A\} = \frac{N}{2g} \int_0^\infty d\tau \int d^2x \sum_{j=1}^{N} |(\partial_\mu + A_\mu)z_j|^2, \tag{3}$$

where the spin field S_β^α relates to the z-field by the equation

$$S_\beta^\alpha = -z_\beta^* z_\alpha, \tag{4}$$

and the z-fields satisfy the constraint

$$\sum_{j=1}^{N} z_j^* z_j = 1. \tag{5}$$

The CP^{N-1} model is well known in the field theory due to its nontrivial physics[5], and was discussed recently in the above mentioned context[6]. The CP^{N-1} model has a disordered ground state if the bare coupling constant exceeds some critical value. We shall show that the disordered spin state of the CP^{N-1} model may be described by the 2 + 1 QED with the Chern-Simons term. This theory arises as an effective theory for the $\vec{A}$-fields after integration over z-quanta in the partition function of action (3). The Chern-Simons term cannot be obtained by the $1/N$-expansion, It means that physical quantities are nonanalytical function of $1/N$, and $N = \infty$ is the phase transition point.

The CP^{N-1} model has $SU(N)/U(1)$ symmetry. At $N = 2$ the CP^{N-1} model is equivalent to the $O(3)$ nonlinear σ model, which has a disordered ground state at large coupling constant [7]. The basic properties of the model, namely the gauge symmetry and existence of topologically nontrivial classical solutions persist for any N.

The most interesting property of the disordered phase which we shall study in detail is that the $\vec{A}$ field being nondynamical on the classical level acquires a nontrivial dynamics due to quantum fluctuations.

Note also that there were some attempts to find the Chern-Simons term as an additional term to action (3) after transition from the lattice model to its continuous version. However as was shown recently[7] this assumption is wrong, thus we shall propose another mechanism of appearance of the Chern-Simons term in the effective action.

$N = \infty$ LIMIT OF THE CP^{N-1} MODEL

The partition function for the CP^{N-1} model with the constraint (5) is

$$Z = \int D\lambda Dz^* DA_\mu \exp\left[-S\{z, \vec{A}\} - i\int d\tau d^2x \lambda(z_j^* z_j - 1)\right]. \tag{6}$$

The mean field approximation corresponds to the steepest descents method of evaluating the integral. After extremizing the action $\vec{A}$ and λ acquire static uniform values. The Green's function of z-fields within the mean field approximation in the momentum space is

$$G(\vec{p}, \omega_n) = \frac{g/N}{(p_\mu + A_\mu)^2 + m^2}, \tag{7}$$

where $p_0 = \omega_n, \vec{p} = (p_x, p_y), m^2 = g\lambda/N$.

Extremizing the free energy with respect to $\vec{A}$ and λ we get the following self-consistency conditions

$$1 = NT \sum_n \int \frac{d^2p}{(2\pi)^2} G(\vec{p}, \omega_n) a^2, \quad a^{-1} = \frac{1}{2\pi}\int dp, \tag{8}$$

$$A_\mu = NT \sum_n \int \frac{d^2p}{(2\pi)^2} G(\vec{p}, \omega_n)(p_\mu + A_\mu) a^2. \tag{9}$$

After substituting expression (7) into equations (8) and (9) we get for $T = 0$:

$$m = \frac{2\pi}{ag}(g - 2a) << 1. \tag{10}$$

Equation (10) is satisfied trivially for any constant field $\vec{A}$, and this manifests the gauge invariance of the saddle point. Later we will work in the Euclidean space-time and assume $T = 0$.

According to Ref. 6 the effective action for $\vec{A}$ and z-fields is

$$S_{eff} = \frac{1}{2}\delta\lambda(q)\Pi_1(q)\delta\lambda(-q) + \frac{1}{2}A_\mu(q)\Pi_{\mu\nu}(q)A_\nu(-q), \tag{11}$$

where

$$\Pi_1(q) = \frac{N}{\pi|q|}\arctan\left(\frac{|q|}{2m}\right), \tag{12}$$

and

$$\Pi_{\mu\nu}(q) = \left(\delta_{\mu\nu} - \frac{q_\mu q_\nu}{q^2}\right)\frac{N}{8\pi}\left[\frac{q^2 + 4m^2}{|q|}\arctan\left(\frac{|q|}{2m}\right) - 2m\right]. \tag{13}$$

At small momenta the action for $\vec{A}$-fields is an ordinary Electrodynamics with the Lagrangian

$$L_{eff}\{A_\mu\} = \frac{N}{16\pi m}F_{\mu\nu}^2. \tag{14}$$

TERMS WHICH THE 1/N-EXPANSION CANNOT REPRODUCE

It is clear from the action (11) that fluctuations of λ-fields cannot give rise to any infrared singularities. It is the presence of λ-fields in the partition function (6) that makes it different from the partition function of the complex bosonic fields with gauge forces. But these two models are not equivalent because CP^{N-1} model has excitations which the bosonic model does not have. The CP^{N-1} model with the local constraint (5) has classical solutions with nonvanishing topological invariant

$$Q_\mu = \int_{S_\mu} d^2x J_\mu = \frac{i}{2\pi} \int_{S_\mu} d^2x \epsilon_{\mu\nu\lambda} \partial_\nu z_j^* \partial_\lambda z_j, \qquad (15)$$

where S_μ is a plane orthogonal to the μ-direction. But the model with the global constraint on the average value $< \sum_{j=1}^{N} z_j^* z_j >$ does not have such a topological invariant. Dealing with the quantum partition function one should integrate over vortex lines which may be treated as world lines of massive particles. The particle mass is equal to the energy of the vortex unit length $M \approx 4\pi N/g$. It is well known that a path integral over world line configurations may be represented as a path integral over bosonic fields. Because an elementary vortex can carry positive or negative topological charge, the vortex is represented by a complex bosonic field Φ. It is important that each vortex carries a gauge field flux. Indeed, minimizing the action (3) with respect to $\vec{A}$ we get the equation

$$A_\mu = iz_j^* \partial_\mu z_j, \qquad (16)$$

Then substituting equation (16) into (15) we get

$$Q_\mu = \int_{S_\mu} d^2x \frac{1}{4\pi} \epsilon_{\mu\nu\lambda} F_{\nu\lambda}. \qquad (17)$$

The current $\vec{J} = (1/4\pi)\epsilon_{\mu\nu\lambda} F_{\nu\lambda}$ is associated with a given configuration of the vortex line. Equation (17) corresponds to the following Lagrangian

$$L = \frac{1}{2}|(\partial_\mu + iA_\mu)\Phi|^2 + \frac{1}{2}M^2\Phi^*\Phi + \frac{i}{4\pi}A_\mu \epsilon_{\mu\nu\lambda}\partial_\nu A_\lambda + \frac{N}{8\pi m}F_{\mu\nu}^2, \qquad (18)$$

Lagrangian (18) is invariant under the following transformations:

$$t \to t, x \to y, y \to x, A_0 \to -A_0, A_x \to A_z, A_y \to A_y, \Phi \to \Phi^*. \qquad (19)$$

To get the effective action for the $\vec{A}$-fields one should integrate the Lagrangian over the Φ-fields, which leads to insignificant corrections to the last term in equation (18), and as a result we get

$$L_{eff}(\vec{A}) = \frac{i}{4\pi}A_\mu\epsilon_{\mu\nu\lambda}\partial_\nu A_\lambda + \frac{1}{8\pi e^2}F_{\mu\nu}^2, e^2 = \frac{2m}{N} + 0(N^{-2}). \qquad (20)$$

Thus we obtain the famous Electrodynamics with the Chern-Simons term. As it was proposed in Ref.6, the disordered ground state of the CP^{N-1} model is described by the effective Lagrangian (20). To avoid confusion we emphasize that The Chern-Simons term in the present case does not change statistics of z- and Φ-quanta.

CHERN-SIMONS ELECTRODYNAMICS AND INTERACTION BETWEEN Z-QUANTA

The effective action in the Lorentz gauge ($\partial_\alpha A_\alpha = 0$) takes the form

$$S_{eff} = \sum_{k,\omega} \{\frac{1}{8\pi e^2}\left[k^2 \phi_{k,\omega}\phi_{-k,-\omega} + (1+\omega^2/k^2)H_{k,\omega}H_{-k,-\omega}\right] +$$

$$\frac{i\mu_0}{2}(\phi_{k,\omega}H_{-k,-\omega} + \phi_{-k,-\omega}H_{k,\omega}), \tag{21}$$

where $\mu_0 = 4e^2$, $\phi_{k,\omega}$ and $H_{k,\omega}$ are the Fourier component of the scalar potential and the magnetic field which are related to the vector potential $\vec{A}$. The correlation function of electromagnetic fields for the action (21) is

$$D_{i,j}(\vec{k},\omega) = \begin{pmatrix} D_{\phi\phi} & D_{\phi H} \\ D_{H\phi} & D_{HH} \end{pmatrix} = \frac{4\pi e^2}{\omega^2 + k^2 + \mu_0^2} \begin{pmatrix} 1 + \omega^2/k^2 & -i\mu_0 \\ -i\mu_0 & k^2 \end{pmatrix}, \tag{22}$$

Using this equations we can estimate the interaction effects on z-quanta. At low energies one can neglect dynamical effects and treat the interaction as static. According to (7) the Green's function of a slowly moving z-quanta ($k \ll m$) is

$$G(\vec{p},\omega_n) \approx \frac{Z}{i\omega_n - m - p^2/2m}, Z = g/2Nm. \tag{23}$$

The main contribution to interaction between slowly moving z quanta comes from the scalar potential correlation function $D_{\phi\phi}$. For $\omega = 0$ this interaction is the screened Coulomb potential

$$V(k) = \frac{4\pi e^2}{k^2 + \mu_0^2}. \tag{24}$$

The bound energy of z-quanta is determined by the Schröedinger equation for two particles.

$$\psi(\vec{r}_1,\vec{r}_2)E = \{-\frac{1}{2m}[\partial_1^2 + \partial_2^2] + U(\vec{r}_1 - \vec{r}_2|)\}\psi(\vec{r}_1,\vec{r}_2), \tag{25}$$

where

$$U(\vec{r}) = (mZa)^2 \int \frac{d^2k}{(2\pi)^2} \frac{4\pi e^2}{k^2 + \mu_0^2} \exp(i\vec{k}\vec{r}). \tag{26}$$

As is known, any two-dimensional attractive potential has at least one bound state. The bound state which correspond to equations (25), (26) is

$$E_{bound} = E_0 \exp\left(-\frac{1}{m_{eff}U(k=0)}\right) = const\frac{m}{N} \exp\left(-\frac{16N}{\pi}\right). \tag{27}$$

Here E_0 is a characteristic depth of the potential $U(\vec{r})$, which we assume to be approximately equal to m/N and $m_{eff} = m/2$ is an effective mass of relative motion of two identical particles. As it is seen from equation (26) the bound energy is very small which means that z-quanta may be considered as practically free (itinerant). The spin-spin correlation function in the disordered phase does not have a one magnon pole.

ABSENCE OF MONOPOLES

The theory of the CP^{N-1} model (3) in the present approach is a continuous theory which originates from the lattice model (1). It means that $\vec{A}$-field is compact i.e the action (21) is a periodic functional of the magnetic field $H = F_{xy}$. Periodicity of this type may be taken into account if the path integral contains magnetic field configurations which violates the Bianchi identity

$$\partial_\mu B_\mu = \sum_\alpha (4\pi n_\alpha)\delta^{(3)}(\vec{x} - \vec{x}_\alpha). \tag{28}$$

The partition function is a sum over all integer numbers n_α. However as was shown recently[9], those configurations are wiped out if the Lagrangian contains the Chern-Simons term. The reason is the following. Let us perform the gauge transformation

$A_\mu(xi) = A_\mu + \partial_\mu \xi$ in Lagrangian (21). According to equation (28) this leads to a new effective action

$$S_{eff}(\xi) = S + \frac{i}{4\pi} \int d^3x \partial_\mu \xi F_{\mu\nu} = S + i \sum_{x_\alpha} \xi(x_\alpha) n(x_\alpha). \qquad (29)$$

It is easy to see that all configurations with nonvanishing n_α are wiped out by the integration over gauge transformations.

CONCLUSIONS

Our results may be summarized as follows. The Chern-Simons term arises in the Lagrangian which describes the disordered phase of the CP^{N-1} model and leads to a mass gap in the spectrum of excitations of the RVB fields. The appearance of the mass gap does not break the gauge symmetry (it is a remarkable property of the Chern-Simons term). We would like to emphasize that these results cannot be obtained within the $1/N$ expansion. Actually, we did not calculate the Chern-Simons term, but we have shown that without this term the effective theory cannot be self-consistent. The mass gap weakens confinement of z-quanta, and they behave like almost independent particles as was predicted from qualitative analysis[10].

The authors are very grateful to the Institute for Scientific Interchange for hospitality at Vila Gualino, Torino where part of this work was done. The work at The Ohio State University was supported by DOE contract # DE-FG02-ER45347.

REFERENCES

[1] P.W. Anderson, Science 235:1196 (1987); G. Baskaran and P.W. Anderson, Phys. Rev. B, 37:580 (1988); I. Affleck and B. Marston, Phys. Rev. B 37:3774 (1988).
[2] I. Affleck, T. Kennedy, E.H. Lieb, and H. Tasaki, Commun. Math. Phys. 115:477 (1988).
[3] S. Deser, R. Jackiv, and S. Templton, Phys. Rev. Lett. 48:975 (1982); Ann. Phys. (NY) 140:372 (1982).
[4] A. Auerbach and D.P. Arovas, in "Field Theories in Condensed Matter Physics", Z. Tesanovic, ed. Addison-Wesley, Redwood City (1990).
[5] A. D'Adda, P. di Vecchia, and M.Luscher, Nucl. Phys. B 146:63 (1978); E. Witten, Nucl. Phys. B 149:285 (1979); I. Ya. Aref'eva, Ann. Phys. (NY) 117:393 (1979); G.W. Semenoff, P. Sodano, and Yong-Shi Wu, Phys. Rev. Lett. 62:715 (1989).
[6] X.G. Wen and A. Zee, Phys. Rev. Lett. 62:1937 (1989).
[7] S. Chakravarty, B.I. Halperin, and D. Nelson, Phys. Rev. B 39: 2344 (1989).
[8] A.Muramatsu, Phys. Rev. Lett. 65:2909 (1990).
[9] E. Fradkin, F.A. Schaposnik, Phys. Rev. Lett. 66:276 (1991).
[10] R. Laughlin, Science 242:525 (1988).

SPECIFIC HEAT OF $YBa_2Cu_3O_7$: VOLUME FRACTION OF SUPERCONDUCTIVITY; PARAMETERS CHARACTERISTIC OF THE "IDEAL" SUPERCONDUCTING STATE

Norman E. Phillips and R.A. Fisher

Materials Sciences Division
Lawrence Berkeley Laboratory
Berkeley, CA 94720, USA

INTRODUCTION

Experimentally determined parameters for $YBa_2Cu_3O_7$ (YBCO) are strongly sample dependent. However correlations among parameters derived from the specific heat (C) suggest that the variation reflects a variation in the volume fraction of superconductivity, (f_s). From these correlations f_s can be quantitatively determined and values of parameters characteristic of fully superconducting material can be derived. With additional assumptions, it is also possible to estimate the Sommerfeld constant (γ) for the fully normal state. Among the parameters of particular importance in establishing the correlations are the discontinuity in C [$\Delta C(T_c)$] at the critical temperature (T_c) and concentration (n_2) of localized Cu^{2+} magnetic moments. These are located on the YBCO lattice, at least in substantial measure, and are directly correlated with f_s. From an analysis of C in the vicinity of T_c it is possible to obtain information on the strength of the coupling responsible for the superconductivity; from a comparison of γ with that calculated for the bare density of states (γ_{bs}) the electron-phonon enhancement parameter (λ) can be obtained.

Some data for $(La_{2-x}Sr_x)CuO_4$ suggest similar correlations, but there are insufficient data for other high temperature superconductors (HTSC) to test this possibility.

The total specific heat of a YBCO sample that is reasonably typical of the better polycrystalline samples currently available is shown in Fig. 1. At T_c the lattice specific heat is large compared with the electronic contribution, and the feature associated with the transition to the superconducting state is only 3% of the total. Furthermore, there is no obvious discontinuity in specific heat. Comparison of data for different samples shows that a major part of the apparent breadth of the transition is associated with sample-to-sample differences, presumably inhomogeneities and other imperfections, but the nature of the specific-heat anomaly at T_c for an ideal sample has not yet been unambiguously established. For YBCO there can be inhomogeneities associated with oxygen stoichiometry and with the ordering of the O atoms. The inclusion of impurity phases probably also contributes to the breadth. The effect of small-scale defects of all kinds can be expected to be enhanced in HTSC relative to that in conventional superconductors because the coherence length (ξ) is smaller and is comparable to the lattice parameters. Furthermore, there is the expectation, also based on the small value of ξ, that fluctuation effects should be important in determining the shape of the specific-heat anomaly at T_c.

Figure 1 also shows the low-temperature "upturn" in C/T that is characteristic of virtually all samples of HTSC, at least those that have been studied below 1K. It is associated with electronic magnetic moments that order below 1K as shown by its magnetic-field dependence. After appropriate correction for the upturn there is still a non-zero intercept of C/T at T=0. This "linear term", $\gamma(0)T$, has attracted much attention. It was recognized very early (1) that it could be simply a manifestation of an incomplete transition to the superconducting state. However, independently of, and more or less simultaneously with, its experimental discovery (2), a linear term was predicted theoretically (3) as one of the consequences of the resonant valence bond (RVB) theory. In the RVB model the excitations from the superconducting ground state are qualitatively different from those in a conventional superconductor, and so is the thermodynamics. The development of an understanding of the origin of $\gamma(0)$ has, for these reasons, been a major goal of specific-heat measurements on HTSC.

ANALYSIS OF LOW-TEMPERATURE DATA

Empirically, at low temperatures the specific heat of the HTSC can be represented (4) as a sum of lattice (C_ℓ), hyperfine (C_h) and localized electronic magnetic moment (C_m) contributions, and the linear term (C_e). The latter is field dependent, $C_e = \gamma(H)T$. It is related to the normal state electronic specific heat, but

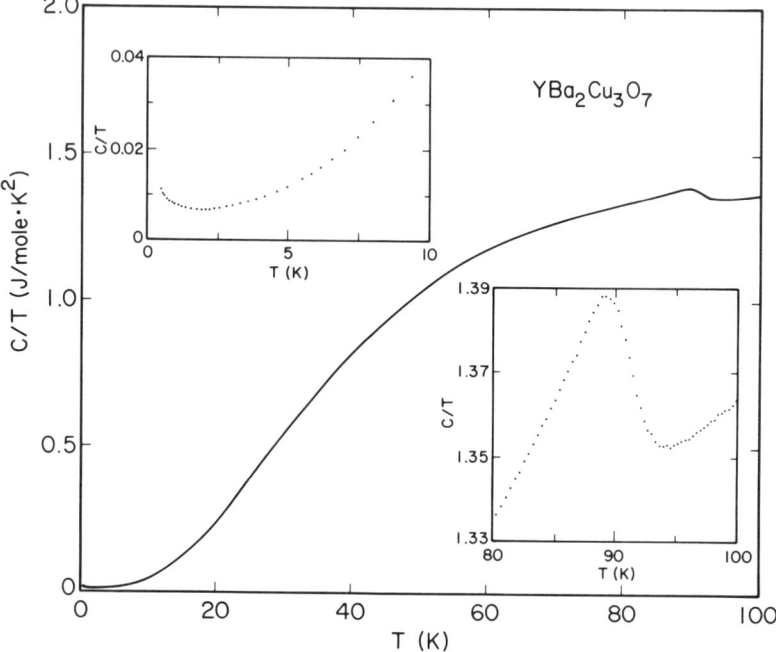

Figure 1. The total specific heat of a typical YBCO sample. The insets show actual data at low temperatures and in the vicinity of T_c

in many cases it includes other contributions. The lattice specific heat is independent of magnetic field (H) and is usually assumed to be the same in the normal and superconducting states. In the low-temperature limit the harmonic lattice expression for C_ℓ is $C_\ell = B_3 T^3 + B_5 T^5 + B_7 T^7 + \cdots$. In the temperature range of interest here, C_h is well represented by the lowest-order term in the high temperature expansion of a Schottky anomaly: $C_h = D(H)T^{-2}$. In the case of a nuclear magnetic moment interacting with the applied field $D(H) \propto H^2$, but this relation would not apply if electric quadrupole interactions or internal magnetic hyperfine fields associated with electronic moments were important.

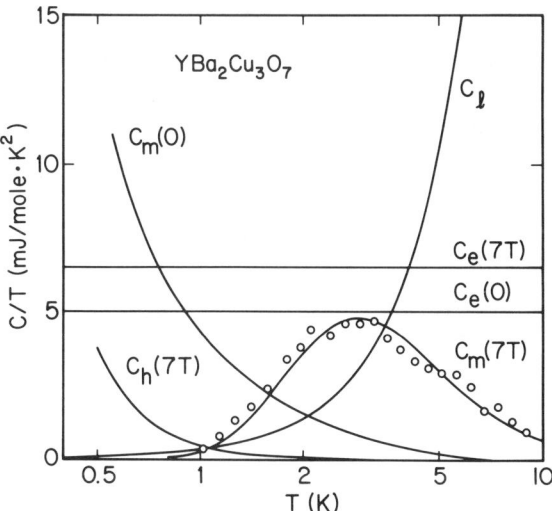

Figure 2. Components of the low-temperature specific heat for a YBCO sample.

The contribution C_m is associated with Cu^{2+} magnetic moments that order at temperatures below 1K, depending on their concentration. In zero applied field C_m is not a simple Schottky anomaly. No doubt this reflects the existence of a distribution of effective magnetic fields associated with disorder in the spatial distribution of the moments and their interactions. Several different approximations to the high-temperature "tail" of the anomaly by expansions in inverse powers of temperature have been used, and they can be represented by the general expression $C_m = \Sigma A_{-n} T^{-n}$. In the presence of a magnetic field of a few T or more the anomaly is shifted to temperatures above 1K. It is the field dependence that suggests Cu^{2+} moments. If the concentration of moments is sufficiently low the anomaly may take the form of the Schottky anomaly expected for Cu^{2+} moments in the applied field (to within the accuracy of the data) but for higher concentrations, and depending on the field, the anomaly may still be broadened by the internal fields. The n_2 concententation of these moments is determined by the Schottky anomaly.

The coefficient of the linear term is approximately linear in H, $\gamma(H) = \gamma(0) + (d\gamma/dH)H$. It is the $H=0$ component, $\gamma(0)$, that is unexpected, by comparison with the superconducting state specific heat of conventional superconductors. The H-proportional term is presumably of the same origin (5) as the analogous term in the mixed-state specific heat of conventional type-II superconductors -- normal-state-like excitations in the vortex cores.

Figure 2 shows an analysis of low-temperature specific-heat data for YBCO into the four components for $H=0$ and 7T. The solid line through the points for $C_m(7T)$ is a Schottky function for Cu^{2+} moments with $n_2 = 0.0044$ moles Cu^{2+}/mole YBCO. The points were obtained by subtracting C_ℓ, C_h and C_e from C. The analysis was made by least-squares fits with the temperature dependences listed above.

ANALYSIS OF SPECIFIC HEAT NEAR T_c

Perhaps the most serious problem associated with specific-heat measurements on HTSC is the impossibility of quenching superconductivity (except very close to T_c) with the magnetic fields available in the laboratory. Thus, the relatively simple methods for separating the lattice and electronic contributions, and determining the zero-field specific-heat anomaly at T_c, which are so important for conventional superconductors, do not work for HTSC. The analysis of data near T_c is further complicated by the broadening of the anomaly by sample inhomogeneities.

In spite of these complications, most specific-heat data for HTSC can be consistently analyzed to obtain the mean-field component of the anomaly, $\Delta C(T_c)$. There are different methods of estimating $\Delta C(T_c)$ from the data, several of which are

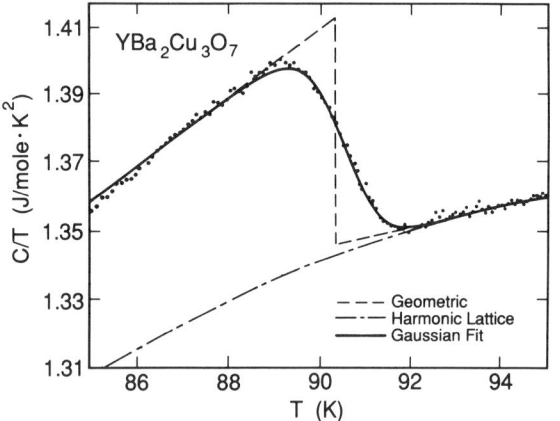

Figure 3. Specific-heat data for YBCO near T_c with several constructions used to determine $\Delta C(T_c)$ as described in the text.

illustrated in Fig. 3, with data obtained on a polycrystalline sample of YBCO. The dot-dash curve is an estimate of $(C_\ell + C_{en})$ based on a harmonic lattice approximation (6) with the addition of a term proportional to temperature which is fitted to the data from 50 to 280K, with the region between 70 and 110K excluded from the fit, and the data below 70K corrected for a small contribution from C_{es} (n and s refer to normal and superconducting, respectively). The dashed lines in the figure represent simple linear extrapolations of the C/T data just above and just below T_c. T_c and $\Delta C(T_c)$ are determined by an entropy-conserving construction that equalizes the two areas between the dashed lines and the data. The height of the vertical dashed line, $\Delta C(T_c)/T_c$, is 66 mJ/mole.K^2 and T_c =90.3K. An entropy-conserving construction that uses the dot-dash curve corresponding to the harmonic-lattice background rather than the straight-line extrapolation above T_c gives $\Delta C(T_c)/T_c$ = 69 mJ/mole.K^2 and essentially the same value of T_c. The smooth curve through the data is the sum of the estimated lattice contribution and a term representing C_{es} calculated for a BCS superconductor with a Gaussian spread of T_c's. The fit was obtained by adjusting the value of γ for C_{es} and the width of the Gaussian distribution. The mean T_c is 91K, γ =45 mJ/mole.K^2 and $\Delta C(T_c)/T_c$=64 mJ/mole.K^2, in reasonable agreement with the other two estimates of $\Delta C(T_c)/T_c$. [γ is here used as a scaling factor to fit the observed specific-heat anomaly. Its high value relative to that determined for the normal state--16 mJ/mole.K^2 (7)-- is an indication of strong-coupling effects.]

The width of $\Delta C(T_c)$ in Fig.3 is probably largely due to sample inhomogeneities which tend to obscure the effects of fluctuations. Figure 4 shows an example of the specific heat for another YBCO sample which has been analyzed using only fluctuations to account for the width of the transition (8). [$\Delta C = C - (C_\ell + C_{en})$ was

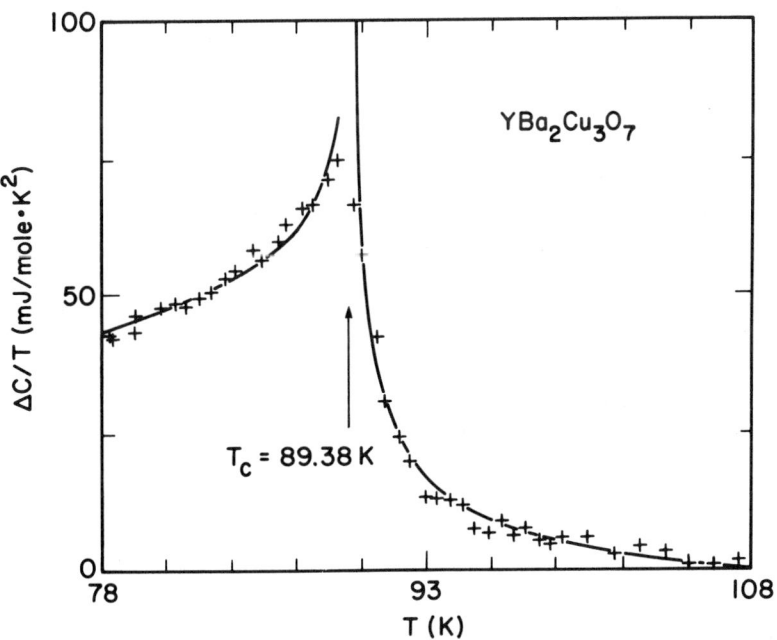

Figure 4. Specific heat of polycrystalline YBCO. The curves represent a fit by the expression for the fluctuation contribution given in the text.

obtained in a similar fashion to that described above for Fig. 3.] The continuous curves represent a BCS like C_{es} with $T_c = 89.38K$; $\gamma = 41$ mJ/mole.K^2 (γ is also used as a scaling factor here); and with an added 3-dimensional Gaussian fluctuation contribution (C_f) given by $C_f = A^{\pm} |T/T_c - 1|^{1/2}$. The values of A^+ ($T > T_c$) and A^- ($T < T_c$) are 0.51 and 0.24 J/mole.K, respectively.

VOLUME FRACTION OF SUPERCONDUCTIVITY

Independently of the sample-to-sample variation of the width of the anomaly, there is a strong sample dependence of $\Delta C(T_c)$. One obvious possibility is that the sample-to-sample variation of $\Delta C(T_c)$ corresponds to a sample-to-sample variation of f_s. This possibility can by tested by a comparison of values of $\Delta C(T_c)$ with the values of two other parameters derived from the specific-heat data that would also be expected to measure f_s, $d\gamma/dH$ and ΔS (defined below). If $d\gamma/dH$ corresponds to the mixed-state electronic specific heat, it should clearly be proportional to f_s. ΔS is defined in Fig. 5 (9). The shaded area represents the entropy decrease produced by the application of a magnetic field between the temperature at which the zero-field and in-field specific heat curves cross (T_x) and T_c. If the zero-field superconducting transition is incomplete, ΔS should also be proportional to f_s. The third law of thermodynamics requires that ΔS be equal to the area between the zero-field and in-field curves for $T < T_x$. The relation $\Delta S/T_x = \Delta \gamma$, where $\Delta \gamma$ is the value measured at low temperatures, has been found to hold to within a few percent for all samples for which it has been tested, suggesting that $d\gamma/dH$ is approximately constant over a wide

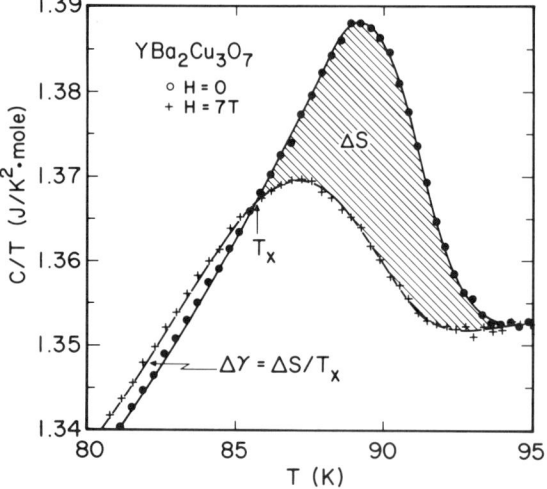

Figure 5. C/T versus T for H=0 and 7T. See text for a discussion of the relation between $\Delta \gamma$ and ΔS.

temperature range extending nearly to T_x. However, it should be noted that although ΔS and $d\gamma/dH$ are not thermodynamically independent, empirically they are independent measures of f_s.

In fact, the three quantities $\Delta C(T_c)/T_c$, $d\gamma/dH$, and ΔS are mutually proportional, as illustrated in Fig. 6, where $d\gamma/dH$ and ΔS are plotted versus $\Delta C(T_c)/T_c$. [Since T_c is essentially constant for these samples, it does not matter whether $\Delta C(T_c)$ or $\Delta C(T_c)/T_c$ is used for this purpose.] The interpretation of all three quantities as measures of f_s is supported by these correlations. [It could be argued that the variations in $\Delta C(T_c)/T_c$, $d\gamma/dH$ and ΔS might be due to variations in the density of electron states. However, the temperature-independent term in the high-temperature susceptibility is constant, suggesting that the electron density of states is also constant.]

Although $\Delta C(T_c)/T_c$, $d\gamma/dH$ and ΔS are all proportional to f_s, there is at this point no means of identifying the values corresponding to $f_s=1$. For that purpose, the correlations of the three quantities with n_2 is useful, as shown in Fig. 7 for $\Delta C(T_c)/T_c$. The fact these parameters correlate with n_2 at all shows that a substantial number of the Cu^{2+} ions that contribute to n_2 must be located on the YBCO lattice. The scatter of the points in Fig. 7 probably reflects not only uncertainty in $\Delta C(T_c)$ but also the possibility that some of the magnetic moments counted in n_2 are in impurity phases and do not contribute to the effect on $\Delta C(T_c)$.

The correlation between $\Delta C(T_c)/T_c$ and n_2 displayed in Fig. 7 implies a maximum value of $\Delta C(T_c)/T_c$ that is approximately 77 mJ/mole.K^2. For other values of $\Delta C(T_c)/T_c$, $f_s=[\Delta C(T_c)/T_c]/77$. Essentially the same conclusion has been reached by the Geneva group (10) on the basis of somewhat different, but related, evidence.

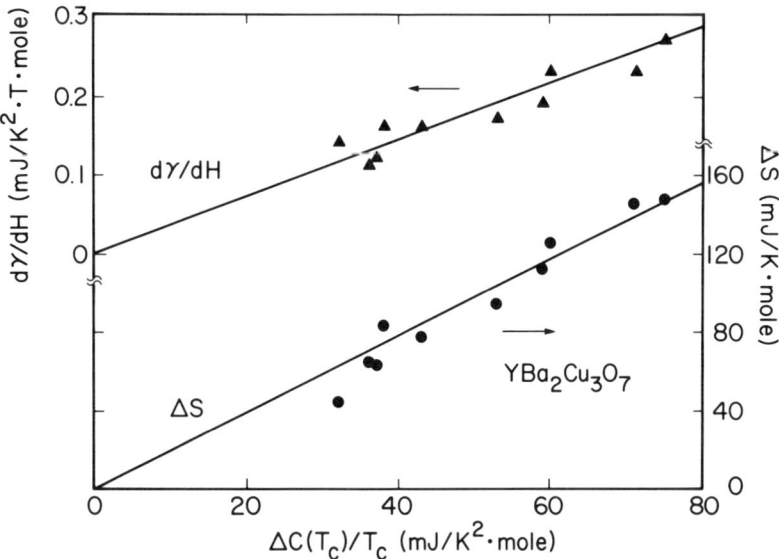

Figure 6. Correlation of $d\gamma/dH$ and ΔS with $\Delta C(T_c)/T_c$.

An interpretation of the above observations, consistent with the small values of ξ, is that defects suppress superconductivity in their immediate vicinity, leaving T_c unchanged elsewhere. The Cu^{2+} moments measured by n_2 may act as pair-breaking centers or they may constitute a measure of the concentration of another defect that has the primary role. Other evidence that supports this interpretation (but does not distinguish the two possibilities for the role of the Cu^{2+} moments) is the existence (7) of a contribution to $\gamma(0)$ that is proportional to n_2.

Another interesting correlation is that of the T=0 Debye temperature $[\theta(0)]$, derived from B_3, with f_s. It is displayed in Fig. 8 and suggests that $\theta(0)=545K$ for fully superconducting material and 345K for non-superconducting material. This correlation may be evidence in support of the conjecture that there is a change in the lattice accompanying the transition to the superconducting state, or it may simply be a direct effect of the defects on the vibration spectrum.

COUPLING STRENGTH AND ELECTRON DENSITY OF STATES

The value of f_s can be used in the analysis of specific-heat data near T_c to obtain information about the strength of the coupling. An example is illustrated in Fig. 9 with data for a YBCO sample prepared by the citrate pyrolysis method. The analysis is based on the α model (11), a model that takes into account the possibility of strong coupling effects. The thermodynamics of the transition corresponds to a gap with the BCS temperature dependence but scaled by a constant factor, represented by $\alpha \equiv \Delta_0/kT_c$. Application of the α model gives the value of α, determined by the shape of the anomaly, and also $f_s\gamma$, determined by the amplitude.

The dotted and solid curves in Fig. 9 represent, respectively, approximations to C_n and C_s. $C_n = C_\ell + \gamma T$, where C_ℓ is assumed to consist of dilatation and harmonic terms (6). C_n was obtained by fitting the data above 96K and between 62-65K, the region expected to include the temperature (T_x) at which $C_n = C_s$. The solid curve is

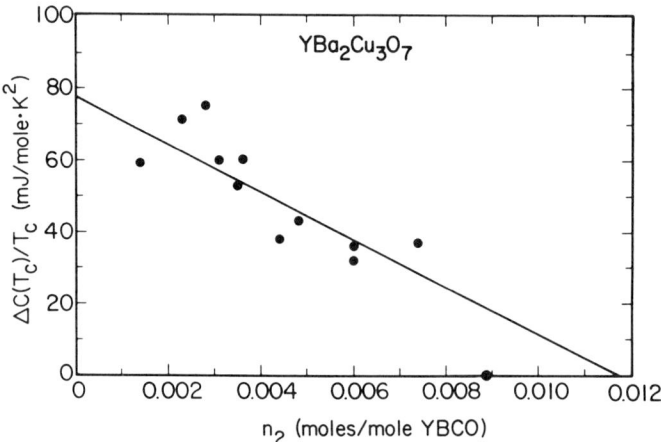

Figure 7. Correlation of $\Delta C(T_c)/T_c$ with n_2.

given by $C_s = C_\ell + f_s C_{es} + (1-f_s)\gamma T$, where C_{es} is calculated by assuming a Gaussian distribution of transition temperatures centered on $<T_c>$ with width ΔT_c. The fit parameters for C_s are given in Fig. 9. (The inclusion of a small fluctuation contribution would slightly improve the fit near 93K, but is not essential.) From the fit: $\gamma = 15 \pm 3$ mJ/mole.K^2 (for this sample $f_s = 0.83$); the gap ratio $2\Delta_0/kT_c = 6.8 \pm 0.6$, 1.9 times the weak-coupling BCS value; $\Delta C(T_c)/\gamma T_c = 5.3$, approximately three times the weak-coupling BCS value; $T_x/T_c = 0.69$ compared with 0.51 for the weak-coupling limit. These comparisons clearly point to the importance of strong-coupling effects.

The value of γ obtained by analysis of the specific-heat anomaly using the α model and the value of f_s is in reasonable agreement with values obtained by several other methods (4): 16 mJ/mole.K^2, from extrapolation of the n_2-proportional term in $\gamma(0)$ to the value of n_2 at which $f_s = 0$; 18 mJ/mole.K^2, from extrapolation of the H-proportional term in $\gamma(H)$ to $H_{c2}(0)$; and 20 mJ/mole.K^2 from an analysis of high-temperature specific-heat data. Comparison of these values with band-structure calculations which give the bare density of states $\gamma_{bs} = \gamma/(1+\lambda)$ provides an estimate of λ. The result, based on $\gamma = 17$ mJ/mole.K^2, and $\gamma_{bs} = 16$ mJ/mole.K^2 (12) to 13 mJ/mole.K^2 (13), is $\lambda \sim 0.1$-0.3. These relatively small values of λ are inconsistent with the strong-coupling effects deduced above if the coupling is via the phonons and the gap has the BCS temperature dependence.

ACKNOWLEDGEMENTS

This work was supported by the Director, Office of Energy Research, Office of Basic Energy Sciences, Materials Sciences Division of the U.S. Department of Energy under Contract No. DE-AC03-76SF00098.

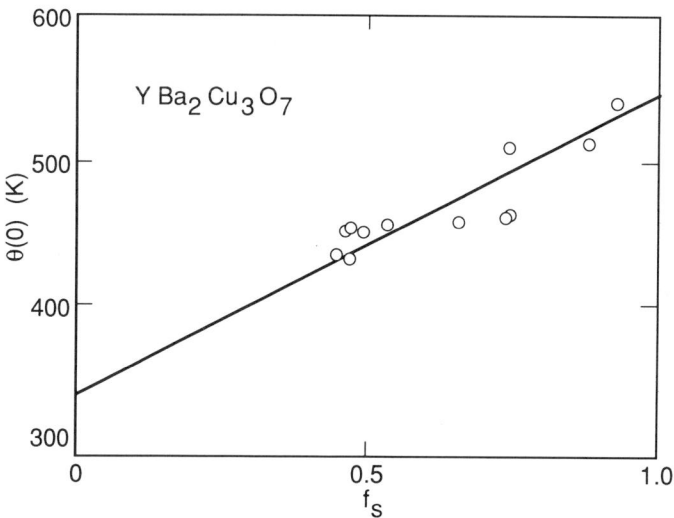

Figure 8. Low-temperature limiting Debye temperature as a function of f_s.

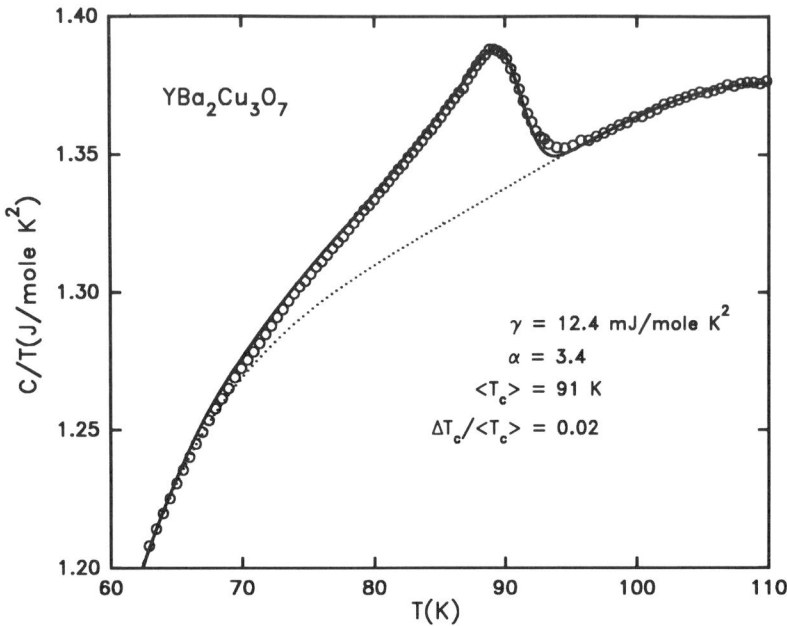

Figure 9. Fit of C to the α model. See text for discussion.

REFERENCES

1. N. E. Phillips et al., in: Novel Superconductivity, eds. S. A. Wolf and V. Z. Kresin (New York, NY: Plenum Press, 1987), 739.
2. The nonzero value of the zero-field "linear term", $\gamma(0)T$, had been observed in measurements at a number of laboratories early in 1987, and reported informally at several meetings (e.g., the 1987 APS March meeting). The first published reports were: M. E. Reeves et al., Phys. Rev. B35, 7207 (1987); L. E. Wenger et al., Phys. Rev. B35, 7213 (1987); B. D. Dunlap et al., Phys. Rev. B35, 7210 (1987).
3. P. W. Anderson et al., Phys. Rev. Lett. 58, 2790 (1987).
4. N. E. Phillips et al., in: Progress in Low-Temperature Physics 13, ed. D. E. Brewer (Amsterdam, The Netherlands: North-Holland, 1991), to be published.
5. K. Maki, Phys. Rev. A3, 702 (1965).
6. J. E. Gordon et al., Solid State Commun. 69, 625 (1989).
7. N. E. Phillips et al., Phys. Rev. Lett. 65, 357 (1990).
8. J. E. Gordon et al., Physica C 162-164, 484 (1989).
9. J. E. Gordon et al., Bull. Mater. Sci., to be published 1991.
10. A. Junod et al., Physica C 162-164, 1401 (1989).
11. H. Padamsee et al., J. Low Temp. Phys. 12, 387 (1973).
12. A. Massidda et al., Phys. Lett. A122, 198 (1987).
13. H. Krakauer et al., J. Superconductivity 1, 111 (1988).

SPECIFIC HEAT AND THERMAL EXPANSION OF THE Bi AND Tl HIGH TEMPERATURE SUPERCONDUCTORS NEAR T_c

D. Wohlleben, W. Schnelle, E. Braun, and H. Broicher
II Physikalisches Institut
Universität zu Köln
Zülpicher Straße 77
5000 Köln 41, Germany

Abstract

We present measurements of the specific heat and of the thermal expansion of the most prominent phases of the Bi and Tl high temperature superconductors (Bi–2212, Bi–2223, Tl–2212, Tl–2223) in the neighbourhood of T_c. In all these systems we observe small but sharp anomalies which have very little similarity with the mean field jump found in the conventional superconductors. The analysis of these anomalies shows clear evidence for the presence of strong fluctuations of the order parameter. For temperatures more than ≈ 5 K away from T_c, 2D Gaussian fluctuations are found, while within ± 5 K of T_c the fluctuation contribution is best fitted by critical fluctuations. The shape of the thermal expansion anomalies is similar to that of the specific heat. Combination of both measurements predicts $dT_c/dp \approx +0.2$ K/kbar for Bi- and Tl–2223 systems.

1. Introduction

The empirical temperature dependence of the electronic specific heat has been of central importance in stimulating and testing theories of superconductivity in the past, and this seems to repeat itself now with the high temperature superconductors (HTSC). In the $REBa_2Cu_3O_{7-\delta}$ system (RE–123) the specific heat anomaly resembles the mean field anomaly of the traditional superconductors at first glance. However, with a resolution of the calorimetry of 0.1 % or better, it was possible to show clear deviations from the step-like mean field behaviour. These deviations are apparently caused by superconducting fluctuations [1]–[3].

For the Bi and Tl HTSC the anomalies are much less prominent than in the RE–123, and they differ qualitatively from the classical mean field behaviour already at first glance: The measurements reveal the form of a broad lambda type anomaly [4,5,2,6,7], as in superfluid He^4.

In this contribution, we present high resolution measurements of the specific heat and of the thermal expansion of a series of well characterized, high quality polycrystalline samples of the four most prominent phases of the Bi and Tl HTSC in the neighbourhood of T_c. They show that the detailed shape of the anomaly depends on the phase, but that the absence of a clear mean field jump as well as the presence of strong superconducting fluctuations in a wide temperature range around T_c are universal features of all these substances. Further away than ≈ 5 K from T_c, the anomalies can be fitted with Gaussian fluctuations, while within 5 K around T_c, critical fluctuations are observed for all Bi and Tl phases. In the thermal expansion data, anomalies can be detected as well. Within the poorer resolution of that measurement, they image those in the specific heat.

Table 1. Overview of characteristic data of the samples ($T_{c,0}$ = temperature of zero resistance, $-4\pi M$ = Meissner fraction measured in DC magnetic field H; a,b,c = lattice constants from powder X-ray diffraction).

substance	sample	$T_{c,0}$ [K]	$-4\pi M$ (H)	a/b [Å]	c [Å]
BiPb–2212	Ha24e	78.5	26 % (25 G)	—	—
	Ha83d	81.6	—	—	—
	Ha83d1	93.3	—	5.382/5.404	30.86
	Ha65c	80.1	83 % (5 G)	5.388/5.402	30.82
	Ha65c1	94.0	65 % (2 G)	5.383/5.405	30.84
Bi–2212	SL S 48	88.1	30 % (5 G) [1]	5.403/5.409	30.84
Bi(+Pb)–2223	Ha14b	107.4	83 % (5 G) [1]	5.402	37.08
Bi–MP	Ha18a	108.5	—	—	—
BiPb–2223	Se58f	105.7	65 % (2 G) [2]	5.405	37.04
Tl–2212	128	107 [3]	15 % (1 G) [4]	3.855	29.31
Tl–2223	12c-2	117.9	27 % (5 G) [1]		

[1] measurement shown in [2] [3] from DC Meissner measurement [10]
[2] measurement shown in [4] [4] relative to shielding effect [10]

2. Sample Preparation and Characterization

Table 1 gives an overview of characteristic data of all samples presented here.

The preparation of all Bi HTSC (except SL S48) is described in [7,8]. None of the BiPb–2212 samples shows any contribution from the Bi–2223 phase (X-ray diffraction, χ_{AC}, Meissner, $R(T)$, $C(T)$). An additional heat treatment in Ar atmosphere transformed sample Ha65c into Ha65c1 and Ha83d into Ha83d1 [7]. It resulted in a small loss of oxygen, in an increase of the orthorhombic splitting and in an increase of the T_c from 80 K to 94 K [8].

The preparation and characterization of the BiPb–2223 sample was described in [4]. According to X-ray diffraction, it contained 6–7 % of Bi–2212. About 20 % of this phase was found in the Bi(+Pb)–2223 sample (Ha14b). Both samples do not show the existence of this phase in any other measured quantity. Both in X-ray diffraction and in specific heat Bi–MP (Ha 18a) showed nearly equal amounts of Bi–2212 and Bi–2223.

The Bi–2212 sample without Pb ($Bi_2Sr_2CaCu_2O_{8+x}$, sample SL S 48) was prepared at Hürth–Knapsack by melt processing [9]. A small contribution of the Bi–

2223 was found by DC Meissner effect (≈ 1 %), by X-ray diffraction (≤ 5 %) and in the electrical resistance. Informations concerning the $Tl_2Ba_2CaCu_2O_{8+\delta}$ sample (Tl–2212, 128) can be found in [10]. Preparation and other measurements of the $Tl_2Ba_2Ca_2Cu_3O_{10+\delta}$ sample (Tl–2223, 12c-2) are reported in [11].

3. Experimental Technique and Results

3.1. Specific Heat

With our calorimeter we can measure the specific heat by a continuous heating method [12] with a sensitivity of $\approx 5 \cdot 10^{-4}$. We emphasize that in all plots shown here the data are *unsmoothed*, and that all observed anomalies must be ascribed to the samples and not to some spurious systematic errors [7].

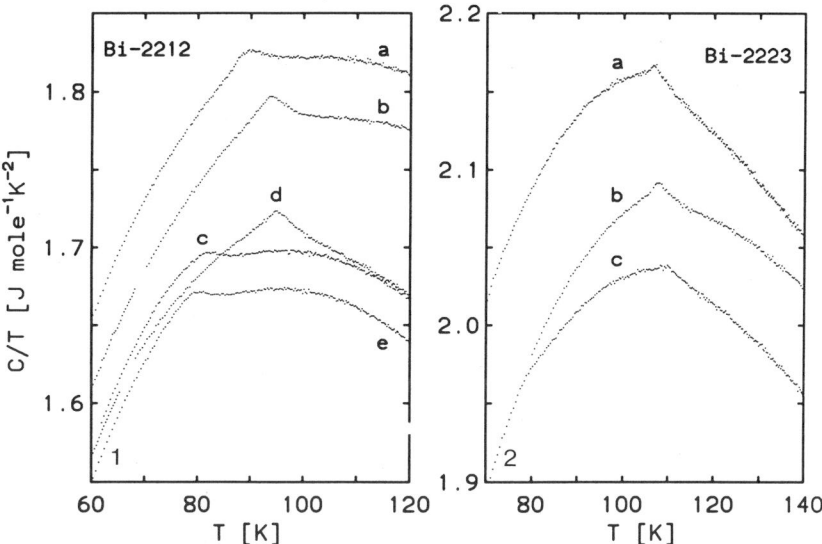

Figure 1. Specific heat C/T vs. T of a pure Bi–2212 (a) and of some lead containing BiPb–2212 samples: b Ha83d1, c Ha65c, d Ha65c1 and e Ha24e. Ha65c and Ha65c1 is the same pellet before and after the heat treatment (see text and [7]).
Figure 2. C/T vs T for a Bi(+Pb)–2223 (Ha14b), b BiPb–2223 (Se58f) and c Bi–MP (Ha18a).

Figures 1–3 show the specific heat of all samples in a representation of C/T vs. T close to T_c. All samples exhibit small but well resolved anomalies around their T_c's. None of the anomalies is reminiscent of the mean field jump of the conventional superconductors. In each case a maximum is located at $T_{c,0}$, the zero resistance transition temperature. This maximum is rounded off within ± 1.5 K. The temperature region where the anomaly grows visibly out of the phononic background reaches not only to 10 K below T_c, but also to at least 10 K *above*. Moreover, note that for the majority of the samples there is a clear *upturn* in C/T vs. T when approaching the critical temperature from below. This occurs in a temperature region where the overall curvature of the phononic background and of the standard mean field contribution is *downwards* [4,3]. Bi–MP (Ha18a) shows two kinks, at $\approx$ 80 K and at $\approx$ 109 K (curve c in fig. 2), consistent with nearly equal amounts of Bi–2212 and Bi–2223 (see section 2).

For the two **BiPb–2212** samples Ha24e and Ha65c (curves c and e in fig. 1) with relatively low T_c and with identical preparation histories, the maximum of the anomalies is much broader than for the other 2212 samples. For these samples the "upturn" below T_c is not visible, while it is well resolved in all the other samples.

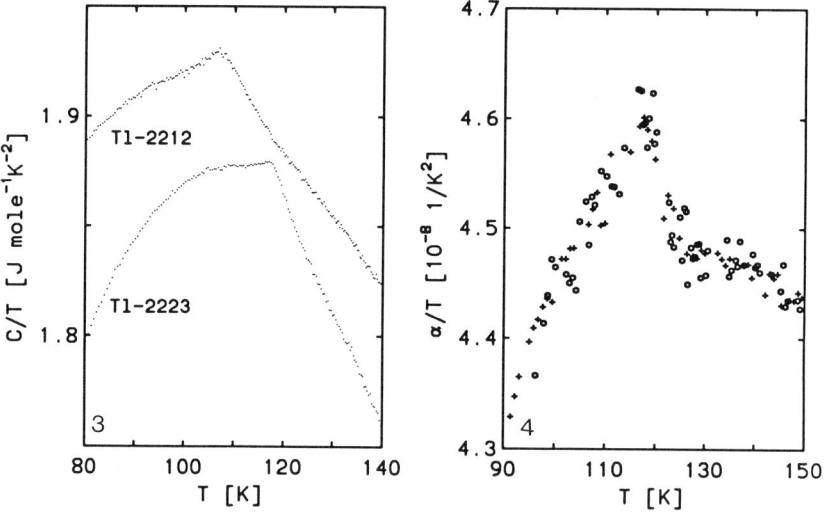

Figure 3. Specific heat C/T as function of T for the samples of Tl–2212 and Tl–2223.
Figure 4. Linear thermal expansion α/T vs. T of the Tl–2223 sample (12c-2).

3.2. Thermal Expansion

The thermal expansion measurements were performed by an improved version of a capacitance dilatometer [13]. The resolution at 100 K is $\Delta\alpha \sim 10^{-8}$ K^{-1}. For polycrystalline samples with their strong anisotropy, the absolute value of α is reduced by intergranular effects to some extent. In the case of Bi and Tl pellets, our α from thermal expansion is 30 % smaller than that from X–ray diffraction. We ascribe this mainly to texture in the pellets, with the a–b direction lying preferentially in the measuring direction. In fig. 4 we present the data for the Tl–2223 sample in the α/T vs. T representation around T_c.

4. Discussion

4.1. Intrinsic Shape of the Anomalies

The anomalies of the specific heat in figures 1–3 and of the thermal expansion in fig. 4 show very little similarity with the standard BCS form (jump of C at T_c). As discussed in previous publications [14,4,3,2], this deformation away from BCS is an intrinsic effect, i. e. it is not due to inhomogeneities within the pellet. The Meissner effect sets in at exactly the temperature of the peak of C/T vs. T, is a linear function of T just below T_c and saturates at values near 100 % at sufficiently low fields ($H < H_m \approx 0.1$–1 Gauß) [15]. The peculiar behaviour of the Meissner effect at intermediate fields ($H_m < H < H_{c,1}$) was explained in [14] and in more detail in [16]. The Meissner flux expulsion at the lowest measuring field is given in table 1 for most samples of this paper.

To a certain extent, the shapes of the anomalies in figures 1–4 speak against inhomogeneous broadening in themselves: The anomalies are quite sharp (with the possible exception of curves c and e in figure 1) and deviate from standard mean field behaviour in two respects: Firstly, there is the upturn below T_c. Secondly, there is entropy belonging to the anomaly *above* T_c (T_c as defined by the onset of the Meissner effect is exactly at the peak in $C(T)$). Further down, we shall explain all these features in terms of intrinsic fluctuations.

There is no universal form to the new shape of the superconducting specific heat anomalies of the Bi and Tl HTSC. Two clearly distinct forms are those of curves c and e in figure 1 on one hand and curves d in fig. 1 and a and b in fig. 2 on the other. Note that both curves can appear in the same pellet (fig. 1, curves c and d; here the T_c of the Bi–2212 pellet was shifted from 80 K to 94 K by a final heat treatment as described in section 2 and [8]). Curves c and e in figure 1 are both found on Bi–2212 with relatively low T_c, whereas curve d in fig. 1 goes with the highest T_c obtainable in this system. Curves a and b in figure 1 belonging to Bi–2212 samples with intermediate T_c's also show intermediate shapes of the anomalies. The anomalies of Bi–2223 and of the Tl HTSC are of the type of curve d in fig. 1.

4.2. Analysis of the Specific Heat Anomaly

In previous publications [4,2,3] we have presented specific heat measurements of various HTSC samples and have analysed them by a superposition of a smooth phonon background, a mean field contribution and a fluctuation contribution. For Y and Dy–123 we found [3] that the fluctuations near T_c can be described by 3D Gaussian fluctuations, i. e. by $\Delta C_{FL} = C_G^{\pm}(|T - T_C|/T_C)^{-\beta}$; $\beta = 1/2$, with certain amplitudes below (−) and above (+) T_c and with a mean field contribution $\Delta C_{MF} = 1.43\gamma T_C$. Figure 5(left) presents a fit, where the ratio C_G^+/C_G^- was fixed to the value corresponding to a two component order parameter. The parameters of the fit are given in the figure. A fit with variable C_G^+/C_G^- is shown in [3]. From these Aslamazov–Larkin type fits we find a coherence volume of $\approx 200\text{Å}^3$. From fits where the anisotropy of Y–123 was taken into account (Lawrence–Doniach type [17], see fig. 5(right)), we find $\xi_{\parallel} \approx 12\text{Å}$ and predict a changeover from 3D to 2D Gaussian fluctuations near $1.25\, T_c$.

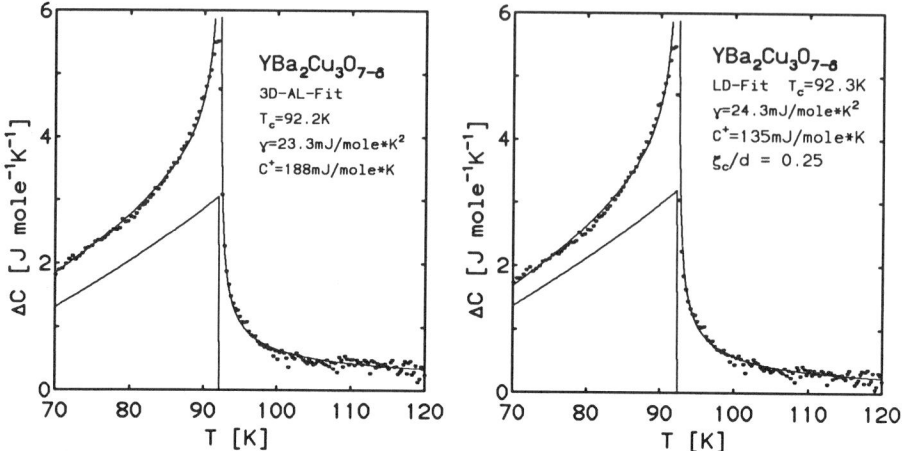

Figure 5. Analysis of the specific heat anomaly of $YBa_2Cu_3O_{7-\delta}$, left: 3D–AL fit (isotropic model), right: LD fit (anisotropic model) (see text).

For Bi and Tl HTSC neither 3D nor 2D Gaussian fluctuations fit, unless one excludes a relatively wide window ΔT_{ex} around T_c from the fit. This window is much wider than the distribution of T_c's from compositional inhomogeneities within the sample. Even when this window is excluded, fits with 3D Gaussian fluctuations do not converge at all. On the other hand fits with a 2D Gaussian contribution ($\beta = 1$) with $C_G^+/C_G^- = 1$ converge well. From this result it can be concluded that the superconductivity in Bi and Tl HTSC is twodimensional for $|T - T_c| > \Delta T_{ex}$, i. e. that $\xi_z(T)$ does not span the distance between adjacent Cu–O planes there. When C_G^+ and C_G^- were varied as individual parameters, the ratio C_G^+/C_G^- showed a wide nonsystematic scatter. For this reason we have restricted ourselves to $C_G^+ = C_G^- = C_G^\pm$, i. e. we assume $N = 2$ for the number of components of the superconducting order parameter. This method of analysis is described in more detail in [7].

Fig. 6 shows the specific heat anomaly of the Tl–2223 sample after subtraction of the phonon background. The anomaly has a clear λ–shape, as observed for He^4. The 2D Gaussian fit and the mean field contribution are shown as drawn out lines. The data begin to deviate from the Gaussian fit at $(T - T_c) \approx \pm 6$ K ($|t| \approx 0.05$). At this temperature the fluctuation contribution has reached the same amplitude as the mean field jump. This coincidence is the definition of the (2D) Ginzburg temperature t_G [18,19], where the changeover from the Gaussian region into the critical region occurs. Therefore it is quite natural that the Gaussian fit greatly exceeds the data near T_c.

The reduction of the fluctuation contribution with respect to the Gaussian approximation in the critical region is due to the growing interaction between the fluctuations. In this region we expect a logarithmic dependence of the fluctuation contribution on $|t|$. That this is indeed the case can be seen from fig. 7, where we find linear sections for $0.015 < |t| < 0.1$ in a plot of ΔC_{FL} vs. $\log |t|$, i. e. within $\pm$ 1.5–10 K around T_c, the fluctuations can be described with a logarithmic divergence [18]. The divergence is rounded off within typically $\pm$ 1.5 K around T_c. This interval of rounding is also observed in the Meissner measurements and is presumably due to sample inhomogeneities. For samples with smaller fluctuation contributions (Bi–2212 with

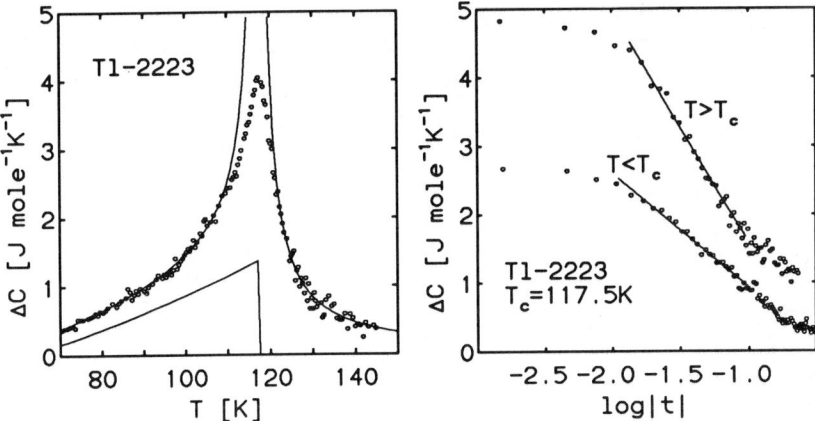

Figure 6. Specific heat anomaly ΔC of the Tl–2223 sample as obtained from the fit after subtraction of a background polynomial. The curved line represents the fit with 2D Gaussian fluctuations, the triangular ramp is the mean field contribution from the same fit (see text).

Figure 7. Fluctation specific heat ΔC_{FL} of the Tl–2223 data for $0.01 < |t| < 0.1$. sample plotted vs. the decadic logarithm of the reduced temperature $|t|$. The data for $T > T_c$ are shifted upwards by one unit. The drawn out lines are linear fits to the

lower T_c) we do not find a clear changeover to logarithmic fluctuations. For these samples the critical region ($|t_G|$) is smaller than the "smearing" by inhomogeneities.

In the temperature interval of $0.05 < |t| < 0.10$ neither kind of fluctuations can describe the data satisfactorily alone. This is the changeover region from the Gaussian approximation (valid only for $|t| \ll t_G$) to the critical region (logarithmic fluctuations, $|t| \gg t_G$).

In table 2 we list the linear specific heat coefficients γ_{BCS}, the amplitudes of the fluctuations $C_G^\pm$ in the Gaussian regime and the amplitudes C_C^+ and C_C^- of the fluctuations in the critical regime for all cases shown in figures 1–3. The ΔT_{ex} was chosen for each sample individually and may represent a crude measure for the Ginzburg temperature t_G (typically $T_{ex} \approx 1.5\, t_G$).

We do *not* claim that these γ_{BCS} values represent the true coefficient γ of the linear electronic specific heat, because the coupling was arbitrarily set to the weak coupling BCS value. The γ_{BCS} represents merely the order of magnitude of the true γ; in principle the true γ can be obtained only by the entropy $S_{el} = \int_0^T C_{el}/T dT = \gamma T$ with $T \gg T_c$.

For the Bi–2212 samples we find rather different values for the parameters, depending on the T_c's of the samples (see table 2). While samples with lower T_c show relatively large γ_{BCS} and $C_{rmG}^\pm$ around 50 mJ/mole·K, the samples with higher T_c show a much smaller γ_{BCS}, a larger fluctuation contribution and a changeover to critical behaviour. For the sample pair Ha65c/c1 the different forms of the anomalies are shown in fig. 8. The drawn out lines again show the mean field contributions. It is know that in the case of Bi and Tl–2212 the T_c is shifted reversibly by changing the hole concentration, e. g. by varying the oxygen content [8,20]. The change of

Table 2. Parameters obtained by an analysis as described in the text. γ_{BCS} is a coefficient of electronic specific heat (triangular construction in figures 6 and 8); $C_G^{\pm}$ is the amplitude of the 2D–Gaussian fluctuations, C_C^+ and C_C^- are the amplitudes of the critical fluctuations. The numbers in brackets show the statistical errors of the last digit.

substance	sample	ΔT_{ex} [±K]	γ_{BCS} [mJ mole^{-1}K^{-2}]	$C_G^{\pm}$ [mJ mole^{-1}K^{-1}]	C_C^- [J mole^{-1}K^{-1}]	C_C^+
BiPb–2212	Ha24e	6	8.0(6)	46(8)	-	-
	Ha83d1	8	6.6(7)	88(12)	2.2	2.6
	Ha65c	7	10.9(7)	57(5)	-	-
	Ha65c1	7	2.9(6)	70(10)	1.6	2.1
Bi–2212	SLS48[1])	7	2.8(4)	80(6)	1.3	1.9
Bi(+Pb)–2223	Ha14b[1])	6	6.4(4)	46(5)	0.8	0.9
BiPb–2223	Se58f[2])	8	4.6(5)	44(9)	0.8	1.7
Tl–2212	128[1])	9	3.7(7)	103(15)	1.8	2.7
Tl–2223	12c-2[1])	8	8.2(7)	90(7)	1.8	3.2

Earlier results are shown in [1]) ref. [2] and [2]) ref. [4].

doping by itself can not explain the enhancement of the fluctuations in Bi–2212 for oxygen deficient samples. However, in a recent paper [21] it was found by angle resolved photoemission that for Bi–2212 with $T_c = 80$ K the Cu–O *and* the Bi–O layers were metallic and superconducting below T_c, while for a sample with $T_c = 89$ K only the Cu–O layers showed superconductivity. The latter could be responsible for the enhancement of the fluctuations with increasing T_c: the structure with insulating Bi–O layers is obviously nearer to 2D than the structure with superconducting Bi–O layers, i. e. the fluctuations in the sample with $T_c = 94$ K are enhanced (Fig. 8). Also the resistivity of the oxygen deficient sample of figure 8 with $T_c = 94$ K departs from linearity at much higher temperatures and shows more rounding at T_c than that of the as-prepared sample with $T_c = 80$ K [8], indicating a stronger contribution of fluctuations to the conductivity [8].

The anomalies of the Tl HTSC are significantly larger than the anomalies of the isomorphic Bi HTSC. This is true for both the mean field contribution and the fluctuations.

The experimental value for the ratio C_C^+/C_C^- (see table 2) scatters between 1 and 2, while it is expected to be one in the 3D critical region [18]. Since only 2D Gaussian fluctuations are detected in our samples, it is possible that the superconductivity in the Bi and Tl HTSC is 2D even in the critical regime. This is very likely for the 2212 phases with their extremely large anisotropy [22,19]. Otherwise the crossover from 2D to 3D superconductivity may occur in addition in the Gaussian to critical changeover region and may therefore complicate the fluctuation behaviour further.

4.3. Magnetic Field Dependence of the Fluctuation Contribution

It is interesting to note that the λ-shaped ΔC_{FL} of the anomaly of the Tl–2223 sample in fig. 6 is congruent with the difference in specific heats in zero field and in magnetic

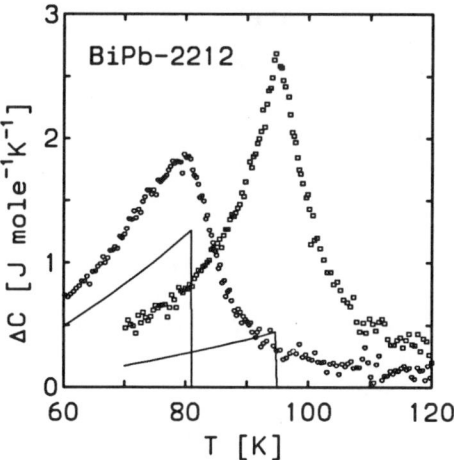

Figure 8. Anomalies of the specific heat ΔC of the **BiPb–2212** sample before ($T_c = 80$ K, Ha65c) and after ($T_c = 94$ K, Ha65c1) the final heat treatment in Argon. In each case, the triangular construction represents the MF term C_{MF} from the fit (see text).

fields of several Tesla, as calculated from reversible magnetization curves by Fang et al. [23] and Gohng et al. [24]. These authors find that $\Delta C(T,H) = C(T,0) - C(T,H)$ has the shape of a nearly triangular peak, with the tip of the triangle at T_c, and without measurable contributions beyond certain temperatures on both sides of T_c, which depend on the field. The width of the magnetically determined peaks is 110 to 123 K for a field of 2 T and 108 to 125 K for 5 T. This width coincides with the temperature region ΔT_{ex}, where critical fluctuations are observed in our specific heat analysis. For our Tl–2212 sample, there is also qualitative agreement with the $\Delta C(T,H)$ of [23]. All this suggests, that only the part of the specific heat anomaly with logarithmic temperature dependence of ΔC_{FL} is measurably affected by fields of order 10 Tesla.

A suppression of the specific heat around T_c in the above described fashion was actually found in direct specific heat measurements of Y–123 [25] and of the multiphase Bi and Tl HTSC [26] in magnetic fields: the anomaly flattens out, while the onset temperature of the anomaly remains nearly the same. Additionally strong anisotropic effects were found even for Y–123 [25].

In the same temperature region, where one observes critical fluctuations, there is also an anomalous contribution to the Hall voltage $U_H(T)$ [4]; in this temperature interval, the inital slope $dU_H/dH|_{H=0}$ drops below the positive slope $dU_H/dH \sim 1/T$ measured at 5 T up to 300 K. This negative contribution leads actually to a reversal of the sign of the low field Hall voltage above T_c. The negative contribution peaks at T_c and decreases again below [4]; altogether this anomalous low field contribution to the Hall effect is again roughly congruent with the specific heat contribution from critical fluctuations and is suppressed by fields of order 10 T.

4.4. The Anomalies in the Thermal Expansion

The shape of the thermal expansion data is similar to the shape of the specific heat data near the phase transition (see fig 4 and Ref. [7]). In particular our Bi–2223 data [7] show an upturn in α/T when approaching T_c from below and in all our data, there is clear evidence for contributions to the anomalies above T_c over a temperature interval of 5–10 K (see fig. 4).

By a simple extrapolation method we have extracted the "jumps" of $\Delta\alpha$ as listed in table 3. Also listed in this table are the values of the pressure dependence of T_c as determined by the Ehrenfest equation $dT_c/dp|_{p=0} = \Delta V\, T_c/\Delta C$. ΔC was taken as the maximum of our ΔC plots (see e. g. fig. 6), which corresponds to the construction used for determining $\Delta\alpha$. These values of dT_c/dp should be taken as rough approximations, since the influence of the fluctuations in ΔC and $\Delta\alpha$ on dT_c/dp is not known. Furthermore the average $\Delta V = 3\,\Delta\alpha V$ taken from textured polycrystals of anisotropic materials give uncertain values for ΔV (For Y–123 see e. g. [27]). In the case of multiphase Tl–2223 a direct measurement gave 0.12 K/kbar [28]. The $dT_c/dp|_{p=0}$ values for both 2223 phases are larger than expected from the systematic variation of $d(\ln T_c)/dp$ vs. T_c (see e. g. [29]), from which we would expect values near zero for HTSC with T_c higher than 100 K. For the Tl–2212 sample (128) a direct measurement gave $dT_c/dp|_{p=0} = 0.08$ K/kbar [30].

Table 3. Extrapolated jump in thermal expansion $\Delta\alpha$ and pressure dependence of T_c for p=0 determined by the Ehrenfest relation.

substance	sample	$\Delta\alpha$ 10^{-7} K^{-1}	dT_c/dp K/kbar
Bi(+Pb)–2223	Ha14b	1.3	0.27
BiPb–2223	Se58f	0.8	0.18
Tl–2223	12c-2	1.2	0.17

5. Conclusions

The anomalies of the specific heat and of the thermal expansion of the Bi and Tl HTSC's with T_c in the 100 K range deviate strongly from the standard mean field behaviour of traditional superconductors. There is no universal shape, but a gradual transition from a more mean field like shape at the lower T_c's to a λ type shape at the higher T_c's. The fluctuation contribution ΔC_{FL} is best fitted by 2D Gaussian fluctuations further than 6–9 K away from T_c, while within ≈ 5 K around T_c critical fluctuations fit well. The critical contribution appears also as an anomaly in the initial slope of the Hall voltage $U_H(B)$ [4] and is strongly affected by magnetic fields [23,24]. The enhancement of the fluctuations with growing T_c in Bi–2212 is possibly due to a gradual transition of the Bi–O planes from metallic (oxygen rich samples) to insulating behaviour [21] (deoxygenated samples). In contrast to the anomalies of the Tl and Bi HTSC, the Y–123 system shows 3D Gaussian fluctuations near T_c and a changeover to 2D behaviour around 1.25 T_c.

Acknowledgements

We thank R. Dömel, S. Ruppel and B. Büchner for Meissner and X-ray measurements. The samples were provided by J. Harnischmacher (Köln), C. Allgeier, W. Reith and J. S. Schilling (München), J. Bock and E. Preisler (Hoechst AG, Hürth–Knapsack) and G. J. Vogt (Los Alamos). The work in Köln was supported by the Deutsche Forschungsgemeinschaft through SFB 341, by the Bundesministerium für Forschung und Technologie (BMFT contract # 13N5494) and by the Ministerium für Wissenschaft und Forschung of the Land Nordrhein-Westfalen.

References

[1] S. E. Inderhees, M. B. Salamon, N. Goldenfeld, J. P. Rice, B. G. Pazol, D. M. Ginsberg, J. Z. Liu, G. W. Crabtree, Phys. Rev. Lett. **60**, 1178 (1988)

[2] D. Wohlleben, E. Braun, W. Schnelle, J. Harnischmacher, S. Ruppel, R. Dömel in: "Proceedings of the International Conference on Superconductivity — ICSC", Jan. 10.–14., 1990, Bangalore, India, Eds. S. K. Joshi, C. N. R. Rao & S. V. Subramanyam, World Scientific 1990, p. 194

[3] W. Schnelle, E. Braun, H. Broicher, R. Dömel, S. Ruppel, W. Braunisch, J. Harnischmacher, D. Wohlleben, Physica **C168**, 465 (1990)

[4] W. Schnelle, E. Braun, H. Broicher, H. Weiss, H. Geus, S. Ruppel, M. Galffy, W. Braunisch, A. Waldorf, F. Seidler, D. Wohlleben, Physica **C161**, 123 (1989)

[5] E. Braun, W. Schnelle, F. Seidler, P. Böhm, W. Braunisch, Z. Drzazga, S. Ruppel, H. Broicher, H. Geus, M. Galffy, B. Roden, I. Felner, D. Wohlleben, Physica **C162–164**, 496 (1989)

[6] N. Okazaki, T. Hasegawa, K. Kishio, K. Kitazawa, A. Kishi, Y. Ikeda, M. Takano, K. Oda, H. Kitaguchi, J. Takada, Y. Miura, Phys. Rev. **B41**, 4296 (1990), T. Atake, H. Kawaji, M. Itoh, T. Nakamura, Y. Saito, Physica **C162–164**, 488 (1989)

[7] E. Braun, W. Schnelle, H. Broicher, J. Harnischmacher, D. Wohlleben, C. Allgeier, W. Reith, J. S. Schilling, J. Bock, E. Preisler, G. J. Vogt, submitted to Z. Phys. B, (May 1991)

[8] N. Knauf, J. Harnischmacher, R. Müller, R. Borowski, B. Roden, D. Wohlleben, Physica **C173**, 414 (1991)

[9] J. Bock, E. Preisler, Sol. State Commun. **72**, 453 (1989)

[10] W. Reith, C. Allgeier, K. Andres, J. Heise, R. Hoben, A.-K. Klehe, R. Kleiner, C. Kowal, A. Moise, P. Müller, J. S. Schilling, Physica **C162–164**, 109 (1989)

[11] G. J. Vogt, D. S. Phillips, in "High Temperature Superconducting Compounds II" edited by S. H. Whang, A. DasGupta, R. Laibowitz; The Minerals, Metals & Materials Society, 1990, p. 105

[12] A. Junod, J. Phys. E — Sci. Instrum. **12**, 945 (1979)

[13] R. Pott, R. Schefzyk, J. Phys. E — Sci. Instrum. **16**, 444 (1983)

[14] F. Seidler, P. Böhm, H. Geus, W. Braunisch, E. Braun, W. Schnelle, Z. Drzazga, N. Wild, B. Roden, H. Schmidt, D. Wohlleben, I. Felner, Y. Wolfus, Physica **C157**, 375 (1989)

[15] S. Ruppel, G. Michels, H. Geus, H. Kalenborn, W. Schlabitz, B. Roden, D. Wohlleben, Physica **C174**, 233 (1991)

[16] D. Wohlleben, G. Michels, S. Ruppel, Physica **C174**, 242 (1991)

[17] K. F. Quader, E. Abrahams, Phys. Rev. **B38**, 11977 (1988)

[18] L. N. Bulaevskiĭ, V. L. Ginzburg, A. A. Sobyanin, Sov. Phys. JEPT **68**, 1499 (1988), translation from Zh. Eksp. Teor. Fiz. **94**, 355 (1988)

[19] A. I. Sokolov, Physica **C174**, 208 (1991)

[20] C. Allgeier, J. S. Schilling, Physica **C168**, 499 (1990)

[21] B. O. Wells, Z. -X. Shen, D. S. Dessau, W. E. Spicer, C. G. Olson, D. B. Mitzi, A. Kapitulnik, R. S. List, A. Arko, Phys. Rev. Lett. **65**, 3056 (1990)

[22] D. E. Farrell, R. G. Beck, M. F. Booth, C. J. Allen, E. D. Bukowski, D. M. Ginzberg, Phys. Rev. **B42**, 6758 (1990)

[23] M. M. Fang, J. E. Ostenson, D. K. Finnemore, D. E. Farrel, N. P. Bansal, Phys. Rev. **B39**, 222 (1989)

[24] Junho Gohng, D. K. Finnemore, Phys. Rev. **B42**, 7946 (1990) (The vertical scale on fig. 4 of this reference should read 6 mJ/cm^3 K instead of 40 mJ/g K; D. K. Finnemore, private communication)

[25] M. B. Salamon, S. E. Inderhees, J. P. Rice, D. M. Ginzberg, Physica **A168**, 283 (1990), E. Bonjour, R. Calemczuk, J. Y. Henry, A. F. Khoder, Physica **B165& 166**, 1343 (1990)

[26] R. A. Fisher, S. Kim, S. E. Lacy, N. E. Phillips, D. E. Morris, A. G. Markelz, J. Y. T. Wei, D. S. Ginley, Phys. Rev. **B38**, 11942 (1988)

[27] C. Meingast, B. Blank, H. Bürkle, B. Obst, T. Wolf, H. Wühl, Phys. Rev. **B41**, 11299 (1990)

[28] I. V. Berman, N. B. Brandt, Yu. P. Kurkin, E. A. Naumova, I. L. Romashkina, V. I. Siderov, A. I. Akimov, V. I. Gapel'skaya, and E. K. Stribuk, JEPT Lett. **49**, 769 (1989)

[29] N. Mori, H. Takahashi and C. Murayama, Supercond. Sci. Technol. **4**, S439 (1991)

[30] C. Allgeier, R. Sieburger, H. Gossner, J. Diederichs, H. Neumaier, W. Reith, P. Müller, J. S. Schilling, Physica **C162–164**, 741 (1989)

CURRENT STATUS OF FERMI LIQUID BASED APPROACHES TO THE CUPRATES

K. Levin, Qimiao Si, Ju H. Kim, and J.P. Lu

Department of Physics and The James Franck Institute, and Science and Technology Center for Superconductivity, The University of Chicago
Chicago, IL 60637

ABSTRACT

A central issue in the field of high temperature superconductivity is the nature of the normal state. Here we review normal state transport, thermodynamic, magnetic and spectroscopic data and discuss their interpretation within a Fermi liquid based framework. It is demonstrated that the cuprate data are not consistent with canonical Fermi liquid behavior. However, if the Fermi liquid is characterized by low energy scales, arising from narrow bandwidths or soft spin fluctuations, then canonical behavior is not expected at the relatively "high" temperatures of the normal state. We discuss two main Fermi liquid based schools, i.e., the "almost localized" and "almost magnetic" descriptions. Both approaches have addressed the data with some success and it is clear that they are, in many respects, closely related. Here it is claimed that the strongest support for a Fermi liquid based approach to the cuprates derives from the comparison with other strongly correlated Fermi liquid systems, which exhibit similar normal state anomalies.

I. INTRODUCTION

The appropriateness of Fermi liquid based schemes for describing the normal state of the cuprates is a subject of considerable controversy. Here we review a variety of Fermi liquid approaches and discuss their applicability to the copper oxides. This paper may be viewed as a kind of status report on this class of theories.

Most Fermi liquid based theories have in common the assumption that there exists a low energy scale ω_c which is of the order of T_c at roughly optimal doping. This energy scale can arise in a number of ways. In the "almost localized " schemes of the present work,[1] of Newns et al,[2] Kotliar and his co-workers[3] as well as of others[4] the energy arises from the assumption that there are narrow bandwidths associated with nearly localized d electrons. This approach is based on analogies with the heavy fermion metals and, in some applications, also builds on the notion that the half filled limit is a Mott insulator, in which the d electrons are *fully* localized Heisenberg spins.

Alternate proposals for the low energy scale correspond to the "almost magnetic" Fermi liquid picture. These have been discussed by Scalapino, Schrieffer and their co-workers[5] as well as by Pines and his collaborators.[6] Here ω_c is the spin fluctuation energy which is assumed to be soft due to proximity with the antiferromagnetic (insulating) phase. Finally, Ruvalds[7] has proposed that another low energy scale may arise away from half filling as a consequence of proximity to the perfect Fermi surface nesting associated with the insulating state. Closely related to this picture are "soft" energies arising from a small separation between the Fermi energy and the Van Hove singularity known to be present in the bandstructure. It is clear that a combination of all of the above phenomena may also be considered to play a role in creating the low energies ω_c.

In all these approaches it is assumed that the $T=0$ phase would be a Fermi liquid if superconductivity did not intervene. However, in the cuprates, the ground state is not particularly relevant from the standpoint of normal state properties. Rather one has to study the *finite temperature* Fermi liquid which contains important correction terms arising from the low characteristic energy scale. It is because of these finite temperature corrections that one can reconcile the various magnetic and transport anomalies observed in the cuprates with a Fermi liquid ground state picture.

The essential physics of these approaches may be summarized by the phase diagram shown in Figure 1. Here temperature is plotted on the vertical axis and hole concentration x (in the CuO_2 planes) is indicated by the horizontal axis. The insulating state is shown at low x by an additional vertical line and the dashed line separates the (unphysically) high temperature regime from the more relevant range of T. The shaded line shows the low energy scale ω_c which is seen to be x dependent. In the nearly localized Fermi liquid theories ω_c becomes "softer" as the Mott localized insulator is approached. This corresponds to an increasingly narrow bandwidth and thus a decrease in ω_c. Similarly in the almost magnetic schemes, the increasing proximity to the antiferromagnetically ordered insulator leads to a decrease in the analogous energy scale. As indicated in the figure, below the shaded line "ground" state Fermi liquid properties will be observed. Above this line, finite T corrections become sufficiently important so that the system will not correspond to a canonical Fermi liquid.

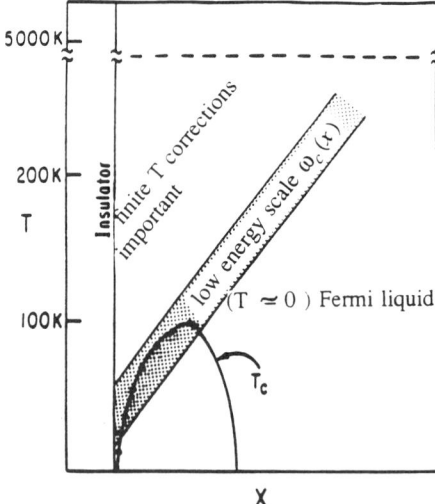

Figure 1   Proposed phase diagram for cuprates in the hole concentration (x), temperature (T) plane. For temperature below the low energy scale ω_c (shaded region) the cuprates would be canonical Fermi liquids, if superconductivity is absent. They show anomalies when temperature is above this energy scale.

Of great importance in this physical picture is the fact that the superconducting phase transition $T_c(x)$ lies very close to $\omega_c(x)$ at and below optimal doping. In this way the system will not exhibit the classical temperature dependences expected of a Fermi liquid for these values of x. On the other hand, in "overdoped" cuprates it is predicted that there will be clear signatures of canonical Fermi liquid behavior. More generally, because the ω_c line represents a *cross-over* rather than a more dramatic "change of state" within the (x,T) plane, these Fermi liquid based theories predict a *smooth* evolution with increasing x and T from the Fermi liquid ground state.

This prediction is an important one which will be addressed experimentally in this paper. One can argue that it is this smooth behavior in the (x,T) plane that distinguishes these theories from alternative schemes such as the gauge field theories. Indeed, it is generally accepted that at sufficiently high x the cuprates are ordinary metals, i.e., Fermi liquids. Presumably, in order for the Luttinger liquid,[8] marginal Fermi liquid,[9] or gauge field theoretic[10] pictures to hold at optimal doping there must be at least one line in the (x,T) plane across which the behavior does not evolve smoothly.

We next discuss the behavior of overdoped cuprates in an effort to ascertain whether these systems appear to evolve smoothly toward classical Fermi liquids. To interpret this data we first review the signatures of canonical Fermi liquid behavior. The T dependence of the spin susceptibility χ, NMR relaxation $1/T_1$, electronic contribution to the resistivity ρ^{el} and Hall coefficient R_H at the lowest temperatures is given by

$\chi \sim constant$

$1/T_1 T \sim constant$

$\rho^{el} \sim T^2$

$R_H \sim constant$

At optimal stoichiometry these properties are *not* observed. In Fermi liquid based approaches this is attributed to the relatively high values of T_c/ω_c. However, in cuprates at higher than optimal x where T_c/ω_c is considerably lower the scenario in Figure 1 suggests that the above T dependences should be observed. It is important to stress that there are well known materials related problems in studying all systems away from optimal stoichiometry. Nevertheless, the trends these data provide seem to be consistent both from one cuprate to another and from one experimental property to another. Because of this consistency and with this caveat we now refer to the data.

The Hall coefficient and resistivity for a Tl 2201 sample from Ref. 11 are shown in Figures 2a and 2b. In this system x is changed by increasing the oxygen concentration above O_6. At optimal x, R_H shown by curve A exhibits the typical temperature dependence seen in other cuprates. In these systems R_H rises rapidly as the temperature is raised above T_c and then falls off after reaching a maximum slightly above T_c. However, as x increases it is clear that the Hall coefficient becomes progressively more T independent and of smaller magnitude. The behavior of curve D essentially coincides with the ideal Fermi liquid behavior. Similarly the resistivity is found[11] to evolve from the ubiquitous linear T dependence at optimal stoichiometry towards a Fermi liquid T^2 behavior. While NMR data are not shown here they, too, exhibit a similar evolution towards the canonical Korringa ($1/T_1 \sim T$) dependence as x increases.[12] In summary, it appears from these and similar data on other cuprates[13] that the **evolution towards a Fermi liquid state at high x is smooth.** This seems to confirm an important prediction of the general Fermi liquid schemes discussed above.

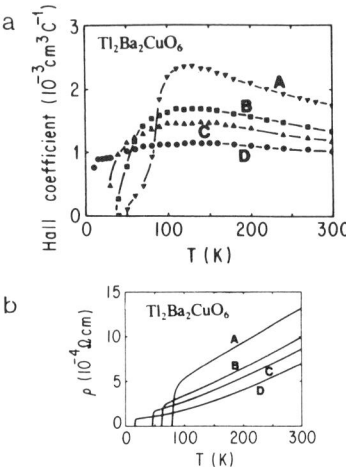

Figure 2 (a) Temperature dependence of Hall coefficient for optimal (A) and higher than optimal hole concentrations (B-D); (b) Temperature dependence of resistivity for various (excess) oxygen concentrations. Data from Ref. 11.

Within these Fermi liquid schemes, how does one distinguish between the various models for low energy scales ω_c? It should be noted at the onset that the different proposals for soft ω_c may not necessarily be incompatible and that because the metallic cuprates seem to be near both magnetic and localization instabilities it is likely that these two effects are contributing simultaneously. Furthermore we will show below that, at least in the 214 system, the Fermi energy is in close proximity to the Van Hove singularity, so that this additional low energy scale may also be playing a role.

It should be stressed that dis-entangling the role of the various soft frequencies is very difficult and that one cannot separate these effects using only one class of measurements. In particular, magnetic measurements (which have provided the motivation for the "almost magnetic" Fermi liquid school) are in general *not* sufficient to separate incipient localization from incipient magnetic order. To emphasize this point one should recall the simple Fermi liquid expression of the $q=0, \omega=0$ susceptibility χ.

$$\chi/\chi^o = (m^*/m)/(1+F_o^a)$$

Thus even this simplest dc measurement involves a convolution of localization (via m^*/m) and spin interaction (via F_o^a) effects. The same result applies to the dynamical susceptibility $\chi(q,\omega)$. These effects can only be separated using a variety of different measurements, just as in the Landau case.

Our own work[1] has led us to conclude that incipient localization effects are playing a role in the cuprates, so that narrow bandwidths are at least partially responsible for low ω_c. Support for this "almost localized" picture derives from the following

(1) The cuprates appear from neutron data[14] to have localized (i.e., Heisenberg) spins in the insulating state. It may be expected that some degree of quasi-localization of the d electrons persists into the metallic phase.

(2) Comparison with the heavy fermion metals (which exhibit a high degree of f-electron localization) shows many similarities with the cuprates.[15] We will present some of these comparisons below.

(3) μsr, x-ray, NMR and neutron spectroscopies[14] have been interpreted as suggesting that even in the metallic state the d electron valence and spin are close to the integer value of the Cu^{+2} state.

Despite this evidence for quasi-localization, we have found that[16] some residual magnetic interaction effects must be invoked to understand NMR and neutron data.[14] It should be emphasized that the softness of the spin fluctuation frequencies needed to explain the data is highly dependent on the degree to which the bands are narrowed by incipient localization. We found elsewhere[16] that the inclusion of band narrowing effects led to somewhat more "robust" parametrizations of these data than in the almost magnetic school. In this way the system need not be on the verge of a magnetic instability in order to explain the low energy $\omega_c \sim 125K$ seen, for example in NMR experiments in fully oxygenated $YBCO$. (See, for example, Figure 5a below).

What is the low energy scale in the almost localized Fermi liquid? As in the heavy fermions, we refer to this energy as the "coherence" temperature T_{coh}. Above this temperature the d electrons of the cuprates begin to dissociate from the Fermi liquid, i.e., to become increasingly more incoherent. We have defined T_{coh} operationally as the temperature above which band edge effects begin to be felt, so that it corresponds to roughly 1/4 the separation between the Fermi energy and the nearest band edge. In our microscopic calculations (to be discussed below) we find that $T_{coh} \sim 150-300K$ depending on the cuprate and its stoichiometry. Thus this energy scale is clearly sufficiently close to T_c so that in the normal state there will be important finite T corrections to the Fermi liquid ground state.

II. COMPARISON WITH HEAVY FERMION METALS

Perhaps the strongest phenomenological evidence for a Fermi liquid T=0 state in the cuprates derives from comparison[1,15] with heavy fermion metals. In these systems the ground state is known to be a canonical Fermi liquid; however, because of low energy scales the temperature dependences of $1/T_1$, R_H, etc. coincide with the canonical state predictions only for $T < T_{coh} \sim 1-10K$. Above this temperature anomalies in magnetic and transport data occur which are very similar to their counterparts in the cuprates.

A comparison of Hall data[17,18] in various heavy fermions and cuprates is shown in Figures 3a and 3b. We choose our temperature scale T* in the heavy fermion plots to be the laboratory temperature multiplied by a factor of 100 for $CeCu_6$ in order to take account the difference of about two orders of magnitude found in the coherence temperature. T* in other heavy fermions is consistently adjusted by the ratio of the (known) Sommerfeld constants γ. In UPt_3 and UAl_2 data it may be seen from Figure 4b that at sufficiently low T, R_H tends towards a T independent value as would be expected in the coherent state.

This comparison provides a "road map" for the cuprates which suggests that the cuprates have a superconducting transition temperature which is slightly above T_{coh}. In this way the normal state properties exhibit important corrections to the T=0 canonical Fermi liquid behavior. Thus, this picture is consistent with the phase diagram of Figure 1.

This "road map" may be further exploited in an attempt to understand the anomalously large values of R_H seen as x decreases *below* optimal stoichiometry. This study thus complements that of Figures 2. In Figure 4a and 4b we again compare heavy fermion and cuprate data.[19] The curves labelled A'-E' correspond to $YBCO$ Hall data at increasingly lower hole concentration. (Here A' represents the optimal stoichiometry). In Figure 4b are plotted Hall coefficient measurements on a heavy fermion alloy $Ce_xLa_{1-x}Cu_2Si_2$. In the limit of x= 0, there

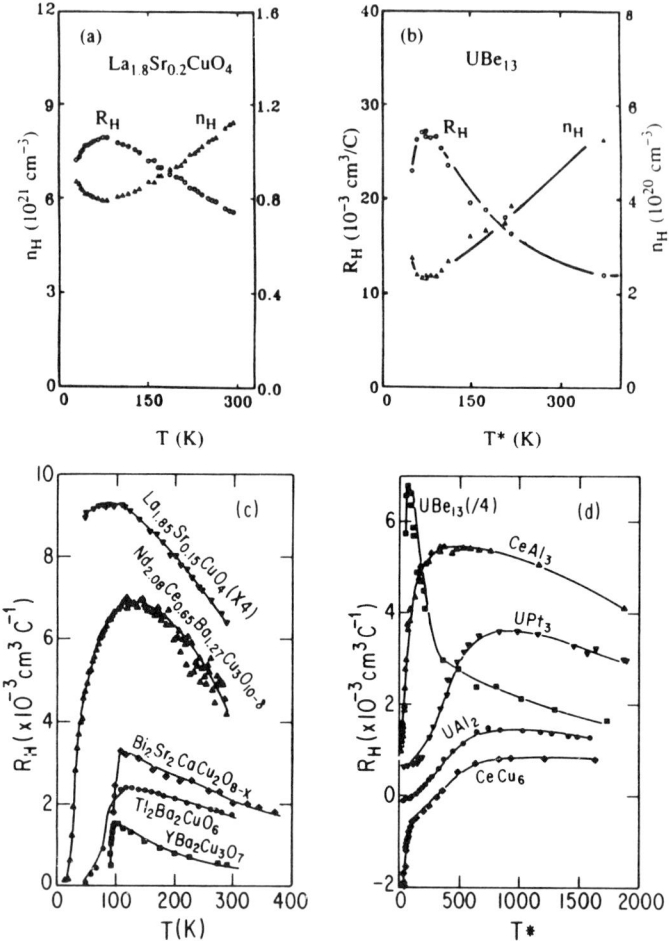

Figure 3   Comparison of Hall coefficient and its inverse (n_H) between cuprate (a) and heavy fermion (b). More comparisons are shown in (c) and (d). Data from Refs. 11, 17, and 18.

are no f electrons in the alloy so that the metal is not "heavy" and R_H is constant in T and of relatively modest size, as expected in a canonical Fermi liquid. However, as Ce is alloyed into the system, it becomes increasingly more localized and associated with this behavior is an increase in the magnitude of R_H, as well as in the degree of T dependence. The comparison of Figures 4a and b suggest that by analogy, in the cuprates as the hole oncentration decreases towards 0, the magnitude of R_H may well be increasing as a result of incipient localization, just as in the heavy fermion system of Figure 4b. It should be stressed, that a full microscopic understanding of the anomalous behavior of the Hall coefficient (also called the "extraordinary" Hall effect) has not been achieved in the heavy fermions. Nevertheless these analogies show how the temperature dependence and large magnitude of R_H can be reconciled with a ground state Fermi liquid picture.

Finally, in Figures 5 we compare the temperature dependence of the NMR relaxation in the cuprates with their heavy fermion counterpart.[12,20] The heavy fermion data show how the Fermi liquid Korringa dependence gives way to a nearly T independent relaxation as the sys-

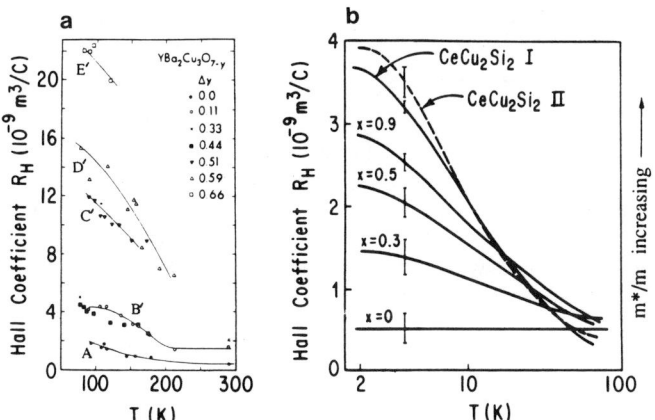

Figure 4 Temperature dependence of R_H (a) under variable oxygen stoichiometry, from Ref. 19; and (b) in heavy fermion alloy $La_{1-x}Ce_xCu_2Si_2$ with variable x. Data from Ref. 19, and the curves I and II refer to different samples of the x=1 material.

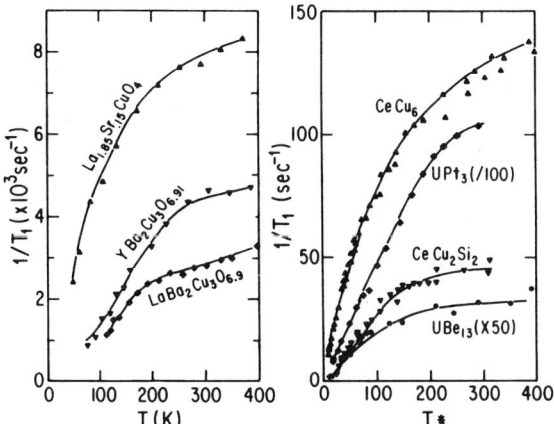

Figure 5 Comparison of Cu NMR relaxation in cuprates with (non f-site) NMR relaxation in heavy fermions. Data from Refs. 12 and 20 respectively.

tem is heated above its coherence temperature. The cuprate data indicates a very narrow Korringa regime (for the Cu spins) also followed by a tendency towards saturation. Thus these data re-enforce the notion that there is a low energy scale (above which Korringa behavior is not observed) in the Fermi liquid. Furthermore, they provide support for the phase diagram which postulates a smooth evolution in temperature from a Fermi liquid ground state.

III. THEORETICAL FRAMEWORK

The theoretical framework[1-3,21,22] for describing an almost localized, or narrow band Fermi liquid is based strongly on that developed for the heavy fermion system.[23,24] While 1/N schemes are commonly used, it should be stressed that there are a variety of other approaches which yield equivalent results. The starting Hamiltonian is an extended Hubbard (for the cuprates) and Anderson lattice (for the heavy fermions) model. These two Hamiltonians are fully equivalent in all essential ways. Both systems consist of basically two hybridizing bands with strong Coulomb U interactions on one component (d electrons) and non zero direct overlap (p electrons) on the other.

It is assumed next that U is infinite so that this parameter does not appear explicitly in the analysis, but rather enters as a constraint that there be no double occupancy on the Cu d site. In this way the hybridization is strongly reduced ($V \to V^*$) and the d level renormalized ($\varepsilon_d \to \varepsilon_d^*$) so as to satisfy the constraint. In 1/N schemes (where N is the spin degeneracy) the leading order or mean field contributions produce a bandstructure which is the same as that of the non interacting, U= 0 system but with the replacement of V and ε_d by their renormalized counterparts. Higher order in 1/N contributions produce the Landau parameters $F_l^{s,a}$ of the Fermi liquid.

This renormalized bandstructure has been used with some success to compute the photoemission and inverse photoemission cross section.[25,4] It has also addressed[21] the ordinary (as distinct from the extraordinary contribution which probably dominates at small x) contribution to R_H. Here it is found that R_H has the correct sign and also undergoes the observed change in sign at high x. This bandstructure has been used to deduce the electron-phonon interaction, within a "frozen phonon" picture.[26] It has also been applied to NMR and neutron data.[16] Finally, in a recent application the electronic contribution to the resistivity has been discussed.[27] Because of space limitations, here we concentrate only on this last contribution. It should be stressed, however, that collectively this body of work by a number of different investigators has addressed as wide or wider a body of experimental data than any other normal state approach to the cuprates.

IV. APPLICATION TO THE RESISTIVITY

In this section we use the almost localized Fermi liquid scheme to calculate the electronic contribution to the resistivity.[27] Elsewhere[26] we have shown that the phononic contribution is probably not sufficient to explain the data, at least below optimal doping. Our approach, like others[28,9] is based on the polarization or dynamical susceptibility $\chi(q,\omega)$ which enters via the "one bubble" self energy. We have shown elsewhere[22,16] that within a 1/N resummation scheme this dynamical susceptibility assumes the RPA form

$$\chi(\mathbf{q},\omega)=\chi_o(\mathbf{q},\omega)/(1-J(\mathbf{q})\chi_o(\mathbf{q},\omega))$$

(We have used a slighly schematic notation in this equation so that the separate Cu and O components are not indicated). Here J(**q**) is the Cu-Cu interaction which contains both superexchange as well as RKKY contributions and is found to be strongly x dependent. This x dependence arises physically from the fact that the Cu-Cu exchange interactions occur via the oxygen states. Since the occupancy of these mediating states is changing with x, this is reflected in a rapid decrease (with increasing x) of the superexchange and a simultaneous increase in the RKKY term. The fact that these interactions are of relatively short range leads to the result that the q dependence of J is roughly of the tight binding ($J_o(\cos(q_x a)+\cos(q_y a))$) form. We use this parametrization in what follows.

There are a number of constraints on a Fermi liquid explanation for the ubiquitous linear resistivity observed in the cuprates. Because this behavior is observed[29] in a 10K superconductor, it is clear that a phononic explanation is inadequate. Thus this contribution must arise from electron-electron scattering. Here we review the constraints on this term:

(1) The canonical T^2 regime must be unobservable (i.e., hidden by the superconductivity). This means that there must be a low energy scale $\omega_c < T_c$, as required for various different reasons throughout this article.

(2) For $T > \omega_c$, it is quite general to expect a linear T dependence, just as one observes a linear "high" temperature regime for phononic scattering above about 1/4 of the Debye energy. **However this linear regime is generally of limited extent** unless the transport spectral function $\alpha^2 F$ is temperature independent. For phonons it is clear that $\alpha^2 F$ *is* T independent. However for electron-electron scattering this is *not* usually the case. The corresponding spectral function is a q-weighted integral of Imχ(**q**,ω). Quite generally, the electronic spectral function will reflect the low energy scale ω_c and thus exhibit a T dependence which varies as ω_c/T, at high T. This T dependence in turn gives rise to a saturation in this (second order) electronic contribution to the resistivity, much as is observed in actinides and heavy fermion metals. A summary of this discussion is that the electronic transport spectral function $\alpha^2 F$ must be essentially T independent to obtain an extended linear resistivity.

(3) The characteristic "soft" NMR and resistivity energy scales must be somewhat different. This is required by the experimental observation that, slightly above T_c, as the Cu NMR begins to deviate from Korringa behavior, there is no structure in the resistivity. Since both measurements depend, in principle, on the same dynamical susceptibility, this provides a reasonably strong constraint on the Fermi liquid picture.

The present calculations are able to satisfy these three constraints. A summary of our results (presented in more detail elsewhere[27]) is that

(1) We find the Van Hove energy scale determines the T^2 to T cross over in the resistivity. This effect has been previously noted by Lee and Read,[28] but the present calculations show that

(2) in more realistic and strongly correlated bandstructure calculations, the Fermi energy is "pinned" near the Van Hove singularity for a range of doping concentrations around optimal x (for the 214 model cuprate).[30]

(3) A very special aspect of the Van Hove low energy scale is the associated *weak* temperature dependence in the transport spectral function. This presumably unique two dimensional feature thus differentiates the heavy fermion behavior (where saturation effects are seen) from that of the cuprates (where ρ remains linear to very high T).

(4) Finally we find that this Van Hove energy is not as relevant in NMR calculations due to its logarithmic integrability. Rather the NMR behavior is more strongly sensitive to T_{coh}, as discussed elsewhere. In summary, we find an extended linear regime for the electronic contribution to ρ with a low T cross-over from the canonical Fermi liquid limit.

Figure 6a shows the q,ω,T dependence of Imχ. As may be seen, along the zone diagonal the marginal ansatz of Ref. 9 works reasonably well for temperatures above the Van Hove

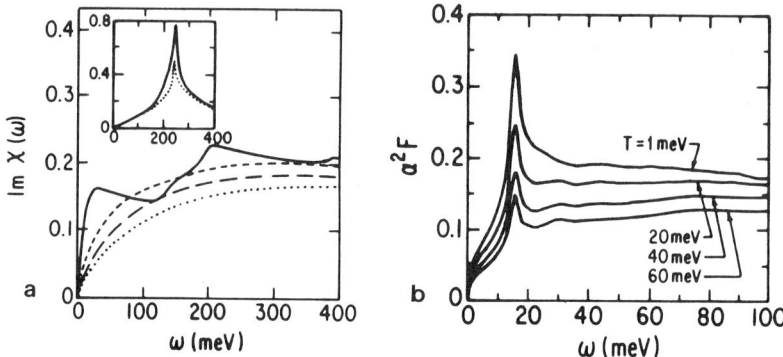

Figure 6 (a) Calculated $\chi''(Q,\omega)$, for x=0.28 and $Q=.85(\pi/a,\pi/a)$, as function of frequency and various temperatures T=1meV (solid curve), and T=20, 40 and 60 meV for remaining three curves from top to bottom. Inset shows analogous results for $(\pi/a,0)$ at T=1 meV (solid curve) and 60 meV (dotted curve); (b) Calculated electronic spectral function (from Ref. 27) for same temperatures and parameters in Fig. 6a. This function results from a weighted q integration of $\chi''(q,\omega)$.

energy. However, as seen in the inset, away from this direction in **q** space the behavior of χ is *not* consistent with this ansatz. The resultant $\alpha^2 F$, shown in Figure 6b, thus does not resemble the marginal ansatz. Rather it appears more characteristic of a phonon-type spectral function in which there is a peak at the characteristic energy (here corresponding to the Van Hove energy scale). There is a very weak (logarithmic) T dependence in the spectral function, which weak T dependence is responsible for an extended linear regime.

Finally in Figure 7a we show the behavior of the calculated resistivity without the inclusion of the Cu-Cu spin exchange. Here two hole concentrations are considered, the lower corresponding to the 214 and upper to the 123 systems, very roughly, at optimal stoichiometry. In the lower x case, the Van Hove and Fermi energies roughly coincide. The cross-over from T^2 to T behavior occurs at around 10 K for the lower and 50K for the higher concentrations.

The effects of including spin exchange interactions J(**q**) into the dynamical susceptibility are shown in Figure 7b for the higher concentration x. In the figure J_c corresponds to the critical value of the exchange strength above which the non-magnetic phase is unstable. Here it may be seen that that these effects generally undermine the linearity in ρ deriving from the Van Hove singularity. This provides a concrete example of a system where the temperature dependence in the spectral function $\alpha^2 F$ is sufficiently strong so as to limit the range of linearity to a narrow interval in temperature. This should serve to underline the unique properties of the particular low energy scale which derives from the Van Hove singularity. Most other models with "soft" energies will not yield an extended linear regime.

V. CONCLUSIONS

In this paper we have demonstrated that a variety of Fermi liquid approaches to the cuprates lead to a similar physical picture: the Fermi liquid must contain low energy scales

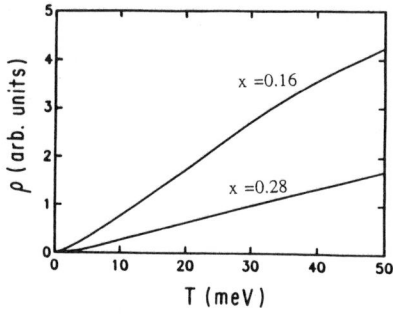 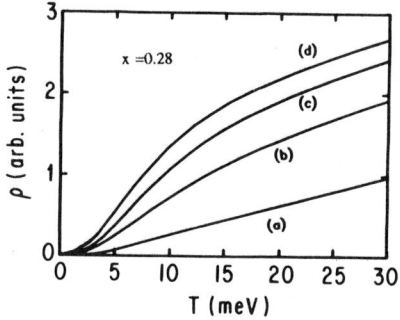

Figure 7 (a) Calculated (Ref. 27) temperature dependence of electron-electron contribution to resistivity, in absence of Cu-Cu spin exchange, for hole concentrations of $YBa_2Cu_3O_7$; (b) Calculated (Ref. 27) temperature dependence of electron-electron contribution to resistivity, for various spin exchange strengths $J_o/J_c=$ (a) 0, (b) 0.7, (c) 0.85 and (d) 0.98. Here J_c represents instability value.

$\omega_c(x)$ which become progressively softer as the insulator is approached. This scenario is best represented by the phase diagram of Figure 1. Fundamental to the picture is the notion that there is a smooth cross-over in the (x,T) plane from a classical Fermi liquid to one in which there are significant correction terms. In this way anomalies in the data can be reconciled with a Fermi liquid ground state (in the absence of superconductivity). The picture is given support by (a) the smooth evolution towards canonical Fermi liquid behavior in overdoped systems. This probes the variation in the x parameter. (b) the similarity between anomalous properties seen in the cuprates and the heavy fermions. These latter metals exhibit smooth temperature variations from a Fermi liquid ground state to a metal with "anomalous" properties. Thus this demonstrates the variation in the T parameter.

When looked at collectively these various (nearly magnetic and nearly localized) Fermi liquid pictures yield reasonable agreement with experiment for transport, magnetic and spectroscopic measurements. But these are "zeroth" order approximations in the sense that the temperature is "dangerously" close to the various low energy scales of the Fermi liquid. Thus finite temperature effects enter only via a thermal smearing. In reality, what is needed is a more complete account of systems which are not in the strict Fermi liquid regime.

Thus we are led to conclude that even if a Fermi liquid ground state is appropriate to the cuprates (in the absence of superconductivity), the material from which high temperature superconductivity evolves may be quite different from this ground state system. In this way the standard BCS or Eliashberg treatment of the pairing is probably inadequate. Ultimately this may be the "bigger" and more relevant question concerning the applicability of Fermi liquid theory for the cuprates.

This work was supported by funds from the NSF-STC Grant No. STC-8809854 and NSF-MRL Grant No. DMR-19860.

REFERENCES

1. For a recent review, see K. Levin, Ju H. Kim, J.P. Lu, and Q. Si, Physica C, *in press*.
2. For a review, see D.M. Newns, and P.C. Pattnaik, in *Strong Correlation* and *Superconductivity*, eds. H. Fukuyama *et al.* (Springer-Verlag, Berlin, 1989).

3. G. Kotliar, P.A. Lee, and N. Read, Physica C 153-155, 538 (1988).
4. C.A.R. Sa de Melo, and S. Doniach, Phys. Rev. B 41, 6633 (1990).
5. N. Bulut, D. Hone, D.J. Scalapino, and N. E. Bickers, Phys. Rev. B 41, 1797 (1990); A. Kampf, and J.R. Schrieffer, Phys. Rev. B 41, 6399 (1990).
6. A.J. Millis, H. Monien, and D. Pines, Phys. Rev. B 42, 167 (1990).
7. A. Virosztek, and J. Ruvalds, Phys. Rev. B 42, 4064 (1990).
8. P.W. Anderson, Phys. Rev. Lett. 64, 839 (1990).
9. C.M. Varma *et al.*, Phys. Rev. Lett. 63, 1996 (1989).
10. N. Nagaosa, and P.A. Lee, Phys. Rev. Lett. 64, 2450 (1990); L.B. Ioffe, and P.B. Wiegmann, *ibid.* 65, 653 (1990); L.B. Ioffe, and G. Kotliar, Phys. Rev. B 42, 10 348 (1990).
11. Y. Kubo *et al.*, Physica C 164-166, 991 (1989); Data were analysed by C.C. Tseui, Physica A, to be published; For resistivity, there presumably is also a Bloch Gruneisen T dependence arising from a phonon background.
12. T. Imai *et al.*, Physica C162-164, 169 (1989); See also K. Asayama, Y. Kitaoka and Y. Kohori, (unpublished).
13. D.B. Mitzi *et al.*, Phys. Rev. B41, 6564(1990); J.B. Torrance *et al.*, Phys. Rev. B 40, 8872 (1989).
14. For reviews see, C.P. Pennington, and C.P. Slichter, in *Physical Properties of High Temperature Superconductors*, Vol. II, ed. D.M. Ginsberg (World Scientific, N.J., 1990); R. J. Birgeneau and G. Shirane, *ibid.*, Vol. I. (World Scientific, Singapore, 1989); For more references, see Ref. 1.
15. Q. Si, Ju.H. Kim, J.P. Lu, and K. Levin, Phys. Rev. B 42, 1033 (1990).
16. J.P. Lu, Q. Si, Ju. H. Kim, and K. Levin, Phys. Rev. Lett. 65, 2466(1990).
17. M. Suzuki, Phys. Rev. B 39, 2312 (1989).
18. T. Penney *et al.*, Phys. Rev. B 34, 5959 (1986); T. Tamegai, Y. Iye, M. Ogata, and J. Akimitsu (unpublished); G. Briceno, and A. Zettl, Phys. Rev. B 40, 11352 (1989); P. Chaudhari *et al.*, Phys. Rev. B 36, 8903 (1987); M. Christen, and M. Godet, Phys. Lett. A 63, 125 (1977); M. Hadzic-Leroux *et al.*, Europhys. Lett. 1, 579 (1986).
19. (a) N.P. Ong, Ref. 14, Vol. II. (World Scientific, N. J., 1990); (b) N.B. Brandt, and V.V. Moschchalkov, Adv. Phys. 33, 373 (1984).
20. K. Asayama, Y. Kitaoka, and Y. Kohori, J. Magn. Magn. Mat. 76&77, 449 (1988).
21. Ju.H. Kim, K. Levin, and A. Auerbach, Phys. Rev. B 39, 11633 (1989).
22. Q. Si, J.P. Lu, and K. Levin, Physica C172, 481 (1991);and (unpublished).
23. N. Read, and D.M. Newns, J. Phys. C 16, 3273 (1983); P. Coleman, in *Theory of Heavy Fermions* and *Valence Fluctuations*, eds. T. Kasuya and T. Saso (Springer-Verlag, New York, 1985).
24. A. Auerbach, and K. Levin, Phys. Rev. Lett. 57, 877 (1986); A.J. Millis, and P.A. Lee, Phys. Rev. B 35, 3394 (1987); A.Houghton, N.Read, and H. Won, *ibid.* 37, 3782 (1988).
25. P.C. Pattnaik, and D. M. Newns, Phys. Rev. B 41, 880 (1990).
26. Ju.H. Kim, K. Levin, R. Wentzcovitch, and A. Auerbach, Phys. Rev. B 40, 11378 (1989).
27. Q. Si, and K. Levin (unpublished).
28. P.A. Lee, and N. Read, Phys. Rev. Lett. 58, 2691 (1987).
29. S. Martin *et al.*, Phys. Rev. B 39, 9611 (1989).
30. See also D.M. Newns, P.C. Pattnaik, and C.C. Tsuei, Phys. Rev. B43, 3075 (1991).

VAN HOVE SCENARIO FOR HITC SUPERCONDUCTIVITY

C. L. Kane, D. M. Newns, P. C. Pattnaik, C. C. Tsuei, and C. C. Chi

IBM Thomas J. Watson Research Center
P.O. Box 218, Yorktown Heights, NY 10598

INTRODUCTION

After many experimental difficulties with high temperature superconductors, the realization is emerging that there is a connection between their anomalous normal state properties, whose origin lies in the anomalously high quasiparticle lifetime broadening, and their superconducting properties [1-3].

What is the first principles origin of these phenomena? In this paper we shall make the case [2,3], which is based on increasingly persuasive evidence, that it is the location of the Fermi level close to the van Hove singularity in the density of states in these materials which is responsible. The energy surfaces around the VHS are quasi-hyperbolic, and both electron-electron [2] and electron-phonon [3] scattering proves to be anomalous when considered on such surfaces.

EXTENDED HUBBARD MODEL

A technical obstacle central to theoretical treatment of the HiTc problem is the strong hole-hole Coulomb repulsion U within the Cu atom. We choose to handle this with the slave boson approach. [4-6] The Hamiltonian for a simple model of the CuO_2 planes is then given by

$$\mathcal{H} = E_1 \sum_{i\sigma} d^+_{i\sigma} d_{i\sigma} + \sum_{ij\sigma} \gamma_{ij} p^+_{i\sigma} p_{j\sigma} + \sum_{ij\sigma} \beta_{ij} d^+_{i\sigma} d_{j\sigma} b^+_j b_i + \sum_{ij} (t_{ij} b^+_j p^+_i d_j + \text{h.c.}) + \lambda \sum_i (\hat{Q}_i - 1) \quad (1)$$

In (1), $p_{i\sigma}$ are hole operators for the p_x and p_y oxygen orbitals of spin σ, and $d_{i\sigma}$ are pseudofermion operators for the d^9 state of the $d_{x^2-y^2}$ orbitals of spin σ. Operators b_i are the slave boson operators for the d^{10} state of these orbitals. The representation (1) exactly corresponds to a $U = \infty$ theory if the constraint

$$\hat{Q}_i = \sum_\sigma d^+_{i\sigma} d_{i\sigma} + b^+_i b_i = 1 \quad (2)$$

is satisfied; it's static component is treated by a Lagrange multiplyer λ. The quantity E_1 is the charge transfer gap, the energy between the $d^9 - d^{10}$ level of Cu $d_{x^2-y^2}$ and the center of the p_x-p_y band (taken as energy zero). E_1 is taken as 5eV. The intersite matrix elements δ_{ij}, β_{ij} and t_{ij} are taken from the NRL band structure parametrization[7] of the 2212 material.

To generate the quasiparticle band structure the mean field approximation [4-6] $b \to <b>$ is taken. Self-consistent calculation of λ and $<b>$ yield the results illustrated in Fig. 1. Comparison with photoemission data [8] shows that the choice of E_1, which controls the effective mass, is good.

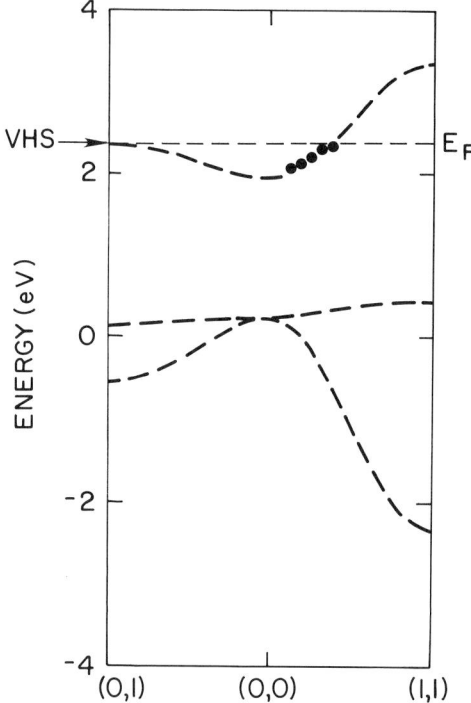

Fig. 1. Band Structure for model of CuO_2 plane in Eq. (1), $x = 0.2$ (dashed curve), compared with angle resolved photoemission data for 2212 material [8] (filled circles). Dotted line indicates Fermi energy E_F, arrow indicates location of van Hove Singularity.

How correlated is the 2212 material according to the slave boson solution? One measure might be the effective mass, which is only $\sim$ a factor of 2 greater than in the LDA band structure. Another measure might be the ratio of spin to charge susceptibility on Cu, which would be 1 in an uncorrelated system. This ratio is found to be $\sim$ 20, showing there is a large degree of correlation hidden in the slave boson solution. A related result [2] is illustrated in Fig. 2, where the weight of added holes on oxygen is plotted vs. doping; since copper is nearly d^9, adding further holes to Cu is impossible for $U = \infty$, and the holes almost all go on to oxygen. This is in agreement with the illustrated spectroscopic data.[9]

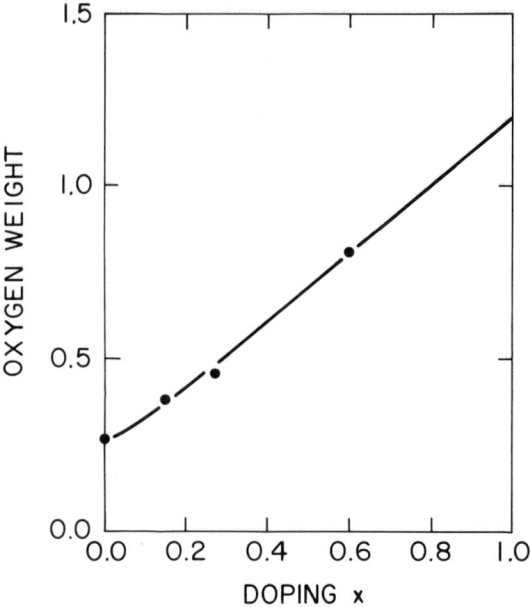

Fig. 2. Weight of holes on oxygen plotted vs. doping for model of CuO_2 plane in Eq. (1). Filled circles, spectroscopic data for 214 material.

The foregoing theory of the normal state is a Fermi liquid one which even seems to describe some properties of the undoped insulator! In Fig. 3 we illustrate a calculation of the optical gap in our model at zero doping for various Cu-O distances. The gap is taken as the energy from the top of the valence band in Fig. 1 to the Fermi energy, while the distance variation is implemented via a scaling of the matrix elements in Eq. (1) according to Harrison's rules. The result, illustrated in Fig. 3, is in good accord with the experimental data of Thomas et al.[11] for various materials, despite the fact that the slave boson solution in the present paper does not provide the supression of the intensity of the intraband transitions at half filling in a Mott insulator.

THE VAN HOVE CONCEPT

The density of states for various dopings is illustrated [2] in Fig. 4 for the 2212 model. It is seen that the Fermi level remains close to the van Hove singularity for a wide range of dopings in the $x = -0.05 - 0.45$ range. This is because added holes, which are mostly located on the oxygen sites (figure 2), are acquired in the oxygen like foot of the band in figure 4, while the van Hove singularity (mostly Cu-like) remains pegged at the Fermi level. The T_c maximum [12] occurs close to $x = 0.3$ in this material, the doping where the Fermi level sits at the VHS in our model. The density of states deduced [2] from specific heat jump and magnetic susceptibility measurements is compatible with the Fermi level lying close to the VHS. Hence we are led to assume that a description of the energy surfaces as hyperbolic may be the appropriate one for high temperature superconductors.

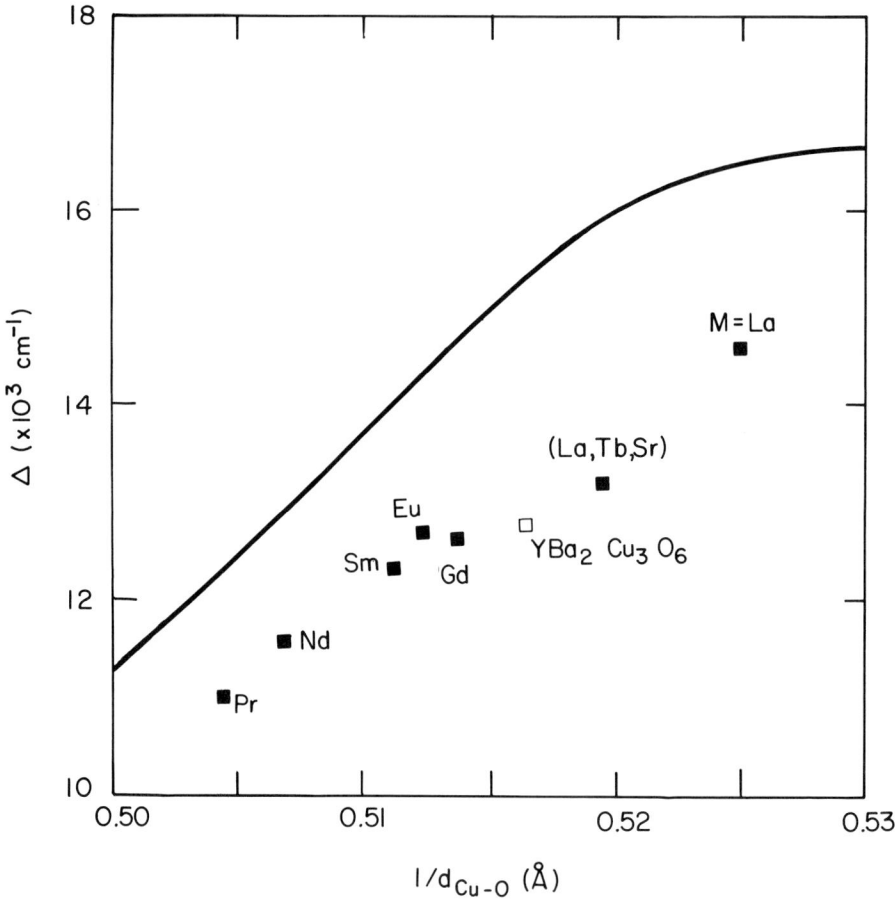

Fig. 3. Interband gap for zero doping plotted vs. Cu - O distance for model of CuO_2 plane (Eq. (1)). Filled squares are data [11] for energy gap Δ in T' - phase M_2CuO_4 (= Pr, Nd, Sm, Eu, Gd), T* phase (La, Tb, Sr)$_2$ CuO_4 and T-phase La_2CuO_4. The energy gap [11] for insulating $YBa_2Cu_3O_6$ is as shown for comparison (open square).

ANOMALOUS SCATTERING

Electron-electron scattering is anomalous when the energy surfaces are hyperbolic as shown by perturbation theory calculations. The result we find from the diagram in figure 5 (insert) for the self energy is

$$\frac{1}{\tau} = 2\mathrm{Im}\,\Sigma(\mathbf{k}, \varepsilon_\mathbf{k}) = C\,\varepsilon_\mathbf{k}, \tag{3a}$$

where

$$C = 2\pi \frac{W^2}{E_F^2}, \tag{3b}$$

when $\varepsilon_\mathbf{k}$ is taken as $\varepsilon_\mathbf{k} = k_a k_b/m$ (k_a, k_b not necessarily orthogonal directions). W is the particle-particle interaction, assumed energy and momentum independent. The Fermi

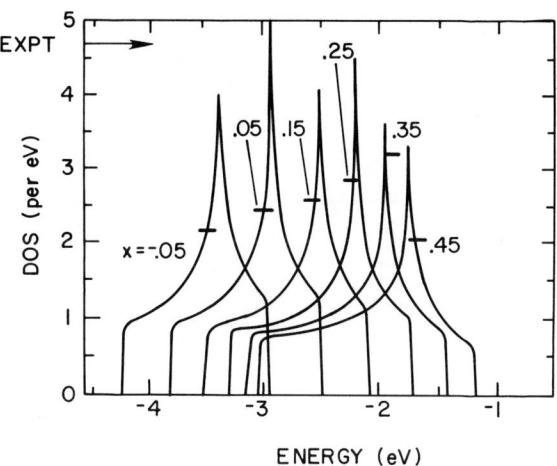

Fig. 4. Mean Field DOS of CuO_2 plane model (Eq. (1)) for a range of dopings. Horizontal bars denote DOS at E_F Hole notation is used. Arrow indicates experimental DOS for 2212 material derived from analysis of Ref. 2.

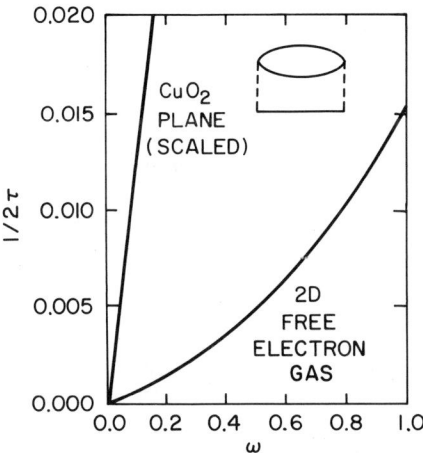

Fig. 5. Comparison of imaginary part of self energy $\Sigma (k, \omega)$ $(= 1/2 \tau)$ vs. energy ω from Fermi level for 2D free electron gas and for CuO_2 model of Eq. (1). Diagram calculated is illustrated in insert. The models have been scaled to have the same Fermi level, which is 1.3 in these units.

energy, E_F is k_c^2/m, where the momentum cutoff k_c for the hyperbolic region is such that its area equals that of the Brillouin zone. The linear dependence of the lifetime broadening in (3) is distinguished from that of the free electron gas in two dimensions, [13] for which the dependence is $\varepsilon_k^2 \log(\varepsilon_k)$, and fulfils the expectations of marginal Fermi liquid theory [1].

The physical origin of the anomalous low energy scattering lies in the additional phase space for scattering nearly parallel to the asymptotes of the hyperbolic surfaces - phase space present irrespective of the angle between the asymptotes. This scattering is quite different from the nesting scattering occurring between nearly parallel sections of a Fermi surface; there is no significant nesting in our model (1) at $x = 0.3$.

In Fig. 5 we compare the numerically calculated $1/2\tau$ for a 2D free electron gas with calculations [2] for our model of the CuO_2 plane, scaled so as to have the same Fermi energy, for a doping $x = 0.3$ at which E_F lies very close to the VHS. The CuO_2 calculation, like the hyperbolic results of Eq. (3), is essentially linear and is also seen to be huge relative to the free electron gas result. The predictions of the CuO_2 theory are in agreement [2] with the photoemission [8] $1/\tau$, and, assuming that (3a) is approximately equal to the transport $1/\tau$, with IR [14] measurements, if W is taken to be $W = 1.3\ E_F$, which seems physically plausible. It follows from the same assumption that linearity of $1/\tau$ with ω implies linearity of resistivity $\rho(T)$, a well-known characteristic of the high T_c cuprate superconductors.

SUPERCONDUCTIVITY

It has long been known[15–18] that the van Hove singularity enhances superconductivity; an approximate formula given early on [16] for T_c is

$$T_c = 1.36 T_F \exp\left\{-\sqrt{\frac{2}{N_0 V}}\right\}; \quad \omega_c << E_F \tag{4}$$

for a density of states of the form $N(E) = N_0 \log[E_F/(E - E_F)]$. Here V is electron-phonon coupling content, and ω_c the Debye temperature. This formula contains two factors which enhance T_c relative to conventional BCS: the presence of T_F in the prefactor rather than the Debye temperature, and the coupling constant in the exponent is square rooted. Moreover, the absence of Debye temperature in the prefactor implies zero isotope shift in the limit (4).

More complete calculations [3] reveal that the characteristic behavior of T_c and the isotope shift in the presence of the VHS is that T_c has a maximum when the Fermi level goes through the singularity, while the isotope shift has a minimum at this point. This is exactly the behavior [3] observed by Crawford et al. [19] for the 214 material. More recent data on the 123 material is compared with numerical integration of the BCS equation in Fig. 6. It is seen to fit the data quite well, especially if μ^* is included.

As a final example of the effect of the VHS on superconductivity, it is interesting to calculate the coherence length, for T near T_c, which may be done by standard procedures. The result for the hyperbolic energy surface model $\varepsilon_k = k_a k_b/m$ is

$$\xi_{ab} = 0.16 \frac{\sqrt{E_F/m}}{kT_c} \sqrt{\frac{T_c}{T_c - T}} \frac{1}{\ln^{1/2}(1.36 E_F/kT_c)}. \tag{5}$$

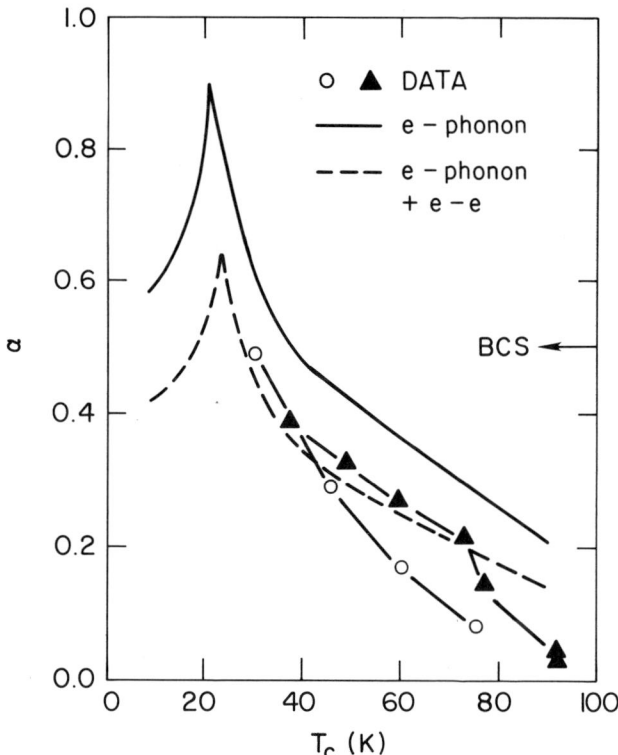

Fig. 6. Plot of isotope shift exponent α vs. T_c for 123 material. Filled triangles and open circles are data obtained by cation doping [19]. Curves are BCS calculation for a model DOS $N(E) = - N_0 \log (E/E_F)$, with $E_F = 0.5$eV, $\omega_D = 0.065$eV.. Full curve, electron-phonon coupling $VN_0 = 0.128$, zero Coulomb repulsion; dashed curve, $VN_0 = 0.202$, Coulomb repulsion V_c given by $V_c N_0 = -0.085$ ($\mu^* \simeq 0.07$).

The coherence length is reduced by an additional factor $\ln^{1/2}(1.36E_F/T_c)$ from the conventional BCS coherence length. The ξ_{ab} calculated from equation (5) is found to be in agreement with the ξ_{ab} for the 123 material[20], obtained from variation of H_{c2} near T_c, to within the accuracy of the parameters.

CONCLUSION

In conclusion, the presence of a van Hove singularity near the Fermi level leads to a problem of scattering on hyperbolic energy surfaces which has not yet been fully explored. In perturbation theory, the quasiparticle lifetime broadening from electron-electron scattering is found to be a linear function of energy ω, instead of the usual $\omega^2 \log \omega$ in 2D. The magnitude of this linear term is of the right order compared with experimental data.

The effects on superconductivity are to enhance T_c, and reduce the isotope shift, when the Fermi level is near the VHS, effects which are observed in both the 214 and 123 materials. The coherence length is reduced by a logarithmic factor.

It seems probable that the VHS provides the missing link needed to understand the special properties of high T_c materials. Future directions we are pursuing involve the possibility of an electronic component to pairing, and the involvement of antiferromagnetism in the van Hove scenario.

REFERENCES

1. C. M. Varma, P.B. Littlewood, S. Schmittrink, E. Abrahams, and A. E. Ruckenstein, Phys. Rev. Lett. **63**, 1996 (1989); Y. Kuroda and C. M. Varma, Phys. Rev. B **42**, 8619 (1990).
2. D. M Newns, P. C. Pattnaik, and C. C. Tsuei, Phys. Rev. B **43**, 3075 (1991); P. C. Pattnaik, D. M. Newns, and C. C. Tsuei, submitted to Phys. Rev.
3. C. C. Tsuei, D. M. Newns, C. C. Chi and P. C. Pattnaik, Phys. Rev. Lett. **65**, 2724 (1990).
4. D. M. Newns, P. C. Pattnaik, M. Rasolt and D. A. Papaconstantopoulos, Phys. Rev. **38**, 7033 (1988).
5. B. G. Kotliar, P. A. Lee, and N. Read, Physica C **153-155**, 538 (1988).
6. J. H. Kim, K. Levin, and A. Auerbach, Phys. Rev. B. **39**, 11633 (1989); Q. Si, J. P. Lu, and K. Levin, Phys. Rev. Lett. **65**, 2488 (1990).
7. M. J. de Weert, D. A. Papaconstantopoulos, and W. E. Pickett, Phys. Rev. B **39**, 4235 (1990). D.A. Papaconstantopoulos, private common.
8. C. G. Olson, R. Kiu, D. W. Lynch, R. S. List, A. J. Arko, B. W. Veal, Y. C. Chang, P. Z. Jiang, and A. B. Paulikas, preprint; R. S. List, A. J. Arko, R. J. Bartlett, S.-W. Cheong, Z. Fisk, J. D. Thompson, C. G. Olson, A.-B. Yang, R. Liu, C. Gu, B. W. Veal, J. Z. Liu, A. P. Paulikas, K. Vandervoort, H. Claus, J. C. Campuzano, J. E. Schirber, and N. D. Shinn, proceedings of the International Conference on the Physics of Highly Correlated Electron Systems, Santa Fe, 1989.
9. A. Santoni, M. Ronay, L. J. Terminello, G. V. Chandrasekhar, M. W. Shafer, Y. Hidecha, and F. J. Himpsel, private common.
10. W. A. Harrison, 'Electronic Structure and the Properties of Solids' (1979).
11. S. L. Cooper, G. A. Thomas, A. J. Millis, P. E. Sulenski, J. Orenstein, D. H. Ropkins, S.-W. Cheong, P. L. Trevor, and B. Batlogg, preprint.
12. Koike et. al., Physica C **159**, 105 (1989).

13. C. Hodges, H. Smith, and J. W. Wilkins, Phys. Rev. **4**, 302 (1971).
14. Z. Schlesinger, R. T. Collins, F. Holtzberg, C. Field, S. H. Blanton, U. Welp, G. W. Crabtree, Y. Fang, and J. Z. Liu, Phys. Rev. Lett. **65**, 801 (1990).
15. J. Friedel, J. Physique **48**, 1787 (1987); ibid **49**, 1435 (1988); J. Phys. Condens. Mater. **1**, 7707 (1989).
16. J. Labbe and J. Bok, Europhysics Lett. **3**, 1225 (1987); J. Labbé, Physica Scripta **T29**, 82 (1989).
17. R. S. Markiewicz, and B. J. Giessen, Physica C **160**, 497, (1989); R. S. Markiewicz, J. Phys. Condens Matter 1, 8911 (1989); ibid 8931 (1989); ibid2 **6223**, (1990).
18. J. E. Dzyaloshinskii, Soviet Physics JETP Lett. **46**, 118 (1987); Soviet Physics JETP **66**, 848 (1987).
19. M. K. Crawford, M. N. Kunchur, W. E. Farneth, E. M. McCarron, III, and S. J. Poon, Phys, Rev. B **41**, 282 (1990): J. P. Franck, J. Jung, M. A. K. Mohamed, S. Gygax, and I. G. Sproule, "Observation of an Oxygen Isotope Effect in Superconducting $(Y_{1-x}Pr_x)$ $Ba_2Cu_oO_{7-\delta}$ (preprint); H. J. Bornemann and D. E. Morris, "Isotope Effect in $YBa_{2-x}La_xCu_3O_7$ — Evidence of Phonon-mediated High-Temperature Superconductivity", (preprint).
20. L. Civale and L. Krusin-Elbaum, private communication.

THE t-J MODEL AT SMALL t/J:

NUMERICAL, PERTURBATIVE AND SUPERSYMMETRIC RESULTS

T.Barnes

Physics Division and
Center for Computationally Intensive Physics
Oak Ridge National Laboratory, Oak Ridge, TN 37831-6373
and
Department of Physics
University of Tennessee
Knoxville, TN 37996-1200

ABSTRACT

We discuss some recent results for one- and two-hole states in the t-J model at small t/J. These include numerical results (bandwidth determinations and accurate t/J values for 4×4 lattice one-hole ground-state level crossings), hopping-parameter perturbation theory (which gives the small-t/J one-hole bandwidth in terms of the static-vacancy ground state), and results at the supersymmetric point $t/J = 1/2$ (exact results for energies and bandwidths). The perturbative results lead us to a new conjecture regarding the staggered magnetization of higher-spin states in the two-dimensional Heisenberg model. We also discuss extrapolation of small-t/J results to high-T_c parameter values; in the two-hole ground states we find $(t/J)^\lambda$ behavior in the rms hole-hole separation, and an extrapolation to $t/J = 3$ gives a bulk-limit rms hole-hole separation of $\approx 7\text{Å}$.

INTRODUCTION: THE t-J MODEL AT SMALL t/J

The "t-J model"[1], which is described by the Hamiltonian

$$H_{tJ} = -t \sum_{<ij>,\sigma} (c^\dagger_{i\sigma} c_{j\sigma} + h.c.) + J \sum_{<ij>} (\mathbf{S}_i \cdot \mathbf{S}_j - \frac{1}{4} n_i n_j) \qquad (1)$$

with an implicit restriction to unoccupied or singly-occupied sites, has attracted considerable interest as a model of the high temperature superconductors. This is in large part due to suggestions that the closely related two-dimensional Hubbard and Heisenberg spin systems might provide useful models of high temperature superconductors[2], and to the close proximity of the disruption of long-range antiferromagnetic order and the onset of superconductivity as hole doping is increased[3]. Although the t-J model is now believed to be unphysical due to its prediction of hole phase separation[4], it nonetheless continues to be of great interest as a prototype high temperature superconductor model, and may require relatively little modification to correct its unphysical features. The parameters appropriate for the high temperature superconductors are $J \approx 125$ meV (which is relatively well established through comparisons of Heisenberg model predictions with neutron scattering[5]), $t/J \approx 3$ (which is estimated in band structure calculations and is much less well established), $a_0 = 3.79$ Å, and fillings close to one electron per site.

The t-J model is unfortunately not amenable to numerical studies using Monte Carlo methods, due to the "minus sign problem" of dynamical many-fermion systems. For this reason numerical studies of energies and other matrix elements have used Lanczos techniques almost exclusively (for a recent review see Dagotto[6]). As Lanczos methods require the storage of several vectors of dimension equal to that of the Hilbert space, which for a given filling fraction increases exponentially with the number of sites N, these studies have been limited to systems of relatively small size, at most $N = 20$ sites[7], and many references consider only the 4×4 lattice.

Although the rather large value of $t/J \approx 3$ is most relevant to the high temperature superconductors, there are several reasons for investigating the small-t/J limit. The first is that at $t/J = 0$ this model is essentially the Heisenberg antiferromagnet with static vacancies, which has been studied using spin-wave theory[8] and Monte Carlo techniques[9]. (These static-hole studies are not very reassuring for Lanczos work, as they find moderately large finite size effects.) The small-t/J limit can also be studied using perturbation theory in the hopping parameter t; this leads to a relation between the structure of the lowest one-hole band and the static hole ground-state wavefunction. Another interesting feature of the small-t/J regime is the occurrence of a supersymmetry at $t/J = 1/2$; this leads to exact relations between the energies of states with different hole number, which are independent of the lattice size. Some of these relations appear in our numerical results as level crossings at $t/J = 1/2$. Finally, some observables such as energies scale quite accurately as powers of t/J; one might hope to establish such behavior at small t/J, where the holes are relatively immobile and finite size effects are less important, and then extrapolate to $t/J \approx 3$ in the bulk limit. We find that this procedure apparently does work for the two-hole ground states of the t-J model, and we estimate the size of these bound states at $t/J = 3$. We shall now discuss these topics (bandwidths, supersymmetry and bulk-limit extrapolations) in more detail.

PERTURBATIVE ONE-HOLE BAND STRUCTURE

The simple $\vec{k}$ dependence of the one-hole band at small t/J can be understood using perturbation theory in the hopping parameter t. The Heisenberg antiferromagnet with static vacancies is taken to be the unperturbed system, and the hopping terms are treated as a perturbation;

$$H_0 = J \sum_{<ij>} (\mathbf{S}_i \cdot \mathbf{S}_j - \frac{1}{4} n_i n_j) , \qquad (2)$$

$$H_I = -t \sum_{<ij>,\sigma} (c^\dagger_{i\sigma} c_{j\sigma} + h.c.) . \qquad (3)$$

One then develops a perturbative expansion in the hopping parameter t, using a basis of static-vacancy states in the Heisenberg antiferromagnet[10]. (For related discussions see Dagotto, Joynt, Moreo, Bacci and Gagliano[11] and Elser, Huse, Shraiman and Siggia[12].) All translations of static one-hole states are degenerate under H_0, so one applies degenerate perturbation theory and finds energy shifts which are proportional to t at leading order. Solution of the one-hole secular equation shows that the resulting linear combinations of static-hole states are momentum eigenstates, with a multiplet structure given by

$$\lim_{t/J \to 0} e_h(\vec{k}) - e_h(t=0) = Z_W \cdot 2t (\cos k_x + \cos k_y) . \qquad (4)$$

The small-t/J "bandwidth renormalization" Z_W (normalized to unity for a free fermion, $W_h = 8t$) is equal to an off-diagonal matrix element of the ground-state wavefunction of a static hole in a Heisenberg background,

$$Z_W = \sum_S \Psi^*_0(S', \vec{j}') \cdot \Psi_0(S, \vec{j}) . \qquad (5)$$

The sum over $\mathcal{S}$ is understood to run over all basis states in the appropriate sector of Hilbert space, for example $S_{tot}^z = 1/2$. $\Psi_0(\mathcal{S}', \vec{j}\,')$ is the amplitude to find a spin configuration $\mathcal{S}'$ in a static-hole ground state with a hole at $\vec{j}\,' = \vec{j} + \hat{x}$, where the pair $(\mathcal{S}', \vec{j} + \hat{x})$ is constructed from $(\mathcal{S}, \vec{j})$ by exchanging the hole at $\vec{j}$ with the spin at $\vec{j} + \hat{x}$. In the basis used to derive (5) the spin-flip terms in $\vec{S}_i \cdot \vec{S}_j$ in the Hamiltonian (2) have positive matrix elements, so that the $\{\Psi_0\}$ are real but do not all have the same sign and hence Z_W in (5) is not positive definite. Negative $W_h = 8 Z_W t$ simply implies an inverted multiplet, with $\vec{k} = (0,0)$ rather than (π, π) at the bottom of the band.

A direct Lanczos evaluation of the $S = 1/2$ static-hole matrix element in (5) on the 4×4 lattice gives a bandwidth renormalization of $Z_W = 0.14880571(1)$ and hence a one-hole bandwidth of $\lim_{t/J \to 0} W_h = 1.1904457(1) \cdot t$, which is consistent with the independent $t/J > 0$ Lanczos results shown in Figure 1. The $\vec{k}$-dependence of the theoretical small-t/J dispersion relation (4) is also consistent with the numerical results shown in Figure 1 at small t/J. (The predictions (4) are shown as solid lines.) This band is shown for a somewhat larger range of t/J in Figure 2.

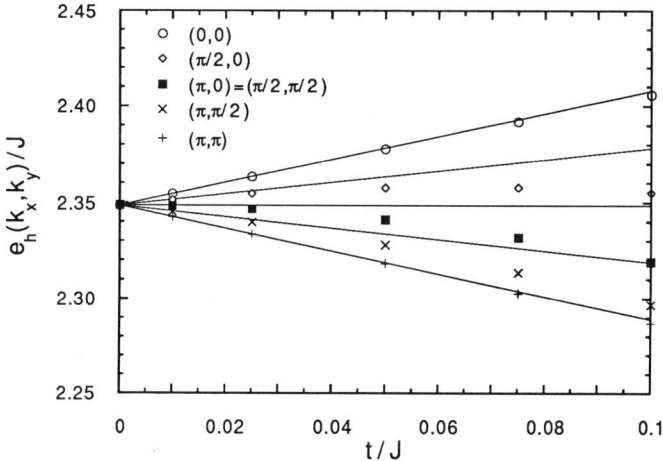

Figure 1. Spin-1/2 one-hole band structure for small t/J.

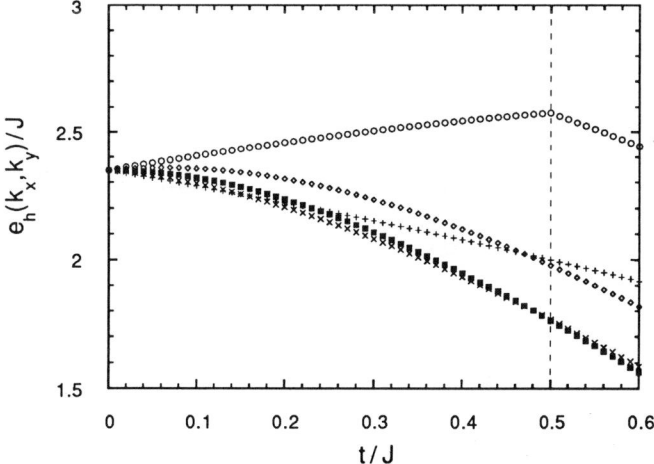

Figure 2. Spin-1/2 one-hole band for 0≤t/J≤0.5 and the supersymmetric point t/J=1/2.

Although Figure 1 and equation (4) appear to confirm that $\lim_{t/J \to 0} W_h = c_1 t$, there is considerable evidence that the coefficient c_1 vanishes in the bulk limit, probably as $c_1 \propto 1/L^2$. Elser, Huse, Shraiman and Siggia[12] argue that c_1 vanishes due to the nonzero bulk-limit staggered magnetization, which acts as a dimerization of the system and reduces the size of the Brillouin zone, resulting in a degeneracy of bulk-limit states which differ by $\Delta \vec{k} = (\pi, \pi)$. In language more appropriate to our bandwidth formula, we would say that the linear-t bandwidth is zero in the bulk limit because the static-hole matrix element in (5) vanishes in this limit. This matrix element is zero because the ground-state staggered magnetizations associated with the initial and final static holes on different sublattices (at $\vec{j}$ and $\vec{j}'$) have opposite signatures. This presumably leaves a bulk-limit one-hole bandwidth of $\lim_{t/J \to 0} W_h = c_2 t^2/J$. Note however that static-hole states with zero staggered magnetization, such as states with sufficiently large total spin, retain a linear-t bandwidth in the bulk limit at small t/J.

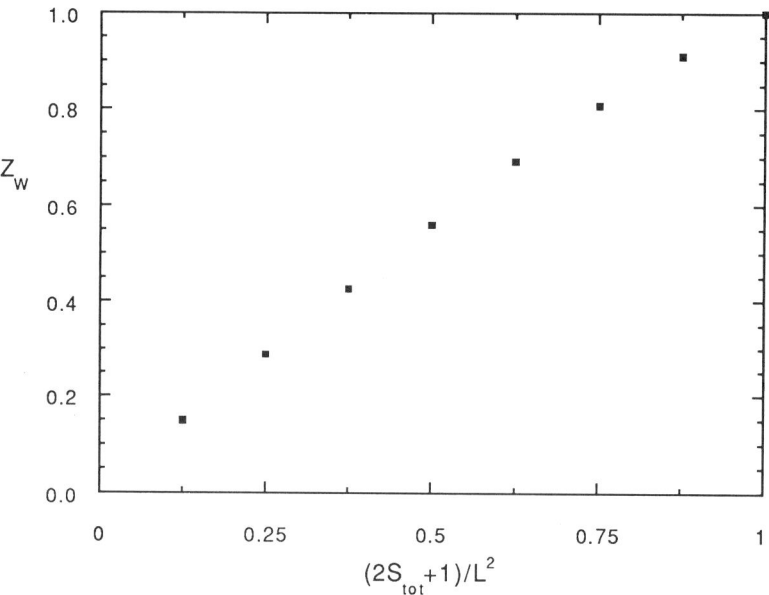

Figure 3. Bandwidth renormalization $Z_W = \lim_{t/J \to 0} W_h/8t$ versus S_{tot}.

An interesting question which has not been widely considered is the dependence of the one-hole bandwidth on the total spin. Previous studies have considered $S_{tot} = 1/2$ almost exclusively, as this is the one-hole ground state on a finite lattice. We can use our small-t/J bandwidth formula (5) to determine the linear-t coefficient of $W_h(S_{tot})$ on the 4×4 lattice from the static-hole ground state of the appropriate S_{tot}. Although we have argued that the linear-t term in the small-t/J bandwidth vanishes in the bulk limit for $S_{tot} = 1/2$, this is clearly not the case for all S_{tot}. For example, in the ferromagnetic state with $S_{tot} = S_{max} = (L^2 - 1)/2$, $W_h = 8t$ and $Z_W = 1$ is an exact result. The modulus of Z_W, determined from (5), is shown in Figure 3.

Some striking features of $|Z_W|$ are immediately apparent; it is approximately proportional to $2S_{tot} + 1$, and an obvious conjecture is that Z_W is nonzero if the total spin is a finite fraction of the maximum allowed spin S_{max}. The approximate linearity of Figure 3 can be summarized by

$$\lim_{t/J \to 0} |W_h(S_{tot})| \approx 8t \cdot \left\{ \frac{2S_{tot} + 1}{2S_{max} + 1} \right\} . \quad (6)$$

In view of the argument of Elser, Huse, Shraiman and Siggia that the linear-t bandwidth vanishes in the bulk limit due to the nonzero staggered magnetization, the nonzero bandwidth in (6) implies a conjecture for the pure Heisenberg antiferromagnet, which is that *the bulk-limit staggered magnetization of the 2D Heisenberg antiferromagnet vanishes in the lowest-lying state in any sector with $S_{tot}/L^2 > 0$.* In other words, there is no ground-state staggered magnetization in the bulk limit of the 2D Heisenberg model if the fraction of up (or down) spins differs from $1/2$. (This assumes that a single static hole does not have an important effect on the modulus of the bulk-limit staggered magnetization.) This conclusion does not require that $Z_W(S_{tot})$ has the particular approximate form suggested in (6), only that it is nonzero in the bulk limit for $S_{tot}/S_{max} > 0$. Another remarkable feature of $Z_W(S_{tot})$ is that it alternates in sign with increasing spin,

$$\frac{Z_W(S_{tot})}{|Z_W(S_{tot})|} = (-1)^{S_{tot}-1/2}. \tag{7}$$

We shall return to this result in our discussion of the supersymmetric point.

We note in passing that degenerate perturbation theory is not similarly applicable to leading-order bandwidth calculations for two or more holes, since these basis states have H_0 eigenvalues which depend on the relative hole locations. For these systems nondegenerate perturbation theory can be used to determine the two-hole bandwidth, for example, which should therefore satisfy $\lim_{t/J \to 0} W_{hh} = \kappa_2 t^2/J$. This form is consistent with our numerical results on the 4×4 lattice[10] with $\kappa_2 = 2.575(4)$.

As we increase t/J, the simple one-hole dispersion relation (4) begins to show higher-order effects due to multiple hops. This is visible in Figures 1 and 2 as a departure from linearity in the $S_{tot} = 1/2$ one-hole band, which becomes quite pronounced for $t/J \gtrsim 0.1$. As t/J increases there are evidently several changes in the ground-state quantum numbers; the t/J values of these changes and the quantum numbers of each ground-state level are given in Figure 4.

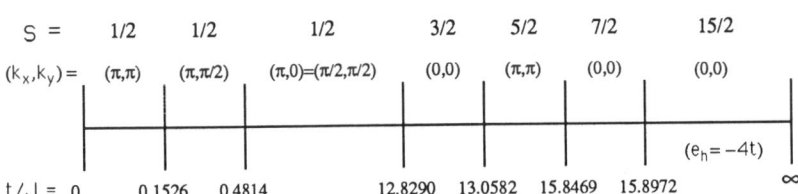

Figure 4. One-hole ground-state level crossings and quantum numbers on the 4x4 lattice.

The $(\pi/2, \pi/2)$ level, which becomes the ground state above $t/J \approx 0.4814$ on the 4×4 lattice, is believed to be the true bulk-limit ground state for t/J values of interest here, as the smaller-t/J crossings recede to $t/J = 0$ in the bulk limit as L^{-2}. (This is because these level crossings are due to the linear-t bandwidth term, which is expected to vanish for $S_{tot} = 1/2$ in the bulk limit.) At large t/J ($\gtrsim 13$), level crossings to ground states of higher total spin occur, as noted in Figure 4. These indicate the generation of a ferromagnetic Nagaoka polaron, and similar ground-state level crossings presumably first appear at a similar t/J value in the bulk limit.

THE SUPERSYMMETRIC POINT $t/J = 1/2$

Returning to Figure 2, a level crossing is apparent in the $\vec{k} = (0,0)$ one-hole states at exactly $t/J = 1/2$, where the $S_{tot} = 1/2$ level crosses a descending $S_{tot} = 3/2$ level. A close investigation of $t/J = 1/2$ reveals more such "accidents"; the energy of the (π,π) one-hole level at $t/J = 1/2$ is exactly $2J$, and the separation between the $(0,0)$ and (π,π) levels is $\approx 0.578529J$, which equals the singlet-triplet gap of the Heisenberg antiferromagnet (without vacancies) on a 4×4 lattice. These simple relations result from an spl(2,1) supersymmetry at $t/J = 1/2$, which has been discussed elsewhere by Förster[13] and Cappon[14].

This supersymmetry can be understood by considering the effect of the off-diagonal terms in the t-J Hamiltonian on two nearest-neighbor sites,

$$H_{tJ}(1,2) = -t\left(c_{1\uparrow}^\dagger c_{2\uparrow} + c_{1\downarrow}^\dagger c_{2\downarrow} + h.c.\right) + \frac{J}{2}\left((c_{1\uparrow}^\dagger c_{1\downarrow})(c_{2\downarrow}^\dagger c_{2\uparrow}) + h.c.\right) + H_{diagonal} . \tag{8}$$

Note that the spin-flip and hopping matrix elements have equal magnitudes at $t/J = 1/2$, which suggests that an additional symmetry might be present. This symmetry can be made more obvious by introducing slave-boson hole operators, which casts the first hopping term in the form

$$-t(c_{1\uparrow}^\dagger h_1)(h_2^\dagger c_{2\uparrow}) . \tag{9}$$

This is now a quartic in electron and hole operators with the same magnitude matrix element as the quartic spin-flip electron operator in (8). A sum of such quartic operators which is invariant under an infinitesimal rotation between electron (Fermi) and hole (Bose) operators (a supersymmetry transformation) is almost identical to the t-J Hamiltonian (1). There is a minor complication due to the $-\frac{1}{4}n_i n_j$ term in (1), which prevents the full t-J Hamiltonian from being supersymmetric. The operator $\tilde{H} = H_{tJ} - 4tN_h$ however is exactly supersymmetric at $t/J = 1/2$, so we find degeneracies between states in an irreducible supermultiplet which have the same hole number, and energy differences of $4t\Delta N_h$ between states in one supermultiplet which have different hole number.

The representation theory of this supersymmetry has been discussed by Cappon[14], who finds three types of irreducible representations. In terms of their total spin and hole content these representations decompose as

$$(S_{tot}, N_h) = (0, n_0) \oplus (1/2, n_0+1) \oplus (0, n_0+2) , \tag{10}$$

$$(S_{tot}, N_h) = (s_0, n_0) \oplus (s_0 - 1/2, n_0+1) \oplus (s_0 + 1/2, n_0+1) \oplus (s_0, n+2) \tag{11}$$

and

$$(S_{tot}, N_h) = (L^2/2, 0) \oplus (L^2/2 - 1/2, 1) . \tag{12}$$

All states within a supermultiplet have the same momentum $\vec{k}$.

The previous "accidents" in the $S_{tot} = 1/2$ one-hole energies on the 4×4 lattice can now be understood as relations between states within a supermultiplet. In particular, $e_h(\pi,\pi)/J = 2$ indicates that this one-hole state belongs to an $(S_{tot}, N_h) = (0,0) \oplus (1/2, 1) \oplus (0,2)$ supermultiplet, with the pure Heisenberg ground state as its spin-zero no-hole partner. Similarly, the degeneracy of $\vec{k} = (0,0)$ spin-1/2 and spin-3/2 levels shows that they belong to a single supermultiplet, necessarily $(1,0) \oplus (1/2,1) \oplus (3/2,1) \oplus (1,2)$, which they share with the lowest Heisenberg triplet level (down in energy by $4t$) and a spin-1 two-hole level (up by $4t$). This also explains why these one-hole levels are separated by the same gap ($\approx 0.578529J$) as the Heisenberg singlet and triplet levels; each one-hole level is $4t$ above its no-hole Heisenberg partner. As the singlet-triplet gap of the Heisenberg model is believed to vanish as L^{-2}, the $e_h(0,0) - e_h(\pi,\pi)$ splitting and hence the linear-t bandwidth coefficient c_1 should also vanish as L^{-2}.

As our attention was originally drawn to this supersymmetry by a degeneracy of $S_{tot} = 1/2$ and $S_{tot} = 3/2$ levels at $t/J = 1/2$, we are naturally led to consider the lowest-lying one-hole bands of higher spin, and to inquire whether or not similar degeneracies occur in these sectors. Rather than tracking each level in detail, as a first indication we simply use the small-t/J bandwidth formula (5), Figure 3, which together with the corresponding static-hole energy gives a linear-t approximate band pattern. The resulting approximate bands on the 4×4 lattice are shown in Figure 5.

Evidently we have found a sequence of level crossings at $t/J = 1/2$, which in order of increasing energy are
$S_{tot} = 1/2 \to 3/2$; $\vec{k} = (0,0)$ (discussed above)
$S_{tot} = 3/2 \to 5/2$; $\vec{k} = (\pi,\pi)$
$S_{tot} = 5/2 \to 7/2$; $\vec{k} = (0,0)$
$S_{tot} = 7/2 \to 9/2$; $\vec{k} = (\pi,\pi)$
and so on until
$S_{tot} = 13/2 \to 15/2$; $\vec{k} = (0,0)$,
which is the bottom of the $S_{tot} = 15/2$ band. This remarkable pattern of level crossings requires the inversion of alternate-spin bands, which was noted in (7). These degeneracies allow us to assign the lowest-lying $\vec{k} = (0,0)$ and $\vec{k} = (\pi,\pi)$ one-hole states to specific supermultiplets; evidently they all belong to multiplets of the type

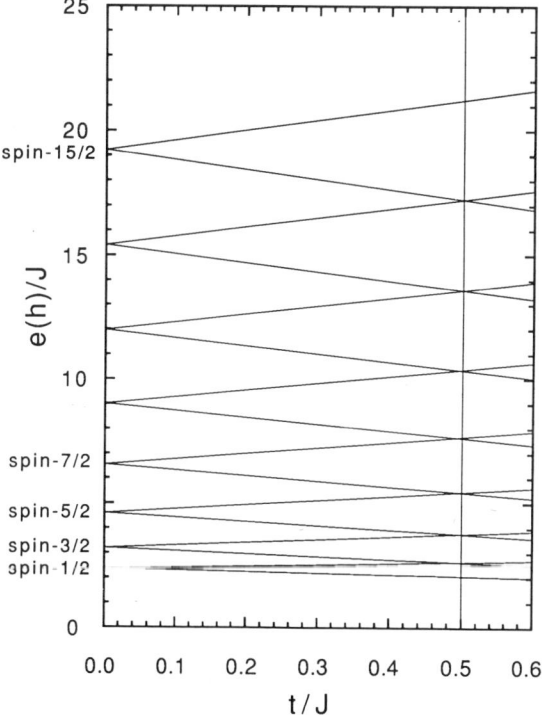

Figure 5. 4x4 higher-spin one-hole bands in the small-t/J approximation.

$$(S_{tot}, N_h) = (s_0, 0) \oplus (s_0 - 1/2, 1) \oplus (s_0 + 1/2, 1) \oplus (s_0, 2), \qquad (13)$$

which implies that each one-hole state is $4t$ higher in energy than the lowest-lying spin-s_0 pure Heisenberg state (in t-J model conventions). Comparison of our results with tabulated energies[15] for the higher-spin Heisenberg states on the 4×4 lattice confirms this identification.

EXTRAPOLATION TO BULK-LIMIT PROPERTIES AT t/J=3

Although we have found several very interesting results for the t-J model at small t/J, one can object that these have little to do with the original motivation for studying this model, which was to learn about a possible magnetic mechanism for high temperature superconductivity. This goal requires that we attempt to learn something about the relevant parameter regime of $t/J \approx 3$.

One might hope that a small lattice such as the 4×4 system discussed here is sufficiently large to learn about the bulk-limit properties of the lightly hole-doped sectors of the t-J model, and the many studies of the t-J model on small lattices tacitly assume this. Of course we cannot conclude with certainty that a 4×4 lattice is too small for this purpose without obtaining accurate results on larger lattices, but there is evidence nonetheless that the 4×4 system does experience large finite size effects for $t/J \gtrsim 1$, which we shall discuss.

Our small-t/J results can be used to estimate bulk-limit physics if we can convincingly demonstrate simple scaling behavior as a function of t/J for physically interesting quantities, which can then be used to extrapolate to the bulk limit at larger t/J. For sufficiently small t/J the holes are relatively immobile, so one expects properties such as the rms hole-hole separation in the two-hole ground state to be insensitive to the finite lattice extent in this limit. As a first test of possible scaling behavior we consider precisely this hole-hole separation, which is defined by

$$r_{rms} = \left\{ \langle \Psi_{0,hh}(\vec{k})| \, r^2 \, |\Psi_{0,hh}(\vec{k})\rangle \right\}^{1/2}. \quad (14)$$

On the 4×4 lattice there are three degenerate two-hole ground states, with $\vec{k} = (0,\pi), (\pi,0)$ and $(0,0)$. The first two have equivalent wavefunctions under rotations, so we consider only $(\pi,0)$ and $(0,0)$.

We anticipate that the hole-hole separation may scale as a power of t/J. This behavior has previously been reported[6,16] for the ground-state energy in various sectors of the t-J model, and can be motivated by the familiar argument that the hole moves in an approximately linear potential due to the trail of energetically unfavorable bonds it creates in hopping through a Néel background. This argument would lead us to expect that the energy of a hole should scale as $e_h/t \propto (t/J)^{-2/3}$, and the rms hole-hole separation as $r_{rms}/a_0 \propto (t/J)^{1/6}$. The approximate t/J dependence actually found for the energy of these d-wave two-hole states[16] is $e_{hh}/t \propto (t/J)^{-0.78}$, which deviates from the naive exponent of $-2/3$ in the direction expected if the linear potential is weakened at large distance by spin flips. A very naive picture of this effect may be obtained by replacing the linear potential by a weaker-than-linear power law[10], $V(r_{hh}) \propto r_{hh}^p$; given the energy exponent of -0.78 cited above we then estimate $p \approx 0.56$ from scaling arguments using the Schrödinger equation, and that the rms hole-hole separation should scale as

$$r_{rms}/a_0 \propto (t/J)^{1/(2p+4)} \approx (t/J)^{0.195}. \quad (15)$$

To test for such simple power-law behavior we show $\ln(r_{rms}/a_0)$ versus $\ln(t/J)$ in Figure 6. At $t/J = 0$ the holes are static and adjacent ($r_{rms} = a_0$), and as $t/J \to \infty$ they approach a limiting rms separation of $r_{rms} \approx 1.987 a_0$ which is fixed by the 4×4 lattice. For an intermediate range of t/J, approximately $0.3 \lesssim t/J \lesssim 1.0$, there is indeed evidence of power-law behavior, with a power remarkably close to the $(t/J)^{0.195}$ predicted by the simple estimate (15). A fit to these two states in this scaling region gives[10]

$$r_{rms}(\pi,0)/a_0 = 1.533(2) \cdot (t/J)^{0.198(4)} \quad (16)$$

and

$$r_{rms}(0,0)/a_0 = 1.423(2) \cdot (t/J)^{0.210(4)}, \quad (17)$$

which are shown as solid lines in Figure 6.

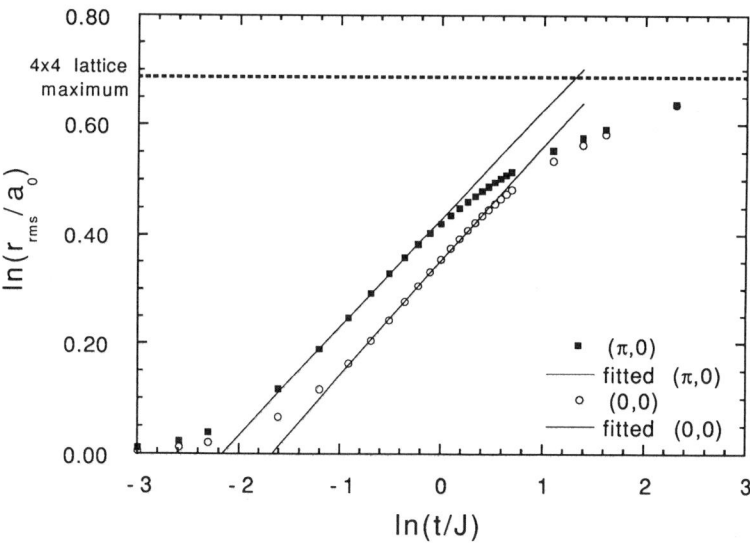

Figure 6. Ground-state rms hole-hole separation versus t/J.

One can see evidence here for finite size effects on the 4×4 lattice; there is a clear departure from the power law (16) for $t/J \gtrsim 1.0$ in the $(\pi, 0)$ state, and a similar departure at somewhat larger t/J in the $(0,0)$ state. This is the origin of our claim that there are important finite size effects in the two-hole ground states on a 4×4 lattice for $t/J \gtrsim 1$. Note however that the characteristic length scales of the two-hole bound states increase very slowly with t/J, so that studies at $t/J = 3$ on only moderately larger lattices would incorporate much smaller finite size effects.

Having found a simple scaling law for the rms hole-hole separation in the small-t/J regime, we can now extrapolate to $t/J = 3$ and estimate properties of the two-hole bound states in the high-T_c regime. In physical units our power-law parametrizations (16) and (17) imply

$$r_{rms}(\pi, 0) = 7.2\text{Å} \tag{18}$$

and

$$r_{rms}(0, 0) = 6.8\text{Å} \tag{19}$$

in the bulk limit. As these numbers are comparable to the in-plane coherence length of $\xi_{ab} \approx 14$Å reported for the high temperature superconductors[17], our results appear to support the identification of hole pairs with the Cooper pairs of high temperature superconductivity. (As the coherence length is only a qualitative length scale of the Cooper pair, it is not clear whether or not the factor-of-two difference between ξ_{ab} and r_{rms} is significant. The fact that we are studying an isolated hole pair in the insulating phase rather than in the metal may also affect the estimated pair size.) Of course the t-J model is incomplete in that it does not incorporate hole-hole Coulomb repulsion and hence suffers hole phase separation; we have studied the effect of such an interaction in the t-J model[9,18], and have concluded that this stops phase separation, results in a hole-hole binding energy of ~ 30 meV, and only slightly increases the rms hole-hole separation to about 8Å.

CONCLUSIONS

We have discussed numerical and analytic results for one- and two-hole energies and rms hole-hole separations in the t/J model at small t/J. An application of perturbation theory in the hopping parameter led to a relation between the linear-t term in the one-hole bandwidth and a static-hole ground-state matrix element. We also discussed evidence that this linear-t bandwidth term actually vanishes in the bulk limit, which leads us to a conjecture for the two-dimensional Heisenberg antiferromagnet: The ground states of the $S_{tot}/L^2 > 0$ sectors all have zero staggered magnetization in the bulk limit. We also gave t/J values and quantum numbers associated with the six one-hole ground-state level crossings. A crossing of $S_{tot} = 1/2$ to $S_{tot} = 3/2$ one-hole levels at exactly $t/J = 1/2$ is due to a supersymmetry at that point, which also leads to exact relations between energies of states having different numbers of holes. We find evidence for a sequence of related degeneracies between levels of higher spin. Finally, we discuss the extrapolation of small-t/J results to the high-T_c regime of $t/J \approx 3$. We find that the rms hole-hole separations in the two-hole ground states scale approximately as $(t/J)^{0.20}$, and an extrapolation to high-T_c parameters gives an rms hole-hole separation of ≈ 7Å in the bulk limit. This and other numerical evidence supports the identification of two-hole bound states with the Cooper pairs of high temperature superconductivity.

ACKNOWLEDGEMENTS

I would like to thank J.Ashkenazi and S.E.Barnes for their kind invitation to the Miami meeting on Electronic Structure and Mechanisms for High Temperature Superconductivity, and for the opportunity to discuss these results with my fellow participants. I would also like to thank S.Bacci, E.Gagliano and W.Stephan in particular for discussions of this work at the meeting, and K.Cappon, A.E.Jacobs, W.G.Macready and especially M.D.Kovarik for discussions and collaboration on many aspects of this work. This research was sponsored by the United States Department of Energy under contracts DE-AC05-840R21400 with Martin Marietta Energy Systems Inc and DE-AS05-76ER03956 with the Physics Department of the University of Tennessee, and by the State of Tennessee Science Alliance Center under contract R01-1061-68.

REFERENCES

[1] J.E.Hirsch, Phys. Rev. Lett. 54, 1317 (1985); C.Gros, R.Joynt and T.M.Rice, Phys. Rev. B36, 381 (1987); F.C.Zhang and T.M.Rice, Phys. Rev. B37, 3759 (1988); C.Gros, Phys. Rev. B38, 931 (1988); J.Bonča, P.Prelovšek and I.Sega, Phys. Rev. B39, 7074 (1989); J.A.Riera, Phys. Rev. B40, 833 (1989); E.Dagotto, A.Moreo and T.Barnes, Phys. Rev. B40, 6721 (1989); Y.Hasegawa and D.Poilblanc, Phys. Rev. B40, 9035 (1989).

[2] P.W.Anderson, Science 235, 1196 (1987); J.R.Schrieffer, X.-G.Wen and S.-C.Zhang, Phys. Rev. Lett. 60, 944 (1988).

[3] R.J.Birgeneau, Am. J. Phys. 58, 28 (1990).

[4] M.Marder, N.Papanicolaou and G.C.Psaltakis, Phys. Rev. B41, 6920 (1990); V.J.Emery, S.A.Kivelson and H.Q.Lin, Phys. Rev. Lett. 64, 475 (1990); J.A.Riera and A.P.Young, Phys. Rev. B39, 9697 (1989).

[5] T.Barnes, Oak Ridge/Caltech report ORNL-CCIP-90-06/C^3P-873, submitted to Int. J. Mod. Phys. C.

[6] E.Dagotto, UCSB preprint NSF-ITP-90-70 (April 1990), to appear in Int. J. Mod. Phys. B.

[7] T.Itoh, M.Arai and T.Fujiwara, Phys. Rev. B42, 4834 (1990).

[8] N.Bulut, D.Hone, D.J.Scalapino and E.Y.Loh, Phys. Rev. Lett. 62, 2192 (1989); N.Nagaosa, Y.Hatsugai and M.Imada, J. Phys. Soc. Jpn. 58, 978 (1989).

[9] T.Barnes and M.D.Kovarik, Phys. Rev. B42, 6159 (1990).

[10] T.Barnes, A.E.Jacobs, M.D.Kovarik and W.G.Macready, Oak Ridge/ Tennessee/ Toronto report ORNL-CCIP-90-08/UTK-90-10/UTPT-90-16 (December 1990), submitted to Phys. Rev. B.

[11] E.Dagotto, R.Joynt, A.Moreo, S.Bacci and E.Gagliano, Phys. Rev. B41, 9049 (1990).

[12] V.Elser, D.A.Huse, B.I.Shraiman and E.D.Siggia, Phys. Rev. B41, 6715 (1990).

[13] D.Förster, Int. J. Mod. Phys. B3, 1783 (1989); "Decoupling of Holes in the tJ Model at $J = 2t$ from spl(2,1) Symmetry", Max Plank Institut für Festkörperforschung report. See also Phys. Rev. Lett. 63, 2140 (1989); "The d=2 tJ model on a triangular lattice at large doping", Max Plank Institut für Festkörperforschung report; S.Sarkar, J. Phys. A23, L409 (1990).

[14] K.Cappon, "Supersymmetric t-J Model: An Exact Result for Two Dimensions", University of Toronto report.

[15] M.Gross, E.Sánchez-Velasco and E.Siggia, Phys. Rev. B40, 11328 (1989).

[16] E.Dagotto, J.Riera and A.P.Young, Phys. Rev. B42, 2347 (1990).

[17] W.C.Lee, R.A.Klemm and D.C.Johnston, Phys. Rev. Lett. 63, 1012 (1989). This reference notes that previous reports of somewhat larger values of $\xi_{ab} \approx 30$Å are inconsistent with recent magnetization measurements near T_c; see U.Welp, W.K.Kwok, G.W.Crabtree, K.G.Vandervoort and J.Z.Liu, Phys. Rev. Lett. 62, 1908 (1989).

[18] T.Barnes, M.D.Kovarik and W.G.Macready, ORNL-UTK-UTPT report (in preparation).

PLASMON EXCHANGE MODEL FOR THE HIGH-T_c SUPERCONDUCTORS

S.M. Bose

Department of Physics and Atmospheric Science
Drexel University, Philadelphia, PA 19104

P. Longe

S.U.P.R.A.S.
Institut de Physique, B5
Université de Liège, Sart-Tilman, B-4000 Liège, Belgium

INTRODUCTION

In this paper we present a plasmon exchange model for the occurrence of superconductivity in the cuprate superconductiors.[1] We first obtain analytic experssions for the effective interaction due to the plasmon exchanges between the charge carriers (holes or electrons) in the CuO layers. Using the Eliashberg model for strong coupling superconductors, we then show that these interactions can lead to superconductivity. Our theory can explain the experimentally observed rise in T_c with the increase in the number of CuO layers per cell in the Tl- and Bi-based compounds.[2,3] We further show that short coherence length is essential for superconductivity at high temperatures.

THE MODEL

We assume that the cuprate superconductor is a layered material consisting of a distribution of CuO layers. The Tl- and Bi-based compounds can have up to L(=1, 2, 3 and 4) CuO layers per cell. The charge carriers in the CuO layers are assumed to form a two-dimensional electron gas (2DEG) and interact with each other via plasmon exchanges in all layers.

We use the following parameters in our theory. The small interlayer distance between two CuO layers within the same cell a=3.2Å and the large interlayer distance between the adjacent

CuO layers in successive cells b=11.55Å, such that the lattice constant is c=b+(L-1)a. The charge carrier density n_s in a CuO layer, the effective mass m* of the charge carriers and the background dielectric constant ε are taken to be 0.004Å^{-2}, 4m_e and 12, respectively.

FORMALISM

The Interaction Potential

We first calculate the effective interaction between the charge carriers. The bare Coulomb interaction between two charge carriers in layers seperated by a height $z-z_0$ is

$$v(q, z-z_0) = v_0(q) e^{-q|z-z_0|} \qquad (1)$$

where

$$v_0(q) = \frac{2\pi e^2}{\varepsilon q}$$

The effective frequency dependent interaction between two charge carriers in the same layer at height z, due to plasmon exchanges among all layers can be written as

$$V(q,\omega) = v_0(q) - \sum_{z_1} v(q,z-z_1) \Pi(q,\omega) v(q,z_1-z)$$

$$+ \sum_{z_1,z_2} v(q,z-z_1) \Pi(q,\omega) v(q,z_1-z_2) \Pi(q,\omega) v(q,z_2-z) + \ldots$$

$$= v_0(q) \sum_{s=0}^{\infty} C_s(q) [-v_0(q) \Pi(q,\omega)]^s \qquad (2)$$

where the plasmon propagator $\Pi(q,\omega)$ in the RPA is

$$\Pi(q,\omega) \approx -\frac{n_s q^2}{m\omega^2} \qquad (3)$$

and the geometric factor

$$C_s(q) = \sum_{z_1 \ldots z_s} \prod_{r=0}^{s} e^{-|z_{r+1} - z_r| q} \qquad (4)$$

with $z_{s+1}=z_0=z$

Both $C_s(q)$ and the summation in Eq. (2) can be calculated in closed form for the cases L=1, .. 4, and the RPA interaction between the charge carriers takes the form

$$V(q,\omega) = \frac{2\pi e^2}{\varepsilon q} + \sum_{\kappa=1}^{L} \int_0^\pi d\kappa' \frac{2w_\kappa(\kappa',q)}{\omega^2 - [\omega_\kappa(\kappa',q)]^2} \qquad (5)$$

While the first term on the right hand side of Eq. (5) corresponds to the bare Coulomb interaction, the second term is the effect of the plasmon (boson) exchange. The functions $w_\kappa(\kappa',q)$ and $\omega_\kappa(\kappa',q)$ are defined elsewhere.[4,5] The second term can obviously become negative providing the attractive interaction for the formation of Cooper pairs.

The poles of $V(q,\omega)$ at $\omega=\omega_\kappa(\kappa',q)$ give the dispersion relations for the plasmon modes. The plasmon modes are obviously distributed into L bands labelled by κ in Eq. (5). These bands are shown in Fig.1 for the case L=3. Note that one of these bands ($\kappa=1$) is the pseudo-optical band and others ($\kappa=2, ...L$) are the acoustic bands. All these bands have a high frequency limit at $\kappa'=0$ and a low frequency limit at $\kappa'=\pi$. As Fig.1 shows, the accoustic bands are very narrow where $\omega \propto q$, but the pseudo-optical band has a width and for small q, $\omega=$ constant at the optical (high frequency) limit, and $\omega \propto q$ at the acoustic (low frequency) limit.

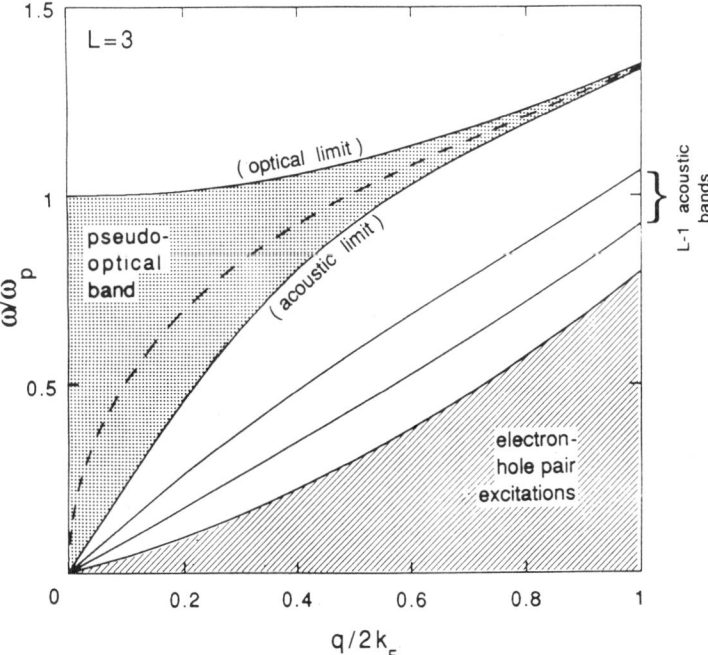

Fig.1. Schematic representation of the plasmon bands. The upper band is the pseudo-optical mode which has the proper optical mode at the upper limit and the acoustic mode at the lower limit. The other bands correspond to the L-1 narrow acoustic bands discussed in the text.

Calculation of T_c

Using the effective interaction given by Eq. (5), we now wish to calculate the T_c using the Eliashberg theory.[6] Because of the complexity of the original theory, some approximate analytic forms have emerged[7,8] and here we use the expressions recently given by Kresin[8] who along with several other authors[9] also considered the contribution of the plasmon to high-T_c superconductivity. He showed that T_c can be obtained from

$$T_c = 0.25\, \tilde{\omega}\, [e^{2/\lambda_{eff}} - 1]^{-1/2} \tag{6}$$

where $\tilde{\omega}$ is the average plasmon frequency and λ_{eff} is the effective interaction strength given by

$$\lambda_{eff} = \frac{\lambda - \mu^*}{1 + 2\mu^* + \lambda\mu^* t(\lambda)} \tag{7}$$

with $t(\lambda) = 0.75 + 0.8(1+\lambda)^{-1} - 0.12(\lambda - 0.5)$ for $0.5 < \lambda < 5$, the range of our interest. Our problem is then to calculate $\tilde{\omega}$, the attractive strength λ and the effective Coulomb repulsion μ^*. Using the effective interaction (Eq. (5)) in the standard expressions[7] for λ and $\lambda\langle\omega^2\rangle$, we find

$$\lambda = \frac{2N(0)}{\pi} \int_{q_m}^{2k_F} dq\, \frac{v_0(q)}{[4k_F^2 - q^2]^{1/2}} \tag{8}$$

and

$$\lambda\langle\omega^2\rangle = \frac{2N(0)}{\pi}\, \sigma \int_{q_m}^{2k_F} dq\, \frac{q\, v_0(q)\, S_L(q)}{[4k_F^2 - q^2]^{1/2}} \tag{9}$$

where $N(0)$ is the density of states at the Fermi curve and $\langle ... \rangle_{FC}$ corresponds to an average over the Fermi curve. Quantities σ and $S_L(q)$ appearing in Eq. (9) have been defined in Refs. 4 and 5.

Then the average frequency is given by

$$\tilde{\omega} = \sqrt{\frac{\lambda\langle\omega^2\rangle}{\lambda}} \tag{10}$$

and the retarded Coulomb repulsion strength is

$$\mu^* = \frac{\mu}{1+\ln(\frac{E_F}{\omega_0})} \quad (11)$$

with the average repulsive Coulomb strength μ calculated from

$$\mu = N(0)\langle v_s(q)\rangle_{FC} \quad (12)$$

where $v_s(q)$ represents the statically screened Coulomb interaction which yields $\mu \approx 1/2$. In Eq. (11) ω_0 is a boson frequency related to the retarding boson fields.

There are two difficulties in our theory which should be discussed. First, we do not know ω_0 and thus μ^*. We just know that we must have $\omega_0 < E_F$ such that $0 < \mu^* < \mu = 1/2$. Note that in the phonon exchange model, μ^* has been quoted to be 0.1.[7] Second, both λ and $\lambda\langle\omega^2\rangle$ diverge at the lower limit but $\tilde{\omega}$ given by Eq. (10) remains finite. Calculation of these intergrals with a non-zero lower limit q_m gives us results for λ and $\tilde{\omega}$ shown in Figs. 2 and 3. Note that λ is independent of L but diverges for $q_m \to 0$ and $\tilde{\omega}$ remains fairly constant for all L, except for $q_m \to 0$. However, as the following argument will show, low q's do not play an important role for superconductivity, as Cooper pairing occurs for $q\sim 2k_F$.

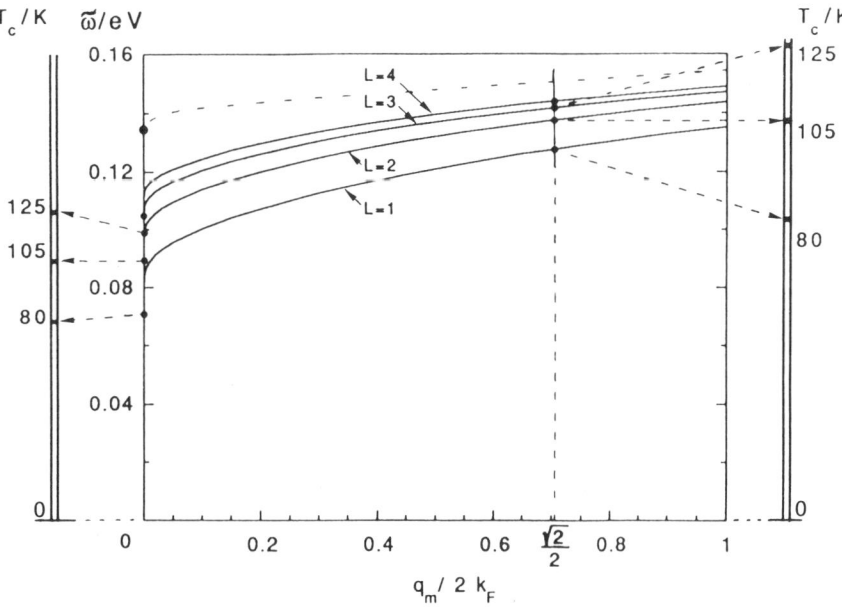

Fig.2. Plot of $\tilde{\omega}$ given by Eqs. (8-10) as a function of the lower limit of integration $q_m/2k_F$. For $q_m/2k_F = 0$ and $\sqrt{2}/2$, the relation $\tilde{\omega} \propto T_c$ is used to fit $\tilde{\omega}$ to T_c scales.

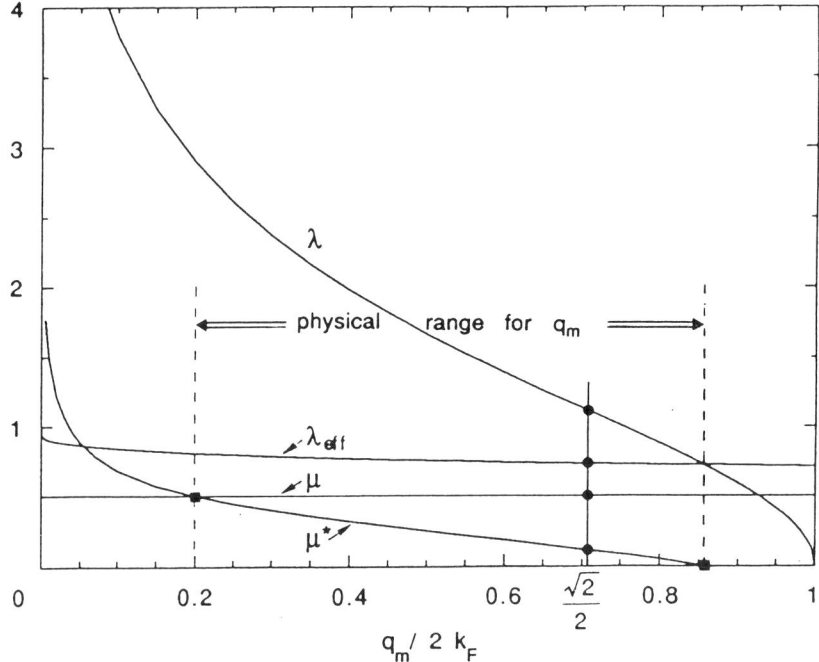

Fig.3. Plot of λ, λ_{eff}, μ and μ^* as functions of $q_m/2k_F$. At $q_m=0$, λ diverges. The physically allowed range of q_m is shown by the double arrow. The values of λ, λ_{eff} and μ^* used in calculation of Table I are shown by black dots.

We resolve the difficulities by noting that λ_{eff} is weakly dependent on L since λ is independent of L and hence we can write

$$T_c = C \tilde{\omega} \qquad (13)$$

where

$$C = \frac{0.25}{\sqrt{e^{2/\lambda_{eff}} - 1}} \qquad (14)$$

Now we calculate the constant $C(q_m)$ using the experimental value of $T_c=105K$ and the value of $\tilde{\omega}$ for L=2 in Eq. (13). Once $C(q_m)$ is known, we can calculate $\lambda_{eff}(q_m)$ and then $\mu^*(q_m)$ using (14) and (7), respectively. These results are shown in Fig. 3.

Note that since both λ_{eff} and $\tilde{\omega}$ are weakly dependent on q_m, T_c is also weakly dependent on q_m except for $q_m \to 0$, but as mentioned before, and for reasons given below, $q_m \to 0$ is not physically acceptable.

If we examine Fig. 3, we note that $\mu^* < \lambda$ for all q_m's, which is a prerequisite for superconductivity. Since we must have $0 < \mu^* < 1/2$, the physically acceptable range of q_m is shown by the vertical lines. Outside this range, our equations are not valid and we will not have superconducitvity.

The final question to answer is what the physically acceptable value of q_m is. If we plot the integrand of λ (Eq. (8)) as a function q, we note that it diverges at both q=0 and $q=2k_F$ and has a minimum at $k_F\sqrt{2}$. So we may decide to choose q_m to be $k_F\sqrt{2}$, since this choice eliminates the region where small q's are dominant and includes the region $q \sim 2k_F$ where the Cooper pairs are expected to be formed. It turns out that such a choice is also physically justified.

Physically, one expects $q_m \sim \xi_0^{-1}$ where ξ_0 is the coherence length. Using the BCS relation $\xi_0 = k_F/\pi m\Delta$ with $\Delta = 1.76 k_B T_c$ in our theory we find that $\xi_0 = 6$Å for $T_c = 105$K or $q_m = 1.04 k_F$, which is on the order of $q_m \approx k_F\sqrt{2}$. For high-T_c superconductors experimental ξ_0 is ~15Å which will lower q_m and will increase the variation of T_c with L. However, too large a value of ξ_0, as in the ordinary superconductors, will make $q_m \sim 0$, and will take us to the unphysical range for superconductivity i.e. there cannot be superconductivity for $q_m \sim 0$. This discussion thus implies that short coherence length is essential for the occurrence of superconductivity at high temperatures.

For numerical computation we have chosen $q_m = k_F\sqrt{2} = 0.224$Å^{-1} which gives us values of $\tilde{\omega}$, λ, λ_{eff} and μ^* indicated by black dots in Figs. 2 and 3. The values obtained are $\lambda = 1.11$, $\mu^* = 0.115$ and $\lambda_{eff} = 0.73$ Note that this μ^* is essentially equal to that quoted by other authors. Using these values in Eq. (6) we can obtain the T_c's for various values of L and the results are presented in Table I.

Table I. The calculated values for $\tilde{\omega}$ as obtained from Eq. (10) and T_c as obtained from Eq. (6) are shown for L=1, 2, 3 and 4.

L	$\tilde{\omega}$/eV	T_c/K
1	0.128	97
2	0.138	105 (exact)
3	0.142	108
4	0.144	110

Note that T_c rises with the number of CuO layers/cell. However, the range of variation of T_c is smaller than the experimental range. This may be primarily due to the choice of our parameters. There is also a saturation effect indicated by $T_{c4}-T_{c3}<T_{c3}-T_{c2}<T_{c2}-T_{c1}$.

CONCLUSIONS

In this paper we have shown that the plasmon exchange model for the interaction between the charge carriers can lead to high-T_c superconductivity in the cuprate superconductors. This theory can explain the rise in T_c with the number of CuO layers/cell in the Tl- and Bi- based compounds. This theory also predicts a saturation effect indicating that T_c cannot be increased indefinitely by increasing the number of CuO layers/cell. Finally, our calculation shows that high-T_c superconductivity is intimately connected with short coherence length.

ACKNOWLEDGMENTS

One of the authors (P.L.) is grateful to the Fonds National de la Recherche Scientifique, Belgium, for financial support. He also thanks Drexel University for its hospitality. The other author (S.M.B.) expresses his thanks to the University of Liège for financial support and hospitality.

REFERENCES

1. J.G. Bednorz and K.A. Müller, Z. Phys. B **64**, 189 (1986); M.K. Wu, J.R. Ashburn, C.J. Torng, P.H. Hor, R.L. Meng, L. Gao, Z.H. Huang, Y.Q. Wang and C.W. Chu, Phys. Rev. Lett. **58**, 908 (1987).
2. Z.Z. Sheng, A.M. Hermann, A.El Ali, C. Almason, J. Estrada, T. Datta, and R.J. Matson, Phys. Rev. Lett. **60**, 937 (1988); Z.Z. Sheng and A.M. Hermann, Nature **332**, 55 (1988); Z.Z. Sheng and A.M. Hermann, Nature **332**, 138 (1988).
3. H. Maeda, Y. Tanaka, M. Fukutomi, and T. Asano, Jpn. J. Appl. Phys. Pt. 2 **27**, 209 (1988); C.W. Chu, J. Bechtold. L. Gao, P.H. Hor, Z.J. Huang, R.L. Meng, Y.Y. Sun, Y.Q. Wang and Y.Y. Xue, Phys. Rev. Lett. **60**, 941 (1988).
4. S.M. Bose and P. Longe, J. Phys.: Condens. Matter **2**, 2491 (1990).
5. S.M. Bose and P. Longe, submitted to Phys. Rev. B; P. Longe and S. M. Bose, submitted to Phys. Rev. B.
6. G.M. Eliashberg, Zh. Eksp. Teor. Fiz. **38**, 966 (1960); **39**, 1437 (1960) [English translation: Soviet Phys. - JETP **11**, 696 (1960); **12**, 1000 (1961)].
7. W.L. McMillan, Phys. Rev. **167**, 331 (1968); P.B. Allen, in *Dynamical Properties of Solids*, edited by G.K. Horton and A.A. Maradudin (North Holland, Amsterdam, 1980) vol. 3, p. 95; P.B. Allen and R.C. Dynes, Phys Rev. B **12**, 905 (1975).
8. V.Z. Kresin, Phys. Lett. A. **122**, 434 (1987); Phys. Rev. B. **35**, 8716 (1987); Solid State Commun. **63**, 725(1987); V.Z. Kresin and H. Morawitz, Phys. Rev. B **37**, 7854 (1988).
9. J. Ashkenazi, C.G. Kuper and R. Tyk, Solid State Commun. **63**, 1145 (1987); J. Ruvalds, Phys. Rev. B **35**, 8869 (1987).

PROPERTIES OF SUPERLATTICES MADE OF HIGH T_c SUPERCONDUCTORS

P. Kumar, R. A. Guyer*, S. Obukhov# and Y. S. Sun+

Department of Physics
University of Florida
Gainesville, FL 32611

INTRODUCTION

A number of experiments[1-3] have reported measurements of the transition temperature and other properties of superlattices made from high T_c superconductors. These superlattices consist of M layers of $YBaCu_3O_7$ (T_c = 90 K), separated by N layers of $PrBaCu_3O_7$, which is a semiconductor. They find[3] that $T_c(M,N)$ for small M decreases with N and then saturates to an N independent value. It increases with increasing M for all N. For large N, when the superconducting layers can be considered independent, the transition temperature $T_c(M)$ increases linearly with M and saturates to the bulk T_c.

The most surprising result[4-5] however is that the superlattices, which are grown in the c direction for which the coherence length is expected to be only approximately 5Å, show T_c variation at length scales that are almost an order of magnitude larger. As the thickness of a superconducting film decreases, the order parameter diminishes due to the surface pairbreaking effects and leads to a suppression of T_c. This change ΔT_c, in a Ginzburg-Landau type description[5] scales with the correspomding coherence length. In high T_c superconductors, this effect is small because the characteristic coherence length is small. When the film thickness is of the order of coherence length, the above considerations break down and a 2–d, Kosterlitz-Thouless (K-T) type[6-8] description is more suitable. In 2–d, the formation energy of vortices is intrinsic and a thermal population of vortices and antivortices is expected to be present. As long as the vortices- antivortices bind to form a pair, the system remains superconducting. However, the pair dissociation leads to dissipation. The corresponding transition temperature scales with the binding energy of a vortex-antivortex pair. When the film thickness increases, this binding energy also increases and leads to a shift in T_c. All of these effects lead to a semi-quantitative agreement with the experiment. The important length scale in the problem involves the London penetration length parallel to the film plane which is large in high T_c superconductors. We also find that in order to fit the experimental data, we have to allow for a dead layer, a critical thickness of the order of the unit cell size in the direction normal to the surface (the thickness d is replaced by d-d_c). While this dead layer is derived from a fit to experiments, it appears reasonable to expect.

High-Temperature Superconductivity
Edited by J. Ashkenazi *et al.*, Plenum Press, New York, 1991

The description of superconducting transition in superlattices in terms of a Kosterlitz-Thouless theory is additionally motivated by the fact that the resistive transition for thin superconducting layers is broad and can be fitted to a K-T[7] behavior. In thin films, the hallmark of a K-T transition, nonlinear current-voltage characteristics has been seen[9]. The observation of this effect in superlattices would support the model presented here.

These considerations do not lead to a satisfactory understanding of the dependence of T_c on the normal layer thickness. The transition temperature of S-N superlattices saturates as a function of the normal metal thickness, both experimentally[3] and also in a Ginzburg-Landau description.[5] In the latter, the length scale for saturation is the coherence length in the semiconductor (defined as $y = \hbar v_F^N/(kT)$) where v_F^N is the Fermi velocity in the normal metal which is expected to be rather small, while in experiments, the saturation length scale is of order of 50Å, clearly an anomaly.

One might expect the 3–d transition, with decreasing normal metal thickness, to appear as a result of interaction between the magnetic fields associated with vortices in adjacent layers. This interaction is (a) rather weak (b) turns on when the separation between superconducting layers (insulator thickness) is of order of the penetration depth which for thin films can be as large as 10μ, and (c) couples the currents in adjacent superconducting layers and not the phases directly and therefore cannot lead to a conventional 3–d phase transition. However, the magnetic interaction between vortices in different planes has a profound effect on vortex interaction in one plane. In a single plane, the vortex energy density (and also the interaction) vanishes exponentially at large distances. As a consequence, single vortices can exist in a 2–d superconductor (in contrast to a superfluid where the London length is effectively infinite and the interaction remains logarithmic at all distances) and leads to dissipation at all temperatures i.e. $T_K = 0$. Because of the inter-plane magnetic coupling, the large scale intraplane interactions becomes logarithmic and the K–T transition occurs at a finite temperature. Also while in a paramagnet, this classical interaction effect is weak, it may be amplified significantly in the presence of an antiferromagnetic coupling between the Pr spins. We return to this subject below.

KOSTERLITZ-THOULESS TRANSITION IN A THIN FILM

For thin superconducting layers (small M), separated by thick non superconducting layers, the resistivity measurements show a broad transition characteristic of a K-T transition. Suppose that the isolated thin layer transition is of the K-T type, then the transition temperature T_K is given[6] by

$$kT_K = \frac{\hbar^2}{4m}\sigma_S(T_K) \qquad (1)$$

where $\sigma_S(T_K)$ is the superfluid areal density (cm^{-2}). The simple, physical derivation of this result compares the binding energy of a vortex-antivortex pair, which is the right-hand side multiplied by $\ln n^{-1}$ where $n^{-1/2}$ is the average distance between the pair constituents, to the entropy of distribution of these particles given by the left-hand side also multiplied by the same logarithmic factor. The free energy F = E - TS changes sign at T=T_K indicating an instability for the unbinding of the pair. The areal density in a thin film is given by $\sigma_S = \int_0^d dz \Delta^2(z)$ where z is the direction normal to the film of thickness d, and Δ is the order-parameter. If Δ is largely z independent but dependent on temperature T, Eq.(1) becomes

$$kT_K = \frac{\hbar^2 d}{4m} n_o \left(1 - \frac{T_K}{T_c}\right) \qquad (2)$$

$$\text{and} \quad \frac{T_K}{T_c} = \frac{d}{d + d_K}, \tag{3}$$

$$\text{where} \quad d_K^{-1} = \frac{\hbar^2 n_o}{4mT_c} = \frac{n}{8}\frac{T_c}{T_F}\xi_\parallel^2 \tag{4}$$

where the last expression for d_K is given in terms of the material parameters (another expression in terms of the London penetration length can be found in ref. (7,8). Here n represents the density of conduction electrons (cm^{-3}), T_c is the bulk transition temperature and $\xi_\parallel$ is the coherence length parallel to the film plane.

There are two additional effects: Δ depends on thickness d for small d and we account for that by imposing a cutoff d_c ($\sigma_s \propto$ (d-d_c)) and the many body problem of the vortices leads to screening of the interaction, normally expressed in terms of an effective dielectric constant. The latter effect can be incorporated via the Kosterlitz scaling equations[6,7] given by

$$\frac{dx}{dl} = 8\pi^2 y^2 \tag{5}$$

$$\frac{dy}{dl} = \frac{xy}{1+x} \tag{6}$$

where x(l) is related to the interaction strength (including screening effects) and y(l) to the pair excitation probability. The transition temperature T_K is determined by the separatrix in Eqs. (4,5) and corresponds to

$$\frac{32\pi^2 \epsilon_c k T_c}{\varphi_o^2 (d-d_c)} \frac{T_K}{T_c} = \frac{1}{\lambda_o^2} \tau \tanh\left[\tau \frac{\Delta(0)}{2kT_c}\frac{T_c}{T_K}\right] \tag{7}$$

where $\tau^2 = \left(1 - \frac{T_K}{T_c}\right)$, ϵ_c is the dielectric constant in the Coulomb gas description of the vortices and φ_o is the flux quantum. All other quantities have their usual meaning. The ratio $\frac{\Delta(0)}{kT_c} = 4$ and the dielectric constant has also been taken to be 4. Fig. (1) shows the variation in T_K with film thickness as derived from Eq. (6). While Eq. (6) is a quantitative description of $T_K(d)$, most physical effects are readily transparent in Eq. (3). If we consider the suppression of the order parameter due to the finite size effect, this can be included in Eq. (3) via the d dependence of the bulk $T_c(d)$. It is clear that if $\xi_\perp$ is small, the order parameter suppression effects are weaker. In an anisotropic superconductor, the entire d dependence of T_c is a consequence of the energetics of vortex-pair binding. It is also the case for small coherence length superconductors. .

DISCUSSION

It can be argued that the Kosterlitz-Thouless transition does not occur in an isolated 2-d superconducting film, contrary to previous suggestions,[7] the reason being that in a 2-d superconductor, the energy of an isolated vortex vanishes exponentially[11] for system size larger than the London penetration length λ. This is in contrast to a neutral superfluid film of ^{4}He for which λ is infinite. As a consequence, there is a finite density of free vortices which lead to dissipation and $T_K = 0$. In a superlattice, the classical-electromagnetic interaction[12,13] between different planes restores the long range logarithmic contribution to the vortex energy, as well as to the vortex-antivortex interaction. The corresponding length scale is governed by the parameter $K = \pi d\omega/\lambda^2$, where ω is the insulator thickness. For $\lambda \sim 1000$ Å and d ~ 10 Å, the length scale for insulator thickness is given by $\lambda^2/d = 10$ μ, an extraordinary length scale. The presence of magnetic moments (Pr) in the insulating

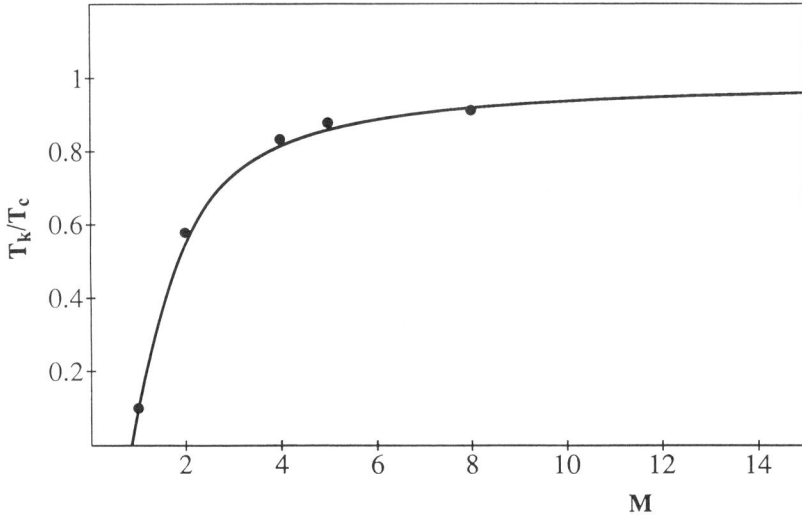

Figure 1. The transition temperature $T_c(M,N)$ as a function of M for large N. The calculations, from Eq. (7), are for independent superconducting layers. The experimental points are from ref. (3).

layer may alter this number significantly. What is required is the susceptibility at short distances (larger wavevectors), which in an antiferromagnet is comparable in magnitude to the uniform susceptibility of a ferromagnet (which typically is of the order of 5000). Below, we also consider some other alternatives.

In the first case, Anderson[11] has argued that in bulk (123) compounds, the resistivity in the c direction is orders of magnitude larger than the Mott's localization limit, the electron motion in the c direction is completely localized. The extended states appear only when the sample becomes superconducting. The effect is similar to the exchange instability in magnetic rare-earths: the ground state is a crystal field singlet with excited states that carry a finite moment and that are occupied because of the gain in the exchange interaction. It is unlikely that the superconducting transition will provide a length scale of 50 Å, given that all coherence lengths are less than 20 Å, albeit at T = 0.

There is however an electronic interaction that operates at a length scale of the order of 10 Å. The carriers in 123 compounds are holes in the CuO planes that are created because of the presence of acceptor atoms with which there is weak, screened Coulomb interaction. The eigenstates between adjacent planes are degenerate and a small splitting in the form of exchange interaction would split them into bonding and non-bonding orbitals. Since the orbitals have length scale of the effective Bohr radius $\sim \epsilon \hbar^2 / m e^2$ where ϵ is the dielectric constant of the background medium, the hopping energy between successive planes which is an overlap between the wavefunctions in the different planes, may be responsible for a long-range interaction between the holes in the adjacent superconducting planes. The effective Bohr radius is still of order 5Å and fails to account for the experiments.

The last possibility is the roughness of the S-N interface. During the epitaxial growth process, the interface fluctuates in units of a monolayer. Thus if the insulator is 2 monolayer thick, there are regions with a non-zero probability where the adjacent superconducting layers touch each other and the coherence between vortices in these layers is complete. When the insulating layer is more than 4 monolayers thick, the probability of touching superconducting layers is vanishingly small. The distance 50 Å translates into a 4 monolayer thickness. This mechanism appears most promising to us, although we cannot rule out some complex charge transfer process entirely.

It is a pleasure to thank D. Tanner and A. Tsvelik for stimulating discussions. This work was supported by a grant from DARPA MDA.972–B5–J–1006, by NSF DMR-9058572 and by the University of Florida, Division of Sponsored Research

REFERENCES

1. Q. Li, X.X. Xi, X.D. Wu, A. Inam, S. Vadlamannati, W. L. McLean, T. Venkatesan, R. Ramesh, D. M. Hwang, J. A. Martinez and L. Nazar, Phys. Rev. Lett. 64, 3986 (1990).
2. J. M. Triscone, O. Fischer, O. Brunner, L. Antognaza, A. D. Kent and M. G. Karkut, Phys. Rev. Lett. 64, 804 (1990); J. M. Triscone et al in Science and Technology of Thin Film Superconductors ed. by R. McConell and R. Naifi, Plenum Press (1990).
3. D. H. Lowndes, D. P. Norton and J. P. Budai, Phys. Rev. Lett. 65, 1160 (1990).
4. Properties of thin film superconductors can be found in reviews by J. P. Burger and D. Saint-James and by G. Deutscher and P.G. deGennes in "Superconductivity" ed. by R. Parks, Marcell Dekker, NY (1969).
5. "Phenomenological Approach to Superconducting Transition in Thin Films and S-N Superlattices" by R. A. Guyer et al, preprint, University of Florida (1991).
6. J. M. Kosterlitz and D. Thouless, J. of Phys. C6, 1181 (1973) and in Progress in Low Temperature Physics ed. by D. Brewer, North Holland, (1978). For a detailed list of references, please see Ref. (7) below.
7. J. E. Mooij, in Nato ASI on Percolation, Localization and Superconductivity ed. by A. Goldman and S. A. Wolf, Plenum Press, NY (1984).
8. L. C. Davis, M. Beasley and D. Scalapino, Phys. Rev. B42, 99 (1990).
9. M. A. Dubson, S. T. Herbert, J. J. Calabrese, D. C. Harris, B. R. Patton and J. C. Garland, Phys. Rev. Lett. 60, 1061 (1988) and N. C. Yeh and C. C. Tsuei, Phys. Rev. B39, 9708 (1989). The effect in Bi (2212) has been reported by S. N. Artemenko, J. G. Orlova and Y. I. Latyshev, JETP Lett. 49, 654 (1989).
10. J. Maps and R. Hallock, Phys. Rev. B27, 5491 (1983).
11. J. Pearl, App. Phys. Lett. 5, 65 (1964).
12. F. Guinea, Phys.Rev. B42, 6244 (1990).
13. R. Guyer, , P. Kumar, S. Obukhov and Y. S. Sun, "Magnetically Coupled Superconducting Layers", Preprint UF (1991).
14. P.W. Anderson, remarks at the ISI Workshop on High T_c Superconductors, Nov. 1990.

---

*:Permanent address: Department of Physics, University of Massachusetts, Amherst, MA 01003, #:Also with the Landau Institute for Theoretical Physics, Moscow, USSR., +:Present address: Department of Physics, Northwestern University, Evanston, Ill. 60208

ELECTRONIC STRUCTURE FERMI LIQUID THEORY OF HIGH T_c SUPERCONDUCTORS

Jaejun Yu and A. J. Freeman

Department of Physics and Astronomy
Northwestern University, Evanston, IL 60208-3112

Abstract

Predictions of local density functional (LDF) calculations of the electronic structure and transport properties of high T_c superconductors are reviewed. As evidenced by the excellent agreement with both photoemission and positron annihilation experiments, the Fermi liquid nature of the 'normal' state of the high T_c superconductors becomes clear for the metallic phase of these oxides. To understand the remaining anomalous features in the observed photoemission experiments, the effect of Coulomb correlations on the electronic structure is determined with the many-body corrections to the known band structure. The effective mass enhancement at E_F, renormalized band dispersions, and the energy dependence of the band state broadening for $Nd_{2-x}Ce_xCuO_4$ are predicted and found to be in agreement with observations in other Cu-oxide superconductors with a choice of the effective Coulomb interaction parameter $U = 3$ eV. In addition, LDF predictions on the normal state transport properties that are qualitatively in agreement with experiments on single crystals are discussed. It is emphasized that the signs of the Hall coefficients for the high T_c superconductors are not consistent with the types of dopants (e.g., electron-doped or hole-doped) but are determined by the topology of the Fermi surfaces obtained from the LDF calculations. Qualitative agreement with experiment is also found for the in-plane anisotropies in both compounds, pointing to the role of the Cu-O chains in the normal state transport properties of these compounds. Finally, with the use of new untwinned single crystal data for $YBa_2Cu_3O_7$, we find good agreement between the experimental and calculated resistivity versus temperature slope, with an estimated value of $\lambda_{tr} \approx 1$.

I Introduction

Ever since the discovery of high T_c Cu-oxide superconductors, there has been controversy and confusion as to whether the "normal" state of the Cu-oxide superconductors is a Fermi liquid — as given by local density energy band calculations — or some other exotic state. Recently, several angle resolved photoemission experiments [1, 2, 3] have provided direct observations of the energy band dispersions in the Cu-oxide superconductors (e.g., $YBa_2Cu_3O_{7-\delta}$ and $Bi_2Sr_2CaCu_2O_{8+x}$). The Fermi surfaces (FS) in $Yba_2Cu_3O_{7-\delta}$ and $Bi_2Sr_2CaCu_2O_{8+x}$ have been mapped by observing the dispersion of the electronic energy bands as they cross the Fermi energy E_F. The observed FS are in good agreement with the predictions of our local density band structure calculations.[4, 5] Moreover, this observation of a FS in $YBa_2Cu_3O_{7-\delta}$ by angle-resolved photoemission experiments is consistent with that of a recent positron annihilation experiment.[6, 7]

Despite the excellent agreement between the calculated[4, 5] and experimental[3, 2] Fermi surfaces as well as the energy bands near E_F for $YBa_2Cu_3O_{7-\delta}$ and $Bi_2Sr_2CaCu_2O_{8+x}$, some features in the photoemission spectra (PES) remain unexplained: (i) a rigid shift to higher binding energies by 0.5 eV of the experimental valence-band spectra compared to the calculated (band theory) spectra for $YBa_2Cu_3O_7$; (ii) a valence-band satellite peak commonly observed at -12 eV for $(RE)Ba_2Cu_3O_7$ samples; (iii) and, of relevance to this paper, a so-called "anomalous" linear energy dependence of the line

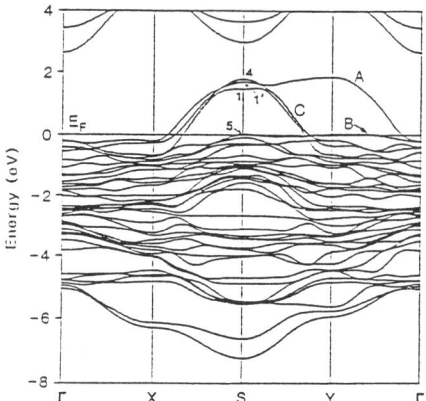

Figure 1. Band Structure of YBa$_2$Cu$_3$O$_7$ along symmetry directions in the $k_z = 0$ plane of the orthorhombic Brillouin zone

width and shape observed [8] in angle resolved photoemission spectra (ARPES) for Bi$_2$Sr$_2$CaCu$_3$O$_{8+x}$.

In attempting to understand the anomalous features observed in the ARPES spectra, new ideas[9, 10] have been proposed recently on the behavior of the normal metallic state of the Cu-oxide superconductors, in which the states are characterized as having a *renormalization factor* $Z_\mathbf{k} \to 0$ for states at E_F. Hence, the behavior of the normal state should be different from that of a conventional Fermi liquid and also the line width and shape observed in ARPES should be "anomalous" as the quasi-particle energy approaches E_F. On the other hand, however, there is an unsolved puzzle as to how such a strongly correlated system can yield the observed FS and energy bands near E_F that are in excellent agreement with band calculations which neglect those same strong correlation effects. Since the Cu d – O p hybrid states are rather itinerant and the single particle excitation spectra are reasonably well described by band calculations in their mean field (i.e., local density) approximation to local Coulomb correlations, many-body corrections based on the best possible band calculations are expected to represent a correct physical picture for these Cu-oxide superconductors.

In this paper, we review the results of our local density functional calculations of the electronic structure of high T$_c$ superconductors and compare them with recent experiments. A Fermi liquid picture of the 'normal' metallic state of the high T$_c$ superconductors is established through a detailed comparison of high-resolution angle-resolved photoemission spectra with our predictions of energy band dispersions and Fermi surfaces. We also discuss the effect of Coulomb correlations on the electronic structure of Cu-oxide superconductors based on the many-body corrections to the known band structure. Further, we present the results of the LDA predictions of the normal state transport properties, which are qualitatively in agreement with experiment.

II LDA Band Structure

For the electronic structure calculations, we used the highly precise full-potential linearized augmented plane wave (FLAPW) method[11] within the local density approximation (LDA) and the Hedin-Lundqvist form for the exchange-correlation potential. In the FLAPW approach no shape approximations are made to either the charge density or the potential. Results obtained on the high T$_c$ Cu-oxides we studied — La$_2$CuO$_4$, YBa$_2$Cu$_3$O$_{7-\delta}$, Bi$_2$Sr$_2$CaCu$_2$O$_8$, Tl$_2$Ba$_2$CaCu$_2$O$_8$ and Tl$_2$Ba$_2$Ca$_2$Cu$_3$O$_{10}$ — indicate a number of common chemical and physical features, especially the role of intercalated layers such as the CuO chains, Bi$_2$O$_2$ and Tl$_2$O$_2$ rock-salt type layers. In the following, we provide a brief summary of the results on the electronic structure of YBa$_2$Cu$_3$O$_7$ as an illustrative example, compare them with experiments, and discuss their implications on the nature of the normal ground state of the high T$_c$ Cu-oxides.

The calculated band structure[4] of stoichiometric YBa$_2$Cu$_3$O$_7$ along high symmetry directions in the bottom ($k_z = 0$) plane of the orthorhombic Brillouin zone is shown in Fig. 1. As seen, a remarkably simple band structure near E_F emerges from this complex set of 36 Cu-O hybrid bands. Four bands — two each consisting of Cu(2) d – O(2) p – O(3) p orbitals and Cu(1) d – O(1) p – O(4) p orbitals — cross E_F. Two strongly dispersed bands labeled C (S$_{1'}$ and S$_4$ in Fig. 1; the labels are given by their character at S) consist of Cu(2) $d_{x^2-y^2}$ – O(2) p_x – O(3) p_y combinations and have the 2D character,

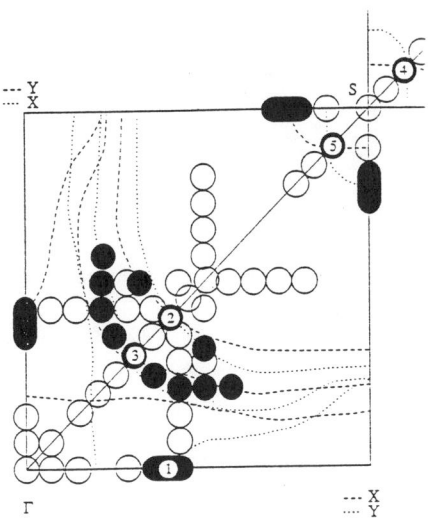

Figure 2. Comparison of the calculated Fermi surfaces with the experimentally determined Fermi surfaces of $YBa_2Cu_3O_7$. See the text for details. (This figure is taken from the paper by Campuzano et al.[3])

which is also common in other high T_c Cu-oxide systems. Significantly, the Cu(1) $d_{z^2-y^2}$ – O(1) p_y – O(4) p_z anti-bonding band labeled A (S 1 in Fig. 1) shows the (large) 1D dispersion expected from the Cu(1)–O(1)–Cu(1) linear chains but is almost entirely unoccupied. This band is in sharp contrast to the π anti-bonding band labeled B (formed form the Cu(1) d_{zy} – O(1) p_z – O(4) p_y orbitals) which is almost entirely occupied in the stoichiometric ($\delta = 0$) compound.

In Fig. 2, our predicted Fermi surfaces (FS) of $YBa_2Cu_3O_7$ determined from our band structure are compared with experimental Fermi surfaces determined by high-resolution angle-resolved photoemission (ARPES) experiments by Campuzano et al..[3] Filled circles indicate the points at which bands are found to cross the Fermi level. The open circles are the points in the Brillouin zone at which bands crossing E_F are not detected. (The size of the circles represents the experimental uncertainty in momentum.) The dashed lines are the predicted FS in Ref. [4]. Two 2D Cu-O $dp\sigma$ bands yield two rounded square FS's centered around S. It is remarkable that these 2D FS's have strong nesting features along the (100) and (010) directions, which are commonly found in other high T_c Cu-oxides except for the case of La-Sr-Cu-O. In addition, the 1D electronic structure also gives a 1D FS with again strong nesting features along the (010) direction. There are additional hole pockets around S(R) which come from the flat $dp\pi$ bands (band B) at E_F. In general, there is a remarkable agreement between the ARPES measurement and our theoretical predictions on the FS of $YBa_2Cu_3O_7$. All of the predicted FS are observed in the experiment. Furthermore, there is also good agreement between the FS measured in the ARPES experiment and the FS obtained by positron annihilation experiment.[6, 7] These results demonstrate that the observed Fermi surfaces in $YBa_2Cu_3O_7$ are in agreement with LDA band calculations and consistent with the Luttinger theorem; the large FS volume, which should be unaffected by the interaction, indicates the Fermi liquid nature of the normal ground state of these high T_c superconductors. It is important to note that the confirmation of the FS results has significant impact on several theories which deny the Fermi liquid nature of the normal ground state in the Cu-oxide superconductors.

Another detailed ARPES study has been carried out on the $Bi_2Sr_2CaCu_2O_8$ system by Olson et al.[8] and leads to a good agreement with the LDA calculated band dispersions and Fermi surfaces. Fig. 3 shows a comparison of the calculated energy bands with the observed bands by Olson et al..[8] (Filled circles in Fig 3 represent the observed band dispersions.) The agreement on the FS dimension between experiment and the LDA band calculations is good, as also expected from the case of $YBa_2Cu_3O_7$. The observed band dispersions along major symmetry lines are large and indicate an enhanced effective mass, which may come from the renormalization effects due to electron-phonon interactions or Coulomb correlations.

In short, the existence of Fermi surfaces and the large band dispersions observed in $YBa_2Cu_3O_7$ and $Bi_2Sr_2CaCu_2O_8$ together with such good agreement between LDA band theory and experiment strongly supports a Fermi liquid description of these high T_c superconductors.

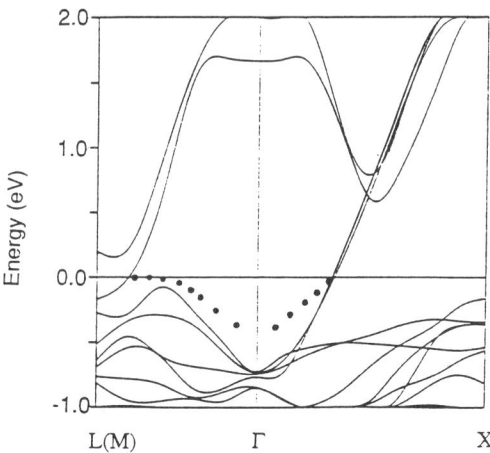

Figure 3. Comparison of the calculated energy bands with the measured bands in the ARPES experiments[8] for $Bi_2Sr_2CaCu_2O_8$.

III Coulomb Correlated Band Structure

Despite the confirmation of our predicted Fermi surface in $YBa_2Cu_3O_7$ and $Bi_2Sr_2CaCu_2O_8$ by angle resolved photoemission experiments, some anomalous features seen in these experiments stimulate questions on the nature of the normal state of the Cu-oxide superconductors. In attempting to understand these anomalous features, we studied the effect of Coulomb correlations on the electronic structure of Cu-oxide superconductors.[12] Many-body corrections to the known band structure are made by calculating perturbatively the self-energy $\Sigma_n(\omega)$ as shown in Fig. 4, where the on-site Coulomb interaction at the Cu sites is given by the effective interaction Hamiltonian

$$H_1 = \frac{U}{2} \sum_{i\nu\sigma\nu'\sigma'} (1 - \delta_{\nu\nu'}\delta_{\sigma\sigma'}) n_{di\nu\sigma} n_{di\nu'\sigma'} \qquad (1)$$

with $n_{di\nu\sigma} = d^+_{i\nu\sigma} d_{i\nu\sigma}$ being an occupation number at the Cu site. As a direct consequence of the self-energy corrections, the energy shift and broadening observed in photoemission spectra are calculated for the $Nd_{2-x}Ce_xCuO_4$ system as an example.

The calculated results of the single particle spectra $n(\varepsilon)$ with the effective Coulomb interaction demonstrate that the major peaks of the valence-band spectra with $U = 3$ eV are shifted by ~ 1 eV toward higher binding energy in comparison to the spectra with $U = 0$ eV. A major contribution to the relative shift of the spectra arises from the first order self-energy term $\Sigma_n^{(1)}$ in Fig. 4(a). The major corrections due to $\Sigma_n^{(1)}$ give rise to the relative shift of the fully occupied valence bands toward lower energies (i.e., higher binding energies), which is attributed to the self-interaction correction (SIC) for the Cu d orbitals in the hybrid band states. The results explain the small difference between the LDA calculated valence band spectra[13] and the experimental spectra measured[14] for $YBa_2Cu_3O_{6.9}$, namely, the residual 0.5 eV rigid shift of the entire valence band spectra except for the bands crossing E_F. When the 0.5 eV rigid shift is employed, there is a one-to-one correspondence of the calculated peaks with those in the experimental spectra. The rigid nature of the shift may come from the strong hybridization of the Cu d – O p orbitals; unless the Cu d and O p orbitals are strongly hybridized, one would expect a much larger shift for the Cu d states than for the O p states and corresponding changes for the states at E_F. (A recent resonant photoemission study[15] showed that the states at E_F have both Cu d and O p character — in agreement with the band theoretical results.[16]) The resulting agreement with PES experiments for $YBa_2Cu_3O_7$, $Bi_2Sr_2CaCu_2O_8$, and $Nd_{2-x}Ce_xCuO_4$ leads to the conclusion that the rigid shift found in the experimental valence band spectra is due to the presence of Coulomb correlations at the Cu-site and that these shifts can be explained by including correlation effects into the strongly hybridized Cu d – O p valence bands.

The second order self-energy correction $\Sigma_n^{(2)}(\omega)$ of Fig. 4(b) for the band crossing E_F is shown in Fig. 5. From the energy dependence of $\Sigma^{(2)}(\omega)$, we can expect few qualitative changes in the band states near E_F due to Coulomb correlation. The real part of the self-energy contributes to the shift of

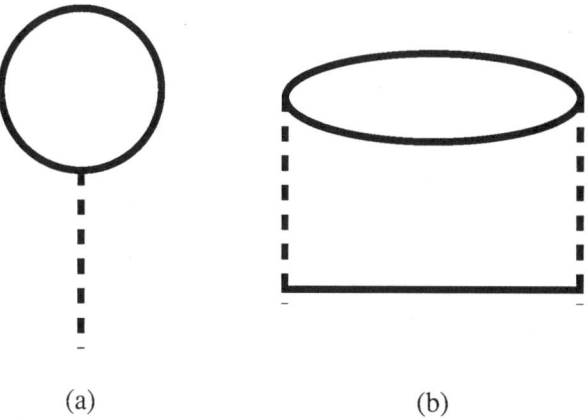

Figure 4. Diagrammatic representations of the perturbation series expansion of (a) the first order self-energy $\Sigma_{n\sigma}^{(1)}(\mathbf{k}\omega)$ and (b) the second order self-energy $\Sigma_{n\sigma}^{(2)}(\mathbf{k}\omega)$.

the original band eigenvalues and the imaginary part of the self-energy corresponds to the broadening (or inverse of the quasi-particle life-time) of the band state.

First, there is a band narrowing due to the energy level shift as $\mathrm{Re}(\partial\Sigma/\partial\omega)|_{E_F} < 0$, as seen, for example, by comparing the calculated spectra in Fig. 5 with $\mathrm{Re}\Sigma > 0$ for $\varepsilon < 0$ and $\mathrm{Re}\Sigma < 0$ for $\varepsilon > 0$. This means that that the renormalized band, $E_{\mathbf{k}} = \varepsilon_{\mathbf{k}} + \mathrm{Re}\Sigma$, has a narrower band width than the original band $\varepsilon_{\mathbf{k}}$. Consequently, the band narrowing will lead to an enhancement of the effective band mass. The calculated renormalized band with $\lambda_{e-e} = 0.28$ for $U = 3$ eV is consistent with the large band dispersion observed in the ARPES experiments for $YBa_2Cu_3O_7$ and $Bi_2Sr_2CaCu_2O_8$. The experimental mass enhancement factor λ_{exp} is estimated to be $0 \lesssim \lambda_{exp} \lesssim 1$ (i.e., $m^* \lesssim 2m_{band}$) from the experimentally observed data[8] in comparison with the LDA calculated band dispersion. This small mass enhancement in the Cu-oxide systems indicates that the effective U of this system is much smaller than that for heavy fermion systems[17] where the typical mass enhancement due to Coulomb correlation is about 10. As a result, we have a large *renormalization factor* $Z_k = 0.78$, which implies that there exists a sharp discontinuity of the momentum distribution (i.e., sharp FS) at $T = 0$ and that the existence of a sharp FS is not affected by the presence of Coulomb correlation. These results are consistent with the observed FS in $YBa_2Cu_3O_7$ and $Bi_2Sr_2CaCu_2O_8$ in agreement with LDA band calculations, where the large FS volume unaffected by the Coulomb correlations indicates the Fermi liquid nature of the normal ground state of these high T_c superconductors. This is also consistent with our starting assumption that the normal metallic ground state of high T_c Cu-oxides can be described by a LDA band structure with perturbative corrections due to Coulomb correlations.

Second, as shown in Fig. 6, the inverse of the quasi-particle lifetime goes to zero as $\omega \to E_F$. According to Landau Fermi liquid theory, the quasi-particle lifetime at E_F should be infinitely long and $\mathrm{Im}\sum(\omega \to E_F)$ should vanish as $\mathrm{Im}\sum \sim (\omega - E_F)^2$ near $\omega = E_F$.[18] The calculated $\Gamma(\omega) = |\mathrm{Im}\Sigma(\omega)|$ shown in Fig. 6 seems to satisfy this $(\omega - E_F)^2$ behavior when ω is very close to E_F, namely, for $|\omega - E_F| < 0.05$ eV. The ω-dependence of $\Gamma(\omega)$ deviates strongly from the $(\omega - E_F)^2$ form for $|\omega - E_F| \gtrsim 0.1$ eV, and becomes linear, i.e., $\sim |\omega - E_F|$ for $|\omega - E_F| \gtrsim 0.25$ eV. As also shown in Fig. 5, the linear behavior of $\Gamma(\omega)$ extends from $|\omega - E_F| \approx 0.25$ eV to almost 0.7 eV, where the range of linearity depends strongly on the band structure features like the density-of-states, band-width, and band-filling. In fact, the linearity of $\Gamma(\omega)$ shown here arises mostly from the highly asymmetric shape of the density-of-states for the band crossing E_F. As a result, the $\mathrm{Im}\Sigma(\omega)$, i.e., $\Gamma(\omega)$, as illustrated in Fig. 5, exhibits a highly asymmetric ω-dependence for $\omega > E_F$ and $\omega < E_F$. Since no experimental data for $\Gamma(\omega)$ are available for $Nd_{2-x}Ce_xCuO_4$, we compare the calculated $\Gamma(\omega)$ for $Nd_{2-x}Ce_xCuO_4$ with the experimental $\Gamma(\omega)$ estimates for $Bi_2Sr_2CaCu_2O_6$.[8] Although the theoretical values are well below the experimental values, the calculated ω dependence of $\Gamma(\omega)$ seen in Fig. 6 is quite consistent with experiment. (After all, it should be kept in mind that the experimental estimates include contributions from final states, limited energy- and angular-resolution, etc., and also depend

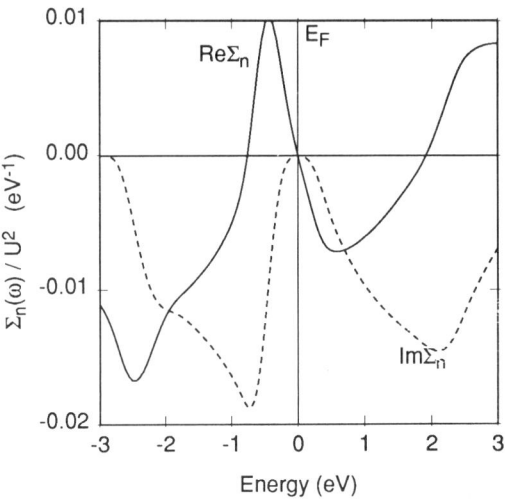

Figure 5. Energy dependence of the second order self-energy $\Sigma^{(2)}(\omega)$ for the band crossing E_F in $Nd_{2-x}Ce_xCuO_4$: the solid curve gives the real part and the dashed curve the imaginary part.

upon the sample quality as seen in many other experiments.[19]) For $U = 3$ eV in $Nd_{2-x}Ce_xCuO_4$, we have $\Gamma(\omega) = 0.1$ eV at $|\omega - E_F| = 0.4$ eV, which is comparable to the experimental fit[8] of $\Gamma(|\omega - E_F| = 0.5) \approx 0.12$ eV in the case of $Bi_2Sr_2CaCu_2O_8$.

Although the linear behavior of the calculated $\Gamma(\omega)$ for $|\omega - E_F| \gtrsim 0.25$ eV is in agreement with the experimental observation of $\Gamma(\omega)$, there exists a large quantitative difference between the theoretical and experimental values of $\Gamma(\omega)$ for $|\omega - E_F| \to 0$. Namely, the extrapolated values of the observed $\Gamma(\omega)$ are estimated to be proportional to $|\omega - E_F|$ even for $|\omega - E_F| \to 0$. Obviously, our Fermi liquid result, $\Gamma(\omega) \propto |\omega - E_F|^2$ for $|\omega - E_F| \to 0$ does not fit the linear $|\omega - E_F|$ behavior very near $|\omega - E_F| = 0$ (e.g. in the region of $|\omega - E_F| < 0.25$ eV). This discrepancy between the Fermi liquid predictions on $\Gamma(\omega)$ and the observed $\Gamma(\omega)$ can be understood in two ways. First, it is possible to consider a background broadening, Γ_b, which may arise from a possible scattering of out-going photoelectrons with phonons, impurities, defect or surface states on the sample surface. In other words, since the surface condition of these Cu-oxides is not ideal, there is a large chance of having surface distortions or impurity phases, which could be an active scattering center for the out-going photoelectrons.

Similar non-lifetime effects in photoemission linewidths have been demonstrated in the high resolution measurements of the intrinsic linewidth for photoemission from bulk and surface states of Cu.[20] This experiment demonstrates that the weakened wave vector conservation may dominate the observed linewidth and the non-lifetime effect is consistent with possible impurity scattering by the extremely small (≈ 0.01 monolayer) surface contamination present.[20] If we assume a background broadening $\Gamma_b \approx 30$ meV, then the resulting line broadening[20] $\Gamma_t(\omega)$ becomes $\Gamma_t(\omega) = \Gamma(\omega) + \Gamma_b$. The total $\Gamma_t(\omega)$, including a background broadening, is shown in Fig. 6 and compared with the calculated and observed $\Gamma(\omega)$; it is seen that $\Gamma_t(\omega)$ seems to be in reasonable agreement with experiment at least for $|\omega - E_F| > 0.1$ eV. Indeed, $\Gamma_t(\omega)$ shows a linear ω-dependence $\sim \alpha|\omega - E_F|$ for $|\omega - E_F| > 0.25$ eV with ($\alpha = 0.25$), while approaching to the saturated value of Γ_b as $|\omega - E_F| \to 0$. Unfortunately, there are only very few measurements of the linewidth and these are not sufficient to resolve this problem due to experimental conditions (or sample problems).

Second, Fermi surface nesting is another possible source of a large contribution to the inverse lifetime of the quasi-particle. As is well-known, the calculated FS of the Cu-oxide superconductors (except for $La_{2-x}M_cCuO_4$) show a large degree of FS surface nesting along the (100) and (010) directions in the Cu-O plane.[4, 16, 21] As previously discussed, the FS nestings observed experimentally for $YBa_2Cu_3O_7$ are in agreement with our predictions. Due to the two-dimensional (2D) nature of the FS, the nesting of the FS reduces to a one-dimensional problem. Some consequences of the nested 2D FS have been extensively studied.[22] Neglecting the wave-vector dependence of the imaginary part of the susceptibility, Virosztek and Ruvalds[23] obtained $\Gamma(\omega) \propto |\omega - E_F|$ for small $|\omega - E_F|$. Further, Levin et al.[24] pointed out that, due to the band narrowing effects, $\Gamma(\omega) \propto |\omega - E_F|$ as $\omega \to E_F$ when the Fermi level is close to the van Hove singularity in the density-of-states. This means that the FS nesting

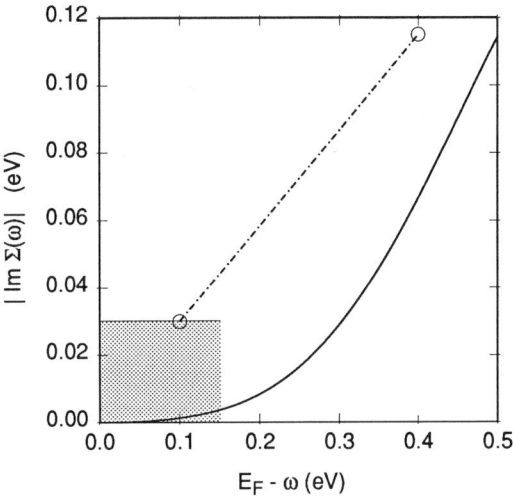

Figure 6. Comparison of the theoretical and experimental broadening of the band states, where the solid line represents our calculated values and the circles and dot-dashed line the experimental estimate from Ref. [8]. The shaded area illustrates a possible background broadening Γ_b discussed in the text.

effect may contribute significantly to the lifetime broadening with a linear energy behavior for small $|\omega - E_F|$, which certainly helps us to understand the qualitative difference between our calculated $\Gamma(\omega)$ and experiments for $|\omega - E_F| \lesssim 0.25$ eV. We should note, however, that the $\Gamma(\omega) \propto |\omega - E_F|$ from the FS nesting does not imply an anomalous behavior of the renormalization constant Z as $|\omega - E_F| \to 0$. When the wave-vector dependence of the imaginary part of the susceptibility is properly taken into account, we obtain a non-singular behavior of Z as $|\omega - E_F| \to 0$, which is again consistent with Luttinger's theorem.

IV Normal State Transport Properties

Normal state transport properties are of course an important test of the band theory of high T_c superconductors. Calculations of the Hall coefficients based on the band theory have previously been performed for $La_{2-x}M_xCuO_4$, $YBa_2Cu_3O_7$,[25] $Nd_{2-x}Ce_xCuO_4$ and $Ba_{1-x}K_xBaO_3$.[26] Among other results, they predicted a change of sign in the current carriers at $x \approx 0.24$ for La-M-Cu-O, and positive and negative signs for the Hall coefficients in Y123 when the magnetic field is aligned parallel and perpendicular to the c-axis, respectively. The latter prediction for Y123 has been confirmed by single crystal experiments[27], while the actual x value at which R_H of $La_{2-x}M_xCuO_4$ changes sign is found to be somewhat larger ($x \approx 0.3$)[28]. As for $Nd_{2-x}Ce_xCuO_4$, we predicted a positive Hall coefficient when the magnetic field $\mathbf{H} \parallel \mathbf{c}$ and negative otherwise; experiments by Takagi et al.[29] show a negative R_H value for small doping, turning positive at $x \approx 0.18$. Recent measurements[28] on single crystals of superconducting $Nd_{2-x}Ce_xCuO_{4-\delta}$ found a positive value of R_{xyz} (R_H with $\mathbf{H} \parallel \mathbf{c}$) at all temperatures up to 300 K (with the exception of one sample where R_{xyz} was positive only below 80 K). R_{xyz} was found to be temperature dependent and to saturate at 300 K at 2×10^{-9} m^3/C, which is about 4 times the value obtained by Hamada et al.[26] based on band theory. The temperature dependence cannot however be recovered by the band approach, at least in the simple approximation used.

A brief summary of both theoretical and experimental results on the Hall coefficients is given in Table I and II. We present the results of a band theory based calculation of the Hall coefficients for cubic perovskite $Ba_{1-x}K_xBaO_3$ and for $Nd_{2-x}Ce_xCuO_4$, $YBa_2Cu_3O_7$, and $YBa_2Cu_4O_8$ together with those of Allen et al.[25] for $La_{2-x}Sr_xCuO_4$ and $YBa_2Cu_3O_7$. Our approach is based on the Bloch-Boltzmann theory, and rests on the Midgal approximation, where the sign of the Hall coefficients is mainly determined by the curvature of FS, i.e., the FS topology. Thus, one expects to have a positive R_H from the hole-like FS and a negative R_H from the electron-like FS, while the open FS can give rise to either positive or negative R_H depending on details of the FS geometry. Note that the "doped-insulator" picture of high T_c superconductors is in disagreement with the experimental observations. On the other hand, the LDA band theory gives correct predictions on the signs of

Table I. Hall coefficients with **H** ∥ **c**

	"doped" carrier	Fermi surface topology	band theory	experiment
$Ba_{1-x}K_xBiO_3$	hole	electron	$-$ [a]	$-$
$Nd_{2-x}Ce_xCuO_4$	electron	hole	$+$ [a]	$+(?)$
$La_{2-x}Sr_xCuO_4$	hole	hole $\to$ electron	$+ \to -$ [b]	$+ \to -$
$YBa_2Cu_3O_7$	hole	hole	$+$ [b]	$+$
$YBa_2Cu_4O_8$	hole	hole	$+$ [c]	$(?)$

[a] Ref. [39]; [b] Ref. [25]; [c] Ref. [38]

Table II. Hall coefficients with **H** ⊥ **c**

	"doped" carrier	Fermi surface topology	band theory	experiment
$Ba_{1-x}K_xBiO_3$	hole	electron	$-$ [a]	$-$
$Nd_{2-x}Ce_xCuO_4$	electron	open	$-$ [a]	$(?)$
$La_{2-x}Sr_xCuO_4$	hole	open	$-$ [b]	$-$
$YBa_2Cu_3O_7$	hole	open	$-$ [b]	$-$
$YBa_2Cu_4O_8$	hole	open	$+/-$ [c]	$(?)$

[a] Ref. [39]; [b] Ref. [25]; [c] Ref. [38]

the Hall coefficients as determined by the FS topologies. For example, although the $Ba_{1-x}K_xBiO_3$ is regarded as "hole-doped", the Hall coefficient is not positive, but negative in agreement with the LDA calculations. Comparing the results of Hall coefficient measurements with the (p-type or n-type) "doped-insulator" picture, we are led to the conclusion that a separation of carriers into positive and negative in the spirit of the effective mass approximation is not possible for these copper oxides.

Together with the Hall coefficients, the linear in-plane resistivity observed in high T_c Cu-oxide superconductivity has been quoted as an "unusual normal state property" of these materials. The high value of the slope $\alpha = d\rho/dT$, and $\beta \approx 0$ ruled out the electron-phonon scattering mechanism as a dominant source of resistance. It was also pointed out that attempts to describe $\rho(T)$ in the framework of electron-phonon scattering lead to internal contradictions[30] and the earlier band theory predictions of $\rho(T)$ for Y123 by Allen et al.[25] turned out to be too small in comparison with the (twinned) single crystal data for $\rho(T)$ available at that time.

Although the observed linearity of the in-plane resistivity, $\rho_{ab}(T)$, in crystals and polycrystals of Bi-Sr-Ca-Cu-O and $YBa_2Cu_3O_7$ is remarkable, the deviation of $\rho(T)$ from linearity in Y124 and in oxygen deficient Y123 is found to be interesting. In particular, a successful fitting of the $\rho(T)$ curve was made for Y124 on the basis of Bloch-Grüneisen-type behavior with a transport Debye temperature $\Theta_D^{tr} \approx 500$ K by Martin et al.,[31] who suppressed superconductivity with a magnetic field along the c axis. This would imply that electron-phonon scattering is sufficient to explain the resistivity behavior in this compound. Another interesting point is the presence of a kink in $\rho(T)$ at ≈ 250 K and interpreted by Schoenes et al. as a tendency to saturation. In addition, recent untwinned single crystal measurements[32] of $\rho(T)$ in $YBa_2Cu_3O_7$ show much smaller values of $\rho(T)$ (by more than a factor of 2) than those for both twinned single crystals[27] and oriented films.[33] These observations suggest that the electron-phonon scattering mechanism may be responsible for the peculiar behavior of $\rho(T)$ in high T_c superconductors.

Since untwinned crystals are available for Y124, the in-plane resistivity anisotropy, ρ_a/ρ_b, of Y124 has been experimentally estimated to be ≈ 3.[34, 35] Also recent experiments on untwinned Y123 single crystals[32] showed $\rho_a/\rho_b \approx 2.2$. This in-plane anisotropy points to the importance of the Cu-O chains in the normal state transport of these systems. The metallicity of the Cu-O chains is in fact considered to be necessary[34, 36] to explain this anisotropy, and arises naturally from band structure calculations.[4, 37] Therefore, along with these data now available with single crystals, it is necessary to examine the band theory predictions of $\rho(T)$ in $YBa_2Cu_3O_7$ and $YBa_2Cu_4O_8$ to try to understand the linearity and magnitude of $\rho(T)$ in terms of the electron-phonon scattering mechanism.

In a recent paper, we have reported[37] on the electronic structure and Fermi surface of Y124 and have compared these results of those obtained previously for Y123.[4] We also reported[38] results of calculations of the resistivity $\rho(T)$ with τ_{e-ph} arising from the electron-phonon scattering of Y123 and Y124.

To understand the origin of the linearity of $\rho(T)$ in high T_c superconductors, we estimated the

Table III. Calculated quantities defining the resistivity of Y123 and Y124 (see text), and comparison of the values obtained with experiment. $\alpha_i = 4.03 \cdot (\lambda_{tr}/\Omega_{pi}^2)$ $\mu\Omega$-cm/K with Ω in eV and $\rho_i(T) = \alpha_i T$.

	Y124		Y123		
	Calc.	Exp.[†]	Calc.	Exp.[*]	Units
$\sum_j (\eta_j/M_j)$	0.086		0.114		eV/(Å^2-amu)
$<\omega^2>_{tr}^{1/2}$	247		247		K
λ_{tr}	0.79		1.05		
α_a^{e-ph}	0.31	0.63	0.58	0.55	$\mu\Omega$-cm/K
α_b^{e-ph}	0.13	0.63	0.22	0.24	$\mu\Omega$-cm/K
α_c^{e-ph}	6.49		4.89	12.5	$\mu\Omega$-cm/K
$l_{ab}(300K)$	18		8		Å
α_a/α_b	2.4	2.8[‡]	2.6	2.2	
α_c/α_b	50	–	22	55	

[†] Ref. [41] (c-axis oriented unfaulted film)

[‡] Ref. [34]

[*] Ref. [32] (untwinned single crystal)

magnitude of $\rho(T)$ for Y123 and Y124 due to electron-phonon scattering following the earlier work by Allen et al.[25] For a given value of $\langle\omega^2\rangle_{tr}^{1/2} = 247$K and the calculated values of $\sum_j (\eta_j/M_j)$, we obtain $\lambda_{tr} = \sum_j (\eta_j/M_j)\langle\omega^2\rangle_{tr}^{-1}$; the results are listed in Table III. For the value of $\lambda = 1.05$ (i.e., $\langle\omega^2\rangle_{tr}^{1/2} = 247$ K chosen), the agreement between theory and experiment for the magnitude of resistivity of Y123 is remarkably close. In particular, the calculated in-plane α_i^{e-ph} ($i = a, b$) values are in excellent agreement (within 10%) with experiment. These α values for Y123, in good agreement with experiment, seem to be consistent (within the frame work of band theory) with the electron-phonon scattering mechanism. This agreement between band theory and experiment is quite plausible, since our estimated value of the mean-free-path $l_{ab}(T = 300K)$ is still larger than the interatomic distance of $a \approx 2$ Å and does not violate the Mott-Ioffe-Regel criterion.[40] However, our results are still in contradiction with the experimental observation that the Bloch-Boltzmann (linear) behavior of the measured resistivity persists up to 600 K, implying $l_{ab}(T \approx 600K) \gg a$ for Y123. In other words, from our estimate of $l_{ab}(T)$, a significant deviation from the Bloch-Boltzmann behavior of $\rho(T)$ should be seen above $T \approx 300$ K, where $l_{ab}(T \approx 300K)/a \approx 4$, whereas the deviation of $\rho(T)$ from linearity has not been observed up to 600 K in experiments.

This contradiction leads to the question of whether the Bloch-Boltzmann theory based on the energy band structure completely breaks down for this problem or if there is any resolution for such a contradiction. For example, Allen et al.[25] suggested the use of the reduced value of Ω_p, admitting a significant discrepancy between LDA energy eigenvalues and the real quasiparticle spectrum. In fact, the use of small value of Ω_p^* and λ_{tr}^* (e.g., $\Omega_p^* = 1.4$ eV and $\lambda^* = 0.27$ as given in the optical measurement[42]) seems to give a reasonable value of $\rho(T)$, which is consistent with experiment. Further, the mean-free-path dilemma looks fairly well resolved if we adopt the relation $\Omega_p^{*2} \sim (n/m)_{eff} \sim k_F^3$ of the free electron-like case with a variable density n. However, this resolution comes from the assumption that the number of carriers scales as k_F^3, but, in the band theory, we should notice that $\Omega_p^{*2} \propto 1/m_{eff}^*$ if the Fermi surface volume is conserved following Luttinger's theorem. Thus, the smaller the Ω_p^* the larger m_{eff}^*, so that $l \approx v_F\tau \propto 1/m_{eff}^*$ reduces to a smaller value unless τ grows as fast as $(m_{eff}^*/m)^{1+\delta}$. Therefore, the renormalized band scheme, having reasonable renormalized parameters like Ω_p^*, λ_{tr}^*, and m_{eff}^*, can not resolve the mean-free-path dilemma unless we have a strongly suppressed electron-phonon scattering for a large value of m_{eff}^* (or small Ω_p^*).

In contrast to the case of Y123, the Y124 system can be well described in terms of Bloch-Boltzmann theory with our band structure. First of all, from our calculations, the average in-plane Fermi velocity, $\langle v_{a,b}^2\rangle^{1/2}$, of Y124 is almost two-times larger than that of Y123. In addition, the calculated λ_{tr} of Y124, using the same $\langle\omega^2\rangle_{tr}^{1/2}$ as that of the Y123 system, is smaller than that of Y123. As a result, the mean-free-path l_{ab} (i.e., $l_{ab}(T \approx 300K)/a \approx 9$) of Y124 turns out to be much larger than that of Y123 (i.e., $l_{ab}(T \approx 300K)/a \approx 4$). (Moreover, if the $\langle\omega^2\rangle_{tr}^{1/2}$ value of Y124 is larger than that of Y123 as suggested by Martin et al.,[31] the l_{ab} of Y124 becomes even larger than the one estimated in Table III.) Hence we find greater consistency in the description of the resistivity in Y124 with the Bloch-Boltzmann theory using the energy band structure than in the cases of $YBa_2Cu_3O_7$ and $La_{2-x}M_xCuO_4$,[30] where the band structure description of $\rho(T)$ was heavily criticized.

In Table III, we also compare our calculated values of $\alpha_{a,b}$ of the resistivity for Y124 with available experimental data. Since, however, the best available resistivity data for Y124 is still on the c-axis oriented unfaulted films,[41] comparisons of the absolute value of α_i^{e-ph} for Y124 with experiment seem not to be appropriate due to the nature of the sample quality (recall that it was seen in Y123 that the factor of two difference in the resistivity can be obtained easily depending on sample characteristics[19]). In this respect, it appears essential to have a high quality single crystal measurement of $\rho_i(T)$ for Y124, in order to clarify the nature of the transport (e.g., dc-resistivity) properties of Y124 in connection with other high T_c superconductors, especially with Y123.

Despite the large uncertainty in the magnitude of λ_{tr} (which again leads to the large uncertainty in α_i^{e-ph}), the in-plane resistivity anisotropy (ρ_a/ρ_b) remains independent of λ_{tr}, i.e.,

$$\frac{\alpha_a}{\alpha_b} = \left(\frac{\Omega_b}{\Omega_a}\right)^2. \qquad (2)$$

As seen in Table III, the calculated in-plane anisotropy of Y123, $\alpha_a/\alpha_b = 2.6$ is in reasonable agreement with the experimental value $\alpha_a/\alpha_b = 2.2$. (Here we emphasize that the resistivity data provided are from the untwinned single crystal,[32] where the in-plane resistivity is very close to the intrinsic value but the c-axis resistivity is still in question.)

V Conclusions

The existence of Fermi surfaces and large band dispersions observed in both $YBa_2Cu_3O_7$ and Bi_2Sr_2-$CaCu_2O_7$ together with such good agreement between LDA band theory and experiment strongly supports a Fermi liquid description of these high T_c superconductors. As a Fermi liquid (metallic) nature of the 'normal' state of the high T_c superconductors becomes clear, these experimental observations have served to confirm the predictions of our local density functional calculations and hence the energy band approach as a valid natural starting point for further studies of their superconductivity.

We have also presented results of many-body corrections to the band structure including Coulomb correlation with an effective interaction parameter $U = 3$ eV, that are in agreement with the experimental PES spectra[43] and that explains the small difference between the LDA calculated valence band spectra[13] and the spectra measured[14] for $YBa_2Cu_3O_{6.9}$; namely, the residual 0.5 eV rigid shift of the entire valence band spectra except for the bands crossing E_F. The small mass enhancement found together with the calculated $Z \approx 0.78$ indicates that the effective Coulomb correlation in the Cu-oxide systems is much weaker than that of the heavy fermion systems. The calculated $\Gamma(\omega) \approx \text{Im}\Sigma(\omega)$ is proportional to $(\omega - E_F)^2$ when ω is very close to E_F, as predicted by Fermi liquid theory; however, such a behavior of $\Gamma(\omega)$ is limited to the very small region close to E_F, i.e., $|\omega - E_F| < 0.05$ eV. It was shown that the calculated $\Gamma(\omega)$ for $|\omega - E_F| \gtrsim 0.25$ eV becomes linear in agreement with the experimental observations.[8] The remaining quantitative differences between theoretical and experimental values of $\Gamma(\omega)$ are discussed and attributed to a possible background broadening, which may arise from surface scattering, and the strong 2D FS nesting present in the Cu-oxide systems.

In addition, LDF predictions on the normal state transport properties that are qualitatively in agreement with experiments on single crystals are discussed. It is emphasized that the signs of the Hall coefficients for the high T_c superconductors are not consistent with the types of dopants (e.g., electron-doped or hole-doped) but are determined by the topology of the Fermi surfaces obtained from the LDF calculations. Qualitative agreement with experiment is also found for the in-plane anisotropies in both compounds, pointing to the role of the Cu-O chains in the normal state transport properties of these compounds. Finally, with the use of new untwinned single crystal data for $YBa_2Cu_3O_7$, we find good agreement between the experimental and calculated resistivity versus temperature slope, with an estimated value of $\lambda_{tr} \approx 1$.

Acknowledgments

We thank S. Massidda, K. T. Park, N. Hamada, and D. D. Koelling for collaboration on the work reported here. Work supported by the National Science Foundation (through the Northwestern University Materials Research Center, Grant No. DMR88-21571) and by a grant of computer time at the NASA Ames Supercomputing Center and the National Center for Supercomputing Applications, University of Illinois at Urbana-Champaign.

References

[1] C.G. Olson, R. Liu, A.-B. Yang, D.W. Lynch, A.J. Arko, R.S. List, B.W. Veal, Y.C. Chang, P.Z. Jiang and A.P. Paulikas, Science **245**, 731 (1989).

[2] C.G. Olson, R. Liu, D.W. Lynch, B.W. Veal, Y.C. Chang, P.Z. Jiang, J.Z. Liu, A.P. Paulikas, A.J. Arko and R.S. List, Physica C**162–164**, 1697 (1989).

[3] J.C. Campuzano, G. Jennings, M. Faiz, L. Beaulaigue, B.W. Veal, J.Z. Liu, A.P. Paulikas, K. Vandervoort, H. Claus, R.S. List, A.J. Arko and R.J. Bartlett, Phys. Rev. Lett. **64**, 2308 (1990).

[4] S. Massidda, J. Yu, A.J. Freeman and D.D. Koelling, Phys. Lett. A**122**, 198 (1987);J. Yu, S. Massidda, A.J. Freeman and D.D. Koelling, *ibid.* **122**, 203 (1987).

[5] S. Massidda, J. Yu and A.J. Freeman, Physica C **152**, 251 (1988).

[6] L.C. Smedskjaer, J.Z. Liu, R. Benedek, D.G. Legnini, D.J. Lam, M.D. Stahulak, H. Claus and A. Bansil, Physica C**156**, 269 (1988).

[7] H. Haghighi, J.H. Kaiser, S.L. Rayner, R.N. West, J.Z. Liu, R. Shelton, M.J. Fluss, R.H. Howell, F. Solal, *The University of Miami Workshop on: Electronic Structure and Mechanisms for High Temperature Superconductivity*, 2-9 January 1991.

[8] C.G. Olson, R. Liu, D.W. Lynch, R.S. List, A.J. Arko, B.W. Veal, Y.C. Chang, P.Z. Jiang and A.P. Paulikas, Phys. Rev. B**42**, 381 (1990).

[9] C.M. Varma, P.B. Littlewood, S. Schmitt-Rink, E. Abrahams and A.E. Ruckenstein, Phys. Rev. Lett. **63**, 1996 (1989).

[10] P.W. Anderson, Phys. Rev. Lett. **64**, 1839 (1990).

[11] H.J.F. Jansen and A.J. Freeman, Phys. Rev. B30, 561 (1984), and references therein.

[12] J. Yu and A. J. Freeman, Physica C **173**, 274 (1991).

[13] J. Redinger, A.J. Freeman, J. Yu and S. Massidda, Phys. Lett. A**124**, 469 (1987).

[14] A.J. Arko, R.S. List, R.J. Bartlett, S.-W. Cheong, Z. Fisk, J.D. Thompson, C.G. Olson, A.-B. Yang, R. Liu, C. Gu, B.W. Veal, J.Z. Liu, A.P. Paulikas, K. Vandervoort, H. Claus, J.C. Campuzano, J.E. Schirber and N.D. Shinn, Phys. Rev. B**40**, 2268 (1989).

[15] J.W. Allen, C.G. Olson, M.B. Maple, J.-S. Kang, L.Z. Liu, J.-H. Park, R.O. Anderson, W.P. Ellis, J.T. Markert, Y. Dalichaouch and R. Liu, Phys. Rev. Lett. **64**, 595 (1990).

[16] S. Massidda, N. Hamada, J. Yu and A.J. Freeman, Physica C**157**, 571 (1989).

[17] H.R. Ott and Z. Fisk, *Handbook on Physics and Chemistry of the Actinide*, vol. 5, p. 85, eds. A.J. Freeman and G.H. Lander (North–Holland, 1987).

[18] M. Luttinger, Phys. Rev. **121**, 942 (1961).

[19] B. Batlogg, *High Temperature Superconductivity: The Los Alamos Symposium – 1989*, eds. K. Bedell, D. Coffey, D. Meltzer, D. Pines and J. R. Schrieffer, (Addison–Wesley, 1990).

[20] S.D. Kevan, Phys. Rev. Lett. **50**, 526 (1983); J. Tersoff and S.D. Kevan, Phys. Rev. B**28**, 4267 (1983).

[21] J. Yu, S. Massidda and A.J. Freeman, Physica C**152**, 273 (1988).

[22] A review article by K. Levin, J.H. Kim, J.P. Lu, and Q. Si, (preprint).

[23] A. Virosztek and J. Ruvalds, Phys. Rev. B (to appear).

[24] Q. Si and K. Levin, (unpublished) cited in Ref. [22]; see also N. Bulut, D.J. Scalapino, and H. Morawitz, (unpublished)

[25] P.B. Allen, W.E. Pickett and H. Krakauer, Phys. Rev. B36, 3926 (1987); P.B. Allen, W.E. Pickett and H. Krakauer, Phys. Rev. B37, 7482 (1987).

[26] N. Hamada, S. Massidda, Jaejun Yu, and A.J. Freeman, Phys. Rev. B**42**, 6238 (1990).

[27] S.W. Tozer, A.W. Kleinsasser, T. Penney, D. Kaiser and F. Holtzberg, Phys. Rev. Lett. **59**, 1768 (1987).

[28] For a review, see for instance N.P. Ong in *Physical Properties of High Temperature Superconductors*, Vol. II, Eds. D.M. Ginsberg, pp. 459-507 (1990), and references therein.

[29] H. Takagi, S. Uchida and Y. Tokura, Phys. Rev. Lett. **62**, 1197 (1989).

[30] M. Gurvitch and A.T. Fiory, Phys. Rev. Lett. 59, 1337 (1988).

[31] S. Martin, M. Gurvitch, C.E. Rice, A.F. Hebard, P.L. Gammel, R.M. Fleming, and A.T. Fiory, Phys. Rev. B**39**, 9611 (1989).

[32] T.A. Friedmann, M.W. Rabin, J. Giapintzakis, J.P. Rice, and D.M. Ginsberg, Phys. Rev. B **42**, 6217 (1990).

[33] Bozovic et al., PRL 59, 2219 (1987).

[34] J. Schoenes, J, Karpinspki, E. Kaldis, J. Keller, P. de la Mora, Physica C **166**, 145 (1990).

[35] B. Bucher, J. Karpinski, E. Kaldis, and P. Wachter, Physica C **167**, 324 (1990).

[36] A. Schilling, A. Bernasconi, H.R. Hott, and F. Hulliger, Physica C **169**, 237 (1990).

[37] Jaejun Yu, K.T. Park, and A.J. Freeman, presented at CAMES, Tokyo, Aug. 28-31, 1990, and Physica C **172**, 467 (1991).

[38] S. Massidda, J. Yu, K.T. Park, and A.J. Freeman, Physica C (submitted).

[39] N. Hamada, S. Massidda, J. Yu and A.J. Freeman, Phys. Rev. B **42**,6238 (1990).

[40] N.F. Mott and E.A. Davis, Electronic Process in Non-crystalline Materials (Clarendon, Oxford, 1979); A.F. Ioffe and A.R. Regel, Prog. Semicond. **4**, 237 (1960).

[41] K. Char, M. Lee, R.W. Barton, A.F. Marshall, I. Bozovic, R.H. Hammond, M.R. Beasley, T.H. Geballe, and A. Kapitulnik, Phys. Rev. B**38**, 834 (1988).

[42] J. Orenstein, G.A. Thomas, A. J. Millis, S.L. Cooper, D.H. Rapkine, T. Timusk, L.F. Schneemeyer, and J.V. Waszczak, Phys. Rev. B **42**, 6342 (1990).

[43] Y. Fukuda, T. Suzuki, M. Nagoshi, Y. Syono, K. Oh-ishi and M. Tachiki, Solid State Comm. **72**, 1183 (1989).

DYNAMICAL SPIRAL STATE IN TWO-DIMENSIONAL HUBBARD MODEL

Z. Y. Weng and C. S. Ting

Department of Physics, University of Houston
Houston, Texas 77204-5504

The existence of the long-range antiferromagnetic (AF) state adjacent to the superconducting (SC) phase is a prominent feature for the copper-oxide high-T_c superconductors. Doping of holes (or electrons) induces the transition from one (AF) into the other (SC). To properly understand the doping effects, many authors[1-4] have studied the problem of a single hole doped into the two-dimensional (2D) AF background by the single-band Hubbard model or $t-J$ model, which has been proposed[5] to describe the basic physics in the CuO_2 layers. The single hole propagator is shown[1-3] to have a quasi-particle-like coherent peak emerges at the bottom of the incoherent band with a small weight, while its incoherent part persists within an large energy scale which could be properly regarded as the local dynamical spin-polaron effect[4]. As doped hole could gain its most important energy through the local spin-polaron effect, it is reasonable to expect the same picture valid through a finite doping regime. On the other hand, the quasi-particle-like coherent part of the doped hole is closely related to the long wave-length behavior of the system with a more sensitive energy scale. Both of them could be changed drastically even at small doping, which will be discussed in the present work.

Recently, we have developed a path integral formalism[6] of the Hubbard model which can be applied to the whole coupling regime. In the strong coupling and half-filling case, within the quadratic fluctuations above the AF saddle-point, we are able to obtain the effective Lagrangian describing the spin, amplitude and charge fluctuations[6] $L_f^{eff} = L_s^{eff} + L_{\Delta,\phi}^{eff}$ where low-lying spin-fluctuation is determined by $L_s^{eff} = \frac{1}{2}\rho \sum_i \left[\frac{1}{c^2}(\partial_\tau \Omega_i)^2 + (\nabla \Omega_i)^2\right]$ + the short-wavelength part, in which the coefficients are in agreement with that of the non-linear σ model derived in strong-coupling at large S expansion.[7] The amplitude and charge fluctuations, described by $L_{\Delta,\phi}^{eff}$, have a higher energy scale. The original electron operator $c_{i\sigma}$ will be expressed in this formalism through a SU(2) transformation U_i as

$$c_{i\sigma} = \sum_{i\sigma} U_{i\sigma\sigma'} a_{i\sigma'} \tag{1}$$

where U_i is determined by $U_i^+ \Omega_i \cdot \sigma U_i = \sigma_z$. The fermion $a_{i\sigma}$ defined in Eq.(1) always sees a fictitious long-range AF order and its band will be split by a Hubbard charge gap $2\Delta \sim U$ within strong coupling, small doping regime. At half-filling, the lower band is filled and holes will only go to this band upon doping, which is equivalent to the non-double-occupancy constraint. The lower band is described by the fermion $\alpha_{k\sigma}$ which is defined by $\alpha_{k\sigma} = u_k a_{k\sigma} + \sigma v_k a_{k+Q\sigma}$ where the momentum $\mathbf{k}$ is restricted in the reduced (magnetic) Brillouin zone and $\mathbf{Q} = (\pm\pi, \pm\pi)$.[6]

Near the AF saddle-point, the Lagrangian for the doped holes is given by[6]

$$L_h = \sum_{\mathbf{k}\sigma}{}' \alpha^+_{\mathbf{k}\sigma}(\partial_\tau - E_\mathbf{k} + \mu)\alpha_{\mathbf{k}\sigma} + \frac{1}{2}\sum_{\mathbf{k},\mathbf{q}\sigma}{}' [\nabla \epsilon_\mathbf{k} \cdot \mathbf{D}_{\mathbf{q}\sigma-\sigma}(\theta_{\mathbf{k}+\mathbf{q}} + \bar{\theta}_{\mathbf{k}+\mathbf{q}\sigma})$$
$$\times \alpha^+_{\mathbf{k}+\mathbf{q}\sigma}\alpha_{\mathbf{k}-\sigma} + H.c.] \qquad (2)$$

in which the upper Hubbard-band has been projected out and the coupling with the amplitude and charge fluctuations is also neglected. In Eq.(2), $\theta_\mathbf{k}$ and $\bar{\theta}_\mathbf{k}$ are step-functions restrict $\mathbf{k}$ only within and out of the reduced Brillouin zone respectively and $\epsilon_\mathbf{k} = -2t(\cos k_x a + \cos k_y a)$. At small doping, doped holes will be located close to the reduced Brillouin zone boundary such that only the leading coupling matrix, $\mathbf{D}_\mathbf{q}$, is retained in the Lagrangian (2) which is defined in the lattice representation by

$$\mathbf{D}_i = \frac{1}{2}(\nabla \mathbf{\Omega}_i \times \mathbf{\Omega}_i) \cdot U^+_i \sigma U_i \qquad (3)$$

Shraiman and Siggia[2,8] have derived a phenomenological Lagrangian from $t - J$ model which is similar to $L_h(\tau)$.

The Lagrangian L_h in Eq.(2) is obtained close to the AF saddle-point with a small amplitude of $|\mathbf{D}_i|$, which will require a strong local AF correlation (i.e., $|\nabla \mathbf{\Omega}_i| \ll 1$). But in general $\mathbf{\Omega}_i$ itself could have an arbitrary long wave-length behavior. If there still exists a global symmetry-broken in z-direction at small doping, $\mathbf{\Omega}_i$ could be expanded around this direction and one finds $\mathbf{D}_\mathbf{q} \simeq i\mathbf{q}\sigma \cdot (\mathbf{\Omega}_\mathbf{q} \times \mathbf{z})$ and the propagator $iD^s = <T(T_r\mathbf{D}_\mathbf{q}(t)\mathbf{D}_{-\mathbf{q}}(t'))>$ reduces to the conventional transverse spin susceptibility function $\frac{1}{2}\mathbf{q}\mathbf{q} < T(\Omega^x_\mathbf{q}(t)\Omega^x_{-\mathbf{q}}(t') + \Omega^y_\mathbf{q}(t)\Omega^y_{-\mathbf{q}}(t'))>$. In this case L_h simply describes the coupling of hole with the conventional spin wave, with the bare kinetic part of hole identified as the sublattice hopping with spectrum $E_\mathbf{k} = J(\cos k_x a + \cos k_y a)^2$. Coupling with doped holes will lead to a renormalization to the spin-waves, which is determined by the following correction to the effective Lagrangian of spin-wave within the Gaussian fluctuations:

$$L'_s(\tau) = \frac{1}{2}\int \int_0^\beta d\tau d\tau' \sum_\mathbf{q} Tr \mathbf{D}_\mathbf{q}(\tau) \cdot \mathbf{\Pi}(\mathbf{q}, \tau' - \tau) \cdot \mathbf{D}_{-\mathbf{q}}(\tau') \qquad (4)$$

where $\mathbf{\Pi}$ is given by

$$\mathbf{\Pi}(\mathbf{q}, \tau' - \tau) = \sum_\mathbf{k}{}' \nabla \epsilon_\mathbf{k} \nabla \epsilon_{\mathbf{k}+\mathbf{q}} G(\mathbf{k}+\mathbf{q}, \tau'-\tau) G(\mathbf{k}, \tau-\tau') \qquad (5)$$

In the long-wavelength, low-energy limit, the right hand-side of Eq.(4) could be reduced to $\int_0^\beta \frac{1}{4}\Pi_0 \sum_i (\nabla \Omega_i)^2$ where Π_0 is the static susceptibility, and then one gets a renormalized spin-wave stiffness $\tilde{\rho} = \rho + \frac{1}{2}\Pi_0$ at $q \ll k_f$ where k_f is the Fermi momentum of holes. For $q > k_f$, $\mathbf{\Pi}(\mathbf{q})$ in Eq.(4) will become negligible and thus the spin-wave will not be essentially renormalized by the doped holes in this region. It is easy to check that $\Pi_0 \sim -t^2/\delta J$ which will lead to $\tilde{\rho} < 0$ as $\rho \sim J$. But here one should be careful in discussing the long wave-length behavior as the coupling of the doped hole with the whole spin-wave excitations is so strong ($\sim t \gg J$) that hole's propagator could be renormalized dramatically and it would be proper for one to use the renormalized Green's function in calculating the long wave-length, low-energy susceptibility Π_0 (the vertex correction will be still neglected here). The renormalized hole propagator is found by $G(\mathbf{k},\omega) = a_\mathbf{k}/(\omega - \omega_\mathbf{k}) + G_{inc}$, where $\omega_\mathbf{k}$ is renormalized downward to $\sim -t$ with a coherent bandwidth of $\sim J$ and a small coherent weight $a_\mathbf{k} \sim J/t$. All of these could be obtained in a similar way as in Ref. 3. The incoherent G_{inc} is shown giving a vanishing contribution to Π_0 but the coherent part of G leads to a finite Π_0 as $\Pi_0 \sim -J^2 a^2 N_f$. The density of states $N_f \sim J^{-1}(m_\parallel/m_\perp)^{\frac{1}{2}}$ where $m_\parallel$ and $m_\perp$ are the anisotropic masses near $(\pm\pi/2, \pm\pi/2)$. The large ratio[1,3] of $m_\parallel/m_\perp$ will still result in $\tilde{\rho} < 0$.[8] Therefore, the quadratic fluctuation of $\mathbf{D}_\mathbf{q}$ at $q < k_f$ will become unstable at finite doping, as driven by the coherent part of the doped holes.

In the first place one would naturally try to look for some stable saddle-point near the AF saddle-point $< \mathbf{D_q} >= 0$ with $< \mathbf{D_q} > \neq 0$. For example, if $\mathbf{\Omega}_i$ is always confined in a plane, say x-z plane, then $\mathbf{D}_i$ is given by $\nabla \theta_i \frac{\sigma_y}{2}$ with θ_i as a polar angle of the vector $\mathbf{\Omega}_i$ in the plane. The uniform plane spiral state $< \mathbf{D_q} >= \nabla \theta \frac{\sigma_y}{2} \delta_{\mathbf{q},0}$ has been first proposed[8] as a new mean-field state away from half-filling. In general case, one will have the chiral spin structure with $\mathbf{\Omega}_i \cdot (\mathbf{\Omega}_{i+\hat{x}} \times \mathbf{\Omega}_{i+\hat{y}}) \neq 0$. A example is the double spiral mean-field state[9] with $< \mathbf{D_q} > \cdot \hat{x} = \Delta \theta \frac{\sigma_y}{2} \delta_{\mathbf{q},0}$ and $< \mathbf{D_q} > \cdot \hat{y} = \Delta \theta \frac{\sigma_x}{2} \delta_{\mathbf{q},0}$. But in a similar argument as discussed above for the AF state $< \mathbf{D_q} >= 0$, the uniform spiral states with $< \mathbf{D_q} >= \delta_{\mathbf{q},0} < \mathbf{D_0} > \neq 0$ could be shown to still have a negative stiffness as against the quadratic fluctuation of $\mathbf{D}_q - < \mathbf{D_q} >$ at $0 < q \ll k_f$.

We do not expect this instability is towards a phase separation if the doped-hole's most important energy could be obtained through the local incoherent process. It has been shown above that the coherent motion of holes is the driving force of the long wave-length instability at $q < k_f$. However the quasi-particle-like coherent part of the hole propagator is obtained under the assumption of the existence of a long-range AF or spiral state, where long wavelength spin-wave has a vanishing coupling with the doped holes. These imply that a type of hole-driven spiral fluctuations would be dominant at $q < k_f$, leading to the disappearance of the long-range order, while interacting with this long-wavelength fluctuating spiral field will essentially change the property of the coherent part of single hole's propagator. This problem will be considered self-consistently in the following.

The propagator of spin-fluctuations is defined by $\mathbf{D}(\mathbf{q}, t - t') = -i < T_t T_r (\mathbf{D_q}(t) \mathbf{D}_{-\mathbf{q}}(t')) >$. At the long-range AF state, this propagator is just the conventional spin-wave propagator $\mathbf{D}^s$ as discussed before. When the hole-induced spiral fluctuation of $\mathbf{D_q}$ is dominant over the conventional spin-wave in the long wave-length regime, it is $\nabla \mathbf{\Omega}_i$ instead of $\mathbf{\Omega}_i$ itself will be meaningful.[10] In this case, the quantity $U_i^+ \partial_\tau U_i$ which would lead to the dynamical part of the non-linear σ model[6] in the long-range state is found no longer well-defined and we shall simply neglect its contribution in this long wave-length regime. Then the propagator $\mathbf{D}(\mathbf{q}, \omega)$ will be determined totally by the dynamics of the doped holes as $\mathbf{D}(\mathbf{q}, \omega) \sim \mathbf{D}^l(\mathbf{q}, \omega) \equiv -1/(\mathbf{\Pi}(\mathbf{q}, \omega) + 2\rho \hat{\mathbf{I}})$, in a long-wavelength, low-energy regime: $q \leq q_c$, $|\omega| \sim \omega_l$. At $q > q_c$, as the contribution from $\mathbf{\Pi}(\mathbf{q}, \omega)$ will no longer be dominant, $\mathbf{D}(\mathbf{q}, \omega)$ reduces to the conventional renormalized spin-wave propagator $\mathbf{D}^s(\mathbf{q}, \omega)$.

First we consider the hole propagator. The self-energy of the hole could be written as

$$\sum(\mathbf{k}) = i \sum_{\mathbf{q}} \int \frac{d\omega'}{2\pi} \nabla \epsilon_{\mathbf{k}} \cdot \mathbf{D}(\mathbf{q}, \omega') \cdot \nabla \epsilon_{\mathbf{k}-\mathbf{q}} G(\mathbf{k} - \mathbf{q}, \omega - \omega') \qquad (6)$$

in the non-crossing approximation[3]. One can write $\sum(\mathbf{k}, \omega) = \sum^s(\mathbf{k}, \omega) + \sum^l(\mathbf{k}, \omega)$ where $\sum^l(\mathbf{k}, \omega)$ is determined by the coupling with the long-wavelength spiral fluctuation $\mathbf{D}^l(\mathbf{q}, \omega)$ at $q < q_c$, while $\sum^s(\mathbf{k})$ is contributed by the local spin-wave fluctuation $\mathbf{D}^s(\mathbf{q}, \omega)$ at $q > q_c$. Defining $G^s \equiv (\omega - E_{\mathbf{k}} + \mu - \sum^s)^{-1}$ and taking it as the "free" propagator, one needs only to consider the spiral fluctuations in Eq.(6), i.e., $\sum^l(\mathbf{k}, \omega)$. If the characteristic momentum q_c and energy ω_l for $D^l(\mathbf{q}, \omega)$ tend to zero (i.e., approaching to the static uniform spiral limit), the self-energy $\sum^l(\mathbf{k}, \omega)$ will reduce to a simple form as $\sum^l(\mathbf{k}, \omega) = \tilde{d}_{\mathbf{k}}^2 / 4 G(\mathbf{k}, \omega)$. In this case, $G = 2\tilde{d}_{\mathbf{k}}^{-2}(G^{s-1} - \sqrt{G^{s-2} - \tilde{d}_{\mathbf{k}}^2})$. If G^s has a coherent peak $\sim \frac{a_{\mathbf{k}}}{\omega - \omega_{\mathbf{k}}}$ at $|\omega - \omega_{\mathbf{k}}| < J$, then G becomes

$$G(\mathbf{k}, \omega) \simeq a_{\mathbf{k}} \frac{\omega - \omega_{\mathbf{k}} - \sqrt{(\omega - \omega_{\mathbf{k}})^2 - d_{\mathbf{k}}^2}}{d_{\mathbf{k}}^2 / 2} \qquad (7)$$

where there is a new characteristic energy scale $d_{\mathbf{k}} \equiv a_{\mathbf{k}} \tilde{d}_{\mathbf{k}} < J$ with $\tilde{d}_{\mathbf{k}}^2 = 4i \sum_{\mathbf{q}} \int \frac{d\omega}{2\pi} \nabla \epsilon_{\mathbf{k}} \cdot \mathbf{D}^l(\mathbf{q}, \omega) \cdot \nabla \epsilon_{\mathbf{k}}$. When $|\omega - \omega_{\mathbf{k}}| > J$, $G \sim G_{inc}^s \sim \frac{1}{t}$. Therefore even at the static uniform limit q_c and $\omega_l \to 0$, we have found a quite different solution for the spiral state. In the case of finite q_c and ω_l, as long as $|\nabla \omega_k| q_c$, $\omega_l \ll d_{\mathbf{k}}$ (the broadening width of the coherent spectral function in Eq.(7)), we expect Eq.(7) still to be a good approximation for $G(\mathbf{k}, \omega)$ at $|\omega - \omega_{\mathbf{k}}| < J$

except $\omega \ll \omega_l$[11]. On the other hand, as $G(\mathbf{k},\omega)$ appears also in the expression for $\sum^s$, with the broadening of its coherent spectral function by Eq.(7) one can show that $G^s(\mathbf{k},\omega)$ will become totally incoherent (i.e., $a_\mathbf{k} \sim 0$) when $|\omega_\mathbf{k}| > \Omega_c$ as $\Gamma_\mathbf{k} = |1/\pi Im \sum^s(\mathbf{k},\omega_\mathbf{k})| > |\omega_\mathbf{k}|$. Ω_c is the low-energy cut-off of the spin-wave spectrum at $q \sim q_0$, i.e. $\tilde{c}q_0 = \Omega_c$ where $\tilde{c}$ is the renormalized spin-wave velocity near $q \geq q_0$. But when $|\omega_\mathbf{k}| < \Omega_c$, one still finds a quasi-particle-like coherent peak in G^s with $\Gamma_\mathbf{k} = 0$ and a weight $a_\mathbf{k} \sim J/t$. Therefore, if there exists a long wave-length, low energy spiral fluctuations, G's coherent part will no longer be quasi-particle-like as Eq.(7) shows.

Next we determine the spiral fluctuation $D^l(\mathbf{q},\omega)$ through the susceptibility $\Pi(\mathbf{q},\omega)$. Without including the vertex correction, $\Pi(\mathbf{q},\omega)$ is easy to calculate by the coherent part of $G(\mathbf{k},\omega)$ in Eq.(7) as follows

$$Im\Pi(\mathbf{q},\omega) \simeq -b|\omega|\theta(\omega_c - |\omega|)\theta(q_c - q) \qquad (8)$$

where $\omega_c \simeq d_f \equiv d_{k=k_f}$ and $q_c = \Omega_c/v_f$ with $v_f = |\nabla \omega_{k_f}|$. We shall identify $q_c = q_0$, that is, the upper cut-off momentum for spiral fluctuation is the same as the lower cut-off momentum for the renormalized spin-wave. Then one finds $\tilde{c} = v_f$. The spectral function for the spiral fluctuations is given by

$$Im\mathbf{D}^l(\mathbf{q},\omega) \simeq \frac{-b^{-1}|\omega|}{\omega_l^2 + \omega^2}\theta(d_f - |\omega|)\theta(q_c - q) \qquad (9)$$

where $\omega_l = 2\tilde{\rho}/b$ with $\tilde{\rho} = \rho + \frac{1}{2}Re\Pi_0$ which should be positive as required by the stability. In the above spectral function, the main weight is located around its maximum peak at $\omega_l \simeq (k_f a)^2 d_f$. The characteristic momentum scale for the spiral fluctuation is found by $q_c \simeq \kappa/\ln[1+(k_f a)^{-4}]k_f$. The energy scale d_f in the hole propagator (7) is determined by $d_f \simeq \kappa^2/\ln[1+(k_f a)^{-4}]\epsilon_f$ where the hole's Fermi energy $\epsilon_f = m_\perp v_f^2/2$ and $\kappa = 1/(4\pi)(m_\parallel/m_\perp)^{\frac{1}{2}}$. We note that $q_c v_f = \kappa^{-1} d_f \ll d_f$ and $\omega_l \ll d_f$ at small doping. Such a long wave-length, low-energy scale of the spiral fluctuation justifies our approximation in obtaining the coherent part of G in Eq.(7).

The true physical propagator is related to the original electron operator $c_{i\sigma}$ as $G^p_{\sigma,\sigma'}(i,j;t-t') = -i<Tc^+_{i\sigma}(t)c_{j\sigma'}(t')>$. Through the transformation (1), the zeroth-order approximation, with the spin part and the charge part being treated independently, will lead to the following expression of G^p

$$G^p(i,j;t-t') \simeq <U_j(t)U_i^+(t')> \frac{1}{N}\sum_\mathbf{k} e^{i\mathbf{k}\cdot(\mathbf{R}_i-\mathbf{R}_j)}(\theta_\mathbf{k} u_\mathbf{k}^2 + \bar{\theta}_\mathbf{k} v_\mathbf{k}^2)G(\mathbf{k},t-t'), \qquad (10)$$

Generally, the long wave-length behavior of $<U_j(t)U_i^+(t')>$ is very difficult to be calculated by $D^l(\mathbf{q},\omega)$, because there are a lot of almost degenerate spin configurations which can not be distinguished by $D^l(\mathbf{q},\omega)$. But this problem could be greatly simplified if $\mathbf{\Omega}_i$ is assumed to be always confined in a plane, say z-x plane, with $\mathbf{D}_i = \nabla \theta_i \sigma_y/2$. In this case, for $t = t'$, $<U_jU_i^+>=<\exp -i\sigma_y/2 \int_i^j d\mathbf{R}\cdot \nabla\theta(\mathbf{R})> = \exp -2\sum_\mathbf{q} iD^l(\mathbf{q},t=0)\sin^2(\mathbf{q}\cdot\mathbf{R}_{ij}/2)/q^2$ which $\sim (1/q_c R_{ij})^{\kappa^2\mu_f/J}$ at the limit $R_{ij} \equiv |\mathbf{R}_i - \mathbf{R}_j| \gg q_c^{-1}$. Similarly, one finds the instantaneous spin-spin correlation $<\mathbf{S}_i\cdot \mathbf{S}_j> \sim \pm(1/q_c R_{ij})^{4\kappa^2\mu_f/J}$ in the same limit. We note that when the local fluctuations have a chiral spiral structure with $\mathbf{\Omega}_i \cdot (\mathbf{\Omega}_{i+\hat{x}} \times \mathbf{\Omega}_{i+\hat{y}}) \neq 0$, the bare spin-stiffness ρ is found having an additional increase with $\Delta\rho \sim \delta J$, which could lead to its fluctuation energy $\omega_l \sim (k_f a)d_f$ as compared to the plane spiral fluctuation energy $\omega_l \sim (k_f a)^2 d_f$. Thus at least in low temperature and small doping, one may reasonably assume the plane spiral fluctuations dominant in a scale $R_{ij} \sim q_c^{-1}$, but in the scale $R_{ij} \gg q_c^{-1}$ the fluctuation of the spiral planes will certainly lead to a more rapid decay of $<U_jU_i^+>$ as well as $<\mathbf{S}_i\cdot\mathbf{S}_j>$.

Thus, the hole propagator G^p_{ij} gets a prefactor $<U_jU_i^+>$ which decays when $R_{ij} > q_c^{-1} \sim k_f^{-1}$. But if we consider a singlet pair of holes, i.e., $b_s^+ = \frac{1}{\sqrt{2}}\sum_\sigma \sigma c_{l\sigma}^+ c_{m-\sigma}^+$, it is found

$$b_s^+ \simeq <U_lU_m^+> \frac{1}{\sqrt{2}}\sum_\sigma \sigma a_{l\sigma}^+ a_{m-\sigma}^+. \qquad (11)$$

Then in the singlet channel of two-particle Green's function, the prefactor $< U_j U_i^+ >$ as a function of the distance will be replaced by $< U_l U_m^+ >$ as a function of the distance between two holes. Therefore, two holes will become more coherent when they form the singlet pair than when they are "free". On the other hand, the true superconducting condensation will be possible when the inter-layer hopping is taken into account. At small doping, the spiral fluctuation could be suppressed by the interlayer coupling $J_\perp$ up to a critical doping concentration δ_c. At $\delta > \delta_c$, the long-range order disappears and the interlayer AF correlation presumably becomes very weak ($J_\perp/J \sim 10^{-5}$ for La_2CuO_4[12]). In this case, we find the charge carrier, described by a_i, hops in the c-direction almost incoherently. But the hoping of a singlet pair of a_i, setting in the same layer, is found still being coherent along c-direction by using Eq.(11). Such an coherent process along c-direction will result in an effective attraction between the charge carriers a within layers which would lead to the superconducting condensation $< b_s^+ > \neq 0$. A similar interlayer-hopping-induced superconducting mechanism was first discussed in Ref.13.

In conclusion, in the finite doping the coherent motion of the doped holes drives a dynamical spiral fluctuation in the long-wavelength, low-energy regime, leading to the decay of the long-range spin correlation. In such a spin liquid background, the electron could be regarded as a composite particle with charge and spin degrees of freedom separated and the true single hole propagator loses its coherence in a scale larger than the spin-spin correlation length, but in the singlet-pair channel holes could acquire a coherence amplitude again. The real superconducting condensation of these singlet pairs will occur when the interlayer hopping is included and only after the interlayer AF correlation has been suppressed beyond some critical doping concentration.

ACKNOWLEDGMENTS

The author would like to thank T.K. Lee, L.Y. Chen, G. Reitor, Z.B. Su, D.N. Sheng and D.C. Mattis for helpful discussions. The present work is supported by a grant from the Robert A.Welch Foundation and also by the Texas Center for Superconductivity at the University of Houston under the Prime Grant No.MDA-972-88-G-0002 from Defence Advanced Research Project Agency.

REFERENCES

[1] S.A. Trugman, Phys. Rev. B**37**, 1597(1988); S. Schmitt-Rink, C.M. Varma, and A.G. Ruckenstein, Phys. Rev. Lett. **60**, 2793(1988); Z.B. Su, Y.M. Li, W.Y. Lai, and L. Yu, ibid. **63**, 1318(1989).

[2] B.I.Shraiman and E.D.Siggia, Phys. Rev. Lett. **61**, 467(1988);

[3] C.L. Kane, P.A. Lee, and N. Read, Phys. Rev. B**39**, 6880(1989).

[4] Z.Y. Weng, C.S. Ting and T.K Lee, Phys. Rev. B**41**, 1990(1990); W.P. Su and X.Y. Chen, ibid. **38**, 8879(1988).

[5] P.W.Anderson, Science **235**, 1196(1987); Proceedings of the Enrico Fermi International School of Physics, *Frontiers and Borderlines in Many Particle Physics*, North-Holland Publ. Co., Varenna, July 1987.

[6] Z.Y. Weng, C.S. Ting and T.K. Lee, Phys. Rev. B1 (February) (1991).

[7] F.D.M. Haldane, Phys. Lett, 93**A**, 464(1983), and Phys. Rev. Lett., **50**, 1153(1983).

[8] B.I.Shraiman and E.D.Siggia, Phys. Rev. Lett. **62**, 1564(1989).

[9] C.L. Kane, P.A. Lee, T.K. Ng, B. Chakraborty and N. Read, Phys. Rev. **B41**, 2653(1990).

[10] Z.Y. Weng and C.S. Ting, Phys. Rev. **B42**, 803(1990).

[11] When $|\omega_\mathbf{k}| \ll \omega_l$, $G(\mathbf{k},\omega)$ could still have a δ-like peak emerging from the broadened coherent part, but its weight ($\sim \omega_l/d_{k_f} a_{k_f} \simeq (k_f a)^2 J/t$) is negligible.

[12] T. Thio et al. Phys. Rev. **B38**, 905(1988).

[13] J.M. Wheatley, T.C. Hsu and P.W. Anderson, Phys. Rev. B37, 5897(1987); P.W. Anderson, Kathmandu lectures (1989).

VALENCE-FLUCTUATION SCENARIO FOR CUPRATE

SUPERCONDUCTIVITY: THE FINITE-U PAIRING MECHANISM

Baird Brandow

Theoretical Division
Los Alamos National Laboratory
Los Alamos, NM 87545

INTRODUCTION

The idea that the normal state of cuprate superconductors is a valence-fluctuation (VF) state is appealing for a number of reasons. We are exploring this scenario quantitatively,[1] focussing especially on the finite-U mechanism[2] for s-wave pairing. The calculations employ a many-body variational technique developed previously for VF and heavy-fermion materials.[3] We are presently exploring the effects of (a) varying the Hamiltonian parameters away from the values obtained from photoemission data analysis,[4] (b) various simple assumptions for the band structure and hybridization k-dependence, and (c) corrections to the Gutzwiller approximation. These are the topics to be discussed in this report.

ARGUMENTS FOR A VALENCE-FLUCTUATION PHASE

There are several good reasons to expect that the electronic state of a cuprate superconductor above T_c may be a valence-fluctuation state. At the outset, we note an obvious lesson from band theory and general solid-state chemistry: The 3d and 2p electrons are *both* close to the Fermi level, so it is appropriate to consider both types of orbitals explicitly within the model Hamiltonian. This leads to an Anderson lattice model (or multi-band Hubbard model), similar to the usual starting point for VF or heavy-fermion studies. This immediately suggests the VF phase as a candidate to be explored. (It should be noted that the heavy-fermion state is basically the same as the VF state, but in a region of parameter space where the effective mass enhancement is considerably larger.)

The VF picture leads directly to a *normal Fermi liquid*, for the state above T_c, with quasiparticles satisfying the Luttinger sum rule. Furthermore, this involves a "renormalized band" or "renormalized hybridization" type of effective band structure, for the quasiparticle spectrum. Many physical properties should therefore be interpretable via the conventional band-theoretic recipes, perhaps with minor modifications due to the renormalization.

There are two ways to qualitatively understand this normal-Fermi-liquid result. One way is to argue that a large U generates local moments on the copper ions, and then the oxygen 2p electrons cancel these moments by Kondo screening. One then obtains a normal Fermi liquid because there are, effectively, no local moments to disrupt the elementary Bloch periodicity. Although quite correct, this description has the disadvantage of suggesting that we are dealing with an exotic scenario. We now want to argue that this is not so.

An equivalent but very pedestrian way to visualize this result is to start with the U=0 solution of the Anderson lattice Hamiltonian. This gives an elementary band-theoretic state, which is obviously a Fermi liquid. We then invoke the idea of Luttinger continuity, i.e., we suppose that U is switched on adiabatically, and that the system evolves continuously without any phase change. The result is therefore a strongly-correlated normal Fermi liquid -- the valence-fluctuation state. The large number of known VF and heavy-fermion materials demonstrate that this scenario is physically reasonable. It must be admitted, however, that there are no well-established previous examples of the VF state in which 3d electrons are the active ingredient.

With either version of the argument, one can see that the 2p orbitals play an essential role in promoting the assumed normal Fermi liquid state. The itinerancy or metallic aspect is provided by the 2p's, because the various models typically neglect any direct hopping between the 3d Wannier orbitals. On the other hand, the Anderson lattice model is flexible enough to permit formation of local moments, and even a Mott-insulator state (at proper stoichiometry), if this is what the system would prefer.

Experimentally, one of the most striking features of the cuprate superconductor materials is the existence of a Fermi surface in close agreement with the prediction of conventional band theory, as revealed by angle-resolved photoemission. Band theory is not fully successful, however, because this data also indicates weaker dispersion in the vicinity of ε_F, i.e. some "heaviness." The band-theoretic Fermi surface, together with some heaviness, is a typical signature for the known VF and heavy-fermion materials. In addition, there are a number of other physical properties for which striking parallels have been found between cuprate superconductors and heavy-fermion materials. These have recently been reviewed by Levin and co-workers.[5] All together, this experimental evidence supports the VF picture very strongly. Recent calculations by Newns and co-workers[6] have also demonstrated quantitative consistency of this picture with a number of cuprate properties, in a fairly *ab initio* manner.

The assumed absence of local moments should be contrasted with the physics of the t-J model, where the use of the J parameter presumes the existence of local moments. This contrast leads to a further implication of the present scenario: There should probably be a first-order phase boundary between the local-moment phase of $La_{2-x}Sr_xCuO_4$ at small doping, and the VF phase at higher doping. This is not seen experimentally, however, at least not clearly. We presume that this is due to local inhomogeneity in the Sr concentration, leading to a range of doping x with both phases coexisting on a microscopic scale. This might, for example, be the explanation for the so-called spin glass region. We note that other speakers at this workshop have also suggested coexisting phases on a microscopic scale, although with differences in the details.[7] Our present scenario envisages just one phase, essentially homogeneous, throughout the region of well-developed superconductivity.

THE FINITE-U PAIRING MECHANISM

Possible mechanisms for pairing within the VF phase have long been studied, with the goal of understanding the heavy-fermion superconductors. Apart from some phonon proposals, most of this work has been based on the 1/N or slave-boson expansion. However, the resulting mechanisms involving exchange of one slave boson[8] [$O(N^{-1})$] or two slave bosons[9] [$O(N^{-2})$] can provide d or p-wave pairing, but *not* s-wave pairing. There is only one known electronic mechanism which can produce

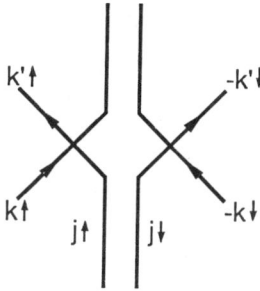

Figure 1. Diagram illustrating the finite-U mechanism for s-wave pairing.

s-wave pairing within the VF phase, the "two-hole resonance" mechanism of Newns,[2] so this is the one we are studying. This arises because of the *finiteness* of U, i.e., the fact that in a real material $U^{-1} \neq 0$. (In contrast, nearly all of the 1/N expansion work has assumed that $U=\infty$, $U^{-1} = 0$.) This Newns mechanism also has a very convenient feature -- it arises already at the "mean field" or "renormalized band" stage of approximation, corresponding to $(1/N)^0$ or no slave-boson exchanges. This is a welcome simplification, which in practice should enhance the reliability of the numerical results.

This finite-U pairing mechanism is illustrated in the diagram of Fig. 1. Here the vertical lines represent occupied $d(x^2-y^2)$ Wannier orbitals (of spin $\uparrow$ and $\downarrow$) at a copper site j. (The other eight d orbitals are assumed to be inert and fully occupied.) The sloping lines represent "conduction" orbitals, in the language of VF theory. These are Bloch functions constructed from the oxygen 2p orbitals. In the present case these 2p Bloch functions are components (projections) of Landau-Luttinger quasiparticle states, so this diagram represents an interaction between quasiparticles. The quantum numbers shown for these quasiparticles correspond to annihilation of one Cooper pair $(k\uparrow, -k\downarrow)$, and creation of another pair $(k'\uparrow, -k'\downarrow)$ so this process constitutes a pairing interaction $V_{k'k}$.

One can see here that the d^8 intermediate state plays the role of an exchanged boson. From another point of view this might be described as a charge-transfer mechanism, because it involves p-d transitions, but it is essential to recognize that this involves the *correlated* action of two electrons (of spin $\uparrow$ and $\downarrow$) doing this simultaneously. Hence the name "two-hole resonance."

Letting M denote the effective matrix element for the two-electron dd→(k,−k) transition, the net effect of the second-order process shown is $|M|^2$/(negative denominator) = *attractive*, as desired. This is quite appealing. But this also presents a problem. The mechanism appears to work "too well," suggesting that high-temperature superconductivity should be a common phenomenon. We shall see, however, that this process is accompanied by an opposing mechanism, and that it is not at all easy for the Newns pairing to win in the resulting competition.

RENORMALIZED HYBRIDIZATION

Our present version of the Anderson lattice Hamiltonian is

$$H = \sum_{k\sigma} \varepsilon_k \, \hat{n}_{k\sigma} + \varepsilon_d \sum_{j\sigma} \hat{n}_{j\sigma} + U \sum_j (1 - \hat{n}_{j\uparrow})(1 - \hat{n}_{j\downarrow})$$
$$+ \sum_{kj\sigma} (v_{kj} \, \eta_{k\sigma}^\dagger \, \eta_{j\sigma} + h.c.) \,, \qquad (1)$$

where $\hat{n}_\alpha = \eta_\alpha^\dagger \eta_\alpha$ is a number operator. Here $\alpha = k\sigma$ refers to a "conduction" Bloch orbital composed of oxygen 2p orbitals, with a dispersion width for ε_k amounting to several eV due to the considerable direct pp orbital overlap. On the other hand, $\alpha = j\sigma$ refers to a $d(x^2-y^2)$ Wannier orbital for a copper ion at site j. The direct dd overlap is surely small, and is neglected. Note that the Hubbard U is assumed to act between 3d *holes*, so it is the d^8 (two-hole) configuration energy which is raised by U. (We start with a Wannier representation for the d's to facilitate the treatment of U, but the resulting recipe is later transformed to the Bloch representation.)

As already mentioned, the solution of this Hamiltonian is elementary when U=0. This becomes a simple two-band hybridization problem, in which the 3d's constitute a band of zero width. The hybridization matrix element connecting the p and d Bloch basis states is $V_k = N^{-1/2} \sum_j e^{-ik \cdot R_j} v_{kj}$, where this N is the number of copper sites. In the mean-field approximation, the effect of replacing U=0 by U=∞ is to leave this simple picture intact, but to renormalize the input parameters:

$$\varepsilon_d \to \tilde{\varepsilon}_d \,, \qquad (2)$$

$$V_k \to \tilde{V}_k = V_k \, Z^{1/2} \,, \qquad (3)$$

where $0 < Z < 1$.

Unfortunately, the VF literature has presented two recipes for the renormalization parameter Z, without providing a compelling reason for choosing either one over the other. These rivals are the so-called mean field recipe,

$$Z_{mf} = 1 - n_f \,, \qquad (4)$$

and the Gutzwiller recipe,

$$Z_G = (1 - n_f)/(1 - n_{f\sigma}) \,. \qquad (5)$$

(Here we use the familiar notation of the VF literature, referring to the f electrons of a cerium compound. For cuprates, the average f *electron* number n_f, at a single site, should be replaced by the d *hole* number.) We have studied this problem in considerable detail,[10] and have concluded that the Gutzwiller version (5) is physically more correct. This conclusion was obtained, independently, from a variety of separate studies, including (a) exactly-solvable test cases, (b) few-site systems, (c) combinatoric analysis, (d) diagrammatic analysis, and (e) the requirement of correct behavior in the U → 0 limit. Since we are now concerned with the effects of finite U, the last of these considerations seems particularly significant. In terms of actual calculations, however, the diagrammatic approach has proved to be most useful.

Using a linked-cluster diagrammatic analysis,[11] we had previously obtained the mean-field version (4) by ignoring the restriction that in a diagram with several sites, all of the site indices (j, j', j", ---) must be distinct. We later managed to incorporate this site-exclusion feature, and thereby obtained the extra factor $(1-n_{f\sigma})^{-1}$ seen in (5).[10] Still later, this recipe was extended to the case of finite U.[1] For finite U this recipe for Z differs somewhat from previous Gutzwiller treatments of the Anderson lattice, but ours has the virtue of summing a well-defined class of diagrams. Examination of the diagrams still omitted shows that this recipe is generally still inexact, although this does become exact in some special cases.

The choice between recipes (4) and (5) has profound consequences for the understanding of cuprate superconductivity. In the mean-field version (4), there is *no* interaction between the quasiparticles when U=∞, and finite U then provides superconductivity. [However, the 1/N correction (exchange of one slave boson) opposes s-wave superconductivity, so the final outcome is still not trivial.] In the Gutzwiller version (5), the explicit spin dependence in the $(1-n_{f\sigma})^{-1}$ factor leads to a very strong Stoner enhancement of the magnetic susceptibility, and this "Gutzwiller magnetism" tendency now strongly opposes the "Newns pairing" tendency. (This agrees, of course, with the general experience that magnetism is bad for s-wave superconductivity.) The result is that we now find it rather difficult to obtain superconductivity.

STRATEGY AND CALCULATION TECHNIQUE

We are considering only isotropic band models, in which the ε_k and V_k depend only on the magnitude $|k|$ and not the direction $\hat{k}$. This is convenient here because the "Gutzwiller magnetism" and "Newns pairing" aspects then *both become purely s-wave effects*, in the sense of Landau. The net result of their competition can therefore be described by means of a single parameter, the Landau F_o^a. The static magnetic susceptibility takes the form

$$\chi = \frac{\mu_B^2 \rho_{qp}(\varepsilon_p)}{1 + F_o^a}, \qquad (6)$$

and thus through study of χ we can determine F_o^a. Now suppose that F_o^a turns out to be positive. If we further assume that the effective pairing matrix element $V_{kk'}$ is constant throughout the Brillouin zone, and likewise that the quasiparticle state density is independent of energy, we obtain the simple result

$$k_B T_c \approx \frac{1.14}{\rho_{qp}(\varepsilon_F)} e^{-\frac{1}{F_o^a}}. \qquad (7)$$

This recipe for T_c is admittedly very crude, and of course it can be refined. (We have made some progress here.) This does show, however, that for initial exploration it makes sense to merely calculate χ and $\rho_{qp}(\varepsilon_F)$, and see whether the resulting F_o^a is positive. This F_o^a should be reliable as a guide to track whether various parameter changes or other modifications are beneficial or harmful for the superconductivity. A lot can be learned by guiding F_o^a from an initially large negative value towards a positive value.

The actual calculations are based on a many-body variational technique.[1,3] This is patterned after the variational approach of BCS, although we adapt and apply the technique to the assumed *normal* Fermi liquid VF ground state. A trial many-body wavefunction Ψ is constructed in which there is one free parameter for each Bloch state $k\sigma$ within the Brillouin zone, plus one extra parameter to control the amount of d^8 component in the ground state.[1] We then obtain an analytic expression for $\langle H \rangle$ via partial summation of a linked-cluster diagrammatic expansion.[11] Although there are approximations in the choice of Ψ and in the evaluation of $\langle H \rangle$, the subsequent steps of optimizing the variational parameters, followed by the evaluation of χ, are carried out exactly, i.e. with no further approximations. This aspect is significant in view of the strong-coupling nature of the present problem.

RESULTS AND DISCUSSION

The F_o^a results in our initial publication[1] are, unfortunately, invalid due to a bug in this part of the program. When corrected, the F_o^a remained strongly negative (and usually <1, implying magnetic instability) for all reasonable choices of input parameters.

This could of course be a sign that the present approach is simply wrong or inappropriate for the cuprate superconductors. We doubt this conclusion, for several reasons: (1) For the conceptual and experimental reasons already discussed, it appears most likely that the normal state (above T_c) is actually a VF state. (2) The many-body variational technique is quite refined, and appears to be adequate for the task at hand. (3) If one grants that we are dealing with a VF state, which arises from a substantial U parameter, then $(-F_o^a) > 0$ corresponds to the effective or screened Coulomb parameter μ^* of the McMillan formula. Thus, a large negative F_o^a would constitute a serious obstacle which any other mechanism (e.g. phonons) would be required to overcome. It is therefore quite inappropriate to simply give up and ignore this problem.

A further motivation is the fact that rare-earth intermetallic compounds typically have local moments, which is consistent with a large negative value for F_o^a. Nevertheless, the nonmagnetic (Pauli paramagnetic) VF and heavy-fermion materials demonstrate that this is not always the case. One of the main problems remaining for VF theory is to explain how this magnetic-nonmagnetic phase boundary is determined, or in terms of the present study, what it is that can counteract the Gutzwiller magnetic tendency. The previous understanding of this phase boundary has been based on the competition between two effects[12]: the Kondo screening (included here), which opposes magnetism, and the RKKY coupling (omitted here), which promotes magnetism. [An RKKY interaction arises from the model Hamiltonian (1), via the $(1/N)^2$ correction.[9] In reality, however, part of the effective coupling between the 3d's and 2p's is due to the Fock exchange between these orbitals, which is omitted from (1).] We now emphasize that the Gutzwiller magnetism and Newns pairing tendencies are *also* very important for this issue, as well as emphasizing that this issue is related to cuprate superconductivity.

Our first stratagem to resolve the numerical discrepancy was to study corrections to the Gutzwiller approximation (5). We identified the leading diagrams omitted from our finite-U version of the Gutzwiller renormalization, and included them in the parameter self-consistency and the evaluation of χ. Their effect was rather small, and of the wrong sign.

We next observed that the assumed band structure model for these studies was extremely crude: V_k independent of k, and $\rho(\varepsilon_k)$ = constant (rectangular state density for the ε_k's). This model has often been used in VF studies, because the required Brillouin-zone integrals become elementary. It turns out that a number of other simple band models can also be integrated analytically, and we are presently exploring some of these. In an actual CuO_2 layer, most of the antibonding band dispersion is due to the k-dependence of V_k, rather than to the ε_k dispersion. Some of the simple models are obtained by attributing *all* of the quasiparticle dispersion to a sinusoidal k-dependence for V_k, using ε_k = constant. [There is more than one model of this type, due to different possible assumptions for the density of states $\rho(k)$.] This modification greatly improved the results. Further refinement in this vein is possible, including for example a logarithmic singularity in the state density to mimic the van Hove singularity. We expect this to make the pairing interaction matrix element a bit more attractive (or less repulsive), besides increasing its state-density prefactor. Our general variational method can in principle deal with a real band structure, to be integrated numerically (as for example in Ref. 6), but it seems premature to take this step.

We have also varied the Hamiltonian parameters away from the values obtained by analyzing CuO photoemission data.[13] It is very helpful to make both U and Δ much smaller. ($\Delta = \varepsilon_d - \varepsilon_p$ is the "charge-transfer energy," where ε_p is the centroid of the hybridizing 2p band states.) Our present inclination is that U $\approx$ 3.5 eV and $\Delta \approx$ 4 eV may be reasonable values here, even though our photoemission analysis[13] gave U = 7.0 eV, Δ = 7.55 eV. We attribute most of this major difference in U to the effect of metallic screening in the VF phase. (CuO is a Mott insulator.) We attribute this also, to a much lesser extent, to a screening renormalization from the eight other d orbitals which are omitted from the model of Eq. (1). Metallic screening should also tend to reduce Δ, but it is difficult to estimate by how much. The present guess for Δ is simply the value obtained phenomenologically by Newns and co-workers,[6] and is not much greater than values obtained from band-theoretic supercell calculations.[14] (We measure Δ from the centroid of the $b_{1g} = x^2 - y^2$ hybridization strength distribution for the oxygen 2p's, about 1 eV above the 2p band center, while other authors refer Δ to the center or to the top of the 2p band.)

Another parameter change appears to be essential here. To obtain sufficient satellite intensity in the CuO photoemission analysis, we found it necessary to assume a stronger hybridization interaction ($t_{pd\sigma} = -$ 1.9 eV) for the d^8-d^9 transitions than for the d^9-d^{10} transitions ($t_{pd\sigma} = -$ 1.45 eV). Sawatzky and co-workers[15] had earlier found photoemission evidence for such a charge-dependence of $t_{pd\sigma}$, and Martin[16] had also found this in his *ab initio* cluster calculations. By taking Martin's value for d^8-d^9 ($t_{pd\sigma} = -2.3$ eV), the U and Δ values just mentioned, and a band model with V_k dispersion, we have finally obtained attractive pairing of a reasonable magnitude. This work is still in a preliminary stage, however, and we are not ready to claim that these values are realistic.

Two other issues should be mentioned. In Ref. 1 we mentioned the finding of Houghton and Sudbø,[17] that coupling to the *bonding* quasiparticle band influences the pairing interaction, namely, the effective interaction between the quasiparticles of the *antibonding* band (the band which intersects ε_F). It turns out that our variational treatment takes care of this automatically, so no correction is needed on this account. Secondly, the previous Anderson-lattice studies[8,9] have found that the $(1/N)^1$ correction (from exchange of one slave boson) has a very significant effect on χ. Specifically, this 1/N correction amounts to $\delta F_o^a \approx -$ 0.5. Such a large and negative shift of F_o^a could be very difficult to cope with in the present scenario. We note, however, that this result was obtained for U=∞, with a highly simplified band structure, and without considering any Gutzwiller factors of the $(1-n_{f\sigma})^{-1}$ type. A more realistic evaluation of the 1/N correction could conceivably give a quite different result. We intend to explore this.

This work was supported by the U.S. Department of Energy.

REFERENCES

1. B. H. Brandow, Solid State Commun. **69**, 915 (1989).
2. D. M. Newns, Phys. Rev. B **36**, 2429 and 5595 (1987); D. M. Newns, M. Rasolt, and P. C. Pattnaik, *ibid.* **38**, 6513 (1988).
3. B. H. Brandow, Phys. Rev. B **33**, 215 (1986) and **37**, 250 (1988).
4. B. H. Brandow, J. Solid State Chem. **88**, 28 (1990).
5. K. Levin, Ju H. Kim, J. P. Lu, and Qimiao Si, to appear in Physica C.
6. D. M. Newns, P. C. Pattnaik, and C. C. Tsuei, Phys. Rev. B **43**, 3075 (1991).
7. See the reports of R. S. Markiewicz, J. C. Phillips, and J. M. Tranquada, this proceedings.

8. A. Auerbach and K. Levin, Phys. Rev. Lett. **57**, 877 (1986); M. Lavagna, A. J. Millis, and P. A. Lee, *ibid.* **58**, 266 (1987).
9. A. Houghton, N. Read, and H. Won, Phys. Rev. B **37**, 3782 (1988).
10. B. H. Brandow, unpublished.
11. B. H. Brandow, J. Magn. Magn. Mat. **63-64**, 264 (1987).
12. S. Doniach, Physica **91B**, 231 (1977).
13. B. H. Brandow, J. Solid State Chem. **88**, 28 (1990).
14. A. K. McMahan, R. M. Martin, and S. Satpathy, Phys. Rev. B **38**, 6650 (1988); A. K. McMahan, J. F. Annett, and R. M. Martin, *ibid.* **42**, 6268 (1990); M. S. Hybertsen, M. Schluter, and N. E. Christensen, *ibid.* **39**, 9028 (1989).
15. J. Ghijsen *et al.*, Phys. Rev. B **42**, 2268 (1990).
16. R. L. Martin, Physica B **163**, 533 (1990), and private communication.
17. A. Houghton and A. Sudbø, Phys. Rev. B **38**, 7037 (1988).

THE VAN HOVE SINGULARITY AND HIGH-T_c SUPERCONDUCTIVITY: THE ROLE OF (NANOSCOPIC) DISORDER

R.S. Markiewicz

Physics Department and Barnett Institute,
Northeastern University, Boston, MA 02115

INTRODUCTION

In any two-dimensional band structure, there must be a saddle-point at which the bands intersect the Brillouin zone boundary and the conduction crosses over from hole-like to electron-like. At this van-Hove singularity (vHs), the density-of-states (dos) has a logarithmic singularity. It has often been suggested that this enhanced dos is responsible for high-T_c superconductivity. Theories of the CuO_2-plane vHs can be separated into two groups: in the earlier theories[1-3] the vHs is assumed to occur exactly at half-filling of the band (square Fermi surface); in this case, there is direct competition with the Mott (or charge transfer insulator) transition, and the optimum T_c occurs away from the vHs. In the other group of theories[4-8] the vHs is assumed to occur away from half-filling, as found in band-structure calculations, and hence well separated from the Mott transition. The optimum T_c occurs exactly at the hole filling appropriate to the vHs. Extensive reviews of these vHs models have recently appeared[3,8]. It is the latter category of vHs models which is the topic of this paper.

An important aspect of the vHs is an intimate association with disorder. Naively, this can be expressed as an intense competition for the large dos, associated with strong electron-phonon coupling. There are two separate routes to this disorder. First, as the material is doped away from the vHs, the holes can bunch up, pinning the Fermi surface in part of the sample to the vHs and leading to a heterogeneous phase. Since this phase separation is electronically driven, it can have strikingly different properties from ordinary spinodal decomposition. In particular, charging of domains can lead to nanoscopic-scale disorder. Secondly, even at the concentration of the vHs, strong electron-phonon coupling can lead to short-range charge-density wave (CDW) disorder, possibly consistent with bipolaron formation. Long-range CDW order is found to quench superconductivity, just as in the A15 compounds.

MOTT TRANSITION

In the original picture of the Mott transition, the transition to an insulating state is expected to occur at exactly half-filling, *independent of the band structure*, due to strong on-site Coulomb repulsion. Thus, regardless of the normal state properties of the material – normal metal, itinerant ferromagnet, high-T_c superconductor – as soon as the doping approaches half-filling there should be a transition to an insulating, antiferromagnetic state where none of the atomic orbitals is doubly occupied. The Hubbard model provides an oversimplified model of the normal state, which nevertheless preserves the physics of the Mott transition *very close to half filling*. It was not designed to give an adequate picture of the normal state away from half-filling; indeed within the spirit of the Mott transition, there can be no unique description since virtually any normal state can undergo a Mott transition.

However, looked at as a model for this normal state, the Hubbard model is decidedly peculiar. It leads to a square Fermi surface at exactly half filling. This means that there is not one but two separate electronic instabilities associated with half filling: an ordinary nesting instability, due to parallel sections of Fermi surface, and the additional singularity associated with the divergent dos due to the vHs. Adding these to the Mott instability, the Hubbard model has three independent singularities all of which are forced to occur at precisely half filling, thereby greatly enhancing the stability of the insulating phase, and complicating understanding of what happens when the material is doped away from half filling.

In contrast, band structure calculations consistently show that the Fermi surface has considerable curvature at half filling, due mainly to the direct oxygen-oxygen hopping matrix element, t_{OO}. This shifts the vHs away from half filling, and eliminates most of the flat-band sections associated with ordinary nesting. However, this reduces the parameter range in which the Mott insulating phase is stable, and the first question which must be asked is whether the model still has a Mott transition, when realistic input parameters are used in the calculations. This question was addressed in a number of calculations[9-11], which used a slave boson technique, which treats the limit of infinitely strong on-site correlations. The results are summarized in Fig. 1, which shows the various calculations of the phase boundary between the metallic (M) state and the charge-transfer insulator (CTI). While Sa de Melo and Doniach[9] used a simplified picture of the oxygen band structure, there is good agreement between the other two calculations. The figure also shows the results of a number of first-principles estimates[12-14] of these band parameters for $La_{2-x}Sr_xCuO_4$ (LSCO). As Hybertsen, et al.[12] pointed out, the earlier calculations (e.g., Ref. 13) seriously underestimated the bare Cu-O energy splitting, ΔE. Using the newer estimates, it is seen that LSCO is right on the border of the Mott transition. A number of points should be made: (1) The transition occurs when the effective Cu-O energy separation $\Delta E'$ is large enough. Near half filling, an important contribution to this separation comes from the Cu-O nearest-neighbor Coulomb repulsion, $V^{10,11,15}$: $\Delta E' = \Delta E_0 + 2V$. The effective V is doping dependent, decreasing as x increases[11], so the transition to the metallic state will occur at even lower doping than expected from a constant-V calculation; however, this will not modify the phase boundary at exactly half filling. (2) The phase boundaries in Fig. 1 correspond to a diverging susceptibility, or vanishing Cu-band bandwidth. However, the transition will actually occur at somewhat smaller values of $\Delta E'$, when the bandwidth becomes comparable to the exchange constant, $J^{7,16,17}$. The dotted line in Fig. 1 indicates when the susceptibility begins to sharply increase, in the calculation of Ref. 10. (3) While there are large error bars on the $\Delta E'$ parameter of Martin, et al.[14], the larger values are favored, as giving better agreement with measured J values.

In conclusion, the slave boson calculations, using realistic band parameters, find that LSCO is on the edge of the transition to a CTI state. Very modest doping pushes it over to the metallic state[10,17].

PHASE SEPARATION AWAY FROM THE MOTT INSULATOR

Since the Mott transition can involve many different normal states with a wide variety of properties, the transition back to this normal state, as a function of doping, should be first order. This phase transition, being electronically driven, has a number of unusual features. Thus, in principal the holes could phase separate by themselves, without accompanying ionic motion. However, when Coulomb

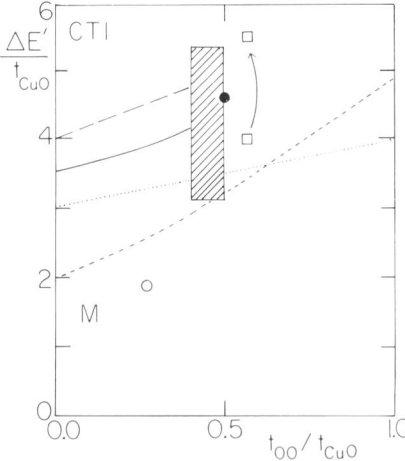

Fig. 1 Metal (M) - Charge-transfer Insulator (CTI) phase diagram. Opening of CTI gap, as calculated by Ref. 9 (short-dashed line), 10 (long-dashed), and 11 (solid), and onset of susceptibility increase, from Ref. 10 (dots). Symbols denote calculated parameter values for La_2CuO_4, open circle = typical of early parameter values (small ΔE), from Ref. 13; others = more recent: filled circle = Ref. 12, open square = Ref. 6 (lower is without $2V$ correction to ΔE), shaded area = Ref. 14.

effects are incorporated, an ordinary macroscopic phase separation becomes energetically impossible. Instead, there can be a microscopic phase separation, where the dense hole phase may contain only a few particles per domain – in an extreme but by no means improbable case, the stable phase corresponds to real-space pairing of holes.

This 'mesoscopic' (more accurately, nanoscopic) phase arises in a pure Hubbard model, and was first noted in the case of three-dimensional doped antiferromagnetic insulators[18]. In three-dimensions, there is strong pinning of these phases to impurities, thereby obscuring their direct observation. In the layered cuprates, the CuO_2-planes are modulation doped, so free hole-domains may be expected to form. Similar domain phases exist in the two-dimensional Hubbard and tJ-models[19], and it has been suggested that these real-space pairs are precursors of superconductivity. A problem with this interpretation arises, since the pair-phase is often found to be metastable with respect to a phase in which the dense holes form 'grain-boundaries' between antiferromagnetic domains[10,20].

There is a much more serious problem with this picture, however. Within the domains, the holes should be at a density close to the normal phase, and it seems highly improbable that the Hubbard model can provide an adequate description of the normal state properties. At this Workshop, Stephan[21] has shown that 10% hole doping is sufficient to eliminate virtually all antiferromagnetic correlations beyond nearest neighbor, so that the system is far from the regime where the Hubbard model can be trusted. It has been proposed that the normal end-phase is exactly at the vHs, this phase being stabilized by strong electron-phonon interactions (a giant Kohn anomaly)[10]. Even if it is not precisely at the vHs, slave boson calculations[6,7] suggest that it must be very close: there is an 'electronic pinning' of the Fermi level near the vHs, so that the Fermi level remains close to the renormalized vHs for a much greater doping range than would be expected using the bare band parameters. Levin[7] has proposed a number of measurements which could in principle determine the energy separation between the Fermi level and the vHs. For instance, the normal-state resistivity should vary linearly with temperature, extrapolating to zero at a finite temperature T^*, which is $\sim 1/4$th of the splitting between E_F and the vHs; the susceptibility should also have a strong peak at this energy.

Simple, essentially classical calculations[10,22] of the nanoscopic phase lead to the phase diagram of Fig. 2. There are three transitions as a function of doping: first, from a (probably incommensurate) antiferromagnetic insulating phase to a spin-glass-like phase, with antiferromagnetic domains surrounded by domain walls of the higher-hole-density phase; secondly, a metal-insulator transition from this phase to a metallic phase, with intrinsic weak links (corresponding to domain walls of the antiferromagnetic phase); and finally to a uniform phase at or near to the vHs. High-T_c superconductivity is associated with this latter phase, and T_c is suppressed by magnetic pair breaking in the mixed phase domain. (A similar phase separation can also occur for hole overdoping.)

There is considerable experimental evidence for this scenario, much of it summarized in Ref. 8. Perhaps the clearest example arises in LSCO. There are two ways to hole-dope La_2CuO_4, either by substituting some divalent ion, such as Sr, for La, or by incorporating interstitial O. The Sr is tightly bound in the La-layer, leading to the nanoscopic phase separation described above. However, the interstitial O remains highly mobile at low temperatures, and it is found that when the holes bunch

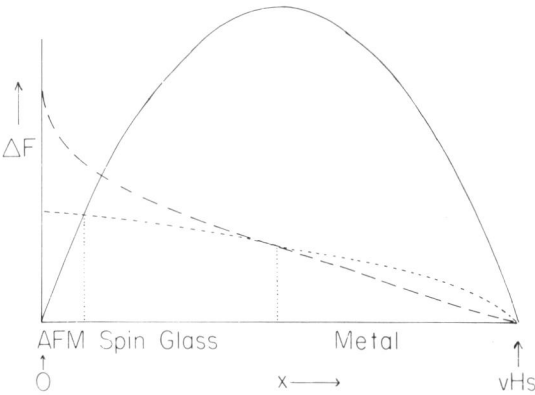

Fig. 2 Schematic free energy (ΔF) diagram for LSCO, showing two-phase coexistence regime for intermediate doping between the Mott insulator (x=0) and the vHs (x$\simeq$0.17). Solid line = uniform phase; short dashed line = islands of antiferromagnetic (AFM) insulator; long-dashed lines = islands of metallic phase at vHs. Lowest energy phases are labelled on horizontal axis [after Refs. 10,22].

up, the O follows, leading to charge-neutral domains and macroscopic phase separation[23]. In this case, it is easy to identify the normal end-phase: it corresponds almost exactly to the phase of highest T_c in the Sr-doped materials – i.e., it is the doping at which the vHs coincides with the Fermi level. It should be noted that the phase transition occurs at $\sim 265K$[24], so evidence for the nanoscopic phase in LSCO should be sought at low temperatures, probably below $150K$[25].

CDW FLUCTUATIONS AT THE vHs

Thus, a true, single phase composition exists only when the hole doping exactly corresponds to the vHs being at the Fermi level. However, at this concentration, a new type of disorder can exist. This disorder can best be understood by comparison with the A15 compounds, with which there is a strong analogy. The large dos leads to strong electron-phonon coupling, and hence to a competition between superconductivity and CDW formation. This CDW is driven by a special form of nesting: by splitting the two vHs peaks the large dos can be driven below the Fermi level, thereby significantly reducing the electronic energy, even though large parts of the Fermi surface remain ungapped. In mean-field, CDW formation wins out, and there can only be a low superconducting T_c associated with the ungapped portions of the Fermi surface[26]. When fluctuation effects are included[4], long-range CDW order is suppressed, but there should still be large two-dimensional short-range CDW order.

Experimental evidence for these effects is found in $La_{2-x}Ba_xCuO_4$. Very close to the vHs, there is a phase transition to a low-temperature tetragonal phase[27]. This phase is associated with an in-plane CDW[28], and strongly suppresses superconductivity. In this Workshop, Cohen[29] has demonstrated that this transition splits the two vHs, thereby greatly reducing the dos at the Fermi level. In LSCO, there are strong fluctuations into this CDW phase, but no long-range order above T_c[27]. Similar fluctuations are found in a Tl-based superconductor, with a local structural change occuring at T_c^{30}. In $YBa_2Cu_3O_{7-\delta}$, this transition is further suppressed, since the two vHs are already split in the orthorhombic phase.

The above considerations suggest that the electron-phonon coupling is highly anisotropic within the basal plane, with strong coupling near the vHs and relatively weak coupling further away. This has led to a 'two-fluid' model[4], similar to that of Bilbro and McMillan[31] for the A15 compounds. It is therefore very significant that Shen[32] reports a large superconducting gap anisotropy within the basal plane of $Bi_2Sr_2CaCu_2O_x$, determined by photoelectron spectroscopy. While the interpretation is complicated by hybridization with a Bi band, the largest gap occurs near the $\bar{M}$-point, very close to the expected vHs.

The 'two-fluid' picture can explain a number of anomalous features of the cuprates. Thus, the vHs-fluid, associated with the large dos, is nearly localized by strong electron-electron and electron-phonon scattering. It may be identified with the 'phenomenological polarizability' of Varma, et al.[33,4] The normal-fluid, associated with the low-dos parts of the Fermi surface (which would remain after CDW formation), is much less localized, and dominates the transport properties. It is these carriers which contribute to the small observed Hall density[4,34] and the Drude part of the frequency-dependent conductivity. The small electron-phonon coupling parameter derived from resistivity measurements is associated with this normal-fluid component. At T_c, both components become superconducting, but only the vHs-fluid is strongly coupled.

These short-range CDW fluctuations are due to the large peak in the *charge* susceptibility at the q-vector connecting the two vHs peaks - i.e., at $(\pi/a, \pi/a)$. There will be comparable peaks in the *spin* susceptibility at the same q-vector. These are precisely the form postulated by Millis, et al.[35] to account for the anomalous nuclear relaxation rates in YBCO. (A similar interpretation of the spin susceptibility is given by Lu, et al.[36].) Within the vHs model, the small value of the cutoff parameter $T_x \simeq 115K$ is readily understood: it is due to a weak interlayer coupling t_z (estimated at $\leq 10meV$) which cuts off the susceptibility divergence. The value of T_x is in excellent agreement with transport measurements[37], which find a change in slope of the linear resistivity in YBCO below about 115K. Clearly, removing oxygen from the chains should reduce t_z, and Millis, et al. find a smaller T_x in $YBa_2Cu_3O_{6.63}$.

CONCLUSION

The vHs model provides a framework for describing many of the phenomena observed in the new layered cuprates, including the high T_c values. Moreover, the underlying mechanism is the same one found in the A15 compounds, and the Bi-oxides, enhanced by lower dimensionality (which suppresses CDW formation). Finally, the model predicts that there is an intrinsic disorder involved with the vHs. The importance of strong microscopic disorder has recently been stressed by a number of experimental groups[38,39].

POSTSCRIPT: WHERE IS THE vHs?

The vHs is a topological singularity (saddle point) of the band structure, and can readily be identified from calculated energy dispersion curves; it is generally found to occur very close to the Fermi level. It should also show up in the calculated dos, but this is a very subtle point. In the first place, a very fine mesh is required to pick out a logarithmic divergence. In the second place, the calculation must be highly converged to properly describe interlayer hybridization, which is the only effect that could wash out the singularity. Under these circumstances, it is easier to find the vHs directly from the band dispersions. This has been discussed in Ref. 8, where it is shown that the vHs occurs very close to the compositions of optimum T_c in most of the cuprate superconductors.

As an example, the Fermi surfaces of $YBa_2Cu_3O_7$, recently calculated by Pickett, et al[40], are shown in Fig. 3. The overall appearance is similar to earlier calculations, but differs in many fine details. In particular, a CuO_2 plane band crosses the CuO chain band, and strongly hybridizes with it near the X-point. Mazin[41] reported that in his work[42] a similar crossing is also found, but with substantially less hybridization: the two segments of the CuO-like bands join in a nearly straight line, producing the appearence of an almost flat band, in good agreement with the photoemission data reported by Haghighi, et al.[43]. In the Pickett, et al., data, the integrated area of the two CuO_2 plane Fermi surface sections, averaged over the Γ and Z planes, corresponds to 1.24 holes per plane, in good agreement with experimental estimates[44]. The most important feature of their data is that *the vHs falls precisely at the Fermi level*, near the center of the X-U-line (see arrows), just as in LSCO.

This work was supported by the Department of Energy under subcontract from Intermagnetics General Corporation. Publication 471 from the Barnett Institute.

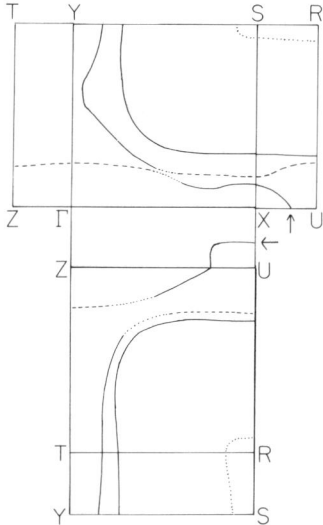

Fig. 3 Fermi surface of YBCO, from Ref. 40. Solid lines = predominantly CuO_2 plane-like sections; dashed lines = predominantly chain-like sections. Arrows indicate position of the vHs, along the X-U-line.

REFERENCES

1. J. Labbe and J. Bok, Europhys. Lett. **3**, 1225 (1987).
2. I. E. Dzyaloshinskii, Zh. Eksp. Teor. Fiz. **93**, 1487 (1987) [Sov. Phys.-JETP **66**, 848 (1987)].
3. J. Friedel, J. Phys. Condens. Matt. **1**, 7757 (1989), and unpublished.
4. R.S. Markiewicz, J. Phys. Condens. Matt. **1**, 8911,8931 (1989), Physica C**168**, 195 (1990), **169**, 63 (1990); R.S. Markiewicz and B.C. Giessen, Physica C**160**, 497 (1989).
5. C.C. Tsuei, Physica A**168**, 238 (1990); C.C. Tsuei, D.M. Newns, C.C. Chi, and P.C. Pattnaik, Phys. Rev. Lett. **65**, 2724 (1990); D.M. Newns, this Workshop.

6. D.M. Newns, P.C. Pattnaik, and C.C. Tsuei, to be published.
7. K. Levin, J.H. Kim, J.P. Lu, and Q. Si, unpublished; Q. Si and K. Levin, unpublished; K. Levin, this workshop.
8. R.S. Markiewicz, unpublished.
9. C.A.R. Sa de Melo and S. Doniach, Phys. Rev. B41, 6633 (1990).
10. R.S. Markiewicz, J. Phys. Condens. Matt. 2, 665 (1990).
11. M. Grilli, B.G. Kotliar, and A.J. Millis, Phys. Rev. B42, 329 (1990).
12. M.S. Hybertsen, M. Schluter, and N.E. Christensen, Phys. Rev. B39, 9028 (1989).
13. D.M. Newns, M. Rasolt, and P.C. Pattnaik, Phys. Rev. B 38, 6513 (1988); D.M. Newns, P. Pattnaik, M. Rasolt, and D.A. Papaconstantinopoulos, Phys. Rev. B38, 7033 (1988).
14. R.M. Martin, S. Bacci, and E. Gagliano, this Workshop.
15. C.A. Balseiro, M. Avignon, A. G. Rojo, and B. Alascio, Phys. Rev. Lett. 62, 2624 (1989).
16. G. Kotliar, P.A. Lee and N. Read, Physica C 153-55, 538 (1988).
17. R.S. Markiewicz, Physica C170, 29 (1990).
18. E. L. Nagaev, "Physics of Magnetic Semiconductors", (Mir, Moscow, 1983); P.B. Visscher, Phys. Rev. B10, 943 (1974).
19. S.A. Trugman, Phys. Scripta T27, 113 (1989); V.J. Emery, S.A. Kivelson, and H.Q. Lin, Phys. Rev. Lett. 64, 475 (1990).
20. J. Zaanen and O. Gunnarsson, Phys. Rev. B40, 7391 (1989); H.J. Schulz, Phys. Rev. Lett. 64, 1445 (1990).
21. W. Stephan, this Workshop.
22. R.S. Markiewicz, Int. J. Mod. Phys. B4, 1551 (1990).
23. J.D. Jorgensen, B. Dabrowski, S. Pei, D.G. Hinks, L. Sonderholm, B. Morosin, J.E. Schirber, E.L. Venturini, and D.S. Ginley, Phys. Rev. B38, 11337 (1988).
24. P.C. Hammel, A.P. Reyes, Z. Fisk, M. Takigawa, J.D. Thompson, R.H. Heffner, S.-W. Cheong, and J.E. Schirber, Phys. Rev. B42, 6781 (1990).
25. J. Saylor and C. Hohenemser, Phys. Rev. Lett. 65, 1824 (1990).
26. C.A. Balseiro and L.M. Falicov, Phys. Rev. B20, 4457 (1979).
27. J.D. Axe, et al., Phys. Rev. Lett. 62, 2751 (1989).
28. R.S. Markiewicz, J. Phys. Condens. Matt. 2, 6223 (1990).
29. R.E. Cohen, H. Krakauer, and W.E. Pickett, this Workshop.
30. B.H. Toby, T. Egami, J.D. Jorgensen, and M.A. Subramanian, Phys. Rev. Lett. 64, 2414 (1990).
31. G. Bilbro and W.L. McMillan, Phys. Rev. B14, 1887 (1976).
32. Z.X. Shen, B.O. Wells, D.S. Dessau, W.E. Spicer, A.J. Arko, R.S. List, C.G. Olson, D.B. Mitzi, and A. Kapitulnik, this Workshop.
33. C.M. Varma, P.B. Littlewood, S. Schmitt-Rink, E. Abrahams, and A.E. Ruckenstein, Phys. Rev. Lett. 63, 1996 (1989).
34. R.S. Markiewicz, unpublished.
35. A.J. Millis, H. Monien, and D. Pines, Phys. Rev. B42, 167 (1990).
36. J.P. Lu, Q. Si, J.H. Kim, and K. Levin, Phys. Rev. Lett. 65, 2466 (1990).
37. T.P. Orlando, K.A. Delin, S. Foner, E.J. McNiff, Jr., J.M. Tarascon, L.H. Greene, W.R. McKinnon, and G.W. Hull, Phys. Rev. B36, 2394 (1987).
38. J.D. Jorgensen, P. Lightfoot, and S. Pei, Supercond. Sci. Tech., to be published; J.D. Jorgensen, S. Pei, P. Lightfoot, B. Dabrowski, D.G. Hinks, and D.R. Richards, Third Int. Symp. on Supercond., Sendai, Japan, Nov., 1990, to be published (Springer, Tokyo).
39. A.W. Sleight, quoted by D.P. Hamilton, Science 250, 375 (1990).
40. W.E. Pickett, R.E. Cohen, and H. Krakauer, Phys. Rev. B42, 8764 (1990).
41. I.I. Mazin, this Workshop.
42. E.T. Heyen, S.N. Rashkeev, I.I. Mazin, O.K. Andersen, R. Liu, M. Cardona, and O. Jepsen, Phys. Rev. Lett. 65, 3048 (1990).
43. H. Haghighi, J.H. Kaiser, S.L. Rayner, R.N. West, J.Z. Liu, R. Shelton, J. Fluss, R.H. Howell, and F. Solal, this Workshop.
44. Y. Tokura, J.B. Torrance, T.C. Huang, and A.I. Nazzal, Phys. Rev. B38, 7156 (1988).

TWO-COMPONENT THEORY AND DYNAMIC STRUCTURAL CORRELATIONS

Y. Bar-Yam*

Dept. of Chemical Physics
Weizmann Institute of Science
Rehovot 76100, ISRAEL

Abstract

In two-component theory[1,2] pairing arises from localized negative-U states and mobility arises from extended single particle states. Signatures of this theory range from distinctive superconductive T_c, Δ, H_c, ξ to linear T dependence of the normal state resistance, and linear voltage dependent normal state tunneling conductance. A unique signature is the prediction of structural correlations which are local above T_c and extended below T_c. Experiments provide direct evidence for such dynamical correlations: neutron diffraction "thermal ovals",[2] channeling experiment cross section changes as a function of temperature near T_c,[3,4] pair-distribution-function neutron diffraction including inelastic and elastic scattering showing direct evidence for dynamic correlations which change at T_c,[5] and EXAFS showing oxygen atom tunneling between sites separated by 0.13Å.[6] In two-component theory the implied strong lattice coupling is consistent with a low isotope shift since the tunneling occurs by a virtual Franck-Condon transition.

1. Background

In recent years a large body of theoretical work has been dedicated to understanding the theory of High-T_c superconductors. No generally accepted theory has emerged as yet to describe the existing body of experimental literature or predict new phenomena. While much of the effort has been dedicated to theories where the origin of superconductivity lies in the antiferromagnetic properties of the materials, such theories have as yet not provided definitive predictions which establish their relevance to the Cu-O based class of high-T_c materials.

The two-component theory[1,2] suggests that high-temperature superconductivity can be achieved by resolving a basic conflict between pairing and mobility. The resolution arises by making use of two kinds of electronic states in the same material. Localized states provide the pairing and extended states provide the mobility. A small hybridization is necessary for the pairing and mobility to work together for superconductivity.

A summary of properties of the two-component theory follows. Details can be found in references 1 and 2. Let μ be the Fermi energy, D_c be the single particle density of states, w be the hybridization of single particles and pairs, M_B be the density of paired states. The theory then describes or predicts:

- functional dependence of $T_c(\mu, D_c, w, M_B)$
- a separation between the condensate evaporation T_c and the resistive transition T_c'
- pressure dependence of T_c. $T_c(P)$
- anomalies in sound velocity and bulk modulii at T_c
- functional dependence of $2\Delta/kT_c$ (μ, D_c, w, M_B)
- a broadening of the fermionic onset Δ
- functional dependence of $H_c^2/8\pi$ (μ, D_c, w, M_B) which is first order in kT_c.
- existence of fermion and boson excitations
- low energy boson excitations in the fermionic gap
- pair boson excitations in the energy range ~4-8 kT_c
- neutral fermion excitations (when μ is set for optimal T_c)
- linear temperature dependent normal state resistance $R(T) \sim T$
- frequency dependence of the conductivity $\sigma(\omega), \tau(\omega)$
- linear voltage dependence of the normal state tunneling conductance $G(V) \sim V$
- non-zero superconducting state zero-bias conductance $G(0)$
- s-wave pairing
- small isotope shift
- oxygen vacancy concentration dependence of superconductivity
- 2e charge carriers in normal state
- dynamic structural coherence
- single particle excitations including photo-induced excitations
- μ dependence of the single particle excitations
- identification of electronic states in calculations and experiment
- compatibility with antiferromagnetism with competition of order parameters
- reasons for advantage of 2-d materials but compatibility also with 3-d materials

Of the properties which are described many are compared favorably with experimental results.[1,2] Others have not yet been tested. In only one case there appeared to be a conflict between theory and experiment and this has been resolved by a new experiment in agreement with the theory. [In the pressure dependence of T_c as a function of doping; the theory predicted that in overdoped $La_{2-x}Sr_xCuO_{4-\delta}$ $dT_c(P)/dP$ is negative. The single existing experiment at the time of writing was in conflict with the theoretical predictions. A new experiment (unpublished report) properly controlling the oxygen vacancy concentration appears to confirm the theoretical prediction.]

Direct evidence for the distinctive signature of structural coherence has been very recently reported in two articles.[5,6] In the first, the pair distribution function measured by both inelastic and elastic neutron diffraction was used to measure the relative positions of atoms rather than their crystallographic positions. It was found that dynamical correlations which change at T_c exist in all of the high-T_c materials studied. In the second, extended X-ray absorption fine structure (EXAFS) was used to show that the oxygen atoms bridging between the planes and the chains of $YBa_2Cu_3O_7$ tunnel between two distinct sites separated by 0.13Å. The tunneling of this particular oxygen atom is in direct correspondence with the predictions of the two-component theory. Specific predictions about the doping dependence of this tunneling are now being tested. Earlier experiments using ion channeling[3,4] also provide evidence for presence of structural correlations and their change at T_c.

Two additional specific predictions of the theory - 2e charge carriers in the normal state (this is one of a few theories which make this prediction), and distinctive Fermi energy dependence of the single particle excitations (Ref. 2) are currently being tested.

II. Dynamical Structural Correlations

In two-component theory two kinds of states are responsible for superconductivity. One provides pairing and the other mobility. The pairing states are localized states known as negative-U centers.[7] Negative-U centers are common at defects in semiconductors. High-T_c Cu-O based materials are formed of layers of semiconductor and metal. They have the proper environment for both extended states associated with Cu-O planes and localized states associated with a common defect - the oxygen vacancy. Band structure calculations[8] place a localized state associated with isolated vacancies near the Fermi energy. In $Y_1Ba_2Cu_3O_7$ the often noted Cu-O "chains" are parallel to chains of vacancies. A flat band of localized states is associated with the chains similar to the states associated with isolated vacancies. It resides on the chain-to-plane bridging oxygen atoms O(4) and on the chain oxygen atoms O(1).[9]

Localized pairing arises from a structural relaxation accompanying the occupation of the localized level, similar to the electron-phonon coupling in Cooper pairing. The real atomic displacement changes the relative positions of nuclei (Figs. 1 and 2). There are two sets of nuclear positions corresponding to a doubly occupied and empty electronic state. Since the position of each nucleus is related to the positions of other nuclei this is a local structural coherence. The singly occupied state is not favored energetically because of the strong relaxation and is not thermodynamically stable - this is the pairing. In two-component theory, the electrons undergo transitions between the localized and extended states thus inheriting both mobility and pairing and the transition rate determines T_c. These transitions lead to dynamical structural coherence both in the normal state and superconducting state.

The quantum state of the negative-U center is described by a superposition:

$$\Psi_1 + e^{i\phi}\Psi_2$$

where Ψ_1 describes the electronically occupied site including the positions of the nuclei and Ψ_2 describes the electronically unoccupied site including the positions of the nuclei. In the normal state of the superconductor the phases ϕ between different sites are uncorrelated. Since this is the phase of the electronic state it becomes coherent in the superconducting state. Thus, the nuclear positions also become coherent between different sites leading to macroscopic quantum structural coherence.

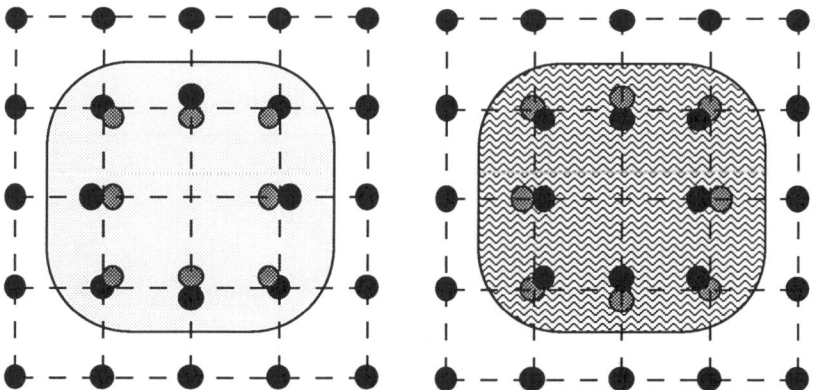

Fig. 1. Schematic illustration of the motion of nuclei associated with a negative-U center. Localized states associated with defects (typically in semiconductors), may have a large lattice coupling. When the localized electronic state is unoccupied (Left) the atoms, shown by solid circles, are in different positions than when the state is doubly occupied (Right). The large lattice relaxation leads to pairing so only unoccupied and doubly occupied cases are usually present. When the transition between occupied and unoccupied sites is responsible for superconductivity, the correlated motion of atoms leads to local structural coherence in the normal state and macroscopic structural coherence in the superconducting state.

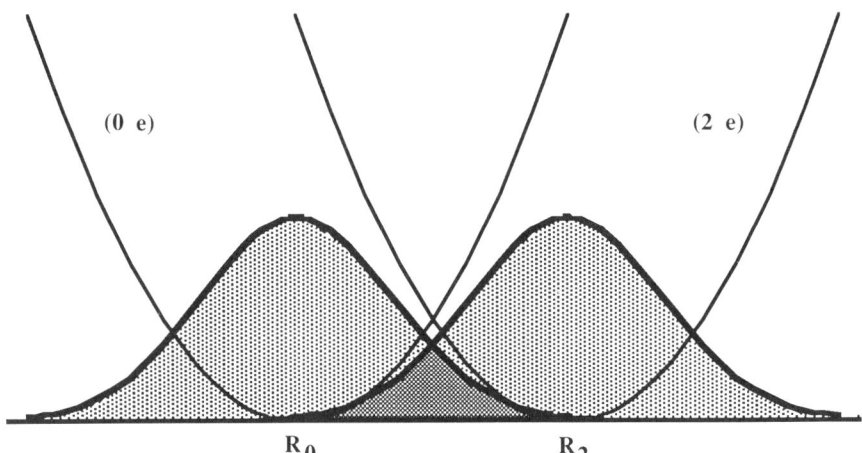

Fig. 2. Illustration of the two potential wells experienced by nuclei neighboring a negative-U center. The two wells correspond to electronically unoccupied (0) and occupied (2) centers. The shaded Gaussians are the zero-point nuclear motions.

Historically, the first evidence for such correlations may be traced to neutron diffraction data which shows large "thermal ovals" associated with oxygen atoms in Cu-O high-T_c materials. In neutron diffraction experiments[10] for $Y_1Ba_2Cu_3O_7$ it is the neighboring oxygen atoms to the vacancies O(4) and O(1) which have large ovals precisely where electronic structure calculations show the localized electronic states.

Evidence for a change in atomic correlations at T_c was first found in ion channeling experiments. In these experiments a beam of ions is sent near to a crystallographic direction down a "channel" of the material. The cross section is measured by back-scattering. In two separate experiments the ion channeling cross section measured as a function of temperature was found to change dramatically at T_c. The first experiment measured the temperature dependence of the smallest channeling cross section,[3] and the second measured the width of the channel.[4] The channeling observations are consistent with the development of atomic correlations which would lead to either a larger cross section if atomic positions are anti-correlated or smaller cross section if they are correlated in the channeling direction.

III. Neutron Diffraction and Dynamical Structural Correlations

In a revealing experiment[5] Toby et al (TEJS) have recently shown direct evidence using neutron diffraction for the presence of *dynamic local-structural-correlations which change at* T_c in a High-T_c Cu-O based material. In their experiment the radial distribution function (RDF), as is typically measured for a liquid, was obtained from a powdered superconductor. The RDF provides information about relative distances between nuclei. TEJS found their data could not be fit by crystallographic distances, requiring correlated nuclear displacements. Further measurements[11] of elastic and inelastic neutron scattering indicate that the correlations are dynamic rather than static. The RDF shows substantial changes near T_c thus associating the dynamical correlations with superconductivity. The experiment of TEJS may be a direct measurement of the dynamic local-structural-coherence of negative-U centers and their change at T_c.

The results of diffraction experiments can be directly compared with theory by calculation of the density-density correlation function of nuclei controlled by the x-y model component of the two-component theory. There are two contributions to the phenomenology - (a) the coherence between two different sites coupled to the electronic occupation of localized states and thus the superconductivity, (b) the motion of the nuclei between the two different states.

The relevant nuclear density for neutron diffraction is given by the expression:

$$\rho(R,t) = \Sigma \, \delta(R-R_i(t))$$

where $R_i(t)$ is the position of the ith nucleus. Considering the coherence between two different sites coupled to the electronic state occupation this can be written as:

$$\rho(R,t) = \Sigma \, B_i^\dagger B_i \, \delta(R-R^1{}_i) + B_i B_i^\dagger \delta(R-R^2{}_i)$$

$$= \Sigma \, \sigma_i^z \, (\delta(R-R^1{}_i) - \delta(R-R^2{}_i)) + 1/2 \, (\delta(R-R^1{}_i) + \delta(R-R^2{}_i))$$

where $B_i^\dagger$ creates an electron pair on site i. $R^1{}_i$ is the position of the atom at the ith site when the localized electronic state is occupied and $R^2{}_i$ is the position of the atom when the electronic state is unoccupied.

$$B_i^\dagger B_i = (\sigma_i^z + 1/2)$$

in the analogous spin x-y model.

Thus, for example, the equal time density-density correlation function can be written as

$$\langle\rho(R,t),\rho(R',t)\rangle = \Sigma \, \langle\sigma_i^z\sigma_j^z\rangle(\delta(R-R^1{}_j) - \delta(R-R^2{}_j))(\delta(R-R^1{}_i) - \delta(R-R^2{}_i))$$

$$+ \langle\sigma_i^z\rangle(\delta(R-R^1{}_j) - \delta(R-R^2{}_j))(\delta(R-R^1{}_i) + \delta(R-R^2{}_i))$$

$$+ (\delta(R-R^1{}_j) + \delta(R-R^2{}_j))(\delta(R-R^1{}_i) + \delta(R-R^2{}_i))$$

or:

$$\langle\rho(q,t)\rho(-q,t)\rangle =$$

$$\{ \langle\delta\sigma_q^z\delta\sigma_{-q}^z\rangle |S^1(q)-S^2(q)|^2 + |S_0(q)[n_B \, S^1(q)+(1-n_B)S^2(q)]|^2 \}$$

This expression describes the effects of the two sets of nuclear positions. $S_0(q)$ is the structure factor of the lattice periodicity, $S^1(q)$ and $S^2(q)$ are the structure factors of nuclei within a cell for the occupied and unoccupied electron state respectively. The second term represents the average scattering from nuclei including their relative probability of being in the different sites, n_B is the relative density of electron occupation. The first term gives the effects of correlations where $\delta\sigma_q^z = \sigma_q^z - \langle\sigma_q^z\rangle$, σ_q^z is the z component of a quantum x-y spin model which is used to describe superfluid correlations in the two-component theory. The results can be analyzed using the properties of quantum x-y model correlation functions and compared directly with the temperature and Fermi energy dependence (μ plays the role of the magnetic field in the x-y model) of experimental results.

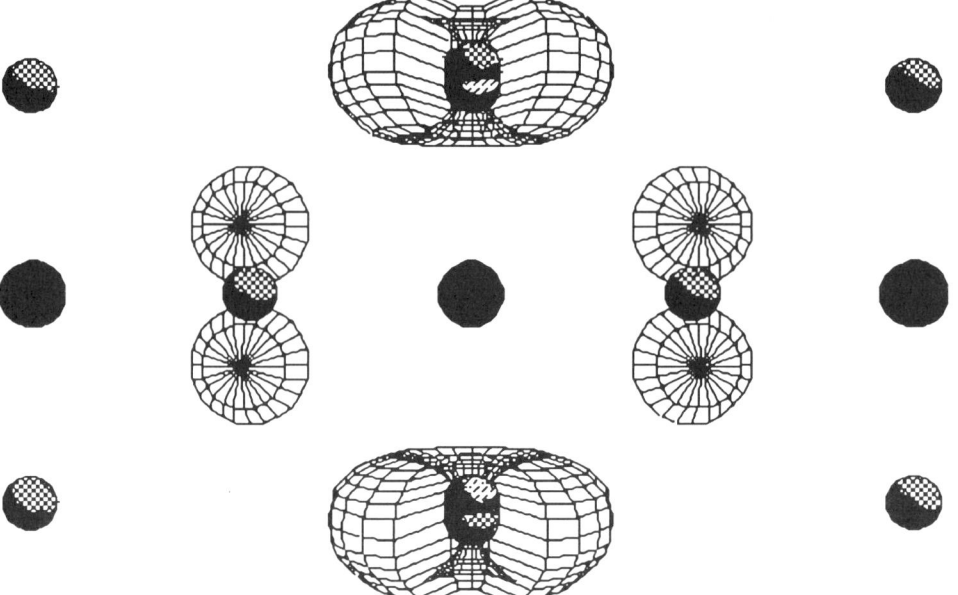

Fig. 3. Illustration of lattice fluctuations in YBa$_2$Cu$_3$O$_7$. The y-z plane is shown, Cu is black and O is shaded. The Cu-O chain lies horizontally at the center, the O4 atoms above and below the chain are also indicated. Two positions of the central O(4) atoms (checkerboard / dashed shading) are shown at tunneling sites according to experiment (Ref. 6). *There is no exaggeration of scale in the displacement of the oxygen atoms compared with other distances.* In two-component theory the oxygen motion is due to occupation of the localized electronic state shown in wire frame by two paired electrons (negative-U center). Since the electronic state is antibonding, when the electrons are present the oxygen positions are shown in checkerboard shading – farther from the Cu, when not present they are shown in dashed shading – closer to the Cu.

IV. EXAFS and Dynamical Structural Correlations

Very recently Mustre de Leon et al[6] (MCBB) reported a new analysis of polarized extended x-ray absorption fine structure (EXAFS) of YBa$_2$Cu$_3$O$_7$. In their analysis the EXAFS spectrum is studied to obtain the distribution of positions of the bridging oxygen O(4). The analysis extends the usual neutron diffraction result showing a broad distribution for the positions of O(4) along the z-axis. MCBB's results indicate O(4) atoms are tunneling between two sites separated by 0.13 Å (Fig. 3). By analyzing the effective potential in which the oxygen nuclei are found, they discover that the best fit is obtained by assuming a double harmonic potential with nearly equal occupation of the two wells (Compare Ref. 8 Fig. 2 with Fig. 2). The temperature dependence of the EXAFS near T_c implies the tunneling is coupled to the superconductive order parameter.

The direct analogy between MCBB's results and theoretical predictions leads to an additional untested prediction. If the two atomic positions are associated with different electronic occupations, their relative probability will be directly controlled by doping. In the measured sample the occupations are similar. This corresponds to half-filling of the negative-U states which gives the highest T_c in two-component theory. Doping by praseodymium should enable a systematic variation of the occupation of the negative-U centers and a test of this prediction in YBa$_2$Cu$_3$O$_7$.

The possibility that some Mössbauer-active nuclei may be used to probe the dynamics of nuclear motion in these materials should also be explored.

V. Dynamical Structural Correlations and External Magnetic Fields

One of the most intriguing possibilities arising from the macroscopic structural coherence is observing the effect of external manipulations of the superconducting order parameter on structural probes. Near T_c correlations have their greatest effect on the structural probes. Thus, for example, magnetic fields should have direct effects on the structural measurements described above, particularly near T_c.

VI. Isotope Shift

Even prior to the direct experiments, it has been generally acknowledged that Infrared, Raman, and Photo-Induced absorption experiments imply the importance of electron-phonon interactions. The only (though key) measurement which appears to be contrary is isotope shift experiments[12] indicating a lack of isotope shift for all atom types. The assumption that an isotope shift must exist for electron-phonon-interaction-induced superconductivity is based on BCS theory. When pairing arises from negative-U centers the strong electron-phonon (or electron-structure) coupling is consistent with a lack of isotope shift.[1]

In two-component theory, T_c is controlled by the transition of electrons from pairing states to mobile states. The matrix element for the transition includes the change in nuclear positions when the electronic occupation of the negative-U center changes. The direct transition between these positions is exponentially suppressed due to the overlap of zero-point-motion wavefunctions (Fig. 2). While the direct transition would give rise to an isotope shift, suppression leads to dominance of a virtual Franck-Condon transition - an electronic transition followed by structural relaxation - which would give rise to *no isotope shift* when the structural relaxation rate is shorter than the electronic transition rate. This shows that a two-component description of negative-U superconductivity is consistent with pairing due to a very strong electron-phonon coupling and a small isotope shift as has been found for high-T_c materials.

References

[*] Current address: ECS, 44 Cummington St. Boston University, Boston, MA 02215
[1] Y. Bar-Yam, Phys. Rev. B 43, 359 (1991)
[2] Y. Bar-Yam, Phys. Rev. B 43, 2601 (1991)
[3] R. P. Sharma, L. E. Rehn, P. M. Baldo and J. Z. Liu, Phys. Rev. Lett. 62, 2869 (1989)
[4] T. Haga, K. Yamaya, Y. Abe, T. Tajima, and Y. Hidaka, Phys. Rev. B 41, 826 (1990)
[5] B. H. Toby, T. Egami, J. D. Jorgensen, and M. A. Subramanian, Phys. Rev. Lett. 64, 2414 (1990)
[6] J. Mustre de Leon, S. D. Conradson, I. Batisti´c and A. R. Bishop, Phys. Rev. Lett. 65, 1675 (1990)
[7] P. W. Anderson, Phys. Rev. Lett. 34, 953 (1975)
[8] R. V. Kasowski, W. Y. Hsu and F. Herman, Phys. Rev. B 36, 7248 (1987)
[9] W. E. Pickett, Rev. Mod. Phys. 61, 433 (1989)
[10] J. D. Jorgensen, et al Phys. Rev. B 41, 1863 (1990)
[11] T. Egami (private communication)
[12] L. C. Bourne, A. Zettl, T. W. Barbee III, and M. L. Cohen, Phys. Rev. B 36, 3990 (1987)

SEARCH FOR THE CORRECT MICROSCOPIC THEORY FOR THE HIGH TEMPERATURE CUPRATE SUPERCONDUCTORS

J. Ashkenazi,[a] D. Vacaru[a] and C. G. Kuper[b]

[a]Physics Department, University of Miami, Coral Gables, FL 33124, U.S.A.

[b]Physics Department and Crown Center for Superconductivity
Technion — Israel Institute of Technology, 32000 Haifa, Israel

INTRODUCTION

It was apparent in this workshop that the scientific community has not yet reached a consensus concerning a microscopic theory for the high-T_c cuprates. The BCS theory assumes a model of free electrons plus a pairing interaction; it contains the essential physics necessary to understand conventional superconductors (SC's). However, few theorists (if any) will claim that this simple three-dimensional free-electron model contains the essential physics of the high-T_c cuprates. Much controversy still surrounds the question which theoretical model is most appropriate. And even for a given model, there is often disagreement about the validity of the various methods of approximation.

The free-electron model disregards structural, orbital and exchange-correlation effects. An important structural aspect of the cuprates is their quasi-two-dimensionality. One approach has been to study them within a two-dimensional renormalized free-electron model ("Fermiology").[1] Another approach (e.g. Anderson[2]) emphasizes not only the role of two-dimensionality but also that of strong electron correlations, but treats other structural and orbital effects only within the framework of a single-band Hubbard model. A number of "band-structure" groups[3,4] treat exchange and correlations within the mean-field local-density approximation (LDA), while taking account fully of structural and orbital effects (assuming that the lattice is perfect). Yet another school believes that lattice defects play a major role.[5,6] It has even been suggested[7] that the theory of conventional SC's needs to be revised. Views intermediate between the above pictures have also been expressed.

Not only is there no agreement as to the correct theoretical description of the SC state of the cuprates; even the specification of their *normal* state is controversial. While many of the papers presented in this workshop (*e.g.* Refs. 1,3,4,7–9) support the Fermi liquid picture, typical of metals, others favor an *exotic* normal state. The various proposals include an "anyon" liquid,[10] a Luttinger liquid,[2] a "marginal" Fermi liquid,[11] bipolaronic electron pairs,[6,12,13] quantum percolation,[5] *etc.*

The diversity in theories of high-T_c reflects the diversity in experimental data on the cuprates, reported in this workshop. For example, photoemission spectroscopy (PES) and inverse PES (IPES),[14–18] de Haas-van Alphen data,[19] and positron annihilation[20] results apparently confirm the LDA band-structure picture (with plausible mass renormalization).[3,4]

Thus these data tend to support a Fermi-liquid picture†. Results of certain other measurements are very anomalous, compared to conventional Fermi liquids. Phenomenological Fermiology[1] yields anomalously low Fermi energy and velocity parameters. Results on mid-infrared[21,22] and Raman[23] continuum spectroscopy, NMR relaxation[24,25] and neutron scattering[26,27] seem to support marginal-Fermi-liquid phenomenology (MFLP), characterized by the following asymptotic behavior of the imaginary part of the charge and spin polarizability:[11]

$$\mathrm{Im}\tilde{P}(\mathbf{q},\omega) \sim \begin{cases} -N(0)\,(\omega/kT) & \text{for } |\omega| \ll kT, \\ -N(0)\,\mathrm{sgn}(\omega) & \text{for } |\omega| \gg kT. \end{cases} \quad (1)$$

Pulsed neutron scattering,[28] EXAFS,[29] and tunneling data[30] show anomalies. But of all the properties of these materials, the most anomalous are the transport coefficients.[31-34]

The low Fermi energy and velocity found by a Fermiology analysis,[1] are consistent with a narrow conduction band. In the following sections, we shall discuss to what extent a narrow-band model can account for the transport properties of the cuprates, and see that, specifically, at high temperatures, the thermoelectric power (TEP) fits such a model very well. We shall show that this model is also consistent with the μSR result[35] that T_c is asymptotically $\propto n/m^*$ (where m^* is an effective mass, and n the carrier concentration). We will then demonstrate that the MFLP arises from transitions between this narrow band and a wide one.

We will subsequently present an explanation of such a band structure, proposing that the narrow band represents singlet polarons of Cu($d_{3z^2-r^2}$) holes surrounded by O(p) holes. Bianconi et al.[36] have suggested that such polarons are more stable than Zhang-Rice singlets.[37] We will argue that the existence of these polarons can account for dynamic structural irregularities,[28,29] "strange" angular-resolved PES (ARPES) and IPES (ARIPES) results[17] near the Fermi level (E_F), and anomalous oscillatory behavior below the SC gap, seen in PES,[14] and can answer the question why the MFLP[11] needs a cutoff frequency ω_c to fit the mid-infrared data, but does not need it to fit the Raman continuum.

In the cuprates, the Hubbard parameter U/W is intermediate in size and our wide band will represent some sort of interpolation between the LDA bands in the small U/W limit and the "spinon" bands[2] in the large U/W limit. (Here U is the intra-atomic Coulomb parameter, and W is the total "bare" bandwidth.) We argue that the number of charge carriers in this band at low and intermediate temperatures is of order $2kT/W_W$, where W_W is the renormalized bandwidth, but that coherence in it is lost at high temperatures.[8] The joint effects of these two bands will generate anomalies in the temperature dependence of the resistivity, the Hall coefficient and the TEP, of the type observed experimentally.

In our final section, we propose a mechanism for high-T_c SC (HTSC). However, the main emphasis of this article is the *search* for the correct theoretical model of the cuprate SC's ; we do not offer a rigorous solution of any specific model.

THE NARROW-BAND MODEL

The normal-state transport properties of the high-T_c cuprates, and especially the dependence of these properties on temperature, are anomalous compared to "good" metals. In samples showing good SC properties, the resistivity (ρ_{ab}) in the ab plane above T_c has metallic behavior (often close to $\rho \propto T$, at moderate temperatures and possibly changing to $\propto T^2$ at high temperatures.[31,38,39]), while ρ_c is much higher.‡ The Hall coefficient R_H, should be independent of temperature for metals — but in at least some of the high-T_c cuprates, the Hall resistance R_H is $\propto 1/T$ in the ab-plane.[31] The TEP (S), which should be a few μV/K in "good" metals, is much higher here, and its temperature dependence is very puzzling. In good

---

† Note, however, that it had been claimed that some details are better explained by a Luttinger-liquid[2] or a marginal-Fermi-liquid[11] approach.

‡ Both metallic and nonmetallic behaviors have been reported[32] for ρ_c.

SC samples, and under conditions of constant stoichiometry, the TEP in the ab plane S_{ab} is very sensitive to the stoichiometry, and at low temperatures has a complicated T-dependence. But in every case, the TEP saturates[33,34,40–43] around 400° C. The saturation value varies with stoichiometry from a few μV/K to more than a hundred — much higher than for a typical metal, even when low T SC has still not been quenched.

A useful feature of the YBa$_2$Cu$_3$O$_{7-\delta}$ (YBCO) system is that one can study the effect of varying two independent parameters, the temperature T and the stoichiometry δ. The normal-state transport properties of this system have been interpreted in terms of the "narrow-band" model.[38,43] This model assumes a narrow conduction band (NCB), with two states per formula unit (i.e. one state per planar copper atom) containing n effectively noninteracting holes per planar Cu atom. The saturation value of S is:

$$S = (k/e)\ln[(1-n)/n], \qquad (2)$$

where k is Boltzmann's constant ($k/e = 86\ \mu$V/K). Saturation should occur when kT is not too smaller than the bandwidth W_N. Thus W_N should be $\lesssim 0.2$ eV — much smaller than in typical metals. With the plausible assumption that $n = \frac{1}{2}(1-\delta)$, (i.e. that the $(1-\delta)$ chain oxygen atoms create, approximately, $(1-\delta)$ holes in the NCB), Eq. (2) fits the high-temperature TEP in YBCO,[38,43] for $0 \lesssim \delta \lesssim 2/3$. When $\delta \simeq 0$ the NCB is close to half full, and S_{sat} is small (and can be either positive or negative), in agreement with experiment.[33] At low temperatures, the fit of Eq. (2) to experiment is no longer good — perhaps due to contributions from other bands[33] (see discussion below). The model is also consistent with the observed high-temperature TEP in Pr-doped[34] and Ca-doped[42] YBCO. Attempts have also been made[44] to apply the model to the transport anomalies at low and intermediate temperatures, but additional assumptions were needed, and the agreement with experiment was not very good.

A priori, the existence of a very narrow conduction band in the normal state would seem to be unphysical because it would tend to cause anomalies and instabilities. Such anomalies are indeed observed in the A15 materials,[45] but not in the cuprates. However, let us consider an NCB which is so narrow that the differences $f(\epsilon_1 - \mu) - f(\epsilon_2 - \mu)$, which appear in the expressions for the anomalies and instabilities, become small for both ϵ_1 and ϵ_2 within the NCB. (This means that in the normal state this band is well away from the low-T limit.) Here μ is the chemical potential, and $f(E) \equiv 1/[\exp(E/kT)+1]$ is the Fermi distribution. Note also that the contribution of the NCB to the specific heat vanishes in the high-T limit. If a low-T limit approach were still valid around T_c, the jump in the specific heat at T_c would have been considerably larger than observed.[46,47]

At low T, one of the instabilities disabling the existence of a normal state with an NCB is the instability against an SC transition; it must indeed be the dominant one below T_c, although we do not rule out (and in some cases we even predict) the possibility that other instabilities may occur between the high-T regime and the SC state. In order to study the SC instability, let us assume for simplicity that the pairing interaction parameter V is constant. Whatever the source of this interaction, the most natural assumption for the NCB is that V is constant throughout the band, and that we can ignore retardation. Thus we will not need to use Eliashberg formalism; the BCS gap equation will suffice:

$$1 = \frac{V}{2}\int_{\epsilon_B - \mu}^{\epsilon_T - \mu} d\epsilon\ N_N(\epsilon + \mu)\frac{1 - 2f(\sqrt{\epsilon^2 + \Delta^2})}{\sqrt{\epsilon^2 + \Delta^2}}. \qquad (3)$$

Here ϵ_T and ϵ_B are, respectively, the top and the bottom of the NCB, and $N_N(E)$ is the density of states (DOS) of the NCB, per planar Cu atom. The gap 2Δ is assumed (for simplicity) to be isotropic, and its value in the SC phase is determined by Eq. (3). The transition temperature T_c occurs when $\Delta \to 0$ and the gap equation (3) reduces to:

$$1 = \frac{V}{2}\int_{\epsilon_B - \mu}^{\epsilon_T - \mu} d\epsilon\ N_N(\epsilon + \mu)\frac{1 - 2f(|\epsilon|)}{|\epsilon|}. \qquad (4)$$

In conventional SC's, the main influence of temperature in Eq. (4) is the temperature dependence of $f(|\epsilon|)$ within the Θ_{Debye} cutoff range. Here, in contrast, the *whole* band is involved, and another effect of temperature becomes significant. Unless the NCB is exactly half full, as we ramp the temperature up from the low-T to the high-T limit, the chemical potential μ will cross the band edge (which is ϵ_{T} for $n < 1/2$). Where this occurs, there will be a sharp drop in the right hand side of Eq. (4), because at lower temperatures the denominator $|\epsilon|$ has a zero within the range of integration, while at higher temperatures it does not.† Thus for the NCB, it is reasonable to assume that the right-hand side of (4) passes the value 1 close to the temperature T_μ where $\mu = \epsilon_{\text{T}}$. This means that T_c is close to T_μ.‡ The chemical potential μ is determined here (in the normal state) through the equation:

$$n = \int_{\epsilon_{\text{B}}-\mu}^{\epsilon_{\text{T}}-\mu} N_{\text{N}}(\epsilon+\mu)\ f(-\epsilon)\ d\epsilon. \tag{5}$$

In the small-n limit, the main contribution to the integral in Eq. (5), when $\mu \simeq \epsilon_{\text{T}}$, comes from the close vicinity of $\epsilon_{\text{T}} - \mu$. In this part of the domain of integration, the effective mass approximation is valid: $m^* \propto N_{\text{N}}(E)$. For the almost-tetragonal quasi-two-dimensional cuprates, with a lattice constant a, and effective m^*, one can then insert in Eq. (5):

$$N_{\text{N}}(E) \cong a^2 m^* / 2\pi \hbar^2. \tag{6}$$

For $\mu = \epsilon_{\text{T}}$, Eq. (5) becomes:

$$n \cong \frac{a^2 m^*}{2\pi \hbar^2} \int_{-\infty}^{0} f(-\epsilon)\ d\epsilon, \tag{7}$$

and if we interpret T_μ as T_c, we get, on integrating (7):

$$T_c(n \ll 1) \simeq \frac{2\pi \hbar^2}{k a^2 \ln 2} \cdot \frac{n}{m^*}. \tag{8}$$

The bandwidth W_{N} corresponding to Eqs. (6) and (8) in the YBCO system is of the order of room temperature. This is somewhat less than would be needed to fit the TEP saturation temperature. However, as we shall see below, there are additional factors affecting the transport properties. These complicating factors disappear at high temperatures.

If n is identified with the concentration of carriers in the SC phase, Eq. (8) is consistent with the μSR-based observation of Uemura et al.[35] that $T_c \propto n/m^*$, with the same proportionality factor for all the cuprates. Note that they all have about the same lattice constant a, though not necessarily the same V; however, V does not appear in (8). As we move away from the small-n limit, further analysis shows that T_c has a lower value than predicted by Eq. (8),‡ in agreement with Ref. 35. A similar mechanism for determining T_c might also occur in the bismuthate, organic, Chevrel-phase and heavy-fermion SC's, making a slight modification to Eq. (6) for isotropic three-dimensional bands (*cf* the conclusion of Uemura et al.[35])

If the above analysis is correct, the behavior of Δ below T_c is expected to be non-BCS like; it is expected to rise much faster than BCS with decreasing temperature to its $T=0$ value and moreover $\Delta(0)$ will be larger than the BCS value $1.75\ kT_c$. Both these deviations from BCS

---

† When the denominator has a zero, the numerator has a zero too, and the integral does not diverge; however it is still considerably larger than it is when $|\epsilon|$ does not have a zero.

‡ When the band is close to half full (*e.g.* in YBCO close to $\delta = 0$), it would appear that μ would not cross the band edge up to very high temperatures. However, in this case we predict that a slight instability, *e.g.* a Peierls charge-density wave instability, will occur above T_c. The instability will split the NCB into subbands, which will still be very close to μ. The high-T saturation value of the TEP will be unaffected. We expect that this instability might disappear below T_c. A specific heat anomaly seen in certain YBCO samples[48] around 220 K may possibly be due to such an instability.

behavior agree with experiment. The "two-fluid model" behavior of the gap[49,50] indicates that Δ is constant over almost the whole SC range. If there is a coherence peak, due to quasiparticle DOS maxima near the gap edges, it should be limited to a much narrower temperature range below T_c than the BCS prediction. A coherence peak, with such a narrow temperature range, has been observed recently in the surface impedance of $Bi_2Sr_2CaCu_2O_8$ single crystals,[51] but no coherence has been observed in NMR relaxation rate.[24,25] In a subsequent section, we will show that the NCB carriers are spinless; and an additional study will be necessary to understand the absence of a coherence peak in NMR relaxation.

INCLUSION OF A WIDE BAND IN THE MODEL

Experimental evidence shows that the NCB is not the only band around E_F. If it were, we would not see a conventional Fermi surface (FS) around room temperature, because then all states in the band would have finite occupation. PES and IPES[14-18] measurements and FS determination[19,20] confirm the existence of wide bands, qualitatively similar to those predicted by LDA,[3,4] although the bandwidth may be renormalized by many-body effects. Let us study the effect on the electron polarizability of transitions between the NCB and a wide conduction band (WCB), whose DOS is $N_w(E)$. Here "wide" means $\gg kT$, so that the conventional low-T approach is valid. The justification of this band structure, and its implications, will be discussed in the subsequent sections.

Let $G_N(\mathbf{k},\epsilon)$ and $G_W(\mathbf{k},\epsilon)$ be the single-particle Green functions for n-type carriers in the NCB and WCB respectively. Here $\mathbf{k}$ is the quasimomentum and ϵ is the energy with respect to the chemical potential μ. Interband transitions contribute to the imaginary part of the polarizability as follows:

$$\text{Im}\,\tilde{P}_{NW}(\mathbf{q},\omega) \propto \int [\text{Im}\, G_N(\mathbf{k},\epsilon) \cdot \text{Im}\, G_W(\mathbf{k}+\mathbf{q},\epsilon+\omega) \\
- \text{Im}\, G_W(\mathbf{k},\epsilon) \cdot \text{Im}\, G_N(\mathbf{k}+\mathbf{q},\epsilon+\omega)] \,[f(\epsilon+\omega)-f(\epsilon)]\, d\epsilon\, d^3\mathbf{k}. \quad (9)$$

To study the general features of $\text{Im}\,\tilde{P}_{NW}$, and specifically to get simple analytic expressions without going into structural details, we will assume that the NCB can be treated in the high-T limit, and the WCB in the low-T limit. Let us model their respective DOS's by a δ-function and a constant:

$$N_N(\epsilon) = \delta(\epsilon - \epsilon_N)\,, \quad N_W(\epsilon) = N(0) = \text{constant}. \quad (10)$$

This gives, for $\text{Im}\,\tilde{P}_{NW}$:

$$\text{Im}\,\tilde{P}_{NW}(\mathbf{q},\omega) \propto \int [N_N(\epsilon)N_W(\epsilon+\omega) - N_W(\epsilon)N_N(\epsilon+\omega)]\,[f(\epsilon-\mu+\omega)-f(\epsilon-\mu)]\, d\epsilon$$
$$= N(0)\,[f(\epsilon_N - \mu + \omega) - f(\epsilon_N - \mu - \omega)]. \quad (11)$$

To find the asymptotic behavior of $\text{Im}\,\tilde{P}_{NW}$ for $|\omega| \ll kT$ and for $|\omega| \gg kT$, we approximate:
(a) for $|\omega| \ll k_B T$, put $[f(\epsilon_N - \mu + \omega) - f(\epsilon_N - \mu - \omega)] \simeq 2\omega[df(\epsilon_N - \mu)/d\epsilon] = -2f(\epsilon_N - \mu)[1 - f(\epsilon_N - \mu)](\omega/k_B T)$;
(b) for $|\omega| \gg k_B T$, put $[f(\epsilon_N - \mu + \omega) - f(\epsilon_N - \mu - \omega)] \simeq -\text{sgn}(\omega)$.
Inserting the model DOS (10) in Eq. (5), we get $f(\epsilon_N - \mu) = 1 - n$. The resulting expressions are:

$$\text{Im}\,\tilde{P}_{NW}(\mathbf{q},\omega) \propto \begin{cases} -2N(0)n(1-n)(\omega/k_B T) & \text{for } |\omega| \ll k_B T, \\ -N(0)\text{sgn}(\omega) & \text{for } |\omega| \gg k_B T, \end{cases} \quad (12)$$

which is close to MFLP[11] (cf Eq. (1)). Note that the $n(1-n)$ factor in Eq. (12) is $\sim 1/4$ for $\delta = 0$ and $\sim 3/16$ for $\delta = 0.5$ (from our analysis of the TEP data for YBCO; see the previous section).

A more realistic DOS than that of Eq. (10) will modify the picture somewhat, but should maintain the main MFLP features. However, since we are considering inter- and not intraband transitions, the effect on the self energy will differ from that of Varma et al.[11] Before discussing it we have to understand the physical origin of these bands.

PHYSICAL ORIGIN OF THE MODEL

The band model introduced above may still look a little puzzling. In particular, it raises the following questions (which we will answer below): (i) What further signatures of the NCB can we expect? (ii) What is its physical origin? (iii) Why does the NCB determine the high-temperature transport properties, and can one see the effect of the WCB on transport?

Many theories of the cuprates concentrate on the CuO_2 planes. A popular model is a three-band one, based on the $Cu(d_{x^2-y^2})$, $O_x(p_x)$ and $O_y(p_y)$ orbitals, where O_x and O_y are planar oxygen atoms, whose nearest Cu neighbors are respectively in the x and y directions. A key question is the importance of electron correlations. Both theoretical estimates[4,52] and Auger spectroscopy data[53] suggest that the intra-atomic Coulomb parameters U for these orbitals are comparable to the LDA bandwidth[3,4] W. If U is sufficiently large,[37,52] the model reduces to a one-band Hubbard model, which can be transformed in the large-U/W limit to the t–J model. This reduction follows from the tendency of the orbitals of a $Cu(d_{x^2-y^2})$ hole and of the hybrid of the $O_x(p_x)$ and $O_y(p_y)$ to form singlet ("Zhang-Rice") pairs around the Cu atom.

When $U \simeq W$, both the small and large U/W approximations fail, and the rigorous theory of the electronic structure becomes complicated. One approach is to perform extended numerical calculations on finite clusters.[52,54,55] However, it is useful to compare approximate solutions for low and high U/W, in order to get some idea about the intermediate range. For large U/W, using the the t–J model, Anderson[2] argues that there is separation of the charge and spin degrees of freedom. The charge is carried by "holons", while the spin is carried by (Fermion) "spinons", which satisfy Luttinger's theorem[56] for the FS of the non-interacting electrons. Barnes,[57] using the "slave boson" formalism that he invented in the seventies, shows that though such a separation is not rigorous in two dimensions (see also Ref. 54), one can still work out an approximate mean-field theory of the charge–spin separation. In it, not only the spinons, but also the holons are fermions, but with one state (and not two) per planar Cu atom. As far as the number of states and their contribution to high-temperature charge transport is concerned, these spinon and holon fields resemble, respectively, the WCB and NCB of our model.

Let us try to build a realistic picture of the electronic structure of the cuprates. This picture should include both the low- and the high-U/W concepts, and the lattice dynamics, and should fit the experimental evidence. The holons of the t–J model represent the Zhang-Rice singlet pairs of the original three-band model. However, the t–J model ignores the effect of the $Cu(d_{3z^2-r^2})$ orbitals; the LDA calculations[3,4] predict that they will contribute close to the FS. On the basis of his spectroscopic-theoretical analysis, Sawatzky[58] doubts the stability of Zhang-Rice singlets against the effect of these $d_{3z^2-r^2}$ orbitals. Using polarized X-ray absorption spectroscopy Bianconi et al.[36] find that the number of $Cu(d_{3z^2-r^2})$ holes is close to the number of $O(p)$ holes. They note that the interatomic Coulomb repulsion integral of an $O_x(p_x)$ or $O_y(p_y)$ with a neighboring $Cu(d_{3z^2-r^2})$ hole is smaller (due to smaller overlap) than the integral with a $Cu(d_{x^2-y^2})$ hole. They therefore suggest that a Zhang-Rice singlet pair is unstable against the creation of a similar pair, but with a $Cu(d_{3z^2-r^2})$ instead of a $Cu(d_{x^2-y^2})$ hole. (Of course the opposite sign must appear in the hybrid between the $O_x(p_x)$ and $O_y(p_y)$ orbitals.) In such a singlet pair there will also be some contribution from the apical $O_z(p_z)$ orbital; this will increase the bonding between the apical oxygen atom and the CuO_2 plane. They further suggest that the lattice responds to the creation of this pair by local displacements of the apical oxygen atom towards the plane, and of the four neighboring planar oxygen atoms perpendicular to the plane. They call the singlet pair, with its surrounding lattice deformation, a "$d_{3z^2-r^2}$ polaron".

There is strong experimental support for the existence of these $d_{3z^2-r^2}$ polarons. Pulsed neutron scattering[28] and EXAFS[29] measurements show dynamic structural irregularities consistent with the assumption that neighboring apical and planar oxygen atoms vary their positions between two short-time-scale equilibria, which should occur when such a $d_{3z^2-r^2}$ polaron is part of the time in that vicinity. We suggest that our NCB is formed by these spinless polarons. It has one state per planar Cu atom, and each state can be only singly occupied, resulting in Fermi statistics.† Polaron formation can yield mass enhancement large enough to lead to bandwidths comparable to room temperature. We can also understand the instability of YBCO for $\delta < 0$: our analysis of the TEP led us to conclude that the NCB cannot hold more than one hole per four planar oxygen atoms. Since each $d_{3z^2-r^2}$ polaron involves four planar oxygen atoms around a copper atom (with possible minor displacements of further neighbors), two polarons cannot overlap. Thus the size of these polarons is on the the borderline between "small" and "large" polarons. Small polarons[12,59] are self-trapped, while large polarons are not localized. Devaux et al.[41], on the basis of TEP results, also postulate polaron formation, but only at high temperatures, where they claim that the polarons will be self-trapped. The $d_{3z^2-r^2}$ polarons have an orbital aspect, missing in conventional polarons, which militates against self-trapping: the atomic displacements are due to a Cu($d_{3z^2-r^2}$) hole, but because of the nonspherical environment of a Cu atom, the $d_{3z^2-r^2}$ nature of the orbital is not conserved. After a while it switches into a Cu($d_{x^2-y^2}$) hole, avoiding self-trapping. This means that the d hole is in the state $d_{3z^2-r^2}$ for part of the time, and $d_{x^2-y^2}$ for the rest of the time. We suggest that there are lattice displacements which follow the motion of the charge carriers in the NCB coherently. Since the bandwidth is of the order of phonon energies, these polarons can constitute fairly stable quasiparticles, which behave like spinless fermions.

Although these quasiparticles are composed of local pairs of a Cu(d) and O(p) holes, we argue that if the electron correlations are strong enough (even for moderate U/W) the NCB quasiparticles only carry charge e, while the WCB quasiparticles carry small current,† which would have almost vanished at $T \to 0$, if SC did not exist. To see this, we note that in the ground state of the system there should be coherence between the particles such that each planar copper atom is occupied most of the time by one hole, which may be in either of the orbitals $d_{3z^2-r^2}$ or $d_{x^2-y^2}$. These d orbitals are, of course, somewhat hybridized with the O(p) orbitals of neighboring atoms. Because of the Coulomb correlations, both zero and double occupancies should be rare (in the ground state). Electric current flows when a $d_{3z^2-r^2}$ polaron moves from one site to another leaving behind a $d_{x^2-y^2}$ hole; thus only charge e has been transferred. The WCB mainly consists of $d_{x^2-y^2}$ orbitals, but also has some $d_{3z^2-r^2}$ character.† This WCB is intermediate in character between the LDA bands of low U/W, and the spinon band in the high U/W (which is expected to carry no current). It is responsible for effects dependent on spin, e.g. the Pauli spin susceptibility.

Particle or hole excitations in the WCB must violate the above single-occupancy condition, causing zero or double occupancies of sites; thus they do carry current. At finite temperature, the number of such induced carriers is about $2kT/W_w$ per planar copper atom. We argue that the induced band carriers lose mobility at high temperatures, since the Fermi-liquid-like character of this band must arise from a high degree of coherence (to achieve optimal balance between the opposing effects of hopping and Coulomb repulsion); this coherence will be lost at high temperatures[8]. If this argument is valid, the high-temperature transport properties will be dominated by the NCB. But at low temperatures both effects are present, and the WCB could even dominate.

---

† In a rigorous treatment of what has been introduced here as a $d_{3z^2-r^2}$ polaron, one should consider it as a pair of fermions, one of which is mobile, and determines the NCB (consisting mainly of O(p) orbitals), while the immobile one is hybridized in the WCB (consisting mainly of Cu(d) orbitals). This introduces an unconventional type of coupling between the two bands.

FURTHER EXPERIMENTAL JUSTIFICATIONS OF THE MODEL

When transport is dominated by the WCB, we expect that the Hall number $1/R_H e$ will correspond to the number of induced carriers in it, namely about $2kT/W_w$ per planar copper atom, which seems to explain the anomalous behavior of the Hall number[31]. The sign of R_H is then expected to be as in the LDA prediction[3,4], which is positive. As temperature rises above room temperature, this linear behavior of the Hall number seems to approach saturation[31] — consistent with a crossover to the high-T regime where transport is dominated by the NCB. In the "n-type" HTSC, $Nd_{2-x}Ce_xCuO_4$, the crossover temperature seems to be considerably lower. In this system we interpret the NCB carriers as polarons around sites where the Cu d-shell is full (*i.e.* with no d-holes); thus their contribution to R_H is expected to be negative unless they fill more than half a band. The WCB, following the LDA prediction, should again give positive R_H. Experiment[60] shows positive R_H at low T, which might change sign around 100 K, and saturates to a constant around 200 K.

The resistivity ρ behaves similarly: the WCB dominates at low T, and there is a crossover to domination by the NCB at high T in the same temperature ranges, under the assumption that in the NCB,[31,38,39] $\rho \propto T^2$, and in the WCB $\rho \propto T$. Since the number of mobile carriers in the WCB is $\propto T$, the $\rho \propto T$ behavior for this band is consistent with either particle-particle scattering or scattering by spin excitations; both possibilities are plausible here. In the limit of zero bandwidth, the phonon-scattering contribution to resistivity crosses over from $\rho \propto T$ to $\rho \propto T^2$ (see Ref. 44), suggesting that NCB transport may be dominated by phonon scattering; this is plausible in view of the polaronic nature of the NCB carriers.

As we have seen in a previous section, the TEP behavior is also dominated by the NCB at high temperatures (where it saturates to a constant value). The crossover between the low-temperature and the high-temperature regimes is reminiscent of the crossover between two plateaux in $S(T)$.[40,41] Further study is required to understand this behavior.

Naturally, we expect that the number of temperature-induced mobile carriers in the WCB, will depend on their energy. This should result in anomalous tunneling[30] conductance and energy dependence of the quasiparticles lifetime (observed by PES[16]). Further study is needed to see whether the predicted anomalous behavior coincides with the observed one.

PES or IPES should excite carriers in both the WCB and the NCB. The experimental spectra correspond to the renormalized LDA bands, and thus also to the WCB of our model. On the other hand, all states in the NCB will already be partly occupied above room temperature (high-T limit), and at first sight we should expect only very-low-energy spectra in both PES and IPES (including all angles in ARPES and ARIPES). However, this is a polaron band, and the lattice cannot respond on the time scale of the PES and IPES processes. Hence PES and IPES processes will not produce quasiparticle excitations in the NCB, but rather excitations of somewhat higher energy (because the lattice does not have time to adjust, to form the NCB polaron states). The true NCB excitations must be produced subsequently by multi-phonon processes. The energy difference between the "unrelaxed" states formed in PES or IPES, and the true NCB quasiparticles should be **k**-dependent, and vary between zero and several tenths of an eV. Thus the signature of the NCB in ARPES and ARIPES should show excitations at all angles, with energies between zero and several tenths of an eV. This will simulate a much wider band than the NCB (although it may still be quite narrow). Such excitations have been seen by the experimental groups, but the main tendency has been to regard them as a kind of noise; indeed some groups excluded them from their plotted curves. The best reports of such excitations have come from Takahashi's group.[17] They observe *two* band crossings of E_F along the ΓX direction in $Bi_2Sr_2CaCu_2O_8$, contrary to the LDA prediction that there should only be one. Furthermore, they find excitation points at all angles at the energy range 0–0.5 eV, both in ARPES and in ARIPES. In the newer ARIPES paper they call this effect: " 'Fermi edge' appearing in all the spectra"

PES measurements in the SC state[14] are anomalous. Not only is some spectral weight transferred from the gap to higher energies, but some oscillatory behavior is also seen, varying with the crystallographic direction. Two interesting explanations to this anomalous behavior have been suggested in this workshop,[2,61]. However, in our approach, it follows naturally from the fact that the PES sees transitions from states below the gap to final unrelaxed states (*i.e.* no adjustment of the lattice). Since the relaxation energy will vary from one NCB state to another, it is not surprising that some apparent spectral transfer is seen in the "wrong direction", and that the effect depends on the crystallographic alignment.

The low-temperature DC resistivity arises mainly from intraband transitions. However, at finite frequencies one expects both intraband transitions and interband ones (between the NCB and the WCB). The former lead to a narrow Drude peak, and the latter to a peak in the mid-infrared. This is in agreement both with the optical experiments,[21,49,50] and with the MFLP predictions. However, at higher frequencies (too high for the lattice to respond), we expect to see transitions to unrelaxed states (*cf* the above discussion of PES and IPES). We should therefore expect deviations from the MFLP behavior at high enough frequencies. This may explain why Varma *et al.*[11] required a cutoff frequency ω_c, in their analysis of the optical data. In the SC state the mid-infrared threshold should be shifted to higher energies by the energy of the gap. This seems to reconcile the apparent differences between Tanner's group and Schlesinger's.

Interband transitions between the NCB and the WCB will influence charge and spin fluctuations strongly, and should cause these fluctuations to exhibit MFLP-type behavior, in agreement with experiment. These transitions should also contribute to Raman spectroscopy[23], explaining why the Raman continuum exhibits MFLP-type behavior. Note, however, that for Raman transitions, the above problem of transition to unrelaxed states does not exist. Thus the Raman continuum should continue to very high frequencies with no ω_c cutoff (from transitions between the NCB and other LDA wide bands), in agreement with experiment. We can also understand the strong influence of the Raman continuum on certain phonon peaks — because of the polaronic nature of the NCB. Interband interactions will also influence the self energies of both the NCB and the WCB. However, this is a complicated problem, which will not be discussed here.

We note that our proposed band structure — a narrow band within a wide band — is rather similar to another popular model: a van Hove singularity within a wide band. Several groups have analyzed the experimental data in terms of a van Hove singularity[8,9,62,63] in the electronic structure. This singularity can give results similar to those of the present analysis. There is also considerable overlap between the predictions of the present model and the models which combine a narrow band of bipolaron-type pairs, with a wide electron band.[6,13,64] However, in these bipolaron models, the pairs are assumed to be bosons, while here their existence introduces a band of spinless fermions. In some respects the present approach is also parallel to the RVB theory,[2] by virtue of the similarity between the WCB and the NCB quasiparticles of the present approach to the RVB spinons and holons respectively. Thus some of our predictions will be similar to those of the RVB model.

INTERLAYER MECHANISM FOR HIGH T_c

Our success in deriving the Uemura[35] relation $T_c \propto n/m^*$ from the NCB model indicates that we should consider the SC charge carriers to be primarily NCB polarons.† Since they are spinless fermions, the NCB carriers in one CuO_2 layer do not have Kramers degeneracy. However, this degeneracy has been shown to be necessary for forming Cooper pairs in conventional SC's. For this reason, the present authors have previously argued[65] that *two* CuO_2 layers are necessary, to restore the Kramers degeneracy, and that HTSC is due to interlayer pairing. We propose a similar hypothesis for an SC condensation by interlayer pairing of polarons.

Because of their coupling to the WCB, the NCB carriers are, for part of the time, in a state with spin. The interplanar layers (the chains in the case of YBCO) are highly polarizable;

---

† Although the WCB may also play some role.

i.e. they have states very close to E_F. They can therefore induce superexchange between NCB carriers, via the virtual spins on opposite sides of the polarizable layer. Denoting the annihilation operators for corresponding NCB carriers in planes "1" and "2" by c_1 and c_2 respectively, we get that interplane exchange (unlike interplane hopping) yields degenerate states $c_\pm$, related to c_1 and c_2 (further details will be published elsewhere).‡ In a coherent state, the degenerate carriers exchange planes simultaneously (through the interplane exchange term), so that they are not simultaneously on the same plane. Hence the Coulomb repulsion between them is interplanar, and thus very small.

Since Coulomb repulsion has been reduced to a small interplanar value, the attractive interaction mediated through boson fields could dominate without the usual requirement that $k\Theta_{Debye}/E_F \ll 1$, (necessary to reduce the Coulomb parameter μ, through retardation, to a much smaller effective μ^*). We do not express any view here as to which is the most important boson field for pairing in the cuprates. Possible candidates include phonons, plasmons and excitons, both within the planes and in the interplanar polarizable medium. We shall not discuss the nature of the SC state, nor what happens to the WCB or to the bands of the interplanar medium, *etc.* The WCB may play a role in the creation of the spin gap, whose nature remains puzzling.[24,25]

We have a comment concerning SC superlattices.[66] In YBCO/PBCO (P≡Pr) superlattices, T_c is reduced. In particular, for a single YBCO layer in PBCO it drops to 12 K. According to the interlayer pairing model presented here, the pairing takes place across the polarizable layer, and not across the Y (or Pr) layer. Thus, for a single YBCO layer we should consider two CuO_2 double layers, separated by the Y layer. Each of these double layers has Pr on one side and Y on the other. Hence, unless effects of three-dimensionality are important, their T_c should be roughly the same as that of randomly Pr-doped YBCO, where 50% of the Y has been replaced by Pr, which is[67] very close to 12 K. In $Bi_2Sr_2CuO_6/Bi_2Sr_2CaCu_2O_8$ superlattices[68] the higher T_c of $Bi_2Sr_2CaCu_2O_8$ is maintained even for ultrathin slabs containing half a unit cell of $Bi_2Sr_2CaCu_2O_8$. This is again consistent with the mechanism we suggest here, if three-dimensionality is not significant. The higher-T_c component of this superlattice has a natural double-CuO_2-layer structure, while the lower-T_c component does not have one (so that unit-cell doubling is necessary for our type of interlayer pairing to be possible). On the other hand, if even half a slab of the higher-T_c component is present, one has a natural double-CuO_2-layer arrangement with the polarizable layer between them. Thus the higher T_c can be achieved.

Because of the polaronic nature of the carriers in the NCB, there should be an anomalous isotope effect, whether or not the pairing interaction is phononic, and the effect should depend on the specific cuprate, and on its stoichiometry. The observed isotope effect in the cuprates[69] is indeed anomalous and depends on stoichiometry; however further study is necessary before we can make any predictions concerning this effect.

SUMMARY AND CONCLUSIONS

In our search for the correct theory of the cuprate HTSC's, we have proposed a model for their electronic structure, and a mechanism for their SC. Both of these are consistent with the rich experimental data, and with basic theoretical requirements. The microscopic origin of the model is not simple: it must include orbital, structural (including dynamic) and electron-correlation effects, characteristic of the cuprates. But the cuprates are not simple either!

Again, we emphasize that this article concentrates on a search for the correct model, including only simple estimates. Rigorous calculations of the fine details are in progress, and we hope to make predictions regarding experiments still to be performed.

---

‡ The states annihilated by c_+ and c_- will be degenerate even when the degeneracy between the states annihilated by c_1 and c_2 is removed by lattice imperfections.

In our discussion, we have emphasized the cuprates, since all the really high T_c's currently known belong to this class. However, as we have pointed out above, the analysis of Uemura et al.[35] reveals the presence of aspects common to the cuprates, bismuthates, organic SC's, Chevrel-phase and heavy-fermion SC's. According to our interpretation, the common feature is that their transition temperatures are determined by the narrow-band limit of the BCS equation. Thus all these classes of SC's appear to be characterized by the presence of a narrow conduction band — too narrow to be treated in the low-T limit in the normal state. We will not be surprised if the newly discovered "buckyball" SC's also fall into this category. However, the physical origin of the narrow band could vary from one class of exotic SC's to another.

Is there an upper limit to T_c? According to our interpretation of the narrow-band limit of the BCS equation, the maximal T_c which could be achieved in principle will occur for a half-full band. However, when the narrow band is close to half full, slight instabilities apparently set in, and prevent our attaining ultra-high T_c. There have been persistent, though still unconfirmed, reports of unstable filamentary superconductivity, even close to room temperature.[70] We speculate that perhaps, in certain cases, and close to boundaries, the unstable state may be metastable, and the transition may not set in immediately. If a method can be found to stabilize this state, room temperature superconductivity might be attainable.

ACKNOWLEDGEMENTS

The authors acknowledge the contribution of B. Fisher, J. Genossar, S. E. Barnes and D. van der Marel to the development of the ideas presented here. JA thanks the participants of the workshop for their important contributions, and the other organizers and editors S. E. Barnes, Fulin Zuo, G. C. Vezzoli and B. M. Klein for their combined effort.

REFERENCES

1. V. Z. Kresin, H. Morawitz and S. A. Wolf, these proceedings.

2. P. W. Anderson, these proceedings; Science **235**, 1196 (1987); Phys. Rev. Lett. **64**, 839 (1990).

3. R. E. Cohen, H. Krakauer and W. E. Pickett, these proceedings; W. E. Pickett, Rev. Mod. Phys **61**, 433 (1989).

4. J. J. Yu and A. J. Freeman, these proceedings.

5. J. C. Phillips, *Physics of High-T_c Superconductors* (Academic, Boston, Diego, 1989).

6. Y. Bar-Yam, these proceedings.

7. J. E. Hirsch, these proceedings.

8. K. Levin, Qimiao Si, Ju H. Kim and J. P. Lu, these proceedings; Physica C **175**, 449 (1991).

9. C. L. Kane, D. M. Newns, P. C. Pattnaik, C. C. Tsuei and C. C. Chi, these proceedings.

10. Z. Zou, these proceedings.

11. C. M. Varma, P. B. Littlewood, S. Schmitt-Rink, E. Abrahams and A. E. Ruckenstein, Phys. Rev. Lett. **63**, 1996 (1989); C. M. Varma, these proceedings.

12. David Emin, these proceedings.

13. J. Ranninger, R. Micnas and S. Robaszkiewicz, Ann. Phys. (Paris) **13**, 455 (1988).

14. Z.-X. Shen, D. S. Dessau and B. O. Wells, these proceedings; B. O. Wells, Z.-X. Shen, D. S. Dessau, W. E. Spicer, C. G. Olson, D. B. Mitzi, A. Kapitulnik, R. S. List and A. J. Arko, Phys. Rev. Lett. **65**, 3056 (1990); D. S. Dessau, B. O. Wells, Z.-X. Shen, W. E. Spicer, A. J. Arko, R. S. List, D. B. Mitzi and A. Kapitulnik, Phys. Rev. Lett. **66**, 2160 (1991).

15. J. G. Tobin, C. G. Olson, F. R. Solal, C. Gu, J. Z. Liu amd M. J. Fluss, these proceedings.

16. J. C. Campuzano, G. Jennings, M. Faiz, L. Beaulaigue, B. W. Veal, J. Z. Liu, A. P. Paulikas, K, Vandervoort, H. Claus, R. S. List, A. J. Arko and R. J. Bartlett, Phys. Rev. Lett. **64**, 2308 (1990).

17. T. Takahashi, H. Matsuyama, H. Katayama-Yoshida, Y. Okabe, S. Hosoya, K. Seki. H. Fujimoto, M. Sato and H. Inokuchi, Phys. Rev. B **39**, 6636 (1989); T. Watanabe, T. Takahashi, S. Suzuki, S. Sato, H. Katayama-Yoshida, A. Yamanaka, F. Minami and S. Takekawa, submitted to Physica C.

18. G. Mante, R. Claessen, T. Buslaps, S. Harm, R. Manzke, K. Skibowlski and J. Fink, Z. Phys. B **80**, 181 (1990).

19. J. L. Smith, C. M. Fowler, B. L. Freeman, W. L. Hults, J. C. Kings and F. M. Mueller, these proceedings.

20. H. Haghighi, J. H. Kaiser, S. L. Rayner, R. N. West, J. Z. Liu, R. Shelton, R. H. Howell, P. A. Sterne, F. Solal and M. J. Fluss, these proceedings.

21. D. B. Tanner, D. B. Romero, K. Kamarás, G. L. Carr, L. Forro, D. Mandrus, D. Mihaly and G. P. Williams, these proceedings.

22. J. H. Kim, I. Bozovic, J. S. Harris, Jr., W. Y. Lee, C.-B. Eom and T. H. Geballe, these proceedings.

23. F. Slakey, M. V. Klein, J. P. Rice and D. M. Ginsberg, Phys. Rev. B **42**, 2643 (1990).

24. R. E. Walstedt, R. F. Bell and D. B. Mitzi, submitted to Phys. Rev. B; these proceedings.

25. P.C. Hammel, M. Takigawa, R. E. Heffner, Z. Fisk and K. C. Ott, Phys. Rev. Lett. **63**, 1992 (1989). R. E. Walstedt, R. F. Bell and D. B. Mitzi, submitted to Phys. Rev. B; these proceedings.

26. G. Aeppli and D. R. Harshman, these proceedings; S. M. Hayden, G. Aeppli, H. Mook, D. Rytz. M. F. Hundley and Z. Fisk, Phys. Rev. Lett. **66**, 821 (1991).

27. J. M. Tranquada, these proceedings.

28. T. Egami, B. H. Toby, S. J. L. Billinge, Chr. Janot, J. D. Jorgensen, D. G. Hinks, M. A. Subramanian, M. K. Crawford, W. E. Farneth and E. M. MaCarron, these proceedings.

29. J. Mustre de Leon, S. D. Conradson, P. G. Allen, I. Batistić and A. R. Bishop, these proceedings.

30. L. H. Greene, J. Lesueur, W. L. Feldmann and A. Inam, these proceedings.

31. N. P. Ong, T. R. Chien, T. W. Jing, T. V. Ramakrishnan and Z. Z. Wang, these proceedings; T. R. Chien, D. A. Brawner, Z. Z. Wang and N. P. Ong, Phys. Rev. B **43**, 6242 (1991).

32. M. Affronte and D. Pavuna, these proceedings.

33. J. L. Cohn, S. A. Wolf, V. Selvamanickam and K. Salama, these proceedings.

34. B. Fisher J. Genossar, L. Patlagan, G. Koren, J. Ashkenazi and C. G. Kuper, these proceedings.

35. Y. J. Uemura et al, Phys. Rev. Lett. **66**, 2665 (1991); these proceedings.

36. A. Bianconi, A. M. Flank, P. Lagarde, C. Li, I. Pettiti, M. Pompa and D. Udron, these proceedings.

37. F. C. Zhang and T. M. Rice, Phys. Rev. B **37**, 3757 (1988).

38. J. Genossar, B. Fisher, I. O. Lelong, J. Ashkenazi and L. Patlagan, Physica C **157**, 320 (1989).

39. C. C. Tsuei, A. Gupta and G. Koren, Physica C **161**, 415 (1989).

40. A. B. Kaiser and C. Uher, in *Studies of High Temperature Superconductivity*, Vol. 7, edited by A. V. Narlikar (Nova Science Publishers, New York, 1990).

41. F. Devaux, A. Manthiram and J. B. Goodenough, Phys. Rev. B **41**, 8723 (1990).

42. C. Legros-Glédel, J.-F. Marucco, E. Vincent, D. Favrot, B. Poumellec, B. Touzelin, M. Gupta and H. Alloul, Physica C **175**, 279 (1991).

43. B. Fisher, J. Genossar, I.O. Lelong, A. Kessel and J. Ashkenazi, J. Supercond. **1**, 53 (1988); B. Fisher, J. Genossar, L. Patlagan, I.O. Lelong and J. Ashkenazi, Physica C **162-164**, 1207 (1989).

44. S. Bar-Ad, B. Fisher, J. Ashkenazi and J. Genossar, Physica C **156**, 741 (1988).

45. M. Weger and I. B. Goldberg, Solid State Phys. **28**, 1 (1973).

46. N. E. Phillips and R. A. Fishe, these proceedings.

47. D. Wohlleben, W. Schnelle, E. Braun and H. Broiches, , these proceedings.

48. T. Lægreid, K. Fossheim, E. Sandvold and S. Julsrud, Nature **130**, 637 (1987).

49. D. van der Marel, A. Wittlin, H.-U. Habermeier and D. Heitmann, these proceedings.

50. Z. Schlesinger, R. T. Collins, L. D. Rotter, C. Feild, U. Welp, G. W. Crabtree, J. Z. Liu and Y. Fang, these proceedings.

51. H. Holczer, L. Forro, L. Mihály and G. Grüner, preprint.

52. S. B. Bacci, E. R. Gagliano and R. M. Martin, these proceedings.

53. R. Bar-Deroma, J. Felsteiner, R. Brener, J. Ashkenazi and and D. van der Marel, these proceedings.

54. J. W. Serene and D. W. Hess, these proceedings.

55. T. Barnes, these proceedings.

56. J. M. Luttinger, Phys. Rev. **119**, 1153 (1960); L.D. Landau, Sov. Phys. JETP, **3**, 920 (1957).

57. S. E. Barnes, these proceedings; J. Phys. F **26**, 1375 (1976).

58. G. A. Sawatzky, private communication.

59. David Emin, Adv. Phys. **24**, 305 (1975); N. F. Mott, *Metal-Insulator Transitions* (Taylor & Francis, London, 1974), p. 55.

60. Z. Z. Wang, T. R. Chien, N. P. Ong, J. M. Tarascon and E. Wang, Phys. Rev. B **43**, 3020 (1990).

61. F. M. Mueller, G. B. Arnold and J. C. Swihart, these proceedings.

62 R. S. Markiewicz, these proceedings.

63. J. Labbé and J. Bok, Europhys. Lett., **3**, 1225 (1987).

64. A. Kallio, these proceedings.

65. J. Ashkenazi and C. G. Kuper, Physica C **153-155**, 1315 (1988); *ibid*, **162-164**, 767 (1989); Ann. Phys. (Paris) **13**, 407 (1988); in *Studies of High Temperature Superconductivity*, Vol. 3, edited by A. V. Narlikar (Nova Science Publishers, New York, 1989), p. 1; D. Vacaru and M. Crişan, Phys. Rev. B **42**, 4767 (1990).

66. D. H. Lowndes and D. P. Norton, these proceedings.

67. Y. Gao, A. Kebede, P. Pernambuco-Wise, M. Kuric, J.E. Crow, R.P. Guertin, T. Mihalisin, N.D. Spencer, and D.W. Cooke, these proceedings.

68. I. Bozovic, J. N. Eckstein, M. E. Klausmeier-Brown and G. Virshup, preprint.

69. J. P. Franck, S. Gygax, J. Jung, M. A.-K. Mohamed and G. I. Sproule, these proceedings.

70. J. T. Chen, L.-X. Qian, L.-Q. Wang, L. E. Wenger and E. M. Logothetis, Mod. Phys. Lett. B **3**, 1197 (1989).

RE-ANALYSIS OF PHOTOEMISSION DATA FOR CuO: REVISION OF THE CONFIGURATION-ENERGY SCHEME FOR CUPRATE MATERIALS

B. H. Brandow

Theoretical Division
Los Alamos National Laboratory
Los Alamos, NM 87545

INTRODUCTION

We have recently carried out a very careful analysis of photoemission and inverse photoemission (BIS) data for CuO,[1] probably the most refined and thorough such analysis for any transition-metal compound to date. This has led to a configuration-energy scheme for cuprate materials which is quite different from the generally-accepted one, due largely to Sawatzky and co-workers.[2-4] For example, it is commonly believed that the "doping holes" in $La_{2-x}Sr_xCuO_4$ are hosted almost entirely within oxygen 2p orbitals. We find instead that these holes are shared roughly equally by 2p and 3d orbitals. In this report we shall sketch the motivations for this study, the main results, and some of the lessons and surprises encountered along the way.

Valence-band photoemission and BIS data provide the main phenomenological access to parameter values for the model Hamiltonians used in studying cuprate superconductivity. The material CuO is particularly convenient for detailed study here. It contains only the elements of most interest (copper and oxygen), without any chain coppers to complicate the interpretation. Furthermore, the ratio of copper to oxygen is much larger than in most cuprate materials, so details of the 3d photoemission can be more clearly resolved. Each copper is square-planar coordinated with four nearby oxygens, as in the superconducting materials. On the other hand this is a magnetic insulator (Mott insulator), like the parent compound La_2CuO_4, which simplifies the data analysis. Thus, CuO deserves to be carefully studied.

Photoemission data for CuO and other cuprates have already been presented and analyzed by a number of workers.[2,5,6] We were skeptical that any of the previous results were sufficiently reliable, however, for several reasons: (1) Previous workers have typically focussed on XPS data, where the high energy of the photons (>1000 eV) suggests that the photoemission is due almost entirely to 3d electrons. Data at lower photon energies reveals additional features, which can be attributed to 2p photoemission, but this 2p information had not been utilized. (2) BIS data is essential for a reliable determination of the Hubbard parameter U, but this data was unavailable for most of the earlier studies. BIS data was used by Eskes, Tjeng, and Sawatzky[2] (hereafter ETS), which is the most refined of the previous analyses, but, as we shall argue below, this data was used incorrectly. (3) The resulting energy-level scheme for CuO was quite different from what had been obtained in the 1970's for the classic Mott insulators NiO, CoO, etc. Although this older scheme[7] was never universally accepted, and was not worked out as thoroughly as is now possible, it did have the support of several additional lines of evidence, including 2p photoemission.[8,9] (We have reviewed this other evidence in Ref. 1.) One should expect CuO to be quite similar to its neighbor NiO. On the other hand, the "new" scheme for CuO (which has also been applied to NiO etc.[3,4]) appears to be nicely confirmed by evidence from resonant photoemission.[10,11]

We were thus faced with an intriguing problem -- a major inconsistency between the old and new pictures, each of which could claim support from a number of other kinds of evidence, but neither of which was based on sufficiently thorough analysis. The most troubling problem, for us, was a paradoxical inconsistency between the 2p data, which supported the old picture, and the resonant photoemission, which favored the new one.

THE MODEL AND ANALYSIS

Our analysis is based on the most refined previous work, by ETS,[2] together with a number of further refinements. Thus, we calculate the full multiplet structure for two holes in D_{4h} symmetry, as appropriate for square-planar oxygen coordination. The Coulomb (direct minus exchange) interactions between the various 3d orbitals are represented by the Racah parameters A, B, C, using the matrices of Zaanen.[2,12] (B and C are obtained from free-ion data, leaving only A to be fitted to the CuO data.) All of the charge-transfer configurations (d^8, $d^9\underline{L}$, $d^{10}\underline{L}^2$) are included [$\underline{L}$ = "ligand" (oxygen 2p) hole]. These features are the same as in ETS.

We use an Anderson-impurity model, rather than the CuO_4 cluster of ETS. This model has only one Cu ion, the "impurity." However, the oxygens are assumed to form a simple cubic lattice of infinite extent. (This idealization is obtained by "undistorting" the monoclinic CuO structure, which makes it essentially the same as the PdO structure.) Furthermore, on *every* Cu site (the empty Cu sites as well as the impurity) we place 4s orbitals. Bloch states are then calculated for this combined 2p-4s tight-binding basis. We then return to the simple four-oxygen cluster of the corresponding CuO_4 cluster model, and combine the twelve 2p spatial orbitals into twelve symmetry-adapted orbitals ϕ_γ for the D_{4h} point group. Then for each ϕ_γ we calculate its overlap with the Bloch 2p-4s eigenstates $\psi_{k\nu}$ (ν = band index) and evaluate the projected state density

$$\rho_\gamma(\varepsilon) = \sum_{k\nu} |\langle \phi_\gamma | \psi_{k\nu} \rangle|^2 \, \delta(\varepsilon - \varepsilon_{k\nu}) . \qquad (1)$$

Five of these symmetries γ are the same as those of the 3d orbitals, and thus can hybridize with the corresponding 3d's, while the remaining seven are nonbonding with respect to the 3d's. For each of the hybridizing γ's we discretize $\rho_\gamma(\varepsilon)$ into 10 states, so that a hole of symmetry γ is then represented by 11 states, 10 for the projected 2p band and 1 for the 3d orbital. It is then a straightforward matrix problem to find the two-hole eigenstates resulting from photoemission. The spectral weight for 2p removal and for 3d removal are then constructed according to the sudden approximation, equivalent to the usual Green's function expression. These results are then Lorentzian broadened, for the sake of illustration. To this "hybridizing" 2p removal spectrum we then add the sum of the seven nonhybridizing densities $\rho_\gamma(\varepsilon)$, similarly broadened, to get the total 2p removal spectrum.

Further refinements include a parameter to allow for the *nonhybridizing component* of the phenomenological crystal field splitting of the 3d orbitals. Such a contribution was found to be significant in *ab initio* cluster calculations,[13] where it was found to arise mainly from the extra kinetic energy due to constructing orthogonal Wannier orbitals out of overlapping ionic orbitals. (ETS assumed that the 3d splitting comes entirely from the 2p-3d hybridization.) The use of projected state densities $\rho_\gamma(\varepsilon)$ likewise provides an effective crystal field splitting for the 2p's, as is apparent if one takes the centroid of each $\rho_\gamma(\varepsilon)$ to represent the energy of the associated four-oxygen cluster orbital ϕ_γ. This feature is similar to the 2p splitting in the cluster model of ETS, although the 4s-2p hybridization now substantially increases this effective splitting. We also further modify the effective crystal field for the $a_{1g} = z^2$ channel, to allow for a further strong effect of the 4s orbital in this symmetry. This refinement was necessary to fit the multiplet structure of the main peak region, and is consistent with previous experience with other square-planar complexes.[14]

In order to obtain sufficient intensity in the satellite region, we found it necessary to allow for a charge dependence of the hybridization strength, i.e., we assume that the magnitude $|t_{pd}|$ of the transfer integral is larger for d^8-d^9 transitions than for d^9-d^{10} transitions. There is independent evidence for this in an *ab initio* cluster calculation,[15] and in a previous photoemission study.[11]

We want to emphasize that each one of these added refinements was motivated in two different ways: (a) by the empirical fitting process, i.e., it was indicated by the data, and (b) it had precedents, usually from previous *ab initio* studies.

RESULTS AND DISCUSSION

We have used the same data[6] as ETS. However, in addition to the XPS spectrum (1478 eV photon energy), which is the one that they analyzed, we have also analyzed the 70 eV (synchrotron radiation) and 21.2 eV (HeI lamp) spectra from Ghijsen *et al*. The main features of the XPS data are a "main peak" with a narrow high-intensity maximum and some multiplet structure, and a considerably weaker "satellite" centered at about 8 eV higher binding energy, also with some multiplet structure.

In comparing the lowest-energy (HeI) and highest-energy (XPS) data, three features can be identified as due to 2p photoemission. There is a strong shoulder on the high-binding-energy side of the main peak, and at higher binding there are two weaker features in the energy range of the satellite. The spacing and relative intensities of these 2p features are well explained by our model. Also, the required 2p-2p and 2p-4s transfer integrals agree well with those deduced from band calculations for La_2CuO_4,[16] especially so for the 2p-2p transfer. It turns out that the calculated 2p removal intensity spectrum has almost the same shape as the state density of the 2p-4s tight-binding band.

The charge-transfer energy parameter Δ, nominally defined as $E(d^{10}\underline{L})-E(d^9)$, is determined mainly by the position of the strongest 2p feature with respect to the main peak, which comes from 3d removal. This determination is as straightforward and unambiguous as one could reasonably hope for, so the resulting Δ value should be reliable. Similarly, the Racah A parameter is determined mainly by the total energy separation (2.9 eV) between the threshold photoemission feature (a rather weak bump at 1.1 eV binding energy, with $^1A_1$ symmetry), and the BIS peak, at 1.8 eV above the "ε_F" reference energy. The A value should therefore also be reliable. The various hybridization strengths (for different symmetries γ, and different charge states, three free parameters) are determined by the position, shape, and intensity of the satellite, including some very weak but apparently genuine features seen in the 70 eV data (which has the highest resolution) at around 16 eV binding energy, which ETS had also fit. The quality of the fit in the satellite and 16 eV regions is very good, provided that we also include a significant amount of 2p intensity. We return to the latter issue below.

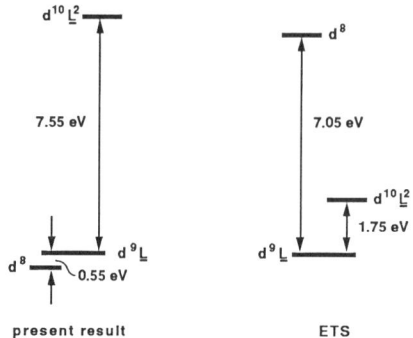

Figure 1. The present configuration-energy scheme for two holes of $x^2 - y^2$ symmetry, compared with the result of ETS, Ref. 2.

The Hubbard U parameter is not uniquely defined, but depends on the quantum numbers of the two-hole eigenchannel. The $^1A_1$ eigenchannel is the one of most interest for understanding superconductivity, since this is the one containing two d holes of $b_{1g} = x^2 - y^2$ symmetry. (As expected, this channel contains the photoemission threshold state.) It would be pointless to present all of the fitting parameters of our model without carefully defining them, so we shall now focus on only the Δ and U obtained for this $^1A_1$ eigenchannel. Here the effective U is

$$U_{eff}(^1A_1) = A + 4B + 3C, \qquad (2)$$

for which we find 7.0 eV. (ETS found 8.8 eV.) Our charge-transfer energy, $\Delta = 7.55$ eV, is now very different from the corresponding ETS value of 1.75 eV (by our definition -- we measure from the *centroid* of $\rho_\gamma(\varepsilon)$ for $\gamma = b_{1g}$). Using the definitions of the configuration-energy differences (for two $b_{1g} = x^2 - y^2$ holes),

$$E(d^{10}\underline{L}^2) - E(d^9\underline{L}) = \Delta, \qquad (3)$$

$$E(d^9\underline{L}) - E(d^8) = \Delta - U, \qquad (4)$$

the present configuration-energy scheme is compared with that of ETS in Fig. 1. The difference is quite drastic, and this deserves an extended discussion.

For a qualitative understanding of the main-peak and satellite states, according to our scheme, one can visualize these as the bonding and antibonding combinations of d^8 and $d^9\underline{L}$, ignoring the high-lying $d^{10}\underline{L}^2$ configuration. The latter is important only for high-lying states which have negligible intensity. This picture is valid for every one of the two-hole eigenchannels. The two-hole ground state (photoemission threshold state = present "bonding" state) corresponds to the state of a doping hole in $La_{2-x}Sr_xCuO_4$. Since the energy separation of d^8 and $d^9\underline{L}$ is far smaller than their very strong hybridization matrix element, it is clear that the doping holes should have roughly equal amounts of 3d and 2p character. This is quite contrary to the generally accepted picture, where these holes are nearly pure 2p, but this is consistent with the near-threshold intensity behavior (*vs.* photon energy) in photoemission from metallic cuprates.[17]

All previous analyses of cuprate photoemission, done to obtain effective Hamiltonian parameters, have been based on the premise that the necessary information can be obtained from the 3d photoemission alone, where the 3d removal spectrum is assumed to be the XPS spectrum. In a simplified model[18] containing only the holes of b_{1g} symmetry, the final states are obtained from an energy matrix of the form

$$\begin{pmatrix} \varepsilon_8 & T & 0 \\ T & \varepsilon_9 & T \\ 0 & T & \varepsilon_{10} \end{pmatrix}.$$

We now observe that the diagonal elements ε_8 and ε_{10} (energies of the d^8 and $d^{10}\underline{L}^2$ configurations) can be *interchanged* without affecting the final-state eigenvalues. This indicates that *two different solutions* are possible, when attention is restricted to the customary choice of data. The two schemes in Fig. 1 do indeed differ by a rough interchange of the d^8 and $d^{10}\underline{L}^2$ energies, although this interchange is imperfect because of multiplet effects.

One might suppose that a careful treatment of the two-hole multiplet structure would suffice to resolve this ambiguity. We found that this is not so, at least not definitively, because of the extra free parameters we have introduced to treat the crystal-field aspect. (Nevertheless, we believe that these extra parameters have more reasonable values for our solution.) By appealing to the 2p spectrum, however, our solution is selected quite unambiguously.

Our result is qualitatively the same as the old picture for NiO, CoO, and MnO, obtained in the early 1970's.[7] This picture has been buttressed by several independent observations.[1] For example, the strong 2p feature has an essentially constant energy separation from the oxygen 2s peak, independent of the material, for a series of transition-metal oxides.[8] Likewise, the main 2p signal always falls between the "main peak" and "satellite" of the 3d spectrum.[9] Our interpretation of CuO is consistent with these old observations, provided we identify this "main 2p feature" with the strong state-density peak at the *top* of our 2p band.

In this original picture, the insulating gap (identified with the threshold for strong optical absorption) was assigned to transitions from 3d orbitals to 4s/4p band states. We believe that this conclusion remains correct, for *all* of the late 3d transition-metal oxides including CuO, and that the more recent 2p → 3d assignment[2-4,19] is wrong. In other words, we are arguing that these materials are *not* "charge-transfer insulators." This can be deduced from Fig. 1, with allowance for the energy shifts from p-d hybridization, and with a band-theoretic estimate for the 2p-4s band gap.

ETS were quite convinced that the photoconductivity threshold (insulating gap) of 1.35 eV[20] is due to 2p → 3d transitions. They therefore used this gap value (raised to 1.8 eV to allow for some bandwidth of the relevant 2p state) in place of the experimental d-d gap value of 2.9 eV, the separation between the centers of the BIS and photoemission threshold features. We consider this illogical, because one ought to *deduce* the assignment of the insulating gap transitions from the results of the data analysis, rather than presuming a particular choice as input. Furthermore, the calculated d-d gap value should be made to agree with its experimental value, 2.9 eV, instead of the value for a different property. Using the 3d-removal spectrum (XPS data) and their value of 1.8 eV for the d-d gap, but *without* considering the available 2p data, we obtained two solutions, as expected from the argument above, with one of these solutions closely matching their solution. But when we raised the dd gap value to 2.9 eV, this sufficed to eliminate the Sawatzky-type (small Δ) solution, *even without* constraining Δ to match the 2p spectral information. (At small Δ but large d-d gap, the required U or A became unphysically large.) This provides further evidence against their assignment.

In contrast, the resonant photoemission[11] appears to be very strong evidence *for* the Sawatzky (ETS) solution in Fig. 1. Our response to this is necessarily rather subtle. At 70 eV photon energy, which is just outside the copper 3p → 3d resonance region centered around 74 eV, the satellite intensity (relative to the main peak) is only about 1/4 of its intensity in the XPS data. This loss of satellite intensity is surprising. The only explanation we could find, which is consistent with similar data over a broad range of photon energies,[11] is that the sudden approximation is failing quite seriously here. Using a semiclassical analysis, we have found[1] that the adiabatic corrections to the "direct" and "resonant" channels are sufficient to qualitatively account for the fact that the resonance is strongly concentrated in the satellite region. Thus, a high energy for the d^8 configuration is *not* demanded by this resonance data.

In matching our calculations to the spectra for 1478 eV, 70 eV, and 21.2 eV photon energies, in each case we treated the orbital photoemission cross section ratio σ_{2p}/σ_{3d} as a free parameter. That is, we added the 3d and 2p removal spectra with an adjustable intensity ratio. At 1478 eV, our empirical value for this ratio was an order of magnitude larger than the very small theoretical value,[21] based on neutral-atom orbitals. This surprising result is consistent with the appearance of rather prominent 2p-like features in XPS spectra of some other 3d-metal oxides.[8] To account for this, we came up with two explanations which are presumably working together: (a) The removal of a d electron allows the orbitals of the remaining occupied d electrons to collapse suddenly (in XPS), thus reducing the "passive electron" overlap and suppressing the d removal intensity with respect to the 2p intensity. This is analogous to the band-narrowing in polaron theory. (This happens also for 2p's, but less strongly.) (b) Since the theoretical analysis presumes orthogonal orbitals, our "2p" orbitals are actually Wannier functions containing bits of 3d and several of the other copper-ion orbitals. At 1478 eV these "Wannier tail" components can contribute substantially to the photoemission cross section, and can thus enhance the apparent 2p intensity.

An important lesson here, as in the cases of the weakened satellite intensity and the resonant photoemission, is that one should be prepared for surprises in intensity behavior, and not rely too heavily on the standard recipes for intensities.

A few more remarks are needed to put these results in perspective. The present model has no Coulomb interactions involving 2p electrons (no U_p or U_{pd}), so the results should not be compared directly with models which do include such interactions. The present parameter set gives a Heisenberg magnetic coupling J which is too small, by about a factor of two, so there is room for some modest revision of the parameters. Finally, it should be emphasized that our results are specific for the *insulating state* of CuO. In applying these parameters to superconducting cuprate materials, one should be wary about possibly important changes due to metallic screening, especially for U and Δ. Also, Δ can be significantly affected by the Madelung potential effects of the ions between the CuO_2 layers, and especially by the presence or absence of apical oxygens.[22]

This work was supported by the U.S. Department of Energy.

REFERENCES

1. B. H. Brandow, J. Solid State Chem. **88**, 28 (1990).
2. H. Eskes, L. H. Tjeng, and G. A. Sawatzky, Phys. Rev. B **41**, 288 (1990).

3. G. A. Sawatzky and J. W. Allen, Phys. Rev. Lett. **53**, 2339 (1984); J. Zaanen, G. A. Sawatzky, and J. W. Allen, *ibid.* **55**, 418 (1985).
4. J. Zaanen and G. A. Sawatzky, Canad. J. Phys. **65**, 1262 (1987); J. Solid State Chem. **88**, 8 (1990).
5. A. Fujimori, E. Takayama-Muromachi, Y. Uchida, and B. Okai, Phys. Rev. B **35**, 8814 (1987); Z. Shen *et al.*, *ibid.* **36**, 8414 (1988); D. D. Sarma, *ibid.* **37**, 7948 (1988); F. Mila, Phys. Rev. B **38**, 11358 (1988); D. E. Ramaker, *ibid.* **38**, 11816 (1988); Γ. Okada and A. Kotani, J. Phys. Soc. Japan **58**, 1095 (1989).
6. J. Ghijsen *et al.*, Phys. Rev. B **38**, 11322 (1988).
7. R. J. Powell and W. E. Spicer, Phys. Rev. B **2**, 2182 (1970); D. Adler and J. Feinleib, *ibid.* **2**, 3112 (1970); L. Messick, W. C. Walker, and R. Glosser, *ibid.* **6**, 3941 (1972).
8. G. K. Wertheim and S. Huffner, Phys. Rev. Lett. **28**, 1028 (1972).
9. D. E. Eastman and J. L. Freeouf, Phys. Rev. Lett. **34**, 395 (1975).
10. S.-J. Oh, J. W. Allen, I. Lindau, and J. C. Mikkelsen, Phys. Rev. B**26**, 4845 (1982); M. R. Thuler, R. L. Benbow, and Z. Hurych, *ibid.* **27**, 2082 (1983).
11. J. Ghijsen *et al.*, Phys. Rev. B **42**, 2268 (1990). See also M. R. Thuler, R. L. Benbow, and Z. Hurych, *ibid.* **26**, 669 (1982); Okada and Kotani, in Ref. 5.
12. J. Zaanen, Ph.D. thesis, Groningen (unpublished).
13. T. F. Soules, J. W. Richardson, and D. M. Vaught, Phys. Rev. B **3**, 2186 (1971); A. J. H. Wachters and W. C. Nieuwpoort, *ibid.* **5**, 4291 (1972).
14. P. Ros and G. C. A. Schuit, Theor. Chim. Acta (Berl.) **4**, 1 (1966); B. Roos, Acta Chem. Scand. **20**, 1673 (1966); H. Basch and H. B. Gray, Inorg. Chem. **6**, 365 (1967); F. A. Cotton and C. B. Harris, Inorg. Chem. **6**, 369 (1967); J. Demuynck, A. Veillard, and U. Wahlgren, J. Am. Chem. Soc. **95**, 5563 (1973).
15. R. L. Martin, Physica B **163**, 533 (1990), and private communication.
16. A. K. McMahan, R. M. Martin, and S. Satpathy, Phys. Rev. B **38**, 6650 (1988); M. J. DeWeert, D. A. Papaconstantopoulos, and W. E. Pickett, *ibid.* **39**, 4235 (1989); L. F. Mattheiss and D. R. Hamann, *ibid.* **40**, 2217 (1989); M. S. Hybertsen, M. Schluter, and N. E. Christensen, *ibid.* **39**, 9028 (1989); A. K. McMahan, J. F. Annett, and R. M. Martin, *ibid.* **42**, 6268 (1990).
17. R. S. List *et al.*, Phys. Rev. B **38**, 11966 (1988) and Physica C **159**, 439 (1989).
18. Z. Shen *et al.*, in Ref. 5.
19. S. Hüfner, Z. Phys. B **61**, 135 (1985).
20. F. P. Koffyberg and F. A. Benko, J. Appl. Phys. **53**, 1173 (1982).
21. J. J. Yeh and I. Lindau, At. Data Nucl. Data Tables **32**, 1 (1985).
22. J. B. Torrance and R. M. Metzger, Phys. Rev. Lett. **63**, 1515 (1989).

SELF-ENERGY CORRECTIONS FOR NiO[*]

Barbara Szpunar and Vedene H. Smith Jr.

Department of Chemistry, Queen's University
Kingston Ont., K7L 3N6

Noriaki Hamada

NEC Fundamental Research Laboratories, 34 Miyukigaoka
Tsukuba 305, Japan

INTRODUCTION

The discovery of high transition temperature superconductors has redirected attention to older materials, the so - called Mott insulators[1] which are related to the undoped phases of the superconducting cuprates. Due to strong electronic correlations, one can not describe this class of materials within the one - electron approximation[1]. The local spin density (LSD)[2] approximation, based on density functional theory (DFT), was a big step forward in describing such materials since it includes correlation effects for the homogeneous electron gas in contrast to the Hartree Fock approximation which gives incorrect results for metals. A big success of the LSD approximation was that it predicted the Mott insulators, NiO and MnO, to be antiferromagnetic band insulators[3]. However, the calculated moment was too small and the band gap was an order of magnitude smaller than experiment. It was suggested[3] that the use of the spherical potential approximation might be a cause of the underestimation of the magnetic moment but full potential linearized augmented plane wave method (FLAPW) calculations[4] discussed herein indicate that the moment on nickel is about the same ($1.03\mu_B$) as in previous calculations[3] and the band gap (0.02 Ry) is one order of magnitude smaller than experiment.

It is well known that for insulators and semiconductors the LSD functional method underestimates the insulating gap[5]. It is believed that the LSD approximation describes ground state properties well but fails to describe properly experimental gaps where quasiparticle excitations should be taken into account[5]. One of the difference between metals and insulators[5] is that in metals the electron affinity and the ionization potential are the same and equal to the chemical potential $\mu = \partial E/\partial N$, where E is the total energy and N is the total number of electrons. In nonmetals the energy difference between electron affinity and ionization potential is nonzero[6]. The DFT band gap (the difference between the lowest unoccupied

---

[*] Supported by NSERC and the OCMR High Tc Consortium (CRAY Research, General Electric Canada and Ontario Hydro).

and the highest occupied orbital energies of the N electron system) is not necessarily the same as a quasiparticle gap between an excited state of N+1 electrons and the ground state of N electrons. In fact it differs by the discontinuity in the exchange - correlation potential when one electron is added. We consider this point further in this report on the so-called GW corrections within the LSD approximation[7]. Although the GW formalism[7] has been known for a long time it has not been used previously for transition metal oxides due to large - scale computer requirements[4].

Since the present GW correction calculations are based on the formalism described previously for paramagnetic materials[8], only the most important features and approximations will be noted here.

DIELECTRIC MATRIX

The microscopic screening effect which results in a reduced effective Coulomb interaction between electrons is fully described by the dielectric matrix. To calculate this matrix, it is convenient to use plane waves due to the simplicity of the Coulomb potential in this representation:

$$v_{G,G'}(q+G) = 4\pi \mid q+G \mid^{-2} \delta_{G,G'} \qquad 1$$

where G and G' are vectors in reciprocal space, q is a vector in the first Brillouin zone and $\delta_{G,G'}$ is the Kronecker delta. The number of vectors needed in reciprocal space is dependent on the degree of localization of the electron wave functions, e.g. more points are required for the more localized d electrons than for the more extended p electrons. An alternative procedure has been proposed recently for localized basis sets[9].

The dielectric function is calculated in the random phase approximation: $\xi = 1 - vP$ where P is the bare polarizability to be calculated with the use of the eigenstates and eigenenergies generated by the band structure calculations with the FLAPW method[10] within the LDA. For convenience in numerical calculations, the quantity: $u(q+G) = (v(q+G))^{1/2}$ is introduced. Then, the dielectric function is represented by the Hermitian matrix:

$$\varepsilon_{GG'}(q,\omega) = \delta_{GG'} + u(q+G) \sum_{v,c,k,\sigma} <v,k,\sigma \mid e^{-i(q+G)r} \mid c,k+q,\sigma>$$

$$\times <c,k+q,\sigma \mid e^{-i(q+G')r} \mid v,k,\sigma> u(q+G')/(E_{c,k+q,\sigma} - E_{v,k,\sigma} - \omega - i0^+) \qquad 2$$

where the c and v indices correspond to the conduction band (zero occupation) and valence band (full occupation) respectively, σ is a spin index and E represents the corresponding eigenvalues calculated within the FLAPW method ($\hbar \equiv 1$). The number of eigenvalues used was limited to forty, since it has been shown[8] that the error introduced should not exceed a few percent. Since these calculations are very extensive, further approximations are needed. First, the above formula is used only to calculate the static dielectric matrix $\varepsilon_{GG'}(q,\omega=0)$.

To calculate the dynamical dielectric matrix ($\omega \neq 0$ and real) a generalized plasmon pole model is used[8]:

$$1/\varepsilon_i(q,\omega) = 1 + C_i(q)/(\omega - \omega_i(q) + i0^+) - C_i(q)/(\omega + \omega_i(q) + i0^+) \qquad 3$$

where $\epsilon_i(q,\omega)$ is the i th eigenvalue of the dynamical dielectric matrix, and $\omega_i(q)$ is found from the required static limit, i.e. $\epsilon_i(q,\omega) \to \epsilon_i(q,0)$ when $\omega \to 0$. The requirement of the proper high frequency limit leads to the expression[8] $\omega_p^2/(2\text{Re}\omega_i(q))$ for $C_i(q)$ where the plasma frequency $\omega_p^2 = 4\pi n$ for the uniform valence electron density (n). Since electrons in NiO behave quite differently from a free electron gas, the calculations were tested by assuming two different plasma frequencies, once with all - valence electron contributions and once with only d electron contributions taken into account. There was not much effect on the final value of GW corrections due to the choice of the plasma frequency and in the following we present results with the plasma frequency corresponding to the all - valence electron contributions.

The next approximation is to calculate the full dynamical dielectric matrix from the above eigenvalues using eigenvectors ($U_{Gi}(q)$) of the static dielectric matrix. This means that the off - diagonal elements will not be correct for higher frequencies but for calculations of the self - energy the most important contribution comes from $\omega < \omega_p$.[8]

In Fig.1 the real part ($\epsilon_1(\omega) \equiv \text{Re}(\epsilon_0(0,\omega))$) and imaginary part ($\epsilon_2(\omega) \equiv \text{Im}(\epsilon_0(0,\omega))$) of the calculated dielectric function are shown as a function of frequency for NiO. We applied at this stage a correction where all conduction bands are shifted by a constant amount in order to correct the LSD gap. Without doing this the calculated dielectric constant would be much too large (40). The experimental dielectric constant is 5.4[11]. When using a gap shift (0.24 Ry; upper curve in Fig.1) corresponding to the experimental value of the gap the dielectric constant equals 12. A larger gap shift (0.40 Ry; lower curve in Fig.1) results in a better value of the dielectric constant: 8.7.

GW APPROXIMATION

The next task is to calculate the corrections to the LSD gap. To do this, the self - energy, calculated by including the first diagram of the expansion in terms of the screened Coulomb operator (W), is of the following form ($\delta \equiv 0^+$):

$$\Sigma(r,r';\omega) = i/4\pi \int_{-\infty}^{\infty} d\omega' e^{i\omega'\delta} G(r,r';\omega-\omega') W(r,r';\omega') \qquad 4$$

where the Green function is calculated using the FLAPW eigenvalues and eigenvectors;

$$G(r,r';\omega) = \sum_{n,k\sigma} \phi_{nk\sigma}(r) \phi^*_{nk\sigma}(r')/(\omega - E_{nk\sigma} + i\delta_{nk\sigma}) \qquad 5$$

$\delta_{nk\sigma}$ is 0^- for valence band energies and 0^+ for conduction band energies. The screened Coulomb interaction, W, is;

$$W(r,r';\omega') = \sum_{q,G,G'} e^{i(q+G)r} \epsilon^{-1}_{GG'}(q,\omega') v(q+G) e^{-i(q+G')r'} \qquad 6$$

After substituting (5) and (6) into (4) the real part of the self - energy can be written as:

$$\text{Re}\Sigma = \Sigma_{SEX} + \Sigma_{COH} \qquad 7$$

where the notation due to Hedin[6] is used. The first part comes from the poles of the Green function and the second part, the so-called Coulomb hole part, arises from the poles of the screened interaction.

If enough G points are used and the same approximations are made in the calculation of the LSD exchange - correlation potential and the self-energy, the self - energy corrections should be calculated as;

$$\Delta_{nk\sigma} = \Sigma_{nk\sigma}(E_{nk\sigma}) - V^{XC}_{nk\sigma} \qquad 8$$

where $\Sigma_{nk\sigma}(E_{nk\sigma}) \equiv \langle nk\sigma|\Sigma(E_{nk\sigma})|nk\sigma\rangle$ and $V^{XC}_{nk\sigma} \equiv \langle nk\sigma|V^{XC}(\rho(r))|nk\sigma\rangle$. Since we use only 111 G points and a very crude approximation for the self - energy, we will make use of the fact that the exchange correlation energy within the LSD approximation is equal to the self-energy calculated at the Fermi energy[12]. In this way, the corrections can be calculated more accuratelly[13] as;

$$\Delta_{nk\sigma} = \Sigma_{nk\sigma}(E_{nk\sigma}) - \Sigma_{nk\sigma}(E_{n_F k_F \sigma}) \qquad 9$$

To simplify the calculations we used for the second term in expression (9) the self - energy of the valence band which lies closest to the Fermi energy. This approximation is probably not bad for d electrons which have energies close to the Fermi energy. For p electrons it means just adding (renormalised by a constant value) the self energy onto the LSD exchange - correlation energy. In Figure 2, the d bands calculated with and without corrections are presented. A significant gap increase can be seen in Fig. 2b.

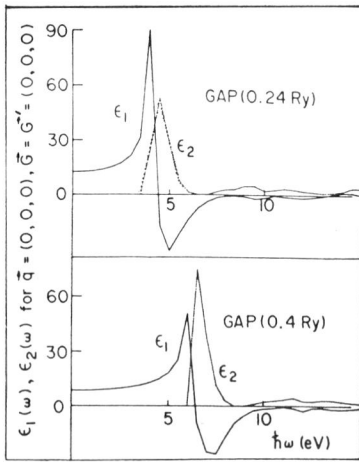

Fig.1 Dielectric function for 0.24 Ry and 0.4 Ry gap shifts.

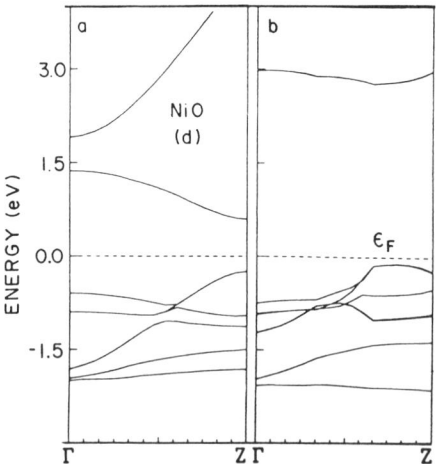

Fig.2 d electron FLAPW bands for antiferromagnetic NiO in z direction; a. without and b. with GW corrections.

It is interesting to compare LDA and Hartree Fock (HF) results[14] for materials with poor screening such as NiO, since the HF method does not take screening into account (but includes self - interaction corrections exactly). In Fig.3 we present calculations[15] in the Γ-Z direction for NiO where we compare at the left, paramagnetic Hartree Fock and LDA - Linear Muffin Tin - Atomic Sphere Approximation (LMTO-ASA)[16] results. On the right, FLAPW spin polarized calculations are shown where the opening of the gap due to antiferromagnetic ordering is observed. On the paramagnetic plots, we represented with broken lines, the image of the bands in the reduced Brillouin zone for the antiferromagnetic state in order to make easier the comparison between spin polarized and non-spin polarized calculations. The last plot on the right of Figure 3 shows our FLAPW calculations with GW corrections included. It is interesting that the conduction bands in the HF approximation at the Γ point are at almost the same energy as the corrected GW-LSD bands. The principal difference between the HF and LDA results occurs in the d electron bands of t_{2g} symmetry. The position of these bands lies below the p electron bands due to the presence of self - interaction corrections in the HF method and its absence for localized electrons in the LDA method[17]. It is obvious that this shift of the t_{2g} bands is probably too large since there is no screening effect in HF. The same shift in the GW approximation is much smaller. There is some indication from a high resolution photoemission study[18] that there is a feature in the density of states, which is not present in bare calculations which could be a shifted d band (t_{2g}).

The above discussion leads us to propose a modification of self - interaction corrections used previously for LSD exchange correlation potentials[17] by taking screening into account. This correction has a good limiting behaviour since it disappears for metals where there is full

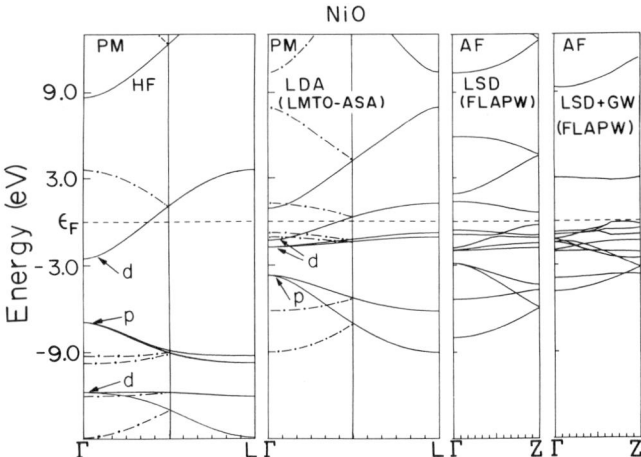

Fig.3 The band structure for NiO in the z direction starting from the left as indicated: paramagnetic HF, paramagnetic LMTO - ASA LSDF, spin polarized FLAPW LSDF calculations; bare bands and with GW corrections included.

screening ($\varepsilon = \infty$) and becomes significant for materials with poor screening. This correction will be discussed in details elsewhere[15].

The p bands corrected with our GW procedure are included for completeness in Fig.3. However, the relative shift of d and p bands is not expected to be very reliable as discussed above.

SUMMARY

Self-energy corrections (SEC) within the GW approximation have been calculated for d electrons of NiO. Due to the extensive computer require-ments of this method for d electrons, the calculations of SEC were restricted to only four k points for each band, 40 eigenstates and eigenvalues generated by the FLAPW method and 111 points in reciprocal space. The SEC corrections are not very sensitive to the plasma frequency values used in the plasmon-pole model. It is important to use the same model for both the self-energy and exchange correlation potential so that the accumulated error cancels in the calculation of SEC. These preliminary results are very encouraging.

ACKNOWLEDGEMENTS

We would like to thank A. J. Freeman for close collaboration at the beginning of this project and C. Dharma - Wardana, W. E. Pickett, M. R. Norman, J. Yu, C.-J. Mei and B. Brandow for useful discussions.

REFERENCES

1. N. F. Mott, Proc. Phys. Soc. **A62**, 416 (1956).
2. W. Kohn and L. J. Sham, Phys. Rev. **140**, A1133 (1965).
3. T. Oguchi, K. Terakura and A. R. Williams, Phys. Rev. **B28**, 6443 (1987).
4. B. Szpunar and V. H. Smith, Jr., Physica **B163**, 29 (1990).
5. J. P. Perdew and A. Zunger, Phys. Rev. **B23**, 5048 (1981); L. J. Sham and M. Schlüter, Phys. Rev. Lett. **51**, 1888 (1983).
6. M. S. Hybertsen and S. G. Louie, Phys. Rev. **B34**, 5390 (1986), R. W. Godby, M. Schlüter and L. J. Sham, Phys. Rev. **B37**, 10159 (1988).
7. L. Hedin, Phys. **139**, A796 (1965); L. Hedin and S. Lundquist, Sol. State Phys. **23**, 1 (1969).
8. N. Hamada, M. Hwang and A. J. Freeman, Phys. Rev. **B41**, 3620 (1990).
9. O. Gunnarson, P. Gies, W. Hanke and O. K. Andersen, Phys. Rev. **B40**, 12140 (1989).
10. H. J. F. Jansen and A. J. Freeman, Phys. Rev. **B30**, 561 (1984).
11. R. Newman and R. M. Cherenko, Phys. Rev. **114**, 1507 (1959).
12. H. A. Bethe, Phys.Rev. **103**, 1353 (1956).
13. W. E. Pickett and C. S. Wang, Phys. Rev. **B30**, 4719 (1984).
14. R. Dovesi, C. Pisani, C. Roetti, M. Causa' and V.R. Saunders, CRYSTAL 88, QCPE program No.577; C. Pisani, R. Dovesi and C. Roetti, Hartree - Fock ab Initio Treatment of Crystalline Systems, Lecture Notes in Chemistry, v. **48** (Springer, Heidelberg, 1988).
15. B. Szpunar, C.-J. Mei and V. H. Smith, Jr., to be published
16. O. K. Andersen, O. Jepsen and D. Glötzel, Highlights of Condensed Matter Theory, Proceedings of the International School of Physics, Enrico Fermi, Varenna, July 1983, F. Bassani et al., Ed. (Elsevier, Amsterdam, 1985).
17. A. Svane and O. Gunnarson, PRL **65**, 1148 (1990).
18. Z.-X. Shen, private communications.

2D ONE-BAND HUBBARD MODEL FOR THE CUPRATES

Silvia B. Bacci, Eduardo R. Gagliano
and Richard M. Martin

Physics Department, Material Research Laboratory and
Science and Technology Center for Superconductivity
University of Illinois at Urbana-Champaign
1110 W. Green Street, Urbana, IL 61801

INTRODUCTION

Experiments on the Cu-O superconductors and related materials show that the electronic states near the Fermi energy have dispersion similar to simple bands, yet have significant electron-electron interactions.[1,2,3,4,5,6] Low energy excitations and transport - namely linear resistivity vs. temperature, linear Raman scattering, and other anomalous low energy behaviors - in these materials reveal behavior different from that expected for normal Fermi liquids.[7] To address whether or not these anomalous behaviors signify a fundamental breakdown of Fermi Liquid Theory[8,9] or are instead manifestations of FLT for the special Hamiltonians that describe these materials,[10] we believe it is important to have some idea of the nature of the interacting- electron Hamiltonian appropriate for CuO planes in the actual cuprate materials.

Here we describe results of studies of several years by ourselves and others to attempt to calculate the appropriate interactions and to compare the results quantitatively with experiments. At this point in time the principle conclusion is that the appropriate Hamiltonian is a one-band Hubbard model, with intermediate U of order the bare band width.[11] This is our principle result and we argue here that it is in accord with known experimental results. In particular, we note that it predicts a spin Hamiltonian for the insulator which has terms beyond the usual Heisenberg form. We emphasize that additional terms may also be important, e.g., extended range hopping and interactions in the one-band model, phonons, or other terms. Our work is not sufficient to firmly address the role of any such additional terms.

One question immediately arises: Is the magnitude of the interaction U important, or are the key results universal and thus independent of any quantitative considerations? The latter point of view has been argued by Anderson[8]; however, others[12] have argued that there is no universal breakdown of Fermi Liquid Theory in 2d as there is in 1d. Others have explicitly used expansions expected to be valid only for small or moderate U.[13] Studies[14,15] of the 2d Hubbard model by QMC have shown a crossover behavior from small U to large U in the range U of order W=8t.

A small number of groups have taken the challenge to determine realistic Hamiltonians. Many are cited in reviews[4,5] and we will give a very incomplete list here. Some have used experimental data to determine parameters in models;[3,2] others have attempted to calculate the parameters a priori[16,17,18,19,20,21]. Since no present exact method can treat complex many-body systems like Cu-d electrons in condensed matter, the calculations of parameters have been based upon local density (LDA) or Hartree Fock (HF) approximations. The approach is, however, different from usual LDA and HF and the results are predictions of parameters in many-body models. In section II, we will review the resulting model Hamiltonians, which appear to be "realistic" and to be supported by experiments. Nevertheless, they are still complex, so that it is essential to carry out very difficult many-body calculations to attempt to distill their essence. In section III we will discuss results of exact calculations on clusters[11,21,22] to answer the questions of whether these "realistic" models reduce to simpler one-band or t-J type models.

CONSTRAINED DENSITY FUNCTIONAL CALCULATIONS

Density Functional Theory is by far the best-developed method for quantitative calculations of electronic structures of solids. In the CuO materials, usual calculations in the local density approximation (LDA) predict metallic behavior in all cases, even those which are in fact antiferromagnetic insulators.[4,5] Thus, the LDA certainly does not describe all the properties of these materials; furthermore, such one-electron methods have nothing to say about the question of whether or not the Fermi liquid is stable in the presence of electron-electron interactions. Another way to use LDA calculations has been developed for strongly correlated systems and applied previously to anomalous rare earths like Ce and CeO,[23] and to transition metal oxides like NiO.[24] In the "constrained LDA", electronic energies are calculated as a function of occupations of localized wannier functions and matrix elements between sites are determined for these functions. This leads to interacting-electron Hamiltonians which then must be solved by many-body techniques. The efficacy of this approach has been established by carrying out calculations on atoms where the energies are directly comparable with experiments. For example, the second and third ionization energies (averaged over multiplets) of Cu atoms have been calculated and shown to agree well with experiments; their difference, 17eV, is the average U_{dd} in the atom, which is reduced by screening in the solid.[16] In addition, the Slater parameters describe the different Coulomb interactions between the different d orbitals; these are expected to carry over to the solid with little change.[2]

The constrained LDA method has been applied to solids, e.g. rare earth systems with highly correlated 4f states.[23] Self-consistent LDA calculations are done on supercells in which the occupations of chosen orbitals are constrained and the energies of that state as well as other nearby states are determined as a function of the occupation. Detailed calculations have been reported for the f electron energies e_f and U_{ff} in CeO and PrO as well as the wide O bands. The resulting many-body problems can be solved exactly in the impurity case (the Gunnarsson-Schonhammer method[25]) giving the O-band width, f energies, and U_{ff} in agreement with photoelectron spectra. The ground states are correctly predicted to be non-magnetic for CeO and insulating magnetic for PrO_2; electronic gaps are somewhat low indicating an error of order 1eV in the f energy relative to the energies of the O bands.[23]

Similar calculations have been carried out for La_2CuO_4 by McMahan, et. al.[16,18] and by Hybertsen, et. al.[20] This work involves self-consistent LDA calculations for supercells in which the charge on a single Cu or O site is constrained. The most complete work is that of McMahan, Annett, and Martin [18], which has considered many different O states on and off the planes and different symmetry Cu d states. These authors carefully defined properly orthogonal Wannier states and found the set of parameters e_p, e_d, U_{dd} (d's on same site), U_{pp}, U_{pd}, and the hopping matrix elements t_{pd}, and t_{pp}. The most important states are the Cu $d_{x^2-y^2}$ and the two bonding O p states per cell. For these bands, the values found by McMahan, Annett, and Martin with estimated uncertainties are shown in Fig. 1. Values which are very similar but different in some quantitative details have been found by Hybertsen, et. al.[20] We will focus upon these states in the present work; however, we note that McMahan, et. al., have emphasized that there are also other states very near the Fermi energy, such as the p states on the out-of-plane apical oxygens, which mix with the Cu $d_{3z^2-r^2}$ states. We will not consider these states here because they are not the main object of our present work; we emphasize, however, that these states cannot be ruled out on present theoretical grounds.

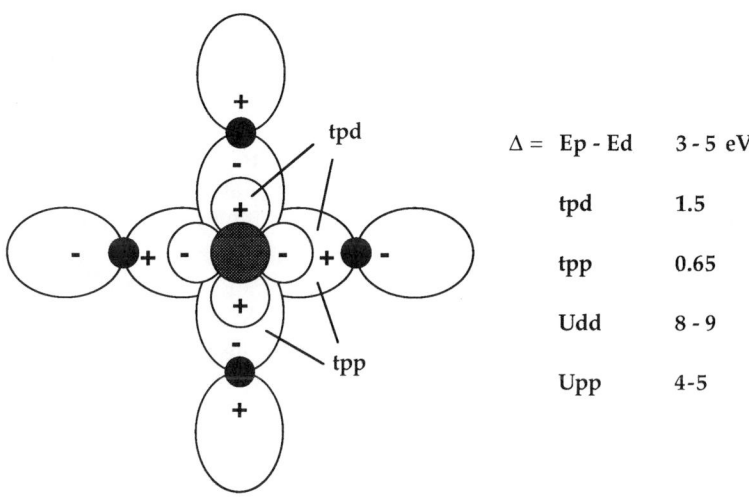

Fig. 1. Orbitals of the Three band model and calculated parameters.

The parameters calculated from the CDF theory are in rather amazing agreement with experiments. This may be summarized by noting that extensive analysis of experiments has been done by others, such as Sawatzky and coworkers,[2] which find empirical values very close to the results given in Fig. 1. This is not totally unexpected based upon the previous results for the rare earth oxides. There the predicted 4f screened U's are in excellent agreement with experiment and the 4f energy relative to the oxygen states agrees with experiment to within about 1 eV. Furthermore, this

1 eV error is just the magnitude expected for LSD errors in exchange energies[26] Based upon this reasoning, we estimate the uncertainty in our least-well-determined parameter, the energy difference $e_p - e_d$, to be around 1 eV.

The states shown in Fig. 1 are the well-known three-band model which is generally accepted as the starting point for calculations (possibly with other bands added). The model is defined by the Hamiltonian $H = H_d + H_p + H_{pd}$,

$$H_d = \epsilon_d \sum_{i,\sigma} d_{i\sigma}^\dagger d_{i\sigma} + U_d \sum_i n_{di\uparrow} n_{di\downarrow}$$

$$H_p = \epsilon_p \sum_{l,\sigma} p_{l\sigma}^\dagger p_{l\sigma} + \sum_{\langle l,l'\rangle,\sigma} t_{pp}^{ll'}(p_{l\sigma}^\dagger p_{l'\sigma} + h.c.)$$

$$H_{pd} = \sum_{\langle i,l\rangle,\sigma} t_{pd}^{il}(d_{i\sigma}^\dagger p_{l\sigma} + h.c.)$$

Here i (l) denotes a Cu (O) site. The operator $d_{i\sigma}^\dagger$ ($p_{l\sigma}^\dagger$) creates a $Cu3d$ ($O2p$) hole with spin σ at site i (l). Here t_{pp} is the $O-O$ hopping matrix elements and t_{pd} is the $Cu-O$ hybridization. The matrix elements $t_{pp}^{ll'} = \pm t_{pp}$ take into account the sign due to the symmetry of the $O(2p)$ states, and $t_{pd}^{il} = \pm t_{pd}$ of the $Cu(3d)$.

This model is the basis for many models. For example, the model proposed by Emery[27] is the three band model assuming the large interactions cause the states to separate into Cu d spins with added holes on the O states. On the other hand, Zhang and Rice[28] proposed that this same model in the strong interaction limit leads to d spins tightly bound to O holes in singlet states, so that the combination acts as a spinless object that can hop in the spin background. This leads to the "t- J model", probably the most widely studied model for the high Tc materials, because it is the simplest version of a strongly interacting one-band model. Thus a key question is: Does the three band model reduce to a one- band Hubbard type model (as strongly argued by Anderson[8]) and does it further reduce to the t-J model (as proposed by Zhang and Rice[28])? Or is there some important feature that is not included in either of these models?

FINITE CLUSTER CALCULATIONS

In this section we discuss results concerning a *numerical mapping* between the 3- and 1-band Hubbard models.[11] To this end, we have made small cluster calculations on periodic cells. We have considered two cluster sizes, 2×2, Cu_4O_8 and $\sqrt{8} \times \sqrt{8}$, Cu_8O_{16}. (In addition, we have carried out some calculations on the 2-band model of Annett and Martin [17], for which calculations can be done on larger clusters than can be done for the full 3-band model. Although we have not done complete studies, the results support our conclusions for the 3-band model.)

In order to make a numerical mapping values of the parameters must be chosen. We have considered values of the parameters obtained from constrained density functional theory given in Fig. 1. The on-site Coulomb repulsion $U_p \sim (4-6)eV$ at an O site, has not been taken into account explicitly since, due to the large O bandwidth

($\sim 5eV$), these holes will be delocalized so that their effective repulsion will be reduced. As we mentioned above the charge-transfer energy Δ is not well-determined by the theoretical results. Yet it is known that this is a key parameter. For $U_d \gg \Delta$ a charge-transfer regime is obtained while for $U_d \ll \Delta$ the system behaves as a Mott insulator.[29] Zhang and Rice[28] used the large Δ limit (but still $U_d \gg \Delta$) in their derivation of the t-J model.

Fixing the value of parameters to the ones mentioned above, except the charge-transfer energy Δ which is the least well-established value, we have analyzed the evolution of the properties as a function of Δ. The appropriate regime for the ceramic oxides was estimated by considering the experimental value of the insulating gap. In this regime we compare the excitations of multiband models with those of the 1-band Hubbard and $t-J$-like models taking into account: a) the *quantum numbers* of the energy levels and b) the *overall behavior* of the low energy levels with the parameters.

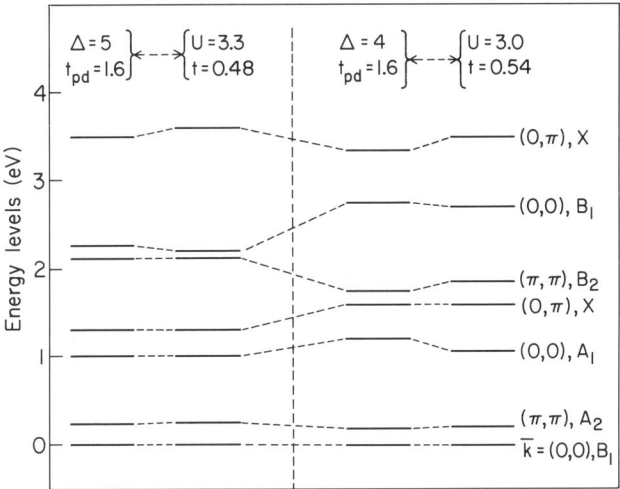

Fig. 2. Mapping of the low energy levels of the stoichiometric compound for different values of the multiband parameters.

Let us first consider the undoped case. By analyzing the energy spectrum as a function of Δ, we can distinguish two regimes. When Δ is large, a low energy band is well separated from upper bands which correspond essentially to states with double occupied sites. The excitations in the low energy band are spin-waves well described by the antiferromagnetic Heisenberg model. For intermediate values of $\Delta \approx (4-5)$ eV that fit the observed energy gap in the insulator, charge excitations coming from the upper bands appear at energies of order of the band gap, mixing with spin excitations. A similar behavior is obtained for the Hubbard model as a function of t/U.

In Fig. 2, we present the fitting of the low energy levels of the three and one band Hubbard models. All the states, except the one with quantum numbers $\vec{k} = (\pi, \pi)$ B_2, correspond to spin excitations. In table I, we show the estimated single-band parameters (U, t) for $\Delta = 4$ and 5 eV. These parameters have been obtained considering the magnitude of the gap in the one particle spectral function (which is mainly determined by U), and the dispersion of the band (from which we obtain

t). Similar calculations for the Hubbard model on bigger lattices show that the overall behavior of the energy levels is not very sensitive to the cluster size, although slightly higher values of U/t are obtained. This finite size consideration suggests that $U \approx W(= 8t)$ is an appropriate estimation of the thermodynamic limit. Note that our value of U/t is smaller than the calculated $U \sim 5.4 eV$ (which gives $U/t \sim 10-12$) found by Hybertsen, et. al.[21], which was obtained using different clusters and which *can not* be used to fix the gap. The value of U/t which we find to fit the gaps is similar to that obtained by Chen *et al*,[30] $U = 4.1 eV$, which has been used to explain x-ray absorption in $La_{2-x}Sr_xCuO_{4+\delta}$.

Table 1. Effective parameters obtained from fitting the 1-band Hubbard model for different sets of the 3-band Hubbard parameters. All energies are in units of eV

t_{pd}	$\Delta = 3$	$\Delta = 4$	$\Delta = 5$
1.6	$U = 2.4, t = 0.57$	$U = 3.0, t = 0.54$	$U = 3.4, t = 0.48$
1.3	$U = 2.3, t = 0.50$	$U = 2.7, t = 0.43$	$U = 2.9, t = 0.37$

Now, we turn to the behavior of hole-doped systems. Similar calculations to the undoped parent compound, show that in the presence of carriers there is also a one-to-one correspondence between the low energy levels of the 3-band model with those of the one-band Hubbard model. Even more important, we obtained the *same* Hubbard model with an estimated thermodynamic limit $U/t \sim 8$. Finally, starting from the one-band Hubbard Hamiltonian, we also studied the reduction to the $t-J$ model, finding a correspondence between the low energy structure of both models. As expected when the hole concentration is increased, the double occupancy decreases and therefore the $t-J$ approximation becomes even more appropriate. Although, this agreement between the low-lying excitations, the $t-J$ and Hubbard models may be very different for the study of bound states of two-holes, at the intermediate regime $U/t \sim 8$. Indeed in this range, the Hubbard model presents a small binding energy[31], while the $t-J$ leads to phase separation.

COMPARISON WITH EXPERIMENT

Many experimental methods - photoemission, inverse photoemission, optical conductivity, neutron scattering, Raman spectroscopy, and other techniques - provide measurements of different excitations, which can be compared quantitatively with our 3-band and reduced 1-band models.

a) One particle spectral function

First let us consider addition/removal spectra for electrons, which we can calculate exactly for small clusters - including all effects of electronic correlations - using exact diagonalization methods. In principle, in order to calculate the spectral density of one-particle excitations, $A(\omega) = \frac{1}{N}\sum_n |<\psi_n^{N-1}|a_i|\psi_0^N>|^2\delta(\omega - E_0^N + E_n^{N-1})$, we need all eigenvectors which are connected with the ground state through the application of the desired operator, in this case a_i, the annihilation of one particle at site i. Here $|\psi_0^N>$ is the ground state of the system with N-particles and E_0^N its energy, with n we label the eigenstates of the system with $N-1$ particles. Clearly for this procedure becomes impractical as the size of the system increases, and one needs alternative approaches. The method we have used is essentially a Lanczos algorithm based in the relation between the spectral function and the imaginary part of the one- particle Green function,[32] $A(\omega) = \frac{-1}{\pi}ImG(\omega + E_0 + i\delta)$ where $G(x) = <\psi_0^N|a_i^\dagger(x-H)^{-1}a_i|\psi_0^N>$ is a resolvent operator which can be evaluated recursively. We have included a small parameter δ to give a finite width to the delta-functions appearing at each pole of G.

In Fig.3a, we show the PES/IPES spectra for the undoped and one hole doped systems for $\Delta = 5$ eV in the Cu_4O_8 cluster. For the stoichiometric compound the insulating gap is $E_g \sim (1.75 - 2.2)eV$ for $\Delta = (4-5)eV$. These results are

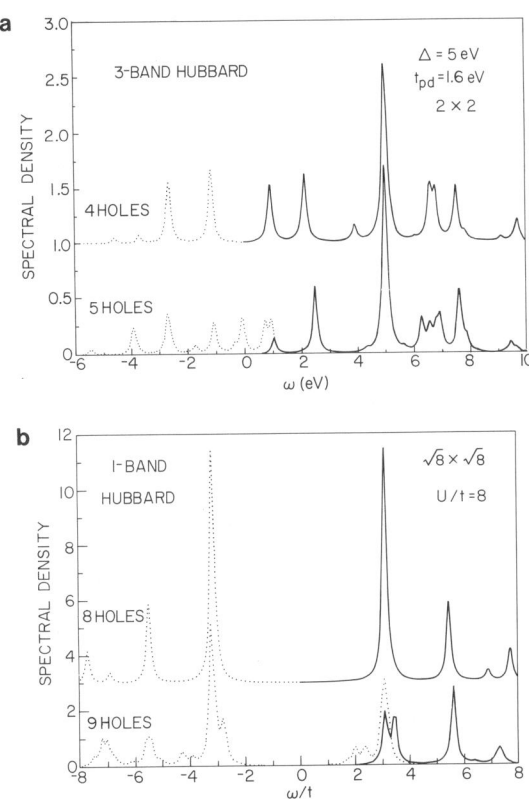

Fig. 3. a) One-particle spectral function of the 3BHM for the Cu_4O_8 cluster. b) One-particle spectral function of the one-band Hubbard model for the $\sqrt{8} \times \sqrt{8}$ cluster at $U/t = 8$. The upper figure correspond to the insulating state while the other to the one-hole doped system ($S = 1/2$ subspace).

close to the experimental value for La_2CuO_4, $E_g \sim (1.65 - 2.0)eV$.[33] This is the primary evidence that $\Delta = (4 - 5)eV$ is the "realistic" range for these materials. The chemical character of the holes in the insulating ground state is Cu dominated (60 − 70%), although there exists a strong $Cu - O$ hybridization. When holes are introduced to the system by doping (see lower part of the figure), the d spectral weight decreases while the p weight increases indicating that the added hole weight goes mainly to the O_{2p} states. We also note a negligible gap, which indicates that the system has crossed the metal-insulator boundary.

A similar behavior is obtained for the one band Hubbard model as is shown in Fig.3b. In the stoichiometric regime the insulating gap is $E_g \sim U$, while for the 1 hole-doped case, a metallic phase is obtained for $U/t \leq 8$, but not for higher values ($U/t \geq 12$), for which the doped system still behaves as an insulator.

b) Magnetic excitations

Spin excitations in AF insulators have been measured by Neutron[34] and Raman[35] scattering experiments, which provide evidence of quantum spin fluctuations as well as a quantitative determination of the exchange parameter J.[36] They are in agreement, giving an exchange parameter, quite large, $J \sim (0.10 - 0.15)eV$. In order to obtain J from our cluster calculations, it is necessary first to analyze the Hubbard-Heisenberg reduction for the intermediate range $U/t \sim 8$.

In Fig.4, we make a comparison between the Hubbard and Heisenberg model for the $\sqrt{8} \times \sqrt{8}$ cluster. We consider two cases $U/t = 8$ and $U/t = 12$ which represent the ranges obtained here and in Hybertsen et al.[21] For $U/t = 12$, the fit to the Heisenberg model is rather good considering the lowest (5-6) energy levels

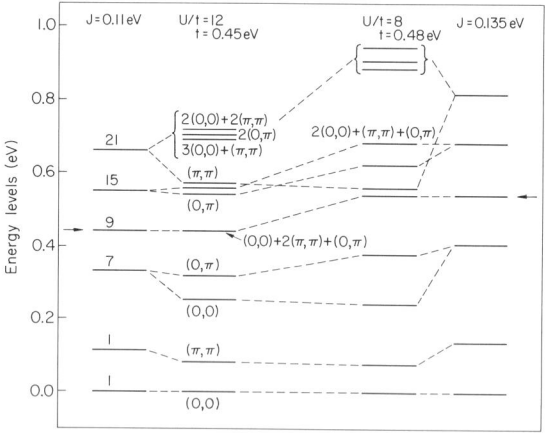

Fig. 4. Comparison of the low energy levels of the one band Hubbard and Heisenberg models for $U/t = 8$ and $U/t = 12$, for the $\sqrt{8} \times \sqrt{8}$ cluster with periodic boundary conditions. Each state with $\vec{k} = (0, \pi)$ has a degeneracy of 6.

and is much poorer for higher states. In the intermediate regime $U/t = 8$ only the lowest 3 states are approximately described. Fitting the lowest excited states (shown in the figure by an arrow) or the highest state of the low energy band we obtain $J = (0.08 - 0.13)$ eV and $J = (0.08 - 0.18)$ eV for $U/t = 12, 8$ respectively. Considering the uncertainties on the multiband parameters and that the fit is not perfect in any case, our estimation $J \sim 0.1$ eV is in good agreement with the values obtained from the analysis of neutron and Raman experiments.

Although within the CuO_2 layers, the leading term in the effective spin Hamiltonian is the antiferromagnetic Heisenberg interaction, there is not a perfect fitting of all low energy levels which indicates that other terms in the effective low energy spin Hamiltonian may be required. These higher-order terms appear at fourth order in the effective Hamiltonian obtained by a canonical transformation of the Hubbard model.[37] Aside from a constant term, the obtained effective spin Hamiltonian for the 2D square lattice is, $\mathcal{H} = \mathcal{H}_J + \mathcal{H}_K$, where

$$\mathcal{H}_J = \sum_i \sum_{n=1,2,3} J_n \mathbf{S_i} \cdot \mathbf{S_{i+\vec{\delta}_n}},$$

and

$$\mathcal{H}_K = K \sum_i (\mathbf{S_i} \cdot \mathbf{S_{i+\hat{x}}})(\mathbf{S_{i+\hat{y}}} \cdot \mathbf{S_{i+\hat{x}+\hat{y}}}) + (\mathbf{S_i} \cdot \mathbf{S_{i+\hat{y}}})(\mathbf{S_{i+\hat{x}}} \cdot \mathbf{S_{i+\hat{x}+\hat{y}}})$$

$$-(\mathbf{S_i} \cdot \mathbf{S_{i+\hat{x}+\hat{y}}})(\mathbf{S_{i+\hat{y}}} \cdot \mathbf{S_{i+\hat{x}}}).$$

where $J_1 = 4t^2/U - 24t^4/U^3$, $J_2 = J_3 = 4t^4/U^3$ and $K = 80t^4/U^3$. Here $\vec{\delta}_1 = (\hat{x}, \hat{y})$ denotes unit vectors in both directions of a square lattice, while i denotes the sites. $\vec{\delta}_2$ and $\vec{\delta}_3$ correspond to second and third near-neighbors respectively. As is shown in Fig. 5, this effective model provides a ground state energy in remarkably good agreement with the corresponding to the Hubbard model at intermediate values of U/t. In fact, for $U/t = 4$ (12) the ground state energy per site of the Hubbard model differs from the corresponding to the Heisenberg model with multiple-spin interactions by 6%(1%) (4 × 4 cluster).

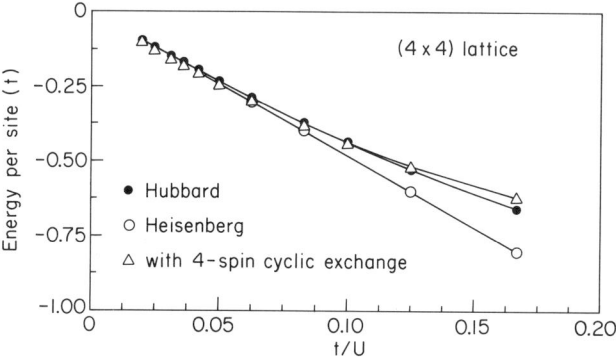

Fig. 5. Comparison of the ground state energy of the Hubbard, Heisenberg and the effective spin Hamiltonian up to $\sigma(4)$ as a function of t/U.

It has been argued based upon experimental spectra that the line shape of the B_{1g} Raman spectra can not be explained by a pure Heisenberg model and it has been suggested[38] that 4- spin processes are required to explain the great spectral weight appearing on the high frequency side of the two magnon peak. This is consistent with our present results and we are investigating in more detail the Raman spectra of the Hubbard and multiple-spin models. Preliminary results using the Parkinson Raman Hamiltonian up to first-nearest-neighbor and $U/t \sim 8$, indicate appreciable intensity in the A_{1g} symmetry, absent in the Heisenberg model (with this Raman operator), as well as an enhancement of the 4-magnon processes in the B_{1g} geometry. [Note added: After the conference we received a preprint[39] from Y. Kuramoto describing calculations on 3-band models which conclude the 4-spin cyclic terms are important and affect the shape of the Raman spectra.]

CONCLUSIONS

We have reviewed recent theoretical calculations which have attempted to derive - at least semi-quantitatively - essential features of the Hamiltonian describing electrons in the actual CuO planar materials. From constrained density functional (CDF) calculations[16,18,20] have been found realistic model parameters for a 3-band Hubbard model in general agreement with high energy spectroscopies and low energy magnetic excitations. Exact calculations on clusters show that the low energy charge and spin states are well-described by a 1-band Hubbard model with U of order 8t = band width.[11] One consequence of such intermediate values of U are spin interactions beyond the usual Heisenberg form, e.g., 4-spin cyclic terms, which may be important for understanding the spin spectra of the insulator.[38,39] We emphasize that in addition to t and U, there may also be other terms in the Hamiltonian, e.g., extended range hopping and interactions in the one-band model, other bands such as out-of-plane p states, phonons, etc.; our conclusion is that the 1-band model with intermediate U is the appropriate starting point for realistic theoretical models.

ACKNOWLEDGEMENTS

This project was supported by the NSF grants DMR- 8920538 and STC-8809854. The computer simulations were done on a CRAY2s at NCSA and MIT. We acknowledge NCSA and Cray Corporation for a grant of time on the MIT Cray.

REFERENCES

1. C. G. Olson, et. al., Phys. Rev. **B42**, 381 (1990).

2. L. H. Tjeng, H. Eskes, and G. A. Sawatzky in Proceedings of Symposium on "Strong Correlations and Superconductivity" (1989); H. Eskes, and G. A. Sawatzky, Phys. Rev. **B43**, 119 (1991).

3. A review of photoemission experiments is given by P. A. P. Lingren, et. al., Surface Science Reports **111**, 1 (1990).

4. K.V. Hass, in Solid State Physics vol. 41, p. 213. (1989).

5. W. E. Pickett, Rev. Mod. Phys. **61**, 433 (1989).

6. E.Fradkin,"Field Theories of Condensed Matter Systems" (Addison Wesley, 1991), Chap.3.

7. See papers in MMS-HTS Conferences: Interlaken, Switzerland, published in Physica C 153-55 (1988); Stanford, CA, published in Physica C **162-64** (1989).

8. P.W. Anderson, Science **235**, 1196 (1987); Phys. Rev. Lett. **64**, 1389 (1990) and **65**, 2306 (1990).

9. R.Laughlin, Science **243**, 525(1988)

10. K.Levin, *et al*, Preprint (1990).

11. S.Bacci, E. R. Gagliano, R.M.Martin, and J. F. Annett, Phys.Rev.B**44**,7504 (1991).

12. J. R. Engelbrecht and M. Randeria, Phys. Rev. Lett. **65** 1032 (1990).

13. A. Kampf and J.R. Schrieffer, Phys.Rev. **B41** 6399 (1990).

14. S.R.White, *et al*, Phys.Rev.B **40**,506 (1989).

15. D. J. Scalapino, in High Temperature Superconductivity, Proc. of Los Alamos Symposium, ed. by K. Bedell, et. al. (Addison-Wesley, 1990).

16. A.K. McMahan, R.M. Martin, S. Satpathy, Phys. Rev. **B38**, 6650 (1988).

17. J. F. Annett and R.M. Martin, Phys. Rev. **B42**, 3929 (1990).

18. A. K. McMahan, J. F. Annett and R.M. Martin, Phys. Rev. **B42**, 6268 (1990).

19. E. B. Stechel and D. R. Jennison, Phys. Rev. **B38**, 4632 (1988).

20. M. S. Hybertsen, M. Schluter, and N. E. Christensen, Phys. Rev. **B39**, 9028 (1989).

21. M. S. Hybertsen, E. B. Stechel, M. Schluter, and D. R. Jennison, Phys. Rev. **B41**, 11068 (1990).

22. T. Tohyama and S. Maekawa, J. Phys. Soc. Japan **59**, 1760 (1990).

23. A.K. McMahan and R.M. Martin, in Narrow Band Phenomena, ed. by J. C. Fuggle, G. A. Sawatzky, and J. W. Allen (Plenum, New York, 1988), p 133.

24. M.R. Norman and A.J. Freeman, Phys. Rev. **B33**, 8896 (1986).

25. O. Gunnarsson and K. Schonhammer, Phys. Rev. **B28**, 4315 (1983). The present work differs in that exact solutions can be found without the need for the 1/N expansion, since for undoped La2CuO4 the oxygen bands are fully occupied.

26. R. O. Jones and O. Gunnarsson, Rev. Mod. Phys. **61**, 689 (1989).

27. V.J.Emery, Phys.Rev.Lett **58**, 2794 (1987).

28. F.C. Zhang and T.M. Rice, Phys. Rev. **B37**, 3757 (1988).

29. C.Balseiro, *et al*, Phys. Rev. Lett.**62**, 2624 (1989).

30. C.T.Chen *et al*, Phys. Rev. Lett. **66**, 104 (1991).

31. J.Riera and A.Young, Phys.Rev.B **39**, 9697 (1989); E.Dagotto *et al*, Phys. Rev. **B41**, 811 (1990); E.Dagotto *et al*, Phys. Rev. **B41**, 9049 (1990); G.Fano *et al*, Phys. Rev. **B42**, 6877 (1990).

32. E.R.Gagliano and C.A.Balseiro, Phys. Rev. Lett. **59**, 2999 (1987); Phys. Rev. **B38**, 11766 (1988)

33. T.Thio *et al*, Phys. Rev. **B42**, 10800 (1990); S.L.Cooper *et al*, Phys. Rev. **B42**,10785(1990).

34. Shirane *et al*, Phys. Rev. Lett.**59**, 1613 (1987); J.M. Tranquada *et al*, Phys. Rev. **B40**, 4503 (1989); G. Shirane et al, Phys. Rev. Lett.**63**, 330 (1989); G. Aeppli, et al, *ibid* **62**, 2052 (1989).

35. K.B.Lyons *et al*, Phys. Rev. Lett. **60**, 732 (1988); S.L. Cooper *et al*, Phys. Rev. B **37**, 5920 (1988); *ibid* **38**, 11934 (1988); S. Sugai *et al*, *ibid* **38**, 6436 (1988); W.H. Weber and G.W. Ford, *ibid* **40**, 6890 (1989); R.R.P. Singh *et al*, Phys. Rev. Lett. **62**, 2736 (1989).

36. A review is given by E.Manousakis, Rev.Mod.Phys.**63**, 1 (1991).

37. A.H.MacDonald *et al*, Phys. Rev. B **41**, 2565 (1990).

38. S. Sugai et al, Phys. Rev. B **42**, 1045 (1990); E.Gagliano *et al*, Europhysics Lett., **12**, 259 (1990); H.J.Schmidt and Y.Kuramoto, Physica C (1990).

39. Y. Kuramoto, "Exact Dynamics of Highly Correlated' Electrons in Two Dimensions," preprint (1991).

INTERLAYER PAIRING AND c-AXIS VERSUS ab-PLANE GAP ANISOTROPY IN HIGH T_C SUPERCONDUCTORS

R. A. Klemm and S. H. Liu*

Materials Science Division, Argonne National Laboratory
Argonne, IL 60439

ABSTRACT

Interlayer BCS-like pairing in nearly two-dimensional superconductors with $N=1,2,3$ conducting layers per unit cell is investigated. For $N=1$, the singlet Δ_s and triplet Δ_t order parameters have identical T_c's ($T_{cs} = T_{ct}$) and magnitudes, and the gap is isotropic. For $N=2$, $T_{cs} \geq T_{ct}$ and $|\Delta_s| \geq |\Delta_t|$. In region I of the parameters, the gap is anisotropic and nodeless for $T < T_{cs}$. In region II, the gap has nodes near to T_{cs}, but is nodeless at low T. For $N=3$, $T_{cs} >> T_{ct}$, and the gap is anisotropic but nodeless on two bands, but has a node on the third band. For $N=2,3$, the k_z-dependence of the gap and the quasiparticle density of states are found. For $N=1$ (2,3), intra- and interlayer pairing are completely (essentially) incompatible.

INTRODUCTION

Recently, there have been a number of measurements of the high T_c superconductors that have the potential of giving information regarding the superconducting energy gap. Measurements[1] of the penetration depth $\lambda(T)$ in $YBa_2Cu_3O_{7-\delta}$ (Y123) for magnetic fields $B||\hat{c}$ and $B\perp\hat{c}$ gave results consistent with BCS behavior in both directions, suggesting that the gap is unlikely to exhibit nodes at low temperatures T, although it could be anisotropic. Far infrared[2], Raman[3], and tunneling[4] data on Y123 suggest that the gap is different for measurement wavevectors $k||\hat{c}$ and $k\perp\hat{c}$, but isotropic within the ab plane. The lack of any appreciable azimuthal dependence of the upper critical field measurements[5] parallel to the ab plane in Y123 are consistent with no pair effective mass anisotropy and a nodeless gap within the ab planes. Recent SNS' tunneling data on Y123 films[6] resulted in a Josephson effect when the tunneling currents were into the ab plane, but not in the c-axis direction. Tunneling measurements[4] on $La_{2-x}Sr_xCuO_4$ were interpreted in terms of an isotropic gap, however. Photoemission results[7] on $Bi_2Sr_2CaCu_2O_8$ are only sensitive to the gap within the ab planes, for which it is found to be isotropic.

In this letter, we present the essential results of a detailed calculation, which will be published elsewhere[8]. We demonstrate that for intralayer pairing of the BCS type, the resulting gap $2|\Delta_0|$ cannot exhibit any anisotropy between the c-axis direction and the ab planes, even in the presence of single particle interlayer hopping processes, which includes Josephson tunneling. The paired quasiparticles must be spatially separated in the $\hat{c}$ direction for such gap anisotropy to occur. This fact has important consequences regarding the mechanism for high T_c superconductivity, which have not been addressed in the literature.

For such gap anisotropy to exist, it is further necessary that the number N of conducting layers per unit cell edge s be ≥ 2. For N=1, the interlayer pairing singlet and vector triplet order parameters (OP's) Δ_s and Δ_t have identical transition temperatures T_{cs} and T_{ct}. While their associated normalized gap functions are highly k_z-dependent, exhibiting nodes, the actual gap $2|\Delta|$ is found to be isotropic, as $|\Delta_s| = |\Delta_t|$. However, the gaps $2|\Delta|$ and $2|\Delta_0|$ are completely incompatible. For $N \geq 2$, $T_{cs} \geq T_{ct}$. The anisotropic gap $2|\Delta(k_z)|$ is either completely or predominantly singlet, and is nearly incompatible with $2|\Delta_0|$.

THE MODEL

We assume the normal state (above T_c) is a Fermi liquid[7] with Fermi energy E_F. The quasiparticles (or quasiholes) are assumed to propagate with two-dimensional fermion band dispersion $\xi_0(\mathbf{k})$ and bandwidth $W_\parallel$ within the N conducting layers in each of the M unit cells. Between neighboring layers within a unit cell and in adjacent unit cells, they tunnel through the separations d and $d' = s - (N - 1)d$ with matrix elements J_1 and J_2, respectively. We assume the interlayer hopping bandwidth $W_\perp \ll W_\parallel$, as in band structure calculations[9] and photoemission experiments[7].

Within each layer, the quasiparticles pair via the BCS interaction with strength λ_0. Quasiparticles on neighboring layers within a unit cell and in adjacent unit cells pair via BCS-like interactions with strengths λ_1 and λ_2, respectively, as pictured in Fig. (1). Since the Pauli exclusion principle is not applicable to pairing at a distance, all interlayer pair spin configurations are equally probable. For simplicity, we treat these interactions in the BCS weak-coupling approximation.

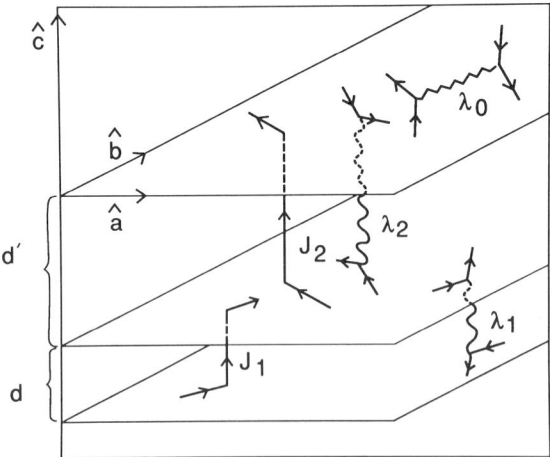

Fig. 1. Cross-sectional view of an $N = 2$ crystal section. The quasiparticles hop with matrix elements J_1 and J_2 between neighboring layers separated by d and $d' = s - d$, respectively. The intra- and interlayer pairing interactions λ_0, λ_1, and λ_2 are also pictured.

The Hamiltonian is thus taken to be $H = H_0 + V$, where

$$H_0 = s\int d^2r\, \Psi^\dagger(\mathbf{r})\hat{\xi}\Psi(\mathbf{r}) \tag{1a}$$

and

$$V = \frac{1}{2}s\int d^2r\, \rho(\mathbf{r})\hat{V}\rho(\mathbf{r}), \tag{1b}$$

where $\Psi(\mathbf{r})$ is a 2MN-dimensional Nambu spinor with spinor component $\Psi_{jn}(\mathbf{r})$ which annihilates a quasiparticle at position $\mathbf{r}$ on the nth layer in the jth unit cell, $\rho(\mathbf{r})$ is an MN-dimensional 'densitor', with component $\rho_{jn}(\mathbf{r}) = \Psi^\dagger_{jn}(\mathbf{r})\Psi_{jn}(\mathbf{r})$,

$$\hat{\xi} = \xi_0\hat{1} + J_1\hat{\delta}_1 + J_2\hat{\delta}_2, \tag{2a}$$

$$-\hat{V} = \lambda_0\hat{1} + \lambda_1\hat{\delta}_1 + \lambda_2\hat{\delta}_2, \tag{2b}$$

$$-\xi_0(\mathbf{r}) = \nabla^2/2m_0 + E_F, \tag{2c}$$

and $\hat{1}$, $\hat{\delta}_1$, and $\hat{\delta}_2$ have matrix elements given by

$$\left(\hat{1}\right)^{nn'}_{jj'} = \delta_{jj'}\delta_{nn'}, \tag{3a}$$

$$\left(\hat{\delta}_1\right)^{nn'}_{jj'} = \delta_{jj'}(\delta_{n',n+1} + \delta_{n',n-1}), \tag{3b}$$

and

$$\left(\hat{\delta}_2\right)^{nn'}_{jj'} = \delta_{j',j+1}\delta_{n'1}\delta_{nN} + \delta_{j',j-1}\delta_{n'N}\delta_{n1}, \tag{3c}$$

respectively. We use units in which $\hbar = c = k_B = 1$. For N=1, we set $J_1 = \lambda_1 = d = 0$.

QUASIPARTICLE AND PAIR EIGENSTATES

H_0 is easily diagonalized by Fourier transformation, as required by Bloch's theorem. The resulting quasiparticle dispersion is simply $\xi(\mathbf{k}) = \xi_0(\mathbf{k}) + 2J_2\cos k_z s$. The Fermi surface is a corrugated cylinder, given by $\xi(k_F) = 0$. For N=2, H_0 is diagonalized[8,10] by Fourier transformation plus a rotation by $\pi/4$ of the operators within a unit cell. There are two quasiparticle energy bands, $\xi_\pm(\mathbf{k}) = \xi_0(\mathbf{k}) \pm \epsilon_\perp(k_z)$, where

$$\epsilon_\perp(k_z) = [J_1^2 + J_2^2 + 2J_1J_2\cos(k_z s)]^{1/2}. \tag{4}$$

There are two Fermi surfaces, given by $\xi_\pm(k_{F\pm}) = 0$. For N=3, H_0 is diagonalized by Fourier transformation plus a unitary transformation of rank 3. There are three quasiparticle energy bands, $\xi_n(\mathbf{k}) = \xi_0(\mathbf{k}) + \epsilon_n(k_z)$, where

$$\epsilon_n(k_z) = 2\bar{J}\cos[\eta(k_z) - 2\pi(n-1)/3], \tag{5a}$$

$$\cos[3\eta(k_z)] = [J_1^2 J_2/\bar{J}^3]\cos(k_z s), \tag{5b}$$

$$\bar{J}^2 = (2J_1^2 + J_2^2)/3, \tag{5c}$$

and n =1,2,3. There are three Fermi surfaces, given by $\xi_n(k_{Fn}) = 0$.

The fact that the single quasiparticle states are diagonal in the band (or momentum) representation, but not in the layer index (or real space) representation requires that we perform the same transformations upon the pairing interaction V. This is necessary in order to assure that the resulting gap functions are commensurate with the single quasiparticle bands, which are periodic in k_z with period $2\pi/s$. This is particularly important for interlayer pairing, but is also necessary for intralayer pairing, as the quasiparticles are not localized to individual layers, but propagate from layer to layer in their respective bands.

We then apply the above transformations to the operators in the interaction V. The resulting mean-field interlayer interaction strengths $\tilde{V}_N(k_z, k_z')$ exhibit the lattice periodicity, and are separable in k_z and k_z'. These features allow us to write

$$\tilde{V}_N(k_z, k_z') = \sum_{S=\pm} \sum_{m=1}^{L_{NS}} \lambda_{Nm}^S |u_{Nm}^S(k_z)\rangle \langle u_{Nm}^S(k_z')| \qquad (6)$$

for $N = 1, 2$, where the even ($S = +$) and odd ($S = -$) eigenfunctions $|u_{Nm}^S(k_z)\rangle$ are periodic in $k_z s$ and orthonormal over the first zone $-\pi/s \leq k_z \leq \pi/s$. The $\lambda_{Nm}^S \geq 0$ are the eigenvalues, and the L_{NS} are the numbers of linearly independent even and odd eigenfunctions (indexed by m), which are independent of the band index n. For inequivalent conducting layers, the eigenvalues λ_{Nnm}^S also depend upon n. For $N = 1$, it is easy to show[11] that $\tilde{V}_1(k_z, k_z') = \lambda_2 \cos(k_z - k_z')s$. This implies $L_{1+} = L_{1-} = 1$, $|u_{11}^+(k_z)\rangle = \sqrt{2} \cos k_z s$, $|u_{11}^-(k_z)\rangle = \sqrt{2} \sin k_z s$, and $\lambda_{11}^+ = \lambda_{11}^- = \lambda_2/2$.

For $N = 2$, $L_{2-} = 1$ and $L_{2+} = 2$. We write the eigenvalues and eigenfunctions for $J_2 \geq J_1 \geq 0$. The eigenfunctions are also independent of the band index n. The eigenvalues are readily obtained, and found to satisfy $\lambda_{21}^+ \geq \lambda_{21}^-$ and $\lambda_{21}^+ > \lambda_{22}^+$. The orthonormal eigenfunctions are

$$|u_{21}^-(k_z)\rangle = \sqrt{2} J_2 \sin k_z s / \epsilon_\perp(k_z), \qquad (7a)$$

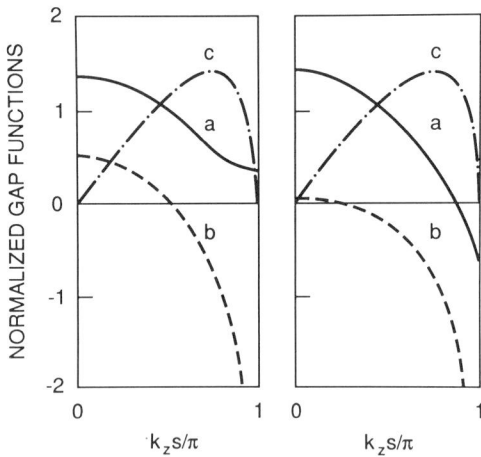

Fig. 2. Plotted are the orthonormal eigenfunctions $|u_{2n}^\pm(k_z)\rangle$ as functions of $k_z s/\pi$ for the parameters $\lambda_1/\lambda_2 = 0.9$, $J_1/J_2 = 0.7$, and $\lambda_2/\lambda_1 = 0.4$, $J_1/J_2 = 0.8$ in the left and right figures, respectively. The curves (a), (b), and (c) refer respectively to $|u_{21}^+(k_z)\rangle$, $|u_{22}^+(k_z)\rangle$, and $|u_{21}^-(k_z)\rangle$. The relative eigenvalues $\lambda_{2n}^\pm/\lambda_2$ for the left figure are (a) 0.9679, (b) 0.2371, and (c) 0.6950. In the right figure, the relative eigenvalues $\lambda_{2n}^\pm/\lambda_1$ are (a) 0.6635, (b) 0.1085, and (c) 0.6280.

$$|u^+_{2m}(k_z)\rangle = J_2(a_m + b_m \cos k_z s)/\epsilon_\perp(k_z), \qquad (7b)$$

where a_m and b_m are appropriately chosen to insure orthonormality. For $J_1 = J_2$, $\tilde{V}_2(k_z,k'_z)$ reduces to the form of $\tilde{V}_1(k_z,k'_z)$, with s and λ_2 replaced by $s/2$ and $\lambda_1+\lambda_2$, respectively. The forms for $J_1 \geq J_2 \geq 0$ are obtained from the above by interchanging J_1 with J_2 and λ_1 with λ_2. Plots of eigenfunctions for two particular choices of J_1/J_2 and λ_1/λ_2 are shown in Fig. (2).

For $N = 3$, the eigenfunctions depend upon the band index n. On each band, there are two even and one odd orthonormal eigenfunctions, as for $N = 2$. Because the 3 bands are inequivalent, there are 2 additional degrees of freedom, taking account of the band sum rule. These additional degrees of freedom are evident in the SU(3) symmetry of the matrix representing the band index: there are two traceless diagonal matrices (which we denote μ_1 and μ_2), whereas for $N = 2$, there is only the single traceless diagonal matrix μ_3 (a Pauli matrix). This complication changes the form of Eq. (6) to

$$\tilde{V}_N(k_z,k'_z) = \sum_{S=\pm} \sum_{m,m'=1}^{L_{NS}} \sum_{p,p'=1}^{N-1} V^S_{Nmp,m'p'} |u^S_{Nmp}(k_z)\rangle \langle u^S_{Nm'p'}(k'_z)|, \qquad (8)$$

where p and p' index the projections of the gap functions onto the traceless diagonal representations of SU(N). For $N = 3$, $L_{3+} = 2$ and $L_{3-} = 1$. The $V^S_{3mp,m'p'}$ thus comprise a 4×4 and a 2×2 matrix, for states even and odd in k_z, respectively. The

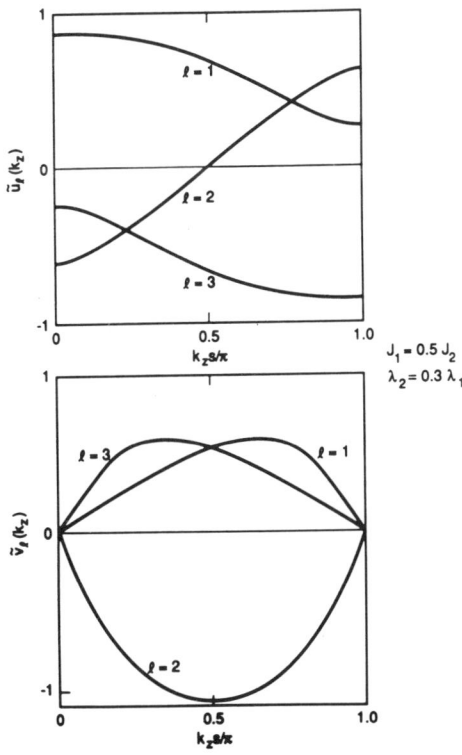

Fig. 3. Normalized gap functions $|u_{31n}(k_z)\rangle$ corresponding to the largest eigenvalues for the singlet (top) and triplet (bottom) states on each band.

613

diagonalization of the interaction is then performed after writing the self-consistency (gap function) equations, leading to four and two real eigenvalues λ_{3q}^S for even and odd states, respectively. We rank order the λ_{3q}^S for each S value, with λ_{31}^S being the largest. In addition, we find $\lambda_{31}^+ \geq \lambda_{31}^-$, as for N = 2.

The orthonormal eigenfunctions which diagonalize the gap function equations have the form

$$|u_{3q}(k_z)\rangle = \sum_{n=1}^{3} |u_{3qn}(k_z)\rangle, \qquad (9a)$$

$$|u_{3qn}^-(k_z)\rangle = \frac{c_{qn}\bar{J}^2 \sin k_z s}{(\epsilon_n^2(k_z) - \bar{J}^2)}, \qquad (9b)$$

$$|u_{3qn}^+(k_z)\rangle = \frac{\bar{J}(d_{qn}\epsilon_n(k_z) + \bar{J}e_{qn}\cos k_z s)}{(\epsilon_n^2(k_z) - \bar{J}^2)}, \qquad (9c)$$

where the c_{qn}, d_{qn}, and e_{qn} are appropriately chosen. Plots of the eigenfunctions on each band corresponding to the largest (q = 1) even and odd eigenvalues are given in Fig. (3).

For N = 1, 2, Eq. (6) for $\tilde{V}_N$ applies for all types of interlayer pairing, as the eigenfunctions and eigenvalues are independent of the band index. It is shown elsewhere[8] for N = 2 that intraband pairing dominates completely over interband pairing below the maximum T_c value, so that interband pairing can be neglected. For N = 3, Eq. (8) for $\tilde{V}_3$ applies only to intraband pairing. However, interband pairing can be neglected below the maximum T_c value, so Eq. (8) contains all of the important interlayer pairing interactions.

ORDER PARAMETERS AND GAP FUNCTIONS

For intralayer intraband pairing alone, the interaction λ_0 leads to a constant singlet OP Δ_0, the T dependence of which is obtained from the self-consistency equation. For N = 1, 2, 3 Δ_0 is defined by

$$\Delta_0 = T \sum_{|\omega| \leq \omega_{||}} \int \frac{d^3k'}{(2\pi)^3} \lambda_0 \text{Tr}[\mu_0 \sigma_2 F_{\sigma\sigma'}^{\mu\mu'}(k',\omega)], \qquad (10)$$

where $\int d^3k \equiv \int d^2k \int_{-\pi/s}^{\pi/s} dk_z$, $\omega_{||}$ is the BCS cutoff for intralayer pairing, and F is the usual Fourier transform of the pairing function, which is a 2 × 2 matrix in spin space and an N × N matrix in the band space (indexed by μ). The quantities σ_j are Pauli spin matrices, and the μ_j are the orthonormal band matrices of rank N, plus the identity matrices for j = 0. For equal intralayer pairing on each layer within the unit cell (as above), Δ_0 is independent of k_z, has a singlet spin configuration (represented by σ_2), and arises from equal pairing in each of the bands, even for N = 3. If the intralayer pairing on the different layers in the unit cell is inequivalent, the dominant OP is still given by Eq. (10) for N = 2, but for N = 3 involves inequivalent intraband pairing in the various bands, and an effective intralayer pairing interaction $\lambda_0(k_z, k_z')$ analogous to Eq. (8), resulting in a k_z-dependent gap function $\Delta_0(k_z)$.

For interlayer pairing, $\tilde{V}_N$ forces the four-vector gap functions $\tilde{\Delta}_N$ appearing in the self-consistency equations to be k_z-dependent. Each $\tilde{\Delta}_N$ contains one singlet component, $\tilde{\Delta}_{N2}$, with an even k_z dependence, and three triplet components $\tilde{\Delta}_{Nj}$ for $j = 0, 1, 3$, which are odd in k_z. These functions are defined by

$$\tilde{\Delta}_{1j}(k_z) = T \sum_{|\omega| \leq \omega_\perp} \int \frac{d^3k'}{(2\pi)^3} \tilde{V}_1(k_z, k'_z) \text{Tr}[\sigma_j F_{\sigma\sigma'}(k', \omega)], \tag{11a}$$

$$\tilde{\Delta}_{2j}(k_z) = T \sum_{|\omega| \leq \omega_\perp} \int \frac{d^3k'}{(2\pi)^3} \tilde{V}_2(k_z, k'_z) \text{Tr}[\mu_3 \sigma_j F^{\mu\mu'}_{\sigma\sigma'}(k', \omega)], \tag{11b}$$

and

$$\tilde{\Delta}_{3mpj}(k_z) = T \sum_{|\omega| \leq \omega_\perp} \int \frac{d^3k'}{(2\pi)^3} \sum_{\substack{m,m'=1 \\ p,p'=1}}^{2} V^\pm_{3mp,m'p'}(k_z, k'_z) \text{Tr}[\mu_{p'} \sigma_j F^{\mu\mu'}_{\sigma\sigma'}(k', \omega)], \tag{11c}$$

where $\omega_\perp$ is the effective BCS cutoff for interlayer pairing, and the upper (lower) sign in Eq. (11c) corresponds to $j = 2$ ($j \neq 2$), respectively. In Eq. (11c), the matrices $\mu_{p'}$ are the two traceless diagonal representations of SU(3). For $N = 2$, $\tilde{V}_2$ is independent of the band index, and that the gap function $\tilde{\Delta}_2$ is proportional to the difference between the pair functions $F^{++}_{\sigma\sigma'}$ and $F^{--}_{\sigma\sigma'}$, brought about by the interlayer pairing interaction $\tilde{V}_2$.

We then expand the $\tilde{\Delta}_{Nj}(k_z)$ in the appropriate eigenfunction basis (corresponding to the largest eigenvalues), letting $\tilde{\Delta}_{Nj}(k_z) = \Delta_{tm}|u^-_{N1}(k_z)\rangle$, for the odd functions, where $m = +, 0, -$ for $j = 0, 1, 3$, respectively. For the even functions, $\tilde{\Delta}_{N2}(k_z) = \Delta_s|u^+_{N1}(k_z)\rangle$. Δ_s is a singlet OP, and the Δ_{tm} are the components of a vector triplet OP Δ_t.

The self-consistent gap function equations may be obtained to all orders in Δ_0, Δ_s, and Δ_t. To linear order in these OP's, we may write

$$\Delta_0 = \lambda_0 a_\parallel(T) \Delta_0, \tag{12a}$$

$$\Delta_s = \lambda^+_{N1} a_\perp(T) \Delta_s, \tag{12b}$$

and

$$\Delta_t = \lambda^-_{N1} a_\perp(T) \Delta_t, \tag{12c}$$

where $a_\parallel(T) = N(0) \ln(2\gamma\omega_\parallel/\pi T)$, $a_\perp(T) = N(0) \ln(2\gamma\omega_\perp/\pi T)$, $N(0) = m_0/(2\pi N s)$ is the single quasiparticle density of states per band, and $\gamma \approx 1.78$.

The bare T_c values are obtained when each OP becomes non-vanishing in the absence of the others. T_{c0}, T_{cs}, and T_{ct} are given by $1 = \lambda_0 a_\parallel(T_{c0})$, $1 = \lambda^+_{N1} a_\perp(T_{cs})$, and $1 = \lambda^-_{N1} a_\perp(T_{ct})$ respectively. For $N = 1$, $T_{cs} = T_{ct}$, because[11] $\lambda^+_{11} = \lambda^-_{11}$. For $N = 2, 3$, $T_{cs} \geq T_{ct}$, the equality occuring only for the special cases of the parameters J_1, J_2, λ_1, and λ_2 for which those cases reduce to the $N = 1$ case.

GINZBURG-LANDAU FREE ENERGY

The Ginzburg-Landau free energy F relative to the normal state is obtained by expanding the gap function equations to cubic order in Δ_0 and $\tilde{\Delta}_N$, integrating over k'_z, and functionally integrating the resulting equation for each OP with respect to its complex conjugate. For $N \leq 3$, we obtain $F = \frac{1}{2} N(0) f_{30}$, where

$$f_{30} = |\Delta_s|^2 \ln(T/T_{cs}) + |\Delta_t|^2 \ln(T/T_{ct}) + |\Delta_0|^2 \ln(T/T_{c0}) + b_s|\Delta_s|^4$$
$$+ b_t\left(2(|\Delta_t|^2)^2 - \Re(\Delta_t^{*2}\Delta_t^2)\right) + 2c\left(2|\Delta_s|^2|\Delta_t|^2 - \Re(\Delta_s^{*2}\Delta_t^2)\right)$$
$$+ b_0\left(|\Delta_0|^4/2 + 2|\Delta_0|^2(|\Delta_s|^2 + |\Delta_t|^2) + \Re[\Delta_0^{*2}(\Delta_s^2 - \Delta_t^2)]\right), \tag{13}$$

where $b_0 = 7\zeta(3)/[8(\pi T)^2]$, and $b_t = \frac{3}{4}b_0$. For $N = 1$, we also have $b_s = b_t = 3c$, and $T_{cs} = T_{ct}$. For $N = 2, 3$, the quantities b_s, c, and T_{cs}/T_{ct} are functions of J_1/J_2 and λ_1/λ_2.

We first consider the $N = 1$ case with $\Delta_0 = 0$. Equation (13) is phase-minimized when Δ_s and the Δ_{tm} are in phase, or π out of phase. The free energy is then only a function of $|\Delta_s|$ and $|\Delta_t|$, and is minimized when $|\Delta_s| = |\Delta_t| = |\Delta|/\sqrt{2}$. The resulting gap $2|\Delta|$ is isotropic for $N = 1$, since

$$|\tilde{\Delta}_1|^2/2 = |\Delta_s \cos k_z s|^2 + |\Delta_t \sin k_z s|^2, \tag{14}$$

which equals $|\Delta|^2/2$, independent of k_z. This prediction is consistent with tunneling measurements on the $N = 1$ compound $La_{2-x}Sr_xCuO_4$ by Bulaevskii et al.[4] Although intralayer impurity scattering will suppress T_{cs} and T_{ct} from their clean limit values, the result $T_{cs} = T_{ct}$ still holds in the self-consistent Born approximation[12], so that Δ remains isotropic.

Including Δ_0, we see that the phases of Δ_s, Δ_t, and Δ_0 are mutually incompatible. For $T_{c0} \neq T_{cs}$, whichever gap has the higher T_c value destroys the other gap completely. These results contradict those of previous workers[13], who neglected Δ_t.

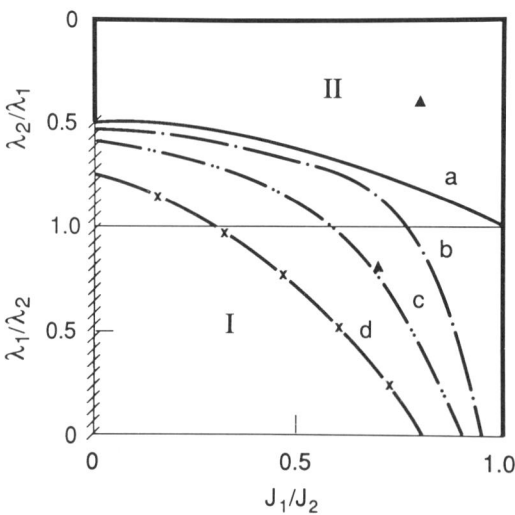

Fig. 4. Regions of different gap $2|\Delta|$ behavior for $N = 2$ are shown. Regions for $1 \geq J_2/J_1 \geq 0$ are obtained from $J_1 \leftrightarrow J_2$ and $\lambda_1 \leftrightarrow \lambda_2$. On the dark exterior line, $T_{cs} = T_{ct}$, and Δ is isotropic. On the cross-hatched line, Δ is isotropic and pure singlet. Curve (a) divides regions I and II, where the singlet gap function is nodeless or has a pair of nodes, respectively. Curves (b), (c), and (d) denote $|\Delta_t| = 0.01|\Delta_s|$ at $T = 0$ for $\lambda_{21}^+ N(0) = 0.2, 1.0$, and 2.0.

For $N \geq 2$, we always have $T_{cs} \geq T_{ct}$. There are three regions of T_c values relevant to the compatibility of Δ_s, Δ_t, and Δ_0. For $T_{c0} > T_{cs}$, $\Delta_s = \Delta_t = 0$, and the gap is just $2|\Delta_0|$. Conversely, $\Delta_0 = 0$ for $1 \geq T_{ct}/T_{cs} > (T_{c0}/T_{cs})^\kappa$, where $\kappa = (b_s - c)/(b_s - b_0/2)$. There can be a second order phase transition at $\tilde{T}_{ct} < T_{ct} < T_{cs}$ below which Δ_t becomes non-vanishing, where $\tilde{T}_{ct}/T_{cs} = (T_{ct}/T_{cs})^\mu$ and $\mu = b_s/(b_s - c)$. In the (rare, for $N = 2$) case $1 \geq (T_{c0}/T_{cs})^\kappa > T_{ct}/T_{cs}$, $\Delta_t = 0$, and Δ_0 can coexist with Δ_s, as discussed elsewhere[8].

DISCUSSION

In Figs. (4) and (5), the case $\Delta_0 = 0$ for $N = 2$ is illustrated in detail. The gap $2|\Delta(k_z)| = 2|\tilde{\Delta}_2(k_z)|$. In Fig. (4), we have plotted the regions of different gap behavior for $0 \leq J_1/J_2 \leq 1$ and $1 \geq \lambda_2/\lambda_1 \geq 0$ ($0 \leq \lambda_1/\lambda_2 \leq 1$) in the upper (lower) half. Results for $1 \geq J_2/J_1 \geq 0$ are obtained by setting $J_1 \leftrightarrow J_2$ and $\lambda_1 \leftrightarrow \lambda_2$. On the dark solid line on the right, top, and the top part of the left edges of Fig. (4), $T_{cs} = T_{ct}$, $|\Delta_s| = |\Delta_t|$, and the gap is isotropic. On the cross-hatched part of the left edge, the gap is also isotropic, but is pure singlet, as $|\Delta_t| = 0$ and the singlet gap function $|u_{21}^+(k_z)\rangle = 1$. Below the solid line (a) (region I), $|u_{21}^+(k_z)\rangle$ is nodeless. In the remaining region (region II), $|u_{21}^+(k_z)\rangle$ has a pair of nodes. Since the nodes in $|u_{21}^\pm(k_z)\rangle$ occur for different k_z values, when $|\Delta_s| \geq |\Delta_t| > 0$, the gap is nodeless. Curves (b), (c), and (d) in Fig. (4) are obtained when $|\Delta_t| = 0.01|\Delta_s|$ at $T = 0$ for the effective interaction strengths $\lambda_{21}^+ N(0) = 0.2, 1.0$, and 2.0, respectively. Below these curves, Δ_t can be neglected. We see from Fig. (4) that region II lies above these curves. Hence, even for weak coupling, the gap will *always* be nodeless at low T. We note that the gap anisotropy is greatest in the central region of Fig. (4).

In Fig. (5), we have plotted $|\Delta(k_z,T)|/|\Delta(0,0)|$ for $N = 2$ as a function of $k_z s/\pi$ for $T/T_{cs} = 0, 0.6$, and 0.9. In Fig. (5a), $\lambda_1/\lambda_2 = 0.9$ and $J_1/J_2 = 0.7$, which is in region I, and corresponds to the parameter choice of the left figure in Fig. (2). In this

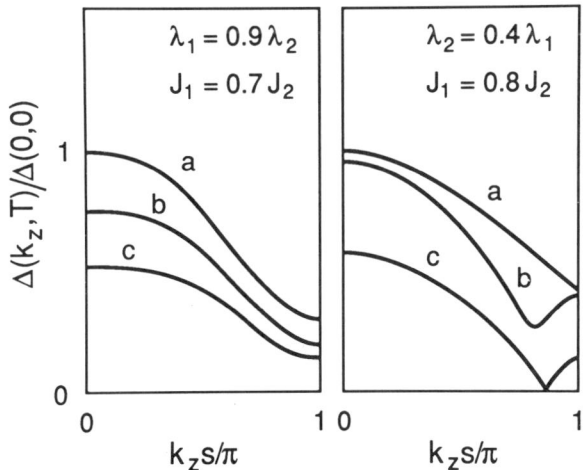

Fig. 5. Plots of $|\Delta(k_z,T)|/|\Delta(0,0)|$ versus $k_z s/\pi$ for T/T_{cs} values of $0, 0.6$, and 0.9 in curves (a), (b), and (c), respectively. Left plot: $\lambda_1/\lambda_2 = 0.9$, $J_1/J_2 = 0.7$. Right plot: $\lambda_2/\lambda_1 = 0.4$, $J_1/J_2 = 0.8$.

case, the gap is nodeless for $0 \leq T < T_{cs}$ and anisotropic. In Fig. (5b), $\lambda_2/\lambda_1 = 0.4$ and $J_1/J_2 = 0.8$, which is in region II, and corresponds to the parameter choice of the right figure in Fig. (2). There is a pair of nodes at $k_z s/\pi = \pm 0.86$ for $T/T_{cs} = 0.9$, at which $|\Delta_t| = 0$. For $T/T_{cs} \leq 0.6$, the nodes have been removed, as $|\Delta_t| \neq 0$.

In Fig. (6), we have plotted $|\Delta(k_z, T)|/|\Delta(0,0)|$ for one $N = 2$ example for which $\Delta_0 \neq 0$ below $\tilde{T}_{c0}$. In this figure, $\Delta_t = 0$. The gap anisotropy remains nearly T-independent for $T > \tilde{T}_{c0}$, but is reduced in magnitude below $\tilde{T}_{c0}$.

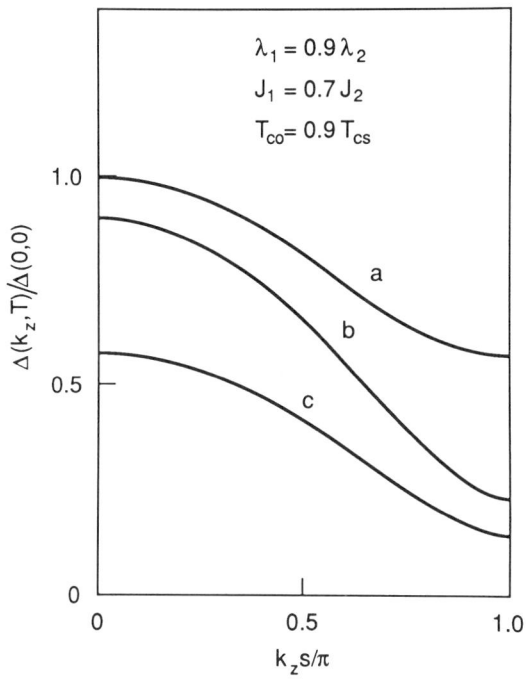

Fig. 6. Plotted is $|\Delta(k_z, T)|/|\Delta(0,0)|$ versus $k_z s/\pi$ for $\lambda 1/\lambda_2 = 0.9$, $J_1/J_2 = 0.7$, and $T_{c0}/T_{cs} = 0.9$. Curves (a), (b), and (c) are for T/T_{cs} values 0, 0.5, and 0.9, respectively. In these curves, $\Delta_t = 0$, and $\Delta_0 \neq 0$ in curve (a).

We note that for $N = 2$, the gap is the same on both bands. This leads to a quasiparticle density of states for region I such as that pictured in Fig. (7). For this case (with $\Delta_0 = 0$), the density of states has a shoulder below the maximum density of states, due to the gap anisotropy. For the parameters appropriate for region II, there would be a substantial amount of T-dependence to the density of states, as follows: just below T_{cs}, the node in the gap would contribute greatly to the density of states inside the gap region, varying linearly from the center, but smeared by the finite T. As T is lowered below $\tilde{T}_{ct}$, the gap would open up, resulting in a density of states that is similar to that pictured in Fig. (7).

For $N = 3$, the situation is rather different. As the gap functions on the different bands are inequivalent, with one gap function always having a pair of nodes at $k_z s = \pm \pi/2$, as pictured in Fig. (3), the density of states will generally appear as in Fig. (8). For $N = 3$, usually $T_{cs} \gg T_{ct}$, so the nodes on one of the bands are not so easy to remove, unless T_{c0} is close to T_{cs}. Hence, the $N = 3$ case with interlayer pairing is likely to exhibit a linear density of states inside the 'gap'.

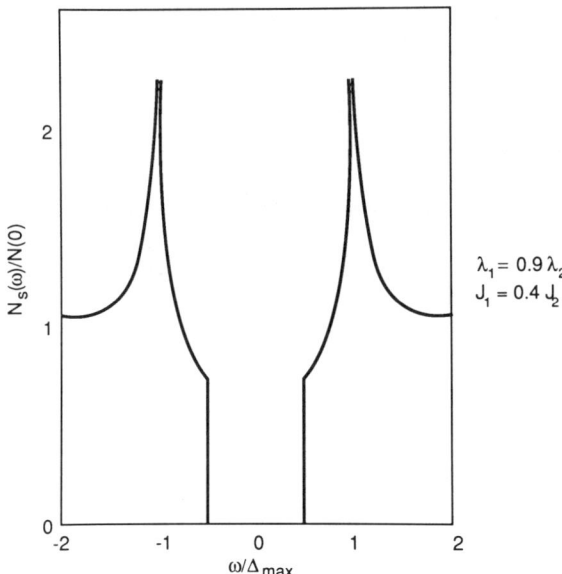

Fig. 7. Thermodynamic density of states at $T = 0$ for the $N = 2$ case whose gap functions are pictured in the left figure of Fig. (2) (region I), as a function of energy.

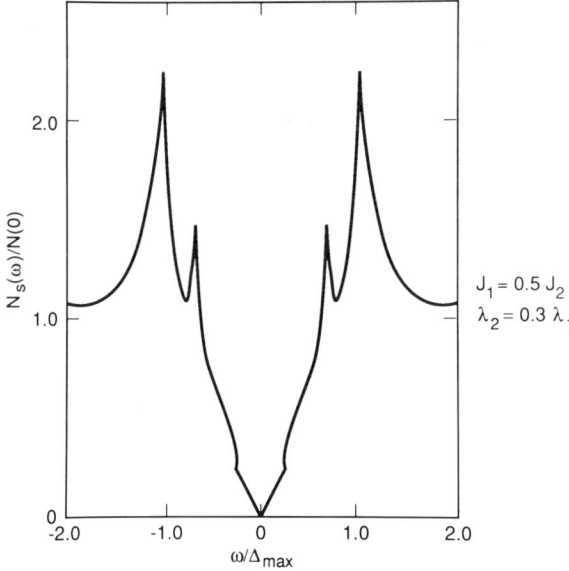

Fig. 8. Thermodynamic density of states at $T = 0$ for the $N = 3$ case whose gap functions are pictured in Fig. (3), as a function of energy.

Region II and parts of region I for $N = 2$ are consistent with penetration depth[1], infrared reflectance[2], Raman[3], SNS' tunneling[6], and NMR[15] experiments on Y123. It would be interesting to determine if the phase transition at $\tilde{T}_{ct}$ exists in that material. In $Bi_2Sr_2CaCu_2O_8$, it is likely that $J_2 \ll J_1$, but the conducting, electron-like Bi-O layers form a band at the Fermi energy[9], so that system is likely an example of $N = 3$. If so, tunneling experiments should give evidence of a line node, with a density of states similar to that pictured in Fig. (8).

We urge further experiments on Y123 and other materials to confirm or contradict the evidence for c-axis versus ab plane gap anisotropy. In addition, we urge experiments to explicitly demonstrate the directional and polarization dependence of the pairing interaction. Neutron scattering experiments could be a useful tool for investigating the T-dependence near to T_c of the phonon modes, and we suggest looking at the high energy optical phonon modes at the c-axis zone boundary.

ACKNOWLEDGMENTS

This work was supported by the U. S. Department of Energy, Division of Basic Energy Sciences, under contracts no. W-31-109-ENG-38 and no. DE-AC005-84OR21400 with Martin Marietta Energy Systems.

REFERENCES

* Solid State Division, Oak Ridge National Laboratory, Oak Ridge, TN 37831-6032
1. S. Mitra et al., Phys. Rev. B**40**, 2674 (1989).
2. R. T. Collins et al., Phys. Rev. Lett. **63**, 422 (1989); Z. Schlesinger et al., Phys. Rev. Lett. **65**, 801 (1990).
3. K. McCarty et al., Phys. Rev. B**42**, 9973 (1990).
4. M. Gurvitch et al., Phys. Rev. Lett. **63**, 1008 (1989); J. S. Tsai et al., Physica C**157**, 537 (1989); G. Briceno and A. Zettl, Sol. State Commun. **70**, 1055 (1989); L. N. Bulaevskii et al., Supercond. Sci. Technol.**1**, 205 (1988).
5. U. Welp et al., Physica C **161**, 1 (1989).
6. L. H. Greene et al., Physica C **162-164**, 1069 (1989) and unpublished; H. Akoh et al., Jpn. J. Appl. Phys. **27**, L519 (1988); M. R. Beasley et al., unpublished.
7. A. Arko et al., Phys. Rev. B**40**, 2269 (1989); C. G. Olson et al., Science **245**, 731 (1989); C. G. Olson et al., Sol. State Commun. (to be published).
8. R. A. Klemm and S. H. Liu, Physica C**175** (to be published); Phys. Rev. B (to be published).
9. Z. -X. Shen et al., Phys. Rev. B**38**, 7152 (1988).
10. R. A. Klemm, Phys. Rev. B**41**, 2073 (1990).
11. R. A. Klemm and K. Scharnberg, Phys. Rev. B**24**, 6361 (1981).
12. Y. Suwa, Y. Tanaka, and M. Tsukada, Phys. Rev. B **39**, 9113 (1989).
13. T. Schneider, H. De Raedt, and M. Frick, Z. Phys. B **76**, 3 (1989).
14. K. Maki, in Superconductivity, ed. by R. D. Parks (Marcel Dekker, Inc., New York, 1969), p. 1035.
15. P. C. Hammel et al., Phys. Rev. Lett. **63**, 1992 (1989).

MUON SPIN RELAXATION STUDIES OF THE LAYERED COPPER OXIDES

G. Aeppli and D. R. Harshman

AT&T Bell Laboratories
Murray Hill, New Jersey 07974

ABSTRACT

We give a brief review of the major contributions of the muon spin relaxation (μSR) technique to the science of layered copper oxides. These include (i) establishment of the conventional nature of the gap function, unlike that found for heavy fermion systems such as UPt_3 (ii) absolute values for the magnetic penetration depths (iii) identification of the inhomogeneous nature of the superconductivity in ceramic $La_{2-x}Sr_xCuO_4$ with x away from "optimal" value (0.15), and (iv) demonstration of local moment ordering at low temperatures near the metal-insulator transition in $La_{2-x}Sr_xCuO_4$. We also sketch the results of complementary neutron scattering and transport investigations on a sample near the metal-insulator transition. These experiments provide good evidence for a strong coupling between charge and spin degrees of freedom, as well as the marginal Fermi Liquid hypothesis in the normal state.

INTRODUCTION

Muon spin relaxation (μSR) was applied to the cuprous oxide superconductors almost immediately after their discovery. This technique[1] measures a time-dependent asymmetry function $A(t)$ which is directly proportional to the autocorrelation function, $<\vec{S}_\mu(t)\cdot\vec{S}_\mu(0)>$, for the spins of muons implemented at time t = 0 with known $S_\mu(0)$. Thus, μSR is a local probe more akin to nuclear magnetic resonance than to neutron scattering, which measures the space and time Fourier transform of the correlations of the electronic moments belonging to the solid under investigation. The primary contribution of μSR to the study of cuprous oxides has been to measure the temperature-dependent magnetic penetration depth $\lambda(T)$, and thereby to provide the first demonstration of the conventional (s-wave) nature of the pairing in the oxide superconductors.[2] The technique has also provided values of $\lambda(T=0)$ with greater absolute accuracy than more traditional methods.[3] A second and little known application of μSR to the copper oxide problem has been its use to examine phase purity with greater sensitivity than more common techniques, notably x-ray powder diffraction.[6] Finally, μSR has been a helpful survey tool[7] for establishing microscopic magnetic correlations,[7,8] especially near the metal-insulator transition. Here, the technique's advantage has been to allow relatively rapid data acquisition for polycrystalline samples. However, because the muon site is generally unspecified and the technique is a local probe, μSR has not been as prominent as NMR and neutron scattering.

The purpose of the present report is to review the three major applications described above, of μSR to the cuprous oxides. In particular, the remaining three sections of this paper deal with magnetic penetration depths, phase purity, and the metal-insulator transition.

MAGNETIC PENETRATION DEPTHS

A static field applied perpendicular to the initial muon spin direction $\vec{S}_\mu(0)$ results in Larmor procession of the muon spin implanted in the solid. The procession frequency is simply $\nu = \gamma_\mu H$, where $\gamma_\mu = 135.5$ MHz/T is the gyromagnetic ratio of the muon. For general distributions $\rho(H)$ of static fields, the asymmetry function A(t) is simply proportional to the Fourier transform of $\rho(H)$. The local fields H can either be intrinsic to the solid, as they are for ordered magnets and spin glasses, or induced by an external field, as for the vortex lattice of a type-II superconductor. In the latter case, the width of $\rho(H)$, as found from A(t), is related[9] to a length Λ_{eff} determined by the flux lattice constant d, the magnetic penetration depth λ, and pair coherence length ξ:

$$<|\Delta H|^2> = 3.706 \times 10^{-3} \Phi_0^2 \Lambda_{\text{eff}}^{-4} \qquad (1)$$

In (1), $\Phi_0 = 2.068 \times 610^{-7}$ G $-$ cm^2 is the magnetic flux quantum. As long as the applied H_{ext} satisfies the condition that

$$\xi << d << \lambda, \qquad (2)$$

Λ_{eff} is simply λ. We note parenthetically that while (2) is usually obeyed in investigations of the high-T_c materials, it is not always satisfied in systems with smaller upper critical fields H_{c2}.[10]

A primary motivation for studies of $\lambda(T)$ is to determine the nature of the superconducting energy gap, and also to obtain an absolute measure of the electromagnetic response of the condensate. As is well known, for good bulk superconductors, $\lambda(T=0)$ is simply c/ω_p where ω_p is the normal state plasma frequency. Figure 1 shows the original data[2] which demonstrated that $\lambda(T)$ for YBa$_2$Cu$_3$O$_7$ is of the conventional s-wave type. The important point is that $\lambda(T)$ depends only weakly on T between T = 0 and T = 0.7 T$_c$ (note that the horizontal axis in Fig. 1 corresponds to $(T/T_c)^2$), in contrast to what is found for unconventional superconductors such as UBe$_{13}$[11] (solid lines in Fig. 1) and UPt$_3$[12] (see Fig. 2). In the latter case, $\lambda_\parallel$, measured in the basal plane, actually rises *linearly* with T from its T = 0 value; the anisotropic nature of this superconductor is apparent from the difference between the temperature dependences of λ measured parallel and perpendicular to the c-axis of the hexagonal close-packed compound.

There are three CuO compounds of which single phase samples (see below) are available to us and where the interpretation of the μSR results is not complicated by vortex motion. Table I lists these materials with their respective transition temperatures and magnetic penetration depths. We consider the base of reliable data too small to determine whether there is a systematic variation between T$_c$ and λ.

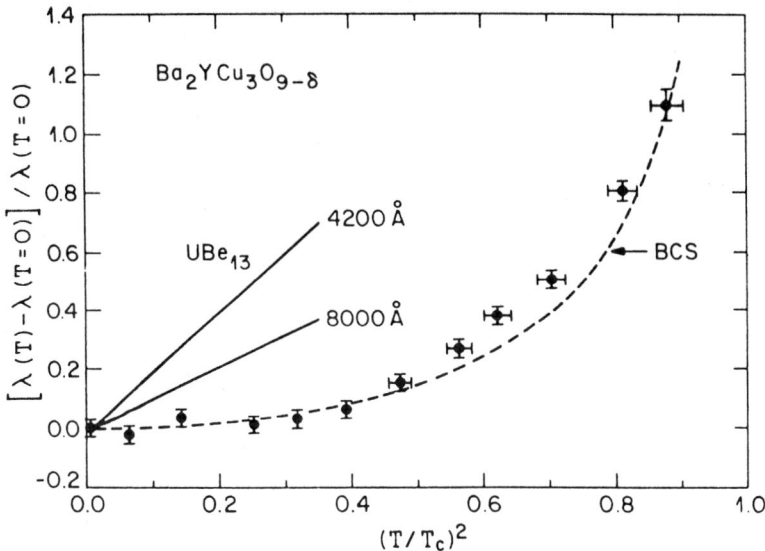

Fig. 1. Relative incremental penetration depth plotted against $(T/T_c)^2$ for a ceramic sample of $YBa_2Cu_3O_7$. Dashed line represents two-fluid model, while solid lines represent UBe_{13} dataref(11) assuming $\lambda(T=0) = 4200$ and 8000Å. From ref. 2 (1987).

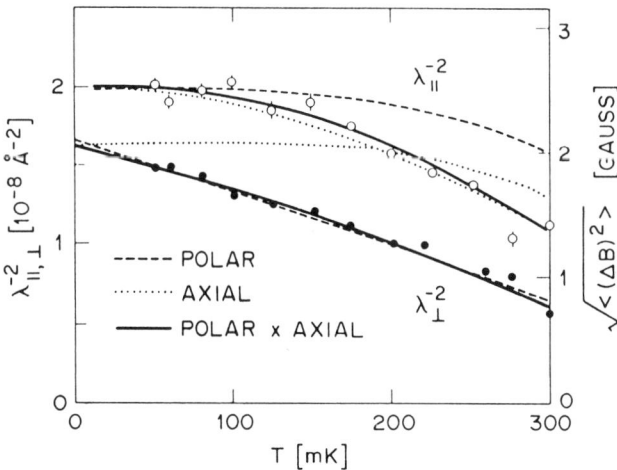

Fig. 2. Temperature dependence of λ^{-2} measured parallel and perpendicular to c-axis of UPt_3. Lines are fits assuming various nodal structures in the superconducting gap and $T_c = 474$ mK. From ref. 12 (1990).

Table I

Basal plane magnetic penetration depth λ_{ab} and superconducting transition temperature T_c for three single phase cuprous oxides.

Material	T_c (K)	λ_{ab} (Å)	Ref.
$YBa_2Cu_3O_7$	82–90	1400	[2,4]
$YBa_2Cu_3O_{6.7}$	65	2550	[4]
$La_{1.85}Sr_{0.15}CuO_4$	37	2400	[3]

PHASE INHOMOGENEITY IN $La_{2-x}Sr_xCuO_4$

Since the original investigations, multiphase coexistence has been known to be a serious problem in this series of compounds, with magnetic measurements[13,14] showing high Meissner fractions and narrow transition widths only for samples with compositions close to the "optimal" value, $x = x_c = 0.15$. Figure 3 shows some of these data, which make it clear that superconducting samples with large Meissner fractions exist only for $x \approx x_c$. A strong indication that inhomogeneties rather than flux pinning effects cause the smaller Meissner fractions for x away from x_c are tails in the temperature-dependent diamagnetism which extend to the maximum $T_c = 38K$ (see Fig. 3(b)). Transverse field μSR measurements[6] also show that the reduced Meissner fractions are indeed due to reduced volume fractions of the superconducting phase. In such experiments, type II superconductors yield an asymmetry function A(t) which is oscillatory, with a frequency shifted (to the negative) with respect to its normal state value and a decaying envelope due to the inhomogeneous internal field associated with the flux lattice. Because the magnetic penetration depth and flux lattice constant are generally much larger than the spacing between different possible muon stopping sites, a single-phase type II superconductor is characterized by a single decaying precession signal. By the same token, the presence of an additional slowly decaying, unshifted component is an unambiguous indication of a non-superconducting fraction in the sample. Figure 4 shows low-temperature transverse-field ($H_{ext} \approx 0.3$ T) data, for $x = 0.075$, 0.15, and 0.25, along with fits to Larmor signals with Gaussian envelopes (top frames) and residuals representing the difference between the data and the fits (bottom frames); the data are presented in the reference frame rotating at the fitted Larmor frequency. For the material ($x = 0.15$) with the highest T_c (and full Meissner signal), A(t) has only a single component, as expected for a single-phase superconductor, while for samples with compositions well below or above the optimum, the fits to a single Larmor signal are very poor, and the residuals contain several zero crossings. The presence of these other signals clearly demonstrates that $La_{2-x}Sr_xCuO_{4-\delta}$ samples fabricated by standard methods with x far from 0.15 are electronically inhomogeneous. Thus, although the materials form a continuous chemical solid solution at least to $x = 0.25$, the correct composition for bulk superconductivity occurs only for x near 0.15. For x far from 0.15, an electronically two-phase mixture exists due to short-range chemical inhomogeneties within the "single-phase" solid solution.

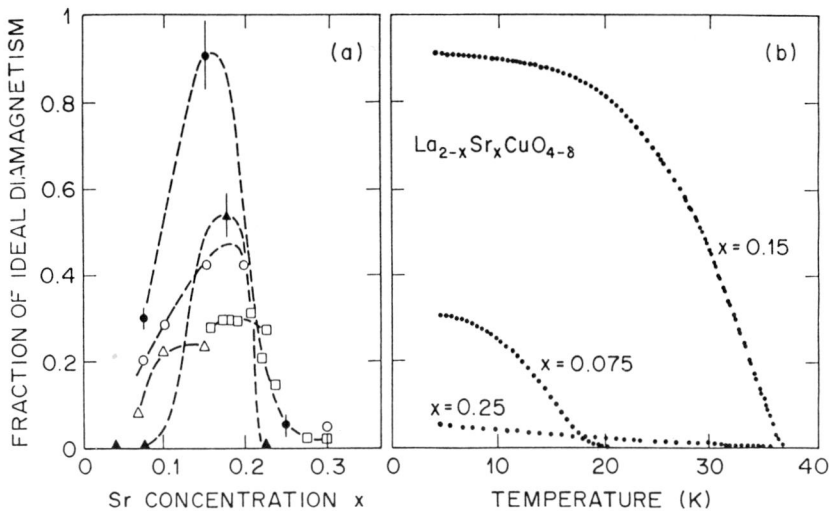

Fig. 3. The (a) diamagnetic fraction for various $La_{2-x}Sr_xCuO_{4-\delta}$ samples as a function of x. The filled circles correspond to our recently prepared high-density samples for which μ^+ SR data are shown in Fig. 2. The other symbols correspond to samples made by various groups at other timesref(13,14). Frame (b) shows $\chi_{Meissner}(T)$ for our three high-density samples (field-cooled in 25 Oe). From ref. 6 (1989) and B. Batlogg, private communication.

In summary, the magnetization and μSR measurements demonstrated that homogeneous bulk superconductivity exists only for a narrow range of compositional in $La_{2-x}Sr_xCuO_4$. This implies that when considering the systematics[14] of how superconducting properties evolve with x, one needs to be aware of the multiphase nature of $La_{2-x}Sr_xCuO_4$. While easily found in the μSR experiments, the inhomogeneties cannot be identified in routine x-ray powder diffraction studies.

MAGNETIC CORRELATIONS NEAR THE METAL-INSULATOR TRANSITION

We have used μSR to examine a series of $La_{2-x}Sr_xCuO_4$ ceramics with $x \leq 0.05$.[8] These non-superconducting samples show an oscillatory A(t) in zero applied fields at low temperature. Figure 5 shows the corresponding temperature-dependent procession frequencies ν for three such samples, as well as a single crystal of La_2CuO_4 from PbO flux. These frequencies measure the internal fields generated by the frozen electronic (Cu^{2+}) moments at the muon sites. An interesting feature of the data is that even with modest doping, the onset temperature (T_f) for the muon precession decreases precipitously, while the size of the moment, measured by ν, depends only weakly on x. Thus, at low T, doping does not introduce more quantum fluctuations for samples which ultimately insulate.

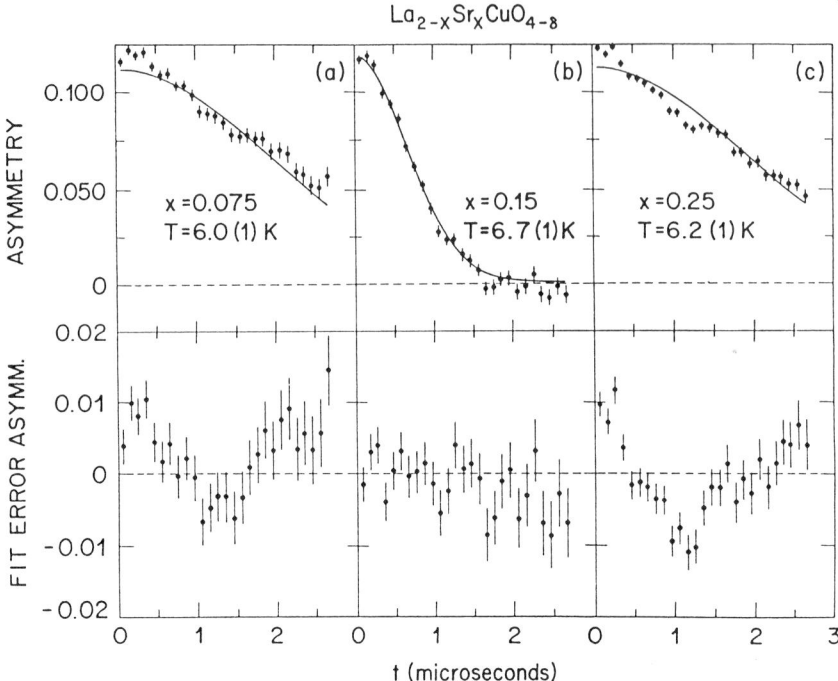

Fig. 4. The muon asymmetry (top frames) and associated residual spectrum (bottom frames) for $La_{2-x}Sr_xCuO_{4-\delta}$ with x equal to (a) 0.075, (b) 0.15, and (c) 0.25. Note that the data (plotted in arbitrary units) show a full initial amplitude, with no background contributions. From ref. 6 (1989).

Because μSR is a local probe, non-zero ν is an indication only of magnetic freezing on the time scale of the experiment, and does not imply long-range (in space) correlations between the frozen moments. Indeed, neutron scattering measurements[15,16] on $La_{2-x}(Ba,Sr)_xCuO_4$ with x = 0.05 demonstrate the absence of true long range antiferromagnetic order as seen for pure La_2CuO_4.[17] Instead, small antiferromagnetic microdomains with a typical extent of five Cu-Cu spacings characterize the frozen state. The freezing of the microdomains can be monitored using the triple-axis neutron scattering technique.[18] The relevant quantity is the nominally elastic magnetic scattering I, which corresponds to the longest-lived correlations and actually represents a measure of the Edwards-Anderson order parameter. The inset in Fig. 6(a) shows the T-dependent I for a $La_{1.95}Ba_{0.05}CuO_4$ single crystal. Magnetic freezing on the time scale (6×10^{-10} sec) of this experiment occurs at $\sim 8K$. Not surprisingly, μSR experiments, with their much longer time scale ($\sim 10^{-6}$ sec), reveal a lower freezing temperature ($\sim 4K$) for comparable levels of doping.[7,8] At the same time, these carriers "freeze" out at the considerably higher temperature ($\sim 50K$) defined by the resistance minimum (see Fig. 6(a)) and linear behavior typical of metallic layered cuprates. The observation of the resistance minimum occuring well above the spin freezing temperature is also found for other samples.[7] Thus, localization of the carriers is a precondition for spin freezing.

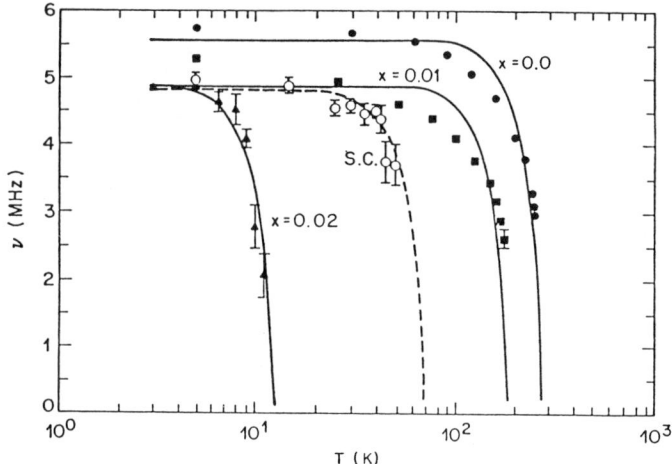

Fig. 5. Muon Larmor precession frequency ν versus temperature for sintered $La_{2-x}Sr_xCuO_4$ and single crystal (s.c.) La_2CuO_4.

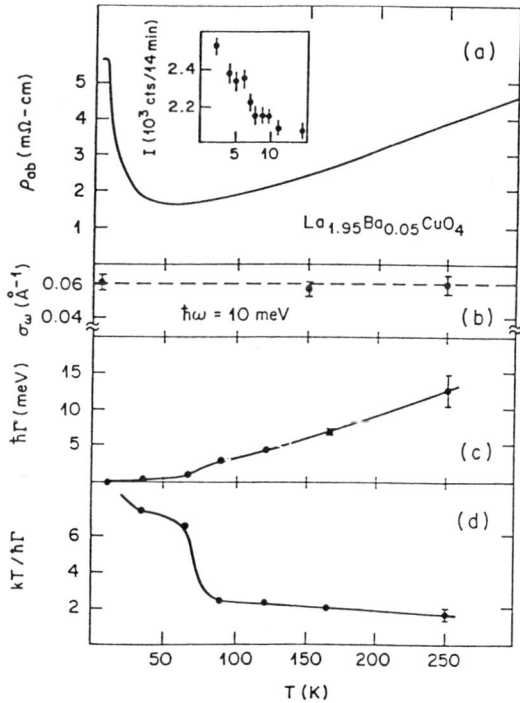

Fig. 6. Temperature dependence of (a) basal plane resistivity (inset: elastic signal proportional to the spin glass order parameter, (b) the Gaussian width parameter σ_ω (in reciprocal space) measures the characteristic inverse length scale for fluctuations with 10 meV energy, (c) magnetic relaxation rate $\hbar\Gamma$, and (d) dimensionless ratio $k_B T/\hbar\Gamma$, proportional to the nuclear spin-lattice relaxation rate. From ref. 15 (1991).

Further evidence for the coupling between spin and charge degrees of freedom is revealed by comparison of frames (c) and (d) to (a) in Fig. 6. The quantities plotted are the characteristic rate Γ and the dimensionless relaxation time, $k_B T/\hbar\Gamma$, for the magnetic fluctuations measured by inelastic neutron scattering. It should be obvious that the magnetic fluctuations slow down considerably near the resistance minimum, while for $T \gtrsim 80K$, $k_B T/\hbar\Gamma$ is an essentially temperature-independent constant of order unity. This result is in remarkable accord with the marginal Fermi liquid hypothesis,[19] where charge and spin response functions are generally constant for energies $\hbar\omega > \hbar\Gamma \sim kT$ and linear in ω for $\omega < \Gamma$, for the metallic state of the cuprates.

Acknowledgements

We are grateful to the many collaborators whose participation in the research described above was indispensable, and the institutions (TRIUMF, Risø, and ILL) which hosted the μSR and neutron and experiments.

References

1. An introduction is given by A. Schenck, "Muon Spin Rotation Spectroscopy Principles and Applications in Solid State Physics" (Adam Hilger, Bristol, 1985).
2. D. R. Harshman et al., Phys. Rev. B 36:2386 (1987).
3. G. Aeppli et al., Phys. Rev. B 35:7129 (1987).
4. D. R. Harshman et al., Phys. Rev. B 39:851 (1989).
5. W. J. Kossler et al., Phys. Rev. B 35:7133 (1987); F. N. Gygax et al., Europhys. Lett. 4:473 (1987); R. Wäppling et al., Phys. Lett. A 122:209 (1987); D. W. Cooke et al., Phys. Rev. B 37:9401 (1988).
6. D. R. Harshman et al., Phys. Rev. Lett. 63:1187 (1989).
7. J. Budnick et al., Europhys. Lett. 5:651 (1988).
8. D. R. Harshman et al., Phys. Rev. B 38:852 (1988); G. Aeppli et al., Physica C 153-155:1111 (1988).
9. E. H. Brandt, Phys. Rev. B 37:2349 (1988); W. Barford and J. M. F. Gunn, Physica 156C:515 (1988).
10. For example, neglecting that condition (2) was not met in their experiments, Y. J. Uemura et al., Phys. Rev. Lett. 66:2665 (1991) erroneously conclude that λ is unmeasurably large in the heavy fermion superconductors.
11. D. Einzel et al., Phys. Rev. Lett. 56:2513 (1986).
12. C. Broholm et al., Phys. Rev. Lett. 65:2062 (1990).
13. R. B. van Dover et al., Phys. Rev. B 35:5337 (1987); R. M. Fleming et al., Phys. Rev. B 35:7191 (1987); J. Orenstein et al., Phys. Rev. B 36:8892 (1987).
14. See, e.g. J. B. Torrance et al., Phys. Rev. Lett. 61:1127 (1988).
15. S. M. Hayden et al., Phys. Rev. Lett. 66:821 (1991).
16. T. R. Thurston et al., Phys. Rev. B 40:4585 (1989).
17. D. Vaknin et al., Phys. Rev. Lett. 58:2802 (1987).
18. For descriptions of how this technique can be applied to spin glasses, see G. Aeppli et al., Phys. Rev. Lett. 54:843 (1985), H. Mook et al. in "Dynamics of Magnetic Fluctuations in High T_c Materials" (Crete, 1989), edited by S. Reiter et al. (Plenum, N.Y., to be published); B. J. Sternlieb et al., Phys. Rev. B 41:8866 (1990).
19. C. M. Varma et al., Phys. Rev. Lett. 63:1996 (1989).

NEUTRON SCATTERING STUDIES OF SPIN CORRELATIONS IN METALLIC $YBa_2Cu_3O_{6+x}$

J. M. Tranquada

Physics Department
Brookhaven National Laboratory
Upton, NY 11973

Introduction

Electron-electron correlations appear to play an important role in determining many properties of the copper oxide superconductors. One significant consequence of the electronic Coulomb interactions are spin correlations. In the insulating phases of the layered cuprates, where the antibonding band due to Cu $3d_{x^2-y^2}$–O $2p_\sigma$ hybridization in the CuO_2 planes is half filled, long-range antiferromagnetic order is observed. As the layers are doped with holes, the Néel order is rapidly destroyed, but dynamical antiferromagnetic correlations survive.[1] In this paper, I will review some inelastic neutron scattering studies of the spin fluctuations in metallic $YBa_2Cu_3O_{6+x}$.

Much of the experimental work that I will discuss is the outgrowth of a collaboration between the neutron scattering group at Brookhaven and Professor Sato and his student Dr. Shamoto, both now at Nagoya University in Japan. Because the High Flux Beam Reactor at Brookhaven has been without neutrons for the past two years, it has been necessary to perform most of our measurements on metallic $YBa_2Cu_3O_{6+x}$ crystals at other institutions. As a result, we have had the pleasure of collaborating with groups at Risø National Laboratory in Denmark[2]; AECL Research in Chalk River, Ontario, Canada[3,4]; Laboratoire Léon Brillouin in Saclay, France[5]; and the National Institute of Standards and Technology in Gaithersburg, Maryland.[6] Of course, quite similar studies have been carried out by Rossat-Mignod's group, working primarily at the Institut Laue-Langevin in Grenoble, France.[7–9]

The paper is organized as follows. In the first section I briefly discuss some background information concerning the phase diagram and spin waves in the insulating phase. Experimental results on metallic $YBa_2Cu_3O_{6+x}$ samples are presented in the second section. The interpretation of these results and their relationship to nuclear magnetic resonance (NMR) studies and to theory are discussed in the final section.

Background

The characteristics of the crystals that we have studied are indicated in the schematic phase diagram for $YBa_2Cu_3O_{6+x}$ shown in Fig. 1. Each of these crystals has a scattering volume of roughly 1 cm^3. With a typical mosaic spread of $\sim 2°$, each sample is best described as a well-oriented collection of many small crystals having a typical dimension of 0.1 mm. This polycrystalline nature is beneficial for the purpose of controlling and homogenizing the oxygen content. The widths of superconducting transitions measured by ac susceptibility (e.g., see Ref. 2) are in the range of 5–10°, and are comparable to those of sintered samples. Measurements of the Meissner fraction are difficult because of the large sample size; nevertheless, field-cooled dc magnetization measurements on a piece broken from crystal #29 clearly indicated bulk superconductivity.[4] The oxygen contents of the crystals are typically determined by comparing the measured c lattice parameter with data in the literature for sintered samples. The uncertainty in x is roughly 0.05–0.10. It is important to note that no sign of magnetic Bragg peaks has been observed in any of the superconducting crystals. This result implies that there is no significant volume fraction of tetragonal, insulating phase present which might contribute to the inelastic magnetic scattering that is of interest here.

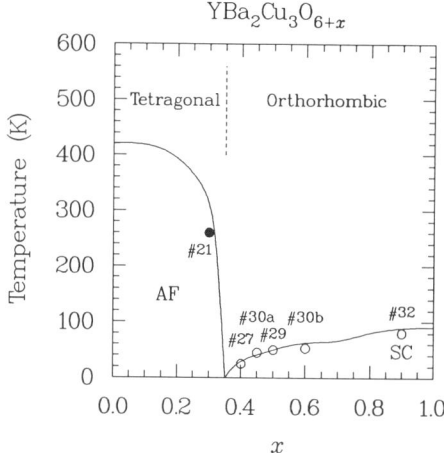

Fig. 1. Schematic phase diagram of $YBa_2Cu_3O_{6+x}$ showing the relative positions of the crystals that have been studied.

All of our studies of inelastic magnetic scattering in metallic $YBa_2Cu_3O_{6+x}$ crystals have been performed in the (h, h, l) zone of reciprocal space illustrated in Fig. 2(b). In this zone, the 2D antiferromagnetic Bragg point, which I will denote by Q_{AF}, extends along the rod $(\frac{1}{2}, \frac{1}{2}, l)$. If the magnetic correlations were purely 2D, then the scattered intensity would be independent of l (except for the Q dependence of the magnetic form factor). However, the coupling $J_{\perp 1}$ between nearest-neighbor CuO_2 planes [planes A and C in Fig. 2(a)] causes the 2D-like spin waves in the insulating phase to be split into acoustic and optical branches. The scattered intensity from acoustic modes is modulated by the factor $\sin^2(\pi z l)$, where zc is the bilayer separation and l is measured in reciprocal lattice units, $2\pi/c$. (Similarly, we specify h in units $2\pi/a = 1.63$ Å^{-1}.) A measurement of the modulation in an antiferromagnetic crystal above T_N is shown in Fig. 3(b). The energy gap for the optical modes depends on the product of $J_{\perp 1}$ and $J_\parallel$, the in-plane exchange energy. Since $J_\parallel \approx 100$ meV, even a small value of $J_{\perp 1}$ results in a large optical gap. $J_{\perp 1}$ is not known because the optical modes have not yet been observed, but its value is $\gtrsim 2$ meV.[10,7] In the absence of long-range order, one would expect the optical gap to decrease as the in-plane correlation length gets smaller and the effective interplanar coupling is reduced. However, the bilayer modulation is still observed in the $x = 0.5$, $T_c = 50$ K sample, as shown in Fig. 3(a). As a consequence, measurements must be peformed at values of l corresponding to maxima in the modulation ($l = 1.8, 5.2, \ldots$). This restriction makes it difficult to probe the $(h, k, 0)$ zone in search of possible incommensurate scattering.

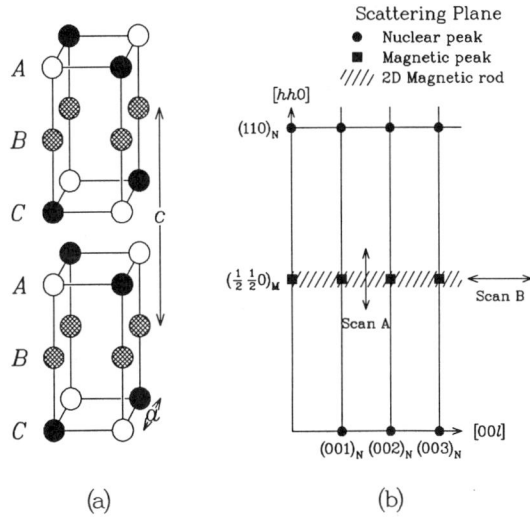

Fig. 2. (a) Antiferromagnetic structure of $YBa_2Cu_3O_6$. Open and solid circles denote Cu^{2+} sites with antiparallel spins. (b) $[hhl]$ zone of the reciprocal lattice for $YBa_2Cu_3O_{6+x}$.

Experimental Results

Doping the CuO_2 planes with holes destroys long-range magnetic order. We can obtain a measure of the dynamical spin-spin correlation length by measuring inelastic magnetic scattering for a fixed energy transfer ΔE and scanning the momentum transfer perpendicular to the 2D rod $\mathbf{Q}_{AF}$, as indicated by scan A in Fig. 2(b). Several such scans, measured in crystals with varying oxygen content, are shown in Fig. 4. In the superconducting crystals, we always observe a single peak with a width greater than resolution. For low enough excitation energies, we expect the width of the peak to be proportional to κ, where $\kappa = 1/\xi$. From various fitting analyses of the data, it appears that $\xi/a \sim 2$ for $x = 0.5$.

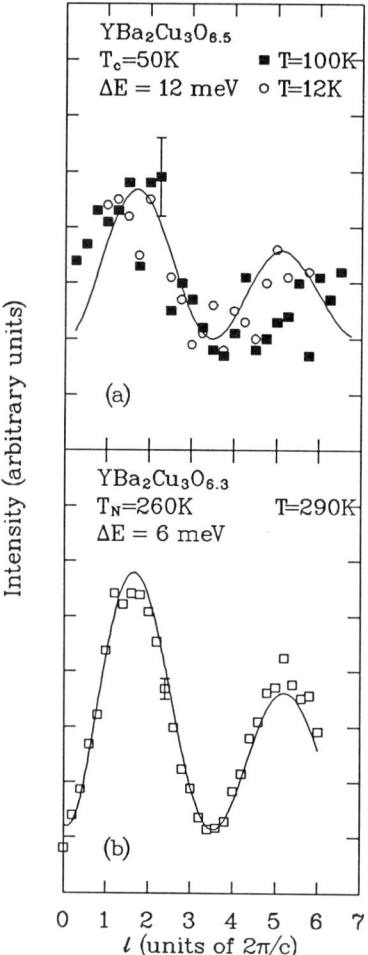

Fig. 3. (a) Scan along the 2D rod $(\frac{1}{2},\frac{1}{2},l)$ at energy transfer $\Delta E = 12$ meV for the $x = 0.5$, $T_c = 50$ K sample. Little change in the modulation is observed between 12 and 100 K. The solid line is calculated with $z = 0.286$, taking into account the magnetic form factor of Cu, but neglecting resolution effects. Note that the curve in the original figure of Ref. 4 was calculated with the incorrect value of z. (b) A similar scan at $\Delta E = 6$ meV for an insulating sample above its Néel temperature. From Ref. 4.

Another useful measurement is to sit at $\mathbf{Q}_{AF}$ and measure the magnetic scattering as a function of ΔE. Figure 5 shows such measurements for several different crystals at low temperature.[5] The solid lines are fits to the data.[3,4] The fitted cross section is essentially $\omega\Gamma/(\omega^2 + \Gamma^2)$, corresponding to damped spin fluctuations. The energy width Γ increases from 3 meV for the $x = 0.45$ sample to 30 meV for $x = 0.5$. Taking $\Gamma \sim \kappa^n$, with $n = 1$ or 2, the increase in Γ is consistent with the large change in κ indicated in Fig. 4.

The large change in Γ and κ for such a small change in x and T_c is rather surprising. However, it can be rationalized if we take into account the results of a μSR

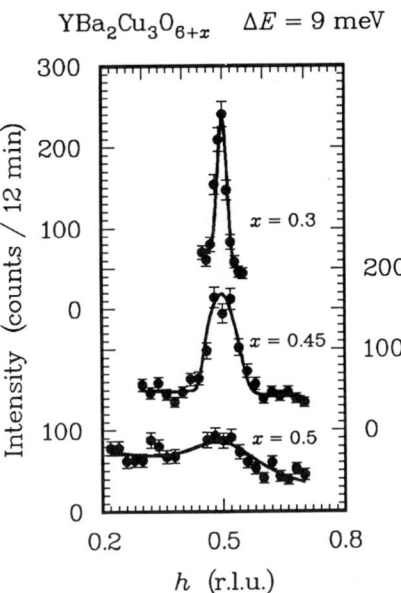

Fig. 4. Constant-E scans measured at Chalk River (Ref. 3) along $(h, h, 1.8)$ at $\Delta E = 9$ meV for crystals with $x = 0.3$ ($T_N = 260$ K), $x = 0.45$ ($T_c = 45$ K), and $x = 0.5$ ($T_c = 50$ K). The $x = 0.3$ scan was measured at room temperature; the other two were measured at 12 K.

study on a series of sintered samples by Kiefl et al.[11] In that study, a static internal magnetic field was observed at very low temperature ($T < 0.1$ K) for samples with $x \lesssim 0.5$ and $T_c \lesssim 50$ K. The coexistence of superconductivity and static magnetic order (presumably short range) in such samples suggests that they are electronically inhomogeneous. Such an inhomogeneity may be an intrinsic property related to ordering of the oxygen vacancies in the basal plane and related problems of charge transfer to the planes.[12] In any case, the strong damping observed in our $x = 0.5$ sample clearly reveals the dramatic effects of hole doping and suggests homogeneity.

Fig. 5 also indicates that the low-energy magnetic cross section at $\mathbf{Q}_{AF}$ continues to decrease on increasing x from 0.5 to 0.9. The dashed line indicates the expected cross section if Γ increases by a factor of 2, a not unlikely possibility. Thus, the search for the magnetic scattering for $x \sim 1$ is quite a challenge, one that is compounded by the possible existence of large energy gap below T_c. The measurements so far are not inconsistent with the changes in spin susceptibility near $\mathbf{Q}_{AF}$ extracted from NMR measurements by Imai.[13]

Our initial measurements of the magnetic scattering were a bit too noisy for a proper analyis of the temperature dependence of Γ and κ in the $x = 0.5$ crystal.[4] The recent measurements at Saclay are a significant improvement.[5] Constant-E scans for the $x = 0.5$ sample were fit with a Gaussian plus a background linear in q. At $\hbar \omega = 8$ meV, where the resolution contributed just 25% of the Gaussian line width, the width was found to be independent of temperature between 10 and 300 K. If the q dependence of $S(\mathbf{q}, \omega)$ is of the form $(q^2 + \kappa^2)^{-1}$, where $\mathbf{q} = \mathbf{Q} - \mathbf{Q}_{AF}$, then the

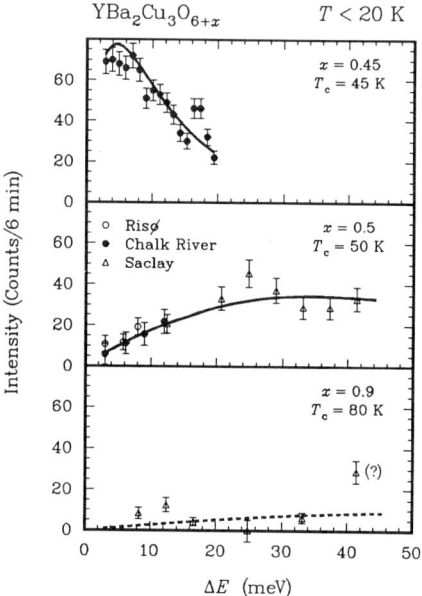

Fig. 5. Constant-**Q** scans measured at $\mathbf{Q} = (\frac{1}{2}, \frac{1}{2}, 1.8)$ and $(\frac{1}{2}, \frac{1}{2}, 5.0)$ at $T \leq 20$ K for $x = 0.45$ and 0.5 are fitted with the solid lines. Open triangles for $x = 0.5$ and 0.9 represent constant-E scans taken at Saclay. Note that these data have been scaled to match the data taken at Chalk River. All data have been corrected for a constant background. From Ref. 5

observations directly indicate that the inverse correlation length κ is independent of temperature. Such a conclusion is less obvious if one assumes the dynamical susceptibility to have the form proposed by Millis, Monien, and Pines,[14] and also by Moriya, Takahashi, and Ueda[15]:

$$\chi''(\mathbf{q}, \omega) = \chi_0 \xi^2 \frac{\omega \Gamma}{\omega^2 + \Gamma^2(1 + q^2 \xi^2)^2}, \quad (1)$$

where

$$S(\mathbf{q}, \omega) = \frac{1}{1 - e^{-\hbar\omega/k_B T}} \chi''(\mathbf{q}, \omega). \quad (2)$$

For fixed ω, the half width at half maximum (HWHM) of S as a function of q is then

$$\text{HWHM} = \kappa \left\{ [2 + (\omega/\Gamma)^2]^{1/2} - 1 \right\}^{1/2}. \quad (3)$$

The width, in general, is a function of ω. However, we have found that at low temperature the ω-dependence of the data requires $\hbar\Gamma \sim 30$ meV, and, as I will discuss later, if Γ changes with temperature it only gets larger. Thus, for $\hbar\omega =$

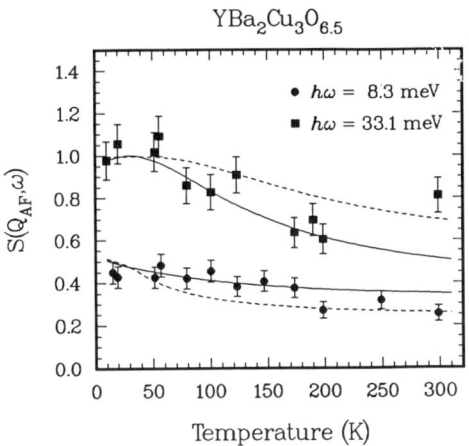

Fig. 6. Temperature dependence of the intensity measured at $\mathbf{Q} = \mathbf{Q}_{AF}$ for the $x = 0.5$, $T_c = 50$ K sample. Because the measurements at the two energies were performed under different conditions, their relative intensities have been scaled using the low temperature data in Fig. 4. The lines are model calculations which are discussed in the text. Data from Ref. 5.

8.3 meV, $(\omega/\Gamma)^2 \sim 0.08$, and the HWHM is essentially independent of ω. Therefore, we can still conclude that κ has relatively little temperature dependence below 300 K. Rossat-Mignod et al.[7] have reported a similar result. The direct observation that κ is independent of temperature is in conflict with the *assumption* of a temperature dependent ξ made in some analyses of NMR relaxation rate measurements.[14,16,17]

The fitted Gaussian amplitude, which is essentially equal to $S(\mathbf{Q}_{AF}, \omega)$, is plotted as a function of temperature in Fig. 6. I will discuss the temperature dependence of the data in the next section.

Besides the strong damping of spin fluctuations by the added holes, Rossat-Mignod et al.[7,9] have reported a gap in the magnetic scattering at $\mathbf{Q}_{AF}$ which appears at low temperature. The size of the gap grows rapidly from about 4 meV at $x = 0.51$ to 16 meV at $x = 0.69$. No such gap was apparent in measurements on our $x = 0.5$ sample. However, in recent measurements at NIST, we observed a gap in a crystal with $x = 0.6$.[6] Typical constant-E scans are shown in Fig. 7(a). The integrated peak intensities as a function of ΔE are plotted in Fig. 7(b), where they are compared with the data of Rossat-Mignod et al.[9] Because of the time-consuming nature of these measurements, we do not have a good characterization of the temperature dependence of the gap; however, we do know that it is not present at 150 K. The implications of these data will be discussed in the next section.

Discussion

Normal-State Temperature Dependence

The Saclay measurements[5] on the $x = 0.5$ sample indicate that the correla-

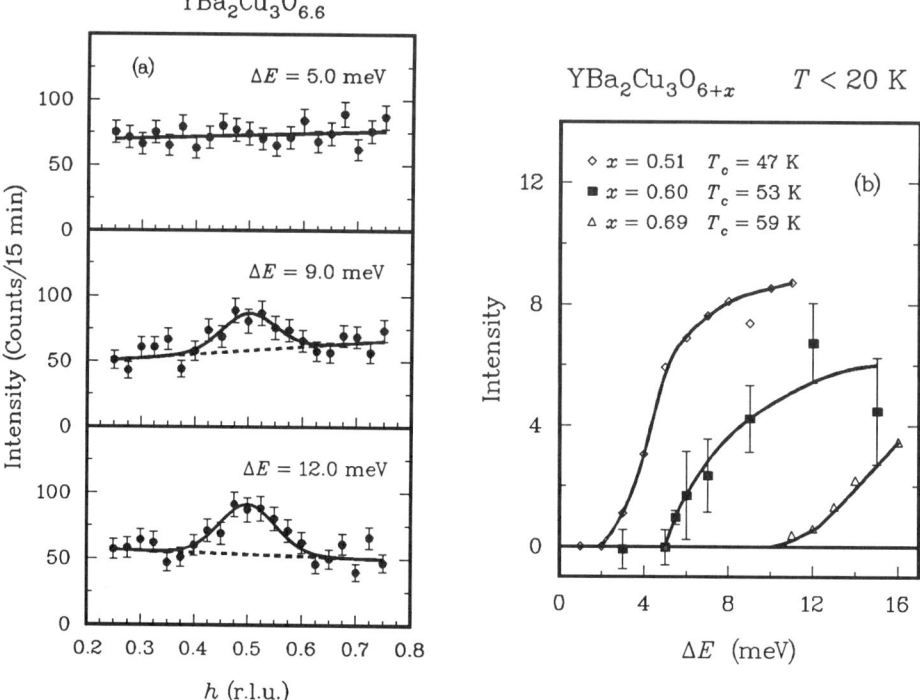

Fig. 7. (a) Constant-E scans measured along $(h, h, 1.8)$ at $\Delta E = 5$, 9, and 12 meV and $T = 10$ K, using a fixed incident neutron energy of 30.5 meV. Solid lines represent best fits to a Gaussian lineshape plus linear background. (b) A composite figure showing the frequency dependence of the magnetic scattering cross section measured for $x = 0.6$ relative to those for $x = 0.51$ and $x = 0.69$ reported by Rossat-Mignod et al. (see Ref. 9). The data for $x = 0.6$ are integrated intensities of constant-E scans such as those shown in (a). The other data sets are background-corrected constant-$\mathbf{Q}$ scans. The solid lines are guides to the eye. From Ref. 6

tion length is essentially independent of temperature below 300 K. On the other hand, $S(\mathbf{Q}_{\mathrm{AF}}, \omega)$ decreases slightly with increasing temperature, which tells us that $\chi''(\mathbf{Q}_{\mathrm{AF}}, \omega)$ must be decreasing significantly. So where is the temperature dependence coming from? Before attributing it entirely to behavior in the individual CuO$_2$ planes, we must first consider the interplanar coupling within the bilayers. With the very short correlation length in the $x = 0.5$ crystal, it seems likely that the gap is on the order of thermal energies. Thus, some temperature dependence at a given point along the 2D rod is expected due to the disappearance of the modulation, with a maximum possible decrease of 50% due to this effect. Rossat-Mignod[9] has mentioned some weakening in the interplanar coupling above 150 K in an $x = 0.51$ sample, but did not give details. This problem is currently being investigated.

For the sake of argument, let us assume that all of the temperature dependence is due to intraplanar effects. One possible model is to assume that Γ has temperature dependence independent of κ. We can model the data of Fig. 6 using

$$\chi''(\mathbf{Q}_{\mathrm{AF}}, \omega) = \chi_0 \frac{\omega \Gamma}{\omega^2 + \Gamma^2}, \tag{4}$$

with
$$\Gamma = \Gamma_0 \sqrt{1 + (T/T_0)^2}. \tag{5}$$

Eq. (4) is consistent with both overdamped spin waves and Eq. (1). Using the value $\hbar\Gamma_0 = 30$ meV, determined from the ω dependence at low temperature,[3,4] and setting $T_0 = 53$ K one obtains the solid lines shown in Fig. 6.

This parametrization gives a quite decent fit to the data; however, it is not completely satisfying from a theoretical point of view. It is not at all clear why Γ should have a strong temperature dependence when κ is temperature independent. Also, at high temperature we find that $\hbar\Gamma \approx \alpha k_B T$ with $\alpha = 6.6$, which is quite different from the result $\alpha \approx 0.5$ found in two recent studies of lightly doped La_2CuO_4.[18,19] [The latter experimental works evaluated the ω and T dependence of χ'' after integrating out the **q** dependence. They found that the q-integrated dynamical susceptibility is reasonably well described by the function $\tan^{-1}(\omega/\Gamma)$. It should be noted that one obtains the same form by integrating Eq. (1).]

Both the renormalization-group analysis[20] and the Schwinger boson mean field theory[21] for the square lattice Heisenberg antiferromagnet find that $\Gamma \approx \kappa c$, where c is the spin-wave velocity. In the antiferromagnetic Fermi liquid theories,[14,15] the relationship becomes $\Gamma = \Gamma_0(\kappa/q_0)^2$. In either case, the temperature dependence of Γ comes from κ. In a heavily doped CuO_2 plane one expects the correlation length to be controlled by the hole concentration. The kinetic energy of the holes is quite large compared to the temperature, and with a sufficiently short correlation length there is no problem with divergences due to thermal excitation of long-wavelength spin fluctuations. Thus, it is reasonable that the correlation length be temperature independent, but it is not clear from the theories mentioned what the source of the temperature dependence of Γ would be.

In the nested-Fermi-liquid theory of Virosztek and Ruvalds,[22] χ'' has a temperature dependence of the form $\tanh(\hbar\omega/\gamma k_B T)$, with γ in the range 2 to 4, which comes from Fermi occupation factors. There are problems with the theory, since, as emphasized by Levin's group,[23] the Fermi surface for $YBa_2Cu_3O_{6+x}$ found in LDA calculations and photoemission experiments would yield nesting at $\mathbf{Q} = (\pi, 0)$ instead of (π, π). Nevertheless, it suggests the following alternative parametrization for χ'':

$$\chi''(\mathbf{Q}_{\mathrm{AF}}, \omega) = \chi_0 \frac{\omega \Gamma_0}{\omega^2 + \Gamma_0^2} \tanh\left(\frac{\hbar\omega}{\gamma k_B T}\right). \tag{6}$$

This form is roughly consistent with the phenomenological model of Varma et al.[24] The dashed lines in Fig. 6 correspond to Eqs. (2) and (6) with $\gamma = 2.3$ and $\Gamma_0 = 30$ meV. The agreement with the data is qualitatively quite reasonable.

Comparison with NQR/NMR

Several groups have performed NQR and NMR studies of $YBa_2Cu_3O_{6+x}$ samples with $T_c \sim 60$ K.[26] In particular, Takigawa et al.[25] have carefully analyzed measurements of the Knight shifts and nuclear-spin-relaxation rates, $1/T_1$, for both ^{63}Cu and ^{17}O sites in the CuO_2 planes of a sample with $x = 0.63$. They found that various components of the Knight shift tensors for Cu and O share an identical temperature-dependent spin component. From these measurements they extracted the static planar spin susceptibility $\chi_s = \chi_0/\mu_B^2$ shown in Fig. 8(a).

From arguments based on the commonly used hyperfine form factor model (see, for example, Ref. 14) combined with the assumption of antiferromagnetically-correlated Cu spins, one expects[25] for the planar O sites that

$$1/{}^{17}T_1 T \sim \chi_0/\Gamma(0), \tag{7}$$

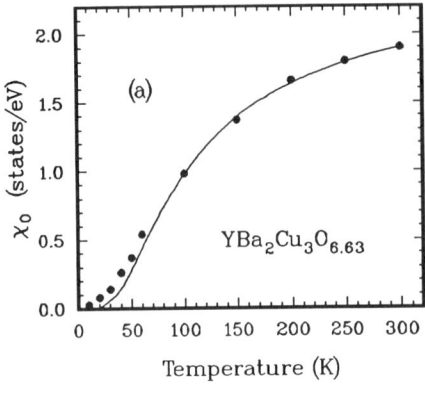
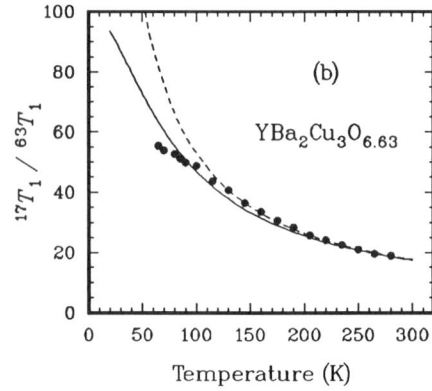

Fig. 8. (a) The static spin susceptibility per μ_B^2 for YBa$_2$Cu$_3$O$_{6.63}$ from Ref. 25. The solid line is a fit disussed in the text. (b) Ratio of nuclear spin relaxation rates for Cu and O in YBa$_2$Cu$_3$O$_{6.63}$, from Ref. 25. The solid and dashed lines are discussed in the text.

where $\Gamma(\mathbf{0})$ is a spin fluctuation frequency characteristic of $\mathbf{Q} = \mathbf{0}$. Takigawa et al.[25] showed that their measurements of $^{17}T_1$ have this form with $\Gamma(\mathbf{0})$ nearly temperature independent. On the other hand, one expects for the planar copper sites that

$$1/^{63}T_1T \sim \chi(\mathbf{Q}_{AF})/\Gamma(\mathbf{Q}_{AF}), \qquad (8)$$

where $\Gamma(\mathbf{Q}_{AF})$ is the damping factor for antiferromagnetic spin fluctuations. By making the somewhat bold assumption that $\chi(\mathbf{Q}_{AF}) = \chi_0$, they arrived at the result

$$^{17}T_1/^{63}T_1 \sim \Gamma(\mathbf{0})/\Gamma(\mathbf{Q}_{AF}). \qquad (9)$$

Since $\Gamma(\mathbf{0})$ has very little temperature dependence, the experimentally determined ratio, shown in Fig. 8(b), can be compared with the temperature dependence of $\Gamma(\mathbf{Q}_{AF})$ determined by neutron scattering in our $x = 0.5$ sample. The solid line in Fig. 8(b) is proportional to the reciprocal of Eq. (5), with $T_0 = 53$ K. Expanding Eq. (6) for $\omega \ll T, \Gamma_0$, one finds $\Gamma(\mathbf{Q}_{AF}) \sim T$, and this model is indicated by the dashed line. While the former model appears to be in better agreement with the relaxation rate ratio at low temperatures, the important point is that the temperature dependence of $\Gamma(\mathbf{Q}_{AF})$ determined at relatively large frequencies by neutron scattering is consistent with the $\omega \to 0$ result from NMR.

But what about the assumption $\chi(\mathbf{Q}_{AF}) = \chi_0$? First of all, the temperature dependence of χ_0 suggests that an energy gap is present. To represent the temperature effects of such a gap, I have considered the simple function

$$f(\omega) = 1 + \tanh[\hbar(\omega - \omega_g)/k_B T]. \qquad (10)$$

The solid line in Fig. 8(a) corresponds to $\chi_0 \sim f(0)$ with $\hbar\omega_g = 6$ meV. Note that in this parametrization the gap is present at all temperatures, and that there is no transition temperature such as T_c or T_K present. Also, the size of the gap is quite similar to the magnitude of the gaps seen by neutron scattering for spin fluctuations near $\mathbf{Q}_{AF}$. The temperature dependence of the gap in an $x = 0.69$ sample reported by Rossat-Mignod et al.[9] shows a similar gradual temperature evolution with no clear onset temperature; however, to model those results quantitatively requires a somewhat more complicated ω dependence than that give in Eq. (10). Thus, it appears that the assumption $\chi(\mathbf{Q}_{AF}) = \chi_0$ is quite reasonable, provided that one allows the factor $\chi(\mathbf{Q})$ entering $\chi''(\mathbf{Q},\omega)$ to have some frequency dependence similar to that in Eq. (10).

Bulut and Scalapino[27] have obtained a pseudo-gap in $\chi''(\mathbf{Q}_{AF},\omega)$ from their RPA analysis of a 2D single-band Hubbard model with a temperature dependence similar to that seen by Rossat-Mignod et al.[9] However, their pseudo-gap appears only at $\mathbf{Q} = \mathbf{Q}_{AF}$, and not for all $\mathbf{Q}$, which is not consistent with the NMR results. Mila[28] has taken a different approach, considering the problem of a local-moment system. He uses the Schwinger-boson mean-field theory (SBMFT) of Auerbach and Arovas[21] for the 2D Heisenberg antiferromagnet, assuming the effects of doping to correspond to the reduction of the effective spin per Cu site to a magnitude below the critical value required for long-range order at $T = 0$. He finds that the static susceptibility goes to zero at $T = 0$ when long range order is absent. Recently Mila, Poilblanc, and Bruder[29] have applied the SBMFT to a frustrated Heisenberg model. They obtain an energy gap in $S(\mathbf{Q},\omega)$ at $T = 0$, in qualitative agreement with experiment. However, the doping dependence of the gap is not explained.

The strong temperature dependence of χ_0 shown in Fig. 8(a) is not observed in $YBa_2Cu_3O_{6+x}$ with $x \sim 1$, suggesting that any spin-gap behavior in that compound is much smaller than in the $T_c \sim 60$ K material.[25] Thus, if the spin fluctuation gap seen in neutron scattering is indeed connected with the temperature dependence of χ_0, it might not have any direct connection to the superconductivity. Certainly the gaps that have been measured are much less than the values of $2\Delta/k_B T_c \sim 6$–8 commonly reported for cuprate superconductors. If the spin-gap behavior is reduced in the $x \sim 1$ material, will the superconducting gap be observable in $S(\mathbf{Q},\omega)$? The first challenge for neutron scattering is to find the magnetic scattering at any energy or temperature in a highly oxygenated crystal. So far the measurements have been limited by background.

Acknowledgments

Besides thanking my collaborators, I would like to acknowledge stimulating conversations with V. J. Emery, K. Levin, A. Auerbach, N. Bulut, M. Takigawa, and R. Walstedt. A major part of the work discussed in this paper was supported by the U.S.-Japan Cooperative Neutron Scattering Program. Research at Brookhaven National Laboratory is supported by the Division of Materials Sciences, U.S. Department of Energy, under Contract No. DE-AC02-76CH00016.

References

1. For a review, see: R. J. Birgeneau and G. Shirane, in *Physical Properties of High Temperature Superconductors I*, edited by D. M. Ginsberg (World Scientific, Singapore, 1989), pp. 151–211.
2. G. Shirane, J. Als-Nielsen, M. Nielsen, J. M. Tranquada, H. Chou, S. Shamoto, and M. Sato, Phys. Rev. B **41**, 6547 (1990).
3. J. M. Tranquada, W. J. L. Buyers, H. Chou, T. E. Mason, M. Sato, S. Shamoto, and G. Shirane, Phys. Rev. Lett. **64**, 800 (1990).
4. H. Chou, J. M. Tranquada, G. Shirane, T. E. Mason, W. J. L. Buyers, S. Shamoto, and M. Sato, Phys. Rev. B **43**, 5554 (1991).
5. P. Bourges, P. M. Gehring, B. Hennion, A. H. Moudden, J. M. Tranquada, G. Shirane, S. Shamoto, and M. Sato, Phys. Rev. B **43**, 8690 (1991).
6. P. M. Gehring, J. M. Tranquada, G. Shirane, J. R. D. Copley, R. W. Erwin, M. Sato, and S. Shamoto, Phys. Rev. B (to be published).

7. J. Rossat-Mignod, L. P. Regnault, J. M. Jurgens, P. Burlet, J. Y. Henry, G. Lapertot, and C. Vettier, in *Dynamics of Magnetic Fluctuations in High T_c Materials*, edited by G. Reiter, P. Horsch, and G. Psaltakis (Plenum, New York, 1991).
8. J. Rossat-Mignod, L. P. Regnault, M. J. Jurgens, C. Vettier, P. Burlet, J. Y. Henry, and G. Lapertot, Physica B **163**, 4 (1990).
9. J. Rossat-Mignod, L. P. Regnault, C. Vettier, P. Burlet, J. Y. Henry, and G. Lapertot, Physica B (to be published).
10. J. M. Tranquada, G. Shirane, B. Keimer, S. Shamoto, and M. Sato, Phys. Rev. B **40**, 4503 (1989).
11. R. Kiefl *et al.*, Phys. Rev. Lett. **63**, 2136 (1989).
12. H. F. Poulsen, N. H. Andersen, J. V. Andersen, H. Bohr, and O. G. Mouritsen, Nature **349**, 594 (1991).
13. T. Imai, J. Phys. Soc. Jpn. **59**, 2508 (1990).
14. A. J. Millis, H. Monien, and D. Pines, Phys. Rev. B **42**, 167 (1990).
15. T. Moriya, Y. Takahashi, and K. Ueda, J. Phys. Soc. Jpn. **59**, 2905 (1990).
16. H. Monien, D. Pines, and M. Takigawa, Phys. Rev. B **43**, 258 (1991).
17. B. S. Shastry, Phys. Rev. Lett. **63**, 1288 (1989).
18. S. M. Hayden, G. Aeppli, H. Mook, D. Rytz, M. F. Hundley, and Z. Fisk, Phys. Rev. Lett. **66**, 821 (1991).
19. B. Keimer, R. J. Birgeneau, A. Cassanho, Y. Endoh, R. W. Erwin, M. A. Kastner, and G. Shirane (preprint).
20. S. Chakravarty, in *High Temperature Superconductivity*, edited by K. S. Bedell, D. Coffey, D. E. Meltzer, D. Pines, and J. R. Schrieffer (Addison-Wesley, Redwood City, CA, 1990), p. 136.
21. A. Auerbach and D. P. Arovas, Phys. Rev. Lett. **61**, 617 (1988).
22. A. Virosztek and J. Ruvalds, Phys. Rev. B **42**, 4064 (1990).
23. J. P. Lu, Q. Si, J. H. Kim, and K. Levin (preprint).
24. C. M. Varma, P. B. Littlewood, S. Schmitt-Rink, E. Abrahams, and A. E. Ruckenstein, Phys. Rev. Lett. **63**, 1996 (1989).
25. M. Takigawa, A. P. Reyes, P. C. Hammel, J. D. Thompson, R. H. Heffner, Z. Fisk, and K. C. Ott, Phys. Rev. B **43**, 247 (1991).
26. See the review by R. E. Walstedt and W. W. Warren, Jr., Science **248**, 1082 (1990).
27. N. Bulut and D. J. Scalapino (preprint); N. Bulut, D. Hone, D. J. Scalapino, and N. E. Bickers, Phys. Rev. Lett. **64**, 2723 (1990).
28. F. Mila, Phys. Rev. B **42**, 2677 (1990).
29. F. Mila, D. Poilblanc, and C. Bruder, Phys. Rev. B **43**, 7891 (1991).

RECENT STUDIES OF CHEMICAL DOPING IN HIGH-T_c SUPERCONDUCTORS

Y.H. Kao

Department of Physics
and New York State Institute on Superconductivity
State University of New York at Buffalo
Buffalo, NY 14260

ABSTRACT

Various experimental techniques using synchrotron radiation, transport and magnetic measurements were employed to investigate the effects of chemical doping in different high-T_c superconductors. These results reveal features in the electronic structure which are closely related to the transition temperature, as well as a universal scaling law of flux-pinning force that greatly simplifies the usual complex description of flux pinning in the mixed state as a function of magnetic field, temperature, and chemical doping.

INTRODUCTION

A most prominent property of the oxide superconductors is that the coherence length ($\xi \sim 10$Å) is much smaller than that of conventional low-T_c superconductors. This implies that real space pairing of electrons plays an important role, and Cooper pairs can be bound by a non-retarded potential; the chemical potential is comparable to the binding energy, pairs could exist at temperatures above T_c, and dynamic charge fluctuations in real space are closely related to the superconducting behavior. It is plausible that a change in the local microstructure within ξ in the material, e.g. by chemical doping, could allow an effective means to control the superconducting properties. This approach will be important for device applications. A complete understanding of the high-T_c phenomena can only be achieved with a clear picture of the electronic structure and local order in real space on the scale of ξ. To this end, detailed information on the short-range structure and development of more effective methods for exploring the relationship between microstructures and superconducting properties would seem highly desirable.

The advent of high intensity polarized x-rays from synchrotron radiation provides unique opportunities to study microscopic structures in multi-component material systems. Several x-ray techniques are well suited for probing the local environment around selected atomic species and the local electronic structure on specific sites.

These methods include x-ray fluorescence, electron yields, x-ray absorption spectroscopy (XAS) and x-ray absorption fine structure (XAFS), they are especially useful for our investigation of the oxide superconductors and chemical doping.

Of special interest is the recent development of a detector for measuring x-ray fluorescence from low-Z elements. This has proved to be particularly useful for studying the local environment and electronic structure about oxygen atoms in the high-T_c superconductors. This study of oxygen is important in light of the fact that oxygen is the only common element found in all the high-T_c superconductors known to date, and the charge carriers (holes) are largely located on the oxygen sites.

Measurements of I-V curves, Hall effect and magnetization M as a function of magnetic field H and temperature T can reveal useful information on physical properties. The broadening in superconductive transition can be used to study the effects of flux pinning and creep. Variations of the critical current density J_c with H can be used to investigate the changes in flux pinning and irreversibility line with imperfections. The Hall coefficient R_H provides useful information on the carrier type and its effective concentration, however, the relationship between R_H and T_c is still not well understood. A detailed comparison of Hall effect with the **local** unoccupied electronic states obtained from the XAS measurements could provide more insight into the correlations between the mobile/local charge carriers and T_c.

Since the activation potential is in general a **nonlinear** function of field gradient and many different flux pinning processes might be present, M and J_c usually show complicated dependence on H, T and doping. We have found in various Y-Ba-Cu-O films, despite of this complexity, that the pinning force density J_cxH (as well as J_c) follows a simple universal scaling behavior in H up to the irreversibility line. This behavior is very useful for evaluating the effects of chemical doping on J_c, and also reveals the conditions required to optimize the pinning effects via doping.

In the following, some recent results in our studies of chemical doping in high-T_c superconductors will be briefly reviewed.

A NEW DETECTOR FOR STUDYING OXYGEN FLUORESCENCE

X-ray fluorescence affords one of the most reliable means for probing the local structure around a specific element in a complex material system. By measuring the fluorescence, XAS and XAFS spectra can be obtained while the integrity of the multi-element sample is preserved. This is due to the fact that the fluorescence technique is bulk sensitive, thus avoiding the problems associated with surface contamination. This consideration is most important for oxygen. The K absorption edge of oxygen is at 543 eV where most of the conventional x-ray detectors are ineffective. In the past, most x-ray absorption measurements in this spectral region were made by using the surface-sensitive electron-detection techniques which usually contain spurious signals arising from unwanted structures or contaminants on the surface. The advent of a parallel plate photon detector[1] designed for high efficiency soft x-ray measurements has now solved this longstanding technical problem. This detector has the capability of single photon counting and has already been successfully used in several experiments on oxygen fluorescence studies of high-T_c superconductors[1]. The probing depth of the oxygen fluorescence measurement is around 200 nm, thus it offers a truly bulk-sensitive technique for investigating the local structures around the oxygen atoms.

X-RAY ABSORPTION SPECTROSCOPY (XAS)

The XAS technique is particularly useful for exploring the electronic structure of specific atoms in a complex material system. The XAS spectrum generally refers to the absorption structure in the energy range around first 15 eV near the onset due to transitions from a localized core level to the unoccupied states above the Fermi energy. Since the core levels due to different atoms are well separated from each other, this technique is therefore element specific. The intensity of XAS structure provides direct information on the density of unoccupied states. We have applied this technique to the investigation of unoccupied oxygen 2p states in several high-T_c superconductors. Results of two recent experiments are summarized here.

A systematic study of the evolution of hole states on the oxygen sites has been pursued with various Sr-doping in $La_{2-x}Sr_xCuO_{4-y}$. This is a result of collaboration between our group at SUNY Buffalo and an AT&T research group[2]. The Sr content x was varied from 0 to 0.15. Two distinct peaks at incoming photon energies around 528.8 and 530.3 eV were observed (labelled A and B, respectively) below the edge. These pre-edge peaks are sensitive to Sr concentration and related to T_c. The peak A grows while peak B diminishes systematically as a function of x. Variations in the spectral weight thus provide direct evidence for changes in the local hole density of states on the oxygen sites. The measured spectra can be compared with calculations based on the Hubbard model and are consistent with a description of electronic structure by a doped charge-transfer insulator[3].

Another example for the useful application of the XAS technique is our recent study[4] of $Tl_2Ba_2Ca_2Cu_3O_{10-y}$. Similar to the two pre-edge peaks found in the previous case, we observed three peaks at 528.3, 529.4, and 530.6 eV in this material. These peaks are ascribed to core-level excitations of oxygen 1s electrons to unoccupied states around the Fermi level which have predominantly oxygen 2p character located in the CuO_2, BaO, and TlO planes, respectively. A typical spectrum obtained with a fully oxidized compound ($T_c = 114K$) is shown in Fig. 1. The solid line represents a quantitative fit to the three peaks. When the oxygen content was reduced and T_c decreased to 92K, the peak at 528.3 eV decreases considerably, while the middle peak remains practically unchanged; the peak at 530.6 eV shows only a moderate decrease. These results lend more support to the description of electronic structure by the charge-transfer insulator model and provide strong evidence for a direct connection between the unoccupied oxygen hole states and T_c.

The behavior of these peaks in Tl-Ba-Ca-Cu-O also implies difficulty in observing the electronic specific heat anomaly at T_c. Since only a small fraction of the electronic density of states changes with T_c, the total electronic specific heat is expected to carry only a weak signal of superconductive transition. This problem could be further complicated by the lattice contributions near the transition.

We have also studied the XAS spectra arising from Cu 2p levels in the same experiment[4], which reveal the different roles played by the Cu^{2+} and Cu^{3+} states.

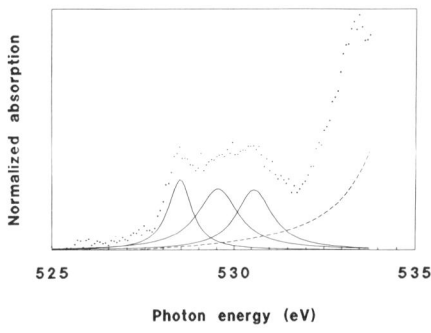
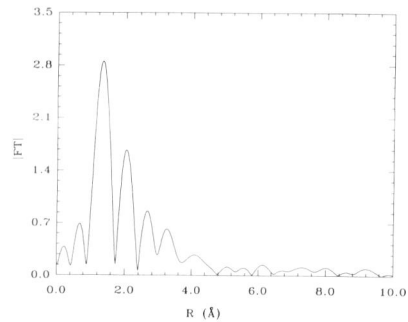

Fig. 1. Normalized x-ray absorption spectrum of $Tl_2Ba_2Ca_2Cu_3O_{10-y}$.

Fig. 2. Fourier transform of the oxygen K XAFS spectrum of PBCO

Moreover, in collaboration with Dr. N. Nücker et al., we have recently investigated the XAS spectra with a twin-free $YBa_2Cu_3O_{7-y}$ (YBCO) single crystal and found a large anisotropy in the XAS with different x-ray polarizations. This result provides a quantitative measure of the contributions by chains to the total hole density of states in YBCO.

X-RAY ABSORPTION FINE STRUCTURE (XAFS)

XAFS contains information on the interatomic distances, coordination numbers, and local disorder about a selected atomic species in multi-element systems. Detailed description of the XAFS techniques can be found in several review articles[5]. In the case of dilute chemical doping, since there is generally no long-range structural order of the impurity atoms, the conventional diffraction techniques are not useful for probing the location of impurities. The attractive features of element selectivity and sensitivity to short-range-order structures make XAFS a particularly useful tool for investigating chemically-doped high-T_c superconductors. This technique has been used by many researchers in previous studies of chemical doping[6-8].

Our new soft x-ray detector offers a unique advantage for studying XAFS around the oxygen K edge by measuring fluorescence. In contrast to the electron-yield detection method used in the past which is sensitive to surface conditions, the fluorescence detector now allows a reliable measurement of **bulk** oxygen K XAFS. An example is given here mainly to demonstrate the advantage this technique can offer to the study of local structure around the Y sites in YBCO. The Fourier transform peak arising from the O-Y pairs in YBCO is well separated from other peaks, this is particularly useful for studying the substitution of Y by other elements, such as Pr. The composite system of Y-Ba-Cu-O/Y-Pr-Ba-Cu-O is a material of great technical and fundamental interest.

A Fourier transform of the XAFS spectra of $PrBa_2Cu_3O_{7-y}$ (PBCO) near the oxygen K edge is shown in Fig. 2. The first strong peak is due to Cu backscatterers and the second peak is due to Pr. By comparison with the similar oxygen K XAFS spectra obtained with YBCO, the location of the second peak and the coordination number provide clear evidence that Pr has indeed substituted for Y. This result can be compared with that of $Y_{1-x}Pr_xBa_2Cu_3O_{7-y}$ (YPBCO) where Y has been partially replaced by Pr. Although the position of Pr atoms can be identified to be largely on the Y site, there are strong local disorder as well as lattice distortion associated with the Pr impurities. Our XAFS data near the Pr K edge of YPBCO show a structure very similar to that of Pr_2O_3, suggesting that the formal valence of Pr in YPBCO is close to +3. These XAFS results suggest that the effects of local disorder play an important role in lowering Tc. (This work is a collaboration between our group at SUNY Buffalo and C.C. Tsuei's group at IBM.) This observation can be compared with a recent EELS experiment by Fink et al.[9] which shows no change in the density of states of holes of oxygen 2p character when Y is systematically replaced by Pr.

A relevant question is whether the decrease in T_c with increasing Pr content is caused by the local disorder or by hole filling. Our XAFS results provide an evidence that the main reason for lowering T_c in YPBCO is caused by cation disorder within the coherence length when Pr is introduced into the YBCO system. The local disorder around the Pr atoms in YPBCO is actually expected in light of the valency difference between Y and Pr. Hybridization of Pr f-orbitals with the neighboring atoms may result in large distortions in the local structure. However, due to the strong bonding between O and Cu in the basal plane, it is probably more difficult to inject electrons into the CuO_2 planes (i.e. hole filling) than to change the effective valency of other atoms such as Ba, which lies right above the Pr(Y) site. The local disorder or structural distortion resulting from mixed valence of cations can suppress the order parameter associated with the mobile holes, thereby leading to a lower value of T_c.

CONCEPT OF VALENCE FRUSTRATION

The foregoing discussion on mixed valence in cations and suppression of T_c can be related to a general mechanism possibly important for superconductivity. It is interesting to note that cations in the high-T_c superconductors, e.g. Y, Bi, and Tl, can assume multivalent chemical states. When one of these elements forms a three-dimensional compound with Cu and O atoms, the degree of freedom associated with the multivalent states could allow different stoichiometry (e.g. variations in the oxygen content) in the material in order to maintain charge balance. Hybridization of the wave functions pertaining to different cation valency could therefore give rise to different charge distributions, bond length variations, and different electronic density of states. However, if the energy difference due to hybridization of different valence states is not sufficiently large compared to kT or is influenced by the presence of static/dynamic fluctuations, the choice of a specific valence state for the cations to take, which in turn determines the effective charge transfer among the atoms or the bonding configuration, cannot be precisely determined; thus the final state valency of the system could become "frustrated".

The valence frustration of cations could lead to three possible consequences: (i) The competition between different valence states could eventually settle to a metastable configuration of the system by choosing a local energy minimum. This can be achieved at the expense of allowing local distortions in the otherwise long-range ordered structural arrangement. For an inhomogeneous multi-element system such as the oxide superconductors, this will most likely give rise to local disorders around the consitutent atoms or in the bond length. The oxide superconductors, even in carefully prepared single crystals, all seem to show a large degree of structural disorder. (ii) The system might choose to create global potential minima so that the energy difference, say between two different valence states, can be shared by two adjacent but slightly different potential minima in each unit cell; i.e. an otherwise stable minimum for the ions may "split" to two wells separated by a local small barrier. This new configuration could have an important effect on electron coupling. If the energy difference associated with valence frustration is large enough, the system may be stabilized by forming even more than two potential wells around the cations, resulting in a superstructure or some type of charge density waves. Another alternative to relax the "frustrated configuration" is the appearance of puckering in the CuO_2 planes. It is possible that superstructures observed in Bi-Sr-Ca-Cu-O (BSCCO) and puckering of CuO_2 planes in YBCO are related to valence frustration of Bi and Y ions, respectively. (iii) The formation of symmetry-breaking new configurations could result in additional states in the band structure which was based on translational symmetry of the unit cells. If oxygen and copper atoms are involved in such a new configuration, it is likely that the density of states on the oxygen and copper sites at the Fermi level could change accordingly. These ideas of valence frustration were discussed before by the author at the 1988 Singapore Conference on Condensed Matter Physics[10].

These possibilities of valence frustration could have different effects on high-T_c superconductivity. Due to the short coherence length, the local cation disorder can seriously suppress the order parameter and lower T_c. The formation of double wells could lead to an enhancement in T_c, as discussed by Dr. Hardy at this conference[11]. It can be speculated that the formation of superstructures in BSCCO and puckering of basal planes in YBCO could be the origin of the double wells giving rise to the high values of T_c. An increased hole density of states on the oxygen sites could also favor higher value of T_c. An interplay between these different mechanisms can have an important effect on the final state of superconductivity. At this point it seems that Pr doping in YBCO results in mainly an increased amount of cation disorder which tends to suppress superconductivity.

SCALING BEHAVIOR OF FLUX PINNING IN THIN FILMS

Thin films of high-T_c superconductors usually exhibit J_c values several orders of magnitude higher than in bulk materials. Flux pinning, as well as the interrelated effects of flux creep and flow, is believed to be mainly responsible for the magnetic-field and temperature dependence of $J_c(H,T)$ in the mixed state.[12-14] Introduction of impurities into the superconducting material could affect the pinning mechanisms, and it might afford a means to control the behavior of J_c for applications in superconductive devices. Very little experimental data based on this approach are presently available. For practical purpose as well as for fundamental understanding of the magnetic and electrical properties, it is essential at first to characterize the effects of flux pinning as a function of H, T, the impurity concentration (x), and the type of dopants or defects (D). A useful physical parameter for characterizing these effects is the pinning force density $F_p = |J_c \times H|$.

We have found a universal scaling law which greatly simplifies the rather complex description of $F_p(H,T,x,D)$, as well as a low-field peak in F_p which has not been reported before. The observed universal behavior indicates the likelihood of similar pinning mechnisms at work in various thin films, and this result can be employed to predict flux pinning under other conditions.

Thin films of YBCO with chemical doping were made by a laser deposition technique described elsewhere[15,16]. Yttrium-stablized ZrO_2 (100) single crystals were used as substrates. Typical thickness of the films is 200nm. The structure of films was checked by XRD, SEM and EDX. X-ray rocking curve on (006) peak showed a half-width less than one degree, indicating a high-quality c-axis orientation. Superconducting transition was examined by a SQUID magnetometer (made by Quantum Design) and by a standard four probe measurement with a $1\mu V$ onset voltage as the criterion for R=0. T_c in zero field was 85-90K for undoped films, and $69 \pm 3K$ with 5% Fe-doping. The SQUID was used for measuring magnetization (M) hysteresis in a field upto 5 Tesla. Intragranular critical current density J_c was obtained from ΔM by applying Bean's critical state model.[17] The validity of Bean's model was also checked by comparison with transport critical current density as a function of temperature; the overall agreement was within a factor of 2. All our measurements were performed with the sample c-axis parallel to the applied field and zero-field-cooled to each measurement temperature.

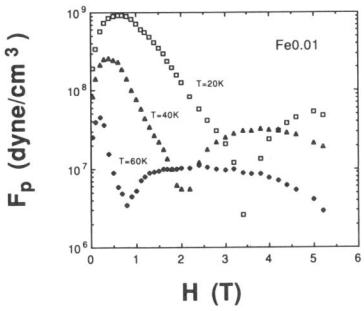
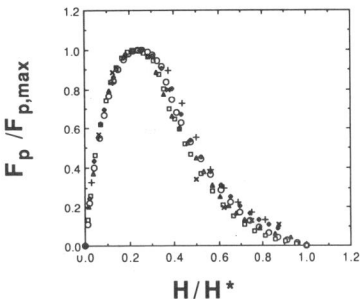

Fig. 3. Flux pinning force density vs. applied field at different temperatures.

Fig. 4. Normalized flux pinning force as a function of scaled variable H/H*. Data points obtained with Fe and Ni doping (0-5%) between 5-60K.

Figure 3 shows F_p as a function of applied magnetic field H for a 1% Fe-doped YBCO film at different temperatures. Although J_c is a monotonic decreasing function of H, two peaks in F_p are generally present, separated by a well-defined minimum. Besides its own physical meaning, the pinning force density also serves as an indication of the rate of J_c variation in different field regions. The broad peak in higher field (designated P_h) is similar to that discussed by Wördenweber et al.[18]; however, the stronger peak in lower field (designated P_l) has hitherto not been reported. Both peaks shift towards lower field with diminishing height when T is increased. The lower field peak seems to be a feature only found in thin films by the magnetization measurements with H perpendicular to the film surface.

The presence of a minimum in F_p at field values much below H_{irr} on the irreversibility line[14] (hence much lower than the upper critical field H_{c2}) suggests that there exists a new pinning mechanism different from what controls the irreversibility line. The field value at this minimum is called H^*, and it signifies a crossover between two different pinning mechanisms. It is plausible from free energy considerations that for H not far below H^* the pinning force density should contain a function of ($H^* - H$) characteristic of the predominant pinning mechanism in this field region.[19] For simplicity, in terms of a reduced field variable $h^* = H/H^*$, we assume this function to be proportional to $(1 - h^*)^q$. Thus H^* represents a certain critical field at which this particular pinning mechanism becomes ineffective, the exponent q will be determined from experimental data. For H below the value at P_l, F_p rises nearly as a linear function of H. Combining these considerations, we write $F_p(H,T,x,D) = F_{p,max}(T,x,D) f(h^*)$, where $f(h^*)$ is proportional to $h^*(1-h^*)^q$. This functional dependence of F_p is compared with our data. It is interesting that all the field dependence of the normalized F_p can be fitted very well to this expression with $q \approx 3$. This result is summarized in Fig. 4, where the F_p data points were obtained from various values of temperature and chemical doping. There is no adjustable parameter in this plot.

The exponent q being equal to 3 is physically plausible for the reasons given below. In the neighborhood of the peak P_l, flux density is high enough that flux line lattice (FLL) starts to form. Those flux lines surrounding a pinned fluxon will undergo plastic shear, the pinning strength should therefore be related to the FLL shear modulus C_{66}.[19,20] By an analogy with the field dependence of C_{66} on $(1-b)^2$ in conventional type II superconductors[20,21], where $b = H/H_{c2}$, we would expect a field dependence of C_{66} on $(1-h^*)^2$ for H below the new critical field H^*. In addition, for H perpendicular to the film plane, the large demagnetization factor should result in a strong distortion in the lines of force across the film.[22] Hence certain amount of energy is associated with the curvature of the flux lines.[23] Furthermore, since the film thickness is somewhat smaller than the penetration depth, the interface between the film and substrate or the surface could also pin the fluxons.[24] This type of pinning leads to a line tension which is proportional to the tilt modulus C_{44}.[23] In analogy with the dependence of C_{44} on (1-b) in conventional type II superconductors,[20,21] we expect that C_{44} should show a field dependence as $(1-h^*)$ for H below H^*. The dependence of F_p on C_{44} is also consistent with our observation that P_l disappears in bulk material or when H is in the film plane[25]. The combined effects of C_{66} and C_{44} thus give the field dependence of Fp on $(1-h^*)^3$. The product of h^* and $(1-h^*)^3$ has a maximum at $h^* = 1/4$, as shown in Fig. 4 for data points obtained under a variety of conditions.

In addition to the simple functional form of $f(h^*)$, another important simplification arises from the fact that the maximum pinning force density $F_{p,max}$ at the maximum P_l is expressable as a simple power of H^*. In all the cases studied, $F_{p,max}$ can be written simply as $C_1 H^{*2}$, where C_1 is a parameter depending on doping, and the exponent is found experimentally very close to 2. Hence, we write

$$F_p(H,T,x,D) = C_1 H^{*2} f(h^*) \qquad (1)$$

This is a useful result in view of the fact that the generally complex variation of F_p can now be described by a simple power law. This expression indicates that F_p/C_1 is practically an implicit function of T,x,D, and it obeys a "law of corresponding state" with H* appearing as the only scaling parameter. Once H* is determined from experiment as an explicit function of T, x, and D, (1) can then be used to evaluate the changes in pinning force under different conditions.

The peak P_h in higher field can be analyzed in a similar manner except that the pinning mechanism is different. The field dependence of F_p around P_h can be scaled by using a different reduced field variable $h = H/H_{irr}$. H_{irr} can be obtained by fitting data points to the expression: $F_p(H,T,x,D) = F'_{p,max}(T,x,D) g(h)$, where $h = H/H_{irr}$. The function g(h) accounts for all the H dependence in this field region. The results are shown in Fig. 5. All the data can be scaled to fit the field dependence of the form $h(1-h)^2$. g(h)=1 at the maximum where h=1/3. From the scaling behavior, H_{irr} can be conveniently obtained from the location of the maximum. This can be useful for the determination of H_{irr} in superconductors when H_{irr} is a very high field difficult to reach.

The field dependence governed by g(h) represents a pinning mechanism different from that associated with P_1. For H>H*, the effects due to line tension become negligible, more collective pinning can take place as the flux density is increased. This leads to an increase in F_p as H is increased above H*. Nevertheless, this process must compete with plastic shearing and thermally activated creep which tend to disorder the FLL. When H becomes high enough, presumably beyond the peak P_h, the FLL may either be depinned or start to "melt".[13,14] Thus Fp decreases monotonically with increasing H in the range $0.33H_{irr} < H < H_{irr}$, and it becomes undetectable as H reaches the apparent critial field H_{irr} under the influence of flux creep. Combining these considerations, the field dependence of F_p should therefore scale with the product of h and $(1-h)^2$. The $(1-h)^2$ term arises from C_{66} but now with H_{irr} replacing H_{c2} in h.[15,18]

The peak height of P_h can be expressed as a simple power of the relevant critical field H_{irr} similar to the peak height scaling of P_1. All of our high field data can be fitted to

$$F_p(H,T,x,D) = C_2 H_{irr}^2 g(h) \qquad (2)$$

Hence a similar type of scaling behavior is also established for the high field peak with g(h) replacing f(h*) and H_{irr} replacing H*.

We have obtained from our Fe-doped YBCO films the temperature dependence of both H* and H_{irr} in the region 5K<T<60K. It is tempting to find a connection between H* and H_{irr}, as measurements of one scaling behavior can then be related to the other. These data can be expressed as: $H^* = A_1(1-t)^{1.2}$, $H_{irr} = A_2(1-t)^{0.8}$, where $t=T/T_c$, and H* is proportional to $H_{irr}^{1.5}$, where A's are constants depending on the amount of doping. All the temperature and doping variations can be fitted to the scaling behavior shown in Figs. 4 and 5. This empirical result at least indicates that the two apparent critical fields H* and H_{irr}, although associated with different pinning mechanisms, are indeed related to each other. The dependence of H_{irr} on (1-t) does not show the 3/2 power as found in YBCO crystals,[14] probably due to fluctuation effects.

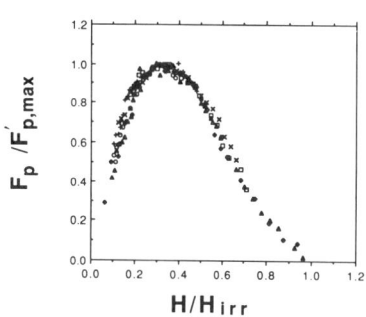

Fig. 5. Scaling behavior around the high field peak with data points similar to Fig. 4.

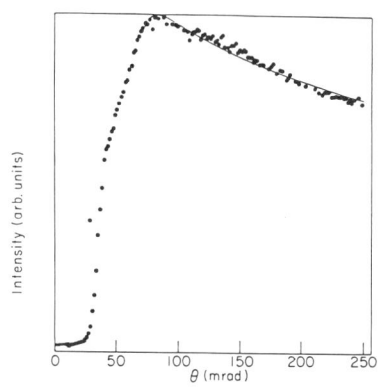

Fig. 6. Angular dependence of oxygen x-ray fluorescence of YBCO.

GRAZING INCIDENCE X-RAY FLUORESCENCE

The angular dependence of x-ray reflectivity, fluorescence, and electron yields contains much useful information on surface and interface structures. This is particularly important for the study of layered materials, such as thin film heterojunctions, composite film-substrate structures, and superlattices. By measuring the reflectivity, fluorescence, and electron yields as a function of the incidence angle of incoming x-rays, the roughness and other microstructures (e.g. interdiffusion of constituent atoms, formation of clusters) on the sample surface as well as the interfaces can be probed nondestructively. The problem of x-ray scattering from a multilayer material was studied earlier by Kiessig[26] and Parratt[27] but without considering the effects of interfacial roughness. By a modification of the Fresnel equations and incorporating a vector theory of scattering, we have developed a method which can now be readily used for the determination of interfacial roughness in multilayer systems.[28,29] An example of our results on the angular dependence of oxygen fluorescence yield from a high-Tc YBCO film is shown in Fig. 6. From this result we estimated that oxygen deficiency of about 5% or more could occur near the surface of YBCO.

CONCLUSIONS

These results of recent experiments, some still in progress, are presented mainly to show the methods currently used by our research group at SUNY Buffalo to investigate the effects of chemical doping on electronic structure, local environment surrounding various constituent atoms as well as impurities, and flux pinning as a function of magnetic field and temperature. As more work is needed, I hope the results presented here at least show that chemical doping, coupled with x-ray investigation using synchrotron radiation and preparation of high-quality thin films, can lead to a very useful direction for future studies of high-T_c superconductivity.

Synchrotron radiation provides many effective means for probing microstructures in the multi-component high-T_c superconductors in ways which were not possible before. This is particularly useful for the study of electronic structure and density of states pertaining to a specific atomic species such as oxygen or copper. The variation with T_c of the available states with oxygen 2p character in the Hubbard band shows a direct connection between the relevant electronic structure and superconductivity. This result should be useful for understanding the connection between Hall coefficient and T_c which is still unclear at the present time. Any successful theory on the electronic structure of high-T_c superconductors may eventually have to confront the type of XAS data as shown by these experiments. Further XAS measurements with different polarizations on single crystal superconductors will be very useful. Although it does not have a good energy resolution as compared to x-ray photoemission spectroscopy (XPS), the XAS technique still offers many unique advantanges for probing the electronic states above the Fermi level. The XAS and XPS results can nicely complement each other for a more complete understanding of the electronic structure. These techniques, together with the element-specific information available from XAFS and angular dependence of x-ray fluorescence, will constitute a main force in the future for probing the short-range-order structures in high-T_c superconductors.

The attempts of chemical doping so far have not resulted in any increase of T_c. However, effects of doping can suggest many useful approaches to understanding the mechanisms for a quantitative control of the critical current density J_c, a physical parameter of paramount importance to technological applications. The discovery of scaling laws could greatly simplify the usually rather complicated description of the dependence of J_c on magnetic field, temperature, defects and chemical doping. An understanding of the pinning mechanism in different field regions could suggest new ways for tailoring J_c values to suit the specific needs in some superconductive devices, such as radiation detectors or switching circuits. For instance, an increase in H* will improve the performance of J_c in a low magnetic field up to about one Tesla. This can be achieved by the formation of interfaces as in the case of multilayer structures. Further studies of correlations between the interfacial structure in the multilayer and superconducting properties will be of great interest.

This research is carried out in collaboration with my research associates at SUNY Buffalo A. Krol and L.W. Song, and graduate students Z.H. Ming, Y.L. Soo, M. Yang. I am indebted to many collaborators at other institutions, their names have appeared in our joint publications cited in this paper. The present research is supported in part by DOE and NYSIS.

REFERENCES

1. G.C. Smith, A. Krol, and Y.H. Kao, Nucl. Instr. Meth. **A291**, 135 (1990); A. Krol, C.S. Lin, Z.H. Ming, C.J. Sher, Y.H. Kao, C.T. Chen, F. Sette, Y. Ma, G.C. Smith, Y.Z. Zhu, and D.T. Shaw, Phys. Rev. **B42**, 2635 (1990); A. Krol, C.S. Lin, Z.H. Ming, C.J. Sher, Y.H. Kao, C.L. Lin, S.L. Qiu, J. Chen, J.M. Tranquada, M. Strongin, G.C. Smith, Y.K. Tao, R.L. Meng, P.H. Hor, C.W. Chu, and Gang Cao, Phys. Rev. **B42**, 4763 (1990).
2. C.T. Chen, F. Sette, Y. Ma, M.S. Hybertsen, E.B. Stechel, W.M. Foulkes, M. Schlutter, S.W. Cheong, A.S. Cooper, L.W. Rupp, Jr., B. Batlogg, Y.L. Soo, Z.H.

Ming, A. Krol, and Y.H. Kao, Phys. Rev. Lett. **66**, 104 (1991).

3. J. Zaanen, G.A. Sawatzky, and J.W. Allen, Phys. Rev. Lett. **55**, 418 (1985).

4. A. Krol, C.S. Lin, Y.L. Soo, Z.H. Ming, Y.H. Kao, J.H. Wang, M. Qi, and G.C. Smith, preprint.

5. P. A. Lee, P. H. Citrin, P. Eisenberger, and B. M. Kinkaid, Rev. Mod. Phys. 53, 769 (1981). T. M. Hayes and J. B. Boyce, in "Solid State Physics" edited by E. Ehrenreich, F. Seitz, and D. Turnbull, Vol 37, p. 173 (Academic Press, New York, 1982). "X-ray Absorption", edited by D.C. Koningsberger and R. Prins (Wiley, New York, 1988).

6. F. Bridges, J.B. Boyce, T. Claeson, T.H. Geballe, and J.M. Tarascon, Phys. Rev. **B42**, 2137 (1990); and references therein.

7. Y. H. Kao, Y. D. Yao, L. Y. Jang, F. Xu, A. Krol, L. W. Song, C. J. Sher, A. Darovsky, J. C. Phillips, J. J. Simmins, and R. L. Snyder, J. Appl. Phys. **67**, 353 (1990).

8. C.Y. Yang, A.R. Moodenbaugh, Y.L. Wang, Y. Xu, S.M. Heald, D.O. Welch, M. Suenaga, D.A. Fischer, and J.E. Penner-Hahn, Phys. Rev. **B42**, 2231 (1990).

9. J. Fink, N. Nücker, H. Romberg, M. Alexander, M.B. Maple, J.J. Neumeier, and J.W. Allen, preprint.

10. Y.H. Kao, Proc. First Asia-Pacific Conf. on Condensed Matter Physics (Singapore, 1988), p.136.

11. J.R. Hardy and J.W. Flocken, Phys. Rev. Lett. **60**, 2191 (1988).

12. J. Mannhart, P. Chandhari, D. Dimos, C.C. Tsuei, and T.R. McGuire, Phys. Rev. Lett. **61**, 2476 (1988).

13. K.A. Müller, M. Takashige, and J.G. Bednorz, Phys. Rev. Lett. **58**, 1143 (1987); R.S. Markiewicz, Physica **C162-164**, 235 (1989).

14. Y. Yeshurun and A.P. Malozemoff, Phys. Rev. Lett. **60**, 2202 (1988).

15. L.W. Song, E. Narumi, F. Yang, H.M. Shao, D.T. Shaw, and Y.H. Kao, Physica C, in press.

16. E. Narumi, L.W. Song, F, Yang, S. Patel, Y.H. Kao, and D.T. Shaw, Appl. Phys. Lett. **56**, 2684 (1990).

17. C.P. Bean, Phys. Rev. Lett. **8**, 250 (1962).

18. R. Wördenweber, G.V.S. Sastry, K. Heinemann, and H.C. Freyhardt, J. Appl. Phys. **65**, 1648 (1989); Cryogenics **29**, 458 (1990).

19. See, for example, E.J. Kramer, J. Appl. Phys. **44**, 1360 (1973).

20. E.H. Brandt, Phys. Stat. Sol. **77**, 551 (1976); J. Low Temp. Phys. **26**, 709 (1977); Phys. Rev. Lett. **63**, 1106 (1989).

21. A.I. Larkin and Yu.N. Ovchinnikov, J. Low Temp. Phys. **34**, 409 (1979).

22. G.M. Stollman, B. Dam, J.H.P.M. Emmen, and J. Pankert, Physica **C159**, 854 (1989).

23. A.M. Campbell and J.E. Evetts, Adv. Phys. **21**, 199 (1972).

24. B. Ross, L. Schultz, and G. Saemann-Ischenko, Phys. Rev. Lett. **64**, 479 (1990).

25. L.W. Song, E. Narumi, M. Yang, F. Yang, D.T. Shaw, and Y.H. Kao, to be published.

26. H. Kiessig, Ann. Phys. **10**, 715 (1931).

27. L.G. Parratt, Phys. Rev. **95**, 359 (1954).

28. A. Krol, C.J. Sher, and Y.H. Kao, Phys. Rev. **B38**, 8579 (1988).

29. A. Krol, C.J. Sher, and Y.H. Kao, Phys. Rev. **B42**, 3829 (1990).

PARTICIPANTS

Dr. M. Abella, Physics Department, University of Miami, Coral Gables, FL 33124.

Dr. G. Aeppli, AT&T Bell Laboratories, Murray Hill, NJ 07974.

Dr. D. Agassi, Naval Surface Warfare Center, Silver Spring, MD 20903-5000, 10901, New Hampshire Ave.

Dr. G. Alexandrakis, Physics Department, University of Miami, Coral Gables, FL 33124.

Dr. P.B. Allen, Department of Physics, State University of New York at Stony Brook, Stony Brook, NY 11794.

Dr. H. Alloul, Laboratoire de Physique des Solides Université Paris Sud, Bat. 510, 91405 Orsay, France.

Dr. P. W. Anderson, Joseph Henry Laboratories of Physics, Princeton University, Princeton N.J. 08540.

Dr. E. J. Ansaldo, Physics Department, University of Saskatchewan, Saskatoon S7N 0W0, Canada.

Dr. J. Ashkenazi, Physics Department, University of Miami, Coral Gables, FL 33124.

Dr. Assa Auerbach, Physics Department, Boston University, Boston, MA 02215.

Dr. C. Balseiro, Centro Atómico Bariloche, 8400 Bariloche, Argentina.

Dr. Silvia Bacci, Physics Department, University of Illinois at Urbana-Champaign, Urbana, IL 61801.

Dr. Richard, D. Bardo, NAVSWC, Code R13, 10901, New Hampshire Avenue, Silver Spring, MD 20903-5000.

Dr. S. E. Barnes, Physics Department, University of Miami, Coral Gables, FL 33124.

Dr. T. Barnes, Center for Computationally Intensive Physics, Bldg. 6003, Oak Ridge National Lab. Oak Ridge, Tennessee 37831-6373.

Dr. Yaneer Bar-Yam, Departmet of Chemical Physics, Weizmann Institute of Science, Rehovot, Israel.

Dr. A. Bianconi, Dipartimento di Fisica, Univ. degli Studi di Roma "La Sapienza", 00185 Rome, Italy.

Dr. S.M. Bose, Department of Physics and Atmospheric Science, Drexel University, Philadelphia, PA 19104.

Dr. B.H. Brandow, Los Alamos National Laboratory, Los Alamos, NM 87545.

Dr. Helmut Brandt, AT&T Bell Laboratories, Room 1E402, Murray Hill, N.J. 07974.

Mr. F. Carvajal, Physics Department, University of Miami, Coral Gables, FL 33124.

Dr. Leslie C. Case, Eltron, Inc., 14 Lockland Road, Winchester, MA 01890.

Dr. Ron Cohen, Carnegie Institution, Geophysical Laboratory, 5251 Broad Branch Road, N.W. Washington, D.C. 20015.

Dr. J. L. Cohn, Naval Research Laboratory, Materials Physics Branch, Washington, D.C. 20375-5000.

Dr. J. E. Crow, Physics Department, The Florida State University, Tallahassee, FL 32306-3016.

Dr. F. de la Cruz, Centro Atómico Bariloche, 8400 Bariloche, Argentina.

Dr. I. Dzyaloshinskii, Institute of Theoretical Physics, University of California, Santa Barbara, CA 93106.

Dr. T. Egami, Department of Materials Science and Engineering, University of Pennsylvania, Philadelphia, PA 19104-6272.

Dr. David Emin, Sandia National Laboratory, Solid State Theory Division, Albuquerque, NM 87185.

Dr. A.J. Fedro, Materials Science and Technology Division, Argonne National Laboratory, 9700 South Cass avenue, Argonne IL 60439-4837.

Dr. M. J. Fluss, L-326, Lawrence Livermore National Laboratory, Livermore, CA 94550.

Dr. J. P. Franck, Department of Physics, University of Alberta, Edmonton, AB T6G 2J1, Canada.

Dr. Eduardo R. Gagliano, Physics Department, University of Illinois at Urbana-Champaign Urbana, IL 61801.

Dr. L. H. Greene, Bell Communications Research, 331 Newman Springs Road, Red Bank N.J. 07701.

Dr. J. R. Hardy, University of Nebraska, Loncoln, NE 68588-0111.

Dr. David G. Hinks, Materials Science and Technology Division, Argonne National Laboratory, Argonne, IL 60439.

Dr. J. E. Hirsch, Physics Department, University of California at San Diego, La Jolla, CA 92093.

Dr. R. H. Howell, L-280, Lawrence Livermore National Laboratory, Livermore, CA 94550.

Dr. O. Jepsen, Max-Planck-Institut für Festkörperforschung D-7000 Stuttgart 80, Federal Republic of Germany.

Dr. Keith H. Johnson, Mass. Inst. of Technology, Room 13-5013, Cambridge, MA 02139.

Dr. M. D. Johnson, Department of Physics, University of Central Florida, Orlando FL, 32816-0385.

Dr. A. Kallio, University of Oulu, Dept. of Theoretical Physics, SF90570 Oulu, Finland.

Dr. Y. H. Kao, Department of Physics and Astronomy, SUNY at Buffalo, 239 Fronczak Hall, Buffalo, NY 14260.

Dr. J. H. Kim, 226 McCullough Bldg., Stanford University, Stanford, CA 94305.

Dr. B. M. Klein, Naval Research Laboratory, Complex Systems Theory Branch, code 4690, Washington, D.C. 20375-5000.

Dr. M. V. Klein, Physics Department, University of Illinois at Urbana-Champaign, Urbana, IL 61801.

Dr. H. Kleinert, Physics Department, University of Miami, Coral Gables, FL 33124.

Dr. R. A. Klemm, Materials Science Division, Argonne National Laboratory, 9700 South Cass Avenue, Argonne, IL 60439.

Dr. Vladimir Z. Kresin, Materials and Molecular Research Division, Lawrence Berkeley Laboratory, University of California, Berkeley, CA 94720.

Dr. P. Kumar, Physics Department, University of Florida, 215 Williamson Hall, Gainesville, FL 32611.

Dr. Charles G. Kuper, Physics Department, Technion, Haifa 32000, Israel.

Dr. F. V. Kusmartsev, Institut für Theoretische Physik, Universitat zu Köln, D-5000 Köln 41, Federal Republic of Germany.

Dr. T. K. Lee, Department of Physics, Virginia Tech., Blacksburg, VA 24061.

Ms. L. Leong, Physics Department, University of Miami, Coral Gables, FL 33124.

Dr. K. Levin, James Franck Physics Institute, University of Chicago, 5640 Ellis Avenue, Chicago, IL 60637.

Dr. Sudan Liang, Physics Department, Penn State University, University Park, PA 16802.

Dr. Douglas H. Lowndes, Solid State Division, Building 2000, Oak Ridge National Laboratory, P.O. Box 2008, Oak Ridge, TN 37831-6056.

Dr. D. van der Marel, Faculty of Applied Physics, Delft University of Technology, Lorentzweg 1, 2628 CJ Delft, The Netherlands.

Dr. R. S. Markiewicz, Physics Department and Barnett Institute, Northeastern University, Boston, MA 02115.

Dr. R. M. Martin, Physics Department, University of Illinois at Urbana-Champaign, Urbana, IL 61801.

Dr. I. I. Mazin, P.N. Lebedev Physics Institute Leninski pr. 53, Moscow 117924, USSR.

Dr. F. M. Mueller, Los Alamos National Laboratory, Physics Division, Los Alamos, NM 87545.

Dr. J. Mustre de Leon, Los Alamos National Laboratory, Physics Division, Los Alamos, NM 87545.

Dr. D. Newns, T.J. Watson Research Center, P.O. Box 218, Yorktown Heights, N.Y. 10598.

Dr. David P. Norton, Solid State Division, Bld. 2000, Oak Ridge National Laboratory, P.O. Box 2008, Oak Ridge, TN 37831-6056.

Dr. N.-P. Ong, Joseph Henry Laboratories of Physics, Princeton Univesity, Princeton, NJ 08544.

Dr. D. A. Papaconstantopoulos, Naval Research Laboratory, Complex Systems Theory Branch, code 4690, Washington, D.C. 20375-5000.

Dr. D. Pavuna, Physics Department, Swiss Federal Institute of Technology (EPFL), PH-Ecublens, CH-1015 Lausanne, Switzerland.

Dr. J. C. Phillips, AT&T Bell Laboratories, Murray Hill, N.J. 07974.

Dr. N. E. Phillips, Department of Chemistry, University of California, Berkeley, CA 94720.

Dr. George F. Reiter, Physics Department, University of Houston, 4800 Calhoun, Houston TX 77004.

Dr. Peter S. Riseborough, Physics Department, Polytechnic University, 333, Jay Street, Brooklyn, NY 11201.

Dr. M. B. Salamon, Physics Department, University of Illinois at Urbana-Champaign, Urbana, IL 61801.

Dr. Z. Schlesinger, T. J. Watson Research Center, Yorktown Heights, N.Y. 10598.

Dr. J. W. Serene, Naval Research Laboratory, Complex Systems Theory Branch, code 4690, Washington, D.C. 20375-5000.

Dr. A. T. Seshadri, Department of Physics, Indian Institute of Technology, Madras–600 036, India.

Dr. Z.-X. Shen, Stanford Electronics Laboratory, Stanford University, Stanford, CA 94305.

Dr. Qimiao Si, James Franck Physics Institute, University of Chicago, 5640 Ellis Avenue, Chicago, IL 60637.

Dr. J. L. Smith, Los Alamos National Laboratory, Physics Division, Los Alamos, NM 87545.

Ms. X. M. Song, Physics Department, University of Miami, Coral Gables, FL 33124.

Dr. W. Stephan, Max-Planck-Institut, für Festkörperforschung, D-7000 Stuttgart 80, Federal Republic of Germany.

Dr. Barbara Szpunar, 12 Lakeview Av., Kingston, Ont. K7M 3S7, Canada.

Dr. D. B. Tanner, Physics Department, University of Florida, Gainesville, FL 32611.

Dr. J. G. Tobin, L-268, Lawrence Livermore National Laboratory, Livermore, CA 94550.

Dr. J.M. Tranquada, Physics Department, Brookhaven National Laboratory, Upton, N.Y. 11973.

Dr. A. M. Tsvelik, Department of Physics, University of Florida, Gainesville, FL 32611.

Dr. Y. J. Uemura, Physics Department, Columbia University, New York, NY 10027.

Mr. D. Vacaru, Physics Department, University of Miami, Coral Gables, FL 33124.

Dr. Z. Vardeny, Department of Physics, 201 JFB, University of Utah, Salt Lake City, UT 84112.

Dr. C. M. Varma, AT&T Bell Laboratories, Murray Hill, NJ 07974.

Dr. Gary Vezzoli, Army Materials Tech. Laboratory, Materials Science Branch, Watertown, MA 02172-0001.

Dr. V. M. Vinokur, Materials Science Division, Argonne National Laboratory, 9700 South Cass Avenue, Argonne, IL 60439.

Dr. Purushotham Voruganti, Physics Department, University of Toronto, 60 St. George Street, Toronto, M5S 1A7, Canada.

Dr. R. E. Walstedt, AT&T Bell Laboratories, Murray Hill, NJ 07974.

Dr. Ziqiang Wang, Department of Physics, Rutgers University, P.O. Box 849, Piscataway, NJ 08854.

Dr. M. Weger, Hebrew University of Jerusalem, Racah Institute of Physics, Jerusalem, Israel.

Dr. Gene L. Wells, Editor, *Physical Review Letters*, 1 Research Road, Box 1000, Ridge, NY 11961-9802.

Dr. Z. Y. Weng, Department of Physics and Texas Center for S/C, University of Houston, Houston TX 77204-5506.

Dr. R. N. West, Department of Physics, P.O. Box 19059, University of Texas at Arlington, Arlington, TX, 76019-0059.

Dr. D. Wohlleben, II. Physikalisches Institut, Universitat zu Köln, D-5000 Köln, Federal Republic of Germany.

Dr. Stuart A. Wolf, Naval Research Laboratory, Materials Physics Branch, Washington, D.C. 20375-5000.

Dr. L. M. Xie, Xsirius Superconductivity, Inc., 7590 E. Gray Road, Suite 103, Scottsdate, AZ 85260.

Dr. Jaejun Yu, Department of Physics, Northwestern University, Evanston, IL 60208-3112.

Ms. S. Zane, Physics Department, University of Miami, Coral Gables, FL 33124.

Dr. Zhou Zou, Stanford University, Physics Department, Stanford, CA 94305.

Dr. Fulin Zuo, Physics Department, University of Miami, Coral Gables, FL 33124.

AUTHOR INDEX

Aeppli, G., 621
Affronte, M., 231
Allen, P.G., 425
Anderson, P.W., 1
Ansaldo, E.J., 447
Arnold, G.B., 417
Ashkenazi, J., 205, 215, 569
Auerbach, A., 53

Bacci, S.B., 597
Bar-Deroma, R., 215
Bar-Yam, Y., 561
Barnes, S.E., 95
Barnes, T., 503
Batistic, I, 425
Bianconi, A., 363
Billinge, S.J.L., 389
Bishop, A.R., 425
Bonfim, O.F.de Alcantara, 45
Bose, S.M., 515
Bozovic, I., 251
Brandow, B.H., 547, 583
Brandt, E.H., 437
Braun, E., 469
Brener, R., 215
Broicher, H., 469
Burke, T., 257

Canright, G.S., 89
Carr, G.L., 159
Chen, M.F., 257
Chi, C.C., 493
Chien, T.R., 131
Clougherty, D.P., 341
Cohen, R.E., 7
Cohn, J.L., 235
Collins, R.T., 147
Conradson, S.D., 425
Cooke, D.W., 119
Crabtree, G.W., 147
Craver, F., 257
Crawford, M.K., 389
Crow, J.E., 119

Dessau, D.S., 113

Egami, T., 389
Emin, D., 319
Eom, C.-B., 251

Fang, Y., 147
Farneth, W.E., 389
Fedro, A.J., 37
Feild, C., 147
Feldman, W.L., 137
Felsteiner, J., 215
Fisher, B., 205
Fisher, R.A., 459
Flank, A.M., 363
Flocken, J.W., 331
Fluss, M.J., 183, 189
Forro, L., 159
Fowler, C.M., 177
Franck, J.P., 411
Freeman, A.J., 529
Freeman, B.L., 177

Gagliano, E.R., 597
Gao, Y., 119
Geballe, T.H., 251
Genossar, J., 205
Girvin, S.M., 89
Greene, L.H., 137
Gros, C., 89
Gu, C., 189
Guertin, R.P., 119
Guyer, R.A., 523
Gygax, S., 411

Habermeier, H.-U., 197
Haghighi, H., 183
Hardy, J.R., 331
Harmon, B.N., 37
Harris, J.S. Jr., 251
Harshman, D.R., 621
Heitmann, D., 197
Hess, D.W., 61
Hinks, D.G., 389
Hirsch, J.E., 295
Holtzberg, F., 147
Howell, R.H., 183
Hults, W.L., 177

Inam, A., 137

Janot, Chr., 389
Jing, T.W., 131
Johnson, K.H., 341
Johnson, M.D., 89
Jorgensen, J.D., 389
Jung, J., 411

Kaiser, J.H., 183
Kamaras, K., 159
Kane, C.L., 493
Kallio, A., 241
Kao, Y.H., 641
Kebede, A., 119
Kim, H., 107
Kim, Jae H., 251
Kim, Ju H., 481
King, J.C., 177
Klemm, R.A., 609
Koren, G., 205
Krakauer, H., 7
Kresin, V.Z., 275
Kumar, P., 523
Kuper, C.G., 205, 569
Kuric, M., 119
Kusmartsev, F.V., 77

Lagarde, P., 363
Larson, B.E., 53
Lee, T.K., 69
Lee, W.K., 251
Lesueur, J., 137
Leung, T.C., 37
Levin, K., 481
Li, C., 363
Liu, J.Z., 147, 183, 189
Liu, S.H., 609
Longe, P., 515
Lu, H.M., 331
Lu, J.P., 481
Lowndes, D.H., 377

Mandrus, D., 159
Markiewicz, R.S., 555
Martin, R.M., 597
McCarron, E.M., 389
McHenry, M.E., 341
Mihalisin, T., 119
Mihaly, T., 159
Mohamed, M. A-K, 411
Moon, B.M., 257
Morawitz, H., 275
Mueller, F.M., 177, 417
Mustre de Leon, J., 425

Newns, D.M., 493

Norton, D.P., 377

Obukhov, S., 523
Olson, C.G., 189
Ong, N.P., 131

Pattnaik, P.C., 493
Pavuna, D., 231
Pernambuco-Wise, P., 119
Pettiti, I., 363
Philips, N.E., 459
Pickett, W.E., 7
Pompa, M., 363

Ramakrishnan, T.V., 131
Rayner, S.L., 183
Reiter, G., 143
Reizer, M.Yu, 453
Remschnig, K., 131
Riseborough, S., 107
Romero, D.B., 159
Rotter, L.D., 147

Salama, K., 235
Schlesinger, Z., 147
Schnelle, W., 469
Selvamanickam, V., 235
Serene, J.W., 61
Seshadri, A.T., 225
Shelton, R., 183
Shen, Z.-X., 113
Si, Q., 481
Smith, J.L., 177
Smith, V.H. Jr., 591
Solal, F.R., 183, 189
Spencer, N.D., 119
Sproule, G.I., 411
Sterne, P.A., 183
Subrahmanyam, B., 225
Subramanian, M.A., 389
Sun, Y.S., 523
Swihart, J.C., 417
Szpunar, B., 591

Tanner, D.B., 159
Tarascon, J.M., 131
Ting, C.S., 541
Tobin, J.G., 189
Toby, B.H., 389
Tranquada, J.M., 629
Tsuei, C.C., 493
Tsvelik, A.M., 453

Uemura, Y.J., 353
Udron, D., 363

Vacaru, D., 569

Van der, Marel, D., 197, 215
Varma, C.M., 19
Vezzoli, G.C., 257

Walstedt, R.E., 401
Wang, Z., 69
Wang, Z.Z., 131
Weger, M., 309
Wehrhahn, R.F., 309
Wells, B.O., 113
Welp, U., 147
Weng, Z.Y., 541
West, R.N., 183
Williams, G.P., 159
Wittlin, A., 197
Wohlleben, D., 469
Wolf, S.A., 235, 275

Yu, J., 529

Zhou, Yu, 37
Zou, Z., 25

SUBJECT INDEX

Anderson model, 1, 547, 549, 569
Anharmonicity, 7, 347, 350, 398, 434
Anisotropy, 186, 404, 609
Antibonding state, 307, 531, 553, 629
Antiferromagnetic interactions, 23, 53, 55, 107, 120, 127, 257, 258, 260, 272, 451, 503, 545, 555, 561, 604, 630, 637
Anyons, 89, 104, 451, 569
Auger spectra, 216-218, 574
Axial oxygen, 425-434

Band structure, 13, 27, 215, 489, 529, 555, 569
Bipolaron, 319-323, 325-328, 363, 373, 374, 397, 398, 569, 577
Bloch functions, 549-551

Charge (and charge transfer) 120, 272, 434, 549
Charge density wave, 272, 555, 558, 572
Chern-Simons treatment, 456-458
Chevrel phase, 355
Chiral spin liquid, 26
Cluster, 600
Coherence length, 279, 500, 622
Conductivity, 107, 149, 161, 163, 165, 315
Coulomb effects, 215, 220, 222, 232, 332, 370, 530, 532, 574, 588
Coupling, 16, 292, 371, 373, 448, 466, 543, 588

Debye temperature, 466, 489, 572

de Haas van Alphen effect, 177, 179, 181, 277, 290
Density fluctuation, 7
Density of states, 194, 218, 286, 466, 497, 555, 591, 573
Dielectric property, 202, 285
Disorder, 306, 453, 458, 555
Doping effects, 119-121, 257-267, 304, 428, 566, 604, 641-649
Drude model, 152, 159, 161, 163, 169, 202, 422, 558, 577
Dynamics, 331-335, 391, 541, 561-568, 578

Electron-phonon interaction, 7, 14, 17, 287, 306, 415, 423, 434, 467, 555, 567
Eliashberg treatment, 1, 291, 292, 314, 331, 332, 417-420, 491
Exciton, 23, 173, 251, 272, 423

Fermi liquid, 19, 67, 103, 136, 160, 168, 173, 481-491, 529-538, 547, 551, 610
Fermi surface, 8, 10, 103, 177, 180, 183, 186, 187, 281, 289, 397, 531, 555
Feynman diagram, 314
Fluctuations, 257, 383, 427, 434-444, 475, 478, 503, 543, 544, 547, 558
Flux, 102, 134, 268, 271, 354, 387, 438, 641, 651
Franck-Condon transition, 567
Frequency-dependence, 107, 163, 173, 199

Gap, 6, 113, 147, 156, 159, 161, 171, 280, 405, 420-422, 609, 614
Gauge invariance, 454
Ginzburg-Landau free energy, 615
Green's functions, 39, 573
Gyromagnetic ratio, 354

Hall effect, 131, 132, 226, 228, 230, 232, 278, 295, 477, 486, 535
Hamiltonian, see Wave function
Hartree-Fock, 89, 594, 598
Heat capacity, see Specific heat
Heavy fermion systems, 357, 484, 489, 547
Holes, 53, 123, 129, 131, 220, 272, 295-300, 307, 341, 342, 363-367, 372, 397, 414, 490, 503, 505, 511, 535, 543, 545, 549, 574, 583, 604, 637
Holon, 28, 75, 95-97, 574
Holstein approach, 160, 161
Hopping, 397, 503
Hubbard model, 1, 2, 37, 41, 43, 61, 63-65, 77, 83, 107, 109, 203, 493, 503, 541, 547, 557, 569, 570, 586, 597-606

Indirect exchange, 262, 265, 266
Infrared studies, 147, 159, 197
Inter-layer, 577, 609-620
Isotope effect, 335, 336, 411-415, 499, 567

Jahn-Teller effect, 272, 341-351, 370, 435
Josephson-Junction, 139, 202

Kondo effect, 45, 547, 552
Kosterlitz-Thouless transition, 377, 380, 524, 525
Kramers-Kronig analysis, 167, 251, 421, 577

Ligand field, 343
Local density approximation, 222, 530, 531, 569, 575, 595, 598
Local field, 272, 335, 354, 481, 622
London penetration depth, 200

Lorentz-Gauge, 456
Luttinger liquid, 5, 68, 574

Magnetic properties, 257-574, 401-407, 447-451, 476, 481, 567, 604, 625
Magnetoresistance, 133, 238
Marginal fermi liquid, see fermi liquid
Mean field, 37, 89, 96, 101, 104, 478, 574
Migdal's theorem, 1
Mixed state, 272, 434, 448
Mobility, 229
Monopole, 457
Mott insulator, 481, 482, 553, 555, 591
Muffin tin approximation, 335 595
Multi-band effects, 306
Muon spin resonance, 127, 621-628

Narrow-band model, 570
Neel temperature, 120, 126, 127

Ordering effects, 119, 127-129, 450, 614, 622
Organics, 355
Oscillations, 144, 257

Pairing, 25, 119, 203, 265, 279, 291, 316, 346, 347, 359, 378, 548, 561, 577, 609
Pair breaking, 119, 241, 257, 354, 466
Penetration depth, 353, 449, 622
Perturbation, 61, 62, 503-512
Phase fluctuations, 91, 93
Phase separation, 556
Phonon, see Electron-phonon interactions
Photoemission, 113, 116, 189, 290, 325, 396, 418, 434, 529, 553, 576, 583-588, 598
Plasmon, 241, 251, 255, 283-286, 290, 292, 596, 622
Polarizability, 23, 434, 556, 558
Polaron, 55, 56, 58, 319, 373, 574, 576, 577
Pre-onset, 267
Pressure dependence, 305

Quantum Hall effect, 77, 89

Radial distribution function, 431, 564
Rare earth series, 159-162, 262-268
Reflectivity, 148, 153, 154, 198
Resistivity, 9, 205, 209, 228, 529
Resonating valence bond, 453, 458

Scattering, 131, 136, 172, 489, 496
Schottky anomaly, 461
Screening, 222, 335, 346, 594, 595
Self energy, 591-596
Slave boson, 69, 95, 104, 574
Slave fermion, 73, 95
Sommerfeld constant, 358
Specific heat, 277, 459-469, 471, 472
Spin liquid, 453
Spinon, 28, 75, 95-97, 257, 272, 574
Spiral state, 541-545
Statistical transmutation, 89
Strong coupling, 291, 305, 315
Structural anomaly, 389
Substitutions, 119-121, 262-268, 413, 451
Superlattice, 381, 382, 385, 449, 523-526

t-J model, 45, 53, 55, 69, 75, 76, 503-512, 598
Thermal expansion, 472
Thermoelectric power, 205, 212, 235-238, 324
Thin films, 137, 205, 524
Tomonaga-Luttinger model, 70
Transmittance, 164-166
Transport properties, 225, 231, 535
Tunneling, 137, 141, 241, 291, 305, 309, 313, 315, 430, 433

Valence, 120, 122, 257, 272, 434, 547-553
van Hove singularity, 116, 480-490, 493-500, 555-559, 577
Variational analysis, 90
Vortex, 102, 134, 268, 271, 354, 387, 437-444, 523, 622

Wannier orbital, 108, 549, 550, 588, 598
Wavefunction, 344, 427, 503, 505, 598, 612, 613
Wide-band, 573